Handbook of Experimental Pharmacology

Volume 125

Editorial Board

G.V.R. Born, London
P. Cuatrecasas, Ann Arbor, MI
D. Ganten, Berlin
H. Herken, Berlin
K.L. Melmon, Stanford, CA

Springer
*Berlin
Heidelberg
New York
Barcelona
Budapest
Hong Kong
London
Milan
Paris
Santa Clara
Singapore
Tokyo*

Physiology and Pharmacology of Biological Rhythms

Contributors

F. Andreotti, J. Arendt, P.M. Belanger, G.D. Block, B. Bruguerolle
J. Camber, G.E. D'Alonzo, L.N. Edmunds, Jr., M.H. Hastings
E. Haus, M. Karzazi, G. Labrecque, B. Lemmer, F. Levi, A. Lluch
K.F. Martin, L. Mejean, H. Merki, S. Michel, J.G. Moore
R.Y. Moore, N. Mrosovsky, G. Patti, M. Pons, F. Portaluppi
P.H. Redfern, A.E. Reinberg, L. Rensing, A.M. Rosenwasser
R. Smaaland, M.H. Smolensky, A. Stricker-Krongrad, Y. Touitou
M.-C. Vanier, A. Wirz-Justice, K. Witte

Editors

P.H. Redfern and B. Lemmer

 Springer

Professor PETER H. REDFERN, BPharm, PhD
School of Pharmacy and Pharmacology
University of Bath
Claverton Down
BA1 7AY Bath
United Kingdom

Professor Dr. B. LEMMER
Direktor des Instituts für Pharmakologie und Toxikologie
Fakultät für klinische Medizin Mannheim
Universität Heidelberg
Maybachstr. 14–16
68169 Mannheim
Germany

With 166 Figures and 43 Tables

ISBN 3-540-61525-3 Springer-Verlag Berlin Heidelberg New York

Library of Congress Cataloging-in-Publication Data. Physiology and pharmacology of biological rhythms/ contributors, F. Andreotti ... [et al.]: editors. P.H. Redfern and B. Lemmer.   p. cm. – (Handbook of experimental pharmacology: v. 125) Includes bibliographical references and index. ISBN 3-540-61525-3 (alk. paper) 1. Chronopharmacology. 2. Chronobiology. I. Andreotti, F. II. Redfern, P. H., 1942- . III. Lemmer, Bjorn. IV. Series. [DNLM: 1. Circadian Rhythm–physiology. 2. Circadian Rhythm–drug effects. 3. Chronobiology. 4. Drug Therapy, W1 HA51L v. 125 1997 QT 167 P578 1997] QP905.H3 vol. 125 [RM301.3.C47] 615'.1 s–dc20 [612'.022] DNLM/DLC For Library of Congress 96-29357

This work is subject to copyright. All rights are reserved, whether the whole or part of the material is concerned, specifically the rights of translation, reprinting, reuse of illustrations, recitation, broadcasting, reproduction on microfilm or in any other way, and storage in data banks. Duplication of this publication or parts thereof is permitted only under the provisions of the German Copyright Law of September 9, 1965, in its current version, and permission for use must always be obtained from Springer-Verlag. Violations are liable for prosecution under the German Copyright Law.

© Springer-Verlag Berlin Heidelberg 1997
Printed in Germany

The use of general descriptive names, registered names, trademarks, etc. in this publication does not imply, even in the absence of a specific statement, that such names are exempt from the relevant protective laws and regulations and therefore free for general use.

Product liability: The publisher cannot guarantee the accuracy of any information about dosage and application contained in this book. In every individual case the user must check such information by consulting the relevant literature.

Cover design: Design & Production GmbH, Heidelberg

Typesetting: Scientific Publishing Services (P) Ltd, Madras

SPIN: 10477021   27/3136/SPS – 5 4 3 2 1 0 – Printed on acid-free paper

Preface

Measured by any criteria, research in chronobiology in general and chronopharmacology in particular has expanded rapidly in recent years. This expansion has been paralleled by an increasing recognition by those outside the field of the relevance and significance of recent developments in chronobiology.

Advances in two areas have been chiefly responsible. First, application of the full range of modern techniques in behavioral, neurochemical, and molecular biology have greatly improved our understanding of basic clock mechanisms. In several species the genetic basis of the circadian clock is being progressively delineated. A complete picture of the neurochemical and neuroanatomical structure of the mammalian clock is emerging and the complex pattern of control mechanisms involving endogenous clock mechanisms and photic and nonphotic zeitgebers is being built up as a result of behavioral studies.

Secondly, in parallel with these exciting developments in basic science, clinical applications are being convincingly demonstrated in the general fields of pharmacology and medicine as well as in specific areas, e.g., jet lag, shiftwork maladaption syndrome, blindness, and cardiovascular system.

It is therefore an opportune time to review progress in the field of chronopharmacology and to introduce some of the exciting developments and prospects to a readership beyond the confines of the chronobiological cognoscenti. This volume is therefore aimed primarily at the pharmacologist – whether basic, applied, or clinical – who is not a specialist in chronobiology. The core of the volume is a systematic consideration of the relevance of chronobiological principles in basic and clinical pharmacology, covered in 23 chapters dealing with topics ranging from pharmacokinetics, the respiratory system, cardiovascular pharmacology, cytotoxic drugs, diabetes, the gastrointestinal tract, the central nervous system, hemopoiesis and hemostasis, allergy, pain and analgesios, local anesthetics, and toxicology.

These chapters are preceded by an overview of clock mechanisms – the localization and connections of the biological clock within the central nervous system, the mechanisms of entrainment, and the molecular and cellular biology of the clock. Also included are chapters dealing in more detail with specific topics such as the pineal gland and melatonin, rhythms in development and aging, and rhythms in retinal mechanisms, in second messengers, and in neurotransmitter turnover.

We hope that this volume presents a balanced picture of the present state of knowledge in chronopharmacology and that it may stimulate a wider interest in what is, to us, a fascinating and rewarding field of research.

Our unreserved thanks go to the individual authors whose work is contained herein. We are grateful equally for the patience of those who assiduously met deadlines and for the tolerance of those who finally succumbed to our persistent chidings. Finally we would also like to acknowledge our debt to Doris Walker, desk editor for bioscience, who has controlled all aspects of the production of this volume, including the scientific editors.

Bath, Great Britain P.H. REDFERN
Mannheim, Germany B. LEMMER

List of Contributors

ANDREOTTI, F., Università Cattolica del Cacro Cuore, Policlinico A. Gemelli, Instituto di Cardiologia, Largo F. Vito 1, 00168 Roma, Italia

ARENDT, J., Department of Biochemistry, School of Biological Sciences, University of Surrey, Guildford, Surrey GU2 5XH, Great Britain

BELANGER, P.M., Université Laval, École de Pharmacie, Cité Universitaire, Sainte-Foy, Quebec, Canada G1K 7P4

BLOCK, G.D., NSF Center for Biological Timing, Department of Biology, University of Virginia, Gilmer Hall, Charlottesville, VA 22901, USA

BRUGUEROLLE, B., Laboratoire de Pharmacologie Médicale et Clinique, Faculté de Médecine de Marseille, 27, Boulevard Jean Moulin, 13385 Marseille Cedex 5, France

CAMBAR, J., Groups d'Etude de Physiologie et Physiopathologie Rénales, Faculté de Pharmacie, 91, rue Leyteire, 33000 Bordeaux, France

D'ALONZO, G.E., Division of Pulmonary Disease and Critical Care Medicine, Temple University School of Medicine, 3400 North Broad Street, Philadelphia, PA 19140, USA

EDMUNDS, Jr., L.N., Molecular and Cellular Biology, Division of Biological Sciences, Graduate Program, State University of New York, Life Sciences Building 370, Stony Brook, NY 11794-5215, USA

HASTINGS, M.H., Department of Anatomy, University of Cambridge, Downing Street, Cambridge CB2 3DY, Great Britain

HAUS, E., Health Partners, St. Paul-Ramsey Medical Center, University of Minnesota, 640 Jackson, St. Paul, MN 55101, USA

KARZAZI, M., Département de Pharmacie, Hôpital du Saint-Sacrement, Cité Universitaire, Sainte-Foy, Quebec, Canada G1K 7P4

LABRECQUE, G., Université Laval, École de Pharmacie, Cité Universitaire, Sainte-Foy, Quebec, Canada G1K 7P4

LEMMER, B., Institut für Pharmakologie und Toxikologie der Fakultät für klinische Medizin Mannheim, Ruprecht-Karls-Universität Heidelberg, Maybachstr. 15–16, 68169 Mannheim, Germany

LEVI, F., Institut du Cancer et d'Immunogenetique (ICIG), Hôpital Paul Brousse, Laboratoire "Rhythmes Biologiques et Chronothérapeutique", 14, avenue Paul Vaillant-Couturier, 94807 Villejuif Cedex, France

LLUCH, A., Equipe de Recherches Aliment et Comportement, INSERM U 308, Unite de Recherches sur les Mecanismes de Regulation du Comportement Alimentaire, 38, rue Lionnois, 54000 Nancy, France

MARTIN, K.F., Biological Research, Boots Pharmaceuticals, Building 3R, Pennyfoot Street, Nottingham NG2 3AA, Great Britain

MEJEAN, L., Equipe de Recherches Aliment et Comportement, INSERM U 308, Unite de Recherches sur les Mecanismes de Regulation du Comportement Alimentaire, 38, rue Lionnois, 54000 Nancy, France

MERKI, H., Gastrointestinal Unit, Inselspital, University of Berne, 3010 Berne, Switzerland

MICHEL, S., NSF Center for Biological Timing, Department of Biology, University of Virginia, Gilmer Hall, Charlottesville, VA 22901, USA

MOORE, J.G., Utah School of Medicine, GI Section, Department of Veterans Affairs Medical Center, 500 Foothill Boulevard, Salt Lake City, UT 84148, USA

MOORE, R.Y., Center for Neuroscience, University of Pittsburgh, Q1656 Biomedical Science Tower, Pittsburgh, PA 15261, USA

MROSOVSKY, N., Department of Zoology, University of Toronto, 25 Harbord Street, Toronto, Ontario, Canada M5S 1A1

PATTI, G., Università Cattolica del Cacro Cuore, Policlinico A. Gemelli, Instituto di Cardiologia, Largo F. Vito 1, 00168 Roma, Italia

PONS, M., Groups d'Etude de Physiologie et Physiopathologie Rénales, Faculté de Pharmacie, 91, rue Leyteire, 33000 Bordeaux, France

PORTALUPPI, F., Hypertension Unit36, University of Ferrara, Department of Internal Medicine, Via Savonarola 9, 44100 Ferrara, Italy

REDFERN, P.H., School of Pharmacy and Pharmacology, University of Bath, Claverton Down, Bath BA2 7AY, Great Britain

REINBERG, A.E., Unité de Chronobiologie, Chronopharmacologie et Chronothérapie, Fondation Adolphe de Rothschild, 29, rue Manin, 75940 Paris Cedex 19, France

List of Contributors

RENSING, L., Institut für Zellbiologie, Biochemie und Biotechnologie, Universität Bremen, Fachbereich 2, Leobener Straße, Gebäude NW 2, Postfach 33 04 40, 28334 Bremen, Germany

ROSENWASSER, A.M., Department of Psychology, University of Maine, Orono, ME 04469, USA

SMAALAND, R., Department of Oncology, Haukeland Hospital, University of Bergen, 5021 Bergen, Norway

SMOLENSKY, M.H., University of Texas-Houston, School of Public Health, Rm. 442, Health Science Center, P.O. Box 20186, Houston, TX 77225, USA

STRICKER-KRONGRAD, A., Equipe de Recherches Aliment et Comportement, INSERM U 308, Unite de Recherches sur les Mecanismes de Regulation du Comportement Alimentaire, 38, rue Lionnois, 54000 Nancy, France

TOUITOU Y., Faculté de Médecine, Pitié Salpètrière, Laboratoire de Biochimie Médicale, 91, Blvd. de l'Hôpital, 75013 Paris, France

VANIER, M.-C., Centre de Recherche, Hôpital du Saint-Sacrement, Sainte-Foy, Quebec, Canada G1K 7P4

WIRZ-JUSTICE, A., Psychiatric University Clinic, Wilhelm-Klein-Str. 27, 4025 Basle, Switzerland

WITTE, K., Centre of Pharmacology, Johann-Wolfgang-Goethe University, Theodor-Stern-Kai 7, 60590 Frankfurt/Main, Germany

Contents

CHAPTER 1

The Vertebrate Clock: Localisation, Connection and Entrainment
M.H. HASTINGS. With 4 Figures 1

A. Introduction 1
    I. Rhythms and Our Lives 1
    II. Circadian Rhythms and Internal Temporal Order 1
    III. Entrainment and the Phase Response Curve 3
B. Localisation of the Circadian Clock in Vertebrates 4
    I. Criteria for Identifying the Clock 4
    II. The Retina as a Circadian Oscillator 5
    III. The Pineal Gland as a Circadian Oscillator 6
    IV. The Suprachiasmatic Nuclei of the Hypothalamus
        as a Circadian Oscillator 8
C. Neuroanatomy of the SCN 9
    I. Internal Structure and Organisation of the SCN 9
    II. Afferent Connections of the SCN 11
    III. Outputs of the SCN 12
D. Entrainment of the SCN 13
    I. Light and Glutamate 13
    II. Glutamatergic Signals and Gene Expression in the SCN . . . 14
    III. Non-photic Entrainment of the Circadian System 17
    IV. Melatonin and Entrainment of the Circadian System 18
    V. Conclusion 20
References 21

CHAPTER 2

**Regulation of Cell Division Cycles by Circadian Oscillators:
Signal Transduction Between Clocks**
L.N. EDMUNDS, JR. With 10 Figures 29

A. Introduction 29
    I. Regulation of the Cell Division Cycle 31
    II. Cell Division Rhythmicity in Wild-Type
        and Mutant *Euglena* 32

B. Coupling Between Oscillator and Cell Cycle: Role of Cyclic AMP ... 33
  I. Perturbation of the Cell Division Rhythm by Cyclic AMP ... 34
  II. Upstream Analysis: Genesis of the Oscillation
      in Cyclic AMP 41
  III. Downstream Pathway: Cyclic AMP-Dependent Kinases ... 43
  IV. Model for Circadian Control of the Cell Division Cycle ... 47
C. Problems and Prospects: Interface with the Cyclin Clock 48
References 50

CHAPTER 3

Genetics and Molecular Biology of Circadian Clocks
L. RENSING. With 6 Figures 55

A. Introduction 55
  I. Oscillations and Genes 55
  II. Analytical Approaches to the Basic Circadian Oscillator ... 56
B. Genetics and Molecular Biology of the Clock 58
  I. Overview and Generalizations 58
  II. *Neurospora* 60
      1. Circadian Rhythm 60
      2. Clock Mutants 60
      3. Molecular Biology of *frq* 62
  III. *Drosophila* 64
      1. Circadian Rhythms 64
      2. Clock Mutants 65
      3. Molecular Biology of *per* 66
  IV. Mammals 69
  V. Conclusions 70
C. Zeitgeber Signal Pathways to the Clock 71
  I. *Neurospora* 72
  II. Molluscs 73
  III. Mammals 73
D. Output Signals from the Clock 73
References 75

CHAPTER 4

Chemical Neuroanatomy of the Mammalian Circadian System
R.Y. MOORE. With 7 Figures 79

A. Introduction 79
B. Entrainment Pathways 80
  I. Retinohypothalamic Tract 80
  II. Geniculohypothalamic Tract 83

III.	Other Inputs to the Suprachiasmatic Nucleus	85
C. Pacemaker-Suprachiasmatic Nucleus		85
I.	Suprachiasmatic Nucleus Organization	85
II.	Suprachiasmatic Nucleus Efferents	87
D. Conclusions		89
References		90

CHAPTER 5

Chronobiology of Development and Aging
E. Haus and Y. Touitou 95

A. Introduction		95
B. Developmental Chronobiology		97
I.	Biologic Rhythms in the Human Fetus	97
II.	Fetal Rhythms of Susceptibility to Toxic and Teratogenic Agents	97
III.	Rhythms in the Newborn and During Infancy	100
C. Chronobiology of Aging		100
I.	Changes in Sleep Patterns During Aging	100
II.	Body Temperature	101
III.	Blood Pressure and Heart Rate	104
IV.	Hypothalamic–Pituitary–Adrenal Axis	106
	1. Dehydroepiandrosterone and Aging	107
V.	Pineal Gland	109
VI.	Growth Hormone and Prolactin	110
VII.	Pituitary–Gonadal Axis	111
VIII.	Catecholamines	112
IX.	Age-Related Changes in the Pituitary–Thyroid Axis During Aging	113
X.	Clinical Chemical Variables and Aging	113
	1. Plasma Proteins and Volemia	114
X.	Hematology	115
D. Conclusions		116
References		122

CHAPTER 6

Rhythms in Second Messenger Mechanisms
K. Witte and B. Lemmer. With 4 Figures 135

A. Introduction		135
B. Principles of Transmembraneous Signal Transduction		135
I.	Adenylyl Cyclase Pathway	135
II.	Guanylyl Cyclase Pathway	137

| III. Phospholipase C Pathway 138
C. Rhythms in Signal Transduction
   in the Cardiovascular System 138
         I. Adenylyl Cyclase Pathway 138
        II. Guanylyl Cyclase Pathway 145
       III. Phospholipase C Pathway 145
D. Rhythms in Signal Transduction
   in the Central Nervous System 146
         I. Adenylyl Cyclase Pathway 146
        II. Guanylyl Cyclase Pathway 147
       III. Phospholipase C Pathway 151
E. Rhythms in Signal Transduction in Other Tissues 151
         I. Adenylyl Cyclase Pathway 151
        II. Guanylyl Cyclase Pathway 152
       III. Phospholipase C Pathway 152
F. Conclusions and Perspectives 153
References 153

CHAPTER 7

**5-Hydroxytryptamine and Noradrenaline Synthesis,
Release and Metabolism in the Central Nervous System:
Circadian Rhythms and Control Mechanisms**
K.F. MARTIN and P.H. REDFERN. With 2 Figures 157

A. Introduction 157
B. Circadian Rhythms in Turnover of Serotonin 158
         I. Tissue Levels of Serotonin 158
        II. Tryptophan Hydroxylase 160
       III. Tryptophan Availability 160
        IV. Tryptophan Uptake 161
         V. 5-Hydroxytryptophan Decarboxylase Activity 161
        VI. 5-Hydroxytryptamine Release and Metabolism 162
       VII. Firing Rate of Serotonergic Neurones 165
      VIII. 5-Hydroxytryptamine Autoreceptor Activity 165
        IX. What Is the Function of the Circadian Rhythm
             in 5-Hydroxytryptamine Turnover? 168
C. Circadian Rhythms in Noradrenaline Turnover 169
D. Conclusions 171
E. References 171

CHAPTER 8

Rhythms in Pharmacokinetics; Absorption, Distribution, Metabolism and Excretion
P.M. BÉLANGER, B. BRUGUEROLLE, and G. LABRECQUE.
With 13 Figures 177

A. Introduction 177
B. Absorption 178
    I. Parameters and Modifying Factors 178
    II. Inorganic Compounds and Macromolecules 179
    III. Organic Compounds 179
    IV. Factors Modifying the Temporal Variation
        in Gastrointestinal Absorption 181
        1. Solubility 181
        2. Pharmaceutical Formulation 182
        3. Secretion and Motility 182
        4. Blood Flow 183
C. Distribution 185
    I. Plasma Protein Binding 185
    II. Binding to Erythrocytes 187
D. Biotransformation 188
    I. Oxidation: The Cytochrome P-450 Monooxygenase 188
    II. Conjugations 192
        1. Conjugations to Acetate, Glucuronic Acid and Sulfate . 192
        2. Conjugation to Glutathione 193
    III. Factors Related to Biotransformation 195
        1. Hepatic Blood Flow 195
        2. First-Pass Effect 196
E. Excretion 196
F. Conclusions and Perspectives 197
References 199

CHAPTER 9

Progress in the Chronotherapy of Nocturnal Asthma
M.H. SMOLENSKY and G.E. D'ALONZO. With 19 Figures 205

A. Introduction 205
B. Day-Night Pattern of Acute Asthma 205
C. Early, Late, and Recurrent Asthma Reactions 207
D. The Chronobiology of Asthma 208
E. Pharmacotherapy of Asthma 211
F. The Chronopharmacology and Chronotherapy
   of Asthma Medications 212

G. β_2-Adrenergic Agonist Medications 213
H. Anticholinergic Agents 219
I. Theophylline 221
J. Chronotherapy of Glucocorticoids 227
K. Mast Cell Stabilizers 239
L. Conclusion 240
References 241

CHAPTER 10

Chronopharmacology of Cardiovascular Diseases
B. LEMMER and F. PORTALUPPI. With 2 Figures 251

A. Historical Background 251
B. Chronobiological Mechanisms of the Cardiovascular System 252
   I. Physiology and Pathophysiology
      of Blood Pressure Regulation 252
   II. Cardiovascular Haemodynamics 258
   III. Electrical Properties of the Heart 260
   IV. Pathophysiology in Coronary Heart Disease 261
C. Chronopharmacology of Hypertension 262
   I. β-Adrenoceptor Antagonists 262
      1. Pharmacological Characterization 262
      2. Clinical Data 262
   II. Calcium Channel Blockers 267
      1. Pharmacological Characterization 267
      2. Clinical Data 267
      3. Chronopharmacokinetics 269
   III. Converting Enzyme Inhibitors 270
      1. Pharmacological Characterization 270
      2. Clinical Data 270
   IV. Other Antihypertensives 271
   V. Chronopharmacology of Blood Pressure
      in Congestive Heart Failure 271
   VI. Conclusion 271
D. Chronopharmacology of Coronary Heart Disease 272
   I. β-Adrenoceptor Antagonists 272
   II. Calcium Channel Blockers 278
   III. Organic Nitrates 280
      1. Pharmacological Characterization 280
      2. Clinical Data 280
   IV. Drugs Affecting Coagulation 281
   V. Conclusion 282
E. Concluding Remarks 282
References 283

CHAPTER 11

Chronopharmacology of Anticancer Agents
F. LÉVI. With 10 Figures 299

A. Introduction 299
B. Experimental Chronopharmacology 299
    I. Toxicity Rhythms 299
    II. Chronopharmacokinetics 301
    III. Rhythms in Susceptibility of Target Tissues 303
    IV. Circadian Rhythms in Antitumor Efficacy 305
C. Clinical Chronopharmacology 307
    I. Rhythms in Target Tissues 307
    II. Chronopharmacokinetics 309
D. Clinical Validation of Chronotherapy in Oncology 314
    I. Phase I Trials of Chronomodulated Chemotherapy ... 315
    II. Phase II Trials 321
    III. Chronotherapy of Metastatic Colorectal Cancer
        with 5-Fluorouracil, Folinic Acid and Oxaliplatin
        (Chrono-FFL) 322
E. Conclusions and Perspectives 324
References 325

CHAPTER 12

The Endocrine System and Diabetes
L. MEJEAN, A. STRICKER-KRONGRAD, and A. LLUCH.
With 4 Figures.. 333

A. Introduction 333
B. Chronobiology of Blood Glucose Levels 334
C. Chronobiology of Glucose Tolerance 335
D. Chronobiology of Plasma Insulin 335
E. Chronobiology of the Corticotropic Axis 337
F. Chronobiology of Feeding Behaviour 338
G. Chronopharmacological Approach to Diabetes 339
References 345

CHAPTER 13

Gastrointestinal Tract
J.G. MOORE and H. MERKI. With 21 Figures 351

A. Introduction 351
B. Gastrointestinal Motility Rhythms 351
    I. Pharmacological and Therapeutic Implications 354

    II. Other Circadian Rhythms
        Influencing Drug Bioavailability 355
C. Rhythms in Gastric Acid Secretion 357
    I. Pharmacological Implications 361
    II. Therapeutic Implications 367
D. Rhythms in Gastric Mucosal Defense 367
E. Circadian Influence
    in Cancer Chemotherapeutic Treatment Regimens 369
References 371

CHAPTER 14

The Pineal Gland, Circadian Rhythms and Photoperiodism
J. ARENDT. With 10 Figures 375

A. Introduction 375
B. Melatonin Production 376
C. Light Control of Melatonin Secretion 380
D. Control of Seasonal Cycles 383
E. Mechanism of Action of Melatonin
   in Control of Seasonal Rhythms 385
F. Human Seasonality 387
G. Melatonin as a Circadian Marker Rhythm 387
H. Effects of Melatonin on Circadian Rhythms 391
I. Melatonin and Core Body Temperature 395
J. Therapeutic Uses of Melatonin 395
K. Mechanism of Action of Melatonin
   in the Control of Circadian Rhythms 398
L. Melatonin Receptors 401
M. Melatonin Antagonists and Agonists 401
References 404

CHAPTER 15

Problems in Interpreting the Effects of Drugs on Circadian Rhythms
N. MROSOVSKY. With 8 Figures 415

A. Multiplicity of Sites for Chronotypic Action 415
B. Alternative Interpretations for Mode of Drug Action ... 416
    I. Gonadal Steroids and Changes in Period 416
    II. Benzodiazepines and Phase Shifts 418
    III. Periodicity After a Single Treatment 420
    IV. Melatonin Injections and Infusions 420

C. Tactics for Research on Behavioural Mediation 422
   I. Measuring Activity 422
   II. Altering Opportunities for Activity 422
   III. Confinement and Restraint 423
   IV. Testing Drugs In Vitro 424
D. Problems in Assessing Behavioural Mediation 424
E. Mimicking and Modulation of Photic Effects 425
   I. The Retinohypothalamic Tract Transmitter 425
   II. Period and Aschoff's Rule 427
   III. Sensitivity to Light and Phase Shifts 427
   IV. Behavioural Inhibition of Photic Shifts 427
F. Saline Injections 428
   I. Peripheral Injections 428
   II. Central Injections 429
G. Concluding Remarks 431
References 431

CHAPTER 16

Rhythms in Retinal Mechanisms
G.D. BLOCK and S. MICHEL. With 7 Figures 435

A. Introduction 435
B. Retinal Rhythmicity in *Xenopus* 435
   I. Pacemaker Localization 436
   II. Rhythm Generation 436
   III. Pacemaker Entrainment 436
   IV. Rhythm Expression 439
C. Retinal Rhythmicity in *Aplysia californica* 440
   I. Pacemaker Localization 441
   II. Rhythm Generation 441
   III. Pacemaker Entrainment 442
      1. Light-Induced Phase Shifts 442
      2. Serotonin-Induced Phase Shifts 444
   IV. Rhythm Expression 446
D. Retinal Rhythmicity in *Bulla gouldiana* 446
   I. Pacemaker Localization 447
   II. Rhythm Generation 447
   III. Pacemaker Entrainment 449
   IV. Rhythm Expression 450
E. Conclusions 451
References 451

CHAPTER 17

Circadian Rhythms and Depression: Clinical and Experimental Models
A.M. ROSENWASSER and A. WIRZ-JUSTICE. With 5 Figures 457

A. Introduction and Scope 457
B. Measurement of Circadian Rhythms 458
C. Circadian Rhythms and Depression: Clinical Research 459
    I. Phase-Advanced Circadian Rhythms 459
    II. Depression, Sleep, and Rhythmicity 460
    III. Seasonal Depression: A Circadian Rhythm Disorder? 461
    IV. Circadian Amplitude and Waveform 464
    V. Instability and Zeitgeber Coupling 465
    VI. Free-Running Rhythms 466
    VII. Activity Feedback 466
    VIII. Summary 467
D. Circadian Rhythms and Depression: Animal Models 468
    I. Psychopharmacology of Circadian Rhythms 468
      1. Phase-Shifting and Free-Running Period 468
      2. Antidepressants 471
      3. A Pharmacogenic-Developmental Model 472
    II. Stress-Induced Depression Models 472
    III. Behaviorally Characterized Inbred Strains 473
      1. Selection for Hypertension 473
      2. Selection for Drug Sensitivity 474
      3. Inbred Mice Differing in Affective Behavior 474
    IV. Ablation-Based Depression Models 475
      1. Olfactory Bulbectomy 475
      2. Thyroidectomy 475
    V. Animal Models of SAD? 476
    VI. Behavioral Feedback Effects 476
    VII. Summary 477
E. Conclusions 479
References 479

CHAPTER 18

Chronobiology and Chronopharmacology of the Haemopoietic System
R. SMAALAND. With 8 Figures 487

A. Outline of Haemopoiesis 487
B. The Haemopoietic System and the Clinician 491
C. Circadian Aspects of the Haemopoietic System 494
    I. Circadian Aspects of Proliferative Activity
       in Murine Bone Marrow 494
    II. Circadian Stage-Dependent Cytotoxicity
       of Murine Bone Marrow 498

III.	Circadian and Circannual Proliferative Activity in Healthy Human Bone Marrow	499
IV.	Circadian Variation in Bone Marrow DNA Synthesis in Cancer Patients	504

- D. Regulation of Haemopoietic Circadian Rhythms 507
- E. Chronopharmacological Aspects of the Haemopoietic System ... 508
- F. Disease-Dependent Chronotherapeutic Effects and Consequences 511
- G. Marker Rhythms for Proliferative Activity and Cytotoxicity of the Bone Marrow 513
- H. Chronobiology of Peripheral Blood in Health and Disease 513
  - I. Total White Blood Cells 514
  - II. Neutrophil Leucocytes 515
  - III. Lymphocytes 515
  - IV. Eosinophil Leucocytes 516
  - V. Natural Killer Cells 516
  - VI. Platelets 517
- I. Cyclic Haemopoiesis 517
- J. Scientific and Clinical Implications 519
- References 522

CHAPTER 19

Chronobiology of the Haemostatic System
F. ANDREOTTI and G. PATTI. With 2 Figures 533

- A. The Haemostatic System and Its Laboratory Evaluation 533
  - I. The Endothelium 533
  - II. Vasomotion 533
  - III. Platelets 534
  - IV. Coagulation and Viscosity 535
  - V. Fibrinolysis 536
- B. Chronobiology of the Haemostatic System 537
  - I. Circadian Variation of Endothelial Cell Function 537
    1. Endothelium-Related Factors in Plasma 537
    2. Effect of Endothelial Removal 539
    3. Endothelial Reaction to Stimuli 539
  - II. Circadian Variation of Vasomotor Function 539
    1. Coronary Artery Tone 539
    2. Forearm Arterial Tone 539
    3. Response to Constrictor or Dilator Stimuli 540
  - III. Circadian Variation of Platelet Function 541
    1. Platelet Adhesiveness and Serotonin Content 541
    2. In Vivo Platelet Activation and Aggregation 541
    3. Platelet Aggregation In Vitro 542
    4. Platelet α_2-Adrenoceptor Activity 543

	IV. Circadian Variation of Coagulation	544
	1. Global Measures of Coagulation	544
	2. Specific Coagulation-Related Factors	545
	3. Blood Viscosity	546
	V. Circadian Variation of Fibrinolytic Activity	546
	1. Global Assays of Fibrinolysis	546
	2. Individual Fibrinolytic Factors	547
	3. Possible Determinants of the Circadian Variation of PAI-1	548
C. Conclusions		549
References		550

CHAPTER 20

New Trends in Chronotoxicology
J. CAMBAR and M. PONS. With 9 Figures 557

A. Introduction	557
B. General Chronotoxicity	558
I. Chemical Toxic Agents	558
II. Nonchemical Toxic Agents	562
C. Chronotoxicological Examples in Nonrodents	563
D. Organ Specific Circadian Chronotoxicity	563
I. Liver Chronotoxicity (Chronohepatotoxicity)	564
II. Ear Chronotoxicity	565
III. Gastric Chronotoxicity	567
IV. Renal Chronotoxicity	568
1. Heavy Metal Chrononephrotoxicity	568
2. Antibiotic Chrononephrotoxicity	570
3. Cyclosporine Chronotoxicity	573
E. Chronesthesy as a Mechanistic Approach to Chronotoxicology	578
F. Circannual Chronotoxicology	580
G. General Conclusions	582
References	583

CHAPTER 21

**Chronopharmacology of H_1-Receptor Antagonists:
Experimental and Clinical Aspects (Allergic Diseases)**
A.E. REINBERG. With 7 Figures 589

A. Histamine and H_1-Receptor Antagonists: The Conventional Pharmacological Approach	589
B. Skin Reactivity to Histamine and Allergens	590
C. Lung Reactivity to Histamine and Allergens	592

D. Dosing-Time-Dependent Changes
in the Acute H_1-Receptor Antagonist Effects
of Four Agents 592
    I. Cyproheptadine 592
    II. Clemastine and Terfenadine 595
    III. Mequitazine 596
        1. Skin Tests 596
        2. Psychopharmacologic Tests and Performance 596
    IV. Remarks 596
E. Chronopharmacokinetics of Mequitazine 597
F. Chronesthesy of Mequitazine with Reference
to Chronic Administration 599
    I. Skin Tests 599
    II. Bronchial Tests and Temperature 599
    III. Psychopharmacological and Performance Tests 599
    IV. Remarks 601
G. Chronoepidemiology of a Population
of 765 Allergic Patients 601
    I. Circadian and Circannual Rhythms
       in Allergic Symptoms 601
H. Chronotherapy of Allergic Rhinitis with Mequitazine ... 603
    I. Chronotolerance 603
    II. Chronoeffectiveness 603
    III. Remarks 603
I. General Comments 604
References 605

CHAPTER 22

Local Anaesthetics
B. BRUGUEROLLE. With 3 Figures 607

A. Introduction 607
B. Experimental Chronopharmacology of Local Anaesthetics . 607
    I. Time Dependency in Toxicity 607
    II. Time Dependency in Kinetics 610
C. Clinical Pharmacology of Local Anaesthetics 611
    I. Basis for Clinical Chronopharmacology 611
    II. Time Dependency in Efficiency 611
    III. Time Dependency of Pharmacokinetics in Man 612
D. Possible Mechanisms Involved 614
E. Conclusions 616
References 616

CHAPTER 23

Biological Rhythms in Pain and Analgesia
G. LABRECQUE, M. KARZAZI, and M.-C. VANIER. With 2 Figures ... 619

A. Introduction	619
B. Physiological Mechanisms of Pain	620
I. Pain Conduction Pathways	620
II. Neurotransmitters and Pain Perception	621
III. Measurements of Pain	622
1. The Pain Questionnaires	622
2. The Rating Scales	623
C. Chronobiology of Pain	624
I. Biological Rhythms in Pain Control Mechanisms	624
II. Biological Rhythms of Pain	628
1. Pain Studies in Experimental Models	628
2. Pain Studies in Patients	632
D. Chronopharmacology of Analgesics	636
I. Animal Studies	636
II. Human Studies	637
1. Patients with Arthritic Pain	638
2. Patients with Postoperative and Cancer Pain	640
E. Conclusions	642
References	644

Subject Index ... 651

CHAPTER 1
The Vertebrate Clock: Localisation, Connection and Entrainment

M.H. HASTINGS

A. Introduction

I. Rhythms and Our Lives

Our lives are based around a series of recurrent cycles. Some of these cycles reflect changes in our physical environment, such as the solar day or the passage of the seasons, whilst others are social, for example the alternation between the working week and the weekend/Sabbath, or the imposed schedules of shift work. We are adapted to these cycles and function effectively because our metabolism, physiology and behaviour undergo regular rhythmic changes which are generated by an internal "body clock" synchronised to our environment (ASCHOFF 1981; ASCHOFF et al. 1971; WEVER 1979; MOORE-EDE et al. 1983; FOLKARD et al. 1985). Should this synchrony break down, our mental and physical performance is impaired and we experience a generalised disaffect (MOORE-EDE et al. 1983; VAN CAUTER and TUREK 1986; EHLERS et al. 1988; MOORE 1991; VAN DEN HOOFDAKKER 1994). Such desynchronisation may occur for two basic reasons. First, the internal processes of the clock which generates and controls bodily rhythms may be disrupted, for example due to disease or as a consequence of old age. Second, when our habits and routines are forcibly altered, for example during jet lag or irregular working schedules, the environment provides conflicting information and the regular progression of internal rhythms may be suspended (Chap. 14, this volume). If we are to understand how we are synchronised to the environment and so identify novel therapies to overcome the problems associated with desynchrony, it is necessary to examine the internal processes which generate biological rhythms and to describe how they are influenced by external cues. The aim of this chapter is to review what is known of the neural basis of the vertebrate clock, how it controls daily rhythms and how it responds to physical and social cues.

II. Circadian Rhythms and Internal Temporal Order

The most obvious daily rhythm is the cycle of sleep and wakefulness, but this is only one of a massive array of physiological, behavioural and metabolic rhythms. When monitored over the 24-h day, it is the exception rather than the rule that a particular variable remains constant. Rather than holding an unchanging set-point, our internal state is permanently in a state of flux (Fig. 1).

For example, prior to falling asleep, core body temperature declines independently of any change in the level of locomotor activity (CZEISLER et al. 1989). This spontaneous decline in body temperature is thought to facilitate the onset of sleep. During nocturnal sleep, the pineal gland is activated to secrete the hormone melatonin (Chap. 14, this volume), which mediates a number of nocturnal processes, including a further suppression of core body temperature

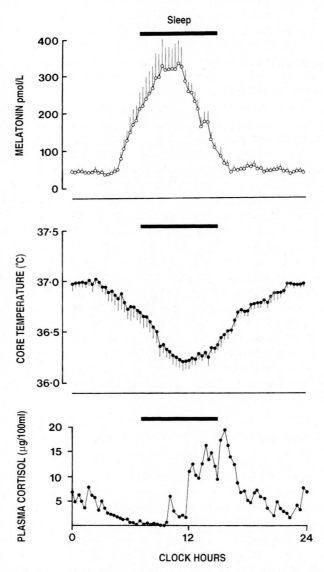

Fig. 1A–C. Examples of circadian rhythms in humans. **A** serum melatonin concentrations; **B** Core body temperature **C** serum cortisol levels. *Shaded area* represents sleep. [**A** and **B** redrawn from CAGNACCI et al. (1993), **C** redrawn from WEITZMAN et al. (1975)]

(CAGNACCI et al. 1992). Naps during the daytime, although they may lead to prolonged periods of sleep, are not associated with the secretion of melatonin, demonstrating the importance of internal timing rather than sleep onset per se in activating the pineal gland. A second endocrine profile with a very pronounced daily cycle is the secretion of corticosteroids by the adrenal gland. In man, late in the night prior to waking the hypothalamo-pituitary-adrenal axis is activated and circulating levels of corticosteroids begin to rise, preparing us for the rigours of the day. Corticosteroid levels peak in the early morning and then decline across the day (MOORE-EDE et al. 1983; LINKOWSKI et al. 1993; RODEN et al. 1993). Corticosteroid levels are also rhythmic in nocturnal rodents but, as might be anticipated, the peak occurs at the beginning of the night, thus preserving the temporal association between adrenocortical function and activity (ALBERS et al. 1985; HASTINGS 1991)

An important feature of all of these daily rhythms is that if subjects are held in temporal isolation, in an unchanging environment, the rhythms do not disappear as would be predicted if they were a direct response to the environmental cycle. Instead, they continue with comparable amplitude and they maintain their temporal relationships to each other and to the cycle of sleep and wakefulness. This observation confirms that they are driven by an internal timing process, an endogenous clock. However, temporal isolation does have one significant effect – the rhythms now run with a period length that is rarely exactly equal to 24 h (it may be slower or faster, depending upon the individual and the experimental context). The rhythms are therefore called circadian to reflect this approximate (i.e. circa-) daily periodicity. During a free-run, the endogenous clock defines within the circadian cycle two major phases: subjective night, when processes normally associated with darkness take place, and subjective day, when typically diurnal activities occur. As a result of the circadian free-run, whenever temporal isolation is maintained the internal time of the subject drifts progressively further away from external time. Under normal, non-experimental conditions, however, this tendency to drift is corrected by environmental cues which reset the clock on every cycle so that it and its dependent rhythms are synchronised (entrained) to external time, running with a period of exactly 24 h.

III. Entrainment and the Phase Response Curve

Entrainment of the circadian system to our external world depends on two events (PITTENDRIGH 1981). First, the period has to be set exactly to 24 h, and this is achieved by small daily readjustments of the clock. Second and as a consequence of this readjustment, the rhythms are held in a particular temporal order or phase relationship, so that nocturnal activities are restricted to the night and diurnal activities to the day. The most potent entraining cue is the light-dark cycle, and the resetting effects of light can be seen when subjects in temporal isolation are exposed to brief pulses of bright light (CZEISLER et al. 1989; ELLIOTT 1976; DAAN and PITTENDRIGH 1976). The effect of light pulses depends on when they are presented. Light delivered during subjective day is without effect, even though an individual may be in continuous darkness. In contrast, light presented at the beginning of subjective night will delay the clock and its rhythms, whereas light pulses given towards the end of subjective night advance the clock. When the magnitude and direction of the shift are plotted as a function of when within the circadian cycle the pulse was presented, a phase response curve (PRC) is produced. For every species the PRC to light has a characteristic shape, a signature (DAAN and PITTENDRIGH 1976). Nevertheless, the general format is as described here with a dead-zone in subjective day and zones of delay and advance shifts in early and late subjective night, respectively (DAAN and PITTENDRIGH 1976; SMITH et al. 1992). Other stimuli capable of entraining the circadian clock, such as arousal or various drugs (see below), can also yield a PRC, but the shapes are often very different to that of the photic PRC, showing shifts during subjective day, implying a very different mode of action on the clock (SMITH et al. 1992; see Chap. 15, this volume). The phase dependence of the response to identical cues demonstrates that the overt rhythms are driven by a true oscillator. Characterisation of the PRC is therefore an essential requirement of any study seeking to investigate the anatomical or neurochemical basis to circadian entrainment.

B. Localisation of the Circadian Clock in Vertebrates

I. Criteria for Identifying the Clock

Circadian rhythms are a ubiquitous feature of eukaryotes, both unicellular and multicellular, and there is now evidence that prokaryotic cyanobacteria can also express circadian cycles (KONDO et al. 1994). It is therefore likely that all cells have some individual capacity to express spontaneous rhythms. Nevertheless, a series of studies over the last 30 years have shown that several structures in vertebrates are specialised to operate as circadian clocks. As pointed out by HERBERT (1991), this ability to represent time has a number of unique features when compared to other sensory systems. There is no external representation of time in the form of chemical, mechanical or electromagnetic energy and so no specialised receptor apparatus can be developed. Nevertheless the body generates an internal representation of time, which has to be cued by sensory events correlated with external time such as changes in light intensity. Second, the clock has to be independent of changes in metabolic rate, body temperature and other functions which vary over the circadian cycle: a clock which ran slower when we became cooler would be useless. The biochemistry of the clock must therefore be protected from the usual changes which lead to a doubling of rate for an increase of 10 °C. A final general point to make is that, given the close association between circadian time-keeping and the light-dark cycle, it is not surprising that in vertebrates (as well as other taxa, e.g. molluscs) there is a close phylogenetic association between photoreceptive structures and the circadian clock.

In seeking to test whether a particular structure is the circadian clock, a number of criteria need to be fulfilled. First, extirpation of the structure should compromise the ability of individuals to express free-running circadian rhythms. However, this criterion may not be adequate in situations where a number of clock structures operate together, such that loss of one alone may not affect circadian function. The second criterion is that some or all of the metabolic activities of the structure should exhibit circadian rhythmicity when isolated from the rest of the animal, both in vivo and in vitro. Third, restoration of the structure by transplantation should reinstate free-running rhythms in lesioned individuals. Fourth, the putative clock should be able to respond to cues known to affect entrainment, and chemical or electrical stimulation of the structure should lead to phase shifts comparable to those observed in the PRC. Several structures in the vertebrate brain have fulfilled some or all of these criteria.

II. The Retina as a Circadian Oscillator

It is quite clear that the retina of vertebrates expresses a number of robust circadian rhythms of very large amplitude and tightly defined phase (McCormack and Burnside 1992; Terman et al. 1991, 1993; Reme et al. 1991). These include changes in cell structure, e.g. shedding of the outer discs of the photoreceptor and accompanying phagocytic activity, and also enzymatic rhythms such as the nocturnal increase in activity of serotonin N-acetyl transferase (SNAT) which leads to a circadian rhythm in the local synthesis of melatonin. When held in isolation in vitro, retinal cups of *Xenopus* continue to express a rhythm of SNAT activity which persists for several days with a clear circadian periodicity, confirming the existence within the tissue of a circadian clock (Besharse and Iuvone 1983; Cahill and Besharse 1991). A similar conclusion has arisen from studies of the eye of quail. Both enucleation and bilateral section of the optic nerve abolish circadian rhythms of activity, indicating an essential role for the eye in generating such rhythms (Underwood et al. 1990a,b). However, the eyes of birds subjected to section of the optic nerve continue to express a circadian rhythm of synthesis of melatonin, and this rhythm can be entrained by light-dark cycles imposed on each eye, indicating that the clock driving the rhythm is present within the eye itself. Furthermore, the avian retina is able to maintain a circadian rhythm of expression of mRNA encoding the iodopsin cone pigment when held in dispersed primary cultures (Pierce et al. 1993). This finding is consistent with the hypothesis that the photoreceptors contain the clock mechanism, a conclusion also reached by Cahill and Besharse (1993) with the retinal clock of *Xenopus*.

The eyes of mammals may also contain a circadian oscillator. In rats, rhythms of disc-shedding and visual acuity survive lesions of the suprachiasmatic nuclei, the principal circadian oscillator in mammals (see below): these lesions disrupt other forms of circadian rhythmicity (Terman et al. 1993). Moreover, the free-running circadian rhythm of visual sensitivity in suprachiasmatic nuclei of the hypothalamus (SCN)-lesioned rats can be phase-shifted by light pulses delivered in subjective night, indicating that the clock in the eyes is photoreceptive (Reme et al. 1991).

Recently, the existence of endogenous circadian time-keeping in the mammalian retina has been demonstrated using *in vitro* cultures of the hamster retina (TOSINI and MENAKER 1996). Under such conditions the production of melatonin follows a very clear circadian cycle. Moreover, the period of the retinal oscillator is temperature-compensated, a cardinal feature of circadian timers in lower species. An even more intriguing observation is that the period of the retinal rhythm is shortened to 20 h in the *tau* mutant hamster, compared to 24 h in wild-type. This indicates that comparable molecular genetic mechanisms underlie the clock of both SCN (see below) and retina, highlighting the conservation of function that appears to be a theme in circadian timing in vertebrates.

III. The Pineal Gland as a Circadian Oscillator

The pineal gland shares many developmental features with the lateral eyes, and phylogenetically can be viewed as an accessory photoreceptive organ (VOLLRATH 1981; OKSCHE 1983, 1991; OKANO et al. 1994). Its metabolic activities are intimately linked to the light-dark cycle, and in lower vertebrates it has cellular components comparable to the photoreceptors and associated neurons and glia of the retina. In higher vertebrates, the exteroceptive sensory function is reduced to a greater or lesser degree, being completely absent in adult mammals, although there may be residual photosensitivity in some species of bird (TAKAHASHI et al. 1989; OKANO et al. 1994). The principal function of the pineal eye of lower vertebrates is probably to monitor changes in light intensity over 24 h rather than to mediate object recognition, and so it has the potential to make an important contribution to circadian entrainment.

As in the lateral eye, the pineal gland secretes melatonin exclusively at night (AXELROD 1974; Chap. 14, this volume). In all species it is therefore the source of an important temporal cue: an endocrine representation of darkness. In some vertebrates, this rhythm of synthesis and secretion is controlled by an endogenous circadian oscillator so that when held in vitro the pineal continues to exhibit robust free-running rhythms (FALCON et al. 1992; TAKAHASHI et al. 1989). There are, however, marked differences in the expression of this pineal clock because although it can be seen in some species of fish, lizard and bird, within these groups the pineal of closely related species fails to exhibit circadian rhythms of secretion of melatonin (FALCON et al. 1992; MENAKER et al. 1981; MENAKER 1985; UNDERWOOD 1989). On transfer to continuous darkness the gland of some species secretes melatonin continuously at a high rate, whereas in others secretion declines rapidly to basal levels. A comparative study of these glands could therefore provide powerful experimental preparations with which to identify the molecular and biochemical components of the clock machinery. This utility of the pineal gland as an in vitro preparation with which to investigate the clock is emphasised by the fact that the gland can be entrained by cycles of light and darkness (confirming the residual photosensitivity of the organ), that it can be phase-shifted by various drugs to generate PRCs and that circadian rhythms of secretion of melatonin persist in dissociated cultures, in which the pinealocytes are isolated from each other (TAKAHASHI et al. 1989; ZATZ 1992). The basis to the circadian clock therefore seems to be contained within individual cells, rather than emerging as a property of a multicellular organisation. In this, the vertebrate clock shares properties of the clock of unicellular organisms.

Consistent with the variable expression of endogenous circadian rhythms in the pineal of lower vertebrates, removal of the gland can have a variety of effects on the circadian system of different species. These range from no effect at all, to impaired entrainment, disorganised rhythms and complete arrhythmicity (UNDERWOOD 1989). A particularly striking effect occurs in sparrows, where removal of the gland obliterates the circadian pattern of rest and activity, the animals remaining active throughout their exposure to continuous darkness (ZIMMERMAN and MENAKER 1979). In this species, a further criterion for identification of the gland as an important circadian oscillator is fulfilled. If the gland is grafted to pinealectomised, arrhythmic recipients, they start to express circadian rhythms of activity and rest. Moreover, the phase of the restored rhythm follows that of the donor animals, confirming that the grafted gland is providing a unique circadian signal, rather than providing a permissive environment, gating the expression of a clock located elsewhere in the recipient (which would not share the phase of the donor). Finally, it is important to note that in the absence of neural reconnection, the transmission of the circadian signal from the grafted pineal must occur by an endocrine pathway, with melatonin being the most likely candidate. Such a role is supported by results in other species, where administration of exogenous melatonin can disrupt circadian rhythms of activity and rest (UNDERWOOD 1989) and the role of melatonin as an effector of the clock is discussed in Chap. 14, this volume. The sites of action of melatonin as an entraining agent are yet to be identified, although autoradiographic studies of melatonin-binding sites in the avian brain have revealed a large number of potential targets, including areas of the brain which receive retinal inputs, midbrain sensory relay areas and the SCN (RIVKEES et al. 1989; CASSONE and BROOKS 1991). The important contribution of the avian SCN to circadian function is underlined by the observation that lesions of this area in the house sparrow and some other species abolish free-running circadian activity rhythms (EBIHARA and KAWAMURA 1981; TAKAHASHI and MENAKER 1982; SIMPSON and FOLLETT 1981), and metabolic activity in the SCN is sensitive to exogenous melatonin. The circadian system of lower vertebrates may therefore be based on an interaction between retinal, pineal and hypothalamic oscillators, linked both by neural and neuroendocrine pathways (MENAKER et al.1981).

In higher vertebrates the principal role of the pineal is as a secretory organ, its major product being melatonin. Nevertheless, the non-sensory pineal of mammals does continue to express a number of molecules associated with the phototransduction pathway, e.g. S-antigen, transducin and receptor kinases, although it is unclear whether these molecules are redundant or serve some other function in the pinealocyte (KORF et al. 1986; SOMERS and KLEIN 1986; FOSTER et al. 1989a,b). In contrast to at least some lower vertebrates, the mammalian pineal gland does not contain an endogenous circadian oscillator. When isolated in vitro the secretion of melatonin remains at basal levels, although it can be activated by adrenergic stimuli (AXELROD 1974; Chap. 14, this volume). In vivo the synthetic and secretory activities of the mammalian pineal are regulated by the circadian system, so that the secretion of melatonin is exclusively nocturnal. The circadian signal controlling the gland is conveyed via its adrenergic, sympathetic innervation, which is regulated ultimately by the SCN.

IV. The Suprachiasmatic Nuclei of the Hypothalamus as a Circadian Oscillator

In a number of experimental animals, including birds and rodents, removal of the lateral eyes and pineal gland does not prevent the expression of free-running circadian rhythms, demonstrating the existence of a circadian clock elsewhere. Early lesion studies in rats indicated that a light-entrained clock was located in the hypothalamus (RICHTER 1967). By using neuronal tracing methods to reveal retinoreceptive areas of the hypothalamus, the SCN (Fig. 2) were identified as this putative clock (MOORE and LENN 1972; MOORE 1973). These ovoid nuclei, which contain about 8000 neurons, sit adjacent to the base of the third ventricle immediately above the optic chiasm (VAN DEN POL 1991a). Their morphology is conserved across mammals, including man (CASSONE and MOORE 1988; CASSONE et al. 1988; SWAAB et al. 1990; MAI et al. 1991; MOORE 1992a), although there is still discussion as to which elements of the avian

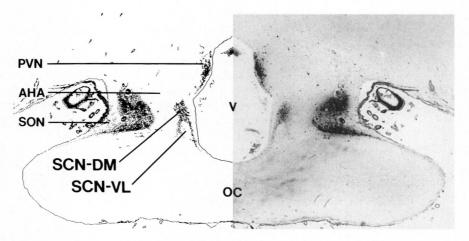

Fig. 2. Coronal section of human hypothalamus immunostained for neurophysin to identify the suprachiasmatic nuclei. Prepared from material generously supplied by Dr. M.V. Sofroniew, Department of Anatomy, University of Cambridge. *OC*, optic chiasm; *PVN*, paraventricular nucleus; *SCN-DM*, dorsomedial suprachiasmatic nucleus (neurophysin-immunoreactive); *SCN-VL*, ventrolateral suprachiasmatic nucleus (not neurophysin-immunoreactive); *SON*, supraoptic nucleus; *V*, third ventricle; *AHA*, anterior hypothalamic area

hypothalamus are the homologue of the mammalian SCN (Cassone and Moore 1987; Cassone 1988; Cassone and Brooks 1991).

The importance of the SCN was illustrated by the observation that their destruction in experimental animals leads to arrhythmicity in a number of circadian functions, including activity/rest cycles, corticosteroid secretion and oestrous cyclicity (Rusak and Zucker 1979). Similarly lesions of the suprachiasmatic region in people can be associated with disturbances of sleep/wake and neuroendocrine cycles (Cohen and Albers 1991). Furthermore, destruction of the retinal pathways beyond the SCN do not affect circadian entrainment, even though such animals are behaviourally blind, indicating that the retinohypothalamic pathway, which enters the SCN from its ventral aspect, is the mediator of photic entrainment of the clock (Morin 1994). The endogenous clock-like properties of the SCN have been examined in several ways. For example, in vivo the nuclei exhibit circadian rhythms of glucose uptake and electrical activity, the latter surviving disconnection from the rest of the brain (Schwartz 1991). These rhythmic properties are also expressed in vitro, the isolated SCN generating free-running rhythms of peptidergic secretion and electrical activity which can be phase shifted by electrical and chemical cues (Earnest and Sladek 1987; Watanabe et al. 1993; Ding et al. 1994). Moreover, dispersed SCN neurons grown on cover slips containing microelectrode arrays have been shown to exhibit autonomous circadian cycles of electrical activity (Welsh et al. 1995). Confirmation of the role of the SCN as a circadian clock in mammals was provided by neural transplantation studies in rodents. When foetal hypothalamic tissue containing the SCN is grafted into the third ventricle of SCN-lesioned adult recipients, the arrhythmic pattern of activity and rest is replaced by a free-running circadian rhythm within a few days (Lehman et al. 1987). Moreover, when the donor and recipient animals are of different genotypes with contrasting circadian periods, the restored rhythm follows the period of the donor animal, demonstrating that the circadian signal being expressed is generated by the grafted tissue block, rather than another site within the brain of the recipient (Ralph et al. 1989; Ralph 1991).

C. Neuroanatomy of the SCN

I. Internal Structure and Organisation of the SCN

The SCN are remarkable structures. Few other parts of the brain exhibit such an obvious relationship between a discrete structure and a single clearly defined function. Moreover, the signal generated by this relatively tiny number of cells controls the most profound alteration of neural function: the interchange between sleep and wakefulness. The development of modern methods of neuroanatomical tract-tracing and the explosion in chemical neuroanatomy have made it possible to identify discrete functional and anatomical domains in the nucleus. In common with other hypothalamic regions, the SCN have a diverse complement of peptidergic neurons, including cells immunoreactive for Arginine vasopressin (AVP), Gastrin-releasing peptide (GRP), Vasoactive intestinal polypeptide (VIP), somatostatin and neurotensin (CARD and MOORE 1984; VAN DEN POL 1991a; MAI et al. 1991; see Chap. 4, this volume). In addition, they have a very dense plexus of astrocytes with levels of immunoreactivity for glial fibrillary protein (GFAP) markedly higher than those of the surrounding hypothalamus (LAVAILLE and SERVIERE 1993). Whereas astrocytes are found in all parts of the nucleus, peptidergic neurons are arranged into specific subdivisions. In the rat and Syrian hamster the principal subdivision is into ventrolateral and dorsomedial SCN, characterised by VIP-immunoreactive (ir) and GRP-(ir) cells, and AVP-ir cells, respectively. In addition to their peptidergic transmitters, the majority of SCN neurons are also GABAergic, and their principal connections are intrinsic to the nucleus, both within and between the subdivisions (MOORE and SPEH 1993). A comparable topography with the same cellular phenotypes is seen in the human SCN (MAI et al. 1991; MOORE 1992a).

It remains to be determined precisely which components of the SCN generate a circadian signal (MILLER 1993; VAN DEN POL and DUDEK 1993). Successful SCN grafts always contain some VIP-ir and AVP-ir perikarya, suggesting that these cells or other cells closely associated with them are critical elements of the clock (LEHMAN et al. 1987). However, it has not been possible to deplete the SCN systematically of any particular cell type prior to grafting to test such a hypothesis. In addition to neurons, astrocytes have also been implicated in control of the clock (VAN DEN POL 1991a; MILLER 1993; MORIN 1994). Indirect evidence comes from studies of the pattern of GFAP-ir in the SCN, which shows a pronounced circadian cycle: during the early subjective night, astrocytic processes are condensed and individual astrocytes clearly defined (LAVIALLE and SERVIERE 1993). However, at other stages of the cycle, GFAP-ir becomes more extensive, the astrocytes forming a network throughout the SCN, clearly delineating its borders. The functional role of such changes is not clear. They may be associated with the well-characterised metabolic cycle of the SCN, the astrocytic network supporting greater levels of oxygen utilisation and glucose uptake during subjective day (SCHWARTZ 1991). Another important point is that the clock is able to continue to oscillate, and also be phase shifted by drugs in the presence of tetrodotoxin, which will block action potentials (SCHWARTZ et al. 1987; SHIBATA and MOORE 1993). Moreover, the clock is established in the foetus prior to synaptogenesis, suggesting that communication between different cellular components of the clock may be mediated by non-synaptic transmission (REPPERT 1992). Alternatively, communication between cells may not be particularly important for circadian timekeeping over the relatively short intervals employed in the studies with tetrodotoxin, and it is an important question as to whether, as appears to be the case in the retina and in the pineal, the clock of the SCN is able to operate autonomously in individual cells. The most recent studies by WELSH et al. (1995) using cultures of dissociated SCN tissue suggest that this is indeed the case.

II. Afferent Connections of the SCN

The principal entraining cue to the SCN, light and darkness, is conveyed by the retinohypothalamic tract (RHT), which emerges from the optic chiasm and enters the ventral and lateral regions of the nucleus (MORIN 1994). Although early studies using less sensitive techniques based on autoradiography of transported tritiated amino acid indicated an exclusive input to the SCN, more sensitive modern tracing methods have shown that the RHT provides an extensive if not abundant input to a number of hypothalamic areas, including the paraventricular nucleus and the anterior hypothalamic area (YOUNGSTROM et al. 1987; CARD et al. 1991). The role of these extensive extra-SCN connections is not clear, although it may be related to some of the masking effects of light on neuroendocrine function, e.g. the suppression of melatonin secretion by the pineal following nocturnal exposure to light (NELSON and TAKAHASHI 1991). Destruction of the retinal input to the SCN by undercutting of the nucleus to interrupt the RHT blocks photic entrainment of the circadian system (JOHNSON et al. 1988; MORIN 1994). Ultrastructural studies have shown that retinal fibres within the SCN synapse on a number of cell types in the ventral SCN, including both VIP-ir and non-VIP-ir perikarya, with glutamate as the principal afferent transmitter, although substance P and the glutamate derivative N-acetyl aspartylglutamate (NAAG) may also be involved in photic input (VAN DEN POL 1991b; VAN DEN POL and DUDEK 1993; CASTEL et al. 1993; TANAKA et al. 1993; DING et al. 1994).

A second major afferent system arises from the raphe nuclei of the midbrain (JACOBS and AZMITIA 1992). Serotonergic fibres and terminals are present in the SCN at a much higher density than in the surrounding hypothalamus. The major, if not exclusive, source of these fibres is from the median raphe nuclei, and they terminate predominantly in the ventral region of the nucleus, partially overlapping the area of retinal inputs (CARD and MOORE 1984; VAN DEN POL 1991a; MORIN 1994). The role of the serotoninergic innervation of the SCN is not yet known, although there is some evidence to associate it with entrainment by non-photic, arousing stimuli (see below) and also as a modulator (probably inhibitory) of photic responses in the SCN (MEIJER and GROOS 1988; MORIN and BLANCHARD 1991; SELIM et al. 1993; REA et al. 1994). Consistent with this latter view is the observation that serotonergic terminals synapse on VIP-ir cells which are known to be retinoreceptive (BOSLER and BEAUDET 1985). In addition, serotonergic agonists can attenuate the electrophysiological responses of the SCN to stimulation of the optic nerve (SELIM et al. 1993), and the light-induced phase advances of hamster activity rhythms are reduced if the animals are active during the interval of illumination (RALPH and MROSOVSKY 1992). Such activity might be expected to increase serotonergic activity in the forebrain, including the SCN.

The third major afferent system to the SCN arises from the intergeniculate leaflet (IGL) of the ventral lateral geniculate nucleus of the thalamus (MOORE 1992a,b; MORIN et al. 1992; see also Chap. 4, this volume). The NPY-ir fibres and terminals are located predominantly in the ventral SCN, overlapping the direct retinal input and some of the serotonergic afferents. Although the IGL receives retinal input, there is strong evidence to suggest that its NPY-ir projections to the SCN are involved in non-photic entrainment (see Chap. 15, this volume).

III. Outputs of the SCN

The major efferent pathways of the SCN have been mapped by using anatomical tracers and by reconstructing the route of neurochemically defined projections (WATTS 1991; KALSBEEK et al. 1993; MORIN et al. 1994). The principal route taken by efferent axons coursing out of the SCN is towards an area ventral to the paraventricular nucleus (PVN), the subparaventricular zone and its more caudal continuation into the retrochiasmatic area. Here, SCN fibres synapse on second-order cells which project extensively to the forebrain and brain stem and it is likely that the circadian signal is conveyed predominantly by these pathways (WATTS 1991; MORIN et al. 1994). The targets of the sub-PVN and retrochiasmatic area also receive a small number of direct inputs from the SCN. The most obvious of these more extensive projections can be traced in the form of AVP-ir, GRP-ir and VIP-ir fibres which run dorsally from the SCN and innervate the paraventricular thalamus, the subparaventricular and paraventricular hypothalamus, and the preoptic and dorsomedial hypothalamus (KALSBEEK et al. 1993). The contribution of individual pathways to the various manifestations of circadian rhythmicity remains an open question, although it does seem clear that dorsal SCN projections convey a circadian signal to the autonomic innervation of the pineal via a paraventriculospinal link (KLEIN et al. 1983; HASTINGS and HERBERT 1986). However, there may well be appreciable redundancy in the efferent pathways because lesions targeted at the sub-PVN and/or retrochiasmatic area do not consistently abolish circadian rhythms of activity and rest. Indeed, it remains to be determined whether neuronal connections are the only way in which the SCN may convey their circadian signal. In several studies, it has been shown that transplanted SCN tissue lodged some distance from the hypothalamus can restore circadian rhythms in lesioned hosts (LEHMAN et al. 1991). Given that few efferent fibres can be traced from such grafts, it has been suggested that the circadian signal is transmitted via paracrine factors in both the intact and the SCN-grafted animal (HAKIM et al. 1991). However, there are problems with such an explanation. First, although only a few fibres can be seen entering host tissue when identified using immunocytochemistry for one or other neuropeptide, this very specific technique will fail to identify the bulk of SCN efferent fibres. Second, a mystery remains about the nature of the circadian function restored. Whereas the graft reinstates a very robust cycle of activity and rest, endocrine rhythms including melatonin and corticosteroid secretion are not restored. If the graft were to use a global paracrine signal, a more general restoration of circadian functions might be anticipated. The time course for restoration is consistent with a neural as opposed to paracrine linkage. In vitro studies have shown that the SCN explants are capable of expressing a robust circadian rhythm of AVP secretion within one or two circadian cycles of culture (EARNEST and SLADEK 1987). In contrast, although foetal SCN grafts restore circadian activity rhythms relatively quickly, it still usually takes about 10 days or so (RALPH 1991; LEHMAN et al. 1991), an interval consistent with the re-establishment of neuronal circuitry. Finally, the success of the graft depends on the production of a lesion to the host brain: in the absence of tissue damage the periodicity of the graft cannot be identified in the circadian activity rhythm of the host animal (VOGELBAUM and MENAKER 1992; MENAKER and VOGELBAUM 1993). Although there are many possible interpretations of this phenomenon, one view is that to be effective, the graft has to make synaptic connections with host tissue, and that the postsynaptic sites are only revealed if some of the efferent fibres of the host SCN are destroyed. This debate has been given a further twist by the recent report that SCN grafts encapsulated in a semi-permeable membrane are capable of restoring circadian activity rhythms in SCN-lesioned hamsters (SILVER et al. 1996). The competence of the graft under circumstances where it would not be expected to be able to establish neural connections is a powerful argument in support of the paracrine hypothesis. Perhaps a factor normally released by synaptic activity is able to operate over longer distances under the conditions of the lesioned and grafted brain. Identification of such a factor is eagerly awaited.

D. Entrainment of the SCN

I. Light and Glutamate

Entrainment of the circadian clock does not require exposure to the full 24-h cycle of light and darkness. Animals held in constant darkness or dim light can be entrained by daily exposure to brief pulses of light, each pulse producing a phase shift which resets the period of the endogenous cycle to exactly 24 h (PITTENDRIGH 1981; SHIMOMURA and MENAKER 1994). The magnitude of the phase-shifting response depends on two factors: the circadian time at which the pulse is delivered, and the total flux of light, which is a product of the duration and irradiance of the pulse (TAKAHASHI et al. 1984; NELSON and TAKAHASHI 1991, 1992). In this way, the circadian system acts as a "photon counter", integrating the photic stimulus over appreciable intervals (microseconds to minutes). The integrative function of the circadian response is also apparent in the electrophysiological responses of SCN neurons to light, which are often of a sustained activation or inhibition rather than a short, phasic response as is so common in other visual areas (RUSAK and ZUCKER 1979; MEIJER 1991; MORIN 1994). In addition, there is a monotonic, non-linear relationship between the amplitude of phase shifts and the light-induced electrophysiological discharge in the SCN of hamsters (MEIJER et al. 1992).

There is now considerable evidence that glutamate is the principal transmitter of the RHT. Stimulation of the optic nerve releases glutamate in SCN slices (LIOU et al. 1986), and putative retinal terminals in the SCN are immunoreactive for glutamate (VAN DEN POL 1991b; CASTEL et al. 1993). Behavioural studies also indicate a critical role for glutamate, in so far as treatment with a variety of glutamatergic antagonists blocks the behavioural phase-shifting responses to light (REA et al. 1993a; COLWELL and MENAKER 1992; VINDLACHERUVU et al. 1992). Both N-methyl-D-aspartate (NMDA) and non-NMDA antagonists are effective, and it is likely that glutamate acts at a non-NMDA, ionotropic receptor (GluR) to depolarise the cell, thereby facilitating its action at the NMDA receptor (NMDAR) (KIM and DUDEK 1991). Immunocytochemical and in situ hybridisation studies have revealed the presence of a number of types and subtypes of glutamatergic receptor in the SCN, including both NMDAR and GluR (MIKKELSEN et al. 1993; VAN DEN POL et al. 1994a,b; HASTINGS et al. 1995). Of particular interest is the expression of the NMDAR2c subunit, which is not detectable elsewhere in the forebrain (MIKKELSEN et al. 1993; HASTINGS et al. 1995). Perhaps the most elegant demonstration of the importance of glutamate as an entraining cue has been provided by in vitro studies using SCN slices (SHIRAKAWA and MOORE 1994; DING et al. 1994). Local treatment of the SCN with glutamate is able to produce dose-dependent phase-shifts of the rhythm of electrical activity in the slice. When the responses to a fixed dose are compared at different circadian phases, the resultant PRC is directly comparable to that observed in vivo in response to light.

II. Glutamatergic Signals and Gene Expression in the SCN

There has been considerable progress in attempts to characterise the signal transduction cascade whereby light, in the form of glutamatergic cues, resets the clock. This is important for at least two reasons: first, elucidation of the pathway will identify possible avenues for intervention in cases of circadian dysfunction. Second, by following the pathway into the cell it may be possible, ultimately, to identify elements of the oscillatory mechanism driving circadian rhythms. In order to study the cascade, it is necessary to have markers for the response of individual SCN cells to light. This was provided by the demonstration that presentation of light pulses to rats and hamsters during the dark phase induced the expression of a number of immediate-early genes (IEGs), including *c-fos*, *egr-1* and *jun-B*, in the SCN (REA 1989; RUSAK et al. 1990; EBLING et al. 1991; VINDLACHERUVU et al. 1992; KORNHAUSER et al. 1992). This effect was spatially specific, in so far as induction occurred only in the retinorecipient zone of the SCN and was not observed in other retinal target sites. Moreover, it was temporally specific because when animals were held in continuous darkness or dim light, IEGs were induced only by light pulses delivered in subjective night (Fig. 3). This correlated well with the behavioural phase-shifting response: light pulses delivered in subjective day do not shift the clock nor do they induce IEGs in the SCN (GROSSE et al. 1995). One interpretation of these data is that the induced IEGs may contribute to re-ordering of the circadian programme, thereby executing a phase shift.

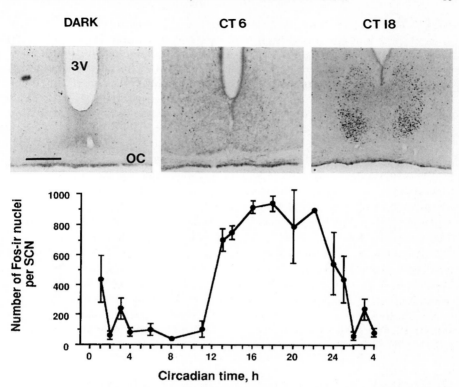

Fig. 3. Photic induction of *Fos* in the hamster SCN depends on circadian phase. *Upper panels*, coronal sections of the hypothalami of Syrian hamsters immunostained for *Fos*. The animals were held in constant dim red light and circadian time defined by activity onset, CT 12. Animals sampled in the dark do not express *Fos-ir* in the SCN. Similarly, there is no *Fos-ir* in the SCN of animals pulsed with light (15 min) in the middle of subjective day (CT 6). However, the presentation of a light pulse at CT 18, the middle of subjective night, induced extensive *Fos-ir* throughout the retinoreceptive, ventrolateral SCN. *OC*, optic chiasm; *3V*, third ventricle. *Lower panel*, the magnitude of *Fos* induction by light varies with circadian phase. Light pulses delivered during subjective day (CT 0–12) do not shift the clock nor do they induce *Fos-ir* in the SCN. In contrast, pulses delivered in subjective night which would shift the clock do induce *Fos-ir*. (Data taken from GROSSE et al. 1995)

The common action of NMDAR is to facilitate the entry of Ca^{2+} into the cell (SEEBURG 1993; BADING et al. 1993), and so the question arises as to how changes in intracellular Ca^{2+} might induce transcription of *c-fos* and other genes. In other cells, the *c-fos* promoter is regulated by a transcription factor which binds to a specific DNA sequence known as the cyclic AMP response element (CRE) (GINTY et al. 1992). The binding protein, CREB, is restricted to the nucleus but is only active when phosphorylated at a particular serine residue (GONZALEZ and MONTMINY 1989; SHENG et al. 1991). Light pulses which induce IEGs also induce the phosphorylation of CREB in the retinorecipient

SCN in vivo (GINTY et al. 1993; SUMOVA et al. 1994). In primary cultures of SCN in vitro glutamate is able to mobilise Ca^{2+} (VAN DEN POL et al. 1992), induce the expression of *c-fos* and *egr-1* and also stimulate the phosphorylation of CREB (Hastings and McNulty, unpublished data). This effect of glutamate on CREB in vitro is blocked by the antagonist dizocilpine, confirming the involvement of NMDAR in the response. The link between NMDAR, Ca^{2+} and CREB and IEGs is probably provided by a Ca-dependent kinase, probably CaM kinase II because local administration into the SCN of an antagonist to CaMK II attenuates the phase-shifting response to light in free-running hamsters (GOLOMBEK and RALPH 1994). A major goal of future studies is to characterise the types of cell in which glutamate is able to activate this signal transduction cascade and thereby identify the anatomical substrate for entrainment of the SCN by light.

A second question is the role of the IEGs in phase-shifting. Treatments with glutamatergic antagonists which attenuate phase shifts to light also block photic induction of IEGs in parts of the retinorecipient SCN (reviewed by HASTINGS et al. 1995). However, this does not imply a causal relationship between the two responses; is *Fos* expression in the retinorecipient SCN necessary and sufficient to cause phase shifts? Local administration of NMDA to the SCN can induce *Fos* in the retinorecipient areas (EBLING et al. 1991; REA et al. 1993a), but such a treatment is not sufficient to cause a phase advance comparable to a light pulse (REA et al. 1993b). A more direct way to test the role of *Fos* in phase-shifting would be to block expression of the gene whilst leaving upstream responses unaffected: if the gene products were necessary, light would not shift the clock under such circumstances. WOLLNIK et al. (1995) have used local infusions into the SCN of a cocktail of antisense oligonucleotides to block the expression of a number of immediate-early genes, including *c-fos*. This treatment attenuated light-induced phase shifts, which indeed indicates a causal role for these proteins in photic entrainment. However, this question has yet to be resolved conclusively. The light-induced protein products of IEGs are detectable 45–60 min after a light pulse, but to be effective they must regulate the expression of a second set of genes. The nature of these target genes in the SCN is unknown, but nevertheless it might be assumed that their protein products would not appear in any abundance until at least 2 h after the light pulse. If it can be shown that the clock is shifted sooner than the time taken for these downstream products to appear, it might be concluded that the expression of IEGs is a delayed correlate of phase-shifting rather than part of a causal link. Indirect support for this is provided by studies of non-photic entrainment in which it was shown that arousal can shift the clock within 1 h, i.e. that non-photic phase shifts are effectively instantaneous (MEAD et al. 1992). Nevertheless, it remains to be determined whether a similar argument can be applied to photic shifts. What is clear from the non-photic studies is that activation of the CREB/IEG cascade is not necessary for all types of phase shift: non-photic cues can entrain the circadian system without activating this glutamate-dependent pathway (MEAD et al. 1992; CUTRERA et al. 1993; SUMOVA et al. 1994).

If the products of IEGs are not causal, how is the photic stimulus transduced further within the SCN? In other tissues, nitric oxide (NO) acts as a diffusible intercellular signal, stimulating the production of cyclic GMP in target cells (LOWENSTEIN and SNYDER 1992). In the brain, the production of NO is known to be sensitive to increases in intracellular Ca^{2+} mediated via activated NMDAR and so it is possible that glutamatergic cues reset the clock by a NO-dependent pathway (BREDT and SNYDER 1992; DAWSON and SNYDER 1994). Evidence for such an effect has been provided from both in vitro and in vivo studies, which have demonstrated, respectively, that glutamatergic phase shifts of the rat SCN slice and photic shifts of free-running activity rhythms in hamsters can be blocked by inhibitors of NO synthesis, whilst activation of the NO pathway can mimic the effects of glutamate in vitro (DING et al. 1994). The observation that phase-shifting can be mediated by non-synaptic communication is very intriguing and echoes the earlier question of whether clock function is at all dependent on synaptic contacts between SCN cells. Given that agonists of cyclic GMP have been shown to phase shift the SCN slice (PROSSER et al. 1989), it may be that this molecule, rather than the products of IEGs, provides the route into the clockwork. Such an interpretation would leave the induction of IEGs as a "house-keeping response", restoring cellular function after the imposition of a metabolic load following photic activation of the SCN (SCHWARTZ 1991).

III. Non-photic Entrainment of the Circadian System

Light is not the only cue to which the circadian system is sensitive (TUREK 1989; SUMOVA et al. 1995; Chap. 15, this volume). A great deal of human behaviour is directed by social cues, and so it is perhaps not surprising to realise that in both experimental or natural settings internal clock time of the subject can be cued by contact with the experimenter, colleagues, family and friends (ASCHOFF et al. 1971; EHLERS et al. 1988). The significance of these cues is easy to overlook, but even in animal studies it is clear that social contact can shift the clock (Chap. 15, this volume). A common feature of all non-photic cues is that they involve arousal of the animal or subject (TUREK 1989; SUMOVA et al. 1995), and it is in this context that the serotonergic input to the SCN from the raphe nuclei assumes importance. The overall level of activity of the serotonergic cells of the raphe varies as a direct function of the state of arousal (JACOBS and AZMITIA 1992) and so cues such as contact with conspecifics, handling stress and enforced exercise which are able to shift the clock are probably also associated with increased serotonergic activation in the forebrain. That serotonin *can* shift the rhythm of the SCN has been confirmed by studies using the SCN slice in vitro (PROSSER et al. 1992, 1993) and treatment of animals with serotonergic agonists in vivo (TOMINAGA et al. 1992; EDGAR et al. 1993). The real question is whether serotonergic input to the SCN is necessary for non-photic entrainment. Depletion of the forebrain serotonin using the neurotoxin 5,7-dihydroxytryptamine has yielded equivocal results, although that may reflect the partial success of the lesion (MORIN 1994). An alternative strategy is to treat animals with serotonergic antagonists and test whether they will respond to an arousing cue. In the Syrian hamster, such antagonism of the endogenous serotonergic activity using the serotonergic antagonists ketanserin or ritanserin can attenuate the phase advances induced by arousal in late subjective day (HASTINGS et al. 1995; SUMOVA et al., in press). Although these drugs are classically antagonists at the $5HT_2$ receptor, it is not yet clear which receptor mediates the effect of serotonin on the clock, although there is strong evidence to implicate the recently cloned $5HT_7$ receptor (LOVENBERG et al. 1993). In other tissues, this receptor is coupled positively to adenylyl cyclase, and so serotonergic activation would be expected to increase levels of cAMP in the SCN during phase shifting. This is consistent with the observation that the circadian rhythm of the SCN slice preparation can be phase advanced not only by serotonergic agents, but also by agonists of cAMP (PROSSER and GILLETTE 1989). However, it is not yet clear which events beyond a rise in levels of cAMP might mediate the phase shift. Arousal-induced shifts are not associated with phosphorylation of CREB (SUMOVA et al. 1994) nor the induction of IEGs in the SCN (MEAD et al. 1992). Non-photic shifting therefore involves pathways different to those addressed by photic (glutamatergic) cues – identification of their point of convergence should reveal the entry point into the mechanism of the clock.

A second afferent to the SCN implicated in non-photic shifting is the NPY-ir geniculohypothalamic tract, which arises from the *IGL (MORIN 1994; see also Chap. 4, this volume). Not only do electrical stimulation of the IGL and local injection of NPY into the SCN shift the circadian rhythm, both in vivo and in vitro, but treatment with antiserum to NPY can block the advances normally produced in hamsters by enforced wheel-running (reviewed in Chap. 15, this volume). Given that the IGL receives a serotonergic innervation from the raphe, it may be that arousal-induced shifts involve simultaneous serotonergic activation of IGL and SCN, and consequent neuropeptide Y (NPY)-ergic activation of the SCN.

IV. Melatonin and Entrainment of the Circadian System

As mentioned above, in lower vertebrates the pineal hormone melatonin can be an important stimulus to the circadian system, providing a link between the oscillator and its effector mechanisms. In mammals, the secretion of melatonin is still exclusively nocturnal and so it provides a potentially important cue to rhythmic physiology and behaviours. This topic is considered in greater detail in Chap. 14, this volume, and so only a few points will be raised here. First, a number of brain and pituitary sites express high-affinity binding sites for melatonin and through these the hormone can exert a direct effect on nocturnal, clock-dependent processes. For example, both sleep latency and core body temperature are reduced in normal human subjects following administration of exogenous melatonin (CAGNACCI et al. 1992; CAGNACCI et al. 1993; DOLLINS et al. 1994). These responses to melatonin may explain some of the reported beneficial effects of the hormone on sleep quality in both normal and

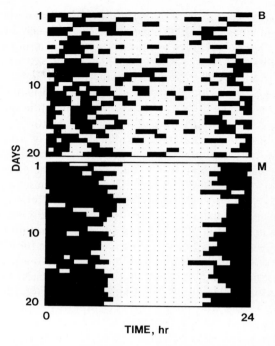

Fig. 4. Schematic representation of sleep patterns in a child suffering from a serious neurological disorder. Prior to therapy the child slept erratically and exhibited no overt circadian pattern. (*B*) Sleep and wakefulness occurred in an essentially random order. (*M*) Following establishment of a regime of daily administration of melatonin, the sleep/wake pattern rapidly stabilised, with consolidated nocturnal sleep and diurnal wakefulness. The beneficial effects of this therapy extend beyond the patient to those charged with his care. (Redrawn from JAN et al. 1994)

brain-damaged individuals (JAN et al. 1994) (Fig. 4). The therapeutic action of melatonin in cases of jet lag may also be associated with this facilitation of sleep (ARENDT et al. 1987; Chap. 14, this volume).

However, it is also clear that the SCN express melatonin receptors, assessed both by autoradiography of ligand binding (BITTMAN 1993; MORGAN et al. 1994) and by in situ hybridisation of mRNA for a recently cloned melatonin receptor (REPPERT et al. 1994). Melatonin therefore has the capacity to regulate clock function directly. This has been confirmed by studies with the SCN slice, which have shown that melatonin applied during late subjective day can advance the circadian rhythm (MCARTHUR et al. 1991). A parallel advance has also been seen in the rat in vivo, albeit at massive, non-physiological doses. Indeed a PRC can be constructed for the effects of melatonin both in vitro and in vivo and the two are remarkably similar (ARMSTRONG 1989). The potential for the use of melatonin and its analogues as chronotherapeutic agents is therefore enormous. However, a biological question remains – when and under what circumstances does melatonin regulate circadian function in a mammal?

The difficulty arises because the sensitive zone of the PRC to melatonin falls at a time in the circadian day when endogenous melatonin is never secreted (McArthur et al. 1991). Under normal circumstances melatonin is only secreted during subjective night, when the SCN is not sensitive to its phase-shifting action. Under what circumstances therefore might melatonin be encountered by the SCN during subjective day? Such a situation might arise in utero when the developing SCN, in the absence of photic input, could establish a circadian phase different to that of its dam. Under these circumstances melatonin derived from the maternal circulation, which is known to cross the placenta, may be important in setting the phase of the foetal clock and thereby synchronising mother and foetus. The presence of melatonin receptors in the foetal SCN, prior to the establishment of retinal input, supports such a role (Reppert et al. 1988), and it has also been shown that exogenous melatonin delivered to pregnant hamsters bearing lesions of the SCN can synchronise the circadian clocks of the litter to a common phase (Davis and Mannion 1988). Melatonin is not, however, the only stimulus to entrain the foetal clock. The foetal SCN expresses high levels of dopamine D_1 receptors, and treatment with a D_1-receptor agonist can also synchronise the litters of SCN-lesioned dams (Viswanathan et al. 1994). It is interesting to note that the phase relationship between the foetal clock and the time of injection of the D_1 agonist is almost opposite to that which would be established by injection of melatonin. The two types of signal are therefore complementary in their effects, and they probably establish circadian day and night, respectively. It is very obvious that the D_1 receptors are rapidly lost after birth probably because photic input assumes control of the clock. Melatonin receptors remain in the SCN for much longer, although there is a trend for them to decline during adult life. The rate of this decline of the (redundant?) signalling pathway probably varies between species, and this might explain why some, e.g. Syrian hamster, are not at all sensitive to entraining effects of exogenous melatonin (Hastings et al. 1992) whilst others, e.g. rat and Siberian hamster, do respond to pharmacological doses (Armstrong 1989; Margraf and Lynch 1993). In the case of the adult Syrian hamster, the melatonin-binding sites which do remain are restricted to the rostral and medial SCN, in a position clearly distinct to that occupied by the photoresponsive areas of the nucleus (Maywood et al. 1995). Should melatonin have any residual action on the SCN in this species, it is likely therefore to be on pathways distinct from those involved in photic entrainment.

V. Conclusion

Endogenous circadian organisation is a basic feature of vertebrate physiology, and is probably expressed in most eukaryotic groups, and indeed in some prokaryotes. An appreciation of circadian rhythmicity is necessary to the understanding of normal human physiology as well as certain pathological conditions. In vertebrates as a group, a number of structures including the retina, the pineal gland and the suprachiasmatic nuclei of the hypothalamus have been identified as locations for a circadian clock or oscillator. In mammals the SCN are the principal oscillator, and considerable progress has been made in understanding the neuroanatomical and neurochemical bases for communication between the clock and the rest of the brain, and the process of clock entrainment. An important future test will be to understand in genetic, molecular and biochemical terms how a repetitive cycle of 24 h can be generated, either by individual cells or by an assemblage of neurons and glia. It is very likely that studies of circadian time-keeping in lower organisms, particularly *Drosophila* and *Neurospora*, will provide a conceptual framework, and possibly some of the components of the clockwork, necessary for resolution of this problem (HASTINGS 1994). The second area of advance is likely to come from the development of "chronotherapeutic" agents, i.e. compounds which by regulating circadian function are able to enhance the quality of life of individuals, e.g. those affected by neurological or psychiatric disturbance, or the aged.

References

Albers HE, Yogev L, Todd RB, Goldman BD (1985) Adrenal corticoids in hamsters: role in circadian timing. Am J Physiol 248:R434–R438
Armstrong SM (1989) Melatonin and circadian control in mammals. Experientia 45:932–938
Aschoff J (1981) Biological rhythms. Plenum, New York (Handbook of behavioural neurobiology, vol 4)
Aschoff J, Fatranska M, Giedke H (1971) Human circadian rhythms in continuous darkness: entrainment by social cues. Science 171:213–215
Axelrod J (1974) The pineal gland: a neurochemical transducer. Science 184:1341–1348
Bading H, Ginty DD, Greenberg ME (1993) Regulation of gene expression in hippocampal neurons by distinct calcium signaling pathways. Science 2260:181–186
Besharse JC, Iuvone PM (1983) Circadian clock in Xenopus eye controlling retinal serotonin N-acetyltransferase. Nature 305:133–135
Bittman EL (1993) The sites and consequences of melatonin binding in mammals. Am Zool 33:200–211
Bosler O, Beaudet A (1985) VIP neurons as prime synaptic targets for serotonin afferents in rat suprachiasmatic nucleus: a combined radioautographic and immunocytochemical sudy. J Neurocytol 14:749–763
Bredt DS, Snyder SH (1992) Nitric oxide: a novel neuronal messenger. Neuron 8:3–11
Cagnacci A, Elliott JA, Yen SSC (1992) Melatonin: a major regulator of the circadian rhythm of core temperature in humans. J Clin Endocrinol Metab 75:447–452
Cagnacci A, Soldani R, Yen SSC (1993) The effect of light on core body temperature is mediated by melatonin in women. J Clin Endocrinol Metab 76:1036–1038
Cahill GM, Besharse JC (1991) Resetting the circadian clock in cultured Xenopus eyecups: regulation of retinal melatonin rhythms by light and D2 dopamine receptors. J Neurosci 11:2959–2971
Cahill GM, Besharse JC (1993) Circadian clock functions localised in Xenopus retinal photoreceptors. Neuron 10:573–577
Card JP, Moore RY (1984) The suprachiasmatic nucleus of the golden hamster: immunohistochemical analysis of cell and fiber distribution. Neuroscience 13:415–431
Card JP, Whealy ME, Robbins AK, Moore RY, Enquist LW (1991) Two alpha-Herpes virus strains are transported differentially in the rodent visual system. Neuron 6:957–969
Cassone VM (1988) Circadian variation of [14C]2-deoxyglucose uptake within the suprachiasmatic nucleus of the house sparrow, Passer domesticus. Brain Res 459:178–182

Cassone VM, Brooks DS (1991) Sites of melatonin action in the brain of the house sparrow Passer domesticus. J Exp Zool 260:302–309

Cassone VM, Moore RY (1988) Retinohypothalamic projection and suprachiasmatic nucleus of the house sparrow, Passer domesticus. J Comp Neurol 266:171–182

Cassone VM, Speh JC, Card JP, Moore RY (1988) Comparative anatomy of the mammalian hypothalamic suprachiasmatic nucleus. J Biol Rhythms 3:71–91

Castel M, Belenky M, Cohen S, Ottersen OP, Storm-Mathisen J. (1993) Glutamate-like immunoreactivity in retinal terminals of the mouse suprachiasmatic nucleus. Eur J Neurosci 5:368–381

Cohen RA, Albers HE (1991) Disruption of human circadian and cognitive regulation following a discrete hypothalamic lesion: a case study. Neurobiology 41:726–729

Colwell CS, Menaker M (1992) NMDA as well as non-NMDA receptor antagonists can prevent the phase-shifting effects of light on the circadian system of the golden hamster. J Biol Rhythms 7:125–136

Cutrera RA, Kalsbeek A, Pevet P (1993) No triazolam-induced expression of Fos protein in raphe nuclei of the male Syrian hamster. Brain Res 602:14–20

Czeisler CA, Kronauer RE, Allan JS, Duffy JF, Jewett ME, Brown EN, Ronda JM (1989) Bright light induction of strong (Type 0) resetting of the human circadian pacemaker. Science 244:1328–1333

Daan S, Pittendrigh CS (1976) A functional analysis of circadian pacemakers in nocturnal rodents. II. The variability of phase response curves. J Comp Physiol [A] 106:253–266

Davis FC, Mannion J (1988) Entrainment of hamster pup circadian rhythms by prenatal melatonin injections to the mother. Am J Physiol 255:R439–R448

Dawson TM, Snyder SH (1994) Gases as biological messengers: nitric oxide and carbon monoxide in the brain. J Neurosci 14:5147–5159

Ding JM, Chen D, Weber ET, Faiman LE, Rea MA, Gillette MU (1994) Resetting the biological clock: mediation of nocturnal circadian shifts by glutamate and NO. Science 266:1713–1717

Dollins AB, Zhdanova IV, Wurtman RJ, Lynch HJ, Deng MH (1994) Effect of inducing nocturnal serum melatonin concentrations in daytime on sleep, mood, body temperature and performance. Proc Natl Acad Sci USA 91:1824–1828

Earnest DJ, Sladek CD (1987) Circadian vasopressin release from perifused rat suprachiasmatic explants in vitro: effects of acute stimulation. Brain Res 422:398–402

Ebihara S, Kawamura K (1981) The role of the pineal and the suprachiasmatic nuclei in the control of circadian rhythms of the Java sparrow, Padda orizivora. J Comp Physiol 141:207–214

Ebling FJP, Staley K, Maywood ES, Humby T, Hancock DC, Waters CM, Evan GI, Hastings MH (1991) The role of NMDA-type glutamatergic neurotransmission in the photic induction of immediate-early gene expression in the suprachiasmatic nuclei of the Syrian hamster. J Neuroendocrinol 3:641–652

Edgar DM, Miller JD, Prosser RA, Dean RR, Dement WC (1993) Serotonin and the mammalian circadian system. II. Phase-shifting rat behavioural rhythms with serotonergic agonists. J Biol Rhythms 8:17–31

Ehlers CL, Frank E, Kupfer DJ (1988) Social zeitgebers and biological rhythms. Arch Gen Psychiatry 45:948–952

Elliott JA (1976) Circadian rhythms and photoperiodic time measurement in mammals. Fed Proc 35:2339–2346

Falcon J, Thibault C, Begay V, Zachmann A, Collin J-P (1992) Regulation of the rhythmic melatonin secretion by fish pineal photoreceptor cells. In: Ali MA (ed) Rhythms in fishes. Plenum, New York, pp 167–198

Folkard S, Hume KI, Minors DS, Waterhouse J M, Watson FL (1985) Independence of the circadian rhythm in alertness from the sleep/wake cycle. Nature 313:678–679

Foster RG, Schalken JJ, Timmers AM, De Grip WJ (1989a) A comparison of some photoreceptor characteristics in the pineal and retina. I. The Japanese quail (Coturnix coturnix). J Comp Physiol 165:553–563

Foster RG, Timmers AM, Schalken JJ, De Grip WJ (1989b) A comparison of some photoreceptor characteristics in the pineal and retina. II. The Djungarian hamster (Phodopus sungorus). J Comp Physiol 165:565–572

Ginty DD, Bading H, Greenberg ME (1992) Trans-synaptic regulation of gene expression. Curr Opin Neurobiol 2:312–316

Ginty DD, Kornhauser JM, Thompson MA, Bading H, Mayo KE, Takahashi JS, Greenberg ME (1993) Regulation of CREB phosphorylation in the suprachiasmatic nucleus by light and a circadian clock. Science 260:238–241

Golombek DA, Ralph MR (1994) KN-62, an inhibitor of Ca^{2+}/calmodulin kinase II attenuates circadian responses to light. NeuroReport 5:1638–1640

Gonzalez GA, Montminy MR (1989) Cyclic AMP stimulates somatostatin gene transcription by phosphorylation of CREB at serine 133. Cell 59:675–680

Grosse J, Loudon A, Hastings MH (1995) Cellular and behavioural responses to light of the circadian system of the tau mutant hamster. Neuroscience 65:587–597

Hakim H, DeBernardo AP, Silver R (1991) Circadian locomotor rhythms, but not photoperiodic responses, survive surgical isolation of the SCN in hamsters. J Biol Rhythms 6:97–113

Hastings MH (1991) Neuroendocrine rhythms. Pharmacol Ther 50:35–71

Hastings MH (1994) Circadian rhythms: what makes the clock tick? Curr Biol 4:720–723

Hastings MH (1995) Circadian rhythms: peering into the molecular clockwork. J Neuroendocrinol 7:331–340Hastings MH, Herbert J (1986) Neurotoxic lesions of the paraventriculo-spinal projection block the nocturnal rise in pineal melatonin synthesis in the Syrian hamster. Neurosci Lett 69:1–6

Hastings MH, Mead SM, Vindlacheruvu RR, Ebling FJP, Maywood ES, Grosse J (1992) Non-photic phase shifting of the circadian activity rhythm of Syrian hamster: the relative potency of arousal and melatonin. Brain Res 591:20–26

Hastings MH, Ebling FJP, Grosse J, Herbert J, Maywood ES, Mikkelsen JD, Sumova A (1995) Immediate-early genes and the neural basis of circadian entrainment. Ciba Found Symp 183:175–197

Herbert J (1991) The brain and interval timing in psychological and photoperiodic time measurement. Adv Pineal Res 5:1–12

Jacobs BL, Azmitia EC (1992) Structure and function of the brain serotonin system. Physiol Rev 72:167–229

Jan JE, Espezel H, Appleton RE (1994) The treatment of sleep disorders with melatonin. Dev Med Child Neurol 36:97–107

Johnson RF, Moore RY, Morin LP (1988) Loss of entrainment and anatomical plasticity after lesions of the hamster retinohypothalamic tract. Brain Res 460:297–313

Kalsbeek A, Teclemariam-Mesbah R, Pevet P (1993) Efferent projections of the suprachiasmatic nucleus in the golden hamster (Mesocricetus auratus). J Comp Neurol 332:293–314

Kim YI, Dudek FE (1991) Intracellular electrophysiological study of suprachiasmatic nucleus neurons in rodents: excitatory synaptic mechanisms. J Physiol (Lond) 444:269–287

Klein DC, Smoot R, Weller JI, Higa S, Markey SP, Creed GJ, Jacobowitz DM (1983) Lesions of the paraventricular nucleus area of the hypothalamus disrupt the suprachiasmatic-spinal cord circuit in the melatonin rhythm-generating system. Brain Res Bull 10:647–651

Kondo T, Tsinoremas NF, Golden SS, Johnson CH, Kutsuna S, Ishiura M (1994) Circadian clock mutants in cyanobacteria. Science 266:1233–1236

Korf H-W, Oksche A, Ekstrom P, Gery I, Zigler JS, Klien DC (1986) Pinealocyte projections into the mammalian brain revealed with S-antigen antiserum. Science 231:735–737

Kornhauser JM, Nelson DE, Mayo KE, Takahashi JS (1992) Regulation of jun-B messenger RNA and AP-1 activity by light and a circadian clock. Science 255:1581–1585

Lavaille M, Serviere J (1993) Circadian fluctuations in GFAP distribution in the Syrian hamster suprachiasmatic nucleus. Neuroreport 4:1243–1246

Lehman MN, Silver R, Gladstone WR, Kahn RM, Gibson M, Bittman EL (1987) Circadian rhythmicity restored by neural transplant. Immunocytochemical characterisation of the graft and its integration with the host brain. J Neurosci 7:1626–1638

Lehman MN, Silver R, Bittman EL (1991) Anatomy of suprachiasmatic nucleus grafts. In: Klein DC, Moore RY, Reppert SM (eds) The suprachiasmatic nucleus: the mind's clock. Oxford University Press, New York, pp 349–374

Linkowski P, van Onderbergen A, Kerkhofs M, Bosson D, Mendlewicz J, van Cauter E (1993) Twin study of the 24th cortisol profile: evidence for genetic control of the human circadian clock. Endocrinol Metab 27:E173–E181

Liou SY, Shibata S, Iwasaki K, Ueki S (1986) Optic nerve stimulation-induced increase of release of ^{3}H-glutamate and ^{3}H-aspartate but not ^{3}H-GABA from the suprachiasmatic nucleus in slices of rat hypothalamus. Brain Res Bull 16:527–531

Lovenberg TW, Baron BM, de Lecea L, Miller JD, Prosser RA, Rea MA, Foye PE, Racke M, Slone AL, Siegel BW, Danielson PR, Sutcliffe JG, Erlander MG (1993) A novel adenylyl cyclase-activating serotonin receptor (5HT-7) implicated in the regulation of mammalian circadian rhythms. Neuron 11:449–458

Lowenstein CJ, Snyder SH (1992) Nitric oxide: a novel biologic messenger. Cell 70:705–707

Mai JK, Kedziora O, Teckhaus L, Sofroniew MV (1991) Evidence for subdivisions in the human suprachiasmatic nucleus. J Comp Neurol 305:508–525

Margraf RR, Lynch GR (1993) Melatonin injections affect circadian behaviour and SCN neurophysiology in Djungarian hamsters. Am J Physiol 33:R615–R621

Maywood ES, Bittman EL, Ebling FJP, Barrett P, Morgan PJ, Hastings MH (1995) Regional distribution of iodomelatonin binding sites in the suprachiasmatic nucleus of the Syrian hamster and the Siberian hamster. J Neuroendocrinol 7:215–223

McArthur AJ, Gillette MU, Prosser RA (1991) Melatonin directly resets the rat SCN circadian clock. Brain Res 565:158–163

McCormack CA, Burnside B (1992) A role for endogenous dopamine in circadian regulation of retinal cone movement. Exp Eye Res 55:1–10

Mead S, Ebling FJP, Maywood ES, Humby T, Herbert J, Hastings MH (1992) A nonphotic stimulus causes instantaneous phase-advances of the light entrainable circadian oscillator of the Syrian hamster, but does not induce the expression of c-fos in the suprachiasmatic nuclei. J Neurosci 12:2516–2522

Meijer JH (1991) Integration of visual information by the suprachiasmatic nucleus. In: Klein DC, Moore RY, Reppert SM (eds) The suprachiasmatic nucleus: the mind's clock.Oxford University Press, New York, pp 107–120

Meijer JH, Groos GA (1988) Responsiveness of suprachiasmatic and ventral lateral geniculate neurons to serotonin and imipramine: a microiontophoretic study in normal and imipramine-treated rats. Brain Res Bull 20:89–96

Meijer JH, Rusak B, Ganshirt G (1992) The relationship between light-induced discharge in the suprachiasmatic nucleus and phase shifts of hamster circadian rhythms. Brain Res 598:257–263

Menaker M (1985) Eyes – the second (and third) pineal glands? CIBA Found Symp 117:78–92

Menaker M, Vogelbaum MA (1993) Mutant circadian period as a marker of suprachiasmatic nucleus function. J Biol Rhythms 8:S93–S98

Menaker M, Hudson DJ, Takahashi JS (1981) Neural and neuroendocrine components of circadian clocks in birds. In: Follett BK, Follett DE (eds) Biological clocks in seasonal reproductive cycles. Wright, Bristol, pp 171–183

Mikkelsen JD, Larsen PJ, Ebling FJP (1993) Distribution of N-methyl D-aspartate (NMDA) receptor mRNAs in the rat suprachiasmatic nucleus. Brain Res 632:329–333

Miller JD (1993) On the nature of the circadian clock in mammals. Am J Physiol 264:R821–R832
Moore RY (1973) Retinohypothalamic projections in mammals: a comparative study. Brain Res 49:403–409
Moore RY (1991) Disorders of human circadian function and the human circadian timing system. In: Klein DC, Moore RY, Reppert SM (eds) The suprachiasmatic nucleus: the mind's clock. Oxford University Press, New York, pp 429–441
Moore RY (1992a) The organisation of the human circadian timing system. Prog Brain Res 93:101–117
Moore RY (1992b) The enigma of the geniculohypothalamic tract: why two visual entraining pathways? J Interdisc Cycle Res 23:144–152
Moore RY, Lenn NJ (1972) A retinohypothalamic projection in the rat. J Comp Neurol 146:1–14
Moore RY, Speh JC (1993) GABA is the principal neurotransmitter of the circadian system. Neurosci Lett 150:112–116
Moore-Ede MC, Czeisler CA, Richardson GS (1983) Circadian time-keeping in health and disease. N Engl J Med 309:469–537
Morin LP (1994) The circadian visual system. Brain Res Rev 67:102–127
Morin LP, Blanchard JH (1991) Depletion of brain serotonin by 5,7-DHT modifies hamster circadian rhythm response to light. Brain Res 566:173–185
Morin LP, Blanchard J, Moore RY (1992) Intergeniculate leaflet and suprachiasmatic nucleus organisation and connection in the golden hamster. Vis Neurosci 8:219–230
Morin LP, Goodles-Sanchez N, Smale L, Moore RY (1994) Projections of the suprachiasmatic nuclei, subparaventricular zone and retrochiasmatic area in the golden hamster. Neuroscience 61:391–410
Morgan PJ, Barrett P, Howell HE, Helliwell R (1994) Melatonin receptors: localisation, molecular pharmacology and physiological significance. Neurochem Int 24:101–146
Nelson DE, Takahashi JS (1991) Comparison of visual sensitivity for suppression of pineal melatonin and circadian phase-shifting in the golden hamster. Brain Res 554:272–277
Nelson DE, Takahashi JS (1992) Sensitivity and integration in a visual pathway for circadian entrainment in the hamster (Mesocricetus auratus). J Physiol (Lond) 439:115–145
Okano T, Yoshizawa T, Fukada Y (1994) Pinopsin is a chicken pineal photoreceptive molecule. Nature 372:94–97
Oksche A (1983) Aspects of evolution of the pineal organ. NATO ASI 65:15–36
Oksche A (1991) The development of the concept of the photoneuroendocrine system: historical perspective. In: Klein DC, Moore RY, Reppert SM (eds) The suprachiasmatic nucleus: the mind's clock. Oxford University Press, New York, pp 5–14
Pierce ME, Sheshberadaran H, Zhang Z, Fox LE, Applebury ML, Takahashi JS (1993) Circadian regulation of iodopsin gene expression in embryonic photoreceptors in retinal cell culture. Neuron 10:579–584
Pittendrigh CS (1981) Circadian systems: entrainment. In: Aschoff J (ed) Biological rhythms. Plenum, New York, pp 95–124 (Handbook of behavioural neurobiology, vol 4)
Prosser RA, Gillette MU (1989) The mammalian circadian clock in the suprachiasmatic nuclei is reset in vitro by cAMP. J Neurosci 9:1073–1081
Prosser RA, McArthur AJ, Gillette MU (1989) cGMP induces phase shifts of a mammalian circadian pacemaker at night, in antiphase to cAMP effects. Proc Natl Acad Sci USA 86:6812–6815
Prosser RA, Heller HC, Miller JD (1992) Serotonergic phase shifts of the mammalian circadian clock: effects of tetrodotoxin and high Mg^{2+}. Brain Res 573:336–340

Prosser RA, Dean RR, Edgar DM, Heller HC, Miller JD (1993) Serotonin and the mammalian circadian system: I In vitro phase shifts by serotonergic agonists and antagonists. J Biol Rhythms 8:1–16

Ralph MR (1991) Suprachiasmatic nucleus transplant studies using the tau mutation in golden hamsters. In: Klein DC, Moore RY, Reppert SM (eds) The suprachiasmatic nucleus: the mind's clock. Oxford University Press, New York, pp 349–374

Ralph MR, Mrosovsky N (1992) Behavioural inhibition of circadian responses to light. J Biol Rhythms 7:353–360

Ralph MR, Foster RG, Davis FC, Menaker M (1989) Transplanted suprachiasmatic nucleus determines circadian period. Science 247:975–978

Rea MA (1989) Light increases Fos-related protein immunoreactivity in the rat suprachiasmatic nuclei. Brain Res Bull 23:577–581

Rea MA, Buckley B, Lutton LM (1993a) Local administration of EAA antagonists blocks light-induced phase shifts and c-fos expression in the hamster SCN. Am J Physiol 265:R1191–R1198

Rea MA, Michel AM, Lutton LM (1993b) Is Fos expression necessary and sufficient to mediate light-induced phase advances of the suprachiasmatic circadian oscillator? J Biol Rhythms 8:S59–S64

Rea MA, Glass JD, Colwell CS (1994) Serotonin modulates photic responses in the hamster suprachiasmatic nuclei. J Neurosci 14:3635–3642

Reme CE, Wirz-Justice A, Terman M (1991) The visual input stage of the mammalian circadian pacemaking system: I. Is there a clock in the mammalian eye? J Biol Rhythms 6:5–30

Reppert SM (1992) Pre-natal development of a hypothalamic biological clock. Prog Brain Res 93:119–131

Reppert SM, Weaver DR, Rivkees SA, Stopa EG (1988) Putative melatonin receptors in a human biological clock. Science 242:78–81

Reppert SM, Weaver DR, Ebisawa T (1994) Cloning and characterisation of a mammalian melatonin receptor that mediates reproductive and circadian responses. Neuron 13:1177–1185

Richter CP (1967) Sleep and activity: their relation to the 24-hour clock. Assoc Res Nerv Ment Dis 45:8–29

Rivkees SA, Cassone VM, Weaver DR, Reppert SM (1989) Melatonin receptors in chick brain: characterisation and localisation. Endocrinology 125:363–368

Roden M, Koller M, Pirich K, Vierhapper H, Waldhauser F (1993) The circadian melatonin and cortisol secretion pattern in permanent night shift workers. Am J Physiol 64:R261–R265

Rusak B, Zucker I (1979) Neural regulation of circadian rhythms. Physiol Rev 59:449–526

Rusak B, Robertson HA, Wisden W, Hunt SP (1990) Light pulses that shift rhythms induce gene expression in the suprachiasmatic nucleus. Science 243:1237–1240

Schwarz WJ (1991) SCN metabolic activity in vivo. In: Klein DC, Moore RY, Reppert SM (eds) The suprachiasmatic nucleus: the mind's clock. Oxford University Press, New York, pp 144–156

Schwartz WJ, Gross RA, Morton MT (1987) The suprachiasmatic nuclei contain a tetrodotoxin-resistant circadian pacemaker. Proc Natl Acad Sci USA 84:1694–1698

Seeburg PH (1993) The molecular biology of mammalian glutamate receptor channels. Trends Neurosci 16:359–370

Selim M, Glass JD, Hauser UE, Rea MA (1993) Serotonergic inhibition of light-induced fos protein expression and extracellular glutamate in the suprachiasmatic nuclei. Brain Res 621:181–188

Sheng M, Thompson MA, Greenberg ME (1991) CREB: a Ca^{2+}-regulated transcription factor phosphorylated by calmodulin-dependent kinases. Science 252:1427–1430

Shibata S, Moore RY (1993) Tetrodotoxin does not affect circadian rhythms in neuronal activity and metabolism in rodent suprachiasmatic nucleus in vitro. Brain Res 606:259–266

Shimomura K, Menaker M (1994) Light-induced phase shifts in tau mutant hamsters. J Biol Rhythms 9:97–110

Shirakawa T, Moore RY (1994) Glutamate shifts the phase of the circadian neuronal firing rhythm in the rat suprachiasmatic nucleus in vitro. Neurosci Lett 178:47–50

Silver R, LeSauter J, Tresco PA, Lehman MN (1996) A diffusible coupling signal from the transplanted suprachiasmatic nucleus controlling circadian locomotor rhythms. Nature 382:810–812

Simpson SM, Follett BK (1981) Pineal and hypothalamic pacemakers: their role in regulating circadian rhythms in Japanese quail. J Comp Physiol 144:381–389

Smith RD, Turek FW, Takahashi JS (1992) Two families of phase-response curve characterise the resetting of the hamster circadian clock. Am J Physiol 262:R1149–R1153

Somers RL, Klein DC (1986) Rhodopsin kinase activity in the mammalian pineal gland and other tissues. Science 226:182–184

Sumova A, Ebling FJP, Maywood ES, Herbert J, Hasting MH (1994) Non-photic circadian entrainment is not associated with phosphorylation of the transcriptional regulator CREB within the suprachiasmatic nucleus, but is associated with adrenocortical activation. Neuroendocrinology 59:579–589

Sumova A, Ebling FJP, Herbert J, Maywood ES, Moore E, Hastings MH (1995) Non-photic entrainment of circadian rhythms. Adv Pineal Res 8:117–132

Sumova A, Maywood ES, Selvage D, Ebling FJP, Hastings MH (1996) Serotonergic antagonists impair arousal-induced phase shifts of the circadian system of the Syrian hamster. Brain Res 709:88–96

Swaab DF, Hofman MA, Honnebeier MBOM (1990) Development of vasopressin neurons in the human suprachiasmatic nucleus in relation to birth. Dev Brain Res 52:289–293

Takahashi JS, Menaker (1982) Role of suprachiasmatic nuclei in the circadian system of the house sparrow, Passer domesticus. J Neurosci 2:815–822

Takahashi JS, deCoursey PJ, Bauman L, Menaker M (1984) Spectral sensitivity of a novel photoreceptive system mediating entrainment of mammalian circadian rhythms. Nature 308:186–188

Takahashi JS, Murakami N, Nikaido SS, Pratt BL, Robertson LM. (1989) The avian pineal, a vertebrate model system of the circadian oscillator: cellular regulation of circadian rhythms by light, second messengers and macromolecular synthesis. Rec Prog Horm Res 45:279–352

Tanaka M, Ichitani Y, Okamura H, Tanaka Y, Ibata Y (1993) The direct retinal projection to VIP neuronal elements in the rat SCN. Brain Res Bull 31:637–640

Terman JS, Reme CE, Terman M (1993) Rod outer segment disk shedding in rats with lesions of the suprachiasmatic nucleus. Brain Res 605:256–264

Terman M, Reme CE, Wirz-Justice A (1991) The visual input stage of the mammalian circadian pacemaking system: II The effect of light and drugs on retinal function. J Biol Rhythms 6:31–48

Tominaga K, Shibata S, Ueki S, Watanabe S (1992) effects of 5HT-1a receptor agonists on the circadian rhythm of wheel-running activity in hamsters. Eur J Pharmacol 214:79–84

Tosini G, Menaker M (1996) Circadian rhythms in cultured mamalian retina. Science 272:419–421

Turek FW (1989) Effects of stimulated physical activity on the circadian pacemaker of vertebrates. J Biol Rhythms 4:135–147

Underwood H (1989) The pineal and melatonin: regulators of circadian function in lower vertebrates. Experientia 45:914–922

Underwood H, Barrett RK, Siopes T (1990a) The quail's eye: a biological clock. J Biol Rhythms 5:257–265

Underwood H, Barrett RK, Siopes T (1990b) Melatonin does not link the eyes to the rest of the circadian system in quail: a neural pathway is involved. J Biol Rhythms 5:349–361

van Cauter EV, Turek FW (1986) Depression: a disorder of timekeeping? Perspect Biol Med 29:511–519

Van den Hoofdakker (1994) Chronobiological theories of nonseasonal affective disorders and their implications for treatment. J Biol Rhythms 9:157–183
van den Pol AN (1991) The suprachiasmatic nucleus: morphological and cytochemical substrates for cellular interaction. In: Klein DC, Moore RY, Reppert SM (eds) The suprachiasmatic nucleus: the mind's clock. Oxford University Press, New York, pp 17–50
van den Pol AN (1991) Glutamate and aspartate immunoreactivity in hypothalamic presynaptic axons. J Neurosci 11:2087–2101
van den Pol AN, Dudek FE (1993) Cellular communication in the circadian clock, the suprachiasmatic nucleus. Neuroscience 56:793–811
van den Pol AN, Finkbeiner SM, Cornell-Bell AH (1992) Calcium excitability and oscillations in suprachiasmatic nucelus neurons and glia in vitro. J Neurosci 12:2648–2664
van den Pol AN, Kogelman L, Ghosh P, Liljelund P, Blackstome C (1994a) Developmental regulation of the hypothalamic metabotropic glutamate receptor mGluR1. J Neurosci 14:3816–3834
van den Pol AN, Hermans-Borgmeyer I, Hofer M, Ghosh P, Heinemann S (1994b) Ionotropic glutamate-receptor gene expression in hypothalamus: localisation of AMP, kainate and NMDA receptor RNA with in situ hybridisation. J Comp Neurol 343:428–444
Vindlacheruvu RR, Ebling FJP, Maywood ES, Hastings MH (1992) Blockade of glutamatergic neurotransmission in the suprachiasmatic nucleus prevents cellular and behavioural repsonses of the circadian system to light. Eur J Neurosci 4:673–679
Viswanathan N, Weaver DR, Reppert SM, Davis FC (1994) Entrainment of the fetal circadian pacemaker by prenatal injections of the dopamine agonist SKF 38393. J Neurosci 14:5393–5398
Vogelbaum MA, Menaker M (1992) Temporal chimaeras produced by hypothalamic transplants. J Neurosci 12:3619–3627
Vollrath L (1981) The pineal gland. In: Oksche A, Volllrath L (eds) Handbuch der mikroskopischen Anatomie des Menschen, vol 4. Springer, Berlin Heidelberg New York, pp 170–190
Watanabe K, Koibuchi N, Ohtake H, Yamaoka S (1993) Circadian rhythms of vasopressin release in primary cultures of rat suprachiasmatic nucleus. Brain Res 624:115–120
Watts AG (1991) The efferent projections of the suprachiasmatic nucleus: anatomical insights into the control of circadian rhythms. In: Klein DC, Moore RY, Reppert SM (eds) The suprachiasmatic nucleus: the mind's clock, pp 77–106
Weitzman ED, Boyar RM, Kapen S, Hellman L (1975) The relationship between sleep and sleep stages and neuroendocrine secretion and biological rhythms in man. Rec Prog Horm Res 31:399–441
Welsh DK, Logothetis DE, Meister M, Reppert SM (1995) Individual neurons dissociated from rat suprachiasmatic nucleus express independently phased circadian firing rhythms. Neuron 14:697–706
Wever RA (1979) The circadian system of man: results of experiments under temporal isolation. Springer, Berlin Heidelberg New York
Wollnik F, Brysch W, Uhlmann E, Gillardon F, Bravo R, Zimmerman M, Schlingensiepen KH, Herdegen T (1995) Block of c-Fos and Jun-B expression by antisense oligonucleotides inhibits light-induced phase-shifts of the mammalian circadian clock. Eur J Neurosci 7:388–393
Youngstrom TG, Weiss ML, Nunez AA (1987) A retinal projection to the paraventricular nuclei of the hypothalamus in the Syrian hamster (Mesocricetus auratus). Brain Res Bull 19:747–750
Zatz M (1992) Does the circadian pacemaker act through cyclic AMP to drive the melatonin rhythm in chick pineal cells? J Biol Rhythms 7:301–312
Zimmerman NH, Menaker M (1979) The pineal gland: a pacemaker within the circadian system of the house sparrow. Proc Natl Acad Sci USA 76:999–1003

CHAPTER 2
Regulation of Cell Division Cycles by Circadian Oscillators: Signal Transduction Between Clocks

L.N. EDMUNDS, JR.

A. Introduction

An important consideration for maximizing the results of radio- and chemotherapy of mammalian cancers is that of host tolerance. It is unfortunate that in most of the earlier cancer work there is little mention of the role of rhythmic variations (particularly circadian periodicities) in the susceptibility of the whole organism to the toxicity of the drug(s) being utilized for treatment (chronotolerance). There now is abundant experimental evidence, however, that properly designed protocols (such as sinusoidally varying drug courses during the 24-h day) can dramatically enhance survival and cure rates by concomitantly maximizing the tolerance of the host to the drug through a temporal shielding of normal, healthy tissues (for recent reviews, see LEMMER 1989; TOUITOU and HAUS 1992; Chap. 11, this volume). Thus, "when" to treat must assume importance together with the "what" and "where" (HALBERG 1975) – a concept embraced by the field of chronotherapeutics.

A significant part of these effects is attributable to rhythms in cell proliferation in mammalian tissues, including bone and blood marrow (see EDMUNDS 1984). There is no lack of awareness that the cell division cycle (CDC) is of paramount importance for cancer treatment, if for no other reason than the fact that studies of synchronized mammalian cells have shown that their sensitivity to a large number of cytostatic drugs (such as cytosine arabinoside), as well as to ionizing radiation, is highly cell cycle-stage-specific (BASERGA 1971). Indeed, some of these drugs may also lead to a partial synchronization of the CDC and have been so employed to obtain synchronized cell populations both in vitro and in vivo. This strategy can maximize the chance for survival and cure by applying the minimum dose of drug necessary to kill the phased, malignant cells at the time of their CDC when they are most susceptible. It is obvious, therefore, that the diurnal and circadian rhythmicities observed in cell flux and in the distribution and duration of CDC phases, as well as the underlying mechanisms that generate these periodicities, must be taken into account in the design of an appropriate chemotherapeutic regimen (SCHEVING et al. 1989).

Among the mechanisms that have been proposed to control microbial cell division cycles, autonomous biological oscillators provide a possible means of integrating the disparate concepts of cell cycle transit, exit, and arrest. An important goal is to determine the nature of the circadian clock that couples to the CDC under certain conditions and generates periodicity of division and other CDC "landmarks" (as well underlying circadian rhythmicity in general). In this overview we are concerned with the means whereby the coupling between the circadian oscillator, on the one hand, and the so-called cyclin clock (see Fig. 1) that constitutes a universal cell-division-cycle "engine," on the other hand, is effected. Cyclic AMP (cAMP), which controls certain rate-limiting steps in the progression of the CDC in many cell types, may play such a coupling role, participating in the gating of CDC events to specific phases of the circadian cycle. Thus, we hope to elucidate the signal transduction pathway whereby information regarding period and phase flows from the circadian clock to the CDC and to determine the links (checkpoints) between the two oscillator systems.

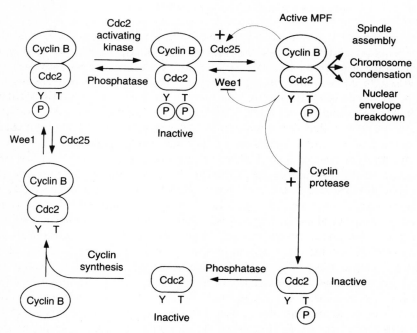

Fig. 1. Biochemical model of the embryonic cell cycle "engine." The different Cdc2/28-cyclin B complexes formed during the embryonic cell cycle and the enzymes that act on them are shown. Wee1, a protein kinase, phosphorylates Cdc2/28 on the tyrosine 15 (*left, Y*) and threonine 161 (*right, T*) residues, while the protein phosphatase Cdc25 removes the phosphate added by Wee1. At the beginning of interphase, Cdc2/28 is unphosphorylated, but then its association with cyclin induces phosphorylation on tyrosine 15 and threonine 161 to produce preMPF. At entry into mitosis, preMPF is converted into its active form by Cdc25, which removes the phosphate from tyrosine 15. Active MPF induces the destruction of cyclin, resulting in the inactivation of MPF. (From MURRAY and HUNT 1993, by permission)

To this end, we have intensively investigated circadian clock control of the CDC in the algal flagellate *Euglena gracilis* Klebs, a highly rhythmic system (EDMUNDS 1982; EDMUNDS and HALBERG 1981) that has been well characterized physiologically and biochemically (EDMUNDS 1965b; BUETOW 1968a,b, 1982). The formal properties of circadian clocks – in particular, entrainability, persistence, susceptibility to, and ability to display phase shifts, and temperature compensation of the free-running period (τ), have been found to characterize the circadian rhythm (CR) of cell division in both the wild type (EDMUNDS and LAVAL-MARTIN 1984) and achlorophyllous, photosynthesis-deficient ZC mutant (CARRÉ et al. 1989a). We now summarize recent results from our laboratory that suggest a key role for cAMP (which exhibits bimodal circadian rhythmicity in *Euglena*) in circadian clock regulation of cell division cycles in this unicell.

I. Regulation of the Cell Division Cycle

The cell division cycle of eukaryotes is a complex cascade of events that culminates in cell duplication (CROSS et al. 1989). Rapid progress has been made within the past several years towards dissecting the regulatory network that provides precise coordination of these processes. A unifying view has emerged: The CDC consists of transitions from one regulatory state to another (LEWIN 1990; MURRAY and KIRSCHNER 1989). These transitions initiate the modification of substrates that determine the physical state of the cell and are themselves feedback controlled. The basic mechanisms, as elucidated by a combination of genetic and biochemical analyses, appear to have been conserved in organisms ranging from unicells such as yeasts, to *Xenopus*, clams, and starfish eggs, and to cultured human cells (MURRAY and HUNT 1993).

Two crucial transition control points exist in the G_1 phase and at the G_2/M boundary. The M phase is characterized by the activation of a kinase [MPF, or maturation (M phase)-promoting factor, so named for its ability to induce cell division when injected into *Xenopus* oocytes], which consists of two component subunits. One is the $p34^{cdc2}$ protein (the product of the *Schizosaccharomyces pombe* gene *cdc2*, or its homologue, *CDC28*, in *Saccharomyces cerevisiae*), which is able to phosphorylate casein and histone H_1 in vitro and has maximal kinase activity at mitosis. The other is cyclin, whose cellular concentration fluctuates during the CDC. It accumulates gradually during interphase, forms a complex with $p34^{cdc2}$, and activates the protein kinase; it is then degraded during mitosis, turning off kinase and MPF activity. The cyclin-*cdc2*/*CDC28* (MPF) system thus behaves like an oscillator, or "clock," which is reset to its interphase state during mitosis (MURRAY and KIRSCHNER 1989), and which appears to constitute a universal cell-division-cycle "engine" (Fig. 1). The first cell divisions in sea urchin eggs, in frog oocytes, and in *Drosophila* embryos are triggered by the accumulation of cyclins above a critical concentration. Less is known about the initiation of DNA synthesis, but a complex comprising p34^{cdc2}, and possibly other cyclin-like molecules, may play a role (CROSS et al. 1989; MURRAY and HUNT 1993).

There is also strong evidence that autonomous oscillators, independent of both p34^{cdc2} and cyclins, are part of, or are coupled to, the CDC (see EDMUNDS 1984, 1988). Thus, when the *S. pombe* cell cycle is arrested by a temperature-sensitive *cdc2* mutation, the activity of nucleotide diphosphate kinase and the rate of CO_2 production continue to oscillate, with a period about that of the CDC (CREANOR and MITCHISON 1986; NOVAK and MITCHISON 1986). The view, therefore, that the duration of the CDC would depend only on the accumulation of cellular components such as cyclins is overly simplistic, particularly in the case of circadian timers, which have been shown to underlie persisting cell division rhythmicity in numerous systems (EDMUNDS and LAVAL-MARTIN 1984).

II. Cell Division Rhythmicity in Wild-Type and Mutant *Euglena*

The CDCs of many algae, fungi, and protozoans exhibit persisting circadian rhythms of cell division, or "hatching" (EDMUNDS 1978; EDMUNDS and ADAMS 1981). Division occurs only at a certain phase of the circadian cycle – often the times ("subjective" nights) in constant darkness (DD) or light (LL) corresponding to the dark intervals in an environmentally synchronizing LD cycle. This "gating" phenomenon, reflecting an interaction between a circadian oscillator and the CDC, has been intensively investigated in the eukaryote *Euglena*. The formal properties of circadian clocks – in particular, entrainment by diurnal and non-24-h light (COOK and JAMES 1960; EDMUNDS 1965a, 1978; EDMUNDS and FUNCH 1969a, b; LEDOIGT and CALVAYRAC 1979) and temperature (TERRY and EDMUNDS 1970a,b), persistence (EDMUNDS 1966), initiation, phase shiftability (EDMUNDS et al. 1982), phase singularity (MALINOWSKI et al. 1985), and temperature compensation (ANDERSON et al. 1985) – have been found to characterize the circadian rhythm of cell division in photoautotrophically cultured wild-type *Euglena gracilis* Klebs (strain Z) (reviewed by EDMUNDS and LAVAL-MARTIN 1984).

In order to eliminate a cellular compartment in which the clock might lie and to obviate possible signaling influences of the LD cycles previously used to elicit rhythmicity and provide energy for growth, we chose the achlorophyllous ZC mutant of *Euglena* – an obligate organotroph – for a series of studies parallel to those carried out in the wild-type strain (CARRÉ et al. 1989a). Division rhythmicity was entrained by LD:12,12 or even by a one-pulse skeleton photoperiod (LD:1,23), and the rhythm free-ran in ensuing DD for at least 8 days with a period (τ) of about 25 h, providing that the overall doubling time (g) was longer than 24 h (see Fig. 3). Similar, though less extensive, results had been obtained earlier with other photosynthesis mutants of *Euglena* (EDMUNDS 1975, 1978; EDMUNDS et al. 1976; JARRETT and EDMUNDS 1970).

A "strong" (type 0) phase-response curve (PRC) for light signals in the ZC mutant strain was generated from 15 phase-shifting experiments (CARRÉ et al. 1989a). Phase delays ($-\Delta\phi$) were obtained between circadian times (CT) 08 and 18, and phase advances ($+\Delta\phi$) between CT 18 and 0: the discontinuity ("breakpoint") inherent to this type of plot occurred at about CT 00. The PRC for the mutant resembled that previously obtained with the wild-type strain (EDMUNDS et al. 1982). These studies, therefore, provide even more conclusive evidence for the role of a circadian oscillator in the control of the CDC and have effectively circumvented the problem of the dual use of imposed LD cycles: as an energy source, or "substrate," for growth, and as a timing cue (zeitgeber) for the underlying clock. They further demonstrate that the presence of a functional chloroplast compartment is unnecessary for circadian timekeeping in *Euglena* and suggest that phase-shifting the clock by light occurs via a similar pathway in photosynthetic and nonphotosynthetic cell types.

Thus, the circadian oscillator(s) plays a key role in the control of the CDC in *Euglena*, taken to be representative of other eukaryotic microorganisms (and, perhaps, multicellular systems). Mitosis would not be an essential part of the oscillator but would lie downstream from it: blockage of cell division should not stop the system from oscillating (EDMUNDS and ADAMS 1981; EDMUNDS and LAVAL-MARTIN 1984). Therefore, cell division would be a "hand" of the underlying clock. We have tested this hypothesis in the wild type in two ways: (1) If the division rhythm (free-running in LD:3,3) was stopped due to low initial levels of vitamin B_{12}, and if this inhibition subsequently was released by readdition of B_{12}, the rhythm started up again in phase with an unperturbed control and (2) if a pulse of lactate was given to a free-running culture, temporarily accelerating the CDC and overriding circadian oscillator controls (JARRETT and EDMUNDS 1970), the phase of the rhythm when it was finally restored after the substrate had been depleted was in phase with that of an unperturbed control. These results are consistent with those found previously for the in-phase restoration of rhythmicity in the P_4ZUL mutant free-running in LL by the addition of sulfur-containing compounds to the medium (EDMUNDS et al. 1976).

B. Coupling Between Oscillator and Cell Cycle: Role of Cyclic AMP

The mechanism whereby the circadian oscillator (and other higher frequency oscillators) interact with the $p34^{cdc2}$/cyclin pathway for the control of cell division is an important question that we now address. Cyclic AMP, which is known to play a pivotal role in cellular regulation (WHITFIELD et al. 1987), if not an element of the clock itself, may effect such a coupling between oscillator and CDC, participating in the gating of CDC events to specific phases of the circadian cycle. Indeed, cAMP seems to have the capacity to control certain rate-limiting steps in CDC progression in many cell types, stimulating the proliferation of some cells, but having the opposite effect, or no effect at all, on others (DUMONT et al. 1989; WHITFIELD 1990). Such contradictory findings might be explained by a variation in different systems of the cAMP concentration required for optimal stimulation. Interestingly, transient increases of cAMP are correlated with cell cycle transitions at both the G_1/S and G_2/M boundaries in many cell types. Genetic experiments in the yeast *Saccharomyces* (MATSUMOTO et al. 1983) and pharmacological studies in mammalian cells (for reviews, see BOYNTON and WHITFIELD 1983; WHITFIELD et al. 1987) have shown that a transient rise and the ensuing fall in the cAMP level are necessary for the initiation of DNA synthesis, and that a second cAMP surge is correlated with the onset of mitosis. These signals coincide with the action(s), respectively, of $p34^{cdc2}p60$ (G_1 cyclin) protein kinase complexes and of $p34^{cdc2}$ cyclin B (mitosis cyclin) protein kinase complexes. Similar changes of cAMP concentration at different phases of the CDC have been shown to occur in *Euglena* (CARELL and DEARFIELD 1982).

In recent studies (CARRÉ et al. 1989b), we have shown that cultures of the achlorophyllous ZC mutant of *Euglena*, displaying free-running, circadian rhythms of cell division in DD (see Fig. 3), also exhibit bimodal circadian variations in their level of cAMP (Fig. 2). These oscillations in cAMP concentration could be phase-shifted by light signals in a manner predicted by the PRC previously derived for the phase-resetting of the cell division rhythm by light pulses (CARRÉ et al. 1989a). These results strongly suggested that cAMP levels are controlled by the same endogenous clock as that regulating the CDC. Maximum cAMP levels occurred at the beginning of the light period (CT 0–2, when cells are in G_1) and at the onset of darkness (CT 12–14, corresponding to the onset of mitosis). At the time, these variations appeared to be to be CDC independent, since they persisted after the cultures had reached the stationary phase of growth (Fig. 2B,C). More recent results indicated, however, that in cell populations maintained for some days in stationary-phase conditions in the virtual absence of division, the oscillations in cAMP level continued only with greatly reduced amplitude or could not be detected. We believe that such a periodic, bimodal cAMP signal (initiated directly by the circadian oscillator, or in conjunction with it) may participate in the gating of CDC events to specific phases of the circadian cycle.

I. Perturbation of the Cell Division Rhythm by Cyclic AMP

Thus, cAMP may play a central role as a coupling link for control of the CDC by the circadian clock, regulating the transition through the G_1/S or the G_2/M boundaries, or both. If this hypothesis holds true, conditions that would disturb or override the control of cAMP level by the oscillator should also cause perturbations of the cell division rhythm.

We first tested for the possible role of these periodic cAMP signals in the generation of cell division rhythmicity in *Euglena* by examining the effects of cAMP signals on the division rhythm (CARRÉ and EDMUNDS 1993). Perturbations of the cAMP oscillation by exogenous cAMP (Fig. 3) resulted in the temporary uncoupling of the CDC from the circadian timer, causing either a shortening or a lengthening of the CDC, depending on the phase of the CDC

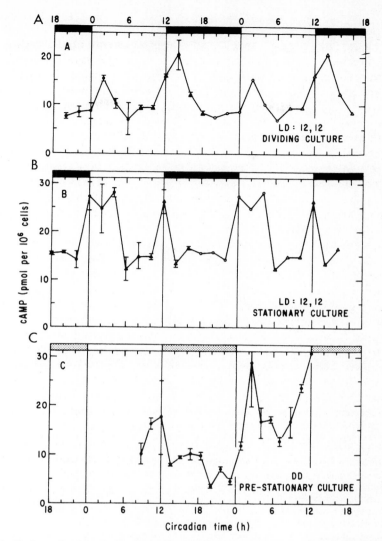

Fig. 2A–C. Entrained and free-running circadian oscillations of the cAMP level in synchronously dividing and in nondividing cultures of the achlorophyllous ZC mutant of *Euglena gracilis* (strain Z), grown at 16.5 °C on a mineral medium supplemented with ethanol (0.1%, v/v). **A** Diurnal variations of the cAMP level in cultures exhibiting rhythmic cell division, entrained to a 24-h period by an LD:12,12 cycle. **B** Persistence of the bimodal 24-h oscillation of cAMP level after the cells had ceased dividing. **C** Variations of cAMP level in a culture displaying a free-running rhythm of division in DD and just entering the stationary growth phase. For experiments **A** and **B**, cell extracts were prepared every 2 h over a 12-h time span from two cultures entrained by out-of-phase LD cycles, thus scanning the entire circadian period. Results were double-plotted (*open symbols*) as a function of the circadian time. In experiment **C**, cell extracts were prepared every 2 h over a 34-h time span. The onset of the last burst of cell division (also corresponding to the first peak on the graph) was used as a reference for CT 12, and the interval between the first and the third peak gave the estimated value (27 h) for τ. In order to facilitate comparison with curves **A** and **B**, results were then normalized to a 24-h period. (From CARRÉ et al. 1989b)

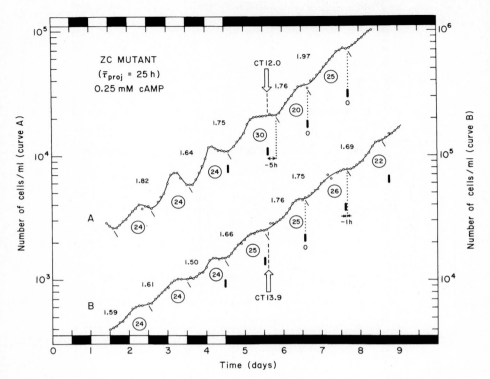

Fig. 3. Perturbation of the circadian rhythm of cell division by cAMP in the achlorophyllous ZC mutant of *Euglena*. Cultures that had been synchronized by LD:12,12 were transferred to DD. Cyclic AMP was injected into the culture during the second circadian cycle in DD at the circadian time (*CT*) indicated by the *open arrow*. The onset of cell division was used as a phase marker (*vertical dotted line*) and compared to the theoretical phase of the rhythm (*black markers*) in an unperturbed control culture, projected from the last division step before the perturbation with a period of 25 h, a value which corresponded to the free-running period of the rhythm in DD. Phase differences are given here in real time. The period length (interval between successive division bursts) is *encircled*. Step size (factorial increase in cell concentration, from plateau to plateau) is given *on the left of the corresponding step*. (From CARRÉ and EDMUNDS 1993)

when the drug was applied (Fig. 4A). Delays of the next synchronous division step (up to 9 h) were obtained when cAMP was given between CT 03 and CT 09, and advances were obtained when cAMP was given between CT 16 and CT 22. Maximum effects were obtained at CT 06–08 and at CT 18–20, corresponding to times when endogenous cAMP levels were minimal (see Fig. 2). Dose-response curves were also derived at CTs when maximum $+\Delta\phi$'s or $-\Delta\phi$'s of the CDC had been obtained. We found that as little as 1–5 nM cAMP was enough to perturb cell cycle transit. The circadian oscillator, in contrast to the CDC cyclin clock, and unlike excitable tissue in which cAMP does phase-shift the output rhythm (ESKIN et al. 1982; PROSSER and GILLETTE 1989), was

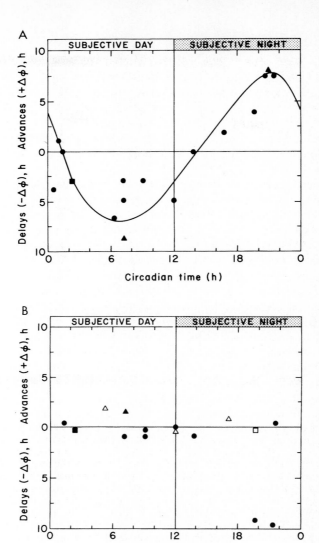

Fig. 4A,B. Phase-response curves for the perturbation of the free-running rhythm of cell division by cAMP in the achlorophyllous ZC mutant of *Euglena*. The curves were derived from 17 experiments similar to those shown in Fig. 3. **A** Advances or delays of the first cell division step following cAMP injection into the culture medium. **B** Steady-state phase shifts, measured (when possible) 3–4 days after the perturbation. *Open symbols*, the cell suspension was diluted 50-fold with fresh medium after a 1-h exposure to cAMP. *Filled symbols*, culture was not diluted. *Triangles*, 500 μM cAMP; *circles*, 250 μM cAMP; *squares*, 100 μM cAMP. (From CARRÉ and EDMUNDS 1993)

not perturbed by the addition of exogenous cAMP, the division rhythm soon returning to its original phase (Fig. 4A).

The advances and delays of the cell division steps observed on the growth curves following cAMP injection reflected real changes in the rate of CDC progression. Measurement of the DNA content of cells by flow cytometry (CARRÉ and EDMUNDS 1993) indicated that cAMP injected between CT 06 and CT 08 delayed progression through S phase, and perhaps also through mitosis (Fig. 5A). When added at CT 18–20, cAMP accelerated the G_2/M transition (Fig. 5B). These effects on the cell division rhythm were only transitory, and after 48 h cell division occurred at the same phase as in unperturbed controls, a finding indicating that the circadian oscillator itself was unaffected (see Fig. 4B).

Thus, these experiments demonstrated that exogenous cAMP signals cause lengthenings or accelerations of the CDC, depending on the CT when the drug is added. We still needed to determine whether the endogenous circadian variations of cAMP level were of sufficient amplitude to have similar effects on cell cycle progression. Our approach was to test the effect of drugs that reduce the amplitude of the cAMP oscillation, or that keep cAMP at a level such that all cAMP receptors should be permanently saturated. We would expect such drugs to prevent the expression of cell division rhythmicity.

Forskolin, which we have shown to stimulate adenylate cyclase in *Euglena* at times when the activity of the enzyme is minimal (TONG et al. 1991), would be expected to permanently elevate the level of cAMP. When forskolin was injected into a culture of the ZC mutant that was maintained in LD:12,12, the cAMP level was increased by as much as threefold, as compared to an untreated control (TONG et al. 1991). Maximum increases of cAMP level corresponded to times when cAMP in the control culture was at its minimum: forskolin had little effect at times when cAMP was at its maximum. As a result, the cAMP level in the presence of forskolin never exceeded the range of cAMP concentrations in control cultures, but oscillated with a reduced amplitude, and at a higher level. When forskolin was added to a free-running culture of the ZC mutant (CARRÉ and EDMUNDS 1993), previously synchronized by LD:12,12, at the time of its transfer to DD (at approximately CT 12), we observed a rapid desynchronization of the cell population (Fig. 6A). The cells appeared to go through one synchronous division step, then resumed nearly exponential growth. Very low amplitude waves of cell division were still visible, however, for a minimum of three more days, a sign that the circadian timer was running unaffected. Such a rapid loss of cell division rhythmicity sometimes occurs after transfer of control cultures to DD, but heretofore has never been

---

Fig. 5A,B. Effects of cAMP on cell cycle progression. Cultures of the ZC mutant synchronously dividing in DD were divided into two halves, one of which received cAMP (*open histograms*), while the other was used as a control (*solid histograms*). Cells were harvested every 4 h following the perturbation and were fixed in 70% ethanol, treated with RNAse A, and then stained with propidium iodide immediately prior to analysis in a fluorescence-activated cell sorter (FACS). **A** The addition of cAMP at CT 06 delayed the progression of cells through S phase. **B** The addition of cAMP at CT 20 stimulated mitosis (note the decrease of the G_2 peak at $t = 12$ h). The time elapsed since the addition of cAMP to the culture medium is indicated *on the right of the corresponding histograms*, and the circadian time is given *on the left*. (From CARRÉ and EDMUNDS 1993)

Regulation of Cell Division Cycles by Circadian Oscillators

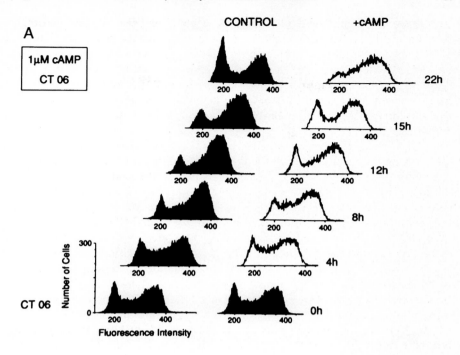

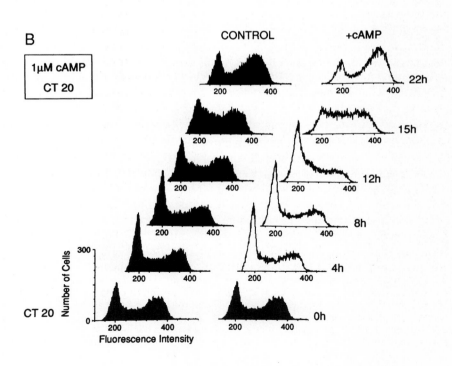

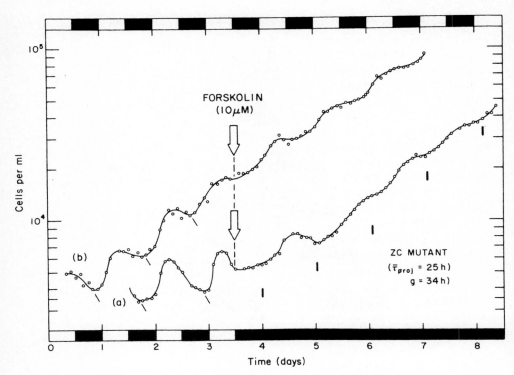

Fig. 6. Effect of forskolin on the cell division rhythm in the ZC mutant. Cell division was synchronized by LD:12,12 prior to the addition of forskolin. *(a)* The culture was transferred to DD at the time of the addition of the drug (indicated by *open arrow*). *(b)* The culture was maintained in LD:12,12. Both experiments show a rapid loss of cell division rhythmicity, but the generation time was unaffected. (From CARRÉ and EDMUNDS 1993)

observed in cultures that had been maintained in 24-h LD cycles. When forskolin was added to a culture that was entrained by LD:12,12, a rapid damping of the cell division rhythm was also observed (Fig. 6B). This effect was not phase dependent: Similar results were obtained when the drug was added at CT 12 (Fig. 6A,B) or at CT 04, 08, 15, or 00 (not shown). These results suggest that the amplitude of the cAMP oscillations determines the amplitude of the cell division rhythm and are consistent with the hypothesis that cAMP acts downstream from the oscillator – perhaps in a parallel pathway, as ZATZ (1992) has suggested for chick pineal cells – and that the cAMP oscillation is an essential component of the signaling pathway for the control of the CDC by the circadian oscillator.

II. Upstream Analysis: Genesis of the Oscillation in Cyclic AMP

What upstream controls regulate the level of cAMP, presumably downstream from the clock? Results from a series of experiments undertaken to elucidate this question for *Euglena* can be summarized as follows. Key factors in the cAMP metabolic pathway are two enzymes responsible for its generation and degradation, namely, adenylate cyclase (AC) and phosphodiesterase (PDE). In LD:12,12, these enzymes were found to undergo bimodal, circadian variation of activity in both dividing and nondividing cultures of the ZC mutant (TONG et al. 1991), although more recent results suggest that the amplitudes of the rhythms are attenuated in later stationary-phase populations. Maximal AC activity occurred 2 h after the onset of the light interval (CT 02) and at the beginning of darkness (CT 12–14); these times corresponded to the acrophase profile for the rhythmic changes in cAMP content (see Fig. 2) that have been previously reported (CARRÉ et al. 1989b). The activity of PDE also exhibited a daily oscillation, but with an inverse phase pattern. Both the AC and PDE activity rhythms persisted in DD (Fig. 7).

The activity of AC was activated significantly in vivo by forskolin (TONG et al. 1991), a highly specific activator of AC that can fully stimulate AC activity directly, independently of G proteins. Forskolin stimulation of AC did not result in uniform effects throughout the 24-h day but varied considerably with circadian time. The degree of AC potentiation by an in vivo forskolin pulse appeared to be inversely dependent on the basal level of AC: at its trough phase (CT 20), AC activity was significantly stimulated by as much as 114% (bringing it to exactly the same level as that found at the peak phase), while at the peak phase (CT 12), only a 10% stimulation was observed. The same result has been demonstrated by LEMMER et al. (1986) in rat heart: The level of cAMP was significantly increased by the addition of forskolin (5 mg/kg) at CT 12 in LD:12,12, but the effect disappeared at CT 20 under the same experimental conditions. Thus, since forskolin appears to stimulate AC activity at different circadian times to the same maximal value regardless of its initial basal level, we conclude that the oscillatory activity of the enzyme derives from modulatory cellular effectors emanating from the circadian pacemaker rather than by changes in the amount of the enzyme protein itself.

While previous studies have emphasized the role of AC, along with cAMP, in signal transduction and neoplastic transformation (WHITFIELD 1990), less attention has been paid to the role of PDE. PDE is an important component of the signal system because it is responsible for the destruction of cAMP after a pulse of synthesis and, thereby, permits the cell to recover from a refractory state induced by cAMP. We found that the addition of 50 μM 3-isobutyl-1-methylxanthine (IBMX) to intact cells at CT 20, corresponding to the peak of the rhythm in PDE activity (see Fig. 7), depressed PDE activity by about 51%, while at CT 12 (the trough phase) IBMX had only a slight (8%) effect (TONG et al. 1991). It is known that the hydrolysis of cAMP by cell extracts is catalyzed by at least five types of PDE isoenzymes: calmodulin-stimulated PDE, cyclic

GMP (cGMP)-stimulated PDE, cGMP-specific PDE, low-K_m PDE, and nonspecific PDE (BEAVO 1988). IBMX is a nonspecific PDE inhibitor, which can affect the activity of calmodulin-stimulated PDE and cGMP-inhibited PDE 20–25 times as much as that of cGMP-stimulated PDE in mammalian cells (BEAVO 1988). Thus, the inhibitory effects of IBMX on PDE observed in *Euglena* extracts may be the resultant of its action on several PDE species. Furthermore, the differing effects of IBMX on PDE activity at CT 12 and CT 20 showed that the degree of inhibition was circadian time dependent and that IBMX had little or no effect on PDE at CT 12 (corresponding to the time when the activity was at a minimum). At CT 20 (when PDE activity was at its

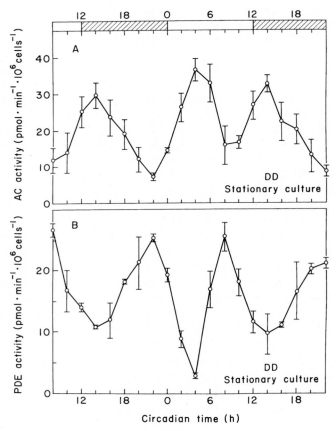

Fig. 7A,B. Free-running circadian rhythms in AC activity (**A**) and PDE activity (**B**) measured over a 40-h time span in stationary cultures of the ZC mutant of *Euglena gracilis* maintained in constant darkness (DD). Cultures, synchronized by LD:12,12, were transferred to DD shortly before the cells had ceased dividing; the onset of the last cell division burst was used as a phase reference point for CT 12. *Hatched bars* subjective night. To facilitate the comparison with the curves in Figs. 2 and 4, free-running periods have been normalized to 24 h. (From TONG et al. 1991)

peak), IBMX brought the activity down to the same level as that observed at CT 12. This suggests that the relative amounts of the different types of PDE vary with circadian time and that a specific PDE (or a specific subset of enzymes that are inhibited by IBMX) is responsible for the oscillation.

In sum, these results indicate that the rhythms of both AC and PDE may be the main factors in generating the circadian oscillations of cAMP content in *Euglena*, which appear to be under control of an endogenous pacemaker. We are now extending this upstream analysis towards the clock, examining the possible roles of Ca^{2+}, calmodulin, inositol 1,4,5-triphosphate (IPE), and cyclic GMP (cGMP) in regulating the activities of AC and PDE (EDMUNDS and TAMPONNET 1990). Thus, the "yin yang," or dualism, hypothesis proposed about 20 years ago (GOLDBERG et al. 1974) provided a theory of biological regulation through the opposing actions of cAMP and another second messenger, cGMP. According to this theory, reciprocal changes in the levels of the two cyclic nucleotides should bring about a maximum expression of the mitogenic signal. An interesting observation was that an increase in tissue cGMP levels was accompanied by either a decrease, an increase, or no change in cAMP concentration. The question then arises as to the origin of the cAMP variation, and whether these changes in cGMP/cAMP ratios are stochastic, or whether, as the hypothesis posits, cGMP may serve as a positive mediator.

To determine if cGMP plays a role in the mediation of circadian rhythmicity of the AC-cAMP-PDE system in the ZC mutant of *Euglena*, the levels of cAMP and cGMP were monitored in synchronized cell populations, and the effects of the cGMP analog 8-bromo-cGMP (8-Br-cGMP) and the cGMP inhibitor LY 83583 (6-anilinoquinone-5,8-quinone) on the activity of AC and PDE, as well as on the level of cAMP, were measured in vivo (TONG and EDMUNDS 1993). A bimodal, 24-h rhythm of cGMP content was found in both dividing and nondividing cultures in either a LD:12,12 or DD (Fig. 8). The peaks and troughs of the cGMP rhythm occurred 2 h in advance of those of the cAMP rhythm (see Fig. 2). The addition of 8-Br-cGMP at different circadian times increased the cAMP level in vivo by two to five times, while LY 83583 reduced the amplitude of the cAMP rhythm so that it disappeared. The effects of 8-Br-cGMP on the activity of AC and PDE were circadian phase dependent and consistent with the changes in cAMP content. These findings suggest that cGMP might serve as an upstream effector that mediates the cAMP oscillation by regulation of the AC-cAMP-PDE system.

III. Downstream Pathway: Cyclic AMP-Dependent Kinases

In summary, we have shown that cAMP acts downstream from the circadian oscillator, on (or in parallel with) the coupling pathway for the control of the CDC. cAMP signals modulate cell cycle progression at restriction points located in both the G_1 and the G_2 phases of the CDC and cause either advances or delays of the CDC depending on the CT at which the signal is given. Whatever the cell cycle regulatory pathways that are affected, it is difficult to

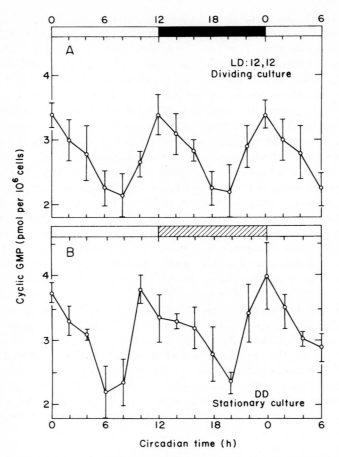

Fig. 8A,B. Circadian rhythms of cGMP in dividing (**A**) and nondividing (**B**) cultures of the achlorophyllous ZC mutant of *Euglena gracilis* Klebs (strain Z), grown in LD:12,12 (**A**) and DD (**B**) at 16.5 °C on a mineral medium supplemented with ethanol (0.1%). The results shown are averages of two independent experiments with duplicate assays; *error bars* the range of the values. *Hatched bar* **B** indicates subjective night in order to facilitate comparision with **A**. (From TONG and EDMUNDS 1993)

explain how identical signals can selectively perturb different pathways depending on the CT at which they are applied, so that cell cycle delays are obtained during the subjective days, and advances during the subjective nights. One possibility is that cAMP acts through different "receptors" that selectively modulate one or the other of the two regulatory pathways.

Indeed, we have recently demonstrated that the ZC mutant of *Euglena gracilis* contains two types of cAMP-dependent kinases (cPKA and cPKB), which have different affinities for cAMP and for several cAMP analogs (CARRÉ and EDMUNDS 1992). Cell extracts were found to contain two cAMP-binding proteins, which bound cAMP with a high affinity (K_d values of 10 nM and 30 nM) and which could be separated by diethylaminoethanol (DEAE)-cellulose chromatography (Fig. 9). Protein kinase activity was assayed using Kemptide as a substrate (specifically phosphorylated by cAMP-dependent kinases from mammalian cells). Stimulation of kinase activity by cAMP was observed after partial purification by DEAE-cellulose chromatography. Two peaks of activity were resolved, corresponding to distinct enzymes with different cAMP-analog specificities (see Table 1). Thus, cAMP signaling in plant cells may proceed by the phosphorylation of target proteins by cAMP-dependent kinases, in a manner similar to that of animal cells.

A correlation between a cAMP-analog's potency in activating one type of kinase and in causing physiological responses can provide evidence for a role of this enzyme in mediating the effect of cAMP. The differential activation of the two kinases identified in *Euglena* extracts by these cAMP analogs (see Table 1) provided us with a tool for the study of their respective roles in the control of cell cycle progression, and we determined the minimum doses of cAMP or of cAMP analogs that caused perturbations of the cell division rhythm during the subjective day, or during the subjective night (CARRÉ and EDMUNDS 1993). Interestingly, different results were obtained at CT 06–08, and at CT 18–20, suggesting that the effect of cAMP at the different CTs was mediated by two different cAMP kinases (Table 1). Thus, 8-benzylamino-cAMP (8-BZA-cAMP), which selectively activates cPKA, induced delays of cell division when added at CT 06–08, but had no effect at CT 18–20. Reciprocally, 8-(4-chlorophenylthio)-cAMP (8-CPT-cAMP), which specifically activates cPKB, induced early cell division and a loss of division rhythmicity when added at CT 18–20 but did not perturb the cell division rhythm when added at CT 06–08. Moreover, there was a correlation between the doses of cAMP, 8-BZA-

Table 1. Cyclic AMP-analog specificities of cAMP-dependent kinases from the ZC mutant of *Euglena* and minimum doses of cAMP analogs that perturb the cell division rhythm when given at CT 06–08 or at CT 18–20. The magnitude of the stimulation (percentage of basal activity) of the cyclic AMP-dependent kinases cPKA and cPKB by the cyclic AMP analogs is given in parentheses (columns 1 and 3). The maximum concentration tested was 1 mM. The lowest concentrations of the cyclic AMP analogs that were found to reproduce the effects of 250 nM cyclic AMP applied at either CT 06–08 (column 2) or at CT 18–20 (column 4) are given. The maximum concentration tested was 1 μM. (From CARRE and EDMUNDS 1993)

Analog	1 cPKA (K_a, nM)	2 CT 06–08 (delays)	3 cPKB (K_a, nM)	4 CT 18–20 (advances)
Cyclic AMP	5 (200%)	1	50 (200%)	50
8-(4-Chlorophenylthio)-cAMP	No effect	No effect	2 (150%)	0.1
8-Benzylamino-cAMP	200 (600%)	100	No effect	No effect
N^6-Monobutyryl-cAMP	5 (130%)	10	50 (130%)	ND

ND, not determined.

cAMP, 8-CPT cAMP, and NY-monobutyryl-cAMP (6-MBT-cAMP) that caused perturbations of the CDC at CT 06–08 and the K_a values of cPKA for these analogs, a result that suggests that cPKA mediates the delaying effects of cAMP at those CTs. Similarly, there was a correlation between the doses of the same nucleotides that caused perturbations of the CDC at CT 18–20 and the K_a of cPKB for these analogs, an observation indicating that cPKB mediates the accelerating effects of cAMP at CT 18–20. A simple explanation for these findings would be that these kinases are expressed at different stages of the CDC, as has been previously described for type I and type II cAMP-dependent kinases from mammalian cells.

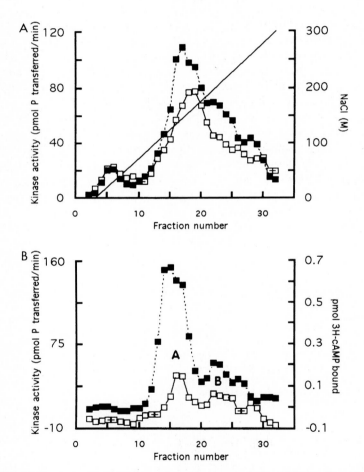

Fig. 9A,B. Separation of cAMP-dependent kinases and of cAMP-binding proteins by DEAE-cellulose chromatography. **A** The basal kinase activity (*open symbols*) was stimulated in the presence of cAMP (*filled symbols*). **B** After subtraction of the basal activity, the cAMP-stimulated kinase activity (*open symbols*) was resolved into two peaks (*A, B*), which overlapped with two peaks of cAMP-binding activity (*filled symbols*). (From CARRÉ and EDMUNDS 1992)

IV. Model for Circadian Control of the Cell Division Cycle

These results can be incorporated into a model (Fig. 10) for the coupling of the CDC to the circadian oscillator (CARRÉ and EDMUNDS 1993). We propose that the cAMP surge at CT 00–02 delays DNA synthesis and holds the cells at a restriction point in G_2 to prevent cell division during the subjective day; the cells are released from this blockage after cAMP levels subside, and the G_2/M transition, or mitosis itself, is accelerated by the second cAMP peak, at CT 12–14, so that mitosis is phased to the subjective night. The delaying effects of cAMP on cell cycle progression during the subjective day would be mediated by the activation of cPKA, and the stimulation of mitosis during the subjective night by the activation of cPKB. Activation of either of these kinases would cause the phosphorylation of a different set of targets and perturb different cell-cycle control pathways. cPKA and cPKB may be expressed at different phases of the CDC. Alternatively, the level of these enzymes might exhibit circadian variations, with cPKAs being expressed during the subjective day and cPKB during the subjective night. Another possibility is that the level of one of their downstream targets oscillates, so that only cPKA activation has an effect on cell cycle progression during the subjective day, and cPKB during the subjective night. Future studies of the regulation of cAMP-dependent kinases during the CDC (or during the circadian cycle) and the identification of targets that are selectively phosphorylated by these enzymes will be necessary to ascertain this part of the model.

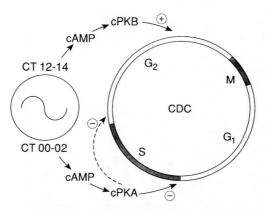

Fig. 10. Model for the "gating" of cell division by the circadian oscillator. We propose that the cAMP surge at CT 00–02 delays DNA synthesis (and perhaps holds the cells at a restriction point in G_2) to prevent cell division during the subjective day. The cells are released from this (these) blockage(s) after cAMP levels subside, and mitosis is stimulated by the second cAMP peak (at CT 12–14) so that cell division is phased to the subjective night. Opposite effects of cAMP on cell cycle progression are explained by its action through two different cAMP-dependent kinases (cPKA and cPKB), which would be expressed at different stages of the CDC (or at different phases of the circadian cycle) and which have different sets of targets. (From CARRÉ and EDMUNDS 1993)

C. Problems and Prospects: Interface with the Cyclin Clock

How does the circadian oscillator, acting through (or in conjunction with) the AC-cAMP-PDE system and the cAMP-dependent kinases (cPKA and cPKB) that we have identified, ultimately couple to the cyclin clock (see Fig. 1) regulating the CDC (see model in Fig. 10)?

Preliminary Western blot analysis of protein extracts from dividing and stationary cultures of the ZC mutant in DD confirms the presence of a homolog of p34^{cdc2}, the catalytic subunit of MPF, reported earlier in wild-type *Euglena* and in other algae (JOHN et al. 1989, 1993) and suggests that its abundance throughout the CDC is invariant, decreasing to a much reduced level when cells cease dividing (EDMUNDS and MOHABIR 1995). Further, a relative shift in electrophoretic mobility of the monomer was observed to occur in a cycle-dependent manner: a more slowly moving band appeared at CT 04 and disappeared at CT 12. Such changes in mobility have been attributed to phosphorylation and dephosphorylation. These initial findings, if replicated, suggest that the activity of this polypeptide depends on post-translational mechanisms, and, thus we have turned our attention to the regulatory subunit(s) of MPF. On the assumption that cyclin B is a likely candidate, we have begun Western blot analyses using monoclonal antibodies and have found that, in contrast to p34^{cdc2}, the abundance of cyclin B in protein extracts varies quantitatively during both the circadian cycle and the CDC, being detected only at CT 20 and CT 24 (00). Further characterization of this key player in CDC regulation would relate to its state of phosphorylation, which can be determined by phosphorylation in vivo with ^{32}P followed by immunoprecipitation, Western blot analysis, and autoradiography. Comparison of extracts from dividing and nondividing cells at different CTs in DD should help dissect CDC-dependent and circadian clock-driven components.

Mitotic kinase activity is dependent on the association of the two subunits of MPF and the phosphorylation state of the complex. Association of cyclin B with the p34^{cdc2} subunit is required for mitotic kinase activity, but only after dephosphorylation of the latter on tyrosine residue 15 by a phosphatase (Cdc25 in fission yeast) is that activity expressed (JESSUS and BEACH 1992). Kinase activity promotes the degradation of cyclin B and thus is itself nullified, thereby preempting catastrophic mitotic arrest (FELIX et al. 1990). Rephosphorylation of tyrosine 15 occurs after mitosis by a tyrosine kinase (TK), the *wee1* gene product of fission yeast. These modulators of MPF kinase activity, TK and tyrosine phosphatase (TP), might very well constitute upstream links to the circadian oscillator. The findings that (1) genistein, a specific TK inhibitor, can reset the clock in the molluscan eye (ROBERTS et al. 1992) and (2) *wee1* mutants of fission yeast, defective in TK, lose circadian rhythmicity of division (F. KIPPERT, personal communication) suggest that the TK of p34^{cdc2} might mediate circadian control of the CDC, or even be a component of both clocks.

Conceptually, circadian clock precision in the gating of cell division perhaps might best be achieved if the coupling point were to occur at the M phase of the CDC, perhaps by the dephosphorylation of tyrosine 15 via circadian-clock-controlled transcription of TP, although it is possible that the S phase of the CDC also is circadian clock controlled, as HOMMA and HASTINGS (1989) have shown for the dinoflagellate *Gonyaulax polyedra*. Transcriptional regulation of the string gene, the *Drosophila* homolog of the Cdc25 product (TP) of fission yeast, is known to control CDC progression after mitosis 13 during embryogenesis (EDGAR and O'FARRELL 1989), and the cyclic accumulation of Cdc25 itself regulates mitosis (MORENO et al. 1990). On the other hand, the abundance of p72, the frog homolog of Cdc25, does not oscillate, although its association with the Cdc2-cyclin B complex does so, limited perhaps by an inhibitor (JESSUS and BEACH 1992). It is possible that this putative inhibitor of TP activity, in turn, is regulated by cAMP-dependent kinases, thereby linking the CDC to the circadian clock via cAMP. This type of negative regulation of $p34^{cdc2}$ has been found in mouse fibroblast cell-free extracts (HOHMANN et al. 1993) and may occur in *Euglena*. Thus, the second peak in cAMP level that occurs in *Euglena* just before division may result in the inactivation of the inhibitor of TP, thereby triggering mitosis by dephosphorylation and activation of MPF. Indeed, phosphatase activity of the frog Cdc25 homolog is dependent on its phosphorylation state (IZUMI et al. 1992).

In conclusion, in this review we have raised the question as to how crosstalk takes place between the circadian oscillator that confers periodicity upon rhythmically dividing cell populations and the cyclin clock that more immediately regulates the cell division cycle. We have traced the flow of information along the signal transduction pathway via the AC-cAMP-PDE system and the cAMP-dependent kinases (cPKA and cPKB) that we have identified in *Euglena*, which we believe ultimately couple to the MPF clock regulating the CDC (see Fig. 10). Indeed, we have confirmed the presence of homologs of $p34^{cdc2}$ and cyclin B, the catalytic and regulatory subunits of MPF, in this unicell. Although the level of the former is invariant across the CDC, we did observe cyclic changes in phosphorylation and dephosphorylation of Cdc2 that persist in nondividing cells under free-running conditions. This finding, if confirmed, raises the possibility that reversible protein phosphorylation might mediate circadian control of the CDC. Indeed, reversible protein phosphorylation recently has been implicated in the mechanism of the circadian clock itself. Thus, EDERY et al. (1994) have reported persistent rhythmic phosphorylation changes in the *PER* protein of *Drosophila* under constant conditions, while COMOLLI et al. (1994) have shown that the inhibitor of protein phosphorylation 6-DMAP (6-dimethylaminopurine), which reversibly blocks cell division in *Gonyaulax polyedra*, also affects the circadian expression of bioluminescence in this marine dinoflaghellate. Recently, cdk- and cyclin-like proteins have been identified in post-mitotic tissues of the isolated eye of the gastropod *Bulla gouldiana* that are affected by treatments that phase shift the circadian rhythm of ocular membrane potential

(KRUCHER and ROBERTS 1994; ROBERTS et al. 1992). Further work may well reveal a close similarity, if not a shared common mechanism, between the eukaryotic CDC and the circadian clock.

Acknowledgments. This review, based on a presentation at the Third Stromboli Conference on Aging and Cancer held in Stromboli, Sicily, Italy, June 4–9, 1993 (see EDMUNDS 1994), includes recent findings from my laboratory. Particular credit is due to Dr. Isabelle A. Carré, Dr. Gangaram Mohabir, and Dr. Jian Tong for the use of their published and unpublished results. The study was supported in part by National Science Foundation grants DCB-8901944 and DCB-9105752 to L. Edmunds.

References

Anderson RW, Laval-Martin DL, Edmunds LN Jr (1985) Cell cycle oscillators: temperature compensation of the circadian rhythm of cell division in *Euglena*. Exp Cell Res 157:144–158
Baserga R (ed) (1971) The cell cycle and cancer. Dekker, New York
Beavo JA (1988) Multiple isozymes of cyclic nucleotide phosphodiesterase. In: Greengard P, Robison GA (eds) Advances in second messenger and phosphoprotein research, vol 22. Raven, New York, pp 1–38
Boynton AL, Whitfield JF (1983) The role of cyclic AMP in cell proliferation: a critical assessment of the evidence. Adv Cyclic Nucleotide Res 15:193–294
Buetow DE (ed) (1968a,b, 1982) The biology of *Euglena*, vols I–III. Academic, New York
Carell EF, Dearfield KL (1982) A relationship between adenosine 3′-5′-cyclic monophosphate levels and deoxyribonucleic acid synthesis in *Euglena*. Life Sci 31:249–254
Carré IA, Edmunds LN Jr (1992) cAMP-dependent kinases in the algal flagellate *Euglena gracilis*. J Biol Chem 267:2135–2137
Carré IA, Edmunds LN Jr (1993) Oscillator control of cell division in *Euglena*: cyclic AMP oscillations mediate the phasing of the cell division cycle by the circadian clock. J Cell Sci 104:1163–1173
Carré I, Oster AS, Laval-Martin DL, Edmunds LN Jr (1989a) Entrainment and phase shifting of the circadian rhythm of cell division by light in cultures of the achlorophyllous ZC mutant of *Euglena gracilis*. Curr Microbiol 19:223–229
Carré I, Laval-Martin DL, Edmunds LN Jr (1989b) Circadian changes in cyclic AMP levels in synchronously dividing and stationary-phase cultures of the achlorophyllous ZC mutant of *Euglena gracilis*. J Cell Sci 94:267–272
Comolli J, Taylor W, Hastings JW (1994) An inhibitor of protein phosphorylation stops the circadian oscillator and blocks light-induced phase shifting in *Gonyaulax polyedra*. J Biol Rhythms 9:13–26
Cook JR, James TW (1960) Light-induced division synchrony in *Euglena gracilis* var. *bacillaris*. Exp Cell Res 28:524–530
Creanor J, Mitchison JM (1986) Nucleotide diphosphokinase, an enzyme with step changes in activity during the cell cycle of the fission yeast, *Schizosaccharomyces pombe*. J Cell Sci 86:207–215
Cross F, Roberts J, Weintraub H (1989) Simple and complex cell cycles. Annu Rev Cell Biol 5:341–395
Dumont JE, Jauniaux J-C, Roger PP (1989) The cyclic AMP-mediated stimulation of cell proliferation. Trends Biochem Sci 14:67–71
Edery I, Zwiebel LJ, Dembinska ME, Rosbash M (1994) Temporal phosphorylation of the *Drosophila* period protein. Proc Natl Acad Sci USA 91:2260–2264

Edgar BA, O'Farrell PH (1989) Genetic control of cell division patterns in the Drosophila embryo. Cell 57:177–187

Edmunds LN Jr (1965a) Studies on synchronously dividing cultures of *Euglena gracilis* Klebs (strain Z). I. Attainment and characterization of rhythmic cell division. J Cell Comp Physiol 66:147–158

Edmunds LN Jr (1965b) Studies on synchronously dividing cultures of *Euglena gracilis* Klebs (strain Z). II. Patterns of biosynthesis during the cell cycle. J Cell Comp Physiol 66:159–182

Edmunds LN Jr (1966) Studies on synchronously dividing cultures of *Euglena gracilis* Klebs (strain Z). III. Circadian components of cell division. J Cell Physiol 67:35–44

Edmunds LN Jr (1975) Temporal differentiation in *Euglena*: circadian phenomena in non-dividing populations and in synchronously dividing cells. In: Les cycles cellulaires et leur blocage chez plusieurs protistes. Centre National de la Recherche Scientifique, Paris, Colloques Intern du CNRS, no 240, pp 53–57

Edmunds LN Jr (1978) Clocked cell cycle clocks: implications toward chronopharmacology and aging. In: Samis HV Jr, Capobianco S (eds) Aging and biological rhythms. Plenum, New York, pp125–184 (Adv Exp Med Biol, vol 108)

Edmunds LN Jr (1982) Circadian and infradian rhythms. In: Buetow DE (ed) The biology of *Euglena*, vol III. Academic, New York, pp 53–142

Edmunds LN Jr (ed) (1984) Cell cycle clocks. Dekker, New York

Edmunds LN Jr (1988) Cellular and molecular bases of biological clocks. Springer, Berlin Heidelberg New York

Edmunds LN Jr (1994) Clocks, cell cycles, cancer and aging: role of the adenylate cyclase-cyclic AMP-phosphodiesterase axis in signal transduction between circadian oscillator and cell division cycle. In: Pierpaoli W, Regelson W, Fabris N (eds) The aging clock: the pineal gland and other pacemakers in the progression of aging and carcinogenesis. The New York Academy of Sciences, New York, pp 77–96 (Ann NY Acad Sci, vol 719)

Edmunds LN Jr, Adams KJ (1981) Clocked cell cycle clocks. Science 211:1002–1013

Edmunds LN Jr, Funch RR (1969a) Circadian rhythm of cell division in *Euglena*: effects of a random illumination regimen. Science 165:500–503

Edmunds LN Jr, Funch R (1969b) Effects of 'skeleton' photoperiods and high frequency light-dark cycles on the rhythm of cell division in synchronized cultures of *Euglena*. Planta (Berl) 87:134–163

Edmunds LN Jr, Halberg F (1981) Circadian time structure of *Euglena*: a model system amenable to quantification. In: Kaiser HE (ed) Neoplasms – comparative pathology of growth in animals, plants and man. Williams and Wilkins, Baltimore, pp 105–134

Edmunds LN Jr, Laval-Martin DL (1984) Cell division cycles and circadian oscillators. In: Edmunds LN Jr (ed) Cell cycle clocks. Dekker, New York, pp 295–324

Edmunds LN Jr, Mohabir G (1995) Circadian regulation of the cell division cycles in Euglena gracilis: role of reversible tyrosine phosphorylation in the timing of mitotic kinase activity. Abstracts, Am Physiol Soc Conf. Understanding the biological clock: from genetics to physiology, 8–12 July, Hanover, New Hampshire

Edmunds LN Jr, Tamponnet C (1990) Oscillator control of cell division cycles in *Euglena*: role of calcium in circadian timekeeping. In: O'Day DH (ed) Calcium as an intracellular messenger in eucaryotic microbes. American Society for Microbiology, Washington, pp 97–123

Edmunds LN Jr, Jay ME, Kohlmann A, Liu SC, Merriam VH, Sternberg H (1976) The coupling effects of some thiol and other sulfur-containing compounds on the circadian rhythm of cell division in photosynthetic mutants of *Euglena*. Arch Microbiol 198:1–8

Edmunds LN Jr, Tay DE, Laval-Martin DL (1982) Cell division cycles and circadian clocks: phase response curve for light perturbations in synchronous cultures of *Euglena*. Plant Physiol 70:297–302

Eskin A, Corrent G, Lin C-Y, McAdoo DJ (1982) Mechanism for shifting the phase of a circadian rhythm by serotonin: involvment of cAMP. Proc Natl Acad Sci USA 79:660–664

Félix M-A, Labbé J-C, Dorée M, Hunt T, Karsenti E (1990) Triggering of cyclin degradation in interphase extracts of amphibian eggs by cdc2 kinase. Nature 346:379–386

Goldberg ND, Haddox MK, Dunham E, Lopez C, Hadden JW (1974) The Yin Yang hypothesis of biological control: opposing influences of cyclic GMP and cyclic AMP in the regulation of cell proliferation and other biological processes. In: Clarkson B, Baserga R (eds) Control of proliferation in animal cells. Cold Spring Harbor Laboratory, Cold Spring Harbor, pp 609–625

Halberg F (1975) When to treat. Indian J Cancer 12:1–20

Hohmann P, DenHaese G, Greene RS (1993) Mitotic CDC2 kinase is negatively regulated by cAMP-dependent protein kinase in mouse fibroblast cell free extracts. Cell Prolif 26:195–204

Homma K, Hastings JW (1989) The S phase is discrete and is controlled by the circadian clock in the marine dinoflagellate *Gonyaulax polyedra*. Exp Cell Res 182:635–644

Izumi T, Walker DH, Maller JL (1992) Periodic changes in phosphorylation of the *Xenopus* cdc25 phosphatase regulate its activity. Mol Biol Cell 3:927–939

Jarrett RM, Edmunds LN Jr (1970) Persisting circadian rhythm of cell division in a photosynthetic mutant of *Euglena*. Science 167:1730–1733

Jessus C, Beach D (1992) Oscillation of MPF is accompanied by periodic association between cdc25 and cdc2-cyclin B. Cell 68:323–332

John PCL, Sek FJ, Lee MG (1989) A homolog of the cell cycle control protein p34cdc2 participates in the division cycle of *Chlamydomonas*, and a similar protein is detectable in higher plants and remote taxa. Plant Cell 1:1185–1193

John PCL, Zhang K, Dong C, Diederich L, Wightman F (1993) p34cdc2 related proteins in control of cell cycle progression, the switch between division and differentiation in tissue development, and stimulation of division by auxin and cytokinin. Aust J Plant Physiol 20:503–526

Krucher NA, Roberts MH (1994) Identification of cdk-like proteins in the eye and brain of the marine snail, *Bulla gouldiana*. Abstracts, Fourth Meeting of the Society for Research on Biological Rhythms, 4–8 May 1994, Amelia Island, Jacksonville, no 128, p 96

Ledoigt G, Calvayrac R (1979) Phénomènes périodiques, métaboliques et structuraux chez un protiste, *Euglena gracilis*. J Protozool 26:632–643

Lemmer B (ed) (1989) Chronopharmacology: cellular and biochemical interactions. Dekker, New York

Lemmer B, Bissinger H, Lang PH (1986) Effect of forskolin on cAMP levels in rat heart at different times of day. IRCS Med Sci 14:1103–1104

Lewin B (1990) Driving the cell cycle: M phase kinase, its partners, and substrates. Cell 61:743–752

Malinowski JR, Laval-Martin DL, Edmunds LN Jr (1985) Circadian oscillators, cell cycles, and singularities: light perturbation of the free-running rhythm of cell division in *Euglena*. J Comp Physiol B 155:257–267

Matsumoto K, Uno I, Ishikawa T (1983) Control of cell division in *Saccharomyces cerevisiae* mutants defective in adenylate cyclase and cAMP-dependent kinase. Exp Cell Res 146:151–161

Moreno S, Nurse P, Russell P (1990) Regulation of mitosis by cyclic accumulation of p80^{cdc25} mitotic inducer in fission yeast. Nature 344:549–552

Murray AW, Hunt T (1993) The cell cycle: an introduction. Freeman, New York

Murray AW, Kirschner MW (1989) Cyclin synthesis drives the early embryonic cell cycle. Nature 339:275–280

Novak B, Mitchison JM (1986) Changes in CO_2 production in synchronous cultures of the fission yeast *Schizosaccharomyces pombe*: a periodic cell cycle event that persists after the DNA-division cycle has been blocked. J Cell Sci 86:191–206

Prosser RA, Gillette MU (1989) The mammalian circadian clock in the suprachiasmatic nuclei is reset in vitro by cAMP. J Neurosci 9:1073–1081

Roberts MH, Towles JA, Leader NK (1992) Tyrosine kinase regulation of a molluscan circadian clock. Brain Res 592:170–174

Scheving LE, Tsai T-H, Feuers RJ, Scheving LA (1989) Cellular mechanisms involved in the action of anticancer drugs. In: Lemmer B (ed) Chronopharmacology: cellular and biochemical interactions. Dekker, New York, pp 317–369

Terry OW, Edmunds LN Jr (1970a) Phasing of cell division by temperature cycles in *Euglena* cultured autotrophically under continuous illumination. Planta (Berl) 93:106–127

Terry OW, Edmunds LN Jr (1970b) Rhythmic settling induced by temperature cycles in continuously-stirred autotrophic cultures of *Euglena gracilis* (Z strain). Planta (Berl) 93:128–142

Tong J, Edmunds LN Jr (1993) Role of cyclic GMP in the mediation of circadian rhythmicity of the adenylate cyclase-cyclic AMP-phosphodiesterase system in *Euglena*. Biochem Pharmacol 45:2087–2091

Tong J, Carré IA, Edmunds LN Jr (1991) Circadian rhythmicity in the activities of adenylate cyclase and phosphodiesterase in synchronously dividing and stationary-phase cultures of the achlorophyllous ZC mutant of *Euglena*. J Cell Sci 100:365–369

Touitou Y, Haus E (eds) (1992) Biologic rhythms in clinical and laboratory medicine. Springer, Berlin Heidelberg New York

Whitfield JF (1990) Calcium, cell cycles, and cancer. CRC Press, Boca Raton

Whitfield JF, Durkin JP, Franks DJ, Kleine LP, Raptis L, Rixon RH, Sikorska M, Roy Walker P (1987) Calcium, cyclic AMP and protein kinase C – partners in mitogenesis. Cancer Metastasis Rev 5:205–250

Zatz M (1992) Does the circadian pacemaker act through cyclic AMP to drive the melatonin rhythm in chick pineal cells? J Biol Rhythms 7:301–311

CHAPTER 3
Genetics and Molecular Biology of Circadian Clocks

L. RENSING

A. Introduction

I. Oscillations and Genes

Oscillations characterize a state of temporal order in nonlinear dynamic systems far from equilibrium. The oscillatory state requires interactive structures (activating or inhibiting interactions) between the system components, energy dissipation and a set of specific conditions. Living systems demonstrate a wide spectrum of oscillations, ranging from action potentials to circadian and annual rhythms (RENSING and JAEGER 1985). An even wider spectrum of oscillations exists in nonliving, i.e., physical and chemical, systems, showing that the oscillatory state as such is not confined to living systems and does not require genetic information. Cells and organisms, however, have made use of the available oscillatory mechanisms in the course of evolution, for example, for signal transmission (action potentials, pulsatile hormone release, intracellular calcium waves), for locomotory or pumping mechanisms (cilia, leg, wing movements, heartbeat, breathing, peristaltic muscle contractions) and for "clock" functions (circadian, lunar and annual rhythms). The term "clock" has been introduced mainly as a metaphor for the basic mechanism of the latter rhythms because it directs a number of processes ("hands" of the clock) in a rhythmic fashion. These rhythms probably allow optimal adaptation of the organism to the periodic changes in the environment. They also establish a state of temporal order that may per se represent a functional advantage. In man, a multitude of circadian rhythms in almost every functional variable have been described, whose maxima and minima map to defined phases of sleep and wakefulness (or night and day, respectively).

In order to utilize oscillations for the purposes listed above, living organisms need oscillatory systems with specific properties. These properties must have evolved under the control of genes which determined the components and interactive structures of the oscillatory systems and the specific conditions necessary for the oscillatory state to exist. The general characteristics of an oscillation are its *frequency* (number of repetitions per unit time) or *period length* (the time between two subsequent repeated events), its *amplitude* (maximal excursions of the oscillatory variable from its zero or mean value) and its *phase* (temporal position of a certain event within a period or with respect to external temporal cues). The term *level* is used to define the mean value of the oscillating variable with respect to the zero line. Genetic control is involved not only in determining these general characteristics, but also in determining more specific properties of the oscillations: circadian rhythms, for example, are characterized by a temperature independence of period length under constant conditions ("temperature compensation"), a property of convincing significance for the functioning of a clock. Another particular property is entrainability by means of light and temperature signals (these and other entraining signals are called zeitgebers).

II. Analytical Approaches to the Basic Circadian Oscillator

The basic circadian oscillator resides within each cell as deduced from the existence of the rhythm in unicellular organisms and from cell cultures of plant and some animal tissues. Possibly, there exist more than one circadian oscillator in a single cell, as has been shown in plants and unicellular organisms. These oscillators are usually coupled, but may uncouple under certain conditions (VON DER HEYDE et al. 1992; ROENNEBERG and MORSE 1993). It is not known whether these multiple circadian oscillators represent slight variations of the same mechanism or whether they are based on different oscillators. In any case, coupling phenomena of oscillators within each cell or between cells of animal pacemaker tissues (eyes, brain regions, see Chap. 1, this volume) may play an important role but are not well understood. In animal organisms it is not clear how many cell types – apart from pacemaker tissue – maintain circadian oscillator(s).

The search for the basic cellular mechanism of the clock began simultaneously with the general establishment of circadian rhythm research ("chronobiology") about 45 years ago. Until the first "clock" gene in *Drosophila* was detected in 1971, this search was mainly based on a classical "pharmacological" approach. It consisted of the experimental manipulation of a process within cells and tissues putatively related to the cellular clock mechanism and the subsequent analysis of the circadian rhythm response (Fig. 1). Most often, a particular cell process was perturbed by applying a pulse of an agent at different phases of the oscillation and determined whether or not these perturbations resulted in phase shifts of the oscillator (WINFREE 1980). A large number of agents have been tested and analyzed in this way.

The conclusions about the underlying clock mechanism were of the following kind: if a process X is activated or inhibited and the circadian rhythm phase-shifted, process X can be assumed to be part of the mechanism. For example: protein synthesis inhibitors were generally effective in phase shifting the clock and their effects were taken to indicate a role of protein synthesis in the mechanism. This conclusion was apparently correct, as shown by recent experiments using molecular biological techniques (see below). In this manner, various processes were postulated to belong to the clock mechanism: RNA synthesis (which was based, however, on ambiguous results), protein synthesis,

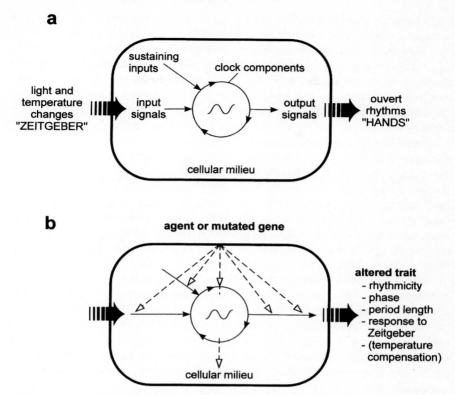

Fig. 1a,b. Functional components of the circadian clock (**a**) and their manipulation (perturbation) by pharmaca or mutations (**b**). For further explanations see text

protein phosphorylation, energy metabolism and concentration changes of some ions, particularly of calcium (EDMUNDS 1988; RENSING and HARDELAND 1990). This analytical approach has to deal with the following problems: (a) the agents often influence large classes of processes such as the synthesis of all proteins, and cause multiple side effects, and (b) they can affect the oscillatory mechanism, the cellular milieu, controlling or sustaining inputs of the oscillatory process. These effects may all lead to phase shifts of the oscillator (Fig. 1b). Therefore, the conclusions from this approach remained rather unspecific.

In addition to or in combination with the above-mentioned approach, oscillatory changes of cellular processes were described, for example, in the amount of RNA and protein species, in the phosphorylation state of proteins, in second messenger concentrations and in photosynthetic and bioluminescence activity. Most or all of these processes do not belong to the oscillatory mechanism, but rather represent "hands" of the clock, i.e., controlled variables. These results thus added much to our knowledge of the many "driven" oscillations in a cell but little to the identification of the basic oscillatory mechanism.

This unsatisfactory situation was dramatically improved by the molecular genetic approach used in recent years. This approach was opened up by detecting single gene mutations in *Drosophila* which differed from the wild-type strain in their circadian period lengths or in the complete absence of circadian rhythmicity (KONOPKA and BENZER 1971). Similar mutants were isolated in the fungus *Neurospora* (FELDMAN and HOYLE 1973) and later in golden hamsters and mice (RALPH and MENAKER 1988; VITATERNA et al. 1994). The great advantage of analyzing the effects of a single gene mutation on an interactive oscillatory system is that only one compound (a protein and its function) is altered. All other parts of the system remain unchanged. Even though a protein may have more than one function, the genetic alteration is much more specific than the usual inhibitor/activator approach described above. Again, it is not a priori clear whether a mutated clock property is caused by an alteration in the oscillatory mechanism, in the cellular milieu or – in the case of an arrhythmic mutant – by uncoupling hands from the clock mechanism (Fig. 1).

In man, knowledge of circadian genetics is very limited: there are no family histories of circadian period lengths in constant conditions. A few drastically altered period lengths of persons living in a bunker may represent variants which are caused by changes in the hormonal milieu or in the coupling of oscillators. The same may apply to patients suffering from so-called winter depression who were successfully treated by additional doses of light. In man, the morning and evening types characterized by different phases of bodily and psychic functions are probably due to genetic differences (J. Zulley, personal communication). There are strong chronopharmacological implications associated with morning and evening types with respect to the timing of therapeutic treatments. Therefore, and in order to understand rhythm disorders in man, knowledge of the molecular basis of the circadian oscillators is especially important.

B. Genetics and Molecular Biology of the Clock

I. Overview and Generalizations

Among the huge number of organisms which have been described with respect to the phenotype "circadian rhythm," only a few were analyzed as to the genetics of this trait: The first milestones were laid by KONOPKA in *Drosophila* (KONOPKA and BENZER 1971) and by BRUCE addressing this problem in the unicellular alga *Chlamydomonas* (BRUCE 1972). KONOPKA described a single locus *period* (*per*) on the X chromosome which was responsible for either different clock period lengths or the absence of rhythmicity. This essential clock gene and its products were characterized by using molecular genetic techniques through the efforts of YOUNG and his group as well as HALL and ROSBASH and their coworkers (YOUNG 1993; HALL 1990). *Neurospora crassa*, a filamentous fungus, was the second organism most successfully introduced into circadian rhythm genetics by FELDMAN and HOYLE (1973). FELDMAN was also able to map several frequency mutants (long and short period length) to a single locus, *frequency (frq)*. The molecular genetics of this gene and its expression was then extensively studied by DUNLAP and his group, who also contributed substantially to the present knowledge of the clock mechanism (DUNLAP 1993). The gene *tau* in the golden hamster was the first clock gene to be detected and further analyzed in mammals by MENAKER (RALPH and MENAKER 1988), a finding recently extended by a *Clock* gene found in mutagenized mice by TAKAHASHI and colleagues (VITATERNA et al. 1994). These mammalian clock genes promise a future extension of the molecular genetic approach to the analysis of the clock mechanism in mammals and man.

Because of the few, presently only two, systems so far analyzed in molecular terms, it may be premature to make generalizations about the genetic and molecular genetics of the clock. However, some common features appear in both organisms – *Drosophila* and *Neurospora* – which are important to emphasize.

1. In each organism, most of the independently isolated clock alleles map to one locus: *per* on the X chromosome of *Drosophila* and *frq* on the linkage group VII of *Neurospora*. The genes *per* and *frq* apparently code for essential parts of the clock mechanism. This observation does not mean that there are no other genes involved in the functioning of the clock, but some of them may be lethal when mutated and some may not be detected in the screening procedure applied.
2. The different *per* and *frq* mutants show different phenotypes: their circadian period length may either be shorter or longer than that of the wild-type strain. This observation already suggests that different mutations of the same gene can affect the activity of the gene product in different ways, e.g., give rise to mutants with more or with less active gene products. These different products apparently cause higher or lower frequencies of the circadian oscillation. This conclusion about different activities of the gene products is corroborated by the observation that the different frequency alleles are *semidominant*: two different *per* alleles in one organism result in a period length which is intermediate between the period lengths of the two alleles in the homozygous state. Most often, mutations lead to a complete loss of a gene function which is overcome by the wild-type allele in the heterozygous state. The wild-type allele is thus dominant and produces a normal phenotype, whereas mutants are recessive. In the case of the clock mutants *per* and *frq*, this is observed only in mutants which give rise to arrhythmic phenotypes (null mutants).
3. In *Drosophila* and *Neurospora* the *per* and *frq* gene products (*per*-mRNA, PER protein as well as *frq*-mRNA) were shown to oscillate with a circadian period. There is good evidence that both proteins act as an inhibitor of their own gene transcription. These results suggest a feedback loop as

the basic oscillator mechanism. By means of fusing clock genes with inducible promoter regions from other genes, it became possible to manipulate the product level of the clock gene in *Drosophila* and *Neurospora*. In both cases it was demonstrated that perturbations of the gene product levels perturb or abolish the clock's function (EDERY et al. 1994b; ARONSON et al. 1994a).
4. Mutations of the *per* or *frq* locus, but also of some other clock genes, often affected not only the frequency (period length) of the circadian oscillation but also its phase, its temperature compensation, its response to entraining signals and the expression of ultradian rhythms. Since the phase with respect to the entraining light-dark cycle generally depends on the frequency of the entrained oscillator, this may explain the covariance of these variables. The effects of clock mutations on temperature compensation and zeitgeber responses point to inbuilt mechanisms (a) of temperature compensation and (b) of responses to light or to neurotransmitters acting within light-transmitting pathways.

II. *Neurospora*

1. Circadian Rhythm

In the filamentous fungus *Neurospora crassa*, a rhythm of asexual spore (conidia) formation is observed which shows all the characteristics of a true circadian oscillation: a period of about 24 h in constant darkness (usually 21.5 h at 25 °C), temperature compensation of the period length and entrainment by light or by temperature changes. In most experiments, the *band* (*bd*) mutant is used, because it expresses a particularly clear periodic banding pattern on agar. The free-running period length of this conidiation rhythm can be registered mainly in constant darkness and not in constant light because light induces the permanent formation of conidia.

The formation of conidia is a morphogenetic process and part of the vegetative life cycle of this organism. Conidia are induced to germinate by contact with water, which leads to the outgrowth of vegetative hyphae. These hyphae form long filaments which branch and contain numerous nuclei in a continuous cytoplasm (syncytium). The growing hyphae develop a network of filaments called a mycelium. At a certain phase of the circadian oscillation or after induction by external stimuli (blue light, heat shock, starvation, oxygen), aerial hyphae branch off from the vegetative hyphae, grow above the surface and form conidia. This morphogenetic process involves many gene activities some of which have been shown to be under direct circadian control (LOROS et al. 1989).

2. Clock Mutants

The available clock mutants were mainly derived by subjecting the fungus to mutagenic treatment (nitrosoguanidine or UV) and a subsequent screening of the mycelia derived from single spores to detect differences in the period length (FELDMAN et al. 1979). Fifteen independently isolated alleles were traced to seven different genes which, when mutated, affected the period length of the conidiation rhythm. The genes were called *frequency* (*frq*), *period–1* (*prd–1*), *period–2* (*prd–2*), *period–4* (*prd–4*), *chrono* (*chr*) and *clock-affecting–1* (*cla–1*). Nine of these alleles map to the *frq* locus, whereas the others map to individual loci. For reasons discussed below, *frq* appears to be of central importance for the functioning of the clock and was, therefore, intensively analyzed by molecular genetic techniques. The main focus of this review, therefore, will be on this gene.

Most alleles mapping to *frq* give rise to altered period lengths. Some cause shorter (frq^1, $frq^2 = frq^4 = frq^6$) and some longer (frq^3, $frq^7 = frq^8$) period lengths. These alleles all show semidominant behavior when combined with a different allele: when heterokaryons are formed by fusion of hyphae then containing two types of nuclei with different alleles, the resulting mycelium expresses a circadian rhythm of conidiation with a period length which is intermediate between the two period lengths coded for by the two alleles. For example, heterokaryons containing equal numbers of nuclei with the wild-type allele (period length of 21.5 h) and of nuclei with the short-period allele (period length of 16.5) show a period length of 19 h. Furthermore, heterokaryons constructed so that they contain different proportions in the number of nuclei from the one type to nuclei from the other type show proportional intermediate values of their period length (FELDMAN et al. 1979). This remarkable gene dosage effect led FELDMAN to suspect "that the *frq* gene or the *frq* gene product is intimately related to, or a direct part of, the basic clock-timing apparatus."

Another surprising feature of *frq* (and *per*) is that some alleles give rise to increased and some alleles to decreased period lengths. This means that the different alleles either slow down or speed up the oscillatory process. It is also interesting to note that the long-period mutant (frq^7) exhibits defects in the temperature compensation of its period length (ARONSON et al. 1994b).

Two recessive alleles of *frq* (frq^9, frq^{10}) have been described, both probably representing null mutants (DUNLAP 1993). Frq^9 was detected after a mutagenizing treatment and later identified to originate from a deletion of a nucleotide pair. frq^{10} was generated by molecular genetic techniques in vitro and transformed into the cells. Under certain conditions no conidial banding pattern at all was observed in these mutants. Under some other conditions a ragged rhythm appeared that was hardly entrainable by a light-dark cycle and devoid of temperature compensation.

Temperature compensation of the period length thus seems to be associated with the *frq* product or rather with the domain affected by the long period mutation. This part of the FRQ protein may be involved in some adaptive control mechanism keeping the activity of the protein temperature-independent. One might assume that both synthesis and degradation of the FRQ protein are equally affected by a temperature change, i.e., that both processes may be enhanced by increasing the temperature. This would amount to a temperature compensation at the level of the FRQ protein. Mutational changes resulting, for example, in an altered sensitivity of FRQ protein to proteinases may upset this balance, prolong the period length and render the oscillation less temperature compensated. Such a mechanism was recently shown in a model oscillator (RUOFF and RENSING 1996). Changes in temperature compensation were simulated by varying the degradation rate of a clock component (RUOFF et al. 1996).

A great number of *Neurospora* mutants defective in metabolic pathways have been screened as to possible changes in circadian rhythm variables. Major changes of these variables were observed in a mutant defective in fatty acid chain elongation (*cel*) (LAKIN-THOMAS et al. 1990). These metabolic mutants, however, also affect growth and differentiation and turn out to be recessive in combination with wild-type alleles – indicating that these mutations influence the clock mechanism indirectly by changing the cellular milieu (Fig. 1).

3. Molecular Biology of *frq*

As was already anticipated from the genetic analysis, *frq* turned out to be essentially involved in the clock mechanism. The gene was cloned and subjected to extensive molecular genetic analysis (DUNLAP 1993). The piece of DNA identified to contain *frq* encoded two transcripts, a shorter (~1500 nucleotides) and a longer transcript (~5000 nucleotides, Fig. 2).

Part of the genomic region for the long transcript as well as the complete cDNA from this transcript have been sequenced and thus allow conclusions

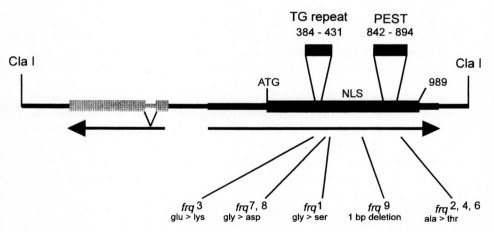

Fig. 2. Structural domains, mutations and transcripts of *frq*. *ATG,* start codon for translation; *NLS,* nuclear localization sequence; *PEST sequence,* amino acid sequence of proline-glutamic acid-serine-threonine; *TG,* threonine, glycine; *ClaI,* recognition cut site for the endonuclease. The different mutated *frq* alleles, the location of their mutated nucleotides and the resulting changes of amino acids are indicated *in the lower panel.* *Arrows* indicate direction of transcription. (After DUNLAP 1993)

about the amino acid sequence of the FRQ protein and its functional domains. The long transcript contains an open reading frame of 989 codons (amounting to a protein of about 108 kDa) and a long untranslated leader region of about 1500 nucleotides. The deduced amino acid sequence contains potential sites for N- and O-linked glycosylation as well as for serine and threonine phosphorylation. It furthermore shows a potential nuclear localization sequence (*nls*), a hyperacidic tail as found in transcription factors and several sequences (PEST) characteristic for high-turnover proteins (Fig. 2).

Sequence homologies to *frq* exist, for example, in fungal species belonging to the same family, such as *Sordaria finicola*. Little homology, however, was found with respect to the *per* gene in *Drosophila*, except for the modification sites and a region of 48 amino acids including threonine-glycine and serine-glycine repeats (TG-repeat, Fig. 2). The functional significance of this region is still unclear and, indeed, it may not be functionally important.

The same laboratory also made successful efforts to identify the sites of mutation in the various alleles of *frq* described above. By sequence analysis of the mutated *frq* DNA they were able to show that the mutations (except frq^9 and frq^{10}) are characterized by single base pair (G to A) changes, albeit at different sites within the DNA sequence (see Fig. 2). The frq^2, frq^4 and frq^6 mutants were due to the same mutation at a single nucleotide pair towards the terminal region of the gene, resulting in a shorter period length. In the FRQ protein, the change of the corresponding amino acid is alanine to threonine. The very short period length of frq^1 is due to a glycine to serine change in a more proximal part of FRQ, whereas the long period lengths result either from a glutamic acid to lysine (frq^3) or a glycine to aspartic acid change (frq^7) in the vicinity of the frq^1 mutation. It is not quite clear at present which property of the protein – its binding affinities to other proteins, its activity as transcription factor, its accessibility to modifications or its turnover rate – is affected by the amino acid alterations. The frq^9 and frq^{10} mutations, in contrast, consist of larger defects which apparently render the FRQ protein inactive [null mutations (DUNLAP 1993)].

A great step forward in understanding the functioning of the clock and the role of *frq* products in the circadian clock mechanism was taken when it became possible to determine the amount of *frq* mRNA and to manipulate its level. The *frq* mRNA revealed a circadian rhythm (ARONSON et al. 1994a). It was shown that *frq* mRNA peaks in the "subjective" morning, i.e., at a time (*ct* o) in constant darkness when in a light-dark cycle the light time would begin. The *frq* mRNA in the long-period mutant frq^7 showed a period of about 28 h and a larger amplitude, indicating that this oscillation may give rise to the conidiation rhythmicity of the same period length. The *frq* mRNA in the null mutant frq^9 did not oscillate regularly.

In order to be able to manipulate the amount of *frq* mRNA, the *frq* open reading frame was fused to a promoter of a quinic acid-inducible gene (*qa*–2). This construct expressed *frq* mRNA, and hence FRQ protein only in the presence of the inducer quinic acid, and was transformed into an frq^+ strain

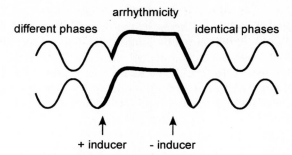

Fig. 3. Schematic picture of two differently phased *frq*-mRNA oscillations interrupted simultaneously by addition of the inducer (quinic acid). The inducer maintains a constant high level of *frq*-mRNA and suppresses the oscillation as observed in the rhythm of conidia formation. After removal of the inducer the oscillations start at identical phases (after Aronson et al. 1994a)

which resulted in a strain containing a wild-type frq^+ gene with its normal promoter and an additional frq^+ gene with the quinic acid-inducible promoter. The conidiation rhythm of this strain was then tested in the absence or presence of the inducer. A normal conidiation rhythm was expressed in the absence of quinic acid, but the rhythm was totally abolished when it was present above a certain threshold concentration (Fig. 3). This and other experiments unequivocally showed that permanent high expression of *frq* abolishes the oscillation. It was further shown that higher levels of *frq* mRNA (induced by quinic acid) inhibited the transcription of the normal frq^+ gene. This result proved a negative feedback loop between *frq* and its products and corroborated earlier speculations that the FRQ protein may have inhibitory effects on its own expression. Finally, it was demonstrated that the time of release from the quinic acid induction sets the phase for the recovering rhythmicity (Fig. 3, Aronson et al. 1994a). These results taken together reveal a feedback loop within a single gene expression as a basic feature of the molecular clockwork.

III. *Drosophila*

1. Circadian Rhythms

In animal organisms, *Drosophila melanogaster* has been the main subject of molecular genetic analyses of circadian rhythms. Mainly two overt rhythmicities have been monitored and used for detecting clock mutants: the eclosion from the pupal case and the locomotory activity of adult insects. Eclosion requires metamorphosis (i.e., a differentiation process) and a neuromotory program in the adult fly in order to open and escape the pupal case. Because eclosion happens only once in the lifetime of a fly, a rhythmic eclosion can be observed only in a population. This fact often makes eclosion unsuitable for longer time series. The circadian rhythm of locomotory activity, in contrast, can be monitored in individual organisms over longer periods of time. There are other behavioral rhythmicities in *Drosophila*, such as courtship, mating and egg-laying, which peak – as the locomotor activity – either around dawn or dusk or at both times. It seems to be of high selective value for the fly to restrict its activity and particularly its eclosion to these times of day of optimal humidity and temperature.

2. Clock Mutants

The most important clock mutants in *Drosophila* result from mutations in the *period* (*per*) locus on the X chromosome (KONOPKA and BENZER 1971). Basically, three phenotypes of the different alleles were observed: (a) shorter periods (per^S), (b) longer periods (per^L) and (c) arrhythmic flies (per^O). In *Drosophila* as well as in *Neurospora* it was surprising to find different alleles with different effects on the clock phenotype mapping to one locus. The shorter period mutants further show an advanced peak of eclosion in light-dark conditions, i.e., a phase alteration predictable from the general behavior of entrained oscillators. A loss of temperature compensation was observed in the per^L mutants – similar to frq^7, but with increasing period lengths when the temperature was raised. The *per* mutants per^S and per^L were shown to be semidominant when combined, for example, with per^+ and to give rise to intermediate-period lengths in heterozygotes. The null mutation per^O is recessive with respect to the expression of rhythmicity. Per^O was shown to exhibit ultradian oscillations of locomotor activity with period lengths between 5 and 15 h (DOWSE et al. 1987).

Apart from the different circadian phenotypes, *per* mutants apparently influence the rhythm of courtship song of the males: the wild-type males (per^+) show ultradian periods of about 1 min, per^S males a period of about 40 s and per^L males a period of about 80 s per cycle. The mechanism of this effect of *per* is still the subject of numerous studies and controversial discussions (HALL 1990, 1995).

A possible involvement of *per* in the timing of the ovarian diapause was analyzed, because circadian oscillators are assumed to be the main components of measuring the day length in the photoperiodic organisms. Surprisingly, per^O females were able to determine the "critical day length" for ovarian diapause, even though at a different day length than wild-type females (SAUNDERS 1990).

Many studies focussed on the localization of *per* products (*per* mRNA, *PER* protein) in different tissues and their expression at different phases of the fly's development. Both products were found in the developing CNS of the embryo, whereas controversial results were obtained with different tissues of the larvae (HALL 1990). There is unambiguous evidence concerning *per* expression in pupal and adult tissues, particularly in the brain. The adult brain has been shown to contain the circadian pacemaker for the locomotor activity rhythm as deduced from transplantation studies. This was corroborated by analyzing the brain of genetically mosaic flies. The results of the latter experiments indicated that many cells – neurons and glia cells – in the brain contribute to the pacemaker, which could not be strictly localized in a certain part (EWER et al. 1992). Some laterally located neurons have been strongly implicated as possessing pacemaker function (HALL 1995). About half of them contain a putative neuropeptide (a so-called pigment-dispersing hormone, PDH, found also in other organisms). In agreement with the pacemaker function of parts of the brain, a high-amplitude oscillation in the expression of *per* is observed (see below). Several other adult tissues were shown to express *per*, such as the photoreceptor cells in the eyes and cells of the ocelli, optic lobes, thoracic ganglion, testes, ovaries, gut, antennae, esophagus, proboscis, rectal papilla and Malpighian tubules. Whether all these tissues are capable of oscillating autonomously is not clear at present. At least the photoreceptor cells have maintained this capability, as shown in the mutant *disconnected* (*disco*), which is arrhythmic in its eclosion and locomotor behavior and defective in its eye-brain connections (HARDIN et al. 1992).

Apart from the *per* locus shown to be of prime importance to clock function, another X chromosomal mutant, *Andante* (*And*), causes a "moderately slow" circadian oscillation and is allelic to a pigment mutation *dusky* (*dy*). A recently isolated mutant named "*timeless*" is arrhythmic in both eclosion and activity, and the mutated gene is localized on the distal end of the second (2L) chromosome (VOSSHALL et al. 1994).

3. Molecular Biology of *per*

The *per* gene was cloned in 1984 and the complete sequence independently determined in 1986/1987 by the group of HALL and ROSBASH at Brandeis and the group of YOUNG at the Rockefeller University (HALL 1990; YOUNG 1993;). The major transcript of this gene consists of eight exons spliced together, the first of which is not translated. The coded protein contains six highly conserved regions which were similar in the several strains sequenced. These regions show functional similarities to the regions found in *frq* (see above): a nuclear localization sequence, a number of potential sites for phosphorylation and glycosylation, and a long 270 amino acid motif found also in transcription factors. This motif (termed PAS) functions in vitro as a protein dimerization motif which can mediate associations between members of the PAS protein families. The dimerization efficiency decreased in the per^L mutation, suggesting that homotypic interactions between PER proteins and a monomer–dimer transition may play an important role in period length and temperature compensation of the clock (HUANG et al. 1993). A region with a threonine-glycine repeat was initially interpreted as an indication that PER protein is a proteoglycan localized in the external periphery of the cell. However, later experiments showed that this region can be deleted without affecting the clock and that it is poorly conserved in different *Drosophila* species.

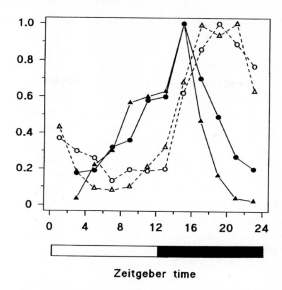

Fig. 4. Oscillation of *per*-mRNA (*solid symbols*) and PER protein (*open symbols*) in the brain (*circles*) and eye (*triangles*) of *Drosophila* under light-dark conditions. *Ordinate*, relative amounts; *abscissa*, zeitgeber time. 0 h is beginning of light time, not midnight. (After ZENG et al. 1994)

PER expression oscillates with a high amplitude particularly in the brain and the eyes of the adult fly: the *per* mRNA shows a peak in the beginning of the night, the PER protein at the end of the night, about 4–6 h later (Fig. 4). These oscillations and phase relationships are also found in constant darkness. The short- and long-period mutants show corresponding period lengths in the oscillations of the *per* products in constant darkness. The lag between the peaks of *per* mRNA and protein may reside in post-transcriptional events. This lag contributes to the exceptionally long period of the circadian oscillation when compared to strictly biochemical oscillations, such as glycolytic oscillations.

There is evidence that the oscillation of *per* expression is due to a feedback between the PER protein and the transcription of the *per* gene: this can be indirectly deduced from the observation that the mutated (per^S, per^L) proteins speed up or slow down the mRNA oscillation, that the PAS motif of the PER protein shows homologies to transcription factors and that – at least in head tissues of the fly – PER protein is translocated into the nucleus. Direct evidence for a negative feedback was recently provided when a *per* gene was fused to the promoter region of the opsin gene. Flies transformed with this construct show a constitutive overexpression of PER in the eyes. This overexpression repressed the endogenous *per* RNA cycling in the eyes but not in other *per*-expressing tissues (ZENG et al. 1994). Since no DNA-binding domain has been found in PER, it may form a feedback loop by binding to other DNA-binding proteins, but may also affect post-transcriptional steps.

PER protein is modified by means of phosphorylation at different sites: a sequential protein phosphorylation takes place in the course of the night (EDERY et al. 1994a). Possibly, the phosphorylated forms of PER serve as a signal for degradation of PER. Thus, the feedback loop may be speeded up or slowed down by affecting the phosphorylation of PER (as for example in per^S). Furthermore, the sequential phosphorylation may represent a process within the oscillatory mechanism which can be affected by various external signals mediated, for example, by cyclic nucleotides and their target protein kinases (see below). PER was found to accumulate in the cytoplasm for several hours before entering the nucleus in a rather short time (CURTIN et al. 1995). The translocation may rely on oligomerization processes, which depend on intermolecular (PER-PER or PER-TIM) and intramolecular binding affinities. The PAS domain and about 40 amino acids at the C-terminus of PER (C domain) have been proposed as intramolecular binding domains (ROSBASH 1995). The per^L mutant rendered the oligomerization process temperature sensitive. Therefore, it is assumed that temperature compensation of PER activity – and hence the period length – is achieved by competition between intermolecular and intramolecular interactions with similar temperature coefficients. The accumulation and nuclear translocation of PER seem to be controlled by binding to another protein (TIM), the product of the gene "*timeless*" (VOSSHALL et al. 1994; PRICE et al. 1995).

The important question here (as in *frq*) is whether or not the feedback interaction between PER protein and *per* transcription represents the basic mechanism for its own oscillation and for the clock that controls rhythmic eclosion and locomotor activity. An initial indication that the PER oscillation is indeed part of the clock stems from studies of the arrhythmic mutant, which does not show such an oscillation of PER. When per^O mutants were transformed with the wild-type gene (per^+), the oscillation reappeared together with the overt rhythmicities. This fact may still be explained by assuming that PER delivers a necessary input to a separate oscillator. If PER were central to the clock mechanism, a transient increase in its concentration should cause a stable phase shift in its own oscillation as well as in the locomotor activity rhythm. This was shown to be the case by using transgenic flies bearing a heat-inducible promoter that controls expression of *per*. When the flies were subjected to heat shock, the *per* products were induced and thus caused permanent phase shifts of the circadian rhythms (EDERY et al. 1994b). These and other experiments with heat-inducible promoters of the *per* gene are more complicated to interpret than the corresponding experiments with substrate-inducible promoters of *frq* (see above) because of the effects of heat shock on gene expression in general. However, both experimental approaches strongly suggest that the oscillations of *per* and *frq* products are essential components of the clock mechanism.

IV. Mammals

Single-gene mutants also exist in mammals which show altered circadian rhythm phenotypes. The best-known example is the *tau* (τ) mutant of the golden hamster detected in a commercial shipment of animals (RALPH and MENAKER 1988). The single male with a period length of 22 h turned out to be heterozygous. When the offspring was crossed the homozygous mutant individuals showed a period length of 20 h (Fig. 5). Thus, the *tau* gene turned out to be semidominant – as had already been shown for the *Drosophila* and *Neurospora* clock alleles (except for the null mutations).

In addition to the period alteration, the *tau* mutant showed a different phase (beginning of locomotor activity) when entrained to a light-dark cycle. In some animals no entrainment was observed at all – particularly in the case of *tau* homozygotes. This altered entrainment might be explained on the basis of the shorter period length. In addition, the mutant individuals showed a different response to light signals, i.e., the resulting phase shifts were much larger than those of the wild type. These pleiotropic effects of the mutation cannot be explained at present. This also applies to the phenomenon of increased "splitting," i.e., the appearance of two peaks of locomotor activity (PITTENDRIGH 1974). Splitting can be interpreted to indicate uncoupling of two (multiple) oscillators or destabilization of the circadian "master" oscillator.

Other pleiotropic effects observed in this mutant were a drastic shortening of the photoperiod required for testis growth, lengthened periods of luteinizing hormone pulsatility, smaller size of the animals, irritability and perhaps a reduced life span. In addition, *tau* mutations have been extensively used for studying the effects of transplanted suprachiasmatic nuclei (SCN): the successful transplantation of *tau* SCN can be readily recognized by the short

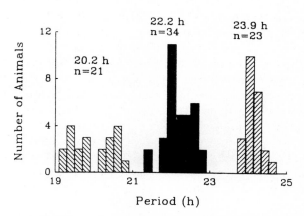

Fig. 5. Frequency distribution of the period length from offspring of heterozygous parents. Three distinct clusters can be observed, with mean periods of 20.2, 22.2 and 23.9 h (Redrawn from RALPH and MENAKER 1988)

period lengths of locomotor rhythmicity expressed by the host animal, whose own SCN was removed prior to transplantation (MENAKER and REFINETTI 1993). In cases of two SCNs in one animal – a normal and a mutant one – both oscillations are expressed, indicating a lack of coupling between the oscillators (VOGELBAUM and MENAKER 1992).

In the progeny of mutagenized mice a semidominant mutation was found that lengthens the intrinsic circadian period length and abolishes the persistence of the rhythmicity in constant darkness (VITATERNA et al. 1994). The mutated allele was termed "*Clock*" and was mapped to the midportion of mouse chromosome 5. Especially homozygous "*Clock/Clock*" animals showed a lengthened and lower-amplitude rhythm of locomotor activity during the first ten cycles in constant darkness. The rhythm gradually deteriorated in the course of further oscillations. Thus, there seem to exist similarities between "*Clock*" in mice, per^L in *Drosophila* and frq^7 in *Neurospora*.

V. Conclusions

The molecular clock mechanism emerging from the analyses of *Neurospora* and *Drosophila* thus consists of the following processes (Fig. 6): (a) transcription of the clock gene (*frq*, *per*), the further processing of the transcript and its nuclear export, (b) translation of the clock gene mRNA into the corresponding clock protein and the degradation of the mRNA, (c) homo- and heterodimer formation of the protein and its nuclear translocation, and (d) interactions between the clock protein and transcription factors, its inhibitory effect on the clock gene transcription, clock protein phosphorylation and the subsequent clock protein degradation.

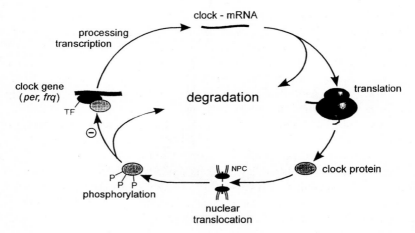

Fig. 6. Model of the basic oscillatory process consisting of a negative feedback loop between transcription of the clock gene and the level of its product (clock protein). For further explanation see text

This scheme, now solidly based on molecular biological experiments, was envisaged earlier on by EHRET and TRUCCO (1967), who called it the "chronon hypothesis." In addition to this – mainly intuitive – vision of the clock mechanism, the pharmacological analyses had led to a principally similar concept (RENSING and HARDELAND 1990). Model oscillators for biological rhythms contained nonlinear components either in the negative or in the positive feedback (or autocatalytic) part. GOODWIN (1963), for example, developed a negative feedback oscillator model with a highly nonlinear transcriptional inhibitor. Other researchers incorporated nonlinear autocatalytic processes, such as product activation and cooperative effects or combined positive (amplifying) and negative (stabilizing) parts into the oscillatory mechanism. When the effects of different temperatures on the kinetics of the amplifying and stabilizing parts were balanced, i.e., when the temperature affected period shortening and period lengthening processes in the same way, as was done in the Goodwin oscillator, this oscillator showed temperature compensation (RUOFF and RENSING 1996).

Ultradian rhythms of various kinds and period lengths within a cell or multicellular organism have often been assumed to couple to each other and thus give rise to the circadian oscillation. In the light of the, now emerging, molecular mechanism of the circadian oscillator, this hypothesis appears rather unlikely. Instead, the reverse may be true: a strong circadian oscillation may "enslave" the independently existing ultradian oscillation. A weak or arrhythmic state of the circadian oscillator as in per^o, tau and $Clock$ might allow expression of the ultradian oscillations.

C. Zeitgeber Signal Pathways to the Clock

The endogenous circadian oscillation is entrained by the periodic environment to a period of exactly 24 h. The main entraining signals (called zeitgebers) are periodic light and temperature changes. In homeothermic organisms, such as birds and mammals, light is the dominant geophysical zeitgeber, but social cues (song, noise) can also act as entraining signals. In lower eukaryotic organisms (unicellular, fungi, plants) zeitgeber signals are perceived by cellular receptors and directly transformed into cellular signals which are then transduced to the intracellular clock mechanism. In animal organisms light signals are mainly perceived by receptor cells and then transmitted to circadian pacemaker cells via neurons and neurotransmitters. Neurotransmitters and neuropeptides are then transformed into intracellular signals and can thus reach the cellular oscillator. The intracellular pathways of light and temperature signals in lower eukaryotes or the pathway of light-dependent neurotransmitters to the circadian oscillating mechanism in animal organisms have not yet been clearly elucidated (see Chaps. 1, 4, this volume).

Approaches to analyze these pathways are as follows: A zeitgeber signal is applied and its effect on putative intracellular signals determined. When intracellular signal pathways were activated after zeitgeber stimuli, these intracellular signals are manipulated experimentally and analyzed as to resulting phase shifts. Manipulations of the putative signal pathways can be achieved by means of inhibitors or activators or by genetic means, i.e., by using mutants of such a pathway.

I. *Neurospora*

In *Neurospora crassa*, light is perceived by a membrane-bound blue-light receptor containing flavine and cytochome$_b$ (BORGESON and BOWMAN 1985). Light-induced changes in the receptor are transduced into the cell by way of signals not yet clearly identified. Changes in cyclic AMP and cyclic GMP levels, protein phosphorylation and the expression of a set of blue-light-inducible genes were observed (RENSING et al. 1993; SOMMER et al. 1989). The cAMP level increased transiently about 30–90 s after the onset of light and then decreased below control levels, kinetics which were paralleled by changes in the amount of the catalytic subunit of the cAMP-dependent protein kinase in the nucleus. The cGMP level, in contrast, immediately decreased after light exposure and slowly increased later (KALLIES et al. 1996). There has been no evidence as yet for a role of calcium and inositol phosphates in light signal transmission. The target of the light signal apparently resides in the transcriptional control of *frq*: a few minutes after the onset of light *frq*-mRNA is synthesized rapidly (CHROSTHWAITE et al. 1995).

How temperature changes are perceived by the cell is not known. There are, however, many effects within the cell which may reset the clock mechanism as discussed above (RENSING et al. 1995): a moderate temperature increase (heat shock) probably affects gene expression and degradation of mRNA and protein positively. However, transcriptional and translational processes of most gene expression pathways, including probably the clock gene, are inhibited by a temperature increase (heat shock). The degradation rate of most long-lived proteins in *Neurospora* is also inhibited by heat shock, whereas the degradation of short-lived proteins is less affected. At the same time the cellular milieu and second messenger level are drastically altered: the level of H^+ and Ca^{2+} increase transiently, as do the levels of cAMP and inositol phosphates. In contrast, the cGMP level decreases after heat shock (GEBAUER et al., submitted for publication). Phase shifting of the oscillator by moderately higher temperatures or by heat shock may thus be the result of multiple alterations. In fact, all variables of the oscillator as well as of the milieu of the cell are altered and shift the oscillator into a different state.

In contrast to *Neurospora* the expression of the *Drosophila* gene (*per*) is apparently suppressed by light (minimal synthesis of *per*-mRNA at the beginning of the light phase and also in constant light; PRICE et al. 1995). This appears to be due to a rapid degradation of TIM (ZENG et al. 1996). Recent experiments further indicated that protein kinase A is involved in the circadian rhythm (LEVINE et al. 1994).

II. Molluscs

The eyes of molluscs contain autonomous circadian clocks. The intracellular pathway of light to the clock has been extensively analyzed by Block and Eskin and their groups and is dealt with in Chap. 16 of this volume.

III. Mammals

In mammals, the most important circadian pacemakers are the suprachiasmatic nuclei (SCN) in the anterior hypothalamus. Light stimuli perceived by the eyes are transmitted to the SCN by way of the retinohypothalamic tract. How the transmitters released from these neurons are actually resetting the intracellular clock mechanism is being actively investigated at present. In the SCN of rat and other rodents early response genes were shown to be stimulated by light (SCHWARTZ et al. 1994). For a detailed review of the anatomical, physiological and molecular aspects of this input pathway (see Chaps. 1, 4, this volume).

In conclusion, one may arrive at the following still very speculative picture: light acts in fungi and plants via cellular receptors and in animals and man via receptor cells, neurons and neurotransmitters on second messenger levels (cyclic nucleotides, calcium) and thus activate protein kinases. Kinases in turn may act on the level of transcription factors or on the level of phosphorylation of a clock protein (and its degradation) and thus reset the clock. The resetting effect of temperature changes on the clock is probably the result of multiple pathways affecting several or all clock processes.

D. Output Signals from the Clock

Gonyaulax polyedra, a unicellular green dinoflagellate, shows circadian oscillations of various processes: for example, bioluminescence, photosynthetic capacity, cell division, protein synthesis and locomotion. These overt oscillations or "hands" are driven by as yet unknown signals. That there is more than one signal pathway from the clock(s) to the "hands" can be deduced from the different processes influenced by the clock including transcription, translation, post-translational modifications, bioluminescence and membrane properties.

A wealth of such clock-controlled intracellular processes are reported in higher plants. In 1985, KLOPPSTECH described for the first time a daily and also a circadian oscillation in the mRNA level of nuclear genes in the pea. This observation was related to genes coding for light-harvesting complexes of the photosystem II which were later extensively analyzed in *Arabidopsis* (KAY and MILLAR 1993) as well as in tomato plants, where 19 genes coding for photosystem proteins were shown to be under the control of a circadian rhythm. Apart from the photosynthetic apparatus, the expression of several other genes was found to oscillate including, for example, nitrate reductase, catalase, phosphoenol-pyruvate carboxylase, chalcone reductase, glutamine synthetase and two heat-inducible proteins (PIECHULLA 1993).

Some of these oscillations were reported to be controlled at the transcriptional level: in the 5'-upstream regions of the genes encoding chlorophyll a,b binding proteins (*cab* 1, *cab* 2), sequences exist that are responsible for the circadian transcription pattern. This sequence was termed "clock responsive element." When this 5' upstream region was fused to a firefly luciferase gene and this construct transformed into *Arabidopsis*, the transgenic plant showed a circadian rhythm of bioluminescence (KAY and MILLAR 1993). These results suggest the existence of trans-active oscillating transcription factors. This function may be performed by the basic clock protein (see above) or as yet unknown signals.

In *Neurospora*, a few genes have been found whose expression is controlled by the circadian clock (LOROS et al. 1989). They seem to be involved in morphogenetic processes leading to asexual spore formation (conidiation).

Apart from transcriptional control exerted by the clock, there is good evidence for circadian translational control: in *Gonyaulax polyedra* the mRNA level of the luciferin-binding protein (LBP) remains constant whereas the protein level oscillates drastically (MORSE et al. 1989). Modifications of proteins, such as phosphorylation changes, are other targets of oscillatory control. A particularly well analyzed example is the phosphorylation rhythm of the phosphoenol pyruvate carboxylase (PEPC) controlling nocturnal CO_2 fixation in *Bryophyllum* (CARTER et al. 1991). The activity changes of this enzyme seem to be controlled by a rhythm in PEPC-kinase activity, which in turn seems to depend on de novo synthesis. Thus, many of the observed rhythmic changes in the cells of plants may be due to output signals affecting gene expression.

In animal organisms, less is known about the intracellular "hands" of the clock. In SCN cells, for example, rhythms in cAMP level and in vasopressin mRNA processing were observed. The protein (CREB) binding to the cAMP-responsive element (CRE) is present in the SCN at all times but its phosphorylation (i.e., activity) appears to be modulated by the clock. Another modulator of CRE, the transcription factor CREM, is expressed rhythmically in pinealocytes, suggesting that cAMP and cAMP-dependent genes are important outputs of the clock (review: SASSONE-CORSI 1994). More is known in animals about the secondary "hands," i.e., the oscillating processes influenced by the neuronal and hormonal outputs of the pacemaker cells. These "secondary" oscillations comprise the whole list of circadian "hands."

From the available evidence about output signals from the clock one may conclude that the clock exerts transcriptional control over a number of genes and seems also to control post-transcriptional processes. It is not clear, however, by means of which signals, either second messenger levels or membrane-associated processes, such as resting and action potentials, are directed by the clock.

References

Aronson BD, Johnson KA, Loros JJ, Dunlap JC (1994a) Negative feedback defining a circadian clock: autoregulation of the clock gene frequency. Science 263:1578–1584

Aronson BD, Johnson KA, Dunlap JC (1994b) Circadian clock locus frequency: protein encoded by a single open reading frame defines length and temperature compensation. Proc Natl Acad Sci USA 91:7683–7687

Borgeson CE, Bowman BJ (1985) Blue light-reducible cytochromes in membrane fractions from Neurospora crassa. Plant Physiol 78:433–437

Bruce VG (1972) Mutants of the biological clock in Chlamydomonas reinhardtii. Genetics 77:211–230

Carter PJ, Nimmo HG, Fewson CA, Wilkins MB (1991) Circadian rhythms in the activity of a plant protein kinase. EMBO J 10:2063–2068

Crosthwaite SS, Loros JJ, Dunlap JC (1995) Light-induced resetting of a circadian clock is mediated by a rapid increase in *frequency* transcript. Cell 81:1003–1012

Curtin KD, Huang ZJ, Rosbash M (1995) Temporally regulated nuclear entry of the Drosophila period protein contributes to the circadian clock. Neuron 14:363–372

Dowse HP, Hall JC, Ringo JM (1987) Circadian and ultradian rhythms in period mutants of Drosophila melanogaster. Behav Genet 17:19–35

Dunlap J (1993) Genetic analysis of circadian clocks. Annu Rev Physiol 55:683–728

Edery I, Zwiebel LJ, Dembinska ME, Rosbash M (1994a) Temporal phosphorylation of the Drosophila period protein. Proc Natl Acad Sci USA 91:2260

Edery I, Rutila JE, Rosbash M (1994b) Phase shifting of the circadian clock by induction of the Drosophila period protein. Science 263:237

Edmunds LE (1988) Cellular and molecular bases of biological clocks. Springer, Berlin Heidelberg New York

Ehret CF, Trucco E (1967) Molecular models for the circadian clock I. The chronon concept. J Theor Biol 15:240–262

Ewer J, Frisch B, Rosbash M, Hall J (1992) Expression of the period clock gene in different cell types within the adult brain of Drosophila melanogaster. J Neurosci 12:3321–3349

Feldman JF, Hoyle M (1973) Isolation of circadian clock mutants of Neurospora crassa. Genetics 75:605–613

Feldman JF, Gardner G, Denison R (1979) Genetic analysis of the circadian clock of Neurospora. In: Suda M, Hayaishi O, Nakagawa H (eds) Biological rhythms and their central mechanism. Elsevier/North Holland, Amsterdam, pp 57–66

Gebauer G, Kallies A, Rensing L (submitted for publication) Heat shock-induced changes in second messenger levels and differentiation of Neurospora crassa

Goodwin BC (1963) Temporal organization in cells. Academic, London

Hall JC (1990) Genetics of circadian rhythms. Annu Rev Genet 24:659–697

Hall JC (1995) Tripping along the trail of the molecular mechanisms of biological clocks. TINS 18:230–240

Hardin PE, Hall JC, Rosbash M (1992) Behavioral and molecular analyses suggest that circadian output is disrupted by disconnected mutants in Drosophila melanogaster. EMBO J 11:1–6

Huang ZL, Edery I, Rosbash M (1993) PAS is a novel dimerization domain shared by the Drosophila period protein and several transcription factors. Nature 364:259–262

Huang ZJ, Curtin KD, Rosbash (1995) Per protein interactions and temperature compensation of a circadian clock in Drosophila. Science 267:1169–1172

Kallies A, Gebauer G, Rensing L (1996) Light signal pathways to the circadian clock of *Neurospora crassa*. Photochem Photobiol 63:336–343

Kay SA, Millar AJ (1993) Circadian regulated cab gene expression in higher plants. In: Young MW (ed) Molecular genetics of biological rhythms. Dekker, New York, pp 73–89

Kloppstech K (1985) Diurnal and circadian rhythmicity in the expression of light-induced plant nuclear messenger RNAs. Planta 165:502–506

Konopka RJ, Benzer S (1971) Clock mutants of Drosophila melanogaster. Proc Natl Acad Sci USA 68:2112–2116

Lakin-Thomas P, Coté GG, Brody S (1990) Circadian rhythms in Neurospora. CRC Crit Rev Microbiol 17:365–416

Levine JD, Casey CI, Kalderson KK, Jackson FR (1994) Altered circadian pacemaker functions and cyclic cAMP rhythms in the Drosophila learning mutant Dunce. Neuron 13:967–974

Loros JJ (1995) The molecular basis of the Neurospora clock. Neurosciences 7:3–13

Loros JJ, Denome SA, Dunlap JC (1989) Molecular cloning of genes under control of the circadian clock in Neurospora. Science 243:385–388

Menaker M, Refinetti R (1993) The tau mutation in golden hamsters. In: Young MW (ed) Molecular genetics of biological rhythms. Dekker, New York, pp 255–269

Morse D, Milos PM, Roux E, Hastings JW (1989) Circadian regulation of bioluminescence in Gonyaulax involves translational control. Proc Natl Acad Sci USA 86:172–176

Pittendrigh CS (1974) Circadian oscillations in cells and the circadian organization of multicellular systems. In: Schmitt FO, Worden FG (eds) The neurosciences third study progam. MIT Press, Cambridge, pp 437–458

Piechulla B (1993) "Circadian clock" directs the expression of plant genes. Plant Molec Biol 22:533–542

Price JL, Dembinska ME, Young MW, Rosbash M (1995) Suppression of PERIOD protein abundance by the Drosophila clock mutation timeless. EMBO J 14:4044–4049

Ralph MR, Menaker M (1988) A mutation of the circadian system in golden hamsters. Science 241:1225–1227

Rensing K, Kallies A, Gebauer G, Mohsenzadeh S (1995) The effects of temperature change on the circadian clock of Neurospora. In: Waterhouse J, Redfern P (eds) Circadian clocks and their adjustment. Ciba Symp 7:26–50

Rensing L, Hardeland R (1990) The cellular mechanism of circadian rhythms – a view on evidence, hypotheses and problems. Chronobiol Intern 7:353–370

Rensing L, Jaeger N (eds) (1985) Temporal order. Springer, Berlin Heidelberg New York

Rensing L, Kohler W, Gebauer G, Kallies A (1993) Protein phosphorylation and circadian rhythms. In: Battey HN, Dickinson HG, Hetherington AM (eds) Post-translational modifications in plants. Cambridge University Press, Cambridge, pp 171–185

Roenneberg T, Morse D (1993) Two circadian oscillators in one cell. Nature 362:362–364

Ruoff P, Rensing L (1996) The temperature-compensated Goodwin model simulates many circadian clock properties J Theor Biol 179:275–285

Ruoff P, Mohsenzadeh S, Rensing L (1996) Circadian rhythms and protein turnover: The influence of temperature on the period lengths of clock mutants simulated by the Goodwin Oscillator. Naturwiss (in press)

Sassone-Corsi P (1994) Rhythmic transcription and autoregulatory loops: winding up the biological clock. Cell 78:361–364

Saunders DS (1990) The circadian basis of ovarian diapause in Drosophila melanogaster. Is the period gene causally involved in photoperiodic time measurement? J Biol Rhythms 5:315–332

Schwartz WJ, Takeuchi J, Shannon W, Davis EM, Aronin N (1994) Temporal regulation of light-induced Fos and Fos-like protein expression in the ventrolateral subdivision of the rat suprachiasmatic nucleus. Neuroscience 58:573–583

Sommer T, Chambers JAA, Eberle J, Lauter FR, Russo VEA (1989) Fast light-regulated genes of Neurospora crassa. Nucleic Acid Res 17:5713–5723

Vitaterna MH, King DP, Chang AM, Kornhauser JM, Lowrey PL, McDonald JP, Dove WF, Pinto LH, Turek FW, Takahashi JS (1994) Mutagenesis and mapping of a mouse gene, Clock, essential for circadian behavior. Science 264:719–725

Vogelbaum MA, Menaker M (1992) Temporal chimeras produced by hypothalamic transplants. J Neurosci 12:3619–3621

von der Heyde, F, Wilkens, A, Rensing L (1992) The effects of temperature on the circadian rhythms of flashing and glow in Gonyaulax polyedra: are the two rhythms controlled by two oscillators? J Biol Rhythms 7:115–123

Vosshall LB, Price JL, Sehgal A, Saez L, Young MW (1994) Block in nuclear localization of period protein by a second clock mutation, timeless. Science 263:1606–1609

Winfree AT (1980) The geometry of biological time. Springer, Berlin Heidelberg New York

Young MW (1993) Molecular genetics of biological rhythms. Dekker, New York

Zeng H, Quian Z, Myers MP, Rosbash M (1996) A light-entrainment mechanism for the Drosophila circadian clock. Nature 380:129–135

Zeng H, Hardin PE, Rosbash M (1994) Constitutive overexpression of the Drosophila period protein inhibits period mRNA cycling. EMBO J 13:3590–3598

CHAPTER 4
Chemical Neuroanatomy of the Mammalian Circadian System

R.Y. MOORE

A. Introduction

The fundamental functional unit of the nervous system is the neuron. Neurons in the central nervous system are organized into networks, or systems, which subserve specialized functions. The communication between neurons within a system, or with other systems, occurs at specialized points of functional contact, synapses. The mechanism of communication is the release of neuroactive substances from the presynaptic element which act on specialized receptor molecules on the postsynaptic element. Thus, neurons may be classified not only with respect to their morphology, location and inclusion in particular functional systems but on the basis of their use of particular neuroactive substances in synaptic transmission. This classification of neurons on the basis of the production of particular neuroactive substances is termed "chemical neuroanatomy." Over the last 25 years, our knowledge of chemical neuroanatomy has become very extensive, a fact exemplified by the existence of a *Journal of Chemical Neuroanatomy* and a long series of volumes which comprise the *Handbook of Chemical Neuroanatomy*. Indeed, at this point we are able to specify at least one neuroactive substance for nearly all of the neuronal components of the mammalian brain. This has been extended recently to the circadian timing system (CTS) as will be described below.

There are three major classes of neuroactive substances which are utilized by neurons in synaptic transmission. The first, and most extensively studied, class of substances is small molecule neurotransmitters (Table 1). Most, if not all, brain neurons produce one small molecule neurotransmitter, often colocalized with one or more members of the second class of substances, neuropeptides. Of the small molecules, γ-aminobutyric acid (GABA) and glutamate are the most common, probably produced by more than 90% of all brain neurons. The remainder are less common and are listed in Table 1 in a descending order of occurrence in the CNS. The final class of substances is gases, of which the only one to be studied extensively is nitric oxide (BREDT and SYNDER 1992). The distribution and function of neuroactive substances in the CTS is the topic of this chapter.

The mammalian CTS has three major components, entrainment pathways, pacemakers and pacemaker output to effector systems that express circadian function. We will discuss the chemical neuroanatomy of the CTS in this context.

Table 1. Neuroactive substances involved in synaptic transmission in the nervous system

Small molecule transmitters	Exemplary colocalized peptides
Glutamate	Substance P
GABA	Neuropeptide Y, somatostatin
Acetylcholine	Galanin, vasoactive intestinal polypeptide
Serotonin	Substance P, thyroid-releasing hormone
Dopamine	Neurotensin, cholecystokinin
Norepinephrine	Galanin, neuropeptide Y
Histamine	Enkephalin, galanin
Glycine	[a]

[a] No known colocalized peptide.

B. Entrainment Pathways

I. Retinohypothalamic Tract

The principal zeitgeber for the mammalian CTS is light. The effects of light on the circadian pacemaker require the lateral eyes and projections through the optic nerve to the brain. The effects of light on pacemaker function have been described in detail elsewhere in this volume. In brief, light produces changes in the period and phase of the pacemaker. The changes in phase are dependent upon the time at which light is present; light early in subjective night produces phase delays whereas light in late subjective night produces phase advances and light during subjective day is largely without effect. What are the pathways through which such light effects are transmitted?

The photic entrainment pathway has three components, photoreceptors, intrinsic retinal pathways and projections of retinal ganglion cells to the brain. Recent data indicate that there may be a specific set of photoreceptors which transduce the changes in luminance that form the basis of photic entrainment (FOSTER et al. 1991, 1993), but the nature of the photoreceptors is not established. Similarly, we do not know the intrinsic retinal pathways involved in photic entrainment. At the level of ganglion cells, however, we now have quite definitive data, at least for the rat (MOORE et al. 1995). These retinal ganglion cells are a subset of class III, or W, cells; small ganglion cells with a mean diameter of 12.8 ± 2.2 µm and a mean area of 81.8 ± 21.8 µm (MOORE et al. 1995). They appear to be dedicated to the CTS, projecting only to its components. The major projections of these ganglion cells are to the circadian pacemaker, the suprachiasmatic nucleus (SCN) of the hypothalamus, and to the intergeniculate leaflet (IGL) of the lateral geniculate complex (PICKARD 1985). The projection to the SCN was demonstrated more than 20 years ago (MOORE and LENN 1972; HENDRICKSON et al. 1972; MOORE 1973) and this direct retinohypothalamic tract (RHT) is both sufficient to maintain entrainment (MOORE and EICHLER 1972; MOORE and KLEIN 1974; KLEIN and MOORE 1979) and necessary for entrainment (JOHNSON et al. 1988).

The neurotransmitter of the RHT is glutamate. RHT terminals contain glutamate (DEVRIES et al. 1993; CASTEL et al 1993). Stimulation of the RHT results in the release of glutamate from RHT terminals in the SCN (LIOU et al.1986). Application of glutamate to the SCN produces depolarization of SCN neurons and an increase in firing rate (MEIJER et al. 1993; SHIRAKAWA and MOORE 1994a). Further, application of glutamate to the SCN in vitro produces phase shifts in the circadian rhythm of neuronal firing rate with a phase response curve that is very similar to that for stimulation of the optic chiasm in vitro (SHIBATA and MOORE 1993) and for light in the intact animal (SHIRAKAWA and MOORE 1994b; DING et al. 1994; Fig. 1). These data fulfill the criteria for establishing glutamate as the small molecule neurotransmitter of the RHT. In its action on SCN neurons, glutamate acts on both N-methyl-D-aspartate (NMDA) and non-NMDA glutamate receptors and requires the production of nitric oxide (DING et al. 1994).

There are only limited data regarding peptide involvement in RHT function. The available information indicates that a subset of retinal ganglion cells projecting in the RHT contain substance P (TAKATSUJI et al. 1991; MIKKELSEN

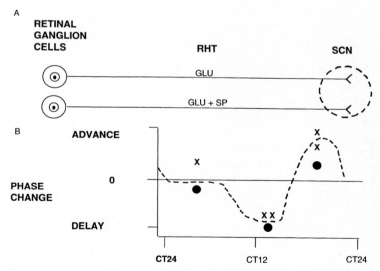

Fig. 1. A Diagram showing the two classes of retinal ganglion cells projecting in the RHT to the SCN. The first is those that contain only glutamate and the second is those that contain glutamate colocalized with substance P. **B** Diagram showing the effects of glutamate and substance P on the firing rate rhythm of SCN neurons in vitro. *Interrupted line* shows the effects of light in the intact animal. *Crosses* are the effects of glutamate and *solid circles* the effects of substance P. See text for further description and references

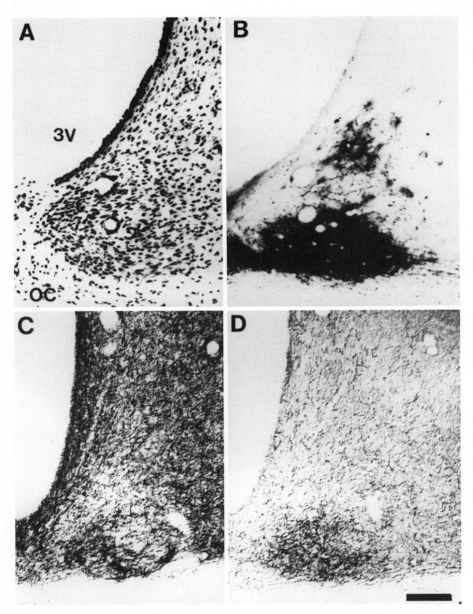

Fig. 2A–D. Photomicrographs of coronal sections through the mid-portion of the rat SCN. **A** Nissl stain. *OC*, optic chiasm; *3*, third ventricle. **B** RHT projection to the core of the SCN shown with cholera toxin (*arrows*). **C** Substance P immunoreactive plexus in the SCN core overlapping the area of the RHT projection. **D** Serotonin plexus in SCN core. *Scale bar*, 100 μm

and LARSEN 1993). Application of substance P to the SCN in vitro also produces changes in the phase of the firing rate rhythm that are similar to the effects of light (SHIBATA et al. 1992), indicating that substance P has effects which would be expected for a colocalized peptide (Fig. 2). The human RHT appears to contain a component of axons which show substance P-like (SP+) immunoreactivity (MOORE and SPEH 1994).

II. Geniculohypothalamic Tract

The GHT arises from a specialized subdivision of the lateral geniculate complex, the IGL, which was first identified by HICKEY and SPEAR (1976) on the basis of its receiving bilateral input from the retina. Subsequently, CARD and MOORE (1982) demonstrated that the IGL contains a population of neuropeptide Y (NPY)-producing neurons that project to the SCN (Fig. 3). In another study, we showed that the IGL also contains a population of enkephalin (ENK)-producing neurons that project in a commissural pathway to the contralateral IGL (CARD and MOORE 1989). The retinal projection to the IGL appears to be from the same set of ganglion cells projecting to the SCN (PICKARD 1985). In addition, there are afferents to the IGL from the locus

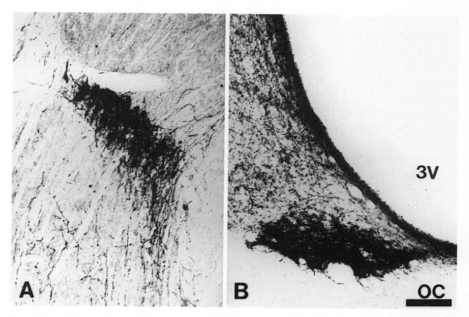

Fig. 3A,B. Photomicrographs showing the NPY+ **(A)** and neurons of the IGL and the NPY+ plexus in the SCN core **(B)**. *Scale bar*, 100 μm

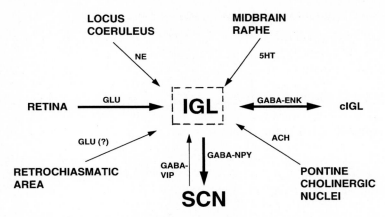

Fig. 4. Diagram illustrating the connections of the intergeniculate leaflet (*IGL*). *Arrows* designate the direction of the projection. *Letters* designate the neuroactive substance(s) produced by the neurons of each projection. *GLU*, glutamate; *ACH*, acetylcholine; *NE*, norepinephrine; *5HT*, serotonin; *ENK*, enkephalin; *NPY*, neuropeptide Y; *VIP*, vasoactive intestinal polypeptide. See text for further discussion

coeruleus (Kromer and Moore 1980; Moore and Card 1994), the midbrain raphe nuclei (Moore and Card 1994), brainstem cholinergic nuclei (Levey et al. 1987), the SCN (Watts 1991) and the retrochiasmatic area (Moore and Card 1994; Fig. 4).

The IGL is made up of small to medium-sized neurons with dendritic arbors that are essentially confined to the IGL. Although we have the most extensive information about IGL organization in the rat, all of the data available indicate that the IGL is quite similar among mammalian species, including primates (Moore 1989). All IGL neurons appear to be GABA producing (Moore and Card 1994). The number of neurons containing the GABA-forming enzyme glutamic acid decarboxylase in the IGL is essentially equal to the number of neurons present in Nissl-stained material (Moore and Card 1994). In contrast, when the number of neurons labeled from retrograde transport from the contralateral IGL and the SCN is compared to the number of ENK- and NPY-immunoreactive neurons, respectively, there is a small population of neurons remaining that projects to the SCN but does not appear to contain NPY (Moore and Card 1994).

There is a very dense SP+ plexus in the IGL (Moore and Card 1994) and SP+ IGL neurons have been reported (Takatsuji and Tokyama 1989), suggesting that a component of the neurons projecting to the SCN is GABA-SP neurons. The function of the IGL-GHT projection appears to be integration of photic and nonphotic information to regulate pacemaker function (cf. Moore and Card 1994; Janik and Mrosovsky 1994, for reviews).

III. Other Inputs to the Suprachiasmatic Nucleus

The two inputs described above, the RHT and GHT, both terminate in a ventral region in the SCN that we will describe below as the core of the nucleus. Three other important inputs terminate predominantly in this area. The first is an input from the pretectal area (MIKKELSEN and VRANG 1994). This overlaps the RHT and GHT inputs as does a nonphotic projection from the serotonin (5-HT)-containing neurons of the midbrain raphe nuclei (MOORE et al. 1978; STEINBUSCH 1981). The final input to the SCN core is from the paraventricular nucleus of the thalamus (PVT). This is a part of a diffuse input over the entire nucleus including both the core and shell regions (MOGA et al. 1995). The functions of the pretectal-SCN projection and that from the PVT are unknown. Recent data indicate that the input from the 5-HT neurons of the raphe acts to modulate the input from the RHT (REA et al. 1994).

There are a number of other nonphotic inputs to the SCN from cerebral cortex, hippocampal formation, basal forebrain, brainstem cholinergic nuclei, medullary noradrenergic nuclei and a number of hypothalamic areas (MOGA and MOORE 1996). All of these terminate in the shell area of the SCN, the zone surrounding the core. Although there have been a few functional studies on some of these, the best assessment of our current knowledge is that we do not have sufficient data to make a definitive statement about the role of any of them in the control of pacemaker function.

C. Pacemaker-Suprachiasmatic Nucleus

I. Suprachiasmatic Nucleus Organization

The SCN is comprised of small neurons which form a compact cell group lying above the optic chiasm and lateral to the third ventricle. Work over the last 15 years has established clearly that the SCN has two subdivisions that are present in all mammals. The first subdivision is found in the ventral part of the nucleus and is the site of termination of the RHT and a secondary visual pathway arising from the IGL, the GHT. It is also the site of termination of another secondary visual pathway arising in the pretectal area (MIKKELSEN and VRANG 1994) and a projection from the midbrain raphe nuclei (MOORE et al. 1978). This region of the SCN, which we designate the core of the nucleus, is characterized by two populations of cells, one containing vasoactive intestinal polypeptide (VIP) and one containing gastrin-releasing peptide (GRP; CARD et al. 1981; VAN DEN POL and TSUJIMOTO 1985; VAN DEN POL 1991; ALBERS et al. 1991). These peptides appear to be colocalized with GABA (OKAMURA et al. 1989; MOORE and SPEH 1993).

The second subdivision of the SCN, which we designate the shell, receives input from a series of nonvisual sources, other hypothalamic nuclei, basal forebrain, cerebral cortex, lateral dorsal tegmental and peripenduncular nuclei and the lateral, medullary tegmentum (MOGA and MOORE 1996). This area is

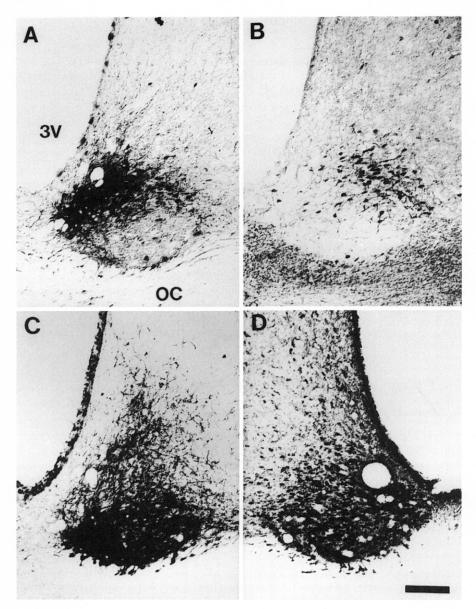

Fig. 5A–D. Photomicrographs showing the distribution of neuroactive substances in the rat SCN. **A** Vasopressin. **B** Calretinin. **C** Vasoactive intestinal polypeptide. **D** GABA. *Scale bar*, 100 μm

characterized by relatively small neurons with smaller dendritic arbors than those in the core (VAN DEN POL 1980). The neurons in the shell contain one or more peptides including vasopressin (VP), angiotensin II (AII), ENK, somatostatin (SS) or calretinin (CAR) (VAN DEN POL and TSUJIMOTO 1985; WATTS and SWANSON 1987; JACOBOWITZ and WINSKY 1991). As with the core, the peptides are colocalized in shell neurons with GABA (Fig. 5; MOORE and SPEH 1993).

The organization of the SCN into two subdivisions has been recognized for a number of years (cf. VAN DEN POL 1980, 1991; MOORE 1983; CASSONE et al. 1988; GILLETTE et al. 1993, for reviews). The subdivisions are evident in all mammals including the human (Fig. 6). In the human brain there is a large population of VP+ neurons that surrounds a zone characterized by VIP+ neurons (MOORE 1992) in which fibers of the RHT terminate (MOORE and SPEH 1994).

In addition to the presence of neurotransmitters and peptides, the SCN has been described to contain nerve growth factor receptor (SOFRONIEW et al. 1989) and to contain VGF, a protein regulated by nerve growth factor in PC12 cells (VAN DEN POL et al. 1989). VGF is found in both shell and core neurons and in the axons of the neurons. These observations raise the possibility that there is a nerve growth factor-containing innervation of the SCN which acts on SCN neurons through regulation of VGF (VAN DEN POL et al. 1989).

II. Suprachiasmatic Nucleus Efferents

SCN efferents project most densely to the hypothalamus, particularly to the zone extending dorsal to the SCN and ventral to the paraventricular nucleus (WATTS and SWANSON 1987; WATTS 1991). All SCN neurons appear to be projection neurons and contribute to the efferents of the nucleus (WATTS and SWANSON 1987). The expression of circadian function in the SCN, and presumably in its efferents, is a circadian rhythm in neuronal firing rate (INOUYE and KAWAMURA 1979; GILLETTE 1991). Although the evidence is incomplete, a high firing rate during subjective day is correlated with a release of peptides (REPPERT et al. 1987) and, presumably, a release of GABA. It is unclear, however, whether release of neuroactive substances at the synaptic cleft with a subsequent action on postsynaptic receptors is the only mechanism through which the SCN acts. Recent data on the effects of fetal transplants to restore rhythmicity in SCN-lesioned hosts has raised the issue of whether rhythmic information may be conveyed by a diffusible substance (LEHMAN et al. 1987; AGUILAR-ROBLERO et al. 1994). Transplants placed in the third ventricle of arrhythmic hosts routinely restore rhythmicity, even in situations in which the transplants appear to establish very minimal connections with host brain. These data, and observations showing preservation of rhythmicity when SCN efferents are sectioned by knife cuts in hamsters (HAKIM et al. 1991), have led to the suggestion that the SCN has redundant outputs, one of which utilizes standard mechanisms of synaptic interaction and the other a diffusible signal which is probably not one of the known SCN neuroactive substances (SILVER and LESAUTER 1993; AGUILAR-ROBLERO et al. 1994).

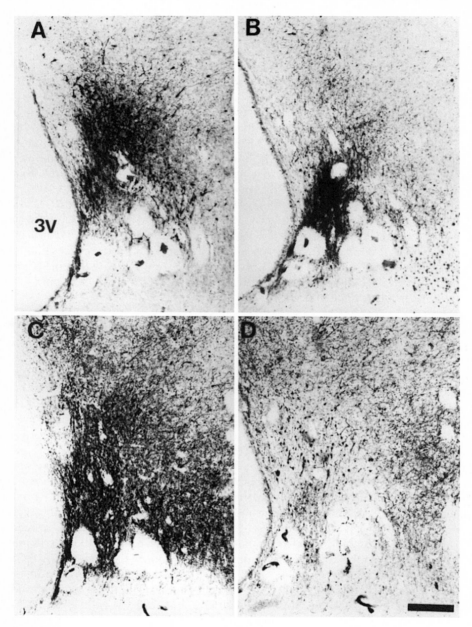

Fig. 6A–D. Photomicrographs showing the distribution of neuroactive substances in the human SCN. **A** Vasopressin. **B** Vasoactive intestinal polypeptide. **C** Neurotensin. **D** Neuropeptide Y. *Scale bar*, 400 μm

D. Conclusions

Our understanding of the chemical neuroanatomy of the CTS is currently well advanced. The CTS has three components, entrainment pathways, pacemakers and pacemaker efferents to effector systems that are under circadian control. The basic organization of the CTS is shown in Fig. 7. The principal entrainment pathway is the RHT, a glutamatergic projection from the retina to the SCN and IGL. There are two other inputs to the SCN core that transmit photic information, the IGL-GHT projection, which is a GABA-NPY projection, and a pretectal-hypothalamic pathway, which also is probably GABAergic. The other inputs to the SCN core are a serotonin-containing projection from the raphe and a glutamate projection from the PVT which terminates in both the shell and core. There is some evidence, albeit not definitive, that intrinsic pacemaker activity is restricted to the core (GILLETTE et al. 1993; AGUILAR ROBLERO et al. 1994). If this is the case, the afferents to the core would be viewed as entrainment pathways whereas those to the shell would serve only to modulate the output of shell neurons. The shell is innervated by glutamate inputs from cortex, basal forebrain, thalamus and hypothalamus, GABA inputs from hypothalamus and cholinergic and noradrenergic inputs from brainstem. Rela-

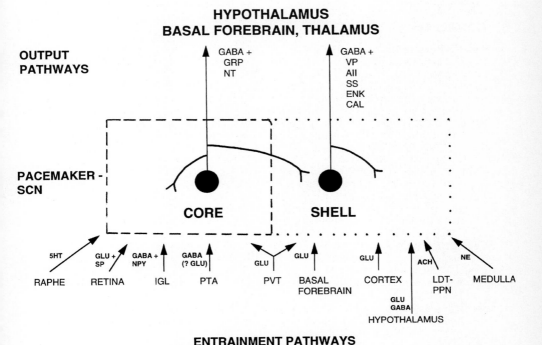

Fig. 7. Organization of the mammalian circadian system with identification of neuroactive substances produced by each component. See text for detailed description

tively little characterization of receptors has been carried out but it is clear that both NMDA and non-NMDA receptors are present in the SCN core and shell (MIKKELSEN et al. 1993; GANNON and REA 1993).

The major input to the SCN is GABAergic. This includes input from intrinsic neurons, the GHT, pretectal area and other hypothalamic areas. The distribution of $GABA_A$ receptors in the SCN and areas receiving SCN projections has been described recently (GAO and MOORE 1996). In the SCN, an unusual subunit organization of $GABA_A$ receptor is expressed, lacking the α_1-subunit. Areas receiving SCN projections express a more usual subunit organization (GAO and MOORE 1996).

The mammalian CTS is dominated by neuronal elements that utilize GABA, colocalized with peptides, as a neurotransmitter. Thus, both intrinsic connections within the system, and projections outside of it, are predominantly inhibitory. How this acts to regulate the effector systems under circadian control remains to be established.

Acknowledgements. The original work from my laboratory here was supported by grants from the National Institutes of Health (NS-16304) and the United States Air Force (NL91-0175). I am grateful to J. Patrick Card, Joan C. Speh, and Nadine Suhan for their contributions.

References

Aguilar-Roblero R, Morin LP, Moore RY (1994) Morphological correlates of circadian rhythm restoration induced by transplantation of the suprachiasmatic nucleus in hamsters. Exp Neurol 30:250–260

Albers HE, Liou SY, Ferris CF, Stopa EG, Zoeller RT (1991) Neurochemistry of circadian timing. In: Klein DC, Moore RY,Reppert SM (eds) The suprachiasmatic nucleus: the mind's clock. Oxford University Press, New York, pp 263–288

Bredt DS, Snyder SH (1992) Nitric oxide, a novel neuronal messenger. Neuron 8:3–11

Card JP, Moore RY (1982) Ventral lateral geniculate nucleus efferents to the rat suprachiasmatic nucleus exhibit avian pancreatic polypetide-like immunoreactivity. J Comp Neurol 206:390–396

Card JP, Moore RY (1989) Organization of lateral geniculate-hypothalamic connections in the rat. J Comp Neurol 284:135–147

Card JP, Brecha N, Karten HJ, Moore RY (1981) Immunocytochemical localization of vasoactive intestinal polypeptide containing cells and processes in the suprachiasmatic nucleus of the rat: light and electron microscopic analysis. J Neurosci 11289–1303

Cassone VM, Speh JC, Card JP, Moore RY (1988) Comparative anatomy of mammalian suprachiasmatic nucleus. J Biol Rhythms 3:71–91

Castel M, Belenkey S, Cohen S, Otterson OP, Storm-Mathisen J (1993) Glutamate-like immunoreactivity in retinal terminals of the mouse suprachiasmatic nucleus. Eur J Neurosci 5:368–381

DeVries MJ, Cardozo BN, van der Waut J, deWolf A, Maijer JH (1993) Glutamate immunoreactivity in terminals of the retinohypothalamic tract of the brown Norwegian rat. Brain Res 612:231–237

Ding JM, Chen D, Weber ET, Faiman LE, Rea MA, Gillette MU (1994) Resetting the biological clock: mediation of nocturnal circadian shifts by glutamate and NO. Science 266:1713–1717

Foster RG, Provencio I, Hudson D, Fiske S, DeGrip W, Menaker M (1991) Circadian photoreception in the retinally degenerate mouse (rd/rd). J Comp Physiol [A] 169:39–50

Foster RG, Argamaso S, Coleman S et al (1993) Photoreceptors regulating circadian behavior: a mouse model. J Biol Rhythms 8:517–523

Gannon RL, Rea MA (1993) Glutamate receptor immunoreactivity in the rat suprachiasmatic nucleus. Brain Res 622:337–342

Gao B, Moore RY (1996) Glutamic acid decarboxylase message isoforms in human suprachiasmatic nucleus. J Biol Rhythms (in press)

Gillette MU (1991) SCN electrophysiology in vitro: rhythmic activity and endogenous clock properties. In: Klein DC, Moore RY, Reppert SM (eds) The suprachiasmatic nucleus: the mind's clock. Oxford University Press, New York, pp 125–143

Gillette MU, DeMarco SJ, Ding JM, Gallman EA, Fairman LE, Liu C, McArthur AJ, Medanic M, Richard D, Cheng TK, Weber ET (1993) The organization of the suprachiasmatic circadian pacemaker of the rat and its modulation by neurotransmitters and modulators. J Biol Rhythms 8:S53–58

Hakim HJ, Philpot A, Silver R (1991) Circadian locomotor rhythms but not photoperiodic responses survive transection of SCN efferents in hamsters. J Biol Rhythms 6:97–113

Hendrickson AE, Wagoner N, Cowan WN (1972) An autoradiographic and electron microscopic study of retino-hypothalamic connections. Z Zellforsch ellforsch 135:1–26

Hickey TL, Spear PD (1976) Retinogeniculate projections in hooded and albino rats. Exp Brain Res 24:523–529

Inouye SIT, Kawamura H (1979) Persistence of circadian rhythmicity in a mammalian hypothalamic "island" containing the suprachiasmatic nucleus. Proc Natl Acad Sci USA 962–5966

Jacobowitz DM, Winsky L (1991) Immunocytochemical localization of calretinin in the forebrain of the rat. J Comp Neurol 304:198–218

Janik D, Mrosovsky N (1994) Intergeniculate leaflet lesions and behaviorally-induced shifts of circadian rhythms. Brain Res 651:174–182

Johnson RF, Moore RY, Morin LP (1988) Loss of entrainment and anatomical plasticity after lesions of the hamster retinohypothalamic tract. Brain Res 460:297–13

Klein DC, Moore RY (1979) Pineal N-acetyltransferase and hydroxyindole-o-methyltranferase: control by the retinohypothalamic tract and the suprachiasmatic nucleus. Brain Res 174:245–262

Kromer LF, Moore RY (1980) A study of the organization of the locus coeruleus projections to the lateral geniculate nuclei in the albino rat. Neuroscience 5:255–271

Lehman MN, Silver R, Gladstone WR, Kahn MR, Gibson M, Brittman EL (1987) Circadian rhythmicity restored by neural transplant. Immunocytochemical characterization of the graft and its integration with the host brain. J. Neurosc., 7: 1626–1638

Levey AI, Hallanger AE, Wainer BH (1987) Choline acetyltransferase immunoreactivity in the rat thalamus. J Comp Neurol 257:317–332

Liou SY, Shibata S, Iwasaki K, Ueki S (1986) Optic nerve stimulation-induced release of ^{3}H-glutamate and ^{3}H-aspartate but not ^{3}H-GABA from the suprachiasmatic nucleus in slices of rat hypothalamus. Brain Res Bull 16:527–531

Meijer JH, Albus H, Weidema F, Ravesloot JH (1993) The effects of glutamate on membrane potential and discharge rate of suprachiasmatic neurons. Brain Res 603:284–288

Mikkelsen JD, Larsen PJ (1993) Substance P in the suprachiasmatic nucleus of the rat: an immunohistochemical and in situ hybridization study. Histochemistry 100:3–16

Mikkelsen JD, Vrang N (1994) A direct pretectosuprachiasmatic projection in the rat. Neuroscience 62:497–505

Mikkelsen JD, Larsen PJ, Ebling FJP (1993) Distribution of N-methyl-D-aspartate (NMDA) receptor mRNAs in the rat suprachiasmatic nucleus. Brain Res 632:329–333

Moga MM, Moore RY (1996) Afferents to the suprachiasmatic nucleus shown by anterograde and retrograde tracing studies. J Comp Neurol (submitted)

Moga MM, Weis RP, Moore RY (1995) Paraventricular thalamic nucleus projections in the rat. J Comp Neurol 359:221–238

Moore RY, Lenn NJ (1972) A retinohypothalamic projection in the rat. J Comp Neurol 146:1–14

Moore RY (1973) Retinohypothalamic projection in mammals: A comparative study. Brain Res 51:403–409

Moore RY (1983) Organization and function of a central nervous system circadian oscillator: the suprachiasmatic hypothalamic nucleus. Fed Proc 42:2783–2789

Moore RY (1989) The geniculohypothalamic tract in monkey and man. Brain Res 486:190–194

Moore RY (1992) The organization of the human circadian timing system. Prog Brain Res 93:101–117

Moore RY, Card JP (1994) The intergeniculate leaflet: an anatomically and functionally distinct subdivision of the lateral geniculate complex. J Comp Neurol 344:403–430

Moore RY, Eichler VB (1972) Loss of circadian adrenal corticosterone rhythm following suprachiasmatic nucleus lesions in the rat. Brain Res 42:201–206

Moore RY, Klein DC (1974) Visual pathways and the central neural control of a circadian rhythm in pineal serotonin N-acetyltransferase activity. Brain Res 71:17–33

Moore RY, Speh JC (1993) GABA is the principal neurotransmitter of the circadian system. Neurosci Lett 150:112–116

Moore RY, Speh JC (1994) A putative retinohypothalamic projection containing substance P in the human. Brain Res 659:249–253

Moore RY, Halaris AE, Jones BE (1978) Serotonin neurons of the midbrain raphe: ascending projections. J Comp Neurol 180:417–438

Moore RY, Speh JC, Card JP (1995) The retinohypothalamic tract originates from a distinct subset of retinal ganglion cells. J Comp Neurol 352:351–366

Okamura H, Berod A, Julien J et al (1989) Demonstration of GABAergic cell bodies in the suprachiasmatic nucleus: in situ hybridization of glutamic acid decarboxylase and immunocytochemistry of GAD and GABA. Neurosci Lett 102:131–136

Pickard GE (1985) Bifurcating axons of retinal ganglion cells terminate in the hypothalamic suprachiasmatic nucleus and the intergeniculate leaflet of the thalamus. Neurosci Lett 55:211–217

Rea MA, Glass JD, Colwell CS (1994) Serotonin modulates photic responses in the hamster suprachiasmatic nuclei. J Neurosci 14:3635–3642

Reppert SM, Schwartz WJ, Uhl GR (1987) Arginine vasopressin: a novel peptide rhythm in cerebrospinal fluid. TINS 10:76–80

Shibata S, Moore RY (1993) Neuropeptide Y and optic chiasm stimulation affect suprachiasmatic nucleus circadian function in vitro. Brain Res 615:95–100

Shibata S, Tsuneyoshi A, Hamada T, Tominaga K, Watanabe S (1992) Effect of substance P on circadian rhythms of firing activity and the 2-deoxyglucose uptake in the rat suprachiasmatic nucleus in vitro. Brain Res 597:257–263

Shirakawa T, Moore RY (1994a) Responses of rat suprachiasmatic nucleus neurons to substance P and glutamate in vitro. Brain Res 642:213–220

Shirakawa T, Moore RY (1994b) Glutamate shifts the phase of the circadian neuronal timing rhythm in the rat suprachiasmatic nucleus in vitro. Neurosci Lett 178:47–50

Silver R, LeSauter J (1993) Efferent signals of the suprachiasmatic nucleus. J Biol Rhythms 8:S89–92

Sofroniew MV, Isacson O, O'Brien TS (1989) Nerve growth factor receptor in the rat suprachiasmatic nucleus. Brain Res 476:358–362

Steinbusch HWM (1981) Distribution of serotonin immunoreactivity in the central nervous system of the rat-cell bodies and terminals. Neuroscience 6:557–618

Takatsuji K, Tohyama M (1989) The organization of the rat lateral geniculate body by immunohistochemical analysis of neuroactive substances. Brain Res 480:198–209

Takatsuji K, Miguel-Hidalgo J-J, Tohyama M (1991) Substance P-immunoreactive innervation from the retina to the suprachiasmatic nucleus in the rat. Brain Res 568:223–229

Van den Pol AN (1980) The hypothalamic suprachiasmatic nucleus of the rat: intrinsic anatomy. J Comp Neurol 191:661–702

Van den Pol A (1991) The suprachiasmatic nucleus: morphological and cytochemical substrates for cellular interaction. In: Klein DC, Moore RY, Reppert SM (eds) The suprachiasmatic nucleus: the mind's clock. Oxford University Press, New York, pp 17–50

Van den Pol AN, Tsujimoto KL (1985) Neurotransmitters of the hypothalamic suprachiasmatic nucleus: immunocytochemical analysis of 25 neuronal antigens. Neuroscience 15:1049–1086

Van den Pol A, Decavel C, Levi A, Paterson B (1989) Hypothalamic expression of a novel gene product, VGF: immunocytochemical analysis. J Neurosci 9:4122–4137

Watts AG (1991) The efferent projections of the suprachiasmatic nucleus: anatomical insights into the control of circadian rhythms. In: Klein DC, Moore RY, Reppert SM (eds) The suprachiasmatic nucleus – the mind's clock. Oxford University Press, New York, pp 77–106

Watts AG, Swanson LW (1987) Efferent projection of the suprachiasmatic nucleus II. Studies using retrograde transport of fluorescent dyes and simultaneous peptide immunohistochemistry in the rat. J Comp Neurol 258:230–252

CHAPTER 5
Chronobiology of Development and Aging

E. Haus and Y. Touitou

A. Introduction

The human time structure consists of a spectrum of rhythms of different frequencies which are superimposed on trends such as development and aging. In some variables and some frequencies, the amplitudes of the rhythmic variations may be larger than the change found in the rhythm-adjusted mean value during a lifetime. Many rhythms are genetically determined (endogenous). Some endogenous rhythms, e.g., in the circadian, circaseptan, or circannual frequency range, are adjusted in time (synchronized) by environmental factors, a process which adapts the human organism to our periodic surrounding. The genetic-environmental interactions in the establishment and the maintenance of rhythms begin in early intrauterine life and continue during infancy and childhood with the development of the mature time structure similar to that seen in the adult during the first 12–24 months of extrauterine life. Optimal functioning of the human organism depends upon an appropriate sequence of metabolic events and related variables within the organism (internal synchronization) and a temporal adjustment of these rhythms to the rhythmic events in our environment (external synchronization) (for review see Haus and Touitou 1994a). The availability of the "right" metabolite at the "right" time allows the orderly sequence of metabolic events required for tissue proliferation and/or other functions. Alterations in the human time structure accompany, and in some instances appear to be responsible for, the decline of many vital functions in the elderly with loss of adaptability to the environmental needs and increased risk of developing and succumbing to disease. However, it remains unclear whether the changes in the human time structure observed in the elderly are a cause or a consequence of the aging process. Changes in the time relation of the elderly to environmental synchronizers and failure to adapt to synchronizer changes may lead to clinical symptomatology and impairment of well-being.

Some rhythms, especially in the infradian (more than 28 h) frequency range, may show a very low amplitude or may not be detectable at rest, but may be induced or triggered by environmental loads such as the introduction of an antigen (Weigle 1975; Levi and Halberg 1982) or exposure to environmental stimuli, e.g., physical therapy (Hildebrandt et al. 1980) or dietary factors (Uezono et al. 1984, 1987). The rhythmic reaction patterns found in some infradian rhythms are essential for the optimal response of the organism to environmental stimuli, and their alteration in the process of aging may lead to functional disturbances and pathology. If the mammalian organism is removed from recog-nizable outside time cues, most if not all endogenous rhythms continue, although often with a frequency slightly but consistently different from the synchronizer schedule, and thus become free running from environmental time cues. Such free-running rhythms have been described for the circadian, circaseptan, and circannual frequency range. In elderly and old subjects and under a variety of pathologic conditions, free-running rhythms in several of these frequencies have also been found under circumstances in which the periodic surrounding persists. "Being out of step" with our periodic surroundings may lead to a loss of adaptation and may be responsible for some of the disturbances observed in the aged.

Changes in the human time structure observed during aging can involve all parameters of biologic rhythms and are found in most frequencies studied. In the circadian range, the most prominent changes in the aging organism are seen in the amplitude and in some instances in the timing of rhythms. The latter may lead to changes in the time relationship of rhythms to each other (desynchronization). Some of these changes may contribute to determine the biologic age of each individual.

A chronobiologic approach to aging has to be extremely broad, and has to consider biochemical, cell, and organ systems as well as the organism as a whole. It cannot be limited to one frequency, but has to consider the entire human time structure, with a wide range of frequencies extending from milliseconds and minutes to many years, in individuals as well as in populations (for reviews, see TOUITOU and HAUS 1994a,b; BROCK 1991; HAUS et al. 1989).

Longitudinal studies in single subjects or in groups of individuals followed regularly over a time span of many years are most likely to answer questions on the relations between changes in time structure and aging. Such studies are possible in laboratory animals with a short life span (e.g., mice or rats), but for obvious reasons are much more difficult to arrange in human subjects. However, in animals with a short life span, it may be difficult to distinguish between aging effects and circannual rhythms. Only recently have some longitudinal studies in human subjects become available (KANABROCKI et al. 1990; SOTHERN and HALBERG 1986). Transverse studies are easier to carry out, but do not follow aging in the strictest sense. The comparison of subjects of different age provides information on differences between age groups, but does not follow the dynamic process of aging. In the upper age groups, we are studying by such an approach a selected population with subjects having developed age-related functional deficits earlier in life, having succumbed to a variety of diseases for which advancing age represents a risk factor. Finally, since the biologic condition of an individual depends both on the genetically fixed aging processes and on the sum of the effects of environmental experience and exposure, there is a very large interindividual variability in aged populations, which must be taken into account in any attempt to characterize aging-related changes in the human time structure.

B. Developmental Chronobiology

The development of the human time structure in children includes two distinct phenomena. One is the process of spontaneous maturation within the framework of the genetic makeup of the individual. Certain genes and gene products determine periodicities of the cell, of organs, and of the organism as a whole. The other is the result of the accumulation of experience by the child.

I. Biologic Rhythms in the Human Fetus

In intrauterine life, the embryo and fetus is exposed to the circadian and circaseptan rhythms of endocrine, metabolic, cardiovascular, and related functions of the mother with rhythmic variations in the passage of hormones and other metabolites through the placental barrier. These maternal factors induce rhythms in the fetus and/or may serve as synchronizers for some fetal rhythms. A number of human fetal rhythms are summarized in Table 1.

II. Fetal Rhythms of Susceptibility to Toxic and Teratogenic Agents

The circadian periodicity of vital organ functions, and apparently also cell proliferation in the fetus, leads to rhythms in sensitivity to drugs with high-amplitude circadian variations of fetal toxicity causing embryopathies and fetal malformations. The rhythms in fetal toxicity have for obvious reasons been studied in experimental animals, predominantly in nocturnal rodents, but it has to be assumed that these findings are also applicable, with differences in timing, to the human situation. Fetal rhythms in toxicity of drugs and other environmental agents pose complex problems since in addition to the stage of gestation, maternal as well as fetal pharmacokinetic factors, maternal drug metabolism, and the extent of placental transfer contribute to the effect of the agent upon the fetus. Susceptibility to teratogenesis varies greatly during gestation. Embryonic tissue is a sensitive target for teratogenic insults. The most critical period in the development of a particular tissue or organ is the time of most rapid cell division. The period of organogenesis ranges in the human embryo from about the 18th to the 60th day of gestation. In mice, this developmental stage corresponds to about gestational days 7–13. Studies with numerous agents have shown that the degree of sensitivity of various embryonic tissues varies markedly, both with the stage of gestation and as a function of the stage of the circadian rhythms of mother and fetus when a toxic agent is administered. Observations on circadian rhythms in fetal toxicity of a variety of agents are summarized in Table 2.

Table 1. Human fetal rhythms

Variable	Frequency	Peak	Trough	Comment – weeks of gestation	Reference
Activity	24-h circadian	During maternal sleep (night)		> 24–30	STERMAN 1967 DEVRIES et al. 1985 NASELLO-PATERSON et al. 1988
	Ultradian				PATRICK et al. 1978 RICHARDSON et al. 1979 CAMPBELL 1980
"Breathing" movements	Circadian	During maternal sleep (0200–0700 hours)		30–40	BODDY and DAWES 1974
	Ultradian (100–500 min)				PATRICK and CHALLIS 1980
Heart rate	Circadian		0200–0600 hours		PATRICK et al. 1982 VISSER et al. 1982a
	Ultradian – multiple frequencies < 1 min to 3–4 h				HAPPENBROWERS et al. 1978 DALTON et al. 1986 VISSER et al. 1982b
Bladder volume	Circadian		0000–0600 hours		CHAMBERLAIN et al. 1984
Estriol	Circadian	Circadian	Midnight (increase in maternal circulation)	Second half of pregnancy, rhythm of fetal-placental unit	TOWNSLEY et al. 1973 – (for review see MEIS 1994)

Seasonal variations in the occurrence of embryopathies and related congenital malformations may be largely caused by environmental exposure of mother and fetus although circannual rhythms in cell proliferation in human tissues have been observed (SMAALAND and SOTHERN 1994), which may suggest differences in the maternal regulation of cell proliferation and its humorally mediated controls, some of which (e.g., corticosteroids) are known to pass the placental barrier. The highest incidence in, e.g., atresia of the esophagus or congenital luxation of the hip is found in children born during winter. In contrast, the peak incidence in congenital stenosis of the aorta and absence of extremities (phocomelia) is observed during summer (for reviews and references see HAUS et al. 1984; SAUERBIER 1994). It is at the present time unknown if these differences in timing may be due to differences in the type of environmental exposure (e.g., seasonal variations in the incidence of certain viral

Table 2. Fetal rhythms of susceptibility to toxic (teratogenic) agents

Agent	Species	Effect	Regimen effect	Minimum effect	Age of gestation (days)	Reference
Cortisone	Mouse	Cleft palate	LD 12:12; 2nd half of dark span		11–15[a]	Isaacson 1959
Dexamethasone	Mouse	Cleft palate	LD 12:12; 2nd half of dark span	End of light span	11–15[a]	Sauerbier 1986
Cyclophosphamide	Mouse	Numerous fetal malformations	LD 12:12; transition from dark to light	Middle of dark span	11–13[a]	Schmidt 1978; Sauerbier 1981, 1983, 1989
5-Fluorouracil	Mouse	Digital defects and kinky tail	LD 12:12; 2nd part of light span	Middle of dark span	11[a]	Clayton et al. 1975
Hydroxyurea	Rat	Digital and limb defects	LD 12:12; early light span		12[a]	Patrick et al. 1982
Ethanol	Mouse, rat	Developmental abnormalities, in utero adsorption, reduction in fetal body weight	LD 12:12; early to middle of dark span	Beginning of light span	7–10[a,b]	Sauerbier 1987, 1988, 1994; Sturtevant and Garber 1985

LD, light-dark
[a] Single exposure.
[b] Single and chronic dosing.

infections) or if differences in maternal-fetal sensitivity to noxious events leading to embryopathies may play a role.

III. Rhythms in the Newborn and During Infancy

At the time of birth, the maternal influences transmitted through the placenta cease and direct environmental stimuli become effective. During the 1st week of life, some of the rhythmicity found in the infant may still in part represent the influence of the mother. With discontinuation of the rhythmic maternal factors transmitted through the placenta, some rhythms detectable in the fetus become less so in the newborn. Ultradian rhythms are predominant in the newborn at the time of birth and shortly thereafter. The development of recognizable circadian periodicity in the infant occurs gradually during the 1st month and, in some variables, may extend over the first 2 years of life, presumably by maturation of the infant and by environmental synchronization of the genetically determined circadian oscillators. After birth, the child is exposed to a large number of new sensorial stimuli which all show marked circadian periodicity, and some of which act as synchronizers of endogenous oscillators. The strongest of these stimuli is the alternation of light/darkness, to which can be added the alternation of noise and silence, heat and cold, and the relations of the neonate to its human environment. Apparently circaseptan rhythms have also been found as group phenomena in the newborn period in some variables with apparently larger amplitudes than the circadian rhythm (HALBERG et al. 1990; WU et al. 1990). Rhythms found in the newborn and infant are summarized in Table 3. During the first 2 years of life a variance transposition takes place, with the circadian rhythm gaining in importance and development of a time structure more and more similar to that seen in the adult.

C. Chronobiology of Aging

I. Changes in Sleep Patterns During Aging

Changes in sleep patterns are a characteristic of the aging process and sleep disturbances are a frequent complaint in the elderly (DEMENT et al. 1982; GOLDENBERG 1991). The regular alternation of sleep and wakefulness is a fundamental endogenous circadian biologic rhythm which under normal conditions is able to entrain (synchronize) other circadian rhythms. Abnormalities of the sleep/wake cycle, with changes in relationships to other endogenous and environmental rhythms, may account in part for some of the disturbances of old age. In experimental animals, the periods of free-running circadian rest/activity cycles decrease with advancing age (PITTENDRIGH and DAAN 1974). Free running of some circadian rhythms in a non-24-h frequency, while others remain environmentally synchronized or may free run with a different period, may change the internal phase relations between circadian rhythms of various physiologic functions, the stable relationship of which is thought to be an important element for physiologic well-being. In some elderly subjects the circadian rhythms of sleep/wake and rest/activity were found to be dissociated, leading to performance decrements (WESSLER et al. 1976; MITLER et al. 1975). A desynchronization between the temperature rhythm and the sleep/wakefulness cycle may lead to sleep disturbances when the time of falling asleep occurs during the ascending branch of the circadian temperature cycle. It was postulated that a diminution of the sensitivity of elderly subjects to the environmental synchronizer, light/darkness, may lead to the appearance of rhythm alterations and desynchronization of the patient. Sleep disturbances related to age in human subjects are summarized in Table 4.

The sleep disturbances in the elderly lead very frequently to medication with hypnotics including barbiturates as well as various benzodiazepines. Some of the rhythm alterations in the elderly suggest that the toxicity profile of a given hypnotic may be different in young adults than in the elderly. Elderly subjects living in a home for the aged, exposed to some of these agents in doses used in geriatric practice, showed few changes in circadian rhythm parameters (see Table 5, HAUS et al. 1988a).

II. Body Temperature

Body temperature in young and mature adults shows ultradian, circadian, and infradian rhythms. Best-known and most extensively studied is the circadian rhythm (HALBERG and PANOFSKY 1961; PANOFSKY and HALBERG 1961; for review see REFINETTI and MENAKER 1992). With advancing age, changes in temperature regulation occur, with a reduction in the circadian amplitude (TOUITOU 1987; VITIELLO et al. 1986; WEITZMAN et al. 1982; and NAKAZAWA et al. 1991) and a phase advance (CAMPBELL et al. 1989; WEITZMAN et al. 1982; ZEPELIN and MACDONALD 1987). WEITZMAN et al. (1982) found in elderly subjects removed from external time cues a shorter free-running period (24.5 h) of the circadian rhythm of rectal temperature than in young adults (25.1 h). In spite of the shorter period of the body temperature rhythm, the sleep/wakefulness cycle did not change its period. Thus, the sleep/wakefulness cycle in the elderly became free running from the body temperature rhythm, which has been found to parallel certain aspects of human performance and to be related to the subject's ability to fall asleep (HAYES and CZEISLER 1994). Internal desynchronization may be responsible for some of the sleep disturbances and performance decrements in many elderly subjects and may also have to be considered in the timing of medication. Some forms of treatment like bright light, melatonin, or a synthetic melatonin receptor agonist may serve to entrain the circadian time structure and may improve, among other factors, the sleep quality in the elderly (ARENDT 1994; see also Chap. 14, this volume).

Table 3. Biologic rhythms in the newborn and infant

Variable	Frequency	Age studied	Peak	Trough	Comment	Reference
Glucose phosphate isomerase	Circadian	At birth			Synchronous with mother (in blood)	Wilf-Miron et al. 1992
Hexoseminidase	Circadian	At birth			Synchronous with mother (in blood)	Wilf-Miron et al. 1992
Body temperature	Circadian	Newborn	Afternoon		In 50% of newborns	Jundell 1904
		Development within first 3–9 months	Afternoon	Night hours		Hellbrügge 1974
		2½ – 4½ years of life	Afternoon	Night hours	Similar to adults	Abe and Fukui 1979
Sleep-wakefulness	Ultradian	Newborn and first weeks of life				Martin du Pan 1974; Menna-Barreto et al. 1993
	Circadian	Development within first 3 months of life, complete during 2nd year of life	Night-time sleep			Kleitman 1963; Kleitman and Engelman 1953; Hellbrügge 1960
Cardiovascular rhythms Systolic BP Diastolic BP	Circadian	Newborn to 48 h			Present in about 35% of individual data series and as group phenomenon	Anderson et al. 1989; Halberg et al. 1991
		1st day	Early in the morning			

	Ultradian	> 2nd day	Shifts to afternoon	ANDERSON et al. 1989 HALBERG et al. 1991
	Circaseptan	Newborn and infant		CORNELISSEN et al. 1987, 1990 HALBERG et al. 1990 WU et al. 1990
Neonatal aortic blood velocity	Circadian	Newborn and infant	As group phenomenon, large circaseptan amplitude, synchronized by process of birth	
	Circannual	Newborn	Circannual as group phenomenon	ZHENGRONG et al. 1993
Endocrine rhythms				
Eosinophils (circulating)	Circadian	Newborn	Absent	HALBERG and ULSTROM 1952
Leukocytes (circulating)	Circadian	15 months	Around noon	HALBERG and ULSTROM 1952
Plasma cortisol	Circadian	3–9 days	Around noon	ZURBRÜGG, 1976
		1–6 months	Night time	
		1.5–11.5 years	Flat or early morning	
Growth hormone	Circadian	Newborn (?) infants	Early morning	Sisson et al. 1974
Urinary sodium$^+$, urinary potassium$^+$, urinary 11-OCS	Circadian	Preterm infants	Early morning Not afternoon	YATSYK et al. 1991
		2–10 weeks	13:00	
		2–10 weeks	10:40	
		2–10 weeks	absent	
TSH, TT$_4$	Circannual	Newborn	Winter	Amplitude increasing with age Fetal-maternal synchronization	ROGOWSKI et al. 1974 ODDIE et al. 1979

Table 4. Age-related sleep rhythm disturbances

Circadian rhythmic variable	Changes observed	Reference
Sleep-wake cycle	Circadian to ultradian frequency transposition	WEBB 1969 For review see DEMENT et al. 1982
	Phase advance	TUNE 1969
	Dissociation from circadian rest-activity, body temperature cycle	WESSLER et al. 1976 MITLER et al. 1975
	Decreased tolerance to phase shifts (e.g., transmeridian flights)	PRESTON 1973
Paradoxical sleep	Phase advance	GOLDENBERG 1991
Urinary water and sodium$^+$ excretion	Shift of urine excretion into night hours (phase delay)	NICOLAU et al. 1985c WEBB and SWINBURNE 1971

III. Blood Pressure and Heart Rate

Human blood pressure shows ultradian variations, a prominent circadian rhythm, and lower-amplitude circasemiseptan, circaseptan, and circannual rhythmicity (for reviews see LEMMER 1989; SMOLENSKY et al. 1976; CORNÉLISSEN et al. 1994).

In clinically healthy adults and in many but not all hypertensive patients, the circadian blood pressure rhythm shows low values during the night. The pressure begins to rise towards morning, reaching a peak during daytime. There may be a drop in the early afternoon with a second, usually lower, peak in the evening hours. With advancing age, the drop around midday becomes more prominent, leading to a double peak within the 24-h span (NICOLAU et al. 1991; CORNÉLISSEN et al. 1994; OTSUKA et al. 1989). The circadian acrophase was reported to show a tendency for phase advance in the elderly (MUNAKATA et al. 1991). In the rhythmometric evaluation of blood pressure data in the elderly, the progressively non-24-h sinusoidal circadian curve shape has to be taken into account. With advancing age, there is a decrease in circadian blood pressure amplitude with appearance of a more prominent ultradian rhythmicity with 12-h and higher frequencies (OTSUKA et al. 1989; CORNÉLISSEN et al. 1994). The ultradian rhythms in systolic and diastolic blood pressure predominate in many subjects during the 10th and 11th decades of life (CORNÉLISSEN et al. 1994). In contrast, in the same subjects, the heart rate remains predominantly circadian.

Similar changes with aging have been found in patients with essential hypertension (ATKINSON et al. 1994; MUNAKATA et al. 1991; SPIEKER et al. 1991). These patients show with advancing age reduced day-night differences especially in systolic blood pressure, due to the early afternoon drop and a rise in night-time values, which becomes more prominent either as an effect of

Table 5. Changes in circadian rhythm parameters of clinical and endocrine variables in elderly subjects (77 ± 8 years of age) under treatment with various commonly used drugs (in doses usually used in geriatric practice; compared to 73 subjects of comparable age without treatment). No changes were found in men and women in growth hormone, insulin, total T_3, or TSH. No changes were found in elderly men in FSH and LH. No changes were found in elderly women in FSH, LH, progesterone, or testosterone. (After Haus et al. 1988a)

Variable	Barbiturate Combined	Barbiturate only	Diazepam Combined	Hydroxizin Combined	Meprobamate Combined	Phenothiazine Combined	Reserpine Combined	Reserpine Only
No. of subjects	63	24	21	11	12	26	62	20
Systolic blood pressure	0	0	0	0	0	0	M+12%	M+13%
Diastolic blood pressure	0	0	M+9%	0	0	0	φ+47° M+11% φ+77°	M+10%
Urine								
Urine volume	0	0	0	0	0	0	M−24%	M−24%
Norepinephrine	0	0	0	0	0	M+37%	M−43%	M−61%
Epinephrine	0	M+128%	0	0	M+344% A+376%	M+191% A+272%	M+112% A+121%	0
Dopamine	0	0	M−29%	0	0	0	0	0
Plasma								
ACTH	0	0	0	0	0	0	0	0
Aldosterone	0	0	0	0	M−32%	A+81%	0	φ+37°
Cortisol	0	M−12%	0	0	0	M+36% A+49%	φ+27°	A+79%
C-peptide	0	0	0	0	0	0	0	0
DHEA-S: men	0	0	0	0	M+79%	0	0	M+160%
DHEA-S: women	M−44%	0	0	0	0	0	0	0
Estradiol: women	0	0	0	0	0	0	M−19%	0
Prolactin	M+26%	0	M+28% A+93%	0	M+48%	M+75% φ+23° A−31%	M+46% A+78%	M+38% A+80%
17-OH progesterone: men	0	0	0	0	0	0	0	0
17-OH progesterone: women	0	0	0	0	0	M+62%	M−27%	0
Testosterone: men	0	A+89%	0	0	0	0	0	0
Total T_4	0	0	0	M+18%	0	0	0	0

Only, drug given alone; Combined, drug given in combination with one or several other agents (e.g., diuretics, laxatives, mild analgesics); M, MESOR; A, Amplitude: +, increase; −, decrease; 0, no change; φ, acrophase: +, phase advance (earlier in the day); −, phase delay (later in the day); 0, no change, 1° = 4 min.

aging or an expression of the longer-standing hypertension (ATKINSON et al. 1994). The afternoon drop in blood pressure in the elderly may in part also be a response to food intake superimposed upon the more prominent ultradian periodicity in this age group. Blood pressure measurements made at this time may lead to "false negatives" in the diagnosis of hypertension (ATKINSON et al. 1994). These rhythms may contribute in part to the time of occurrence of cardiovascular and cerebrovascular pathology in the elderly, including myocardial infarction, sudden cardiac death, and hemorrhagic and ischemic stroke (SMITH 1861; MASTER and JAFFE 1952; REINBERG et al. 1973; SMOLENSKY et al. 1976; MULLER et al. 1985, 1987).

Apparent differences between elderly subjects (77 ± 8 years of age) and children (11 ± 1.5 years of age) were found in the relation between systolic and diastolic blood pressure to plasma aldosterone and urinary norepinephrine excretion (THOMPSON et al. 1987). In children there was a statistically significant positive correlation between the circadian means in systolic and diastolic blood pressure and the norepinephrine excretion, but no significant correlation with the circadian mean of plasma aldosterone. In the elderly subjects, in contrast, there was a weak but statistically significant positive correlation between the circadian mean in diastolic blood pressure and aldosterone, but a negative correlation between the circadian mean of systolic and diastolic blood pressure and norepinephrine excretion. These observations suggest differences in the regulation of blood pressure within the "usual range" between children and elderly subjects and seem to support reports which found a more important role of catecholamines, especially norepinephrine, in blood pressure regulation in young subjects, and in some early stages of essential hypertension as compared with the elderly and with later stages of hypertension (GAVRAS et al. 1982; GOLDSTEIN et al. 1983; NICOLAU et al. 1991).

IV. Hypothalamic–Pituitary–Adrenal Axis

Adrenocorticotropic hormone (ACTH) and adrenal cortisol production show only minor changes during aging. Elderly subjects and young adults have similar circadian patterns of plasma cortisol with a peak between 0600 and 0800 hours, and a progressive decline during the day with a minimum around midnight. This profile was also observed in patients with senile dementia of the Alzheimer type (TOUITOU et al. 1982). An increase of the active free fraction of plasma cortisol was observed in the elderly, which might be related to a decrease in the concentration of binding proteins and/or to a decrease in binding capacity of these proteins (TOUITOU et al. 1982, 1983b). A tendency for a phase advance in plasma cortisol and/or in plasma 18-hydroxy-11-deoxycorticosterone (18-OH-DOC) in the elderly was reported by several investigators (MILCU et al. 1978; TOUITOU et al. 1982; SHERMAN et al. 1985; HAUS et al. 1989), but was not found during all seasons (TOUITOU et al. 1983b) and was observed only in comparison among subjects between the 6th and the 10th decades (HAUS et al. 1989), but not in comparison of elderly subjects to children (HAUS et al.

1988b). A phase advance similar to that seen in cortisol was shown also for dehydroepiandrosterone sulfate (DHEA-S) in elderly subjects between the 6th and the 9th decades (HAUS et al. 1989).

The responsiveness of cortisol production to stimulation by ACTH appears to be unimpaired as is the ACTH response to the infusion of corticotropin-releasing hormone (CRH) (for review see HORAN 1994). The endogenous cortisol to immunoreactive ACTH ratio was actually found to be increased in elderly subjects, suggesting an increase rather than a decrease in the response of the aged adrenal to circulating endogenous ACTH (LAKATUA et al. 1992). An increased responsiveness of the adrenal in the elderly to endogenous ACTH was also suggested by the observation of prolonged elevation of plasma cortisol, but not of plasma ACTH concentrations in elderly subjects after femoral neck fractures (FRAYN et al. 1983; DONCASTER et al. 1993). Also the suppression of ACTH and cortisol secretion by dexamethasone was reported to be less in the elderly (FERRARI et al. 1995) and remained impaired for several weeks after traumatic and surgical stress (FRAYN et al. 1983; ROBERTS et al. 1990).

The human adrenal shows a marked circadian periodicity in its response to endogenous ACTH, with an acrophase at 0700 hours in young adult subjects and with relative resistance to endogenous ACTH stimulation in the evening hours. In the elderly this rhythm shows a marked decrease in amplitude with similar response to ACTH during daytime and evening hours (LAKATUA et al. 1992).

In contrast to the persistence of the circadian periodicity of cortisol into old age, no evidence of seasonal rhythmicity of cortisol was found in the elderly (TOUITOU et al. 1983b; HAUS et al. 1988b).

Aldosterone secretion remains circadian periodic into old age, but with increasing age shows a markedly decreased MESOR (midline estimating statistic of rhythm) and amplitude (CUGINI et al. 1982, 1983b; HAUS et al. 1989). This decrease may be due, at least in part, to a decreased secretion of renin (CUGINI et al. 1982). Aging was accompanied by a large decrease in amplitude of plasma 18-OH-DOC (Touitou et al. 1983b).

1. Dehydroepiandrosterone and Aging

Dehydroepiandrosterone (DHEA) and its sulfate (DHEA-S) are major secretory products of the adrenal, the plasma concentrations of which are circadian periodic. DHEA-S has a longer half-life and is converted to DHEA by steroid sulfatase, which is present in most tissues studied (SONKA 1976). In males, 5%–30% of DHEA-S is of gonadal origin (VERMEULEN 1980). There is a difference in DHEA-S levels between the sexes and there may be different mechanisms of stimulation and metabolism of this hormone between males and females. After administration of DHEA in rats, the highest uptake was found in the pineal within 1 h of administration. DHEA's pineal localization is of interest in view of the pineal's apparent central role in the aging process (see below) and its control of circadian and seasonal rhythms that affect sexual maturity, mating behavior, and sleep patterns, and in view of the recent observation of a stimulation of melatonin production in superfused rat pineals by DHEA-S (San Martin and TOUITOU 1996). Also there appears to be a direct stimulation of DHEA production by mouse adrenals in vitro by melatonin (HAUS et al. 1996a). These observations raise the question of a feedback loop between these two hormones, which appear to play a prominent role in the process of aging.

A decrease of DHEA-S levels with age has been observed in men as well as in women, with elderly women reaching a plateau while elderly men show a negative correlation of DHEA-S levels with age as a continuous variable even after the age of 85 years (BIRKENHAGER-GILLESSE et al. 1994). Some investigators reported a loss in the circadian cyclic production of DHEA in old subjects (REGELSON et al. 1990); others showed persistence of the circadian rhythm of DHEA-S into the 9th decade accompanied by a slight but statistically significant acrophase advance between the 6th and 9th decades (HAUS et al. 1989).

DHEA and DHEA-S appear to play a key role in many processes related to aging. The effects of DHEA and DHEA-S on mechanisms which may contribute to changes found during aging involve a broad range of variables at the cellular level which affect normal and malignant cell proliferation, the integrity of the cardiovascular system, and prominently the immune system. The decrease in DHEA and DHEA-S production with advancing age may lead to the development of some of the degenerative diseases related to aging (REGELSON et al. 1988, 1990, 1994; SONKA 1976) and to favor tumor growth and tumor oncogene expression (SCHULTZ and NYCE 1991; GORDON et al. 1987, 1991, 1993; SCHWARTZ et al. 1988; REGELSON et al. 1988, 1990).

Suggestive evidence for effects of DHEA on the human central nervous system is provided by the observation of reduced plasma concentrations of DHEA-S in patients with Alzheimer's disease (BIRKENHAGER-GILLESSE et al. 1994; SUNDERLAND et al. 1989). DHEA has a profound effect on the immune system. Its action appears to be due to enhancement of host resistance, which limits infection and the related pathology (LORIA and PADGETT 1992). In aging, the decline in host resistance may be related to the decline of DHEA concentrations, which correlates with the general decline in cell-mediated immune response and increased incidence in malignancies. Based on these observations, it was postulated (REGELSON et al. 1994) that the restoration of clinical DHEA values might serve as replacement therapy during aging.

It is not clear which mechanisms are responsible for the decrease of DHEA and DHEA-S with advancing age. Recent in vitro data indicating a stimulation of melatonin production in the pineal by DHEA (San Martin and TOUITOU 1996) and stimulation of DHEA production by melatonin in the adrenal may tie together into a functional relationship, the two hormones most directly changing with age, and presumably playing a role in the aging process.

V. Pineal Gland

Melatonin is the main secretory product of the pineal and its production and blood levels exhibit a high-amplitude circadian rhythm. Extrapineal sources such as the retina (NOWAK et al. 1989) and some cells in the intestinal tract also produce melatonin (HUETHER 1994), which, with the decrease in melatonin production with advancing age, may gain in importance (HUETHER 1994). Although an endogenous relatively low amplitude circadian rhythm of the pineal hormone production is found under constant lighting conditions (e.g., constant dim light), the large-amplitude circadian variation observed is a consequence of the light-dark alternation with acute suppression of melatonin production by bright light in species with diurnal as well as with nocturnal activity patterns. Melatonin interacts with other circadian periodic variables and thus indirectly controls or exerts influence upon a wide variety of physiologic functions such as the sleep-wake cycle, thermal regulation, feeding and sexual behaviors, and certain cardiovascular functions, and through its interaction with serotonin participates in the regulation of the secretion of ACTH, corticosteroids, β-endorphin, prolactin, renin, vasopressin, oxytocin, growth hormone, and luteinizing hormone (LH) (VANDERKAR and BROWNFIELD 1993; EBADI et al. 1993).

In childhood, around the age of 7 years, melatonin shows its maximal amplitude. Its concentration declines at the time of pubertal development (GUPTA et al. 1983). The falling levels of melatonin have been implicated in the onset of puberty. The plasma concentrations of LH at that time were found to increase inversely with the drop in melatonin (WALDHAUSER et al. 1984). In human subjects, plasma melatonin concentrations as well as the excretion of its main urinary metabolite 6-sulfatoxymelatonin decline characteristically with age (TOUITOU et al. 1981, 1984; IGUCHI et al. 1982; SACK et al. 1986). The decrease in the elderly is most prominent at the time of the sharp evening peak of the hormone.

The gradual decline of melatonin which occurs with aging is not caused by calcification of the pineal gland as was previously hypothesized, but is thought to be due to a reduced adrenergic innervation as well as a decrease in β-adrenoceptor density on the surface of the melatonin-producing pinealocytes (REUSS et al. 1990; GREENBERG and WEISS 1978), which decrease in their number and metabolic activity (HUMBERT and PEVET 1993). The positive correlation between plasma melatonin and urinary 6-sulfatoxymelatonin excretion (e.g., MARKEY et al. 1985) indicates that in elderly subjects with normal liver function the low plasma concentrations of melatonin are not a result of more rapid metabolism and clearance (SACK et al. 1986) but of lowered pineal production.

A central role of the pineal in aging has been shown in animal experiments by pineal cross-transplantation from old to young mice and vice versa. Young to old pineal grafting into the thymus, with replacement of the pineal gland of an old mouse with the pineal of a young syngeneic donor mouse, prolonged the life of the old animal, and conversely, the "old" pineal transplanted into a young mouse considerably shortened its life span. Administration of melatonin and pineal extracts or pineal grafting has been reported to prolong the life span of mammals, and to prevent the involution of the gonadal and the thyroid system to stimulate the immune system and to inhibit carcinogenesis (PIERPAOLI et al. 1991; TRENTINI et al. 1991, 1992; ANISIMOV et al. 1992). This apparent role of the pineal gland upon the initiation and progression of senescence and related disorders may offer a novel basis for interventions in the aging process (PIERPAOLI and REGELSON 1994).

Melatonin has been found to be a potent free radical scavenger (REITER et al. 1993) with a protective action on intracellular molecules from oxidative damage. Melatonin seems to scavenge especially the hydroxyl radical (-OH) (TAN et al. 1993), which is considered to be the most damaging of all the free radicals produced (HALLIWELL 1992). With a potent free radical scavenger action, melatonin may be protective against degenerative changes in numerous cell systems. The drop in melatonin during aging appears to be an important factor allowing cumulative oxidative cell damage to occur, which leads to many of the degenerative changes of senescence. With its antioxidant actions, melatonin may, in the opinion of some investigators, have a beneficial effect in the prevention and treatment of disorders related to the aging process and senescence (HARMON 1992; REITER et al. 1994).

Apart from the pineal, melatonin is synthesized in a number of tissues in which oxygen radicals are found in relative abundance, e.g., in the gastrointestinal tract, lungs, liver, skin, and certain nuclear groups in the brain. The beneficial effect of food deprivation on the aging processes and longevity shown in several species including rodents may be related to an increased production of melatonin by the intestinal tract, compensating in part for the decreased pineal secretion (HUETHER 1994).

VI. Growth Hormone and Prolactin

Growth hormone is secreted in a large-amplitude pulsatile fashion. The timing of the main pulse is related to the sleep/wake cycle. Approximately 75% of the pulses occur during the first hours of sleep (during sleep stages 3 and 4). With advancing age, the night-time pulses of plasma growth hormone are markedly decreased in both frequency and amplitude, leading to a decline in secretion to about 20% of the values seen during puberty (FINKELSTEIN et al. 1972; BLICHERT-TOFT 1975; MURRI et al. 1980; HARTMAN et al. 1993), which is comparable to that in patients with organic growth hormone deficiency (Ho and HOFFMAN 1993). The notion that diminished growth hormone action may account for some of the undesirable changes in body composition and function in the elderly is supported by reports on beneficial effects of growth hormone treatment in the clinically healthy elderly (RUDMAN et al. 1990). However, a high incidence of side effects may limit its usefulness (Ho and HOFFMAN 1993). The biologic effects of growth hormone are believed to be mainly mediated via the stimulation of the hepatic production of insulin-related growth factor-1 (IGF-1, somatomedin C), which shows a relatively low amplitude circadian rhythm (NICOLAU et al. 1985b). The concentrations of IGF-1 decline with age probably due to the decrease in the secretion of growth hormone (FLORINI et al. 1985). The night-time reduction of plasma growth hormone concentration appears to be at least in part a consequence of decreased production and decreased responsiveness of the pituitary to the growth hormone releasing factor (GIUSTI et al. 1987; SHIBASAKI et al. 1984).

Prolactin is also secreted episodically. The pulsatile secretions are superimposed upon a circadian pattern characterized by a nocturnal peak (SASSIN et al. 1973; HAUS et al. 1980). Circadian rhythmicity of prolactin continues in the elderly (TOUITOU et al. 1981, 1983a; LAKATUA et al. 1984; HAUS et al. 1988b), with both unchanged and decreased serum concentrations and unchanged or a decreased circadian rhythm amplitude reported (TOUITOU et al. 1983a; HAUS et al. 1989). NELSON et al. (1980) found lowering of the circadian MESOR of prolactin in postmenopausal women. This lowering of the MESOR was associated with changes in pulsatility (HAUS et al. 1996b). The rhythm parameters from post-menopause to old age appear to be unchanged (HAUS et al. 1989). A circannual rhythm in plasma prolactin has been reported in young and in elderly women (REINBERG et al. 1978; TOUITOU et al. 1983a; HAUS et al. 1980, 1989), but could not be found in young and elderly men (TOUITOU et al. 1983a; HAUS et al. 1988b).

VII. Pituitary-Gonadal Axis

The hormones associated with the menstrual cycle undergo characteristic changes with aging. LH shows a marked rise in both sexes while prolactin, estrone, estradiol, and 17-hydroxyprogesterone show a decrease in the circadian mean and in the circadian amplitude, especially in women (NELSON et al. 1980; HAUS et al. 1988b, 1989). Circadian rhythms in plasma testosterone and 17-OH progesterone concentration persist in elderly men and women in spite of markedly decreased circadian mean values. In elderly women circadian rhythms are found in plasma estradiol and progesterone concentration into the 9th decade (NICOLAU et al. 1985a; HAUS et al. 1989). Statistically significant seasonal variations (or if endogenous, circannual rhythms) of the circadian mean were found in elderly women (during the 7th–9th decades of life) in LH, follicle stimulating hormone (FSH), and 17-OH progesterone, and in elderly men in LH, FSH, and testosterone (TOUITOU et al. 1981; NICOLAU et al. 1984). There were significant phase differences between elderly men and women in FSH and LH (NICOLAU et al. 1984). While in seasonally breeding mammals (e.g., hamsters) seasonal variations in the reproductive hormones are thought to be a reflection of light exposure and presumably mediated by melatonin, this is less clear in human subjects as indicated also by the sex difference observed. A discussion of pubertal, menopausal, and related changes in the pituitary gonadal axis is beyond the scope of this review (for reviews see YEN and JAFFE 1991; TIMIRAS et al. 1995).

VIII. Catecholamines

The central and peripheral sympathetic nervous system shows rhythmic variations in its function in various frequencies, but which are especially prominent in the circadian range. The development of the adrenomedullary and sympathetic system is assumed to reach maturity around the 5th year of life (NAKAI and YAMADA 1983). A progressive increase in catecholamine excretion was reported in children from 1 month to 16 years of age (DE SCHAEPDRYVER et al. 1978), which seems to be proportional to the body surface area (PARRA et al. 1980; KIRKLAND et al. 1983). The changes in catecholamine metabolism during the aging process are complex and seem to involve their secretion (RUBIN et al. 1982), blood levels (ROWE and TROEN 1980; GOLDSTEIN et al. 1983), receptors (BERTEL et al. 1980; CONWAY et al. 1971; LAKATTA et al. 1975; VESTAL et al. 1979), peripheral tissue content (MARTINEZ et al. 1981), metabolism (BENDER 1970), and possibly clearance (ESLER et al. 1981; RUBIN et al. 1982). The urinary excretion of norepinephrine increases from 1 to 59 years of age (KARKI 1956), and then decreases during the 7th–10th decades of life. DESCOVICH et al. (1974) found in elderly subjects (66–99 years of age) no difference in the timing of circadian rhythms in catecholamine excretion or a decrease in the MESOR and amplitude of norepinephrine and epinephrine. NICOLAU et al. (1985c) compared the circadian rhythm and the urinary excretion of free epinephrine, norepinephrine, and dopamine between elderly men and women (77 ± 8 years of age) with boys and girls (11 ± 1.5 years of age). The circadian MESOR and the amplitude in epinephrine and dopamine, and to a lesser degree in norepinephrine, were higher in the children than in the elderly subjects. The acrophase of catecholamine excretion remained unchanged in spite of a 4½-h phase shift of the acrophase of the urine excretion in the same elderly subjects into the late night hours (HAUS et al. 1984, 1988b). The excretion of dopamine follows in its timing more closely that of the urine volume.

The decreased excretion of norepinephrine in the elderly (NICOLAU et al. 1985c) contrasts with reports on higher concentrations of plasma norepinephrine concentrations in the aged (ROWE and TROEN 1980; CUTLER and HODES 1983). An alteration in plasma norepinephrine concentrations during aging could be the result of changes at several sites. There is evidence of increased norepinephrine production and release in the aged (WALLIN and SUNDLOF 1979) both in supine and standing positions, while the rate of clearance was found to be the same in young adults and elderly subjects (RUBIN et al. 1982). In contrast, ESLER et al. (1981) suggested a decrease in norepinephrine clearance with aging.

Children showed a circannual rhythm in systolic and diastolic blood pressure and in norepinephrine excretion with peaks during winter (NICOLAU et al. 1985c). In elderly subjects a circannual rhythm in urinary norepinephrine excretion was statistically significant only in women, also with a peak during winter, but was out of phase with the circannual rhythm in blood pressure

(NICOLAU et al. 1986). Absence of a circannual rhythm as a group phenomenon does not necessarily imply absence of such a rhythm in the individual but may be the consequence of a desynchronization within the group.

IX. Age-Related Changes in the Pituitary-Thyroid Axis During Aging

The circadian rhythm in thyroid-stimulating hormone (TSH) in children 5 years of age and older is comparable to that found in adults (NICOLAU et al. 1987a; HAUS et al. 1988b; NICOLAU and HAUS 1989, 1994; ROSE and NISULA 1989). In the elderly, there is a tendency for elevation of plasma TSH concentrations, especially at the trough time of the circadian rhythm (for review see NICOLAU and HAUS 1994). Circadian acrophase differences in total T_4 (TT_4) and total T_3 (TT_3) among subjects of different age groups are independent of the serum protein variations in the same subjects (HAUS et al. 1988b). The circadian acrophase of TSH occurred in elderly subjects only slightly later (40 min) than in children studied in the same geographic region. However, in the elderly, there was a substantial delay in the acrophase of free and of protein-bound thyroid hormones (T_3 and T_4) and of reverse T_3. In spite of the higher circadian mean in TSH, the circadian mean of the circulating thyroid hormones in the elderly was significantly lower than in the children (NICOLAU and HAUS 1994). In the elderly, a clinically apparently euthyroid state is maintained by elevated TSH concentrations, with a lower level of circulating thyroid hormones than in children and a delayed response of the thyroid to the circadian rise in TSH concentrations during the night hours.

The hypothalamic-pituitary-thyroid axis (HPT axis) shows marked seasonal variations which are to a large degree, but probably not exclusively, a response to environmental temperature (LUNGU et al. 1966; LUNGU and NICOLAU 1973; OGATA et al. 1966). In young adult and adult human populations living under climatic conditions with seasonal variations in temperature (irrespective of the yearly temperature average), most investigators reported a seasonal variation in serum or plasma TT_3 and TT_4 concentrations, with highest values during the cold season (for review see NICOLAU and HAUS 1994).

In contrast, in elderly subjects studied in a geographic location with pronounced seasonal variations, no group-synchronized seasonal variation was found in thyroid hormones (HAUS et al. 1988b). This observation raises the question of a defect in cold adaptation in the elderly, but desynchronization of a circannual rhythm within the group including free running of circadian rhythms cannot be excluded.

X. Clinical Chemical Variables and Aging

Few variables frequently studied in clinical chemistry (other than hormonal messengers) show amplitudes in the circadian, circaseptan, or seasonal frequency range which would lead to serious diagnostic problems in laboratory medicine. However, the rhythmicity of some of them may have physiological

Table 6. Extent[a] of circadian variation of biochemical parameters in serum of clinically healthy subjects of different ages

Age (years)	11 ± 2	21 ± 2	76 ± 8
Number of subjects	194	43	189
Variable			
Alkaline Phosphatase (u/l)	8	10	7
CPK (u/l)	28	30	18
Gamma GT (u/l)	50	53	15
LDH (u/l)	5	17	
AST (SGOT) (u/l)	11	15	9
Bilirubin, total (mg/dl)	81	65	42
Cholesterol (mg/dl)	7	6	9
Triglyceride (mg/dl)	82	129	38
Uric acid (mg/dl)	14	10	4
BUN (mg/dl)	30	10	9
Creatinine (mg/dl)	6		7
BUN/creatinine	22	11	9
Glucose (mg/dl)	21	23	48
Calcium (mg/dl)	3	4	4
Chloride (mEq/l)	1	3	2
Iron (µg/dl)	106	54	48
Phosphorus (mg/dl)	24	18	7
Potassium (mEq/l)	8	8	8
Sodium (mEq/l)	1	3	2
Protein, total (g/dl)	5	7	8
Albumin (g/dl)	5	7	7
Globulin (g/dl)	8	6	9
Albumin/globulin	5	6	3

AST, aspartate aminotransferase; CPK, creatinine phosphokinase; Gamma GT, gamma glutamyltransferase; LDH, lactate dehydrogenase; SGOT, serum glutamic oxaloacetic transaminase; BUN, blood urea nitrogen
[a]Difference between highest and lowest value measured over a 24-h span (six samples at 4-h intervals) encountered in groups of subjects expressed as precentage of the lowest value.

importance and, in some instances, may have implications for experimental and clinical pharmacology. The differences between the extent of the circadian change of numerous biochemical variables in children (11 ± 1.5 years of age), young adults (21 ± 2 years of age), and elderly subjects (76 ± 8 years of age) studied at the same geographic location are shown in Table 6.

1. Plasma Proteins and Volemia

Plasma proteins play an important role in pharmacology by binding various molecules including a large number of hormones and drugs. Circadian and seasonal variations of the plasma concentration of plasma proteins can change the proportion and concentrations of the bound and unbound fractions of physiologic or pharmacologic agents, which may be of importance in designing therapeutic protocols that attempt to optimize the tolerance and desired effects of drugs and to diminish their side effects. The plasma protein concentrations in diurnally active elderly human subjects show a circadian rhythm, with a trough around 0400 hours followed by a peak shortly after awakening, around 0800 hours (Touitou et al. 1986). The circadian mean and peak concentrations are lower in the elderly due to a decrease in albumin while the globulin concentrations remain normal or are found to be increased (Haus et al. 1988b). The timing of the circadian periodicity of plasma proteins differs from that of hematocrit, hemoglobin, and red blood cell counts, which are often considered as indexes of the circadian rhythm in blood volume (Touitou et al. 1986; Haus et al. 1988b). Circadian variations were found in the elderly in α_1-acid glycoprotein, α_1-antitrypsin, and C-reactive protein (Bruguerolle et al. 1989), and for immunoglobulins IgA, IgG, and IgM (Casale et al. 1983). Seasonal variations of plasma proteins were documented with a larger peak-trough difference in the elderly than in young adults, with a peak in October and a trough in June (Touitou et al. 1986). The seasonal variation in plasma proteins may explain, at least in part, the higher values of plasma free-cortisol found in June. The circadian and seasonal fluctuations of plasma proteins are of large enough amplitude to produce significant variations in the transport and binding of drugs especially in the aged (Patel et al. 1982; Hequet et al. 1984).

X. Hematology

The circadian rhythms in the number of circulating red blood cells, neutrophils, lymphocytes, eosinophil leukocytes, and platelets persist in the aged (Casale et al. 1982; Haus et al. 1983, 1988b; Touitou et al. 1986). Some rhythm parameters, however, show age-related differences (Haus et al. 1983; Swoyer et al. 1989). In clinically healthy elderly subjects (71 ± 5 years), Swoyer et al. (1989) found the circadian-rhythm-adjusted mean (MESOR) in circulating mature neutrophils in the elderly to be higher than in young adult subjects studied at the same geographic location. The circadian amplitude was unchanged. In contrast, the circadian mean of the "young" neutrophils (band forms) was decreased in the elderly, and a circadian rhythm of these cells was not statistically significant as a group phenomenon. The decrease in the circadian mean of the neutrophil band forms may suggest a lower rate of cell replacement. An increase in the circadian mean in the number of circulating monocytes is of interest since no such change with aging has been found with sampling at single time points chosen without regard to circadian rhythm stage (Dybkaer et al. 1981; Munan and Kelly 1979; Nielsen et al. 1984). Neither MESOR nor amplitude of the circulating number of lymphocytes showed a statistically significant change in the elderly. The circadian acrophase in the number of circulating neutrophils and lymphocytes does occur in the elderly earlier in the day (phase advance) than in young adult and adult subjects, in spite of a comparable time of rising and retiring in the groups examined (Swoyer et al. 1989).

Other findings in the elderly are a decreased MESOR in the number of red cells, and in the mean corpuscular hemoglobin concentration (MCHC), an increased red cell volume (MCV), and an increased amplitude in the circadian variation of mean corpuscular hemoglobin (MCH) and mean corpuscular hemoglobin concentration (MCHC) (Swoyer et al. 1989). In the elderly, serum iron concentrations show persistence of circadian periodicity but, in comparison to children and young adults, with substantially decreased circadian mean and amplitude (Casale et al. 1981; Haus et al. 1988b; Nicolau et al. 1987b).

D. Conclusions

The multifrequency human time structure represents a combination of genetically determined rhythms and trends continuously modified by, and in some frequencies synchronized by, environmental factors. It provides in health for the sequence of metabolic and other physiologic events required for optimal function of the organism and for adaptation to our environment. In clinically healthy and active human subjects, circadian rhythmicity in many variables can be found to be well maintained into very old age. However, observations on such populations may provide a biased view since subjects with genetically determined and/or environmentally induced earlier occurring defects will have succumbed to the disturbances to which aging represents a risk factor.

In the course of the aging process, changes in the human time structure occur which accompany physiologic senescence and may be instrumental in producing some of the performance decrements occurring in old age. These changes may be exaggerated in pathologic aging and contribute to senility.

Age-related changes affect the rhythms of various physiologic variables differently, are found in rhythms of different frequencies, and within a given frequency may be expressed in changes of different rhythm parameters. In the circadian frequency range, the rhythm parameters measured in numerous hormonal and biochemical variables show that the temporal order measured by multiple correlation tests, cluster, and principal coordinates analysis is affected both by gender and age. Up to 76.5% of the variation among the human acrophase dispersion of 39 biochemical and endocrine variables studied by cluster (pattern) analysis could be related to gender and age differences. Acrophase clustering according to gender exhibited a difference greater than did clustering based on age (Ticher et al. 1994). If, in contrast, the circadian amplitude of the same variables expressed as a percentage of the circadian mean is used in a population study, very prominent changes are detected as a function of age. Other factors such as the epidemiologically determined risk to develop breast cancer show clustering according to the acrophase dispersion and not according to the amplitude, although age is regarded as one of the prominent risk factors for the development of breast cancer (Ticher et al. 1994, 1995).

Characteristic changes of the human circadian time structure which are found related to aging are summarized in Tables 7–9. These changes are listed in the order of frequency and consistency with which they have been observed. Age-related changes in circadian rhythm parameters observed in different variables are shown in Tables 7–9. Age-related changes in circadian time adaptations are shown in Table 10.

Among the most prominent changes found in the course of aging is the reduction of the circadian amplitude of many circadian rhythms, which appears to be an important part in the aging process. The measurement of the amplitude of certain circadian periodic variables is regarded by some as a sensitive index of aging and provides a measurable end point to determine its stage and progression. A decrease in amplitude of a physiologic variable during aging may be the expression of a functional decline. Of special interest in this context is the decrease in the circadian amplitude (and especially in the nocturnal rise) of melatonin. The importance of the pineal in the aging process has recently been shown in cross-transplantations of pineals. The antioxidant activity of melatonin, which passes due to its lipophilic nature easily through all cell membranes, could explain the very wide variety of melatonin effects. The decrease in the nightly surge of melatonin may be a factor favoring circadian (external and internal) desynchronization and may lead to a defect in time adaptation, e.g., after transmeridian flights. Of similar interest are the aging changes observed in the concentrations and rhythms of DHEA and DHEA-S. Very recent results suggest in in vitro studies a close functional relationship of these two hormones most prominently changing with age and apparently most intricately in the aging process (HAUS et al. 1996a, San Martin and TOUITOU 1996).

In studies of groups of subjects, the absence of a rhythm as a group phenomenon, as reported occasionally in the elderly, does not necessarily mean absence of a rhythm in single individuals, but rather may be due to a lack of synchronization of the subjects within the group. Rhythms free running from environmental synchronizers and/or among the subjects of a group have been observed in the aged both in the circadian and the circannual frequency range.

In contrast to the majority of circadian rhythms showing a decrease in amplitude, a number of rhythms show an increase in amplitude and/or in circadian mean, which may be interpreted as an adaptive response to some of the changes developing during senescence. The increase in amplitude does not necessarily accompany an increase in circadian mean.

The question has been raised whether a decrease in the circadian amplitude of some variables may play a primary causative role in aging, and if attempts to counteract the decrease of amplitude could delay the aging process. Some attempts along these lines have been made in experimental animals with the administration of melatonin, pineal extracts, and DHEA-S.

Changes related to age have been observed not only in the circadian frequency range but also in ultradian and infradian rhythms including circannual rhythms (HAUS et al. 1988b; HAUS and TOUITOU 1994b). Circannual rhythms of some variables, e.g., thyroid hormones or catecholamine excretion, were not found in groups of elderly subjects, which may represent a lack of adaptation

Table 7. Age-related changes in biologic rhythms in day/night synchronized diurnally active human subjects in order of consistency
1. Reduction in amplitude
2. Increased variability of acrophase
3. Earlier occucrrence of some acrophases (phase advance)
4. Later occurrence of some acrophases (phase delay)
5. Changes in phase relations within the individual (internal desynchronization)
6. Defects in phase adaptation to changes in timing of environmental synchronizers
7. Frequency transposition (e.g., circadian-ultradian)
8. Shorter periods of rhythms studied under "free-running" conditions

Circadian amplitude

Variable	Reference
Decrease	
Blood pressure	OTSUKA et al. 1989
	CORNELISSEN et al. 1994
	ATKINSON et al. 1994
	MUNAKATA et al. 1991
	SPIEKER et al. 1991
Aldosterone	HALBERG et al. 1981
	CUGINI 1984
	CUGINI et al. 1980, 1981, 1982, 1983b
	HAUS et al. 1988b, 1989
Cortisol response to endogenous ACTH	LAKATUA et al. 1992
DHEA, DHEA-S	NELSON et al. 1980
	CUGINI 1984
DHEA-S response to endogenous ACTH	HAUS et al. 1995a
Dopamine	NICOLAU et al. 1985c
	HAUS et al. 1988b
Epinephrine	DESCOVICH et al. 1974
	NICOLAU et al. 1985c
E_2, E_1	NELSON et al. 1980
	HALBERG et al. 1981
Iron (serum)	CASALE et al. 1982
	NICOLAU et al. 1987b
	HAUS et al. 1988b
Melatonin	IGUCHI et al. 1982
	SACK et al. 1986
	SKENE et al. 1990
	CARANI et al. 1987
	TOUITOU et al. 1981, 1984
Norepinephrine	DESCOVICH et al. 1974
	NICOLAU et al. 1985c
17-OH progesterone	NELSON et al. 1980
	HALBERG et al. 1981
Prolactin	CUGINI et al. 1982
	HALBERG et al. 1980
	NELSON et al. 1980
	TOUITOU et al. 1983a
Renin	CUGINI et al. 1981

Table 7. *Contd.*

Circadian amplitude	
Variable	Reference
TSH	Haus et al. 1988b, 1989
	Nicolau and Haus, 1994
	Touitou et al. 1986
	Weitzman et al. 1982
Increase	
Activity rhythm (circadian mean decreased)	Lieberman et al. 1989
C-peptide	Cugini 1984
FSH	Cugini 1984
Insulin	Haus et al. 1989
	Nelson et al. 1980
LH	Haus et al. 1989
	Nelson et al. 1980

Circadian acrophase	
Phase advance	
Sleep-wake cycle	Tune 1969
Peak activity	Lieberman et al. 1989
Blood pressure	Munakata et al. 1991
Body temperature	Campbell et al. 1989
	Weitzman et al. 1982
	Zepelin and MacDonald 1987
Aldosterone	Cugini et al. 1982
Corticosterone	Touitou et al. 1982
Cortisol	Milcu et al. 1978
	Haus et al. 1988b, 1989
	Sharma et al. 1989
	Sherman et al. 1985
DHEA-S	Haus et al. 1988b, 1989
Growth hormone	Cugini and Halberg, 1980
	Cugini et al. 1983a
	Schramm et al. 1980a,b
Iron (serum)	Nicolau et al. 1987b
Lymphocytes	Swoyer et al. 1989
Magnesium	Touitou et al. 1978
Melatonin	Thomas and Miles 1989
Neutrophil leukocytes	Swoyer et al. 1989
Phase delay	
Growth hormone	D'Agata et al. 1974
Melatonin	Sharma et al. 1989
T_3, T_4, fT_3, fT_4, rT_3	Nicolau et al. 1987a
	Nicolau and Haus 1994
Urine excretion	Nicolau et al. 1985c
	Haus et al. 1988b

Table 8. Age-related changes in period in day/night synchronized human subjects

Variable	Observation	Reference
Blood pressure	Circadian to ultradian frequency transposition	CORNELISSEN et al. 1994 HALBERG et al. 1991
Body temperature	Shortening of circadian temperature rhythm in isolation with internal desynchronization	WEITZMAN et al. 1982
Sleep	Circadian to ultradian frequency transposition	DEMENT et al. 1982

Table 9. Age-related internal desynchronization in day/night synchronized diurnally active human subjects

Observation[a]	Reference
Temperature-sleep/activity-rest	ASCHOFF 1969 MOORE-EDE and SULZMAN 1981 WEITZMAN et al. 1982
Sleep-dissociation: sleep/wake vs rest/activity	WESSLER et al. 1976 MITLER et al. 1975
Sleep-wakefulness/temperature in isolation	WEVER 1975, 1979 WEITZMAN et al. 1982
Sleep-polyphasic sleep rhythm circadian to ultradian frequency transposition	WEBB 1969
Temperature/heart rate	NICOLAU et al. 1986, 1991 CAHN et al. 1968
Blood pressure circadian to ultradian frequency transposition	CORNELISSEN et al. 1994
Urine excretion/sleep-wakefulness/catecholamines/electrolytes	HAUS et al. 1988b LOBBAN and TREDRE 1964 NICOLAU et al. 1985c
Thyroid: T_4, T_3, fT_4/TSH ratio	HAUS et al. 1988b
Prolactin/growth hormone	CUGINI 1984

[a] Change in usual phase relation or variance distribution observed in young and middle-aged clinically healthy subjects.

Table 10. Age-related changes in phase shifting in mammalian circadian system

Species	Observation	Reference
Rat	Delay or lack of phase shifing in activity, feeding, water intake	PENG et al. 1980
Mouse	Slower phase adaptation, re-entrainment delayed	BROCK 1991 DAVIS and MENAKER 1981
Human	Inadequate phase adaptation and sleep deficit in airline pilots	PRESTON 1973 DEMENT et al. 1982
	Slower recovery after 6-h phase shift	MONK et al. 1993
	More severe and longer duration of jet lag	MOLINE et al. 1992
	Decrease in sleep time in shift workers	TEPAS et al. 1993

Chronobiology of Development and Aging

to the season-dependent environmental stimuli or a lack of synchronization of circannual rhythms within groups of subjects by those stimuli. In some longitudinally studied subjects, free-running circannual rhythms with periods significantly different from 1 year were found in some variables, e.g., blood pressure (Table 11) (ENGEL et al. 1985; HAUS and TOUITOU 1994b).

Internal desynchronization of circadian rhythms has frequently been documented in older individuals and may lead to disturbances in the sleep-wakefulness pattern (RICHARDSON 1990) and may be responsible for problems in adaptation and resistance to environmental stimuli.

The circadian, circaseptan, and seasonal or circannual variations in human mortality of many causes (REINBERG et al. 1973; SMOLENSKY and SARGENT 1972; MULLER et al. 1985; NICOLAU et al. 1991) indicate transient risk states for many potentially fatal events. Some of these may be related to changes in rhythmic functions (chronopathology), and lack of adaptive capability in the aged, leading to the high mortality noticed during the winter months in human subjects (SMOLENSKY and SARGENT 1972; NICOLAU et al. 1991; REINBERG et al. 1973). In animal experiments in mice (BROCK 1983, 1987a,b), the highest percentage of death of young animals occurred in a circannual rhythmic fashion when the proliferation rates of both T and B cells were lowest. In contrast in old animals, the amplitude of the circannual rhythm of T and B cells was markedly decreased and mortality was consistently higher throughout the year.

In relation to experimental and clinical pharmacology, the differences in human time structure in young children and in the aged are of importance since they may play a role in the pharmacokinetics of drugs as well as in their pharmacodynamics, and in the tolerance of the patient of secondary toxic effects. Medication as used frequently in geriatric practice has been shown to

Table 11. Age-related changes in circannual rhythms in human subjects

Variable	Observation	Reference
Blood pressure	Free-running rhythm (longer than 52-week periods)	HAUS and TOUITOU 1994b ENGEL et al. 1985
Prolactin, aldosterone, LH, 17-OH progesterone, E_2, E_1	Decrease in amplitude	HALBERG et al. 1981
Plasma protein TT_4, TT_3	Increase in amplitude Absence of recognizable circannual rhythm as group phenomenon	TOUITOU et al. 1986 HAUS et al. 1988b NICOLAU and HAUS 1994
Norepinephrine (men only)	Absence of recognizable circannual rhythm as group phenomenon	NICOLAU et al. 1986, 1991
Cortisol	Absence of recognizable circannual rhythm as group phenomenon	TOUITOU et al. 1983b HAUS et al. 1988b

change circadian rhythm parameters in the case of some drugs and some endocrine rhythms as shown in Table 5. Little or no information is available in experimental animals or in human subjects on the relationship of the decreased circadian amplitudes and other changes of the circadian and other rhythm parameters of many variables during aging to changes in pharmacokinetics and pharmacodynamics.

However, the changes in extent and timing, and the potential free running of circadian as well as circannual rhythms, have to be considered in the design of experimental studies as well as in chronotherapy, especially with potent agents used, for instance, in cancer chemotherapy. In the timed treatment with such agents, marker rhythms may have to be used to detect major changes in the elderly patients' time structure, which may require an adjustment in the timing of medication. The question whether aging changes can be delayed or even temporarily reversed by timed treatment with physical or chronopharmacologic agents is intriguing and, at least for the temporary relief of some of the disturbances experienced by the aged, does not seem to be out of reach.

References

Abe K, Fukui S (1979) The individual development of circadian temperature rhythm in infants. J Interdisc Cycle Res 10:227–232
Anderson S, Cornélissen G, Halberg F, Scarpelli PT, Cagnoni S, Germano G, Livi R, Scarpelli L, Cagnoni M, Holte JE (1989) Age effects upon the harmonic structure of human blood pressure in clinical health. Proceedings of the 2nd annual IEEE symposium on computer-based medical systems, Mpls, 26–27 June 1989. Computer Society Press, Washington DC, pp 238–243
Anisimov VN, Bondarenko LA, Khavinson VKM (1992) Effect of pineal peptide preparation (epithalmin) on life span and pineal and serum melatonin level in old rats. Ann N Y Acad Sci 673:53–57
Arendt J (1994) The pineal. In: Touitou Y, Haus E (eds) Biologic rhythms in clinical and laboratory medicine. Springer, Berlin Heidelberg New York, pp 348–362
Aschoff J (1969) Desynchronization and resynchronization of human circadian rhythms. Aerospace Med 40:844–849
Atkinson G, Witte K, Nold G, Sasse U, Lemmer B (1994) Effects of age on circadian blood pressure and heart rate rhythms in patients with primary hypertension. Chronobiol Int 11:35–44
Bender AD (1970) The influence of age on the activity of catecholamines and related therapeutic agents. J Am Geriatr Soc 18:220–232
Bertel O, Buhler FR, Kiowski W, Lutold BE (1980) Decreased beta-adrenoreceptor responsiveness as related to age, blood pressure and plasma catecholamines in patients with essential hypertension. Hypertension 2(2):130–138
Birkenhager-Gillesse G, Derksen J, Lagaay M (1994) Dehydroepiandrosterone sulphate (DHEA-S) in the oldest old, aged 85 and over. Ann NY Acad Sci 719:543–552
Blichert-Toft M (1975) Secretion of corticotrophin and somatotrophin by the senescent adenohypophysis in man. Acta Endocrinol (Copenh) 78:1–157
Boddy K, Dawes GS (1974) Fetal breathing. J Physiol (Lond) 243:599–603
Brock MA (1983) Seasonal rhythmicity in lymphocyte blastogenic responses of mice persists in a constant environment. J Immunol 130:2586–2588
Brock MA (1987a) Age-related changes in circannual rhythms of lymphocyte blastogenic responses in mice. Am J Physiol 252:R299–305

Brock MA (1987b) Temporal order vs. variability in activation of lymphocytes from aging mice. Mech Ageing Dev 37:197–210

Brock MA (1991) Chronobiology and aging. J Am Geriatr Soc 39:74–91

Bruguerolle B, Arnaud C, Bouvenot G (1989) Variations circadienne de l'alpha 1 glycoproteine acid (orosomucoide): consequences therapeutiques? Etude preliminaire. Sem Hos Paris 60:2916–2917

Cahn HA, Folk GE Jr, Huston PE (1968) Age comparison of human day-night physiological differences. Aerospace Med 39:608–610

Campbell K (1980) Ultradian rhythms in the human fetus during the last ten weeks of gestation: a review. Semin Perinatol 4:301–309

Campbell SS, Gillin JC, Kripke DF, Erikson P, Clopton P (1989) Gender differences in the circadian temperature of healthy elderly subjects: relationships to sleep quality. Sleep 12:529–536

Carani C, Baldini A, Morabito F, Resentini M, Diazzi G, Sarti G, Del Rio G, Zini D (1987) Further studies on the circadian rhythms of serum melatonin and testosterone in elderly men. In: Trentini GP, DeGaetani C, Pévet P (eds) Fundamentals and clinics in pineal research. Raven, New York, p 377

Casale G, Migliavacca A, Bonora C, Zurita IE, de Nicola P (1981) Circadian rhythm of plasma iron, total iron binding capacity and serum ferritin in arteriosclerotic aged patients. Age Ageing 10:115–118

Casale G, Emiliani S, de Nicola P (1982) Circadian rhythm of circulating blood cells in elderly persons. Hematologica 67:837–844

Casale G, Marinoni GL, d'Angelo R, de Nicola P (1983) Circadian rhythm of immunoglobulins in aged persons. Age Aging 12:81–85

Chamberlain PF, Manning FA, Morrison I, Lange IR (1984) Circadian rhythm in bladder volumes in the term human fetus. Obstet Gynecol 64:657–660

Clayton DL, Mullen AW, Barnett CC (1975) Circadian modification of drug-induced teratogenesis in rat fetuses. Chronobiologia 2:210–217

Conway J, Wheeler R, Sannerstedt R (1971) Sympathetic nervous activity exercise in relation to age. Cardiovasc Res 5:577–581

Cornélissen G, Halberg F, Tarquini B, Mainardi G, Panero C, Cariddi A, Sorice V, Cagnoni M (1987) Blood pressure rhythmometry during the first week of human life. In: Tarquini B, Vergassola R (eds) Social diseases and chronobiology: proceedings of the IIIrd international symposium on social diseases and chronobiology. Florence, 29 Nov 1986. Esculapio, Bologna, pp 113–122

Cornélissen G, Sitka U, Tarquini B, Mainardi G, Panero C, Cugini P, Weinert D, Romoli F, Cassanas G, Maggioni C, Vernier R, Work B, Einzig S, Rigatuso J, Schuh J, Kato J, Tamura K, Halberg F (1990) Chronobiologic approach to blood pressure during pregnancy and early extrauterine life. In: Hayes DK, Pauly JE, Reiter RJ (eds) Chronobiology: its role in clinical medicine, general biology, and agriculture, part A. Wiley-Liss, New York, pp 585–594

Cornélissen G, Haus E, Halberg F (1994) Chronobiologic blood pressure assessment from womb to tomb. In: Touitou Y, Haus E (eds) Biologic rhythms in clinical and laboratory medicine. Springer, Berlin Heidelberg New York, pp 428–452

Cugini P (1984) Chronobiology and senescence. In: Halberg F, Reale L, Tarquini B (eds) Proceedings of the IInd international symposium on chronobiologic approach to social medicine. Istituto Italiano di Medicina Sociale, Rome, pp 229–268

Cugini P, Halberg F (1980) Age-associated rhythm changes in hormonal levels. All-University Council of Aging: News Walter Library, University of Minnesota, 6:3

Cugini P, Scavo D, Halberg F, Schramm A, Pusch HJ, Franke H (1980) Age and sex difference in circadian amplitude of serum aldosterone. In: Proceedings of the XXVIIIth international congress on physiological sciences. 13–19 July 1980. Budapest, p 1140

Cugini P, Scavo D, Halberg F, Sothern RB, Meucci T, Salandi E, Massimiani F (1981) Aging and circadian rhythm of plasma renin and aldosterone. Maturitas 3:173

Cugini P, Scavo D, Halberg F, Schramm A, Pusch HJ, Franke H (1982) Methodologically critical interaction of circadian rhythms: sex and aging characterize serum aldosterone of the female adrenopause. J Gerontol 37:403–411

Cugini P, Lucia P, Murano G, Scavo D (1983a) Invecchiamento e standards cronobiologici di normalita per i livelli sierici di alcuni ormoni. Rendiconti Soc It Med Int, Roma, pp 183–186

Cugini P, Scavo D, Centanni, Halberg F, Haus E, Lakatua D, Schramm A, Pusch HJ, Franke H, Kawasaki T (1983b) Circadian as well as circannual rhythms of circulating aldosterone have decreased amplitude in aging women. J Endocrinol Invest 6:17–22

Cutler NR, Hodes JE (1983) Assessing the noradrenergic system in normal aging: a review of methodology. Exp Aging Res 9(3):123–127

D'Agata R, Vigneri R, Polosa P (1974) Chronobiological study on growth hormone secretion in man: its relation to sleep-wake cycles and to increasing age. In: Scheving LE, Halberg F, Pauly JE (eds) Chronobiology. Igaku Shoin, Tokyo, p 81

Dalton KJ, Denman DW, Dawson AJ, Hoffman HJ (1986) Ultradian rhythms in human fetal heart rate: a computerized time series analysis. Int J Biomed Comput 18:45–60

Davis FC, Menaker M (1981) Development of the mouse circadian pacemaker: independence from environmental cycles. J Comp Physiol 143:527

Dement WC, Miles LE, Carskadon MA (1982) "White paper" on sleep and aging. J Am Geriatr Soc 30:25–50

DeSchaepdryver AF, Hooft C, Delbeke MJ, den Noortgaete MV (1978) Urinary catecholamines and metabolites in children. J Pediatr 93(2):266–268

Descovich GC, Montalbetti N, Kuhl JFW, Rimondi S, Halberg F, Ceredi C (1974) Age and catecholamine rhythms. Chronobiologia 1:163–171

DeVries JIP, Visser GHA, Prechtl HFR (1985) The emergence of fetal behavior. II. Quantitative aspects. Early Hum Dev 12:99–120

Doncaster HD, Barton RN, Horan MA, Roberts NA (1993) Factors influencing cortisol-adrenocorticotrophin relationships in elderly women with upper femur fractures. J Trauma 34:49–55

Dybkaer R, Lauritzen M, Krakauer R (1981) Relative reference values for clinical, chemical and hematological quantities in "healthy" elderly people. Acta Med Scand 209:1–9

Ebadi M, Samejima M, Pfeiffer RF (1993) Pineal gland in synchronizing and refining physiological events. News Physiol Sci 8:30

Engel R, Sothern RB, Halberg F (1985) Circadian and infradian aspects of blood pressure in a treated elderly mesorhypertensive physician (abstract). Chronobiologia 12:243

Esler M, Skews H, Leonard P, Jackman G, Bobik A, Korner P (1981) Age difference of noradrenaline kinetics in normal subjects. Clin Sci 60(2):217–219

Ferrari E, Magri F, Dori D, Migliorati G, Nescis T, Molla G, Fioraranti M, Solerte SB (1995) Neuroendocrine correlates of the aging brain in humans. Neuroendocrinology 61:464–470

Finkelstein J, Roffwarg H, Boyar R, Kream J, Hellman L (1972) Age-related changes in the twenty-four hour spontaneous secretion of growth hormone. J Clin Endocrinol Metab 35:665–670

Florini JR, Prinz PN, Vitiello MV, Hintz RL (1985) Somatomedin C levels in healthy young and old men: relationship to peak and 24-hour integrated levels of growth hormone. J Gerontol 40:2–7

Frayn KN, Stoner HB, Barton RN, Heath DF, Galasko CS (1983) Persistence of high plasma glucose, insulin and cortisol concentrations in elderly patients with proximal femoral fractures. Age Aging 12:70–76

Gavras H, Hatzinikolau P, North WG, Bresnahan M, Garvas I (1982) Interaction of the sympathetic nervous system with vasopressin and renin in the maintenance of blood pressure. Hypertension 4:400–405

Giusti M, Lomeo A, Marini G, Attanasio R, Baneca A, Camogliano L, Peluffo F, Giordano G (1987) Role of aging on growth hormone and prolactin release after growth hormone releasing hormone and domperidone in man. Horm Res 27:134–140

Goldenberg F (1991) Le sommeil du sujet âgé normal. Neurophysiol Clin 21:267–279

Goldstein DS, Lake CR, Chernow B, Ziegler GM, Coleman MD, Taylor AA, Mitchell JR, Kopin IJ, Keiser HR (1983) Age-dependence of hypertension – normotensive differences in plasma norepinephrine. Hypertension 5:100–104

Gordon GB, Shantz LM, Talalay P (1987) Modulation of growth, differentiation, and carcinogenesis by dehydroepiandrosterone. Adv Enzyme Regul 26:355–382

Gordon GB, Helz L, Sover KJ, Comstock GW (1991) Serum levels of DHEA and its sulfate, and the risk of developing bladder cancer. Cancer Res 51:1366–1369

Gordon GB, Helz L, Sover KJ, Alberg AJ, Comstock GW (1993) Serum levels of dehydroepiandrosterone and dehydroepiandrosterone sulfate and the risk of developing gastric cancer. Cancer Epidemiol Biomarkers Prev 2:33–35

Greenberg LH, Weiss B (1978) b-Adrenergic receptors in aged rat brain: reduced number and capacity of pineal to develop supersensitivity. Science 201:61–63

Gupta D, Riedel L, Frick HJ, Attanasio A, Ranke MB (1983) Circulating melatonin in children: in relation to puberty, endocrine disorders, functional tests, and racial origin. Neuroendocrinol Lett 5:63–78

Halberg F, Panofsky H (1961) I. Thermo-variance spectra; method and clinical illustrations. Exp Med Surg 19:284–309

Halberg F, Ulstrom RA (1952) Morning changes in number of circulating eosinophils in infants. Acta Paediatr 50:160–170

Halberg F, Schramm A, Pusch HJ, Franke H, Cugini P, Scavo D (1980) More prominent circadian amplitude (than any MESOR) – decrease characterizes serum prolactin in human aging. Chronobiologia 7:132–133

Halberg F, Cornelissen G, Sothern RB, Wallach LA, Halberg E, Ahlgren A, Kuzel M, Radke A, Barbosa J, Goetz F, Buckley J, Mandel J, Schuman L, Haus E, Lakatua D, Sackett L, Berg H, Wendt HW, Kawasaki T, Ueno M, Uezono K, Matsuoka M, Omae T, Tarquini B, Cagnoni M, Garcia Sainz M, Perez Vega E, Wilson D, Griffiths K, Donati L, Tatti P, Vasta M, Locatelli J, Camagna A, Lauro R, Tritsch G, Wetterberg L (1981) International geographic studies of oncological interest on chronobiological variables. In: Kaiser HN (ed) Neoplasma – comparative pathology of growth in animals, plants, and man. Wiley, New York, pp 553–596

Halberg F, Cornelissen G, Bakken E (1990) Caregiving merged with chronobiologic outcome assessment, research and education in health maintenance organizations (HMOs). In: Hayes DK, Pauly JE, Reiter RJ (eds) Chronobiology: its role in clinical medicine, general biology, and agriculture, part B. Wiley-Liss, New York, pp 491–549

Halberg F, Halberg E, Halberg J, Ikonomov O, Otsuka K, Holte J, Tamura K, Saito Y, Hata Y, Uezono K, Wang ZR, Xue ZN, del Pozo F, Hillman DC, Samayoa W, Bakken E, Cornelissen G (1991) Womb to tomb blood pressure (BP) monitoring: are single or even 24-hour measurements enough? Proc Assn Adv Med Instr, Washington DC, p 38

Halliwell B (1992) Reactive oxygen species and the central nervous system. J Neurochem 59:1609–1623

Happenbrowers T, Ugartechca JC, Combs D, Hodgman JE, Harpe RM, Sterman MB (1978) Studies of maternal-fetal interaction during the last trimester of pregnancy: ontogenesis of the basic rest-activity cycle. Exp Neurol 61:136–153

Harmon D (1992) Free radical theory of aging. Mutat Res 275:257–266

Hartman ML, Veldhuis JD, Thorner MO (1993) Normal control of growth hormone secretion. Horm Res 40:37–47

Haus E, Touitou Y (1994a) Chronobiology in laboratory medicine. In: Touitou Y, Haus E (eds) Biologic rhythms in clinical and laboratory medicine. Springer, Berlin Heidelberg New York, pp 673–708

Haus E, Touitou Y (1994b) Principles of clinical chronobiology. In: Touitou Y, Haus E (eds) Biologic rhythms in clinical and laboratory medicine. Springer, Berlin Heidelberg New York, pp 6–34

Haus E, Lakatua DJ, Halberg F, Halberg E, Cornelissen G, Sackett LL, Berg HG, Kawasaki T, Ueno M, Uezono K, Matsouka M, Omae T (1980) Chronobiological studies of plasma prolactin in women in Kyushu, Japan and Minnesota, USA. J Clin Endocrinol Metab 51:632–640

Haus E, Lakatua DJ, Swoyer J, Sackett-Lundeen L (1983) Chronobiology in hematology and immunology. Am J Anat 168:467–517

Haus E, Lakatua DJ, Sackett-Lundeen L, Swoyer J (1984) Chronobiology in laboratory medicine. In: Reitveld WT (ed) Clinical aspects of chronobiology. Bakker, Baarn, pp 13–82

Haus E, Nicolau GY, Lakatua DJ, Bogdan C, Popescu M, Sackett-Lundeen L, Fraboni A, Petrescu E (1988a) Circadian rhythm parameters of clinical and endocrine functions in elderly subjects under treatment with various commonly used drugs. In: Reinberg A, Smolensky M, Labrecque G (eds) Annual review of chronopharmacology. Pergamon, New York, pp 77–80

Haus E, Nicolau G, Lakatua DJ, Sackett-Lundeen L (1988b) Reference values for chronopharmacology. Annu Rev Chronopharm 4:333–424

Haus E, Nicolau G, Lakatua DJ, Sackett-Lundeen L, Petrescu E (1989) Circadian rhythm parameters of endocrine functions in elderly subjects during the seventh to the ninth decade of life. Chronobiologia 16(4):331–352

Haus E, Dumitriu L, Nicolau GY, Lakatua, D, Berg H, Petrescu L, Sackett-Lundeen L, Reilly R (1995a) Time relation of circadian rhythms in plasma dehydroepiandrosterone and dehydroepiandrosterone-sulfate to ACTH, cortisol, and 11-desoxycortisol. World Conference on Chronobiology and Chronotherapeutics, Italy, Sept 1995 (abstract)

Haus E, Nicolau GY, Ghinea E, Dumitriu L, Petrescu E, Sackett-Lundeen L (1996a) Stimulation of the secretion of dehydroepiandrosterone by melatonin in mouse adrenals in vitro. Life Sciences 58:263–267

Haus E, Sackett-Lundeen L, Lakatua DJ, Lundeen W, Halberg F, Cornelissen G, Uezono K, Omae T, Kawasaki T (1996b) Circadian rhythm in pulsatile secretion characteristics of plasma cortisol and prolactin in American and Japanese women (in press)

Hayes B, Czeisler CA (1994) Chronobiology of human sleep and sleep disorders. In: Touitou Y, Haus E (eds) Biologic rhythms in clinical and laboratory medicine. Springer, Berlin Heidelberg New York, pp 256–264

Hellbrügge T (1960) The development of circadian rhythms in infants. Cold Spring Harb Symp Quant Biol 25:311–323

Hellbrügge T (1974) The development of circadian and ultradian rhythms of premature and full term infants. In: Scheving LE, Halberg F, Pauly JE (eds) Chronobiology. Igaku Shoin, Tokyo, p 339–341

Hequet B, Meynadier J, Bonneterre J, Adenis L (1984) Circadian rhythm in cisplatin binding of plasma proteins. Annu Rev Chronopharm 1:115–118

Hildebrandt G, Emde L, Geyer F, Weimann H (1980) Zur Frage der periodischen Gliederung adaptives Prozesse. Z Phys Med 9:90–92

Ho KK, Hoffman DM (1993) Aging and growth hormone. Horm Res 40:80–86

Horan MA (1994) Aging, injury and the hypothalamic-pituitary-adrenal axis. Ann N Y Acad Sci 719:285–290

Huether G (1994) Melatonin synthesis in the gastrointestinal tract and the impact of nutritional factors on circulating melatonin. Ann N Y Acad Sci 719:146–158

Humbert W, Pevet P (1993) The decrease of pineal melatonin production with age. Causes and consequences. Ann N Y Acad Sci 719:43–63

Iguchi H, Kato KI, Ibayashi H (1982) Age dependent reduction in serum melatonin concentrations in healthy human subjects. J Clin Endocrinol Metab 55:27–29

Isaacson RJ (1959) Investigation of some of the factors involved in the closure of the secondary palate. Thesis, University of Minnesota, Minneapolis

Jundell I (1904) Über die nykthemeralen Temperatur-Schwankungen im ersten Lebensjahre des Menschen. Jahrb Kinderheilkd 59:521–619

Kanabrocki EL, Sothern RB, Scheving LE, Vesley DL, Tsai TH, Shelstad J, Cournoyer C, Greco J, Mermall H, Ferlin H, Nemchausky BM, Bushnell D, Kaplan E, Kahn S, Augustine G, Holmes E, Rumbyrt J, Sturtevant RP, Sturtevant F, Bremner F, Third JLHC, McCormick JB, Dawson S, Sackett-Lundeen L, Haus E, Halberg F, Pauly JE, Olwin JH (1990) Reference values for circadian rhythms of 98 variables in clinically healthy men in the fifth decade of life. Chronobiol Int 7:445–461

Karki NT (1956) The urinary excretion of noradrenaline and adrenaline in different age groups; its diurnal variation and effect of muscular work on it. Acta Physiol Scand 39 [Suppl 132]:1–96

Kirkland JL, Lye M, Levy DW, Banerjee AK (1983) Pattern of urine flow and electrolyte excretion in healthy elderly people. Br Med J 287:1665–1667

Kleitman N (1963) Sleep and wakefulness. University of Chicago, Chicago

Kleitman N, Engelman TG (1953) Sleep characteristics of infants. J Appl Physiol 6:269–282

Lakatta VG, Gerstenblith G, Angell CS, Shock MW, Weisfeldt ML (1975) Diminished inotropic response of aged myocardium to catecholamines. Circ Res 36(2):262–269

Lakatua DJ, Nicolau GY, Bogdan C, Petrescu E, Sackett-Lundeen L, Irvine PW, Haus E (1984) Circadian endocrine time structure in humans above 80 years of age. J Gerontol 39:654–684

Lakatua DJ, Nicolau GY, Sackett-Lundeen L, Petrescu E, Ortmeier T, Haus E (1992) Circadian rhythm in adrenal response to endogenous ACTH in clinically healthy subjects of different ages. Proceedings of the 5th international conference on chronopharmacology, Amelia Island, FL, USA, abstracts, p IX-1

Lemmer B (1989) Circadian rhythms in the cardiovascular system. In: Arendt J, Minors D, Waterhouse J (eds) Biologic rhythms in clinical practice. Wright, London, pp 51–70

Levi F, Halberg F (1982) Circaseptan (about 7-day) bioperiodicity – spontaneous and reactive and the search for pacemakers. Ric Clin Lab 12:323–370

Lieberman HR, Wurtman JJ, Teicher MH (1989) Circadian rhythms of activity in healthy young and elderly humans. Neurobiol Aging 10:259–265

Lobban M, Tredre B (1964) Diurnal rhythms of renal excretion and of body temperature in aged subjects. J Physiol (Lond) 170:29

Loria RM, Padgett DA (1992) Androstenediol regulates systemic resistance against lethal infections in mice. Arch Virol 127:103–115

Lungu A, Nicolau GY (1973) Circannual rhythm of thyroid function in rat. Rev Roum Endocrinol 10:365–372

Lungu A, Nicolau GY, Cocu F, Teodoru V, Dinu I (1966) Protein-bound iodine variations and spontaneous atmospheric temperature oscillations. Rev Roum Endocrinol 3:279–282

Markey SP, Higa S, Shih S, Danforth DN, Tamarkin L (1985) The correlation between plasma melatonin levels and urinary 6-hydroxymelatonin excretion. Clin Chim Acta 150:221–225

Martin du Pan R (1974) Some clinical applications of our knowledge of the evolution of the circadian rhythm in infants. In: Scheving LE, Halberg F, Pauly JE (eds) Chronobiology. Igaku Shoin, Tokyo, pp 138–144

Martinez JL Jr, Vasquez BJ, Messing RB, Jensen RA, Liang KC, McGaugh JL (1981) Age-related changes in the catecholamines content of peripheral organs in male and female F344 rats. J Gerontol 36:280–284

Master AM, Jaffe HL (1952) Factors in the onset of coronary occlusion and coronary insufficiency. JAMA 148:794–798

Meis PJ (1994) Chronobiology of pregnancy and the perinatal time span. In: Touitou Y, Haus E (eds) Biologic rhythms in clinical and laboratory medicine. Springer, Berlin Heidelberg New York, pp 158–166

Menna-Barreto L, Benedito-Silva AA, Marques N, Morato de Andrade MM, Louzada F (1993) Ultradian components of the sleep-wake cycle in babies. Chronobiol Int 10:103–108

Milcu AE, Bogdan C, Nicolau GY, Cristea A (1978) Cortisol circadian rhythm in 70–100 year old subjects. Rev Roum Med Endocrinol 16:29–39

Mitler M, Sokolove P, Lund R (1975) Activity-inactivity and wakefulness-sleep in mice: induced changes in cyclic relationships by prolonged exposure to constant conditions. Sleep Res 4:267

Moline ML, Pollak CP, Monk TH, Lester LS, Wagner DR, Zendell SM, Graeber RC, Salter CA, Hirsch E (1992) Age related differences in recovery from simulated jet lag. Sleep 14:42–48

Monk TH, Buysse DJ, Reynolds CF III, Kupfer DJ (1993) Inducing jet lag in older people: adjusting to a 6-hour phase advance in routine. Exp Gerontol 28:119–133

Moore-Ede MC, Sulzman FM (1981) Internal temporal order. In: Aschoff J (ed) Biologic rhythms. Plenum, New York, p 215 (Handbook of behavioral neurobiology, vol 4)

Muller JE, Stone PH, Turi SG, Rutherford JD, Czeisler CA, Parker C, Poole WK, Passamani E, Roberts R, Robertson T, Sobel BE, Willerson JT, Braunwald E, MILIS Study Group (1985) Circadian variation in the frequency of onset of acute myocardial infarction. N Engl J Med 313:1315–1322

Muller JE, Ludmer PL, Willich SN, Tofler JH, Aylmer G, Klangos I, Stone PI (1987) Circadian variation in the frequency of sudden cardiac death. Circulation 75:131–138

Munakata M, Imai Y, Abe K, Sasaki S, Minami N, Hashimoto J, Sakuma H, Ichijo T, Sekina M, Yoshizawa M, Sekina H (1991) Assessment of age-dependent changes in circadian blood pressure rhythm in patients with essential hypertension. J Hypertens 9:407–415

Munan L, Kelly A (1979) Age dependent changes in blood monocyte populations in man. Clin Exp Immunol 35:161–162

Murri L, Barreca T, Cerone G, Massetani R, Galamini A, Baldassarre M (1980) The 24 h pattern of human prolactin and growth hormone in healthy elderly subjects. Chronobiologia 7:87–92

Nakai T, Yamada R (1983) Urinary catecholamine excretion by various age groups with special reference to clinical value of measuring catecholamines in newborns. Pediatr Res 17:456–460

Nakazawa Y, Nonaka K, Nishida N, Hayashida N, Miyahara Y, Kotorii T, Matsuoka K (1991) Comparison of body temperature rhythms between healthy elderly and healthy young adults. Jpn J Psychiatry 45:37–43

Nasello-Paterson C, Natale R, Connors G (1988) Ultrasonic evaluation of fetal body movements over twenty-four hours in the human fetus at twenty-four to twenty-eight weeks' gestation. Am J Obstet Gynecol 158:312–316

Nelson W, Bingham C, Haus E, Lakatua DJ, Kawasaki T, Halberg F (1980) Rhythm-adjusted age effects in a concomitant study of twelve hormones in blood plasma of women. J Gerontol 35(4):512–519

Nicolau GY, Haus E (1989) Chronobiology of the endocrine system. Rev Roum Med Endocrinol 27(3):153–183

Nicolau GY, Haus E (1994) Chronobiology of the hypothalamic-pituitary-thyroid axis. In: Touitou Y, Haus E (eds) Biologic rhythms in clinical and laboratory medicine. Springer, Berlin Heidelberg New York, pp 330–347

Nicolau GY, Lakatua D, Sackett-Lundeen L, Haus E (1984) Circadian and circannual rhythms of hormonal variables in elderly men and women. Chronobiol Int 1(4):301–319

Nicolau GY, Haus E, Lakatua DJ, Bogdan C, Sackett-Lundeen L, Popescu M, Berg H, Petrescu E, Robu E (1985a) Circadian and circannual variations of FSH, LH, testosterone, dehydroepiandrosterone-sulfate (DHEA-S) and 17-hydroxy progesterone (17 OH-prog) in elderly men and women. Rev Roum Med Endocrinol 23:223–246

Nicolau GY, Haus E, Lakatua D, Plinga L, Sackett-Lundeen L, Berg H, Petrescu E (1985b) Circadian rhythm in plasma immunoreactive somatomedin-C in children. Rev Roum Med Endocrinol 23:97–103

Nicolau GY, Haus E, Lakatua D, Sackett-Lundeen L, Bogdan C, Plinga L, Petrescu E, Ungureanu E, Robu E (1985c) Differences in the circadian rhythm parameters of urinary free epinephrine, norepinephrine, and dopamine between children and elderly subjects. Rev Roum Med Endocrinol 23:189–199

Nicolau GY, Haus E, Bogdan C, Plinga L, Robu E, Ungureanu E, Sackett-Lundeen L, Petrescu E (1986) Circannual rhythms of systolic and diastolic blood pressure in elderly subjects and in children. Rev Roum Med Endocrinol 24:97–107

Nicolau GY, Dumitriu L, Plinga L, Petrescu E, Sackett-Lundeen L, Lakatua DJ, Haus E (1987a) Circadian and circannual variations of thyroid function in children 11 ± 1.5 years of age with and without endemic goiter. In: Pauly JE, Scheving LE (eds) Progress in clinical and biological research. Liss, New York, pp 229–247 (Advances in chronobiology, vol 227B)

Nicolau GY, Haus E, Lakatua DJ, Bogdan C, Plinga L, Irvine P, Popescu M, Petrescu E, Sackett-Lundeen L, Swoyer J, Robu E (1987b) Chronobiology of serum iron concentrations in subjects of different ages at different geographic locations. Rev Roum Med Endocrinol 25:63–82

Nicolau GY, Haus E, Popescu M, Sackett-Lundeen L, Petrescu E (1991) Circadian, weekly, and seasonal variations in cardiac mortality, blood pressure, and catecholamine excretion. Chronobiol Int 8:149–159

Nielsen H, Blom J, Larsen SO (1984) Human blood monocyte function in relation to age. Acta Pathol Microbiol Immunol Scand [C] 92:5–10

Nowak JZ, Zurawska E, Zawilska J (1989) Melatonin and its generating system in vertebrate retina: circadian rhythm, effect of environmental lighting and interaction with dopamine. Neurochem Int 14:397–406

Oddie TH, Klein AH, Foley TP, Fisher DA (1979) Variation in values for iodothyronine hormones, thyrotropin and thyroxin binding globulin in normal umbilical cord serum with season and duration of storage. Clin Chem 25:1251–1253

Ogata K, Sasaki T, Murakami N (1966) Central nervous and metabolic aspects of body temperature regulation. Bull Inst Const Med Kumamoto Univ [Suppl] 16:1–67

Otsuka K, Kitazumi T, Matsubayashi K, Kawamoto A, Sadakane N, Chikamori T, Kuzume O, Shimada K, Ogura H, Ozawa T (1989) Age-related alterations in the circadian pattern of blood pressure. Am J Noninvas Cardiol 3:159–165

Panofsky H, Halberg F (1961) II. Thermo-variance specta; simplified computational example and other methology. Exp Med Surg 19:323–338

Parra A, Ramirez del Angel A, Cervantes C, Sanchez M (1980) Urinary excretion of catecholamines in healthy subjects in relation to body growth. Acta Endocrinol (Copenh) 94:546–551

Patel IH, Venkataramanan R, Levy RH, Visranathan CT, Ojemann LM (1982) Diurnal oscillations in plasma protein binding of valporic acid. Epilepsia 23(3):283–290

Patrick J, Challis J (1980) Measurement of human fetal breathing movements in healthy pregnancies using a realtime scanner. Semin Perinatol 4:275–286

Patrick J, Fetherston W, Vick H, Voegelin R (1978) Human fetal breathing movements and gross fetal body movements at weeks 34 to 35 of gestation. Am J Obstet Gynecol 130:693–699

Patrick J, Cambell K, Carmichael L, Probert C (1982) Influence of maternal heart rate and gross fetal body movements on the daily pattern of fetal heart rate near term. Am J Obstet Gynecol 144:533–538

Peng MT, Jiang MJ, Hsü HK (1980) Changes in running-wheel activity, eating and drinking and their day-night distributions throughout the life span of the rat. J Gerontol 35:339–347

Pierpaoli W, Regelson W (1994) Pineal control of aging: effect of melatonin and pineal grafting on mice. Proc Natl Acad Sci USA 91:787–791

Pierpaoli W, Dall'ara A, Pedrino E, Regelson W (1991) The pineal control of aging. The effects of melatonin and pineal grafting on the survival of older mice. Ann N Y Acad Sci 621:291–313

Pittendrigh C, Daan S (1974) Circadian oscillations in rodents: a systematic increase of their frequency with age. Science 186:548

Preston F (1973) Further sleep problems in airline pilots on world-wide schedules. Aerosp Med 44:775–782

Refinetti R, Menaker M (1992) The circadian rhythm in body temperature. Physiol Behav 51:613–637

Regelson W, Loria R, Kalimi M (1988) Hormonal intervention: "buffer hormones" or "state dependency". The role of dehydroepiandrosterone (DHEA), thyroid hormone, estrogen and hypophysectomy in aging. Ann NY Acad Sci 521:260–273

Regelson W, Kalimi M, Loria R (1990) DHEA: some thoughts as to its biologic and clinical action. In: Kalimi M, Regelson W (eds) The biologic role of dehydroepiandrosterone (DHEA). De Gruyter, New York, pp 405–445

Regelson W, Loria R, Kalimi M (1994) Dehydroepiandrosterone (DHEA) – the "mother steroid". I. Immunologic action. Ann N Y Acad Sci 719:553–563

Reinberg A, Gervais P, Halberg F, Gaultier M, Poynette N, Abulker C, Dupont J (1973) Mortalité des adultes: rythmes circadiens et circannuels. Nouv Presse Med 2:289–294

Reinberg A, Lagoguey M, Cesselin F, Touitou Y, Legrand JC, Delassalle A, Antreassian J, Lagoguey A (1978) Circadian and circannual rhythms in plasma hormones and other variables of five healthy young human males. Acta Endocrinol (Copenh) 88:417–427

Reiter RJ, Poeggeler B, Tan DX, Chen LD, Manchester LC, Guerrero JM (1993) Antioxidant capacity of melatonin: a novel action not requiring a receptor. Neuroendocrinol Lett 15:103–116

Reiter RJ, Tan D, Poeggeler B, Menendez-Pelaez A, Chen L, Saarela S (1994) Melatonin as a free radical scavenger: implications for aging and age-related diseases. Ann N Y Acad Sci 719:1–12

Reuss S, Spies C, Schroder H, Vollrath L (1990) The aged pineal gland: reduction in pinealocyte number and adrenergic innervation in male rats. Exp Gerontol 25:183–188

Richardson B, Natale R, Patrick J (1979) Human fetal breathing activity during effectively induced labor at term. Am J Obstet Gynecol 133:247–255

Richardson GS (1990) Circadian rhythms and aging. In: Schneider EL, Rowe JW (eds) Handbook of the the biology of aging, 3rd edn. Academic, San Diego, p 275

Roberts NA, Barton RN, Horan MA, White A (1990) Adrenal function after upper femoral fracture in elderly people: persistence of stimulation and the roles of adrenocorticotrophic hormone and immobility. Age Ageing 19(5):304–310

Rogowski P, Siersback-Nielsen K, Hansen M (1974) Seasonal variations in neonatal thyroid function. J Clin Endocrinol Metab 39:919–922

Rose SR, Nisula BC (1989) Circadian variation of thyrotropin in childhood. J Clin Endocrinol Metab 68:1086–1090

Rowe JW, Troen BR (1980) Sympathetic nervous system and aging in man. Endocr Rev 1:167–179

Rubin PC, Scott PJW, McLean K, Reid JL (1982) Noradrenaline release and clearance in relation to age and blood pressure in man. Eur J Clin Invest 12:121–125

Rudman D, Feller AG, Nagraj HS, Gergans GA, Lalitha PY, Goldberg AF, Schlenker RA, Cohn L, Rudman IW, Mattson DE (1990) Effects of human GH in men over 60 years old. N Engl J Med 323(1):1–6

Sack RL, Lewy AJ, Erb DE, Vollmer WM, Singer CM (1986) Human melatonin production decreases with age. J Pineal Res 3:379–388

San Martin M, Touitou Y (1996) Day-night differences in the effects of gonadal hormones on melatonin release from perfused rat pineals. Evidence of a circadian control. Steroids 61:27–32

Sassin J, Frantz AG, Kapen S, Weitzman E (1973) The nocturnal rise of human prolactin is dependent on sleep. J Clin Endocrinol Metab 37:436–440

Sauerbier I (1981) Circadian system and teratogenicity of cytostatic drugs. Prog Clin Res 59C:143–149

Sauerbier I (1983) Embryotoxische Wirkung von Zytostatika in Abhängigkeit von der Tageszeit der Applikation bei Mäusen. Verh Anat Ges 77:147–149

Sauerbier I (1986) Circadian variation in teratogenic response to dexamethasone in mice. Drug Chem Toxicol 9:25–31

Sauerbier I (1987) Circadian modification of ethanol damage in utero in mice. Am J Anat 178:170–174

Sauerbier I (1988) Circadian influence on ethanol-induced intrauterine growth retardation in mice. Chronobiol Int 5:211–216

Sauerbier I (1989) Embryotoxicity of drugs: possible mechanisms of action. In: Lemmer B (ed) Chronopharmacology. Cellular and biochemical interactions. Dekker, New York, pp 683–697

Sauerbier I (1994) Rhythms in drug-induced teratogenesis. In: Touitou Y, Haus E (eds) Biological rhythms in clinical medicine. Springer, Berlin Heidelberg New York, pp 151–157

Schmidt R (1978) Zur zircadianen Modifikation der pränatal-toxischen Wirkung von Cyclophosphamid. Biol Rundsch 16:243–248

Schramm A, Pusch HJ, Franke H, Halberg F, Cugini P, Scavo D (1980a) Circadian exploration of serum cortisol (F) and pituitary growth hormone (GH) as function of age and sex. In: Proceedings of the XXVIIIth international congress of physiological sciences. Budapest, 28:38 (abstract 3076)

Schramm A, Pusch HJ, Müller W, Franke H, Halberg F, Cuginin P, Scavo D (1980b) Circadian exploration of serum prolactin, growth hormone, cortisol, and aldosterone as function of age and sex. In: Proceedings of the XXVth international congress of internal medicine. Hamburg, 15:783 (abstract 3642)

Schultz S, Nyce JW (1991) Inhibition of isoprenylation and p21 membrane association by dehydroepiandrosterone in human colonic adenocarcinoma cells in vitro. Cancer Res 51:6563–6567

Schwartz AG, Whitcomb JM, Nyce JW, Lewbart ML, Pashko LL (1988) Dehydroepiandrosterone and structural analogs: a new class of cancer chemopreventive agents. Adv Cancer Res 51:391–423

Sharma M, Palacios-Bois J, Schwartz G, Iskandar H, Thakur M, Quirion R, Nair NP (1989) Circadian rhythms of melatonin and cortisol in aging. Biol Psychiatry 25:305–319

Sherman B, Wysham C, Pfohl B (1985) Age-related changes in the circadian rhythm of plasma cortisol in man. J Clin Endocrinol Metab 61:439–443

Shibasaki T, Shizume K, Masuda A, Nakahara M, Jibiki K, Demura H, Wakabayashi I, Ling N (1984) Age related changes in plasma growth hormone response to growth hormone releasing factor in man. J Clin Endocrinol Metab 58(1):212–214

Sisson TRC, Root AW, Kendall N (1974) Biologic rhythm of plasma human growth hormone in newborns of low birth weight. In: Scheving LE, Halberg F, Pauly JE (eds) Chronobiology. Igaku Shoin, Tokyo, pp 348–352

Skene DJ, Vivien-Roels B, Sparks DL, Hunsaker JC, Pévet P, Ravid D, Swaab DF (1990) Daily variation in the concentration of melatonin and 5-methoxytryptophol

in the human pineal gland: effect of age and Alzheimer's disease. Brain Res 528:170–174
Smaaland R, Sothern RB (1994) Circadian cytokinetics of murine and human bone marrow and human cancer. In: Hrushesky WJM (ed) Circadian cancer therapy. CRC Press, Boca Raton, pp 119–163
Smith E (1861) Draft of historic review: from periodic fever to chronobiology. P Lavie, personal communication
Smolensky MH, Sargent FS II (1972) Chronobiology of the life sequence. In: Itoh S, Ogata K, Yoshimura H (eds) Advances in climatic physiology. Igaku Shoiu, Tokyo, pp 281–318
Smolensky MH, Tatar SE, Bergman SA, Losman JG, Barnard CN, Dacso CC, Kraft IA (1976) Circadian rhythmic aspects of human cardiovascular function: a review by chronobiologic statistical methods. Chronobiologia 3:337–371
Sonka J (1976) Dehydroepiandrosterone metabolic effects. ACTA Univ Carol 71:9–137
Sothern RB, Halberg F (1986) Circadian and infradian blood pressure rhythms of a man 20 to 37 years of age. In: Halberg F, Reale L, Tarquini B (eds) Proceedings of the 2nd international conference on the medico-social aspects of chronobiology, Florence, 2 Oct 1984. Istituto Italiano di Medicine Sociale, Rome, pp 395–416
Spieker C, Wienecke M, Grotemeyer K-H, Suss M, Barenbrock M, Zierden E, Rahn K-H, Zidek W (1991) Circadian blood pressure rhythms in elderly hypertensive patients. J Int Med Res 19:342–347
Sterman MB (1967) Relationship of intrauterine fetal activity to maternal sleep stage. Exp Neurol [Suppl] 19:98–106
Sturtevant RP, Garber SL (1985) Circadian exposure to ethanol affects the severity of cerebellar cell dysgenesis. Anat Rec 211:187
Sunderland T, Merrill CR, Harrington MG, Lawlor BA, Molchan SE, Martinez R, Murphy DL (1989) Reduced plasma dehydroepiandrosterone concentrations in Alzheimer's disease (letter). Lancet 2(2):570
Swoyer J, Irvine P, Sackett-Lundeen L, Conlin L, Lakatua DJ, Haus E (1989) Circadian hematologic time structure in the elderly. Chronobiol Int 6(2):131–137
Tan DX, Chen LD, Poeggeler B, Manchester LC, Reiter RJ (1993) Melatonin: a potent, endogenous hydroxyl radical scavenger. Endocr J 1:57–60
Tepas DI, Duchon JC, Cersten AH (1993) Shiftwork and the older worker. Exp Aging Res 19(4):295–320
Thomas DR, Miles A (1989) Melatonin secretion and age. Biol Psychiatry 25:363–369
Thompson ME, Nicolau GY, Lakatua DJ, Sackett-Lundeen L, Plinga L, Bogdan C, Robu E, Ungureanu E, Petrescu E, Haus E (1987) Endocrine factors of blood pressure regulation in different age groups. In: Pauly JE, Scheving LE (eds) Advances in chronobiology, part B. Liss, New York, pp 79–95
Ticher A, Sackett-Lundeen L, Ashkenazi IE, Haus E (1994) Human circadian time structure in subjects of different gender and age. Chronobiol Int 11(6):349–355
Ticher A, Haus E, Ron IG, Sackett-Lundeen L, Ashkenazi IE (1995) The pattern of hormonal circadian time structure (acrophase) as an indicator of breast cancer risk. (submitted for publication)
Timiras PS, Quay WD, Vernadakis A (eds) (1995) Hormones and aging. CRC Press, Boca Raton
Touitou Y (1987) Le vieillissement des rythmes biologiques chez l'homme. Pathol Biol 35(6):1005–1012
Touitou Y, Haus E (1994a) Biologic rhythms from biblical to modern times. A preface. In: Touitou Y, Haus E (eds) Biologic rhythms in clinical and laboratory medicine. Springer, Berlin Heidelberg New York, pp 1–5
Touitou Y, Haus E (1994b) Aging of the human endocrine and neuroendocrine time structure. Ann N Y Acad Sci 719:378–397
Touitou Y, Touitou C, Bogdan A, Beck H, Reinberg A (1978) Serum magnesium circadian rhythm in human adults with respect to age, sex and mental status. Clin Chim Acta 87:35–41

Touitou Y, Fevre M, Lagoguey M, Carayon A, Bogdan A, Reinberg A, Beck H, Cesselin F, Touitou C (1981) Age and mental health-related circadian rhythm of plasma levels of melatonin, prolactin, luteinizing hormone and follicule-stimulating hormone in man. J Endocrinol 91:467–475

Touitou Y, Sulon J, Bogdan A, Touitou C, Reinberg A, Beck H, Sodoyez JC, Van Cauwenberge H (1982) Adrenal circadian system in young and elderly human subjects: a comparative study. J Endocrinol 93:201–210

Touitou Y, Carayon A, Reinberg A, Bogdan A, Beck H (1983a) Differences in the seasonal rhythmicity of plasma prolactin in elderly human subjects. Detection in women but not in men. J Endocrinol 96:65–71

Touitou Y, Sulon J, Bogdan A, Reinberg A, Sodoyez JC, Demey-Ponsart E (1983b) Adrenocortical hormones ageing and mental condition: seasonal and circadian rhythms of plasma 18-hydroxy-11-deoxycorticosterone, total and free cortisol and urinary corticosteroids. J Endocrinol 96:53–64

Touitou Y, Fevre M, Bogdan A, Reinberg A, De Prins J, Beck H, Touitou C (1984) Patterns of plasma melatonin with ageing and mental condition: stability of nycotohemeral rhythms and differences in seasonal variations. Acta Endocrinol (Copenh) 106:145–151

Touitou Y, Touitou C, Bogdan A, Reinberg A, Auzeby A, Beck H, Guillet P (1986) Differences between young and elderly subjects in seasonal and circadian variations of total plasma proteins and blood volume as reflected by hemoglobin, hematocrit, and erythrocyte counts. Clin Chem 32:801–804

Townsley JD, Dubin IVH, Grannis GF, Gortman J, Crystle CD (1973) Circadian rhythms of serum and urinary estrogens in pregnancy. J Clin Endocrinol 36:289–295

Trentini GP, DeGaetani C, Criscuolo M (1991) Pineal gland and aging. Aging 3:103–106

Trentini GP, Genazzani AR, Criscuolo M (1992) Melatonin treatment delays reproductive aging of female rat via the opiatergic system. Neuroendocrinology 56:364–370

Tune G (1969) Sleep and wakefulness in 509 normal adults. Br J Med Psychol 42:75–80

Uezono K, Haus E, Swoyer J, Kawasaki T (1984) Circaseptan rhythms in clinically healthy subjects. In: Haus E, Kabat H (eds) Chronobiology 1981–1983. Karger, New York, pp 257–262

Uezono K, Sackett-Lundeen L, Kawasaki T, Omae T, Haus E (1987) Circaseptan rhythm in sodium and potassium excretion in salt sensitive and salt resistant Dahl rats. Prog Clin Biol Res 227A:297–307

Vanderkar LD, Brownfield MS (1993) Serotonergic neurons and neuroendocrine function. News Physiol Sci 8:202

Vermeulen A (1980) Andrenal androgens and aging. In: Thijssen JHH, Siiteri PK (eds) Adrenal androgens. Raven, New York, pp 207–217

Vestal RE, Wood AJ, Shand DG (1979) Reduced beta-adrenoceptor sensitivity in the elderly. Clin Pharmacol Ther 26(2)181–186

Visser GHA, Carse EA, Goodman JDS, Johnson P (1982a) A comparison of episodic heart-rate patterns in the fetus and newborn. Br J Obstet Gynecol 89:50–55

Visser GHA, Goodman JDS, Levine, Dawes GS (1982b) Diurnal and other cyclic variations in human fetal heart rate near term. Am J Obstet Gynecol 142:535–544

Vitiello MV, Smallwood RG, Avery DH, Pascualy RA, Martin DC, Prinz PN (1986) Circadian temperature rhythms in young adult and aged men. Neurobiol Aging 7:97–100

Waldhauser F, Weiszenbacher G, Frisch H, Zeitlhuber U, Waldhauser M, Wurtman RJ (1984) Fall in nocturnal serum melatonin during prepuberty and pubescence. Lancet I:362–365

Wallin BG, Sundlof G (1979) A quantitative study of muscle nerve sympathetic activity in resting normotensive and hypertensive subjects. Hypertension 1:67–77

Webb W (1969) Twenty-four hour sleep cycling. In: Kales A (ed) Sleep: physiology and pathology. Lippincott, Philadelphia, p 53

Webb W, Swinburne H (1971) An observational study of sleep in the aged. Percept Mot Skills 32:895–398
Weigle WD (1975) Cyclical production of antibody as a regulatory mechanism in the immune response. Adv Immunol 21:87–111
Weitzman ED, Moline ML, Czeisler CA, Zimmerman JC (1982) Chronobiology of aging: temperature, sleep-wake rhythms and entrainment. Neurobiol Aging 3:299–309
Wessler R, Rubin M, Sollberger A (1976) Circadian rhythm of activity and sleep-wakefulness in elderly institutionalized patients. J Interdiscipl Cycle Res 7:333
Wever R (1975) The meaning of circadian rhythmicity with regard to aging. Verh Dtsch Ges Pathol 59:160
Wever RA (1979) The circadian system of man. Springer, Berlin Heidelberg New York
Wilf-Miron R, Peleg L, Goldman B, Ashkenazi IE (1992) Rhythms of enzymatic activity in maternal and umbilical cord blood. Experientia 48:520–523
Wu J, Cornélissen G, Tarquini B, Mainardi G, Cagnoni M, Fernández JR, Hermida RC, Tamura K, Kato J, Kato K, Halberg F (1990) Circaseptan and circannual modulation of circadian rhythms in neonatal blood pressure and heart rate. In: Hayes DK, Pauly JE, Reiter RJ (eds) Chronobiology: its role in clinical medicine, general biology, and agriculture, part A. Wiley-Liss, New York, pp 643–652
Yatsyk GV, Syutkina EV, Polyakov YA, Safin SR, Grigoriev AE, Tagbloom M, Halberg E, Halberg F (1991) Circadian variations in urinary Na^+, K^+, and 11-oxycorticoid (11-OCS) excretion by premature infants. Biochim Clin 15:156–157
Yen SSC, Jaffe RB (eds) (1991) Reproductive endocrinology. Saunders, Philadelphia
Zepelin H, MacDonald CS (1987) Age differences in autonomic variables during sleep. J Gerontol 42:142–146
Zhengrong W, Xuechuan S, Cornelissen G, Jinyi W, Ling M, Degni C, Zhennan X, Halberg F (1993) Doppler flurometer-assessed circadian rhythms in neonatal cardiac function, family history, and intrauterine growth retardation. Am J Perinatology 10:119–125
Zurbrügg RP (1976) Hypothalamic-pituitary-adrenocortical regulation: a contribution to its assessment, development and disorders in infancy and childhood with special reference to plasma circadian rhythm. Karger, Basle (Monographs in paediatrics, vol 7)

CHAPTER 6
Rhythms in Second Messenger Mechanisms

K. WITTE and B. LEMMER

A. Introduction

Circadian rhythmicity has been observed in several biochemical, hormonal, and physiological parameters, and evidence exists that disturbances in these rhythms are involved in disease processes such as Cushing's syndrome, secondary hypertension, and affective disorders. A crucial step in physiological regulation and signalling is the transduction of signals from the extracellular to the intracellular space crossing the cell membrane. One of the major pathways involved in transmembraneous signalling is the combination of an extracellular receptor and an intracellular effector molecule responsible for the synthesis of an intracellular second messenger, e.g. the β-adrenoceptor–G-protein–adenylyl cyclase complex or the α-adrenoceptor–G-protein–phospholipase C system. In the case of lipid-soluble mediators such as nitric oxide, an intracellularly located effector molecule, i.e. the soluble guanylyl cyclase, may be the target protein. The present review will focus on circadian rhythms in the above-mentioned second messenger pathways involved in formation of cAMP by adenylyl cyclase, of cGMP by guanylyl cyclases, and of inositol phosphates by phospholipase C. Since the contribution of second messengers to the function of the clock itself has been discussed elsewhere (PROSSER and GILLETTE 1991; TAKAHASHI et al. 1993; TAKAHASHI 1993) and is reviewed in other chapters of this handbook, this point will not be discussed here. We shall also restrict our review to data obtained in humans and in other mammalian species; results concerning unicellular organisms were subject of detailed reviews (EDMUNDS 1988; EDMUNDS et al. 1992) and are addressed in other chapters of the handbook.

B. Principles of Transmembraneous Signal Transduction

I. Adenylyl Cyclase Pathway

Six different forms of adenylyl cyclase have been found in mammalian tissues, called adenylyl cyclase types I-VI. These enzymes differ with regard to tissue distribution, dependence on Ca^{2+}-calmodulin, and their responses to βγ-subunits of G-proteins (TANG and GILMAN 1992). Types I, II, and IV are found in mammalian brain, but only type I depends on Ca^{2+}-calmodulin, whereas the others (II and IV) can be activated by βγ-subunits. Heart tissue contains adenylyl cyclases V and VI, which are neither activated by Ca^{2+}-calmodulin nor by βγ-subunits. All adenylyl cyclases can be stimulated by the α-subunit of the stimulatory G-protein, $G_s\alpha$, which itself is activated by binding of an agonist to its specific G-protein-coupled receptor, e.g. the β-adrenoceptor (Fig. 1). Catalytic activity can be inhibited by activation of the inhibitory G-protein via G_i-protein-coupled receptors, e.g. the muscarinic M_2-receptor. The signal transduction cascade starts with binding of the agonist to the receptor; the resulting agonist-receptor complex facilitates the release of GDP from $G_s\alpha$ and binding of GTP to the $G_s\alpha$-subunit, which then dissociates from the heterotrimeric G-protein and activates the adenylyl cyclase. The cAMP formed activates intracellular effector molecules such as protein kinase A and can, thereby, influence transcriptional activity via cAMP response elements in the

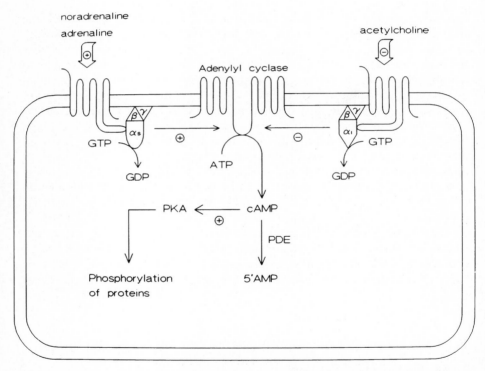

Fig. 1. Summary of signal transduction by the receptor–G-protein–adenylyl cyclase–phosphodiesterase system. The second messenger cAMP is formed by the enzyme adenylyl cyclase, which can be stimulated and inhibited by G_s- and G_i-proteins, respectively. The G-proteins can be activated in the presence of GTP by agonist-bound stimulatory or inhibitory receptors. Effects of cAMP involve intracellular effector molecules such as protein kinase A (*PKA*). Hydrolytic degradation of cAMP is catalysed by phosphodiesterases (*PDE*). For details see text

DNA. Termination of the effects of cAMP is achieved by hydrolysis of the nucleotide by phosphodiesterases, which have been subdivided into at least seven families. These phosphodiesterases differ in substrate preference, i.e. cAMP or cGMP, in their responses to Ca^{2+}-calmodulin and cyclic nucleotides, which can either stimulate or inhibit the catalytic activity of the enzymes (BEAVO et al. 1994).

II. Guanylyl Cyclase Pathway

Two types of guanylyl cyclases are involved in the synthesis of cGMP: the soluble and the particulate guanylyl cyclase (Fig. 2). The soluble guanylyl cyclases, intracellularly located in the cytosol, are heterodimeric proteins composed of α- and β-subunits, both of which are required for enzyme activity and stimulation by nitric oxide (SCHMIDT et al. 1993). The particulate guanylyl cyclases are membrane-bound enzymes consisting of a single protein, which contains both catalytic activity and a specific ligand-binding domain (ANAND-

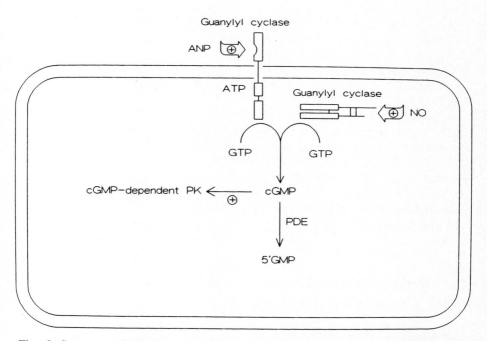

Fig. 2. Summary of signal transduction by the particulate ANP-stimulated guanylyl cyclase and the soluble NO-stimulated guanylyl cyclase. The particulate guanylyl cyclase consists of a single protein with an extracellular agonist-binding domain and intracellularly located domains for binding of ATP, protein kinase-like activity, and catalytic activity. The soluble guanylyl cyclase consists of two subunits, both of which are required for stimulation by nitric oxide and for catalytic activity. For details see text. *PK*, protein kinase; *PDE*, phosphodiesterases

SRIVASTAVA and TRACHTE 1993). Distinct forms (guanylyl cyclases A, B, C, and a human retinal form) have been identified, showing quite similar intracellular regions responsible for catalytic activity, but varying extracellular domains for binding of peptidergic agonists, e.g. atrial natriuretic peptide (guanylyl cyclase A and B), or heat-stable enterotoxin and guanylin (guanylyl cyclase C).

In their intracellular region the guanylyl cyclases contain a protein kinase-like domain, which is required for activiation of the enzyme by atrial natriuretic peptide and ATP. The second messenger cGMP influences intracellular effector molecules such as cGMP-gated channels, cGMP-stimulated or -inhibited phosphodiesterases, and cGMP-dependent protein kinases (ANAND-SRIVASTAVA and TRACHTE 1993). Hydrolysis of the nucleotide by phosphodiesterases (see above) terminates the intracellular effects of cGMP.

III. Phospholipase C Pathway

Several isozymes of phospholipase C have been described and are now divided into three major classes (β, γ, δ). The β-isozyme of phospholipase C can be stimulated via G-protein-coupled receptors (EXTON 1994), e.g. the α_1-adrenoceptor and the angiotensin II AT_1-receptor. These receptors are coupled to a G-protein (Gq and G_{11}), which, after binding of an agonist to the receptor, dissociates into the α- and the $\beta\gamma$-subunits. Both α and $\beta\gamma$ are able to stimulate the hydrolysis of phosphatidylinositol-4,5-bisphosphate by phospholipase C, resulting in the formation of inositol-1,4,5-trisphosphate and 1,2-diacylglycerol (Fig. 3).

Inositol-trisphosphate promotes the release of Ca^{2+} from intracellular stores, while diacylglycerol activates protein kinase C, which can phosphorylate several target proteins, e.g. sarcoplasmatic Ca^{2+}-channels. The effects of inositol trisphosphate are terminated by hydrolysis of one phosphate group yielding inositol-1,4-bisphosphate, which is further degraded to inositol-1-phosphate and, finally, inositol. The last step, mediated by an inositol-1-phosphatase, can be inhibited by lithium (BERRIDGE and IRVINE 1984). Inositol and diacylglycerol are used for recycling of phosphatidylinositol.

C. Rhythms in Signal Transduction in the Cardiovascular System

I. Adenylyl Cyclase Pathway

In healthy humans several groups observed circadian variations in the plasma concentration (MIKUNI et al. 1978; MARKIANOS and LYKOURAS 1981; SCHWARZ et al. 1981; ZIMMERMANN and NAGEL 1987; RICHARDS et al. 1991; LEMMER et al. 1994a) as well as in urinary excretion of cAMP (MURAD and PAK 1972; SAGEL et al. 1973; STONE et al. 1974; KOPP et al. 1977; KÖNIG et al. 1980;

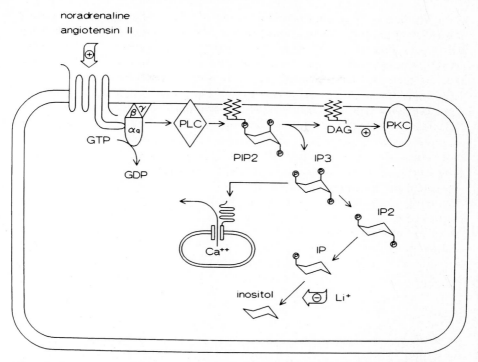

Fig. 3. Summary of signal transduction by the receptor–G-protein–phospholipase C system including stepwise degradation of inositol trisphosphate to inositol. Binding of agonists to their G-protein-coupled receptors activates phospholipase C (*PLC*), resulting in the formation of inositol-1,4,5-trisphosphate (*IP3*) and diacylglycerol (*DAG*) from phosphatidylinositol-4,5-bisphosphate (*PIP2*). IP3 promotes the release of Ca^{2+} from intracellular stores and DAG stimulates protein kinase C. For details see text. *IP2*, inositol-1,4-bisphosphate; *IP*, inositol-1-phosphate

LOGUE et al. 1989). During constant bedrest ENDRES and ROEREN (1979) did not observe circadian variation in plasma cAMP and concluded that a circadian rhythm in cAMP was due to exogenous factors such as posture and muscle activity. However, the comparison of three studies performed in healthy volunteers shows that the rhythm in plasma cAMP was identical, the acrophase ranging from 1200 to 1400 hours, whether subjects got up at 0500 hours (LEMMER et al. 1994a) or at 0800 hours (BEHNE et al. 1990; LEMMER 1989). Moreover, bright light in the morning significantly affected the rhythms in melatonin and cortisol but not in cAMP (LEMMER et al. 1994a). Thus, the extent to which the rhythm in plasma cAMP in humans is endogenous in nature is an open question. Rhythmicities in the formation and degradation of cAMP by adenylyl cyclase and phosphodiesterases, respectively, have not been studied in human tissues, but experiments were performed in myocardial and vascular tissue from laboratory animals.

Circadian and seasonal variation in the adenylyl cyclase pathway in heart ventricles of rats, which has been studied extensively by our group, is summarized in Table 1. LANG et al. (1985) were the first to show that the basal cAMP content in ventricular tissue from Wistar rats depended on the circadian time of sacrifice. The circadian rhythms observed exhibited amplitudes of 12%–28% of the 24-h mean depending on the time of the year. In ventricles from animals sacrificed in summer the lowest amplitude of 12% was found and the circadian variation did not achieve statistical significance. From four separate experiments performed throughout the year, the authors calculated a daily phase-shift of –0.09 up to –0.14 h/day. Furthermore, they reported that the patterns of circadian variation were also influenced by the season, showing sinusoidal profiles in August and May but more complex rhythms with split daytime peaks in October and December. Further studies on cAMP content confirmed its circadian time-dependence, demonstrating circadian amplitudes of up to 30% (LEMMER et al. 1986, 1987a; WITTE and LEMMER 1987). Stimulation of the adenylyl cyclase in vivo by injection of the non-selective β-adrenoceptor agonist isoprenaline increased the 24-h mean in cardiac cAMP but only slightly affected the rhythmic pattern (WITTE and LEMMER 1987). In contrast, injection of forskolin, a direct activator of the catalytic subunit of adenylyl cyclase, resulted in similar cAMP concentrations in ventricles obtained at 1400 hours, the time of the circadian peak, and at 2000 hours, the time of the circadian trough (LEMMER et al. 1986), indicating that cAMP formation by adenylyl cyclase might be the crucial step in the circadian regulation of cAMP content. Therefore, the adenylyl cyclase activity was studied in vitro in ventricular tissue collected throughout 24 h. The basal rate of cardiac cAMP formation was found to depend significantly on the circadian time, amplitudes being 16%–50% in different strains of rats (LEMMER et al. 1987a; LEMMER and WITTE 1989; LEMMER and WALD 1990; WITTE et al. 1995). Interestingly, only in one study was the period length 24 h (LEMMER and WITTE 1989), whereas in the other studies period lengths of 12 h were observed (LEMMER et al. 1987a; LEMMER and WALD 1990; WITTE et al. 1995), with peaks occurring around 0700 hours and 1900 hours, i.e. the onset of light and of darkness. Stimulation of the enzyme via the β-adrenoceptor, via the G-protein, and directly using forskolin (LEMMER and WITTE 1989; LEMMER and WALD 1990; WITTE et al. 1995) amplified the basal rhythmicity in cAMP formation, thus preserving the circadian pattern, i.e. 24-h (LEMMER and WITTE 1989) or 12-h rhythms (LEMMER et al. 1987a; LEMMER and WALD 1990; WITTE et al. 1995). However, activation of adenylyl cyclase by manganese ions, resulting in increased 24-h means in cardiac cAMP formation, abolished the circadian variation in ventricular adenylyl cyclase activity (Fig. 4). Since manganese ions are known to uncouple G-proteins from the adenylyl cyclase, this observation indicates that either the amount of G-proteins or their coupling efficiency could be involved in the circadian regulation of cardiac cAMP (WITTE et al. 1995).

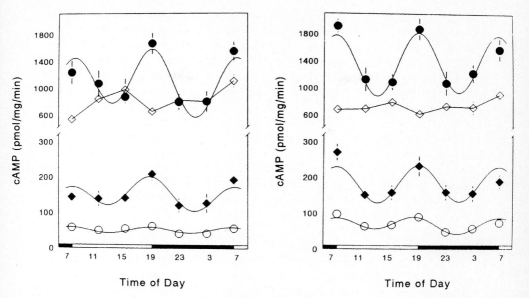

Fig. 4. Circadian variation in basal and stimulated adenylyl cyclase activity in ventricles from Wistar-Kyoto (*left*) and spontaneously hypertensive rats (*right*). Formation of cAMP was measured in vitro under basal conditions (*open circles*) and after stimulation by isoprenaline 1 μM (*closed diamonds*), manganese ions 10 mM (*open diamonds*), and a forskolin derivative 100 μM (*closed circles*). Addition of manganese ions abolished the rhythmicity in cardiac cAMP formation in both strains, whereas stimulation by isoprenaline and forskolin amplified the circadian variation in basal adenylyl cyclase activity. (Redrawn from WITTE et al. 1995 with permission)

In ventricles from Wistar rats aged more than 16 weeks, neither basal nor forskolin-stimulated adenylyl cyclase exhibited circadian rhythmicity. With regard to the results obtained with manganese ions, one might speculate that aging (or maturation) of the animals could affect the myocardial content of G-proteins or their functional activity, resulting in a loss of circadian variation. In support of this hypothesis HARDOUIN et al. (1993) reported on a reduction in mRNA for the $G_s\alpha$-protein in hearts of senescent rats. However, in the study by LEMMER and WALD (1990) a loss of circadian rhythms in cAMP formation was observed in ventricles from 16-week-old rats, while HARDOUIN et al. (1993) used Wistar rats at the age of 22 months.

Future chronobiological studies investigating G-protein content of myocardial membranes as well as G-protein activity are needed in order to elucidate further the role of G-proteins in circadian regulation and in age-dependent changes of the temporal organization.

Since myocardial β-adrenoceptors consist of two subtypes, the β_1- and the β_2-adrenoceptors, it was of interest to investigate whether both subtypes contribute to the circadian rhythm in β-adrenergic stimulation. This question has

Table 1. Circadian variation in the adenylyl cyclase system in heart ventricles from rats and hamsters

Parameter	Species	Age (weeks)	Stimulation	Rhythm	Period length	Amplitude[a]	Reference
cAMP (basal)	Wistar rats	8	None	$P < 0.05$	24 h	15%–28%	Lang et al. 1985
	Wistar rats	8	None	$P < 0.01$	Not calculated[d]	30%	Lemmer et al. 1986
	Wistar rats	10	None	$P < 0.001$	24 h	18%	Lemmer et al. 1987a
	Wistar rats	10	None	$P < 0.05$	24 h	< 10%	Witte and Lemmer 1987
cAMP[b] (stimulated)	Wistar rats	8	Forskolin	NS	Not calculated[d]	—	Lemmer et al. 1986
	Wistar rats	10	Isoprenaline	$P < 0.05$	12 h	< 10%	Witte and Lemmer 1987
AC activity (basal)	Wistar rats	10	None	$P < 0.001$	12 h	50%	Lemmer et al. 1987a
	Wistar rats	10	None	$P < 0.05$	24 h	42%	Lemmer and Witte 1989
	Wistar rats	7	None	$P < 0.05$	12 h	25%	Lemmer and Wald 1990
	Wistar rats	16	None	NS	—	—	
	Wistar rats	40	None	NS	—	—	
	Eur. hamster	???	None	$P < 0.05$	Not calculated[d]	16%	Pleschka et al. 1996
	WKY rats	12	None	$P < 0.05$	12 h	16%	Witte et al. 1995
	SHR rats	12	None	$P < 0.01$	12 h	28%	
AC activity[c] (stimulated)	Wistar rats	10	Isoprenaline	$P < 0.05$	24 h	47%	Lemmer and Witte 1989
			GppNHp	$P < 0.01$	24 h	43%	
			Forskolin	$P < 0.05$	24 h	39%	
	Wistar rats	7	Forskolin	$P < 0.05$	12 h	22%	Lemmer and Wald 1990
		16	Forskolin	NS	—	—	

Eur. hamster	???	Isoprenaline	$P < 0.05$	Not calculated[d]	48%	PLESCHKA et al. 1996
		GppNHp	$P < 0.05$	Not calculated[d]	44%	
		Forskolin	$P < 0.05$	Not calculated[d]	38%	
WKY rats	12	GTP	$P < 0.05$	12 h	30%	WITTE et al. 1995
		Isoprenaline	$P < 0.05$	12 h	24%	
		Forskolin	$P < 0.01$	12 h	38%	
		Manganese ions	NS	–	–	
SHR rats	12	GTP	$P < 0.01$	12 h	30%	
		Isoprenaline	$P < 0.05$	12 h	28%	
		Forskolin	$P < 0.01$	12 h	33%	
		Manganese ions	NS	–	–	
PDE activity Wistar rats	10	None	$P < 0.001$	24 h	32%	LEMMER et al. 1987a

Eur. hamster, European hamster (*Cricetus cricetus*); WKY rats, Wistar-Kyoto rats; SHR, spontaneously hypertensive rats.
[a] If amplitudes had not been calculated in the original publication, they were estimated from the published curves.
[b] After in vivo injection prior to death.
[c] Addition of agonist to the reaction mixture in vitro.
[d] Only two circadian time points studied.
??? Age of animals unknown.

been addressed in our recent study (WITTE et al. 1995) in ventricles from normotensive Wistar-Kyoto and spontaneously hypertensive rats. In both strains the effects of β-adrenergic stimulation of adenylyl cyclase by isoprenaline depended on the circadian time, the period length being 12 h. A similar pattern was observed in the $β_1$-mediated increase in cAMP formation, whereas the stimulation via $β_2$-adrenoceptors did not exhibit circadian rhythmicity. Thus, rhythmic changes in the effect of isoprenaline on adenylyl cyclase were mediated exclusively by a 12-h rhythm in the $β_1$-component. Densities of β-adrenoceptors also varied with time of day and, again, only the $β_1$-subtype contributed to that rhythm (WITTE at al. 1995). Since a number of studies have demonstrated that cardiac $β_1$-adrenoceptors are more severely affected in pathological conditions (BRODDE 1994), it is reasonable to propose that, because of its structure, the $β_1$-adrenoceptor could be more prone to physiological or pathological influences than the $β_2$-subtype. The $β_2$-adrenoceptors could represent a basal receptor population mediating the signals of the hormone adrenaline, whereas the $β_1$-subtype could be involved in rapid adaptive processes elicited by the release of the neurotransmitter noradrenaline from synaptic nerve endings. This hypothesis is supported by observations obtained in European hamsters, which exhibit a marked reduction in the $β_1$-component of the isoprenaline-mediated stimulation of adenylyl cyclase during hibernation, resulting in a relative predominance of the $β_2$-subtype in the hibernating state (PLESCHKA et al. 1996), when hormonal control could be of greater importance than short-term regulation by the sympathetic nervous system.

Summarizing the results obtained in myocardial tissue, it is obvious that the activity of the cardiac adenylyl cyclase system depends on the circadian time and may also be influenced by the season. The physiological role of the circadian rhythm in ventricular cAMP formation remains to be elucidated. Since in most studies 12-h rhythms were observed, it is unlikely that rhythmic changes in myocardial adenylyl cyclase activity contribute to the circadian variation in blood pressure and heart rate, which usually exhibit dominant 24-h rather than 12-h rhythms.

The vascular cAMP system could be another candidate for the circadian regulation of blood pressure. Unfortunately, no data have been published concerning vascular cAMP content and/or formation with regard to the circadian time, with the exception of one study in aortic tissue from Wistar-Kyoto rats and spontaneously hypertensive rats (WITTE et al. 1994). In that study basal and stimulated adenylyl cyclase activity in vitro did not show any circadian variation. Since, however, other parts of the vascular bed, especially the small arteries and arterioles, could be of greater importance for the regulation of peripheral resistance, additional data in these vascular regions are needed in order to elucidate the final contribution of vascular cAMP to the circadian regulation of the cardiovascular system.

II. Guanylyl Cyclase Pathway

Vascular tissue has been found to contain both a soluble guanylyl cyclase, which can be stimulated by nitric oxide, and a particulate guanylate cyclase representing the receptor for atrial natriuretic peptide (ANAND-SRIVASTAVA and TRACHTE 1993; SCHMIDT et al. 1993). Circadian variation in vascular soluble guanylyl cyclase has been studied in aortic tissue from Wistar-Kyoto and spontaneously hypertensive rats (WITTE et al. 1994) under basal conditions and after stimulation by the nitric oxide donor sodium nitroprusside. In both strains formation of cGMP exhibited significant circadian rhythmicity with a period length of 24 h and acrophases in the middle of the animals' daily resting period. Since cGMP is a vasodilating second messenger, an increased formation of cGMP during daytime could result in a lower blood pressure and, thus, contribute to circadian regulation. In order to test the hypothesis that the nitric oxide stimulated guanylyl cyclase could be involved in blood pressure rhythmicity, we studied the effects of the nitric oxide synthase inhibitor Nω-nitro-L-arginine methyl ester after chronic in vivo treatment in Wistar rats (unpublished results). Inhibition of endogenous nitric oxide synthesis resulted in a pronounced increase in systolic and diastolic blood pressure. After 10 days treatment the profiles in systolic and diastolic blood pressure showed an inverse circadian pattern with higher values during the daily resting period, whereas heart rate still peaked at night. In Wistar-Kyoto rats acute application of the nitric oxide synthase inhibitor reversed the circadian blood pressure profile, while chronic application of increasing doses reduced the circadian amplitude (WITTE et al. 1995). This observation could indeed indicate a crucial role of the nitric oxide–cGMP system in the circadian regulation of blood pressure by vascular tone, but further studies are needed to validate these findings.

III. Phospholipase C Pathway

Formation of inositol trisphosphate by phospholipase C would be another possible mechanism involved in the cardiovascular circadian regulation, because several transmitters and hormones elicit their intracellular effects via inositol-trisphosphate-mediated increases in intracellular calcium, e.g. noradrenaline via α_1-adrenoceptors and angiotensin II via AT_1-receptors. Unfortunately, there are no published results on circadian variation in this signal transduction system. Indirect evidence, however, is available from in vivo studies in humans and in laboratory animals. PANZA et al. (1991) observed that circadian variation in forearm vascular resistance in healthy volunteers could be abolished by administration of an α-adrenoceptor antagonist, indicating that either the endogenous transmitter noradrenaline varies with time of day or the same concentrations of the transmitter evoke different intracellular events at different circadian times. Similarly, the converting enzyme inhibitor enalapril (LEMMER et al. 1994b) and the AT_1-receptor antagonist losartan (SCHNECKO et al. 1995) reduced blood pressure in hypertensive transgenic

TGR(mRen-2)27 rats more effectively during daytime than at night, which could be due to circadian rhythmicity in the renin-angiotensin system and/or the AT_1-receptor signalling. The observation that TGR(mRen-2)27 rats exhibit an inverse circadian profile in blood pressure (LEMMER et al. 1993), which is normalized by injection of a converting enzyme inhibitor or an AT_1-receptor antagonist, points to the importance of the renin-angiotensin system and its cellular signal transduction pathway for the circadian rhythmicity in the cardiovascular system.

Therefore, studies on circadian regulation of inositol-trisphosphate-linked signal transduction systems have to be performed in order to obtain more insight into the complex temporal regulation of vascular resistance.

D. Rhythms in Signal Transduction in the Central Nervous System

I. Adenylyl Cyclase Pathway

Several studies in patients with affective disorders demonstrated disturbances in the peripheral β-adrenergic signal transduction, using leucocytes isolated from the patients' blood. Thus, MAZZOLA-POMIETTO et al. (1994) found that the effect of isoprenaline on cAMP formation in mononuclear leucocytes was reduced by 35% in depressed patients and, furthermore, that the degree of reduction was significantly correlated to the severity of depression. Unfortunately, the authors did not investigate circadian time dependency in β-adrenergic signalling in these patients, although in affective disorders several disturbances of the circadian organization have been reported. Whether affective disorders influence circadian rhythms in the cAMP pathway has been studied by MARKIANOS and LYKOURAS (1981), who observed a different circadian pattern and a phase-advance in plasma cAMP in patients compared to healthy controls. These studies indicate an involvement of the adenylyl cyclase system and its temporal organization in depressive disorders, but they do not allow any conclusions to be drawn concerning circadian rhythms of cAMP within the brain.

In laboratory animals circadian rhythms in the formation, concentration, and degradation of cAMP, which have been investigated in different brain regions, are summarized in Table 2.

In cerebellum the concentration of cAMP has been found to depend significantly on the circadian time (CHOMA et al. 1979; KANT et al. 1981), but peaks in cAMP content occurred in the morning in Sprague-Dawley rats (CHOMA et al. 1979) and in the evening in Wistar rats (KANT et al. 1981). In hypothalamic tissue significant circadian rhythms in cAMP content have been observed in some studies (VALASES et al. 1980; KANT et al. 1981; MURAKAMI and TAKAHASHI 1983), whereas other groups did not find circadian variation in hypothalamic cAMP (CHOMA et al. 1979; KAFKA et al. 1986; PROSSER and GILLETTE 1991; YAMAZAKI et al. 1994). However, the circadian patterns observed by VALASES et al. (1980) in 10-week-old Sprague-Dawley rats and by

MURAKAMI and TAKAHASHI (1983) in 5-week-old Wistar rats were almost identical, with peaks in cAMP at the end of the respective activity period. Hippocampal tissue did not show circadian rhythmicity, neither in cAMP content (CHOMA et al. 1979; KAFKA et al. 1986) nor in basal and stimulated cAMP formation (LEMMER et al. 1991). The concentration of cAMP in cerebral cortex did not depend on circadian time in Sprague-Dawley rats aged 6 weeks (PEREZ et al. 1991) and 10 weeks (CHOMA et al. 1979). Since, however, different regions of the cortex have different physiological functions, they could also display specific circadian patterns, which indeed has been observed by KAFKA et al. (1986) in cortical tissue from Sprague-Dawley rats. In that study frontal cortex did not show circadian rhythmicity in cAMP, whereas in parietal, temporal, and occipital cortex the cAMP content significantly depended on the time of day, with peaks always occurring in the resting period but with different circadian amplitudes. Other cortical regions, which have also been studied by KAFKA et al. (1986), displayed either no rhythms in cAMP (insular cortex) or varying circadian patterns with peaks in the activity period (piriform and cingulate cortex). Summarizing these somewhat confusing data, circadian variation in cAMP content is likely to be present in some brain regions, e.g. cerebellum and hypothalamus, can be detected in some cortical areas, and is obviously absent in hippocampus, where neither cAMP formation by the adenylyl cyclase nor cAMP content depended on the circadian time.

II. Guanylyl Cyclase Pathway

The formation or degradation of cGMP in vitro has not been studied with regard to circadian cycles. In some studies, however, the concentration of the second messenger cGMP in different brain regions of rats has been measured at different times of day (Table 3). In most regions cGMP concentrations depended significantly on the circadian time except in parietal cortex (KAFKA et al. 1986). In hippocampus, CHOMA et al. (1979) did not find a time-dependent variation in cGMP, whereas KAFKA et al. (1986) observed significant rhythmicity with an amplitude of approximately 10%. In hypothalamic tissue only one group did not find circadian variation in cGMP (KANT et al. 1981), whereas others reported significant rhythms (CHOMA et al. 1979; VALASES et al. 1980; KAFKA et al. 1986), with amplitudes ranging from less than 10% up to 50% of the 24-h mean. Peaks were observed at the beginning (CHOMA et al. 1979) or in the middle of the activity period (VALASES et al. 1980; KAFKA et al. 1986); only KAFKA et al. (1986) reported an additional peak in the middle of the resting period. As observed for cortical cAMP (see above) different regions of the cortex showed different temporal patterns in cGMP content. Summarizing the data on cGMP in the central nervous system, circadian rhythmicity is likely to be present in hypothalamus and several cortical areas, whereas in hippocampal tissue no or only slight circadian variation was observed, corresponding to the lack of circadian rhythmicity in hippocampal cAMP content.

Table 2. Circadian variation in the adenylyl cyclase system in different brain regions from rats and mice

Parameter	Species	Age (weeks)	Brain region	Rhythm	Period length	Amplitude[a]	Reference
cAMP (basal)	SD rats	10	Cerebellum	$P < 0.01$	24 h	10%	Choma et al. 1979
			Pons medulla	$P < 0.05$	?	<10%	
			Cortex	NS	–	–	
			Hippocampus	NS	–	–	
			Hypothalamus	NS	–	–	
	SD rats	10	Hypothalamus	?	24 h	50%	Valases et al. 1980
	Wistar rats	???	Cerebellum	$P < 0.05$	Not calculated[b]	20%	Kant et al. 1981
			Hypothalamus	$P < 0.05$	Not calculated[b]	20%	
	Wistar rats	5	Hypothalamus (LD)	$P < 0.05$	24 h	20%	Murakami and Takahashi 1983
			Hypothalamus (VMH) (LL)	$P < 0.01$	24 h	20%	
	BALB/C mice	13	Whole brain	$P < 0.01$	24/12/8 h	11%–15%	Joanny et al. 1984
	C57 BL/6 mice	13	Whole brain	$P < 0.05$	24/12/8/6 h	11%–23%	
	SD rats	10	Parietal cortex	$P < 0.01$	24 h	60%	Kafka et al. 1986
			Frontal cortex	NS	–	–	
			Temporal cortex	$P < 0.01$	24 h	20%	
			Occipital cortex	$P < 0.05$	24 h	25%	
			Hippocampus	NS	–	–	
			Hypothalamus	NS	–	–	
	Wistar rats	8	Forebrain	$P < 0.001$	24 h	60%	Lemmer et al. 1987b
	Long-Evans rats	8	Hypothalamus (AH)	$P < 0.1$	12 h	10%	Prosser and Gillette 1991
	SD rats	6	Cortex	NS	–	–	Perez et al. 1991
			Preoptic region	$P < 0.05$	Not calculated[b]	0%–10%	
	Wistar rats	5	Hypothalamus (AH)	NS	–	–	Yamazaki et al. 1994

Activity	Strain	n	Region	Significance	Period	Amplitude	Reference
AC activity (basal)	SD rats	8	Frontal cortex	NS	—	—	Mobley et al. 1983
	Wistar rats	8	Forebrain	$P<0.001$	24 h	32%	Lemmer et al. 1987b
	Long-Evans rats	8	Hypothalamus (AH)	NS	—	—	Prosser and Gillette 1991
	Wistar rats	8	Hippocampus	NS	—	—	Lemmer et al. 1991
AC activity (stimulated)	SD rats	8	Frontal cortex (NE)	?	24 h	20%	Mobley et al. 1983
	SD rats	10	Cortex (IPN)	$P<0.05$	12 h	30%	Kafka et al. 1986
	Wistar rats	8	Hippocampus (IPN)	NS	—	—	Lemmer et al. 1991
			Hippocampus (GPP)	NS	—	—	
			Hippocampus (FOR)	NS	—	—	
PDE activity	Wistar rats	8	Forebrain	$P<0.01$	12 h	11%	Lemmer et al. 1987b
	Long-Evans rats	8	Hypothalamus (AH)	NS	—	—	Prosser and Gillette 1991

SD rats, Sprague-Dawley rats; VMH, ventromedial hypothalamus; AH, anterior hypothalamus; NE, noradrenaline; IPN, isoprenaline; GPP, GppNHp; FOR, forskolin.

[a] If amplitudes had not been calculated in the original publication, they were estimated from the published curves.
[b] Only two circadian time points studied.
??? Age of animals not mentioned in the publication.

Table 3. Circadian variation in cGMP concentrations in different brain regions from rats

Parameter	Species	Age (weeks)	Brain region	Rhythm	Period length	Amplitude[a]	Reference
cGMP (basal)	SD rats	10	Cerebellum	$P < 0.01$	< 12 h	25%	CHOMA et al. 1979
			Pons medulla	$P < 0.05$	< 12 h	40%	
			Cortex	$P < 0.05$	24 h	20%	
			Hippocampus	NS	–	–	
			Hypothalamus	$P < 0.05$	24 h	25%	
	SD rats	10	Hypothalamus	?	24 h	50%	VALASES et al. 1980
	Wistar rats	???	Cerebellum	$P < 0.05$	Not calculated[b]	20%	KANT et al. 1981
			Hypothalamus	NS	Not calculated[b]	–	
	SD rats	10	Parietal cortex	NS	–	–	KAFKA et al. 1986
			Frontal cortex	$P < 0.01$	24 h	40%	
			Temporal cortex	$P < 0.01$	24 h	60%	
			Occipital cortex	$P < 0.01$	24 h	40%	
			Hippocampus	$P < 0.01$	24 h	10%	
			Hypothalamus	$P < 0.05$	?	< 10%	

SD rats, Sprague-Dawley rats.
[a] If amplitudes had not been calculated in the original publication, they were estimated from the published curves.
[b] Only two circadian time points studied.
??? Age of animals not mentioned in the publication.

III. Phospholipase C Pathway

We are not aware of any chronobiological study investigating the circadian time dependency of phospholipase C or inositol trisphosphate in the central nervous system. Since lithium, which is known to affect circadian rhythms, exerts its pharmacological effects by inhibition of enzymes involved in phospholipid turnover (Fig. 3), studies on inositol trisphosphate and other phospholipids with regard to circadian rhythms would be very promising.

E. Rhythms in Signal Transduction in Other Tissues

I. Adenylyl Cyclase Pathway

Circadian variation in cAMP content has been studied in several other tissues such as epidermis (Marks and Grimm 1972; Garte and Belman 1980; Schwarz et al. 1981, 1984), liver (Gagliardino et al. 1978; Tiedgen and Seitz 1980), and adrenals (Moore and Qavi 1971; Guillemant et al. 1980a, b; Guillemant and Guillemant 1986).

In epidermis from mice (Marks and Grimm 1972; Garte and Belman 1980) and man (Schwarz et al. 1981, 1984), rhythmicity in cAMP content has been observed, with peaks occurring in the species' resting period, i.e. the daytime in mice and the night in humans. Since cAMP is involved in the regulation of mitotic activity in the epidermis, a disturbed circadian variation in skin cAMP could contribute to dermatological diseases, but definite data on that issue have not been published.

In hepatic tissue from mice (Gagliardino et al. 1978) and rats (Tiedgen and Seitz 1980), circadian variation in cAMP content has been observed, but the concentration of cAMP in the liver as well as its circadian pattern depended on the composition of the animals' diet (Tiedgen and Seitz 1980). Since the ratio between insulin and glucagon has marked effects on hepatic cAMP content (Seitz et al. 1976) and secretion of both hormones depends on the metabolic requirements, circadian rhythms in hepatic cAMP could reflect the circadian patterns of insulin and glucagon secretion and, thus, the pattern of food intake.

Detailed studies on circadian regulation of tissue cAMP content have been performed in the rat adrenal gland. Since the synthesis of adrenal corticosteroids is controlled by adrenocorticotrophic hormone (ACTH), which is known to stimulate adenylyl cyclase, it was of interest to study whether the pronounced circadian rhythm in plasma corticosteroids is mediated by rhythmic changes in the formation of cAMP. Indeed, Moore and Qavi (1971) observed that, in rats, adrenal adenylyl cyclase activity peaked in the morning when pituitary ACTH content is highest, thus preceding the peak in adrenal corticosterone by approximately 9 h. Transection of a hypothalamic projection pathway, the median forebrain bundle, abolished the circadian rhythmicity in adrenal cAMP formation as well as in adrenal corticosterone concentration.

GUILLEMANT et al. (1978) also found circadian variation in adrenal cAMP and, furthermore, demonstrated that peaks in adrenal cAMP content coincided with peaks in plasma corticosterone. The same authors demonstrated that, even in hypophysectomized rats, circadian variation in adrenal cAMP content was preserved, although the amplitude was reduced, indicating that the rhythm in cAMP could not solely be due to rhythmic release of ACTH (GUILLEMANT et al. 1980a). In another study, it was shown that, in rats, circadian rhythmicity in plasma corticosterone developed simultaneously with the rhythm in adrenal cAMP around the age of 3 weeks (GUILLEMANT and GUILLEMANT 1986); in younger rats neither parameter depended on the circadian time. Thus, the circadian variation in corticosteroids seems to be regulated by rhythms in adrenocortical cAMP, which itself depends only partly on rhythmic release of ACTH. The link between cAMP and the synthesis of corticosteroids could be the steroidogenesis activator polypeptide (SAP), which has been identified in rat adrenal cortex, gonads, and several other tissues (MERTZ and PEDERSEN 1989). Interestingly, rat adrenal SAP exhibited pronounced circadian variation, with 13-fold higher values in the evening than in the morning, and was found to be responsive to cAMP, i.e. to increase after injection of ACTH or a cAMP analogue. In summary, the regulation of corticosteroid rhythmicity seems to involve ACTH as hormonal signal, cAMP as second messenger, which itself depends on the circadian time, and SAP mediating the effects of cAMP on the synthesis of corticosteroids.

II. Guanylyl Cyclase Pathway

Although in mouse epidermis cGMP was found to exhibit significant circadian rhythmicity (GARTE and BELMAN 1980), a study in human epidermis failed to detect any circadian variation (SCHWARZ et al. 1984).

Formation of cGMP seems to be of less importance than cAMP in adrenal steroidogenesis. Although there is significant circadian rhythmicity in adrenocortical cGMP in rats (GUILLEMANT et al. 1978, 1980b), which persists after hypophysectomy (GUILLEMANT et al. 1980a) and can be influenced by ACTH (GUILLEMANT and GUILLEMANT 1981), peaks in cGMP could not be correlated with peaks in corticosteroids. Moreover, circadian variation in adrenocortical cGMP was observed in 2-week-old male rats, which did not show any circadian rhythmicity in plasma corticosterone, supporting the view that cAMP but not cGMP is involved in adrenocortical circadian regulation.

III. Phospholipase C Pathway

No data are available in the literature concerning circadian variation in inositol trisphosphate or other phospholipids of the phospholipase C pathway in epidermis, liver, and adrenals.

F. Conclusions and Perspectives

Circadian rhythmicity in second messenger formation, concentration, and degradation has been studied in several tissues using the adenylyl cyclase pathway as a model. Thus, pronounced and reproducible time-dependent variation has been demonstrated in heart ventricles, in some – but not in all – brain regions, and in the adrenal cortex of rats. The physiological role of these rhythms seems to be obvious in adrenocortical tissue, where rhythmicity in cAMP contributes to the circadian regulation of corticosteroidogenesis, whereas in cardiac and cerebral tissue the functional consequences of circadian time-dependent fluctuations in cAMP remain to be elucidated. Concerning the mechanisms involved in the generation of rhythms in cAMP, it is interesting to note that in adrenocortical tissue rhythmicity persists in the absence of rhythmically released ACTH, indicating circadian variation in the adenylyl cyclase pathway itself. In heart tissue rhythmicity in cAMP formation could not be overcome by receptor-mediated stimulation in vitro, but uncoupling of G-proteins abolished the time-dependent variation. Therefore, density and/or functional activity of G-proteins seem to play a crucial role in the cellular circadian regulation of signal transduction. One possible mechanism of G-protein-dependent rhythmicity could be the variation of their transcription – and synthesis – by cAMP itself, mediated via protein kinase A and cAMP response elements in the DNA. However, definite data on that issue are lacking and further studies, including quantification of G-proteins at different times of day, are required in order to obtain more insight into these intracellular processes.

Less information is available on circadian regulation and possible functional correlates of the guanylyl cyclase pathway. The circadian variation in vascular resistance could partly be mediated by rhythms in cGMP formation, but this hypothesis is based on a single report and has to be confirmed or rejected by further studies. In adrenocortical tissue cGMP – although rhythmic – does not seem to play a role in the circadian regulation of corticosteroids. In the central nervous system rhythmicity in cGMP has been described, but its functional importance is as yet unclear.

Although lithium is known to influence circadian rhythms and has been found to inhibit one enzyme of the inositol-phosphate cascade, the phospholipase C pathway has not been studied with regard to circadian rhythmicity. We, nevertheless, decided to include the phospholipase C pathway in this review in order to promote chronobiological studies investigating this signal transduction pathway.

References

Anand-Srivastava MB, Trachte GJ (1993) Atrial natriuretic factor receptors and signal transduction mechanisms. Pharmacol Rev 45:455–497

Beavo JA, Conti M, Heaslip RJ (1994) Multiple cyclic nucleotide phosphodiesterases. Mol Pharmacol 46:399–405

Behne S, Becker HJ, Liefhold J, Kaiser R, Lemmer B (1990) On the chronopharmacokinetics, effects on cardiovascular functions, and plasma cAMP of oral nifedipine in healthy subjects. In: Lemmer B, Hüller H (eds) Clinical chronopharmacology, vol 6. Zuckschwerdt, Munich, pp 59–63

Berridge MJ, Irvine RF (1984) Inositol-trisphosphate, a novel second messenger in cellular signal transduction. Nature 312:315–321

Brodde OE (1994) Beta-adrenoceptors in cardiac disease. Pharmacol Ther 60:405–430

Choma PP, Puri SK, Volicer L (1979) Circadian rhythm of cyclic nucleotide and GABA levels in the rat brain. Pharmacology 19:307–314

Edmunds LN (1988) Cellular and molecular bases of biological clocks, 1st edn. Springer, Berlin Heidelberg New York

Edmunds LN, Carré IA, Tamponnet C, Tong J (1992) The role of ions and second messengers in circadian clock function. Chronobiol Int 9:180–200

Endres P, Roeren T (1979) Lack of a diurnal plasma adenosine 3',5'-monophosphate rhythm. J Clin Endocrinol Metab 48:872–873

Exton JH (1994) Phosphoinositide phospholipases and G-proteins in hormone action. Annu Rev Physiol 56:349–369

Gagliardino JJ, Pessacq MT, Hernandez RE, Rebolledo OR (1978) Circadian variations in serum glucagon and hepatic glycogen and cyclic AMP concentrations. J Endocrinol 78:297–298

Garte SJ, Belman S (1980) Diurnal variation in cyclic nucleotide levels in normal and phorbol myristate acetate treated mouse epidermis. J Invest Dermatol 74:224–225

Guillemant J, Guillemant S (1981) Effect of exogenous and endogenous ACTH on adrenocortical cyclic GMP in the rat. J Steroid Biochem 14:557–561

Guillemant J, Guillemant S (1986) Development of adrenocortical cyclic nucleotide (cyclic AMP and cyclic GMP) and corticosterone circadian rhythms in male and female rats. Chronobiol Int 3:155–160

Guillemant S, Eurin J, Guillemant J, Reinberg A (1978) Rythmes circadiens des nucléotides cycliques (AMP et GMP cycliques) du cortex surrénalien du rat mâle adulte. Ann Endocrinol 39:49–50

Guillemant J, Guillemant S, Reinberg A (1980a) Circadian variation of adrenocortical cyclic nucleotides (cyclic AMP and cyclic GMP) in hypophysectomized rats. Experientia 36:367–368

Guillemant J, Reinberg A, Guillemant S (1980b) Study of the role of adrenocortical cyclic AMP and cyclic GMP on ascorbic acid depletion and corticosteroidogenesis by analysis of circadian rhythms. Acta Endocrinol 95:382–387

Hardouin S, Bourgeois F, Besse S, Machida CA, Swynghedauw B, Moalic JM (1993) Decreased accumulation of β_1-adrenergic receptor, Gαs and total heavy chain messenger RNAs in the left ventricle of senescent rat heart. Mech Ageing Dev 71:169–188

Joanny P, Chouvet G, Giannellini F, Vial M (1984) Brain diurnal levels of adenosine 3', 5' cyclic monophosphate in C57 BL/6 and BALB/C mice. Chronobiol Int 1:37–40

Kafka MS, Benedito MA, Roth RH, Steele LK, Wolfe WW, Catravas GN (1986) Circadian rhythms in catecholamine metabolites and cyclic nucleotide production. Chronobiol Int 3:101–115

Kant GJ, Sessions GR, Lenox RH, Meyerhoff JL (1981) The effects of hormonal and circadian cycles, stress, and activity on levels of cyclic AMP and cyclic GMP in pituitary, hypothalamus, pineal and cerebellum of female rats. Life Sci 29:2491–2499

König P, Carpenter M, White AA (1980) Urinary cyclic adenosine 3',5'-monophosphate (cAMP) and cyclic guanosine 3',5'-monophosphate (cGMP) in asthmatic and normal children. Eur J Respir Dis 61:218–226

Kopp L, Lin T, Tucci JR (1977) Circadian rhythms in the urinary excretion of cyclic 3',5'-adenosine monophosphate (cyclic AMP) and cyclic 3',5'-guanosine monophosphate (cyclic GMP) in human subjects. J Clin Endocrinol Metab 44:673–680

Lang PH, Bissinger H, Lemmer B (1985) Circadian rhythm and seasonal variations in basal cAMP content of rat heart ventricles. Chronobiol Int 2:41–45

Lemmer B (1989) Circadian variations in the effects of cardiovascular active drugs. In: von Arnim T, Maseri A (eds) Predisposing conditions for acute ischemic syndromes. Steinkopff, Darmstadt, p 1
Lemmer B, Wald C (1990) Influence of age on the rhythm in basal and forskolin-stimulated adenylate cyclase activity of the rat heart. Chronobiol Int 7:107–111
Lemmer B, Witte K (1989) Circadian variation of the in vitro stimulation of adenylate cyclase in rat heart tissue. Eur J Pharmacol 159:311–315
Lemmer B, Bissinger H, Lang PH (1986) Effect of forskolin on cAMP levels in rat heart at different times of day. IRCS Med Sci 14:1103–1104
Lemmer B, Lang PH, Schmidt S, Bärmeier H (1987a) Evidence for circadian rhythmicity of the β-adrenoceptor-adenylate cyclase-cAMP-phosphodiesterase system in the rat. J Cardiovasc Pharmacol 10 [Suppl 4]:S138–S140
Lemmer B, Bärmeier H, Schmidt S, Lang PH (1987b) On the daily variation in the beta-receptor–adenylate cyclase–cAMP–phosphodiesterase system in rat forebrain. Chronobiol Int 4:469–475
Lemmer B, Carlebach R, Stiller M, Ohm TG, Nitsch R (1991) Dose-dependent stimulation of adenylate cyclase in rat hippocampal tissue by isoprenaline, Gpp(NH)p and forskolin: lack of circadian phase-dependency. Brain Res 565:225–230
Lemmer B, Mattes A, Böhm M, Ganten D (1993) Circadian blood pressure variation in transgenic hypertensive rats. Hypertension 22:97–101
Lemmer B, Brühl T, Witte K, Pflug B, Köhler W, Touitou Y (1994a) Effects of bright light on circadian patterns of cyclic adenosine monophosphate, melatonin and cortisol in healthy subjects. Eur J Endocrinol 130:472–477
Lemmer B, Witte K, Makabe T, Ganten D, Mattes A (1994b) Effects of enalaprilat on circadian profiles in blood pressure and heart rate of spontaneously and transgenic hypertensive rats. J Cardiovasc Pharmacol 23:311–314
Logue FC, Fraser WD, O'Reilly DSJ, Beastall GH (1989) The circadian rhythm of intact parathyroid hormone (1–84) and nephrogenous cyclic adenosine monophosphate in normal men. J Endocrinol 121:R1-R3
Markianos M, Lykouras L (1981) Circadian rhythms of dopamine-β-hydroxylase and c-AMP in plasma of controls and patients with affective disorders. J Neural Transmission 50:149–155
Marks F, Grimm W (1972) Diurnal fluctuation and β-adrenergic elevation of cyclic AMP in mouse epidermis in vivo. Nature 240:178–179
Mazzola-Pomietto P, Azorin JM, Tramoni V, Jeanningros R (1994) Relation between lymphocyte β-adrenergic responsivity and the severity of depressive disorders. Biol Psychiatry 35:920–925
Mertz LM, Pedersen RC (1989) The kinetics of steroidogenesis activator polypeptide in the rat adrenal cortex. J Biol Chem 264:15274–15279
Mikuni M, Saito Y, Koyama T, Daiguji M, Yamashita I, Yamakazi K, Honma M, Ui M (1978) Circadian variations in plasma 3′:5′-cyclic adenosine monophosphate and 3′:5′-cyclic guanosine monophosphate of normal adults. Life Sci 22:667–672
Mobley PL, Manier DH, Sulser F (1983) Norepinephrine-sensitive adenylate cyclase system in rat brain: role of adrenal corticosteroids. J Pharmacol Exp Ther 226:71–77
Moore RY, Qavi HB (1971) Circadian rhythm in adrenal adenyl cyclase and corticosterone abolished by medial forebrain bundle transection in the rat. Experientia 27:249–250
Murad F, Pak CYC (1972) Urinary excretion of adenosine 3′5′-monophosphate and guanosine 3′5′-monophosphate. N Engl J Med 286:1382–1387
Murakami N, Takahashi K (1983) Circadian rhythm of adenosine-3′,5′-monophosphate content in suprachiasmatic nucleus (SCN) and ventromedial hypothalamus (VMH) in the rat. Brain Res 276:297–304
Panza JA, Epstein SE, Quyyumi AA (1991) Circadian variation in vascular tone and its relation to α-sympathetic vasoconstrictor activity. N Engl J Med 325:986–990
Perez E, Zamboni G, Amici R, Fadiga L, Parmeggiani PL (1991) Ultradian and circadian changes in the cAMP concentration in the preoptic region of the rat. Brain Res 551:132–135

Pleschka K, Heinrich A, Witte K, Lemmer B (1996) Diurnal and seasonal changes in sympathetic signal transduction in cardiac ventricles of European hamsters. Am J Physiol 270:R304–R309

Prosser RA, Gillette MU (1991) Cyclic changes in cAMP concentration and phosphodiesterase activity in a mammalian circadian clock studied in vitro. Brain Res 568:185–192

Richards AM, Wittert G, Espiner EA, Yandle TG, Frampton C, Ikram H (1991) Prolonged inhibition of endopeptidase 24.11 in normal man: renal, endocrine and haemodynamic effects. J Hypertens 9:955–962

Sagel J, Colwell JA, Loadholt CB, Lizarralde G, Green AS (1973) Circadian rhythm in the urinary excretion of cyclic 3′,5′-adenosine monophosphate in man. J Clin Endocrinol Metab 37:570–573

Schmidt HHHW, Lohmann SM, Walter U (1993) Minireview. The nitric oxide and cGMP signal transduction system: regulation and mechanism of action. Biochem Biophys Acta 1178:153–175

Schnecko A, Witte K, Lemmer B (1995) Effects of the angiotensin II receptor antagonist losartan on 24-hour blood pressure profiles of primary and secondary hypertensive rats. J Cardiovasc Pharmacol 26:214–221

Schwarz W, Schell H, Hornstein OP, Bernlochner W, Weghorn C (1981) Variations of cAMP in epidermis and plasma of male adult subjects. Dermatologica 162:230–235

Schwarz W, Schell H, Bachmann I, Hellmund HW (1984) Cyclic nucleotides in human epidermis – diurnal variations. J Invest Dermatol 82:119–121

Seitz HJ, Müller MJ, Nordmeyer P, Krone W, Tarnowski W (1976) Concentration of cyclic AMP in rat liver as a function of the insulin/glucagon ratio in blood under standardized physiological conditions. Endocrinology 99:1313–1318

Stone JF, Polk ML, Dobbs JW, Graham ME, Scheving LE (1974) Circadian variation in human urinary cyclic AMP and the effect of different diets on this rhythm. Int J Chronobiol 2:163–170

Tang WJ, Gilman AG (1992) Adenylyl cyclases. Cell 70:869–872

Takahashi JS (1993) Circadian clocks à la CREM. Nature 365:299–320

Takahashi JS, Kornhauser JM, Koumenis C, Eskin A (1993) Molecular approaches to understand circadian oscillations. Annu Rev Physiol 55:729–753

Tiedgen M, Seitz HJ (1980) Dietary control of circadian variations in serum insulin, glucagon and hepatic cyclic AMP. J Nutr 110:876–882

Valases C, Wright SJ, Catravas GN (1980) Diurnal changes in cyclic nucleotide levels in the hypothalamus of the rat. Exp Brain Res 40:261–264

Witte K, Lemmer B (1987) Effects of isoprenaline on the daily rhythm in the cAMP content and AC activity of the rat heart. Chronobiologia 14:255–256 (abstract)

Witte K, Zuther P, Lemmer B (1994) Circadian regulation of angiotensin converting enzyme, adenylyl cyclase, and guanylyl cyclase in aortae from normotensive and spontaneously hypertensive rats. Naunyn Schmiedebergs Arch Pharmacol 349 [Suppl]:R59 (abstract)

Witte K, Parsa-Parsi R, Vobig M, Lemmer B (1995) Mechanisms of the circadian regulation of β-adrenoceptor density and adenylyl cyclase activity in cardiac tissue from normotensive and spontaneously hypertensive rats. J Mol Cell Cardiol 27:1195–1202

Witte K, Schnecko A, Zuther P, Lemmer B (1995) Contribution of the nitric oxide-guanylyl cyclase system to circadian regulation of blood pressure in normotensive Wistar-Kyoto rats. Cardiovasc Res 30:682–688

Yamazaki S, Maruyama M, Cagampang FRA, Inouye SIT (1994) Circadian fluctuations of cAMP content in the suprachiasmatic nucleus and the anterior hypothalamus of the rat. Brain Res 651:329–331

Zimmermann T, Nagel M (1987) Untersuchung zur nächtlichen Veränderung von Cortisol, Katecholaminen, c-AMP und Histamin bei Kindern mit nächtlichem Asthma bronchiale und bei gesunden Kindern. Klin Padiatr 199:103–107

CHAPTER 7
5-Hydroxytryptamine and Noradrenaline Synthesis, Release and Metabolism in the Central Nervous System: Circadian Rhythms and Control Mechanisms

K.F. MARTIN and P.H. REDFERN

A. Introduction

The variations in a range of physiological parameters over the course of the daily light: dark cycle have intrigued biologists for many years. The list of physiological functions measured is long and varied, ranging from changes in gross locomotor activity to the functioning of specific ion channels.

The aim of this chapter is to focus on circadian variation in one particular process, that of neurotransmission. Rhythms in neurotransmitter mechanisms are of special interest because they may do more than simply reflect the promptings of the circadian clock. Because of the structure of the mammalian pacemaker and its location within the central nervous system, neurotransmitter mechanisms are intimately involved in zeitgeber inputs to the pacemaker, in the internal workings of the clock and in communicating the efferent signals from the pacemaker.

Circadian rhythms in neurotransmission have been widely investigated by a number of groups in either whole brain or more recently in specific nuclei within the central nervous system. In this chapter we concentrate on neurotransmitter turnover and the factors that control it, since post-synaptic events following receptor activation are addressed elsewhere (LEMMER and WITTE, this volume). Our horizon is further narrowed in that rather than attempting an exhaustive but necessarily cursory listing of all published papers dealing with rhythms in neurotransmitter turnover, we have chosen to focus almost exclusively on one neurotransmitter, namely serotonin (5-hydroxytryptamine, 5-HT). Our intention is that consideration of one neurotransmitter in some detail, with a brief comparison with the noradrenergic system, should serve as a useful model for other transmitters.

To some extent this decision is dictated by the available literature, which provides a wealth of data concerning rhythms in turnover in the serotonergic system and their control, and a relative paucity of information on other neurotransmitters. The reasons for this imbalance are themselves interesting. It is probably a reflection of the fact that serotonin stands at the interface between chronobiological concerns about factors controlling circadian rhythms and the psychopharmacologist's interest in the involvement of rhythm disorders in the aetiology of the affective disorders (see ROSENWASSER and WIRZ-JUSTICE, this volume). For this reason we also consider briefly what is known about circadian rhythms in turnover in noradrenergic neurones in the central nervous system, since noradrenaline is commonly implicated along with serotonin in the aetiology of depressive illness.

B. Circadian Rhythms in Turnover of Serotonin

I. Tissue Levels of Serotonin

Whole brain 5-HT levels were first reported to vary over the course of the day in the mouse (ALBRECHT et al. 1956). Subsequently, similar findings have been reported in numerous other species, including man (e.g. BUCHT et al. 1981; DAVIES et al. 1972; KHAN and JOY 1988; QUAY 1967; SCHEVING et al. 1968; TAYLOR et al. 1982). Rhythms in 5-HT levels have also been described in discrete brain regions. Similar, but not identical, phases have been found in all brain areas studied (KAN et al. 1977; QUAY 1968). In the rat, 5-HT levels are highest during the hours of light and lowest during the hours of darkness. The amplitude of the rhythm in 5-HT tissue concentration has been shown to be largest in the suprachiasmatic nucleus (SCN) (MARTIN and MARSDEN 1985). Daily variations in the serotonin content of specific brain regions of Syrian hamsters have also been reported (OZAKI et al. 1993). Interestingly, these authors also detected daily variations in the concentration of 5-HT in other brain regions thought to play a role in the organisation of circadian time-keeping (e.g. paraventricular nucleus, lateral geniculate nucleus, raphe nuclei).

Over 20 years ago HERY et al. (1972) carried out a detailed examination of the metabolic processes involved in the control of these circadian variations in rat brain 5-HT levels. Interestingly, these workers found that whilst the rate of synthesis of 5-HT was greatest during the light period, breakdown of 5-HT was maximal during the dark period. This observation suggested that, whilst 5-HT synthesis appeared to be lower during the hours of darkness, its utilisation was actually greater at these times. Recent studies, using more sophisticated techniques, have supported these early studies and will be discussed in subsequent sections.

In all of the studies cited thus far, the variations in tissue 5-HT content was determined in animals entrained to a light:dark regimen, typically 12:12 LD. Recently, CAGAMPANG and INOUYE (1994) examined the daily variations in the 5-HT content of the SCN in animals housed under constant darkness. In these experiments, therefore, it was possible to determine whether there is a true circadian rhythm in 5-HT levels. In agreement with previous studies (e.g. MARTIN and MARSDEN 1985), animals housed under 12:12h light:dark cycles showed a peak 5-HT level during the light period and a trough during the dark period. When housed in constant darkness, the rats continued to exhibit a rhythm in 5-HT levels which appeared to free-run. Subsequent exposure to constant light induced a transitory rise in 5-HT levels immediately following lights-on. These data indicate that the circadian rhythm in 5-HT tissue levels in the SCN is generated by an endogenous pacemaker and that it is also influenced by photic cues. Recent reports of direct innervation of raphe nuclei from the retina (SHEN and SEMBA 1994) are pertinent here. By contrast, FERRARO and STEGER (1990) found that the daily variation in the 5-HT content of the anterior hypothalamus, median hypothalamus and olfactory bulbs is abolished following exposure to constant light or constant dark. Clearly, further studies are required to confirm that the circadian rhythm in 5-HT content is driven at least in part by an endogenous oscillator.

Figure 1 shows the various physiological processes which may contribute to the circadian rhythm in serotonergic neuronal function. In the following sections we will consider each of these mechanisms in turn.

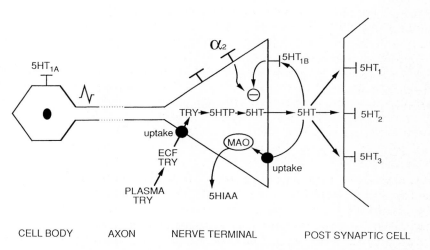

Fig. 1. The potential sites through which control of circadian rhythms in serotonergic neurone function may be exerted. Plasma tryptophan (*TRY*) is transported into the brain by a carrier in the blood-brain barrier (not shown). It then enters serotonergic neurones via a specific, carrier-mediated uptake process. Inside the neurone, TRY is converted to 5-hydroxytryptophan (*5-HTP*) by the enzyme TRY-5-hydroxylase. Conversion of 5-HTP to serotonin (5-hydroxytryptamine, *5-HT*) is facilitated by the enzyme 5-HTP decarboxylase. The rate-limiting step in this pathway is the conversion of TRY to 5-HTP by TRY-5-hydroxylase. The arrival of action potentials at the nerve terminals results in the release of 5-HT into the synaptic cleft, whence it can act upon post-synaptic 5-HT receptors or regulate it's own release by acting upon inhibitory pre-synaptic $5\text{-}HT_{1B}$ autoreceptors. It may also be taken up back into the neurone via an uptake process. Back inside the neurone 5-HT is then metabolised to 5-hydroxy-indoleacetic acid (*5-HIAA*) by monoamine oxidase (*MAO*). The release of 5-HT can also be regulated by other presynaptic receptors such as α_2-adrenoceptors. The frequency with which 5-HT neurones fire is regulated by $5\text{-}HT_{1A}$ autoreceptors located on the cell body

II. Tryptophan-5-Hydroxylase

The rate limiting step in the synthesis of 5-HT is the conversion of the essential amino acid, tryptophan, to 5-hydroxytryptophan (5-HTP). This reaction is mediated by the enzyme tryptophan-5-hydroxylase (ASHCROFT et al. 1965; FRIEDMAN et al. 1972). In any examination of factors responsible for the circadian rhythm in 5-HT levels, modulation of this enzyme is an obvious possibility.

Circadian rhythms in the activity of tryptophan-5-hydroxylase have been reported by several groups (KAN et al. 1977; CAHILL and EHRET 1981; SINEI and REDFERN 1985). All of these studies found that maximal enzyme activity occurred in the middle of the light phase, when 5-HT levels were also at their maximum. However, the area has been confused by other studies that have failed to demonstrate circadian rhythmicity in tryptophan-5-hydroxylase activity (BROWN et al. 1982; MCLENNEN and LEES 1978). The reasons for this diversity are not clear and will not be considered in detail here. They probably relate to the variety of experimental techniques employed. A crucial area of difference is that between in vivo and in vitro studies, which in turn may hinge on the availability of endogenous, and the supply of exogenous, co-factors. In *Xenopus* retina, circadian rhythms in tryptophan-5-hydroxylase activity and in expression of tryptophan hydroxylase mRNA are linked (GREEN et al. 1995). Furthermore, expression of tryptophan-5-hydroxylase mRNA is itself under the control of an endogenous circadian oscillator. It remains to be determined whether a similar situation obtains in mammalian tissue.

III. Tryptophan Availability

Tryptophan concentrations in the brain are rarely sufficient to saturate tryptophan-5-hydroxylase. Thus, considerable effort has been expended in attempts to demonstrate the dependence of brain 5-HT levels upon the availability of tryptophan, as measured by either brain or plasma concentration (CURZON 1979). It is not unexpected, therefore, that similar approaches have revealed circadian rhythms in both plasma and brain tryptophan concentrations which generally parallel those of 5-HT (HILLIER and REDFERN 1977; HUSSEIN and GOEDKE 1979; REDFERN and MARTIN 1985). The obvious inference that has been drawn is that the rhythm in 5-HT levels is directly related to this apparent rhythm in tryptophan availability. However, this seems unlikely to be the case, since it has been shown that the neuronal uptake carrier for tryptophan also has a daily rhythm in its activity (LOIZOU and REDFERN 1986), and it has been argued that this rhythm would effectively buffer the intraneuronal milieu against daily changes in extracellular tryptophan concentrations.

Strong arguments have been presented against a role for tryptophan in controlling the circadian fluctuations in 5-HT concentrations in brain tissue (LEATHWOOD 1989). Several groups have shown that increases in brain tryptophan levels of 80%–100% result in only 10%–15% increases in 5-HT and 5-

hydroxyindoleacetic acid (5-HIAA) concentrations. However, when the change in brain tryptophan is less than two-fold, most workers have failed to demonstrate any changes in 5-HT levels. The normal circadian variation in brain tryptophan concentration is generally less than two-fold, and it therefore follows that it is unlikely that the normal circadian variation in plasma or brain tryptophan concentration will result in a measurable variation in brain 5-HT concentrations. Thus, it is unlikely that tryptophan availability is an important factor in the control of circadian rhythms in 5-HT levels. What is more, the lack of any robust relationship between tissue levels of 5-HT and functional release of 5-HT (see below) further reduces the likely significance of plasma tryptophan levels as a controlling influence on 5-HT concentrations over 24 h.

IV. Tryptophan Uptake

Despite the significant circadian variation in brain tryptophan concentration, it has been argued that the tissue level of tryptophan is unlikely to play a part in the control of 5-HT levels (see above). However, it is possible that large changes in the kinetics of the uptake mechanism responsible for transporting tryptophan into the neurone may be of importance. Indeed, HERY et al. (1972) have claimed a significant role for the tryptophan uptake complex in the control of 5-HT circadian rhythms. However, experiments have shown that the rate of uptake of tryptophan into cortical synaptosomes was maximal in the mid-light phase (LOIZOU and REDFERN 1986) and that the affinity of the carrier for tryptophan varied inversely with the maximum rate of uptake. These data also suggested that the variation in affinity was dependent upon the endogenous extracellular concentration of tryptophan. An inverse relationship between affinity and rate would inevitably result in a fairly stable supply of tryptophan into the neurone that overrides the variation in substrate levels in the extracellular fluid. Whilst these results remain to be confirmed, they do support the view that tryptophan availability does not play an important part in the control of 5-HT neurone function.

V. 5-Hydroxytryptophan Decarboxylase Activity

In the central nervous system the activity of 5-HTP decarboxylase is invariably much higher than that of the rate limiting enzyme tryptophan hydroxylase. In consequence tissue levels of 5-HTP are low, and it is generally considered that the activity of the decarboxylase enzyme plays no part in controlling 5-HT synthesis. One study (HILLIER and REDFERN 1975) has reported a significant circadian variation in 5-HTP decarboxylase activity, and it was suggested that this variation arose from competition for the enzyme between 5-HTP and other amino acids, principally L-Dopa. For the present this remains an isolated observation. In any event the reported changes, though statistically significant, were small, and their influence on the circadian rhythm of 5-HT synthesis is likely to be correspondingly small.

VI. 5-Hydroxytryptamine Release and Metabolism

The early studies of HERY et al. (1972) indicated that 5-HT utilisation in the rat brain was higher during the dark phase when the animals were most active. This led to the suggestion that 5-HT release was also at its maximum during the dark phase. However, their data were not easily interpreted because they did not distinguish between 5-HIAA derived from released 5-HT and that derived from intraneuronal metabolism of newly synthesised 5-HT.

In recent years, new techniques, such as in vivo voltammetry and intracranial dialysis, have become available that allow the measurement of extracellular levels of 5-HT and 5-HIAA. The practicalities of these techniques have been reviewed in detail elsewhere (MARSDEN et al. 1984; MARTIN et al. 1988; UNGERSTEDT 1984).

Using in vivo voltammetry to monitor extracellular levels of 5-HIAA, FARADJI and co-workers (1983) reported a clear circadian pattern of this metabolite in the SCN of the rat. This rhythm was closely correlated to the rhythm in locomotor activity. Peak activity was observed during the dark phase as were the peak levels of extracellular 5-HIAA. These data support the view that functional release of 5-HT was indeed greatest at night. CESPUGLIO et al. (1982) had previously made similar observations in the rat frontal cortex. At the time that these latter experiments were carried out it was not possible to monitor extracellular 5-HT levels using in vivo voltammetry. This situation has now changed with the development of Nafion-coated carbon fibre electrodes that are capable of detecting basal levels of 5-HT in a number of brain regions (CRESPI et al. 1988). The outcome of experiments using these electrodes to study 5-HT release in the suprachiasmatic nucleus and other small areas associated with the control of circadian behaviour is awaited.

Using a miniature push-pull perfusion cannula, RAMIREZ et al. (1987) attempted to measure circadian rhythms in 5-HIAA production in the region of the SCN nuclei. Their experiments were carried out in normally cycling, mature female rats. They showed that the amplitude and characteristics of the circadian rhythm in 5-HIAA production were dependent upon the position of the cannula. Thus, high amplitude rhythms were observed in rostral regions of the SCN compared to low amplitude rhythms observed in caudal regions. However, in both cases the highest 5-HIAA concentrations were found during the hours of darkness. This apparent rostro-caudal variation in rhythm amplitude is difficult to explain, but may be related to the similar rostro-caudal distribution of 5-HT nerve terminals within the suprachiasmatic nuclei (VAN DEN POL and TSUJIMOTO 1985). Nevertheless, these results confirm those obtained using in vivo voltammetry, which showed that 5-HIAA levels were elevated during the dark phase.

For a number of years it has been possible to monitor extracellular 5-HT levels using intracranial microdialysis. However, because of the size of the probes (200 μm diameter × 2–3 mm long), experiments using this technique have been limited to relatively large brain areas. MARTIN and MARSDEN first presented data from experiments in which this technique was used to study circadian rhythms of 5-HT release in 1985. Dialysis probes were implanted into the ventro-medial hypothalamus of male rats under halothane anaesthesia, and the animals were allowed to recover. Twenty-four hours later the extracellular concentrations of tryptophan, 5-HT and 5-HIAA were determined during 20 min epochs throughout the 24-h cycle. The results of these experiments showed that the extracellular concentration of all three compounds varied throughout the light-dark cycle. In agreement with the findings of FARADJI et al. (1983), 5-HIAA levels were found to be elevated during the hours of darkness with rapid changes at the times of transition from light to dark (upwards) and dark to light (downwards). Although these authors were able to show that extracellular 5-HT levels were higher during the hours of darkness, there was considerable 'noise' reflected in unpredictable rapid alterations in dialysate 5-HT concentrations from one epoch to the next. The same pattern was also observed for tryptophan. It is interesting to note that the extracellular concentration of tryptophan was relatively low although the total tissue concentration of tryptophan is reported to be high (REDFERN and MARTIN 1985). Also of note was the fact that the rapid rises in extracellular tryptophan, 5-HT and 5-HIAA concentrations were not in synchrony. As might be predicted, the rise in tryptophan preceded that of 5-HT, which in turn preceded that of 5-HIAA.

These findings have been confirmed by STANLEY et al. (1987) who also used intracranial microdialysis. They found that medial hypothalamic 5-HT followed a similar pattern to that observed by MARTIN and MARSDEN (1985) in the ventro-medial hypothalamus. Close examination of the data of STANLEY et al. (1987) showed that this pattern of 5-HT release persisted when the animals were deprived of food, indicating that the rapid rise in extracellular 5-HT concentrations at the onset of darkness is not simply related to an increased intake of food, and adding support to the argument outlined earlier that dietary intake of tryptophan contributes little to the generation of the circadian rhythms in 5-HT release and metabolism.

KALEN et al. (1989) measured 5-HT release in the hippocampus over 24 h. The hippocampus is not directly associated with the regulation of circadian function but does receive a significant serotonergic input from the same midbrain raphe nuclei that innervate the hypothalamus (AZMITIA and SEGAL 1978). In agreement with the observations of MARTIN and MARSDEN (1985), pronounced fluctuations in the concentration of 5-HT in 30-min dialysate samples were observed in individual animals by KALEN et al. (1989). Nevertheless, these authors were able to show that average 5-HT release was 38% higher during the hours of darkness compared to that during the light period. Interestingly, there was no detectable change in extracellular 5-HIAA concentrations. These authors reported a significant correlation between the behavioural state of the animals, i.e. whether awake or sleeping, and the rate of 5-HT release. Since neither MARTIN and MARSDEN (1985), nor STANLEY et al. (1987), nor KALEN et al. (1989) carried out experiments on animals kept under constant light, it is not possible to draw any conclusion as to whether the variation in the rate of 5-

HT release either in the hypothalamus or the hippocampus is dependent upon an endogenous oscillator or is simply driven either by the behavioural state of the animal or by photic stimuli.

In contrast to the consistency of the findings that in vivo 5-HT release is maximal during the dark phase, contradictory in vitro data have been reported. BLIER and co-workers (1989) studied the release of [^{3}H]5-HT from hypothalamic slices in vitro. They found that the amount of [^{3}H]5-HT released following electrical stimulation was greater in tissue obtained from animals during the light phase, compared to that in tissue extracted in identical fashion during the dark phase. In order to reconcile these in vitro findings with the more physiologically relevant in vivo data, one must examine carefully the methodology used in the in vitro experiments. Animals were killed either at mid-light or at mid-dark. The hypothalamic tissue was then incubated with [^{3}H]5-HT in order to label the intraneuronal stores of 5-HT. Tritium release was determined in superfusion experiments, and the increase in release over basal rates in response to electrical stimulation was determined using standard, well-established methodology. These experiments were carried out at only two time points in the 24-h cycle. This compares with the in vivo experiments which have between 24 and 72 data points in every 24 h. It is clear that for this reason alone the in vivo experiments are likely to provide a much more detailed picture. Equally, it must be conceded that in the in vivo experiments (MARTIN and MARSDEN 1985; STANLEY et al. 1987), although the levels of 5-HT were generally elevated during the dark phase, there was considerable 'noise' in the system. Occasionally the rate of 5-HT release observed in the middle of the dark period dropped to levels seen in the light phase.

In considering the significance of differences in results obtained in vivo and in vitro, a further complication arises because experiments measuring tritium release require the neuronal stores of 5-HT to be labeled. This process first requires uptake of the radiolabel, followed by storage in vesicles. Both of these processes may themselves be subject to circadian variation. Indeed, there is considerable evidence that both the affinity and rate of the 5-HT re-uptake system exhibit a circadian rhythm (MARTIN 1982; BRUNELLO et al. 1987). Indeed, BLIER et al. (1989) themselves reported that the uptake of [^{3}H]5-HT was different at the two time points studied although they did not find any circadian time-dependent differences in the effect of the selective 5-HT re-uptake inhibitor, citalopram, on the uptake of [^{3}H]5-HT, which might be expected to have exerted a greater effect on 5-HT re-uptake at times when the affinity of the carrier complex was greatest.

For these and other reasons, BLIER et al. (1989) acknowledged that their in vitro findings may not be transferable to the in vivo situation. For instance, they pointed out that, in vivo, other factors, such as the firing rate of serotonergic neurones, may be important in regulating the pattern of 5-HT release over 24 h. Furthermore, even in an in vitro slice preparation, much of the interaction between serotonergic neurones and other transmitter systems, depending as it does on maintenance of an intact cytoarchitecture, will be lost.

VII. Firing Rate of Serotonergic Neurones

It is widely accepted that neurotransmitter release is directly related to neuronal firing rate. This immediately suggests itself as a likely mechanism by which the rate of release of 5-HT may be controlled over 24 h.

The evidence in favour of this hypothesis can be briefly summarised as follows. In the cat, Jacobs and co-workers (for a review see FORNAL and JACOBS 1988) have shown that the firing rate of serotonergic neurones is significantly altered only by the sleep-wake cycle. Thus, during REM sleep serotonergic neurones are quiescent, and they fire only very slowly during slow wave sleep (< 1 Hz). During the waking phase firing rates are fairly constant at around 3 Hz. Measurements of 5-HT release in cat forebrain regions indicate that 5-HT release closely follows the rate of neuronal firing; release is minimal during periods of REM sleep (when 5-HT neurones do not fire) and is greatest during the active, waking phase when the rate of neuronal firing is at its highest (WILKINSON et al. 1988).

Since the rat is a nocturnal animal, it follows that slow wave sleep and REM sleep occur during the light phase and the waking state corresponds to the dark phase. Both MARTIN and MARSDEN (1985) and KALEN et al. (1989) showed that the 5-HT concentration in dialysate samples was greatest during the hours of darkness and that the 24-h variation in levels correlated well with behavioural state in a similar manner to that reported in the cat (WILKINSON et al. 1988). Thus, the behavioural state and the rate of firing of serotonergic neurones are closely matched, and it is therefore reasonable to conclude that the circadian rhythm in 5-HT release is normally controlled by the rate at which serotonergic neurones fire.

VIII. 5-Hydroxytryptamine Autoreceptor Activity

It is now well established that the synthesis and release of 5-HT is influenced by the activity of a number of autoreceptors located either in the region of the cell body or on the presynaptic nerve-terminal membrane. In the rat the terminal autoreceptor is believed to be of the 5-HT_{1B} subtype (MORET 1985; MIDDLEMISS 1985). Data obtained from in vivo voltammetric studies have confirmed that in the SCN of the rat, release of 5-HT is controlled by terminal 5-HT_{1B} autoreceptors (MARSDEN and MARTIN 1985a,b). Terminal 5-HT_3 autoreceptors have also been shown to facilitate 5-HT release in the hippocampus in vivo (MARTIN et al. 1992) and in the frontal cortex in vitro (GALZIN et al. 1990).

Activation of 5-HT_{1A} somatodendritic autoreceptors located in the cell body region results in a reduction in firing rate of serotonergic neurones (SPROUSE and AGHAJANIAN 1987). Predictably, the reduced firing rate results in a reduction in 5-HT release. For instance, local injection of the selective 5-HT_{1A} receptor agonist, 8-hydroxy-(di-n-propylamino)-tetralin (8OH-DPAT) into the raphe resulted in decreased 5-HT release in several terminal areas

(GARRATT et al. 1988; HUTSON et al. 1988), but either no change (GARRATT et al. 1988; HUTSON et al. 1988) or an increase (MARSDEN and MARTIN 1985b) in extracellular 5-HIAA levels.

Although it is clear that the predominant somadendritic autoreceptor is of the 5-HT_{1A} subtype, the picture is further complicated by recent reports that functional 5-HT_{1B} and 5-HT_{1D} receptors are also present in the raphe nuclei (DAVIDSON and STAMFORD 1995; MUNDEY et al. 1995). The question that must be addressed is whether these autoreceptors play any part in controlling 5-HT synthesis and release over 24 h, i.e. is the observed circadian variation in 5-HT release mediated by a circadian rhythm in the sensitivity or activation of these autoreceptors?

In the previous section we argued that, under normal conditions, the release of 5-HT is directly related to the rate at which 5-HT neurones fire. It follows, therefore, that the activity of the cell body 5-HT autoreceptor may exert an influence on the circadian activity of these neurones. So far the only attempts to determine whether there is a circadian rhythm in the activity of this receptor have used functional paradigms.

The selective 5-HT_{1A} receptor agonist, 8-OH-DPAT, induces hypothermia in rodents through an action on 5-HT cell body autoreceptors (GOODWIN et al. 1985, 1987; MARTIN et al. 1992). MARSDEN et al. (1985) and MOSER and REDFERN (1985) found that the magnitude of this hypothermic effect did not change over the 24-h light:dark cycle. This suggests that the 5-HT_{1A} cell body autoreceptor does not exhibit a circadian rhythm in its activity. Measurement of the effect of 5-HT_{1A} receptor stimulation on the rate of synthesis of 5-HT in the cerebral cortex of the rat also failed to show any circadian variation (SINGH and REDFERN, unpublished observation). Consequently, it seems unlikely that circadian rhythms in 5-HT synthesis and release are influenced by changes in the activity of the 5-HT_{1A} somadendritic receptor.

The second type of 5-HT autoreceptor that may exert control over rhythms in 5-HT release is the nerve-terminal 5-HT_{1B} autoreceptor. The picture here is less clear-cut. Results from early behavioural experiments suggested that the activity of this receptor did in fact exhibit a circadian rhythm (MARTIN et al. 1986). Although the same doses of the 5-HT_{1B} receptor agonist 5-methoxy-3-[1,2,3,6-tetrahydro-4-pyridinyl]-1H indole (RU 24969) induced the same degree of hyperlocomotion in mice throughout the day, the efficacy of the 5-HT receptor antagonist, metergoline, in blocking this effect was clearly different at different phases of the light-dark cycle.

In contrast to these behavioural experiments, SINGH and REDFERN (1990, 1994a) failed to demonstrate a circadian rhythm in the ability of 5-HT_{1B} receptors to modulate the release of $[^3H]$-5-HT from rat cortical slices. The evidence from these studies appears strongly to support the view that the activity of nerve-terminal 5-HT_{1B} receptors is constant throughout the light-dark cycle in that the pA_2 value for antagonism of 5-HT_{1B}-receptor-mediated inhibition of $[^3H]$5-HT release by methiothepin varied between 6.5 and 6.8 when measured at four evenly-spaced time points in the light-dark cycle.

A similar lack of circadian variation was also reported in terminal 5-HT autoreceptors in the guinea pig, which are of the 5-HT$_{1D}$ type (SINGH and REDFERN 1994b).

These in vitro data, therefore, suggest that nerve-terminal 5-HT$_{1B}$ autoreceptors are unlikely to play a role in controlling circadian rhythms in 5-HT release. However, the problems of comparing the results from in vivo and in vitro experiments have been alluded to earlier. Differences between in vitro and in vivo observations in this area are well documented. For example, the results of in vitro experiments would lead us to believe that α_2-adrenoceptors on serotonergic terminal membranes do not tonically control 5-HT release (GALZIN et al. 1984). By contrast, in vivo experiments have shown that this receptor is indeed tonically activated (MARSDEN and MARTIN 1986). With this in mind, we have carried out a series of in vivo microdialysis experiments to determine whether the function of the hypothalamic 5-HT$_{1B}$ autoreceptor varies with time of day (SAYER et al. 1994). Extracellular 5-HT concentrations were determined following a 15-min infusion of RU 24969, a selective agonist, at the 5-HT$_{1B}$ receptor directly into the anterior hypothalamus via the dialysis probe either at mid-light or end-light. When infused at mid-light the maximal inhibition of 5-HT release was 65%, but at end-light release of 5-HT was completely abolished (SAYER et al. 1994). In addition, at mid-light, 5-HT levels returned to pre-intervention levels within 90 min of termination of drug infusion whereas at end-light, they remained depressed for the time course of the experiment (90 min) (Fig. 2). These data indicate that the function of hypothalamic terminal 5-HT$_{1B}$ autoreceptors does vary with time of day. Receptor binding studies have revealed a 24-h rhythm in the number (B_{max}) of 5-HT$_{1B}$ binding sites in rat cortex (AKIYOSHI et al. 1989). However, 5-HT$_{1B}$-receptor mRNA levels in the SCN were found to remain constant across the light-dark cycle (ROCA et al. 1993). These data are clearly at odds with the in vivo microdialysis data described above. The ligand binding and in situ hybridization studies measure only levels of receptor protein or the message encoding for that protein. The in vivo microdialysis experiments of SAYER et al. (1994) determine receptor *function*. The observed differences reinforce the importance of considering function in vivo before making judgements based on ligand binding studies or mRNA determinations. A further factor that may also explain these reported differences is that ROCA et al. (1993) estimated 5-HT$_{1B}$ mRNA in postsynaptic cells within the SCN whereas the release experiments measured exclusively presynaptic 5-HT$_{1B}$ function; thus it is possible that these two disparate findings are not at odds with each other because rhythms in two different receptor populations have been investigated. It remains to be determined, therefore, whether these functional variations persist through the hours of darkness and in the absence of time cues.

There is to our knowledge no evidence available concerning circadian variations in the activity of terminal 5-HT$_3$ receptors or of somadendritic autoreceptors, beyond the unpublished results referred to above.

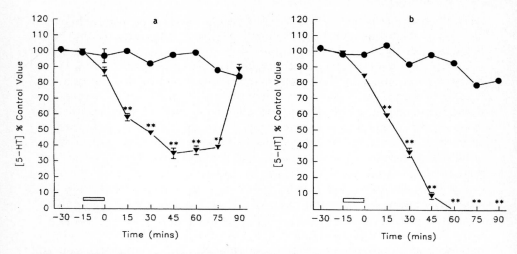

Fig. 2a,b. The effects of a 15-min infusion of 5 μM RU24969 (*open bar*) at mid-light (a) and end-light (b) on 5-HT output in the anterior hypothalamus **a** Control (●, $n=6$) and 5 μM RU 24969 (▼ $n=4$). **$p<0.01$ vs. control value for mid-light (one-way ANOVAR). **b** Control (●, $n=5$) and 5 μM RU24969 (▼, $n=4$). **$P<0.01$ vs. control for end-light (one-way ANOVAR). Data are expressed as a percentage of the concentration of 5-HT in the two dialysate samples taken immediately prior to any intervention (control value); each *point* represents mean with the standard error of mean indicated by *vertical bars*

Thus, on the basis of our limited microdialysis data (SAYER et al. 1994), it is possible to suggest that circadian variations in 5-HT synthesis and release are influenced by changes in 5-HT$_{1B}$ autoreceptor activity. However, the most likely explanation for the circadian rhythm in 5-HT turnover remains the variation in neuronal firing already discussed. Further in vivo studies are clearly required to fully investigate this problem.

IX. What Is the Function of the Circadian Rhythm in 5-Hydroxytryptamine Turnover?

The nerve terminals of serotonergic neurones are widely distributed throughout the central nervous system. Circadian rhythms in turnover in serotonergic neurones have been consistently reported in all brain areas so far examined. These circadian rhythms can be regarded as a tangible response to the promptings of the central circadian oscillator. As such they provide a model of circadian rhythms in other neurotransmitters.

It may be seen that circadian rhythms in 5-HT have an added significance when we consider specifically the serotonergic innervation of the SCN and the intergeniculate leaflet (IGL), both of which are integral to the functioning of the circadian system.

Both SCN and IGL receive a significant serotonergic innervation from midbrain raphe nuclei (AZMITIA and SEGAL 1978; DAHLSTROM and FUXE 1964;

MANTYH and KEMP 1983). Considerable effort has been expended in attempts to delineate the function of the serotonergic innervation of the SCN. The evidence remains somewhat contradictory, although the consensus is that these neurones play a modulatory role, not essential for the generation or perpetuation of the circadian oscillation from the SCN, but affecting phase and amplitude (SMALE et al. 1990). Little is known about the function of the serotonergic innervation of the IGL. Destruction of the raphe-SCN pathway caused a loss of rhythmicity in neuroendocrine marker rhythms (WILLIAMS et al. 1983; BANKY et al. 1986, 1988), but caused phase advances and a decrease in amplitude in behavioural rhythms (TAKAHASHI et al. 1986; SMALE et al. 1990; MORIN and BLANCHARD 1991).

Both in vivo (MARSDEN and MARTIN 1985a; TOMINANGA et al. 1992; EDGAR et al. 1993) and in vitro (MEDANIC and GILLETTE 1992; PROSSER et al. 1993) experiments have shown that pharmacological intervention with selective ligands for 5-HT receptors again produces phase shifts and/or changes in amplitude of the measured rhythm. Similar findings were also reported following depletion of 5-HT by chronic exposure to a diet deficient in the precursor tryptophan (KAWAI et al. 1994).

On the basis of these findings it is clear that serotonergic innervation of the SCN plays an important part in modulating the activity and output of the circadian oscillator. Beyond the physiological significance of this observation, it may also prove crucial in our understanding of the link between affective disorders and rhythmic dysfunction (ROSENWASSER and WIRZ-JUSTICE, this volume), given that most clinically effective antidepressant drugs affect serotonergic neurotransmission.

Two areas require further investigation in order to define this role more precisely: first, the significance of the direct retinal innervation of the midbrain raphe nuclei (SHEN and SEMBA 1994) and the importance of light as a controlling influence on the circadian rhythm of turnover in serotonergic neurones; second, the effect of 5-HT on other neurotransmitter systems within the SCN. Reports of a high density of $5-HT_7$ receptors within the SCN (LOVENBERG et al. 1993) are particularly intriguing here, and should provide a new impetus to research in this area.

C. Circadian Rhythms in Noradrenaline Turnover

The levels of noradrenaline in rodent brain were first reported to vary over the 24-h cycle by MANSHARDT and WURTMAN in 1968. These findings have been confirmed and extended to include specific brain regions and different species. For example, significant daily rhythms in total brain noradrenaline content have been found in the rat, hamster, guinea pig and gerbil with peaks in the dark phase and troughs in the light phase (PHILO et al. 1977). In addition, KEMPF et al. (1982) reported daily variations of hypothalamic noradrenaline content and turnover rate in two strains of mouse. However, during exposure to constant lighting conditions these rhythms persisted only in C57Bl/6 mice.

Furthermore, the magnitude of the day-night differences in noradrenaline content was much less than that reported for 5-HT by these same authors. AGREN et al. (1986) reported that the tissue content of noradrenaline in the locus coeruleus and the dorsal raphe followed similar patterns with peak levels just before the onset of the dark phase and minimum values immediately before the end of the dark period. Interestingly, these authors found similar variations in the 5-HT content of these two brain regions. Whilst there was a significant intercorrelation between noradrenaline and 5-HT levels during the daylight hours, this relationship disappeared during the dark period. As discussed previously, these data confirm the value of taking circadian factors into account when considering the interactions between neurotransmitter systems and the effects of pharmacological interventions.

As with the serotonergic system, the role of the enzymes responsible for the synthesis of noradrenaline in the genesis of the 24-h variations in tissue noradrenaline levels has been studied. NATALI et al. (1980) found that the activity of tyrosine hydroxylase in the mouse locus coeruleus exhibited a marked circadian variation that was paralleled by changes in noradrenaline concentration. Similarly, the existence of circadian variation in the mechanisms involved in terminating the actions of noradrenaline has been investigated. A study by VAN DER GUGTEN and SLANGEN (1975) demonstrated a circadian rhythm in the uptake of noradrenaline by hypothalamic nerve terminals. Maximal uptake rates were recorded at the beginning of the dark phase when food intake is also approaching its maximum. Indeed, using restricted feeding paradigms, these authors clearly demonstrated that the rhythm in noradrenaline uptake was directly related to feeding patterns rather than to the lighting schedule. From this it did not appear that there was an endogenous circadian rhythm in the uptake process.

Although changes in tissue levels of neurotransmitter have been shown to vary over 24 h, it is important to know whether such changes are translated into rhythms of neurotransmitter release. Several authors have shown that the circadian rhythms in tissue levels of, for example, 5-HT are 180 degrees out of phase with the rhythms in extracellular 5-HT (inter alia MARTIN 1991). Using in vivo microdialysis, STANLEY et al. (1989) found a 24-h rhythm in the extracellular levels of noradrenaline in the paraventricular nucleus (PVN) of the hypothalamus of the rat. Their data showed that there was a large but transient increase in release 1 h after lights off and a second, lower amplitude, short lasting increase 2 h prior to lights on. In agreement with the uptake studies of VAN DER GUGTEN and SLANGEN (1975) these increases appeared to coincide with eating behaviour. This pattern disappeared during food deprivation which was associated with a general increase in extracellular noradrenaline levels that declined rapidly on presentation of food but increased again as the animals became satiated (STANLEY et al. 1989). These findings argue that the circadian rhythm in noradrenaline release in the PVN is not endogenously generated and suggest that in the PVN, at least, changes in noradrenaline release are driven by the feeding rhythm rather than by the light-dark cycle. Hippocampal noradrenaline release has also been shown to vary over the normal light-dark cycle (KALEN et al. 1989); pronounced fluctuations over the light-dark cycle were observed, but the mean output during the hours of darkness was 43% higher than during the daylight hours. In addition, it was clear that the levels of noradrenaline were influenced by the behavioural activity state of the animal, i.e. release increased during periods of intensive activity. KALEN et al. (1989) did not, however, investigate the possible link between feeding and noradrenaline release.

It is widely accepted that the release of noradrenaline is controlled by auto- and heteroreceptors in a similar way to the control of 5-HT release. Unfortunately, although there is a wealth of information about the circadian variations in adrenoceptor binding, there are no reports in the literature of circadian changes in the function of these receptors or of their possible role in controlling circadian variations in noradrenergic function.

D. Conclusions

In this chapter we have concentrated on circadian rhythms in the turnover of 5-HT in serotonergic neurones. It is true to say that practically every aspect of serotonergic nerve function, from neuronal firing rate and synthesis to release and metabolism, exhibits some degree of circadian rhythmicity.

The mechanisms responsible for the generation of these rhythms in serotonergic neurones are not clear, and much further work is required to answer many of the important questions that have been alluded to. In addition, the final part of this chapter has illustrated the relative paucity of information concerning the circadian aspects of the noradrenergic system. Since there is even less information about other neurotransmitter systems, a great deal of further work is urgently required fully to understand the mechanisms involved and their relevance to normal physiological function.

References

Agren H, Koulu M, Saavedra JM, Potter WZ, Linnoila M (1986) Circadian covariation of norepinephrine and serotonin in locus coeruleus and dorsal raphe nucleus in the rat. Brain Res 397:353–358

Akiyoshi J, Kuranaga H, Tsuchiyama A, Nagayama H (1989) Circadian rhythm of serotonin receptor in rat brain. Pharmacol Biochem Behav 32:491–493

Albrecht P, Visscher MB, Bittner JJ, Halberg F (1956) Daily changes in 5-hydroxytryptamine concentrations in mouse brain. Proc Soc Exp Biol Med 92:703–706

Ashcroft GW, Eccleston D, Crawford TBB (1965) 5-Hydroxyindole metabolism in rat brain: a study of intermediate metabolism using the techniques of tryptophan loading. J Neurochem 12:483–492

Azmitia EC, Segal M (1978) An autoradiographic analysis of the differential projections of the dorsal and median raphe nuclei in the rat. J Comp Neurol 179:641–668

Banky Z, Halsz B, Nagy G (1986) Circadian corticosterone rhythm did not develop in rats seven weeks after destruction with 5,7-dihydroxytryptamine of the serotonergic terminals in the suprachiasmatic nucleus at the age of 16 days. Brain Res 369:119–124

Banky Z, Molnar J, Csernus V, Halsz B (1988) Further studies on circadian hormone rhythms after local pharmacological destruction of the serotonergic innervation of the rat suprachiasmatic region before the onset of the corticosterone rhythm. Brain Res 445:222–227

Blier P, Galzin A-M, Langer SZ (1989) Diurnal variations in the function of serotonergic terminals in the rat hypothalamus. J Neurochem 52:453–459

Brown F, Nicholass J, Redfern PH (1982) Synaptosomal tryptophan hydroxylase activity in rat brain measured over 24 hours. Neurochem Int 4:181–183

Brunello N, Rovescalli AC, Riva M, Galimberti R, Racagni G (1987) Rhythmic changes in rat hypothalamic 3H-imipramine binding and 5-HT uptake sites: possible biochemical correlate with antidepressant action. Soc Neurosci Abstr 13:200

Bucht G, Adolfsson R, Gottfries CG, Roos B-E, Winblad B (1981) Distribution of 5-hydroxytryptamine and 5-hydroxyindoleacetic acid in human brain in relation to age, drug influence, agonal status and circadian variation. J Neural Transm 51:185–203

Cahill AL, Ehret CF (1981) Circadian variations in the activity of tyrosine hydroxylase, tyrosine aminotransferase and tryptophan hydroxylase: relationship to catecholamine metabolism. J Neurochem 37:1109–1115

Cagampang FRA, Inouye S-CT (1994) Diurnal and circadian changes of serotonin in the suprachiasmatic nuclei: regulation by light and an endogenous pacemaker. Brain Res 639:175–179

Cespuglio R, Faradji H, Crespi F, Jouvet M (1982) Detection by differential pulse voltammetry of 5-hydroxyindoleacetic acid in rostral brain areas: fluctuations occurring during the sleep-waking cycle. 6th European Congress on Sleep Research, Zurich pp 282–284

Crespi F, Martin KF, Marsden CA (1988) Measurement of 5-HT in vivo using Nafion coated carbon fibre electrodes combined with differential pulse voltammetry. Neuroscience 27:885–896

Curzon G (1979) Relationships between plasma, CSF and brain tryptophan. J Neural Transm 15:81

Dahlstrom A, Fuxe K (1964) Evidence for the existence of monoamine containing neurons in the central nervous system. I. Demonstration of monoamines in the cell bodies of brainstem neurones. Acta Physiol Scand 62 [Suppl 232]:1–55

Davidson C, Stamford JA (1995) Evidence that 5-hydroxytryptamine release in rat dorsal raphe nucleus is controlled by $5\text{-}HT_{1A}$, $5HT_{1B}$ and $5\text{-}HT_{1D}$ autoreceptors. Br J Pharmacol 114:1107–1109

Davies JA, Ancill RJ, Redfern PH (1972) Hallucinogenic drugs and circadian rhythms. Prog Brain Res 36:79–95

Edgar DM, Miller JD, Prosser RA, Dean RR, Dement WC (1993) Serotonin and the mammalian circadian system: II. Phase-shifting rat behavioral rhythms with serotonergic agents. J Biol Rhythms 8:17–31

Faradji H, Cespuglio R, Jouvet M (1983) Voltammetric measurements of 5-hydroxyindole compounds in the suprachiasmatic nuclei: circadian fluctuations, Brain Res 279:111–119

Ferraro JS, Steger RW (1989) Diurnal variation in brain serotonin is driven by the photic cycle and is not circadian in nature. Soc Neurosci Abstr 15:293.16

Ferraro JS, Steger RW (1990) Diurnal variations in brain serotonin are driven by the photic cycle and are not circadian in nature. Brain Res 512: 121–124

Fornal CA, Jacobs BL (1988) Physiological and behavioural correlates of serotonergic single unit activity. In: Osborne NN, Hamon M (eds) Neuronal serotonin. Wiley, Chichester, pp 305–346

Friedman PA, Kappelman AH, Kaufman S (1972) Partial purification and characterisation of tryptophan hydroxylase from rabbit hindbrain. J Biol Chem 247:4165–4173

Galzin A-M, Moret C, Langer SZ (1984) Evidence that exogenous but not endogenous norepinephrine activates the presynaptic alpha-2 adrenoceptors on serotonergic nerve endings in the rat hypothalamus. J Pharmacol Exp Ther 228:725–732

Galzin AM, Poncet V, Langer SZ (1990) 5-HT_3 receptor agonists enhance the electrically evoked release of [^{3}H]5-HT in guinea pig frontal cortex slices. Br J Pharmacol 100:307P

Garratt J, Marsden CA, Crespi F (1988) 8OH-DPAT can decrease 5-HT neuronal firing and release but not metabolism. Br J Pharmacol:95 874P

Goodwin GM, DeSouza RJ, Green AR (1985) The pharmacology of the hypothermic response in mice to 8-hydroxy-2-(di-n-propylamino)tetralin (8OH-DPAT). A model of presynaptic 5-HT1 function. Neuropharmacology 24:1187–1194

Goodwin GM, DeSouza RJ, Green AR, Heal DJ (1987) The pharmacology of the behavioural and hypothermic responses of rats to 8-hydroxy-2-(di-n-propylamino)tetralin (8OH-DPAT). Psychopharmacology 91:506–511

Green CB, Cahill GM, Besharse JC (1995) Regulation of tryptophan hydroxylase expression by a retinal circadian oscillator in vitro. Brain Res 677:283–290

Hery F, Rouer E, Glowinski J (1972) Daily variations of serotonin metabolism in the rat brain. Brain Res 43:445–465

Hillier JG, Redfern PH (1975) The 24 hour variation in 5-hydroxytryptophan decarboxylase activity in the rat brain. J Neurochem 27:311–312

Hillier JG, Redfern PH (1977) 24 hour rhythm in serum and brain indoleamines: tryptophan hydroxylase and MAO activity in the rat. Int J Chronobiol 4:197–210

Hussein L, Goedke HW (1979) Diurnal rhythm in plasma level of total and free tryptophan and cortisol in rabbits. Res Exp Med Berlin 176:123–130

Hutson PH, Sarna GS, O'Connell MT, Curzon G (1988) Decrease of hippocampal 5-HT release following infusion of 8OH-DPAT into the dorsal raphe. Br J Pharmacol 94:387P

Kalen P, Rosegren E, Lindvall O, Bjorklund A (1989) Hippocampal noradrenaline and serotonin release over 24 hours as measured by the dialysis technique in freely moving rats: correlation to behavioural activity state, effect of handling and tail pinch. Eur J Neurosci 1:181–188

Kan JP, Chouvet G, Hery F (1977) Daily variations of various parameters of serotonin metabolism in the rat brain. I. Circadian variations of tryptophan-5-hydroxylase in the raphe nuclei and the striatum. Brain Res 123:125–136

Kawai K, Yokota N, Yamawaki S (1994) Effect of chronic tryptophan depletion on the circadian rhythm of wheel-running activity in rats. Physiol Behav 55:1005–1013

Kempf E, Mandel P, Oliverio A, Puglisi-Allegra S (1982) Circadian variations of noradrenaline, 5-hydroxytryptamine and dopamine in specific brain areas of C57B1/6 and BALB/c mice. Brain Res 232:472–478

Khan IA, Joy KP (1988) Seasonal and daily variations in hypothalamic monoamine levels and monoamine oxidase activity in the teleost Channa punctatus (Bloch). Chronobiol Int 5:311–316

Leathwood P (1989) Circadian rhythms of plasma amino acids, brain neurotransmitters and behaviour. In: Arendt J, Minors DS, Waterhouse JM (eds) Biological rhythms in clinical practice. Wright, London, pp 136–159

Loizou G, Redfern PH (1986) Circadian variation in uptake of tryptophan by synaptosomes from rat cortex. J Pharm Pharmacol 38:89P

Lovenberg TW, Baron BM, de Lecea L, Miller JD, Prosser RA, Rea MA, Foye PE, Racke M, Slone AL, Siegel BW, Danielson PE, Sutcliffe JG, Erlander MG (1993) A novel adenylate cyclase-activating serotonin receptor (5-HT_7) implicated in the regulation of mammalian circadian rhythms. Neuron 11:449–458

Manshardt J, Wurtman RJ (1968) Daily rhythm in the noradrenaline content of rat hypothalamus. Nature 217:574–575

Mantyh PW, Kemp JA (1983) The distribution of putative neurotransmitters in the lateral geniculate nucleus of the rat. Brain Res 288:344–348

Marsden CA, Maidment NT, Brazell MP (1984) An introduction to in vivo electrochemistry. In: Marsden CA (ed) Measurement of neurotransmitter release in vivo. Wiley, Chichester, pp 127–151

Marsden CA, Martin KF (1985a) RU 24969 decreases 5-HT release in the SCN by acting on 5-HT receptors in the SCN but not the dorsal raphe. Br. J Pharmacol 86:219P

Marsden CA, Martin KF (1985b) In vivo voltammetric evidence that the 5-HT autoreceptor is not of the 5-HT_{1A} subtype. Br J Pharmacol 86:445P

Marsden CA, Martin KF (1986) Involvement of 5-HT_{1A} and alpha2 receptors in the decreased serotonin release and metabolism in rat suprachiasmatic nucleus after i.v. 8OH-DPAT. Br J Pharmacol 89:277–286

Marsden CA, Martin KF, Webb AJ (1985) Absence in mice of a diurnal variation in 5-HT_{1A} receptor function. J Pharm Pharmacol 37:[Proc Suppl] 155P

Martin KF (1982) Antidepressant drugs and 24-hour rhythms. Ph.D. Thesis. University of Bath

Martin KF (1991) Rhythms in neurotransmitter turnover: focus on the serotonergic system. Pharmac Ther 51:421–429

Martin KF, Crespi F, Marsden CA (1988) In vivo electrochemistry with carbon fibre electrodes – principles and application to neuropharmacology. Trends Anal Chem 7:334–339

Martin KF, Hannon SD, Phillips I, Heal DJ (1992) Opposing roles for 5-HT_{1B} and 5-HT_3 receptors in the control of 5-HT release in rat hippocampus in vivo. Br J Pharmacol 106:139–142

Martin KF, Marsden CA (1985) In vivo diurnal variations of 5-HT release in hypothalamic nuclei. In: Redfern PH, Campbell IC, Davies JA, Martin KF (eds) Circadian rhythms in the central nervous system. Macmillan, London, pp 81–94

Martin KF, Marsden CA (1986a) In vivo voltammetry in the suprachiasmatic nucleus of the rat: effects of RU 24969, methiothepin and ketanserin. Eur J Pharmacol 121:135–140

Martin KF, Marsden CA (1986b) Pharmacological manipulation of the serotonergic input to the SCN – an insight into the control of circadian rhythms. Ann N Y Acad Sci 473:542–545

Martin KF, Marsden CA (1988) In vivo identification of the serotonin (5-HT) autoreceptor in rat suprachiasmatic nucleus (SCN). In: Briley M, Fillion G (eds) New concepts in depression. Macmillan, London, pp 60–68

Martin KF, Marsden CA, Webb AR (1986) The behavioural response to the 5-hydroxytryptamine$_{1B}$ (5-HT_{1B}) receptor agonist, RU 24969, may exhibit a circadian variation in the mouse. Chronobiol Int 4:483–493

Martin KF, Phillips I, Hearson M, Prow MR, Heal DJ (1992) Characterization of 8-OH-DPAT-induced hypothermia in mice as a 5-HT_{1A} autoreceptor response and its evaluation as a model to selectively identify antidepressants. Br J Pharmacol 107:15–21

McLennen IS, Lees GJ (1978) Diurnal changes in the kinetic properties of tryptophan hydroxylase from rat brain. J Neurochem 31:557–559

Medanic M, Gillette MU, (1992) Serotonin regulates the phase of the rat suprachiasmatic pacemaker in vitro only during the subjective day. J Physiol 450:629–642

Middlemiss DN (1985) The putative 5-HT_1 receptor agonist, RU 24969, inhibits the efflux of 5-hydroxytryptamine from rat cerebral cortex slices by stimulation of the 5-HT autoreceptor. J Pharm Pharmacol 37:434–437

Moret C (1985) Pharmacology of the serotonin autoreceptor. In: Green AR (ed) Neuropharmacology of serotonin. Oxford University Press, New York, pp 21–49

Morin LP, Blanchard J (1991) Depletion of brain serotonin by 5, 7-DHT modifies hamster circadian rhythm response to light. Brain Res 566:173–185

Moser PC, Redfern PH (1985) Lack of variation over 24-hours in response to stimulation of 5-HT_1 receptors in the mouse brain. Chronobiol Int 2:235–238

Mundey MK, Fletcher A, Marsden CA, Fone KCF (1995) Effect of iontophoretic application of the 5-HT$_{1A}$ antagonist WAY100635 on neuronal firing in the guinea-pig dorsal raphe nucleus. Br J Pharmacol 115:84p

Natali JP, McRae-Degueurce A, Chouvet G, Pujol JF (1980) Genetic studies of daily variations of first step enzymes of monoamine metabolism in the brain of inbred strains of mice and hybrids. II. Daily variations of tyrosine hydroxylase activity in the locus coeruleus. Brain Res 191:205–213

Ozaki N, Duncan WC, Johnson KA, Wehr TA (1993) Diurnal variations of serotonin and dopamine levels in discrete brain regions of Syrian hamsters and their modification by chronic clorgyline treatment. Brain Res 627:41–48

Philo R, Rudeen K, Reiter RJ (1977) A comparison of the circadian rhythms and concentrations of serotonin and norepinephrine in the telencephalon of four rodent species. Comp Biochem Physiol 570:127–130

Prosser RA, Dean RR, Edgar DM, Heller HC, Miller JD (1993) Serotonin and the mammalian circadian system: I. In vitro phase shifts by serotonergic agonists and antagonists. J Biol Rhythms 8:1–16

Quay WB (1967) Twenty-four hour rhythms in cerebral and brainstem contents of 5-hydroxytryptamine in a turtle, Pseudemys scripta elegans. Comp Biochem Physiol 20:217–221

Quay WB (1968) Differences in circadian rhythms in 5-hydroxytryptamine according to brain region. Am J Physiol 215:1448–1453

Ramirez AD, Ramirez VD, Meyer DC (1987) The nature of in vivo 5-hydroxyindoleacetic acid output from 5-hydroxytryptamine terminals is related to specific regions of the suprachiasmatic nucleus. Neuroendocrinology 46:430–438

Redfern PH, Martin KF (1985) The effect of antidepressant drugs on 24-hour rhythms of tryptophan metabolism in the rat. Chronobiol Int 2:109–113

Roca AL, Weaver DR, Reppert SM (1993) Serotonin receptor gene expression in the rat suprachiasmatic nuclei. Brain Res 608:159–165

Sayer TJO, Hannon SD, Redfern PH, Martin KF (1994) Diurnal variation in nerve terminal 5-HT autoreceptor function in vivo: effects of antidepressant treatment. Br J Pharmacol 112:95P

Scheving LE, Harrison WH, Gordon P, Pauly JE (1968) Daily fluctuation (circadian and ultradian) in biogenic amines of the rat brain. Am J Physiol 214:166–173

Shen H, Semba K (1994) A direct retinal projection to the dorsal raphe nucleus in the rat. Brain Res 635:159–168

Sinei K, Redfern PH (1985) 24-Hour variation in synaptosomal tryptophan-5-hydroxylase activity in rat brain. In: Redfern PH, Campbell IC, Davies JA, Martin KF (eds) Circadian rhythms in the central nervous system. Macmillan, London, pp 193–198

Singh A, Redfern PH (1990) Inhibition of 5-HT$_{1B}$ autoreceptors by methiothepin does not reveal a circadian variation in autoreceptor sensitivity in the cerebral cortex of the rat. Br J Pharmacol 99:242P

Singh A, Redfern PH (1994a) Lack of circadian variation in the sensitivity of rat terminal 5-HT$_{1B}$ autoreceptors. J Pharm Pharmacol 46:366–370

Singh A, Redfern PH (1994b) Guinea pig 5-HT$_{1D}$ autoreceptors do not display a circadian variation in their responsiveness to serotonin. Chronobiol Int 11:165–172

Singh A, Redfern PH (1994c) Lack of circadian variation in the responsiveness of alpha-2-heteroreceptors regulating serotonin release. Chronobiol Int 11:94–102

Smale L, Michels KM, Moore RY, Morin LP (1990) Destruction of the hamster serotonergic system by 5, 7-DHT: effects on circadian rhythm phase, entrainment and response to triazolam. Brain Res 515:9–19

Sprouse JS, Aghajanian GK (1987) Electrophysiological responses of serotonergic dorsal raphe neurons to 5-HT$_{1A}$ and 5-HT$_{1B}$ agonists. Synapse 1:3–9

Stanley BG, Schwartz DH, Hernandez L, Hoebel BG, Leibowitz SF (1989) Patterns of extracellular norepinephrine in the paraventricular hypothalamus: relationship to circadian rhythm and deprivation-induced eating behavior. Life Sci 45:275–282

Stanley BG, Schwartz DH, Hernandez L, Leibowitz SF, Hoebel BG (1987) Diurnal rhythm of medial hypothalamic serotonin metabolism in relation to eating behaviour. Soc Neurosci Abstr 13:9.7

Takahashi K, Shimoda K, Yamada N, Sasaki Y, Hayashi S (1986) Effect of dorsal midbrain lesion in infant rats on development of circadian rhythm. Brain Dev 8:373–381

Taylor PL, Garrick NA, Burns RS, Tamarkin L, Murphy DL, Markey SP (1982) Diurnal rhythms of serotonin in monkey cerebrospinal fluid. Life Sci 31:1993–1999

Tominaga K, Shibata S, Ueki S, Watanabe S (1992) Effects of 5-HT_{1A} receptor agonists on the circadian rhythm of wheel-running activity in hamsters. Eur J Pharmacol 214:79–84

Ungerstedt U (1984) Measurement of neurotransmitter release by intracranial dialysis. In: Marsden CA (ed) Measurement of neurotransmitter release in vivo. Wiley, Chichester, pp 81–106

van den Pol AN, Tsujimoto KL (1985) Neurotransmitters of the hypothalamic suprachiasmatic nucleus: immunocytochemical analysis of 25 neuronal antigens. Neuroscience 15:1049–1086

van der Gugten J, Slangen JL (1975) Norepinephrine uptake by hypothalamic tissue from the rat related to feeding. Pharmacol Biochem Behav 3:855–860

Wilkinson LO, Martin KF, Auerbach SB, Marsden CA, Jacobs BL (1988) Relationship between dialysate serotonin and raphe unit activity in freely moving cats. Br J Pharmacol 95:872P

Williams JH, Miall-Allen VM, Klinowski M, Azmitia EC (1983) Effects of microinjections of 5, 7-dihydroxytryptamine in the suprachiasmatic nuclei of the rat on serotonin reuptake and circadian variation of corticosterone levels. Neuroendocrinology 36:431–435

CHAPTER 8
Rhythms in Pharmacokinetics: Absorption, Distribution, Metabolism, and Excretion

P.M. BÉLANGER, B. BRUGUEROLLE, and G. LABRECQUE

A. Introduction

Pharmacokinetics is the discipline that quantifies the processes concerning the changes of drug concentrations within the body, whereas pharmacodynamics is the study of the biological response to a drug. Hence, the absorption, distribution, metabolism, and excretion of a drug are pharmacokinetic processes that are described in vivo by linear mathematical expressions of drug concentrations related to the times after ingestion. Figure 1 shows the relationship of these events after oral drug administration. Different parameters such as absorption and elimination rate constants, volume of distribution, and clearance are derived from these equations and quantitatively describe each one of these processes for a particular drug under various clinical situations. Over the years, the use of these parameters has provided the basis for the study of the various factors modifying the rates of absorption, distribution, metabolism, and excretion, such as the physicochemical properties of drugs, routes and forms of administration, age, diseases, race, enzyme induction or inhibition, drug interactions, etc. Drugs are usually administered to humans once or many times at different hours of the day, depending on their rate of absorption and elimination, and the times of drug ingestion are randomly selected but convenient to the active period of the patient and to the medical attendants. The aim of the pharmacokineticist is to achieve a steady state blood level of the drug and hence a constant effect by sequential administration of the same dose of the agent at different times of the day.

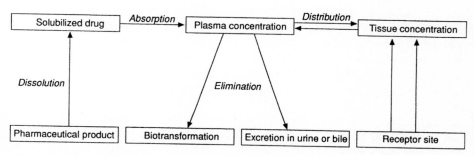

Fig. 1. Schematic representation of drug disposition following oral administration

The various parameters that describe the different pharmacokinetic processes are considered as time invariant under standard conditions, and the models used to derive them do not provide for temporal variations in plasma volume, blood flow rates through various organs, metabolic activities, or other physiologic processes. Obviously, in the presence of some important circadian variations in these variables, steady state levels of drug in the body could not be achieved using the present pharmacokinetic approach. During the past 20 years, numerous studies have shown that the pharmacokinetic properties, therapeutic effects, and toxicity of many drugs vary depending of the hour of the day they are administered in both animals and humans (REINBERG et al. 1990; LABRECQUE and BÉLANGER 1991; BRUGUEROLLE and LEMMER 1993; BÉLANGER 1993; LEMMER and BRUGUEROLLE 1994 and references therein). Moreover, seasonal and monthly variations in the above determinations of drug activity have also been reported in rodents and humans (BÉLANGER 1993). This chapter will deal essentially with the circadian variations in the absorption, distribution, hepatic metabolism, and renal excretion of drugs, and the mechanisms responsible for the time dependency of these pharmacokinetic processes will be emphasized.

B. Absorption

I. Parameters and Modifying Factors

Absorption refers to a series of processes that determine the rate and extent to which a drug is transferred from its pharmaceutical formulation to the circulating blood after administration to animals and humans. Mechanisms of drug absorption may vary depending on the compound, and involve passive or facilitated diffusion, active transport, ion pairing and pinocytosis (SCHANKER 1972). Four parameters are important in the analysis of drug absorption and are derived from a single-dose blood level curve. They are the absorption rate constant (λ_a or ka), the maximal plasma or serum concentration (C_{max}), the time for achieving the C_{max} (t_{max}), and the area under the plasma or serum concentration–time curve (AUC). The first three parameters are indicative of the rate whereas the AUC value estimates the amount of drug absorbed. The AUC value is particularly useful in bioavailability studies for the evaluation of different formulation products of the same compound, but its use in chronopharmacokinetics is limited because few studies, if any, determine the ratio of AUC obtained after intravenous and extravascular administration of a drug at the same time of day. Table 1 lists a series of factors which may affect the rate and/or the extent of drug absorption by extravascular routes. They are classified into two types: the physicochemical characteristics of the drug and its dosage form, and the various biological factors. Any one of these factors may be rate-limiting in the absorption of a particular drug and thus provide the basis for the mechanism(s) of potential temporal variation in the above pharmacokinetic parameters.

Table 1. Factors modifying the extravascular absorption of drugs

Drug and dosage form
Aqueous and lipid solubility
pKa
Stability
Molecular weight
Surface area of particles
Crystal size
Polymorphic form
Solvates
Salt form
Excipient(s) in formulation
Biological function
Blood flow to organ of absorption
Gastric emptying
Intestinal transit
Motility and peristalsis
pH
Enzyme(s)
Bile secretion
Surface area
Permeability of membranes
Specialized transport
Presence of food

II. Inorganic Compounds and Macromolecules

Very few reports have been published on the temporal variation in the absorption of inorganic compounds and macromolecules, and the mechanisms of their absorption is poorly understood. TARGUINI et al. (1979) reported a circadian variation of oral absorption of iron in healthy humans with acrophase value in the evening in both male and female. They suggested that the increased iron absorption in the evening is correlated with a greater iron binding capacity and a reduced sideraemia and hemoglobin levels. WROBEL and NAGEL (1979) described a circadian rhythm of considerable amplitude for the active transport of calcium in rat intestine with peak and trough values at 22.00 h and 09.00 h, respectively. They furthermore reported that both the light-dark cycle and the time of food presentation are important synchronizers of rhythm. On the other hand, AUNGST and FUNG (1981) failed to detect any diurnal variation in intestinal lead absorption in rats. MARKIEWICZ et al. (1981) determined the oral absorption of vitamin B_{12} at different times of day in volunteers and reported a greater absorption at 13.00 h than at 01.00 h, 07.00 h, or 19.00 h.

III. Organic Compounds

Most organic compounds such as drugs, food additives, and toxic agents are absorbed into the blood from their site of administration mainly by passive diffusion based on the pH-partition hypothesis. This mechanism of absorption applies for compounds given through the buccal, dermal, intramuscular, nasal, oral, pulmonary, rectal or vaginal route (DENNIS 1990). Table 2 lists a series of drugs that have a proven effect on the increase in the rate of absorption when administered in the morning, i.e., between 06.00 h and 14.00 h, to humans (BÉLANGER 1993; BRUGUEROLLE and LEMMER 1993; LEMMER and BRUGUEROLLE 1994 and references therein). These studies involved the oral administration of a single dose of the drug given as an immediate-release tablet, capsule, or solution to healthy volunteers after a period of fasting at different times of the day. In all cases the drug led to either an increase in C_{max} or a lower t_{max} value, or both, and in some cases resulted in a greater absorption rate constant or shorter absorption half-life during the morning, which was significantly different ($p < 0.05$) from the parameter determined at any other time of day in the same subject when using crossover design trials. The only exceptions to this rule were hexobarbital and isosorbide dinitrate, showing characteristics of increased absorption at 18.00 h and 02.00 h, respectively (BÉLANGER 1993). Also, there was no temporal variation in the oral absorption of acetaminophen, gentamicin, midazolam, or nortriptyline (BÉLANGER 1993; BRUGUEROLLE and LEMMER 1993). These results emphasize the importance of the time of day in bioavailability studies. The mechanism(s) of the increase rate of drug absorption in the morning has never been characterized, but the knowledge of the physicochemical properties of the drug and of the chronobiological variations of the digestive system could explain many of the results obtained in humans.

Table 2. Drugs known to be absorbed faster during the morning when administered orally as a single dose to young human volunteers in a regular formulation, i.e., nonsustained-release dosage form[a]

Acetylsalicylic acid	
Aminophylline	
Amitriptylline	Lorazepam
Dexamethasone	Mequitazine
Diazepam	Nifedipine
Diclofenac	Nitrendipine
Digoxin	
Dipyridamole	Pranoprofen
	Prednisolone
Enalapril	Propranolol
Ethanol	
	Salicylic acid
Indomethacin	Theophylline
Isosorbide-5-mononitrate	Triazolam
Ketoprofen	Valproic acid

[a] Each drug has at least one parameter of absorption (C_{max}, t_{max}, and λ_a) determined in the morning (06.00 h–12.00 h) and significantly different than at any other time of day, obtained by using a crossover study in the same subjects, i.e., greater C_{max} and λ_a and shorter t_{max} values.

IV. Factors Modifying the Temporal Variation in Gastrointestinal Absorption

1. Solubility

The absorption of lipophilic drugs by passive diffusion from the gastrointestinal tract is based on the concentration gradient of the solubilized drug between the intestinal lumen and the blood. Thus, a solid pharmaceutical formulation must first be desintegrated and the drug particles solubilized in an aqueous mixture at the absorption site to initiate the diffusion process. Table 3 gives the C_{max} and t_{max} values determined in the morning (08.00–09.00 h) and evening (20.00–21.00 h) of four acidic drugs, i.e., furosemide, hydrochlorothiazide, indomethacin, and phenylbutazone, three of which have similar pKa values after administration of a single oral dose in rats at both times of day. Furosemide and hydrochlorothiazide were given as homogeneous aqueous solutions while suspensions of powdered indomethacin and phenylbutazone were used. There was no temporal variation in the C_{max} values of furosemide and hydrochlorothiazide, but the t_{max} values were shorter in the evening.

On the other hand, indomethacin and phenylbutazone have a greater C_{max} and shorter t_{max} values at 20.00–21.00 h. These results are in agreement with those of human studies, considering that rats are nocturnal animals and that their activity period is inversely related to that of humans. An important feature of these results is that the most prominent diurnal variations in C_{max} and t_{max} values were obtained with the drugs given as suspensions, indicating that the rate of solubilization of the two drugs is higher in the evening. SHIGA et al. (1993) tested this hypothesis recently in 13 hypertensive subjects. They studied

Table 3. C_{max} and t_{max} values of some drugs given as single oral dose in rat at 08–09.00 h or 20–21.00 h in an aqueous solution or as a suspension[a]

Drug	pKa	Formulation	C_{max} (µg/ml)		t_{max} (min)	
			08–09.00 h	20–21.00 h	08–09.00 h	20–21.00 h
Furosemide	4.7	Solution	10.15 ± 0.5	10.9 ± 0.5	120	60
Hydrochlorothiazide	7.0 and 9.2	Solution	9.7 ± 0.5	8.8 ± 0.3	180	60
Indomethacin	4.5	Suspension	16.7 ± 4.3[b]	25.8 ± 3.4	120	20
Phenylbutazone	4.4	Suspension	146.7 ± 18.4[b]	190.7 ± 18.0	60	40

[a] Adapted from BÉLANGER et al. (1984). Adult male Sprague-Dawley rats (225–325g) were synchronized on a L:D12:12 lights-on at 07.00 h. The animals were treated orally after a 12-h fast with 30 mg/kg furosemide or 50 mg/kg hydrochlorothiazide given as homogeneous solutions obtained by dissolution of commercial tablets in normal saline or 3 mg/kg indomethacin and 30 mg/kg phenylbutazone as suspensions in 2% acacia solution in normal saline. The serum levels of each drug as mean values ± SEM were determined in at least six rats for each selected time post-drug ingestion. C_{max} and t_{max} values were determined graphically from the serum concentrations-versus-time plot.
[b] $p < 0.01$ from data at 20–21.00 h.

Table 4. C_{max} and t_{max} values of propranolol and atenolol determined after a single oral dose at 09.00 h and 21.00 h in 13 hypertensive subjects[a]

Parameter	Propranolol		Atenolol	
	09.00 h	21.00 h	09.00 h	21.00 h
C_{max} (mg/ml)	17.7 ± 8.6[b]	11.9 ± 4.0	440.0 ± 148.2	391.8 ± 122.9
t_{max} (h)	3.2 ± 1.3[b]	4.0 ± 1.7	3.1 ± 1.1	4.1 ± 1.5

[a] Adapted from SHIGA et al. (1993). Immediate-release tablets of 20 mg propranolol and 50 mg atenolol were used. Results are mean ± SD.
[b] $p < 0.05$ compared with 21.00 h.

the pharmacokinetics of a single oral dose of the lipophilic ß-blocker propranolol, pKa 9.5, and of the hydrophilic agent atenolol, pKa 9.6, given at 09.00 h and 21.00 h to the same subjects in a crossover design. The results of the C_{max} and t_{max} values summarized in Table 4 show that only the absorption of propranolol was greater and faster in the morning, as indicated by the temporal variation in C_{max} and t_{max} values.

2. Pharmaceutical Formulation

An interesting feature reported in many chronopharmacokinetic studies involving many of the drugs listed in Table 2 is that the diurnal variation in their rate of gastrointestinal absorption is abolished when they are administered in a sustained-release formulation or as a rectal suppository. This phenomenon has been shown for indomethacin (TAGGART et al. 1987), isosorbide-5-mononitrate (SCHEIDEL and LEMMER 1991; LEMMER et al. 1991), nifedipine (LEMMER et al. 1991) and valproic acid (YOSHIYAMA et al. 1989). It is not a general rule, as some sustained-release formulations of indomethacin (GUISSOU et al. 1983; BRUGUEROLLE et al. 1986), ketoprofen (REINBERG et al. 1986) theophylline (SMOLENSKY 1989), and verapamil (JESPERSEN et al. 1989) showed temporal variations in the oral absorption in humans which, in many cases, is different from the diurnal pattern of the immediate-release form of the drug. Theophylline is the best-documented drug, as there are many sustained-release formulations marketed in different countries. Some formulations of the bronchodilator agent have characteristic nychtemeral changes in absorption parameters while others do not (SMOLENSKY 1989 and references therein). It appears from these studies that the rate of drug solubilization from the pharmaceutical product is also a determining factor in the temporal absorption of many drugs.

3. Secretion and Motility

Bile secretion and motility of the gastrointestinal tract are two factors known to enhance the rate of drug dissolution, and circadian rhythms have been reported for both in man. VAN BERGE HENEGOWEN and HOFFMANN (1978)

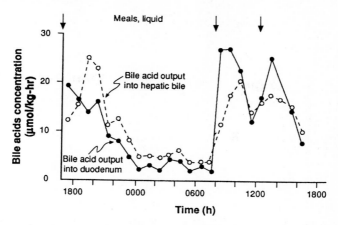

Fig. 2. Temporal variation in the secretion of bile acids into the duodenum (–•–) and into the hepatic bile (--o--) in five healthy male subjects. (Adapted from van Berge Henegowven and Hofmann 1978)

were the first to report that the secretion of bile acids into bile and the duodenum followed a rhythmic pattern related to the intake of a liquid meal in healthy volunteers (Fig. 2). Maximal concentration of bile acids was secreted between 08.00 h and 14.00 h corresponding to the period of day of maximal absorption. Independent of the ingestion of a meal, it was later reported that the synthesis of the various bile acids followed a circadian rhythm in humans with maximal activity in the morning and nadir in late afternoon and evening (Pooler and Duane 1988). Remarkably, similar rhythms in bile flow and secretion have been described in rats whose rates were highest at midnight or during the activity period of the animal and lowest at noon during sleep (Duane et al. 1979; Nakano et al. 1990; Gilberstadt et al. 1991). Also, it has been shown that both the rate of gastric emptying (Fig. 3; Goo et al. 1987) and the motility of the small intestine as determined by the propagation of a migrating myoelectric complex (Fig. 4; Kumar et al. 1986) are greater in the morning. Other factors such as the secretion of acid and gastric fluids and the activity of many enzyme systems of the intestinal membrane have been shown to vary with the time of day (Vener and Moore 1987). Thus, the higher concentration of bile acids secreted into the intestine and the greater rate of gastric emptying and intestinal motility during the morning parallel the time-dependent increase in the absorption parameter of most drugs in humans.

4. Blood Flow

Circadian changes in the blood flow or cardiac output to the gastrointestinal tract could also explain the time-dependent variations in the absorption rate of drugs. Studies carried out in adult rats receiving strontium-labelled microspheres i.v. at four different hours of the day (Labrecque et al. 1988) indicated

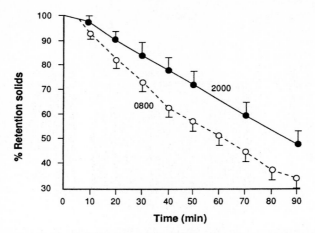

Fig. 3. Gastric emptying determined as the percentage of retention of a solid meal in 16 healthy male subjects at 08.00 h and 20.00 h. (Adapted from Goo et al. 1987)

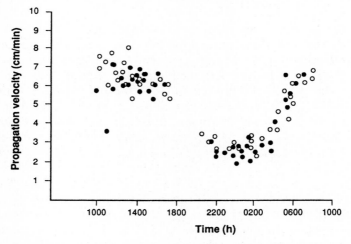

Fig. 4. Temporal variation in the propagation velocity of migrating motor complexes in ten volunteers (o) and in ten patients with irritable lowel syndrome (•). (Adapted from Kumar et al. 1986)

that peak concentration of radioactivity in the intestine was maximal at 21:00 h and 3:00 h, i.e., during the activity periods of the animal, and minimal during day time (Fig. 5). These results showed that the intestine received a higher percentage of the blood supply during the rats' activity period. Also, Larsen and coworkers (1991) reported the circadian rhythm in gastric mucosal blood flow of rat stomach. The temporal pattern was similar to that of the intestine with peak blood flow occurring during the night in the various parts of the stomach and nadir during daytime.

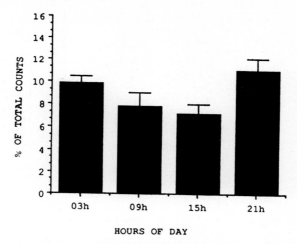

Fig. 5. Temporal variation in the distribution of labelled microspheres to the intestine of rats. Four groups of rats received a bolus dose of Strontium-labelled microspheres four times a day through the aorta. The tissue radioactivity was determined 5 min after injection. Results are mean ± SD

C. Distribution

The factors affecting the distribution of drugs are: body size and composition, blood flow and the cardiac output to various organs, binding to plasma proteins, and membrane permeability to drugs. Of these, the plasma protein binding of a drug is the main factor that controls its diffusion from the plasma to the various organs and tissues. It is the unbound or free fraction of the drug in plasma which diffuses into the tissue compartment, and the volume of distribution of a drug extensively bound (> 95%) to plasma proteins is proportional to its free fraction (BÉLANGER 1993).

I. Plasma Protein Binding

Many studies have described the circadian rhythms of the plasma proteins in humans, especially of albumin and α-1-glycoprotein, the two proteins involved in drug binding (REINBERG et al. 1977); BRUGUEROLLE et al. 1986; HANS et al. 1987; FOCAN et al. 1988; KANABROCKI et al. 1988). In general, peak concentration occurs during the early afternoon and trough during the night. The difference between the diurnal peak and trough amounts to only 10% in young adults but may be as great as 20% in elderly subjects (TOUITOU et al. 1986). Rhythms of these proteins are independent of that plasma hematocrit and probably reflect the temporal variation of hepatic protein synthesis (TOUITOU et al. 1986).

Table 5 summarizes the different studies on the temporal variations in plasma protein binding of drugs in rat and humans. From the limited data available, the temporal variation in drug binding appears to be dependent on

Table 5. Temporal variations in plasma protein binding of drugs[a]

Drug	Species	Temporal variation	References
Carbamazepine	Rat	Lowest free fraction at 04.00 h when albumin concentration is highest	Bruguerolle et al. 1980
Carbamazepine	Human	Highest free fraction between 14.00h and 20.00 h	Riva et al. 1984
Cisplatin	Human	Highest protein binding at 16.00 h	Hecquet et al. 1984
Diazepam	Human	Highest bound fraction at 09.00 h	Naranjo et al. 1980
Diazepam	Human	Highest bound fraction at 09.30 h	Nakano et al. 1984
Disopyramide	Rat	Lowest free fraction at 22.00 h	Bruguerolle et al. 1980
Indomethacin	Human	No temporal change in elderly subjects	Bruguerolle et al. 1986
Lidocaine	Rat	Lowest free level at 22.00 h	Bruguerolle et al. 1982
Phenytoin	Human	No temporal change	Patel et al. 1982
Prednisolone	Human	Highest protein binding at 06.00 h and lowest at 18.00 h	English et al. 1983
Prednisolone	Human	Highest binding at 00.00 h and lowest at 08.00 h	Angeli et al. 1978
Propranolol	Rat	Highest binding at 16.00 h and 00.00 h	Haen et al. 1985
Valproic acid	Human	Highest free fraction between 02.00 h and 08.00 h	Patel et al. 1982
Valproic acid	Human	Unbound clearance is 15% higher during the evening in elderly subjects	Bauer et al. 1985

[a]Adapted from Bruguerolle (1989).

the characteristics of the drug and protein involved. The time dependency of parameters such as the number of drug binding sites and the binding affinity constant has never been studied, nor have the effects of various factors such as age, sex, diet, feeding schedule, and pathological conditions have, rarely been investigated. All these factors are important in drug–protein interactions (Tillement et al. 1984; Lindup 1987). Only two studies established a positive correlation between the extent of drug binding and concentration of total plasma proteins in humans (Patel et al. 1982; Hecquet et al. 1984) although similar correlations have been described for many drugs in rat (Valli et al. 1980; Bruguerolle 1984 a, b). Naranjo et al. (1980) showed that the free fraction of diazepam was higher between 23.00 h and 8.00 h and lower during the day after 09.00 h (Fig. 6). The increase in free fraction was paralleled by a concomitant decrease in the total plasma concentration of diazepam and of its major metabolite, N-desmethyldiazepam. Food intake was associated with increases in total plasma concentrations of diazepam and decreases in its free fraction. Similar findings were reported by Nakano et al. (1984). Patel et al.

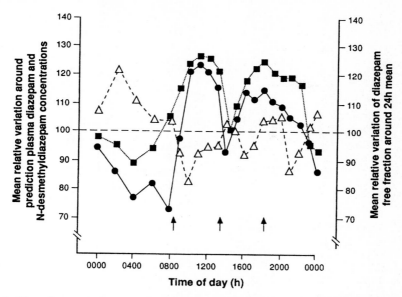

Fig. 6. Circadian variation in the total plasma concentration of diazepam (●) and N-desmethyldiazepam (■), and in the unbound concentrations of diazepam (△) in six healthy subjects. *Arrows* indicate meal intakes. The results are expressed as mean percentage value of relative deviation compared to the 24 h mean setup at 100% (- -). (Adapted from NARANJO et al. 1980)

(1982) showed that the increase in free fraction of valproic acid between 02.00 h and 08.00 h correlated with an increased concentration of free fatty acids known to displace the drug from its binding sites. These findings are clinically relevant since the total plasma concentration of valproic acid is lower and its clearance higher during the evening and night (PATEL et al. 1982; LOISEAU et al. 1982; BAUER et al. 1985). Furthermore, LOCKARD et al. (1985) found that the diurnal fluctuations of valproic acid levels in spinal fluid reflected those of the plasma concentrations after constant infusion of the drug in monkey. ANGELI et al. (1978) describes the circadian rhythm in the plasma binding of prednisolone to corticosteroid-binding globulin or transcortin and showed that the diurnal variability was inversely related to circadian rhythms of serum cortisol (Fig. 7).

II. Binding to Erythrocytes

An interesting model developed by BRUGUEROLLE and coworkers is the use of red blood cells as a tissue compartment for the diffusion of the unbound plasma concentration of drugs at different times of day. They studied the circadian periodicity of drug binding to erythrocytes of many local anesthetics (BRUGUEROLLE and JADOT 1983; BRUGUEROLLE and PRAT 1987, 1988, 1989,

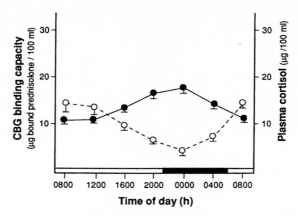

Fig. 7. Mean (±SEM) plasma concentrations of cortisol (--o--) and serum corticosteroid-binding capacity of prednisolone to globulin (*CBG*, —•—) determined at different times of day in ten normal subjects. (Adapted from ANGELI et al. 1978)

1990), indomethacin (BRUGUEROLLE et al. 1986a) and theophylline (BRUGUEROLLE 1987). For example, rats to whom lidocaine was administered at 22.00 h gave a ratio of the drug concentration in erythrocytes to total plasma concentration of 0.74 compared to a ratio of 0.48 when the anesthetic was given at 10.00 h (BRUGUEROLLE and JADOT 1983).

D. Biotransformation

Most drugs and foreign compounds are eliminated from the body by biotransformation or drug metabolism through a variety of chemical reactions catalyzed by enzyme systems predominantly concentrated in mammalian liver. Oxidation and conjugations to endogenous substrate are the two most important reactions, and each one has its characteristic properties (JAKOBY 1980). In particular, each reaction is catalyzed by various isomeric forms of enzyme with different substrate requirements and kinetic properties. Extensive reviews on the chronobiological variations in hepatic drug metabolism and elimination are available (BÉLANGER 1987, 1993; BÉLANGER and LABRECQUE 1989; LABRECQUE and BÉLANGER 1991).

I. Oxidation: The Cytochrome P-450 Monooxygenase

Of all the metabolic reactions involved in drug biotransformations, oxidative reactions are the most common and important and the enzymes involved are the best studied, both in the diversity of their substrates and the biological consequences of their products. Oxidative reactions are in many cases the rate-limiting processes in the elimination or clearance of the drug or toxic agent from the body and they are the locus of many drug interactions of clinical

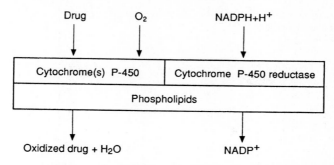

Fig. 8. Schematic representation of the cytochrome P-450 monooxygenase system

relevance. These reactions are mediated by a microsomal monooxygenase or mixed-function system composed of three essential components (Fig. 8): two proteins, cytochrome P-450 and its flavoprotein reductase, and a phospholipid fraction (WINTERS and CEDERBAUM 1993).

RADZIALOWSKI and BOUSQUET (1967, 1968) followed by NAIR and CASPER (1969) were the first two groups to report on the circadian rhythm of hepatic microsomal metabolism in mouse and rat. The highest and lowest values of oxidase activities were obtained during late evening (22.00 h–02.00 h) and morning (10.00 h–14.00 h), respectively, with percentages of variation of about 40%. BÉLANGER and LALANDE (1988) determined the oxidase activities of different substrates as well as the concentration of the microsomal component at 09.00 h and 21.00 h in rat liver (Table 6). These determinations were all performed in purified microsomes isolated from the same rat livers precluding the interindividual variations. The activities of the aminopyrine and p-nitroanisole demethylases and aniline hydroxylase were higher at 21.00 h than at 09.00 h whereas those of the biphenyl and testosterone 7α- and 6β-hydroxylases and of the P-450 reductase were lower in the evening. There was no diurnal variation in testosterone 16α-hydroxylase and in the microsomal concentration of total cytochrome P-450 and protein. Of particular interest, a significant day-night variation in the microsomal concentration of fatty acids was reported for the first time. The levels of the various lipids were much higher at 09.00 h than in the evening except for arachidic acid of which measurable amounts were only obtained at 21.00 h. The activity of each oxidase assayed is mediated selectively by at least one cytochrome P-450 isoenzyme (GUENGERICH 1987; RYAN and LEVIN 1990), and the fact that the temporal variation in activities varied depending on the substrate added in the microsomes strongly indicates different diurnal kinetic properties or concentrations in some P-450 isoenzymes. These results were further confirmed by the recent study of MIYAZAKI et al. (1990) who determined the temporal variations of total P-450 concentration and of some known isoenzymes, IIBl, IIC6, and IIC11, by Western blotting along with the regioselectivity and enantioselectivity of testosterone hydroxylases at the 2α-, 6β-,

Table 6. Diurnal variations in the activity and composition of the microsomal monooxygenase of rat liver[a]

Microsomal activity or component	Time of day		p-value
	09.00 h	21.00 h	
Aminopyrine N-demethylase (nmol/mg 30 min)	29.6 ± 3.4[b]	37.4 ± 0.5	< 0.018
Aniline hydroxylase (nmol/mg 30 min)	6.8 ± 0.3	11.3 ± 0.4	< 0.001
p-Nitroanisole O-demethylase (nmol/mg 30 min)	10.2 ± 0.6	12.2 ± 0.9	< 0.037
Biphenyl 4-hydroxylase (μg/mg 30 min)	3.1 ± 0.1	1.8 ± 0.2	< 0.001
Testosterone hydroxylases (μg/mg 30 min)			
7α-hydroxylase	27.3 ± 5.2	18.2 ± 1.2	< 0.050
6β-hydroxylase	15.0 ± 2.7	10.0 ± 0.4	< 0.042
16α-hydroxylase	14.3 ± 2.5	12.5 ± 1.0	< 0.278 (NS)[c]
Cytochromes P-450 (nmol/mg protein)	0.47 ± 0.03	0.33 ± 0.08	< 0.700 (NS)[c]
Microsomal protein (mg/250 mg of liver)	7.3 ± 0.4	8.5 ± 0.08	< 0.056 (NS)[c]
Cytochrome P-450 reductase (nmol/mg min)	44.0 ± 1.7	33.0 ± 1.5	< 0.001
Fatty acid concentration (μg/g of liver)			
Palmitic acid	571 ± 52	366 ± 36	< 0.004
Stearic acid	935 ± 105	580 ± 74	< 0.010
Linoleic acid	413 ± 53	142 ± 25	< 0.001
Arachidic acid	0	128 ± 5	< 0.001
Arachidonic acid	576 ± 67	281 ± 51	< 0.003

[a] Adapted from BÉLANGER and LALANDE (1988).
[b] Results are expressed as mean value ± SEM of triplicate assays carried out in three animals.
[c] Results are considered nonsignificantly different.

16α- and 16β- positions in rat liver microsomes (Fig. 9). The difference in the temporal patterns of total P-450 concentration and that of the isoenzymes was pronounced. Likewise, typical chronobiological variations were obtained in the rate of the various hydroxylations of testosterone at different sites, i.e., carbon 2, 6 or 16, or on the same carbon molecule but with opposite conformation of the product formed, i.e., 2α versus 2β and 16α versus 16β. The poor correlation between the oxidase activities and the total microsomal concentration of cytochrome P-450 is also evident. More research is needed in this area to further characterize the temporal variation of the cytochrome P-450 isoenzymes and their marker activities.

The fact that most drugs are metabolized to various oxidized products involving many hepatic cytochrome P-450 isoenzymes that exhibit different

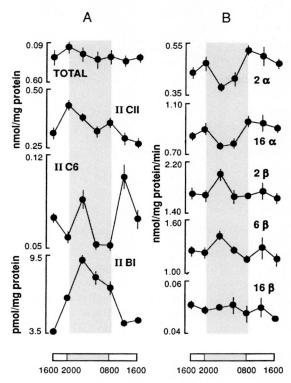

Fig. 9A,B. Circadian variations in the microsomal concentrations of total cytochrome P-450 and of P-450 IIB1, IIC6 and IIC11 isomeric forms (**A**) and the activity of testosterone 2α-, 2β-, 6β-, 16α- and 16β-hydroxylases (**B**) in rat liver. Data are means ± SEM. (Redrawn from MIAZAKI et al. 1990)

and sometimes opposite temporal variations is probably the main reason why so many pharmacokinetic studies failed to show significant diurnal changes in elimination parameters such as plasma half-life and non-renal clearance. However, the in vivo characterization of selected cytochrome P-450 isoenzymes can be studied in animals and humans with a drug that is metabolized to one or two major metabolites in a reaction catalyzed by one specific family of cytochrome P-450 isoenzymes, provided that it has a low hepatic extraction ratio (< 0.3) according to the classification of WILKINSON and SHAND (1975). Known examples of these compounds are antipyrine, carbamazepine, phenylbutazone, phenobarbital, phenytoin, and tolbutamide (ROWLAND and TOZER 1980). In 1980, BÉLANGER et al. reported a 40% decrease in the serum half-life of antipyrine and, inversely, an equivalent increase in the metabolic clearance when the drug was administered at 20.00 h rather than at 08.00 h in rats. LOUBARIS et al. (1984) also determined higher plasma levels of phenylbutazone and its major hydroxylated metabolite at 20.00 h and 01.00 h than at 07.00 h and 13.00 h. Conversely, in humans, SWARSON et al. (1982) determined higher plasma levels of the sulfone metabolite of sulindac when the anti-inflammatory agents were given at 09.00 h rather than 21.00 h. Diurnal variations in the human pharmacokinetics of aminopyrine was reported by POLEY et al. (1978) with a 50% increase in half-life and a 20% decrease in clearance value when the drug was given at 20.00 h rather than at 08.00 h. Furthermore, chronopharmacokinetic variations were also obtained in humans with antipyrine (VESELL et al. 1977), caffeine (LEVY et al. 1984), carbamazepine (HARTLEY et al. 1991), methotrexate (KOREN et al. 1992) and phenytoin (GARRETTSON and JUSKO 1974). All of these studies support the hypothesis that biological rhythms affect the hepatic microsomal metabolism of drugs in both animals and humans.

II. Conjugations

1. Conjugations to Acetate, Glucuronic Acid and Sulfate

Very few studies have dealt with the chronovariation in non-oxidative reactions involved in drug metabolism (BÉLANGER 1987). Only two reports described the diurnal variations in the properties of the transferases involved in the conjugation of acetate, glucuronic acid and sulfate for selected in vitro substrates in rat liver homogenates (Table 7; BÉLANGER et al. 1985, 1989). The rates of acetylation and glucuronide conjugation were much higher at 21.00 h

Table 7. Enzyme/substrate characteristics of the N-acetyltransferase, glucuronide and sulfate congugations of rat liver at 09.00 h and 21. 00 h[a]

Enzyme (cellular homogenate used)	Substrate	Kinetic parameter[a] and amount of protein	Time of day	
			0900h	2100h
N-Acetyltransferase (soluble fraction)	Isoniazid	V (nmol/mg 20 min)	59.0 ± 4.4	135.3 ± 5.5^b
		Protein (mg/g of liver)	138.8 ± 5.2	136.0 ± 4.9
Glucuronosyl-transferase (microsomes)	p-Nitrophenol	Vmax (nmol/g 15 min)	5.6 ± 1.2	8.8 ± 0.8^b
		Km (mM)	0.15 ± 0.05	0.20 ± 0.03
		Protein (mg/g of liver)	31.2 ± 0.8	24.4 ± 2.4^b
Sulfotransferase (soluble fraction)	Phenol	Vmax (μmol/g 5min)	3.5 ± 0.1	1.6 ± 0.2^b
		Km (mM)	0.18 ± 0.01	0.04 ± 0.02^b
		Protein	150.8 ± 8.0	154.0 ± 4.0

[a]Adapted from BELANGER et al. (1985, 1989); V, reaction velocity; Vmax, maximal reaction rate of an enzymatic reaction; Km, Michaelis constant representing the substrate concentration at half the maximal rate of the reaction.
[b]$p < 0.05$.

than at 09.00 h, and inversely the formation of phenol sulfate was twice as high at 09.00 h. These temporal variations in transferase activities were independent of the protein concentration used. The increase in sulfotransferase activity in the morning was associated with a more than fourfold decrease in the apparent affinity of the substrate as estimated by the Km values. The in vitro chronovariation in N-acetyltransferase activity is in agreement with the increased clearance of isoniazid at 21.00 h compared to the value obtained at 09.00 h (BÉLANGER et al. 1989), while those of hepatic glucuronidation and sulfation may explain the chronopharmacokinetics of acetaminophen (paracetamol) in humans and rat (SHIVELEY and VESELL 1975; BÉLANGER et al. 1987). In man, the plasma half-life of acetaminophen was 15% longer at 06.00 h than at 14.00 h, and the mean ratio of the glucuronide conjugate over unchanged acetaminophen excreted in the first 3.5 h urine sample varied from 5.2 at 06.00 h to 7.8 at 14.00 h (SHIVELEY and VESELL 1975). Inversely in rat, the oral clearance and the total metabolism of acetaminophen determined as the extraction ratio were both greater in the evening than in the morning (BÉLANGER et al. 1987).

2. Conjugation to Glutathione

Glutathione is the major nonprotein sulfhydryl compound in the cell (MEISTER and ANDERSON 1983; KAPLOWITZ et al. 1985). The reduced form of this tripeptide has two main cellular functions. First, it serves as a substrate for a peroxidase redox system which removes peroxides formed by other enzyme systems. Secondly, and most important to pharmacologists and toxicologists, it forms an adduct with electrophilic compounds, free-radical species or reactive intermediates of toxic drugs produced by the cytochrome P-450 monooxygenase system. Figure 10 shows the cytochrome P-450 dependent bioactivation of chloroform to free radicals and to the reactive metabolite phosgene that in turn reacts irreversibly with cellular macromolecules, resulting in toxicity, or with reduced glutathione to a nontoxic adduct (DESGAGNÉ et al. 1988). Thus, conjugation to glutathione is a desintoxication process protecting the cellular components from reactive oxygen species or foreign chemicals. LAVIGNE et al. (1983) reported a prominent circadian rhythm in the acute hepatotoxicity of chloroform in rat (Fig. 11) that was found to be inversely related to the circadian rhythm of the hepatic concentration of glutathione (Fig. 12; DESGAGNÉ et al. 1988; BÉLANGER et al. 1991). Maximal and minimal hepatotoxicities were found at 21.00 h and 09.00 h that correlated with the lowest and highest levels of glutathione. Similar findings were obtained for the hepatotoxicity of 1,1-dichloroethylene (JAEGER et al. 1973), allyl alcohol (HANSON and ANDERS 1978), acetaminophen (SCHNELL et al. 1983) and styrene (DESGAGNÉ and BÉLANGER 1986).

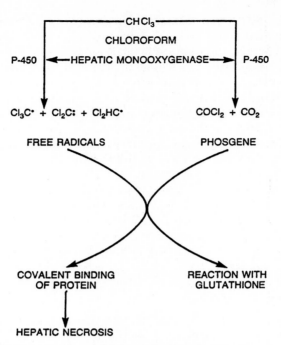

Fig. 10. Metabolic bioactivation and desintoxication of chloroform

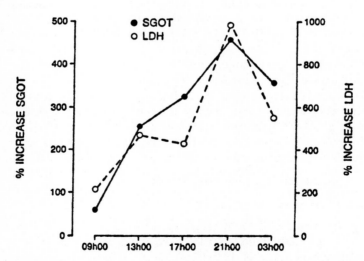

Fig. 11. Circadian hepatotoxicity of chloroform in rat. The hepatotoxicity was determined as the percentage of increase in the serum activity of GOT (•) and LDH (o) compared to the control value determined at the same time of day. Results are mean values of six animals. (Adapted from LAVIGNE et al. 1983)

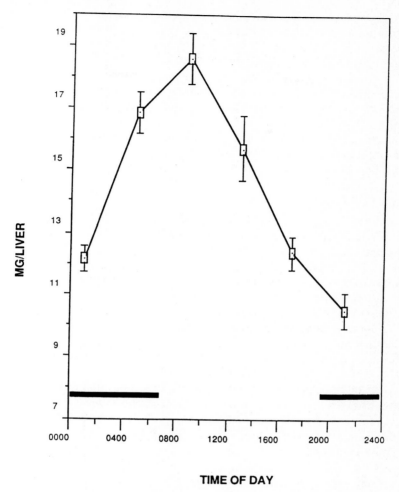

Fig. 12. Circadian variation in the hepatic concentration of reduced glutathione in rats. Results are mean values ± SEM obtained in three rats per time of day. (Adapted from LABRECQUE and BÉLANGER 1991)

III. Factors Related to Biotransformation

1. Hepatic Blood Flow

For drugs having a high hepatic extraction ratio ($E > 0.7$), the hepatic blood flow is the main factor regulating their clearance after intravenous administration (WILKINSON and SHAND 1975). Examples of such drugs are alprenolol, lidocaine, morphine, propranolol, nifedipine, and verpamil. LEMMER and NOLD (1991) determined the time-dependent hepatic blood flow in healthy volunteers and found the highest and lowest values at 02.00 h to 08.00 h and

14.00 h, respectively. The same group reported that the plasma half-life of propranolol was shortest at 08.00 h and longest at 20.00 h (LANGNER and LEMMER 1988). These results are also in agreement with those of chronokinetics of lidocaine in man that showed an increase in the AUC value at 15.30 h compared to those values determined at three other times of day (BRUGUEROLLE and ISNARDON 1985). Similar results were obtained in rats by LABRECQUE et al. (1988) although the hepatic blood flow was not determined directly. These workers determined the fractional distribution of labelled microspheres in different organs following a single cardiac output and found that the lowest and highest distribution of the microspheres in rat liver occurred at 15.00 h and 21.00 h, respectively.

2. First-Pass Effect

The first-pass effect to which a drug is subjected when administered orally is directly related to the hepatic and intestinal metabolizing activity occurring during the absorption process. Therefore, in the case of a drug completely absorbed and eliminated by biotransformation, clearance determined after oral administration is related to the drug metabolizing activity independently of its hepatic extraction ratio (WILKINSON and SHAND 1975). However, the first-pass effect is only significant for drugs having a high hepatic extraction ratio. Assuming complete oral absorption, temporal variations in the first-pass effect and hence in the bioavailability are prominent for many drugs. Examples in rat are acetaminophen (BÉLANGER et al. 1987) and isoniazid (BÉLANGER et al. 1989) in which the first-pass effect was about 70% and 40% higher at 09.00 h than at 21.00 h, respectively. Conversely, the oral bioavailability of both drugs differed markedly as a function of time. Similar effects have been reported in humans for nifedipine (LEMMER et al. 1991), prednisolone (ENGLISH et al. 1993) and propranolol (LANGNER and LEMMER 1988). In all of these cases, the AUC values of the plasma concentrations of the drug determined after oral ingestion were always higher during the morning (08.00 h–12.00 h) than at any other time of day. Thus, the oral bioavailability of drugs subjected to a significant first-pass effect is likely to show a temporal variation in their plasma levels.

E. Excretion

Virtually all drugs and xenobiotics are eliminated from the body either unchanged or as metabolites, and the kidney is the most important organ for their elimination. There are three processes involved in the urinary excretion of chemicals, namely glomerular filtration, passive reabsorption, and, for some compounds, energy-dependent tubular transport (for a review see WATERHOUSE and MINORS 1989). Circadian rhythms of prominent amplitude have been described for glomerular filtration, effective renal plasma flow, urinary volume and pH, and for the urinary excretion of many electrolytes and endogenous compounds in rodents (CAMBAR et al. 1979; PONS et al. 1994) and humans (VAGNUCCI et al. 1969; WISSER and BREUER 1981; ARAKI et al. 1983; KANABROCKI et al. 1988). For all of these rhythms, nadir and trough values occur during the activity and resting periods, respectively. Evidently, these chronovariations will affect the elimination of drugs that are excreted predominantly unchanged in the urine. One rhythm particularly important for the excretion of weakly acidic and basic drugs is the urinary pH fluctuation during the course of 24 h (ELLIOT et al. 1959). In humans, the pH value is usually lower during sleep than during the day. The effect of the temporal variation in urinary pH on the excretion of amphetamines and related compounds has been well characterized (Fig. 13; BECKETT and ROWLAND 1964; WILKINSON and BECKETT 1968). At lower pH, the basic amine is more strongly ionized and readily excreted in the urine and, conversely, as pH increases, the nonionized fraction increases, resulting in stronger reabsorption of the compound and decreased excretion (Fig. 13A). If the pH value is maintained acidic at around 4.5–5.0 by the coadministration of ammonium chloride (Fig. 13A,B), then the ionized forms of amphetamine and norephedrine are readily excreted independently of the urinary output, and the curve of rate of excretion versus time is similar to that of the plasma levels. In contrast, if the pH value is alkalinized at about 8.0 to favor the formation of the nonionized form and the reabsorption process, then the rate of excretion of norephedrine fluctuates as a function of urinary output independently of pH. Other known examples of temporal variation in urinary excretion of drugs related to pH are the sulfonamides (DETTLI and SPRING 1966 and the salicylates (REINBERG et al. 1967; MARKIEWICZ and SEMENOWICZ 1979). Of clinical relevance, the chrononephrotoxicity of drugs such as cisplatin (HRUSHESKY et al. 1982), amikacin (HOSOKAWA et al. 1993), gentamicin (CAL et al. 1985) and tobramycin (LIN et al. 1994) are related to their time-dependent variations in renal excretion in both rodents and humans.

F. Conclusions and Perspectives

During the past 20 years, chronopharmacology and chronopharmacokinetics emerged as the disciplines that explain the temporal variations in the kinetics and effects of drugs. Recent research has provided useful data resulting in a better understanding of the circadian variations in the various pharmacokinetic processes. To some extent, it is now possible to predict certain aspects of the chronopharmacokinetics of drugs but the limited success thus far should lead to more research and generates new hypotheses in the investigation of the mechanisms involved in the circadian variations in the absorption, distribution, metabolism and excretion of drugs. In particular, the role of endogenous and external factors that modulate the temporal variations in these processes have been poorly studied. The results presented in this chapter and in others of this book clearly demonstrate that the conventional presumption that a constant rate of drug delivery will provide constant blood levels and effect(s) over a 24 h

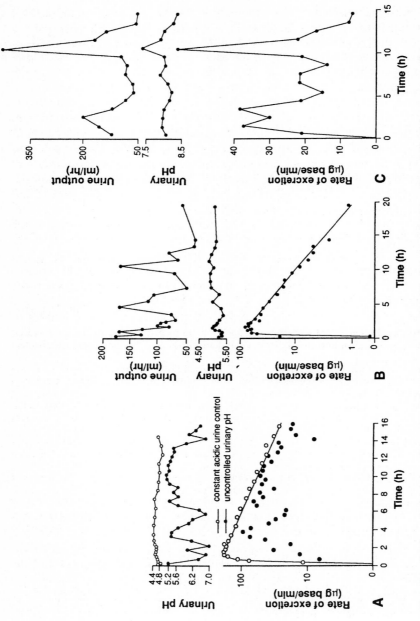

Fig. 13A–C. Time-dependent urinary excretion of amphetamine (A) and norephedrine (B and C) with corresponding urinary pH and volume output after oral administration of 15 mg of S-(+)-amphetamine sulfate or 25 mg of (−) norephedrine hydrochloride under various conditions of urinary pH. (Adapted from BECKETT and ROWLAND 1964 and WILKINSON and BECKETT 1968)

period must be rejected. Furthermore, the application of chronobiological principles to drug therapy will result in a more rational approach to optimizing treatment efficacy and/or reducing drug-induced side or toxic effects.

References

Angeli A, Frajria R, DePaoli R, Fonzo D, Ceresa F (1978) Diurnal variation of prednisolone binding to serum corticosteroid binding-globulin in man. Clin Pharmacol Ther 23:47–53

Araki S, Murata K, Yokoyama K, Yanaghara S, Niinuma Y, Yamanoto R, Ishihara N (1983) Circadian rhythms in the urinary excretion of metals and organic substances in healthy men. Arch Environ Health 38:360–366

Aungst BJ, Fung H-L (1981) Intestinal lead absorption in rats: effects of circadian rhythm, food, undernourishment, and drugs which alter gastric emptying and GI motility. Res Commun Chem Path Pharmacol 34:515–530

Bauer LA, Davis R, Wilensky A, Raisys Y, Levy RH (1985) Valproic acid clearance: unbound fraction and diurnal variation in young and elderly adults. Clin Pharmacol Ther 37:697–700

Beckett AH, Rowland M (1964) Rhythmic urinary excretion of amphetamine in man. Nature 204:1203–1204

Bélanger PM (1987) Chronobiological variation in the hepatic elimination of drugs and toxic chemical agents. Ann Rev Chronopharmacol 4:1–46

Bélanger PM (1933) Chronopharmacology in drug research and therapy. Adv Drug Res 24:1–80

Bélanger PM, Desgagné M, Bruguerolle B (1991) Temporal variations in microsomal lipid peroxidation and in glutathione concentration of rat liver. Drug metab Disp 19:241–244

Bélanger PM, Doré F, Pérusse F, Labrecque G (1980) Temporal variations in the elimination of antipyrine in normal and adrenalectomized rats. Res Commun Chem Pathol Pharmacol 30:243–252

Bélanger PM, Labrecque G (1989) Temporal aspects of drug metabolism. In: Lemmer B (ed) Chronopharmacology: cellular and biochemical interactions, chap. 2. Marcel Dekker, Basel, pp. 15–34

Bélanger PM, Labrecque G, Doré F (1984) Rate-limiting steps in the temporal variations of the pharmacokinetics of some selected drugs. In: Haus E, Kabat H (eds) Chronobiology 1982–1983. Karger, Basle, pp. 359–363

Bélanger PM, Lalande M (1988) Day-night variations in the activity and composition of the microsomal mixed-function oxidase of rat liver. Ann Rev Chronopharmacol 5:219–222

Bélanger PM, Lalande M, Doré F, Labrecque G (1987) Time-dependent variations in the organ extraction ratios of acetaminophen in rat. J Pharmacokin Biopharm 15:133–143

Bélanger PM, Lalande M, Doré F, Labrecque G (1989) Temporal variations in the pharmacokinetics of isoniazid and N-acetylisoniazid in rats. Drug Metab Disp 17:91–97

Bélanger PM, Lalande M, Labrecque G, Doré F (1985) Diurnal variations in the transferases and hydrolases involved in glucuronide and sulfate conjugation of rat liver. Drug Metab Disp 13:386–389

Bruguerolle B, Barbeau G, Bélanger PM, Labrecque G (1986a) Chronokinetics of indomethacin in elderly subjects. Ann Rev Chronopharmacol 3:425–428

Bruguerolle B, Isman R (1985) Daily variations in plasma levels of lidocaine during local anaesthesia in dental practice. Ther Drug Monit 7:369–370

Bruguerolle B, Jadot G (1983) Influence of the hour of administration of lidocaine on its intraerythrocytic passage in the rat. Chronobiologia 10:295–297

Bruguerolle B, Jadot G, Valli M, Bouyard L, Bouyard P (1982) Etude chronocinétique de la lidocaïne chez le rat. J Pharmacol (Paris) 13:65–76

Bruguerolle B, Lemmer B (1993) Recent advances in chronopharmacokinetics: methodological problems. Life Sci 52:1809–1824

Bruguerolle B, Levi F, Arnaud C, Bouvenot G, Mechkouri M, Vannetzel JM, Touitou Y (1986b) Alteration of physiologic circadian time structure of six plasma proteins in patients with advanced cancer. Ann Rev Chronopharmacol 3:207–210

Bruguerolle B, Prat M (1987) Temporal changes in bupivacaine kinetics. J Pharm Pharmacol 39:148–150

Bruguerolle B, Prat M (1988) Circadian phase-dependent pharmacokinetics and acute toxicity of mepivacaine. J Pharm Pharmacol 40:592–594

Bruguerolle B, Prat M (1989) Temporal variations in the erythrocyte permeability to bupivacaine, etidocaine and mepivacaine in mice. Life Sci 45:2587–2589

Bruguerolle B, Prat M (1990) Circadian phase-dependent acute toxicity and pharmacokinetics of etidocaine in serum and brain of mice. J Pharm Pharmacol 42:201–202

Bruguerolle B, Valli M, Jadot G, Bouyard L, Bouyard P (1980) Influence of the hour of administration on free carbamazepine plasma levels. Comm. 14e Rencontre Internationale de Chimie Thérapie, Marseille

Cal JC, Dorian C, Cambar J (1985) Circadian and circannual changes in nephrotoxic effects of heavy metals and antibiotics. Ann Rev Chronopharmacol 2:143–176

Cambar J, Lemoigne F, Toussaint C (1979) Etude des variations nychtémérales de la filtration glomérulaire chez le rat. Experientia 35:1607–1609

Denvis MJ (1990) Absorption processes. In: Hansch C, Sammes PG, Taylor JB (eds) Comprehensive medicinal chemistry vol 5. biopharmaceutics. Pergamon, New York, pp. 1–43

Desgagné M, Bélanger PM (1986) Chronohepatotoxicity of styrene in rat. Ann Rev Chronopharmacol 3:103–106

Desgagné M, Boutet M, Bélanger PM (1988) The mechanism of the chronohepatotoxicity of chloroform in rat: Correlation between binding to hepatic subcellular fractions and histologic changes. Ann Rev Chronopharmacol 5:235–238

Dettli L, Spring P (1966) Diurnal variations in the elimination rate of a sulfonamide in man. Helv Med Acta 33:291–306

Duane WC, Gilberstadt ML, Wiegand DM (1979) Diurnal rhythms of bile acid production in the rat. Amer J Physiol 236:R175–R179

Elliot JS, Sharp RF, Lewis L (1959) Urinary pH. J Urol 81:339–343

English J, Dunne M, Marks V (1983) Diurnal variation in prednisolone kinetics. Clin Pharmacol Ther 33:381–385

Focan C, Bruguerolle B, Arnaud C, Levi F, Mazy V, Focan-Henrard D, Bouvenot G (1988) Alteration of circadian time structure of plasma proteins in patients with inflammation. Ann Rev Chronopharmacol 5:21–24

Garrettson LK, Jusko WJ (1974) Diphenylhydantoin elimination kinetics in overdosed children. Clin Pharmacol Ther 17:481–491

Gilberstadt ML, Bellinger LL, Lindblad S, Duane WC (1991) Liver denervation does not alter the circadian rhythm of bile acid synthesis in rats. Am J Physiol 261:G799–G802

Goo RH, Moore JG, Greenberg E, Alazraki NP (1987) Circadian variation in gastric emptying of meals in humans. Gastroenterology 93:515–518

Guengerich FP (1987) Enzymology of rat liver cytochromes p-450. In: Guengerich FP (ed) Mammalian cytochromes p-450 vol 1, chap. 1. CRC Press, Boca Raton, pp 2–54

Guissou P, Cuisinaud G, Llorca G, Lejeune E, Sassard J (1983) Chronopharmacologic study of a prolonged release form of indomethacin. Eur J Clin Pharmacol 24:678–672

Haen E, Gerdsneir W, Arbogast B (1985) Circadian variation in propranolol protein binding. Naunyn-Schmiedeberg's Arch Pharmakol 329–393

Hanson SK, Anders MW (1978) The effect of diethyl maleate treatment, fasting and time of administration of allyl alcohol hepatotoxicity. Toxicol Lett 1:301–305

Hartley R, Forsythe WI, McLain B, Ng PC, Lucok MD (1991) Daily variations in steady-state plasma concentrations of carbamazepine and its metabolite in epileptic children. Clin Pharmacokin 20:237–244

Haus E, Nicolau GY, Lakatua D, Sackett-Lundeen L (1987) Reference values for chronopharmacology. Ann Rev Chronopharmacol 4:333–424

Hecquet B, Meynadier J, Bonneterre J, Adenis L (1984) Circadian rhythm in cisplatin binding on plasma proteins. Ann Rev Chronopharmacol 3:115–118

Hosokawa H, Nyu S, Nakamura N, Mifune K, Nakano S (1993) Circadian variation in amikacin clearance and its effects on efficacy and toxicity in mice with and without immunosuppression. Chronobiol Int 10:259–270

Hrusheski WJM, Borsch R, Levi F (1982) Circadian time dependence of cisplatin urinary kinetics. Clin Pharmacol Ther 32:330–339

Jaeger RJ, Conolly B, Murphy ID (1973) Diurnal variation of hepatic glutathione concentration and its correlation with 1, 1-dichloroethylene inhalation toxicity in rats. Res Commun Chem Pathol Pharmacol 6:465–671

Jakoby WB (1980) Enzymatic basis of detoxication, vols I and II. Academic, New York

Jespersen CM, Fredericksen M, Fisher Hansen J, Klitgaard NA, Soerum C (1989) Circadian variation in the pharmacokinetics of verapamil. Eur J Clin Pharmacol 37:613–615

Kanabrocki EL, Sothern RB, Scheving LE, Halberg F, Pauly JE, Greco J, Nemchausky BA, DeBartelo M, Kaplan E, McCormick JB, Olwin JH, Marks GE, Bird T, Redmond DP, Graeber RC, Ferrara A, Hrushesky WJM (1988) Ten-year-replicated circadian profiles for 36 physiological, serological and urinary variables in healthy men. Chronobiol Int 5:237–284

Kaplowitz N, Aw TY, Ookhtens M (1985) The regulation of hepatic glutathione. Ann Rev Pharmacol Toxicol 25:715–744

Koren G, Ferrazzini G, Gohl H, Robieux I, Johnson D, Giesbrecht E (1992) Chronopharmacology of methotrexate pharmacokinetics in childhood leukemia. Chronobiol Int 9:434–438

Kumar D, Wingate D, Ruckebusch Y (1986) Circadian variation in the propagating velocity of the migrating motor complex. Gastroenterology 91:926–930

Labrecque G, Bélanger PM (1991) Biological rhythms in the absorption, distribution, metabolism and excretion of drugs. Pharmacol Ther 52:95–107

Labrecque G, Bélanger PM, Doré F, Lalande M (1988) 24-Hour variations in the distribution of labeled microspheres to the intestine, liver and kidneys. Ann Rev Chronopharmacol 5:445–449

Langner B, Lemmer B (1988a) Circadian changes in the pharmacokinetics and cardiovascular effects of oral propranolol in healthy subjects. Eur J Clin Pharmacol 33:619–624

Langner B, Lemmer B (1988b) Circadian phase dependency in pharmacokinetics and cardiovascular effects of oral propranolol in man. Ann rev Chronopharmacol 5:535–538

Larsen KR, Doyton MT, Moore JG (1991) Circadian rhythm in gastric mucosal blood flow in fasting rat stomach. J Surg Res 51:275–280

Lavigne JG, Bélanger PM, Doré F, Labrecque G (1983) Temporal variations in chloroform-induced hepatotoxicity in rats. Toxicology 26:267–273

Lemmer B, Bruguerolle B (1994) Chronopharmacokinetics: are they clinically relevant? Clin Pharmacokinet 26:419–427

Lemmer B, Nold G (1991) Circadian changes in estimated hepatic blood flow in healthy subjects. Brit J Clin Pharmacol 32:627–629

Lemmer B, Nold G, Behne S, Kaiser R (1991a) Chronopharmacokinetics and cardiovascular effects of nifedipine. Chronobiol Int 8:485–494

Lemmer B, Scheidel B, Plume H, Becker H-J (1991b) Clinical chromopharmacology of oral sustained release isosorbide-5-mononitrate in healthy subjects. Eur J Clin Pharmac 40:71–75

Levy M, Granet L, Zylber-Katz E (1984) Chronopharmacokinetics of caffeine in healthy volunteers. Ann Rev Chronopharmacol 1:97–100

Lin L, Grenier L, Bergeron Y, Simard M, Bergeron MG, Labrecque G, Beauchamp D (1994) Temporal changes of pharmacokinetics, nephrotoxicity, and subcellular distribution of tobramycin in rats. Antimicrob Agents Chemother 38:54–60

Lindup WE (1987) Plasma protein binding of drugs - some basic and clinical aspects. Prog Drug Metab 10:142–185

Lockard JS, Viswanathan CT, Levy RH (1985) Diurnal oscillations of CSF valproate in monkey. Life Sci 36:1281–1285

Loiseau P, Cenraud B, Levy RH, Akbaraly R, Brochet-Liermain A, Guyot M, Morselli P (1982) Diurnal variations in steady-state plasma concentrations of valproic acid in epileptic patients. Clin Pharmacokin 7:544–552

Loubaris N, Michel A, Cros G, Serrano JJ, Katz S, Bouchard M (1984) Circadian changes in carrageenin-induced edema: the anti-inflammatory effect and bioavailability of phenylbutazone in rats. Life Sci 34:2379–2384

Maranjo CA, Sellers EM, Giles HG, Abel JG (1980) Diurnal variations in plasma diazepam concentrations associated with reciprocal changes in free fraction. Brit J Clin Pharmacol 9:265–272

Markiewicz A, Gomoluch T, Marek E, Boldys H (1981) Circadian absorption of vitamin B_{12}. Scand J Gastroenterol 16:541–544

Markiewicz A, Senenowicz K (1979) Time-dependent changes in the pharmacokinetics of aspirin. J Clin Pharmac Biopharm 17:409–411

Meister A, Anderson ME (1983) Glutathione. Ann Rev Biochem 52:711–760

Miyazaki Y, Yatagai M, Imaoka S, Funae Y, Motohashi Y, Kobayashi Y (1990) Temporal variations in hepatic cytochromes P-450 isoenzymes in rats. Ann Rev Chronopharmacol 7:149–153

Nair V and Casper R (1969) The influence of light on daily rhythm in hepatic drug metabolizing enzymes in rat. Life Sci 8:1291–1298

Nakano S, Watanabe H, Nagai K, Ogawa N (1984) Circadian stage-dependent changes in diazepam kinetics. Clin Pharmacol Ther 36:271–277

Nakaro A, Tietz PS, Larusso NF (1990) Circadian rhythms of biliary protein and lipid excretion in rats. Amer J Physiol 258:G653–G659

Patel IH, Venkataramanan R, Levy RH, Viswanathan CT, Ojemann LM (1982) Diurnal oscillations in plasma protein binding of valproic acid. Epilepsia 32:282–290

Poley GE, Shiveley CH, Vesell ES (1978) Diurnal rhythms of aminopyrine metabolism: failure of sleep deprivation to affect them. Clin Pharmacol Ther 24:726–732

Pons M, Tranchot J, L'Azou B, Cambar J (1994) Circadian rhythms of renal hemodynamics in unanesthetized, unrestrained rats. Chronobiol Int 11:301–308

Pooler PA, Duane WC (1988) Effects of bile acid administration on bile acid synthesis and its circadian rhythm in man. Hepatology 8:1140–1146

Radzialowski FM, Bousquet WF (1967) Circadian rhythm in hepatic drug metabolizing activity of the rat. Life Sci 6:2545–2548

Radzialowski FM, Bousquet WF (1968) Daily rhythmic variation in the hepatic drug metabolism in the rat and mouse. J Pharmacol Exp Ther 163:229–238

Reinberg A, Levi F, Touitou Y, Le Liboux A, Simon J, Frydman A, Bicakova-Rocher A, Bruguerolle B (1986) Clinical chronokinetic changes in a sustained release preparation of ketoprofen. Ann Rev Chonopharmacol 3:317–320

Reinberg A, Lévi F, Smolensky MH, Labrecque G, Ollagnier M, Decousus H, Bruguerolle B (1990) Chronokinetics. In: Hansch C, Sammes PG, Taylor JB (eds) Comprehensive medicinal chemistry 5: biopharmaceutics. Pergamon, New York, pp 280–295

Reinberg A, Schuller E, Deslanerie N, Clench J, Helary M (1977) Rythmes circadiens et circannuels des leucocytes, protéines totales, immunoglobulines A, G et M. Etude chez neuf adultes jeunes et sains. Nouv Prese Med 6:3819–3823

Reinberg A, Zagulla-Mally ZW, Ghate J, Halberg F (1967) Circadian rhythm in duration of sodium salicylate excretion referred to phase of excretory rhythms and routine. Proc Soc Exp Biol Med 124:826–832

Riva R, Albani F, Ambrosetto G, Contin M, Cortelli P, Perucca E, Baruzzi A (1984) Diurnal fluctuations in free and total steady state plasma levels of carbamazepine and correlation with intermittent side effects. Epilepsia 25:476–481

Rowland M, Tozer TN (1980) Clinical pharmacokinetics. Concepts and applications. Lea and Febiger, Philadelphia, 65–76

Ryan DE, Levin W (1990) Purification and characterization of hepatic microsomal cytochrome P-450. Pharmacol Ther 45:153–239

Schanker LS (1972) Drug absorption. In: La Du BN, Mandel HG, Way EL (eds) Fundamentals of drug metabolism and drug disposition. Williams and Wilkins Co, Baltimore, pp 22–43

Scheidel B, Lemmer B (1991) Chronopharmacology of oral nitrates in healthy subjects. Chronobiol Int 8:409–419

Schnell RC, Bozigian HP, Davies MH, Merrick BA, Johnson KL (1983) Circadian rhythm in acetaminophen toxicity: role of nonprotein sulfhydryls. Tox Appl Pharmacol 71:353–361

Shiga T, Fujimura A, Tateishi T, Ohaski K, Ebihara A (1993) Differences of chronopharmacokinetic profiles between propranolol and atenolol in hypertensive subjects. J Clin Pharmacol 33:756–761

Shiveley CA, Vesell ES (1975) Temporal variations in acetaminophen and phenacetin half-life in man. Clin Pharmacol Ther 18:413–424

Smolensky M (1989) Chronopharmacology of theophylline and beta-sympathomimetics. In: Lemmer B (ed) Chronopharmacology: cellular and biochemical interactions, chap. 5. Marcel Dekker, Basel, pp 65–113

Swanson BN, Boppana VK, Vlasses PH, Holmes GI, Monsell K, Ferguson RK (1982) Sulindac disposition when given once and twice daily. Clin Pharmacol Ther 32:397–403

Taggart AJ, McElnay JC, Kerr B, Passmore P (1987) The chronopharmacokinetics of indomethacin suppositories in healthy volunteers. Eur J Clin Pharmacol 31:617–619

Tarquini B, Romano S, De Scalzi M, de Leonardis V, Benvenuti F, Chegai E, Comparini T, Moretti R, Cognoni M (1979) Circadian variation of iron absorption in health human subjects. Ann Biosci 19:347–354

Tillement JP, Houin G, Zini R, Urien S, Albengres E, Barré J, Lecompte M, D'Athis P, Sébille B (1984) The binding of drugs to blood plasma macromolecules: recent advances and therapeutic significance. Adv Drug Res 13:59–94

Touitou Y, Touitou C, Bogdan A, Reinberg A, Auzeby A, Beck H, Guillet P (1986) Differences between young and elderly subjects in seasonal and circadian variations of total plasma proteins and blood volume as reflected by hemoglobin, hematocrit and erythrocyte counts. Clin Chem 2:801–804

van Berge Henegower GP, Hoffmann AF (1978) Nocturnal gallbladder storage and emptying in gallstone patients and healthy subjects. Gastroenterology 75:879–885

Vegnucci AH, Shapiro AP, McDonald RH (1969) Effect of upright posture on renal electrolyte cycle. J Appl Physiol 26:720–736

Vener KJ, Moore JG (1987) Chronobiologic properties of the alimentary canal affecting xenobiotic absorption. Ann Rev Chronopharmacol 4:257–281

Vesell ES, Shiveley CA, Passananti T (1977) Temporal variations of antipyrine half-life in man. Clin Pharmacol Ther 22:843–852

Waterhouse JM, Minors DS (1989) Temporal aspects of renal drug elimination. In: Lemmer B (ed) Chronopharmacology: cellular and biochemical interactions, chap. 3. Marcel Dekker, Basel, pp 35–50

Wilkinson GR, Beckett AH (1968) Absorption, metabolism and excretion of the ephedrines in man. I. The influence of urinary pH and urine volume output. J Pharmacol Exp Ther 162:139–147

Wilkinson GR, Shand DG (1975) A physiological approach to hepatic drug clearance. Clin Pharmacol Ther 18:377–390

Winters DK, Cederbaum AI (1993) Biochemistry of cytochrome P-450. In: Tavoloni N, Berk PD (eds) Hepatic transport and bile secretion: physiology and pathophysiology, chap. 27. Raven, New York, pp 407–420.

Wisser H, Breuer H (1981) Circadian changes of clinical, chemical and endocrinological parameters. J Clin Chem Clin Biochem 19:323–337

Wrobel J, Nagel G (1979) Diurnal rhythm of active calcium transport in rat intestine. Experientia 35:1581–1582

Yoshiyama Y, Nakano S, Ogawa N (1989) Chronopharmacokinetic study of valproic acid in man: comparison of oral and rectal administration. J Clin Pharmacol 29:1048–1052

CHAPTER 9
Progress in the Chronotherapy of Nocturnal Asthma

M.H. SMOLENSKY and G.E. D'ALONZO

A. Introduction

Bronchial asthma is a chronic airways disease. Symptoms include dyspnea, wheezy chest, cough, and fatigue. It is seldom a problem for those who suffer from the mild form of the disease. However, it is a significant and even life-threatening one for those prone to severe asthma. Asthma is characterized by persistent airways inflammation, airways hyperreactivity, and compromised pulmonary function (MARTIN 1993). All are reversible, at least to some extent, with medication and environmental control. Asthma severity varies between patients and even in the same patient over time. Most patients experience worsening of symptoms during the night; many experience symptoms only at night (DETHLEFSEN and REPGES 1985; TURNER-WARWICK 1988). Successful asthma management must address the major features of the disease – airways inflammation and hyperreactivity, compromised pulmonary function, and temporal patterning.

This chapter addresses the chronotherapy (timed delivery of medication in proportion to biological need during the 24 h) of nighttime asthma. Since an understanding of the disease is central to discussing its chronotherapy with bronchodilator and anti-inflammatory medications, circadian rhythms in the epidemiology and physiology of the disease are first reviewed.

B. Day-Night Pattern of Acute Asthma

The clock-time occurrence of acute asthma is not random during the 24 h, menstrual cycle, and year (DETHLEFSEN and REPGES 1985; ORIE et al. 1964; TURNER-WARWICK 1988; WULFSOHN and POLITZER 1964). Several well-conducted epidemiologic and clinical investigations clearly establish the nighttime prevalence of the disease. Although historical accounts (ADAMS 1856; FLOYER 1698; SMITH 1860; TROUSSEAU 1865) described the nocturnal nature of the ailment, it was DETHLEFSEN and REPGES in Germany and TURNER-WARWICK in the United Kingdom who clearly established the nighttime occurrence of asthma.

The German study involved more than 3,000 presumably diurnally active patients who maintained records of acute asthma during the washout phase of a medication trial. The clock time of the 1,631 recorded bouts of asthma was not random during the 24 h; asthma was several hundredfold more common during the nighttime sleep period than during the middle of the daytime activity span, when it was least common (Fig. 1).

The epidemiologic study by TURNER-WARWICK focused on the extent to which asthma disrupted sleep in a large group of 7,729 noninstitutionalized patients, most of whom were under treatment with equal-interval, equal-dose bronchodilator and anti-inflammatory medications. Some 94% reported their sleep being disturbed by asthma at least one night per month, and 74% indicated that to be the case at least one night per week. Of great significance was the finding that sleep was compromised three nights per week in 64% and nightly in 39%. These findings were unexpected since the majority of patients

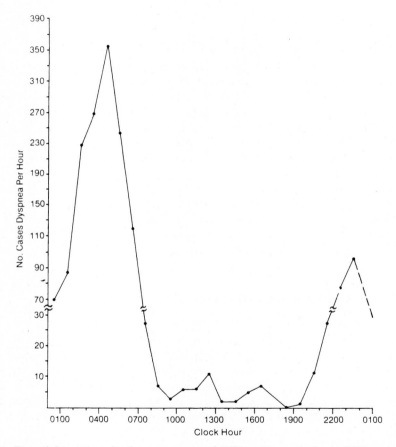

Fig. 1. Day-night pattern in the occurrence of 1631 asthma attacks in 3121 untreated patients. Asthma was most common during the middle of the nighttime sleep span, between 02.00 and 07.00 h, and least during the middle of the diurnal activity span. (Reproduced with permission from DETHLEFSEN and REPGES 1985)

were under treatment with potent antiasthma medications. Clinical studies, such as that conducted by STORMS et al. (1994), confirm the day-night difference in asthma symptoms and its significance to patient assessment and treatment.

The severity of asthma also exhibits day-night differences. HETZEL et al. (1977), based on a 2.5-year survey of 1,169 hospital admissions for asthma in Great Britain, found pulmonary arrest to be relatively uncommon. However, of the ten arrests that were identified, all occurred between midnight and 06.00 h. Death from asthma is also greater at night. COCHRANE and CLARK (1975) found 13 of 19 (68%) asthma deaths took place during the 8 h span between midnight and 08.00 h. Other investigations (ANONYMOUS 1982; BATEMAN et al. 1979; DOUGLAS 1985) also demonstrate greatest risk of death from asthma during the night.

C. Early, Late, and Recurrent Asthma Reactions

Many patients exhibit early, late, and recurrent asthma reactions following clinical challenge with antigens as illustrated in Fig. 2 (NEWMAN TAYLOR et al.

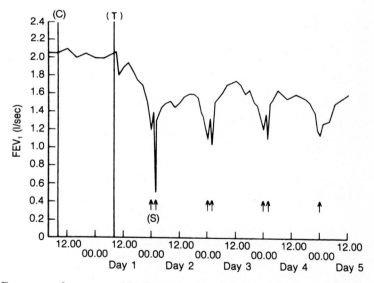

Fig. 2. Exposure of a presumably diurnally active patient sensitized to budgerigar antigen around 11.00 h resulted in an immediate, although minor, early asthma reaction (EAR) confirmed by the temporary dip in FEV_1. A severe episode of nocturnal asthma (late asthma reaction, LAR), verified by profound decline in FEV_1 and need of potent sympathomimetic agents (S) occurred at midnight. Episodes of asthma continued to take place during the subsequent 3 nights (recurrent asthma reaction, RAR) even though no further exposure to the provocative antigen occurred. (Reproduced with permission from NEWMANN TAYLOR et al. 1977)

1977). The initial or early asthma reaction (EAR) occurs within minutes of provocation and is characterized by a decrease of airway tone. The EAR is initiated by antigen-induced release of performed mediators (histamine and specific leukotrienes) from pulmonary mast cells. The late asthma reaction (LAR) occurs 6–12 h after antigen provocation. The LAR is generally more severe than the EAR, with a much greater decrement in airway caliber, due to enhancement of airways inflammation. Cytokines, such as GM-CSF, IL-3, IL-5, and IFN-Gamma, released from the mast cell at the time of the EAR, are believed to recruit, prime, and activate neutrophils, monocytes, eosinophils, macrophages, and lymphocytes, resulting in injury to the epithelial cells of the small airways and enhancement of tissue inflammation (BARNES and PAUWELS 1994; DEMONCHY et al. 1985). The recurrent asthma reaction (RAR) takes place during the nocturnal sleep period for several consecutive nights following daytime provocation (COCKCROFT et al. 1984; DAVIES et al. 1976; NEWMAN TAYLOR et al. 1977; NEWMAN TAYLOR et al. 1979). The mechanisms of the RAR are yet unexplained; however, they are thought to include circadian rhythms in neuroendocrine and other pertinent functions that affect airway status. The biological mechanisms and chemical mediators of the three asthma reactions must be taken into consideration in the clinical management of asthma.

D. The Chronobiology of Asthma

The chronobiology of asthma has been reviewed in great detail (MARTIN 1993; SMOLENSKY et al. 1986). The circadian rhythm-dependencies of the disease are briefly presented here as background and basis for its chronotherapy.

Circadian Rhythm in Pulmonary Function. The airway caliber of normal persons exhibits only slight circadian change. Around-the-clock studies of diurnally active, healthy subjects show that the magnitude of the peak (in the afternoon)-to-trough (in the early morning) variation in flow rate is quite small, amounting to only 5% or so of the 24 h mean level (HETZEL and CLARK 1980; SMOLENSKY et al. 1986). The circadian amplitude of airway patency in asthma patients is much greater than it is in normals. Moreover, the 24 h level is typically greatly reduced compared with predicted levels. In general, the more severe and unstable the asthma, the greater the circadian amplitude and the lower the circadian baseline. Indeed, the increased 24 h variability in self-assessed peak expiratory flow (PEF) values is cited in the *International Consensus Report on Diagnosis and Management of Asthma* (SHEFFER 1992) as an objective indicator of the worsening and growing instability of asthma.

Circadian Rhythm in Airways Inflammation. The mechanisms and signs and symptoms of inflammatory processes and disorders are strongly circadian-rhythmic (HARKNESS et al. 1982; LABRECQUE and REINBERG 1989; LEVI et al. 1985; SMOLENSKY et al. 1995). This is exemplified by the waxing and waning during the 24 h of the symptoms of allergic rhinitis and rheumatoid arthritis.

The symptoms of both conditions intensify during the sleep period, although they are experienced as most severe at the commencement of the daytime activity period. The staging of the circadian rhythm of airways inflammation in patients having a medical history of nocturnal asthma is comparable to that of other inflammatory diseases. MARTIN et al. (1991) conducted bronchoalveolar lavage (BAL) studies at 16.00 h, when asthma risk is low, and at 04.00 h, when it is high. Two groups of patients were involved, one having a history of nighttime asthma and a second with no such history. The yield of the inflammatory marker neutrophil and eosinophil cells in afternoon BAL fluids was very low in both groups. In those without history of nighttime asthma, the yield of anti-inflammatory cells at 04.00 h was low and comparable to that found at 16.00 h. In contrast, in nocturnal asthma patients, the yield was statistically significantly greater at 04.00 h than at 16.00 h (Fig. 3). Other studies by JARJOUR et al. (1992) found BAL histamine and superoxide anion concentration significantly increased in patients with nighttime asthma. CAL-

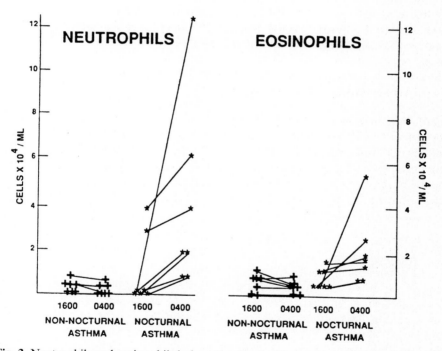

Fig. 3. Neutrophils and eosinophils in bronchoalveolar lavage (*BAL*) fluids in groups of diurnal active asthma patients without and with history of nocturnal symptoms studied at different circadian times. The yield of inflammatory cells in BAL fluids collected in the afternoon at 16.00 h and overnight at 04.00 h in asthmatics without history of nighttime exacerbations was comparable. In contrast, the yield was higher at 04.00 h than 16.00 h in those with a history of nocturnal asthma. These findings indicate the exacerbation of airways inflammation at night contributes to nocturnal asthma. (Reproduced with permission from MARTIN et al. 1991)

HOUN et al. (1992) also demonstrated a greater plasma concentration of activated low density eosinophils at 04.00 h than at 16.00 h. Collectively, these findings infer the day-night difference in airway patency, and asthma risk, is at least in part dependent on the circadian rhythm in airways inflammation.

Circadian Rhythm in Airways Hyperreactivity. The inflamed airways of asthma patients are more reactive to chemical and physical irritants than is the normal lung tissue of non-asthmatics. Based on BAL findings, airways inflammation and hyperresponsiveness are expected to be most intense during the nighttime. Several studies confirm this expectation; the effect of acetylcholine, histamine, and house dust aerosols on airway flow rates is several times greater in tests conducted during the night than during the day (BONNET et al. 1991; DEVRIES et al. 1962; GERVAIS et al. 1977; MOHIUDDIN and MARTIN 1990; REINBERG et al. 1971).

Circadian Rhythm in Autonomic Nervous and Endocrine System Function. Circadian rhythms in the hypothalamic-pituitary-adrenocortical (HPA) and autonomic nervous system contribute to the day-night difference in airway caliber and asthma. In diurnally active patients, serum cortisol is greatest around the commencement of daytime activity and lowest at night during the middle of the sleep period (BARNES et al. 1980; REINBERG et al. 1963; SOUTAR et al. 1975). Serum epinephrine also peaks during the middle of the activity period and dips to a trough during the period of sleep when cholinergic tone predominates (BARNES et al. 1980; MORRISON et al. 1988; SOUTAR et al. 1977). Plasma epinephrine concentration varies threefold during the 24 h; the circadian variation in plasma cortisol concentration is even greater in magnitude. Cortisol exerts potent anti-inflammatory effect and up-regulates ß-receptor number and function; epinephrine induces bronchodilation and modulates mast cell membrane stability. The high-amplitude circadian rhythms in cortisol and epinephrine contribute to the day-night variation in airways inflammation, reactivity, and patency.

Miscellaneous Factors. Nocturnal asthma may be influenced by a variety of factors that vary in a predictable fashion during the 24 h. These include day-night differences in airway secretions, clearance, cooling, and antigen retention as well as body position occurring secondary to sleep and activity and allergen exposure (MARTIN et al. 1990; MARTIN 1993; SMOLENSKY et al. 1986). Since nocturnal asthma oftentimes takes place during sleep, a casual relationship is suspected (BALLARD 1993). However, as yet there is no convincing evidence that nocturnal exacerbations are caused by sleep, per se, or associated with any particular sleep stage. Airway caliber declines nocturnally even in sleep-deprived asthma subjects, although the dip is attenuated in comparison to when sleep is permitted (BALLARD et al. 1989). It is unclear what this finding represents. Since sleep deprivation favors activation of the sympathetic nervous system, some degree of bronchial bronchodilation over basal conditions is expected. Sleep apnea also aggravates nocturnal asthma. Asthmatic patients who exhibit apnea benefit from nasal continuous positive airways pressure (CPAP), suggesting that sleep apnea can contribute to the exacerbation of nocturnal asthma in some patients (CHAN et al. 1988). Gastroesophageal reflux is thought to play a role in nighttime dyspnea in some patients. Certain asthma medications decrease lower gastroesophageal sphincter tone, and this alone or with assumption of the supine position for nighttime sleep facilitates acid reflux. This causes respiratory distress and aggravation of the asthmatic condition either through vagal reflex or by acid aerosol reflux into hyperresponsive airways (BABB et al. 1970; GOODALL et al. 1981; MARTIN et al. 1982). The importance of acid reflux in nocturnal asthma, however, may be overestimated (TAN et al. 1990).

E. Pharmacotherapy of Asthma

Expert consensus panels (BRITISH THORACIC SOCIETY 1993; SHEFFER 1991; SHEFFER 1992; WETTENGEL et al. 1994) have agreed upon a set of goals for asthma therapy. They are: (a) prevention of troublesome acute and chronic symptoms of the disease, (b) maintenance of normal or near normal pulmonary function, life style, activity, and sleep, and (c) avoidance of medication-induced adverse effects. Specific algorithms have been published to guide patient management and to achieve therapeutic goals. Table 1 (SHEFFER 1992), for example, describes the clinical features of patients before treatment, their lung function level and stability, and the medications regularly required to maintain asthma control with reference to its severity. The disease is seldom a problem for mild asthmatics. They can be controlled by intermittent use of short-acting β_2-agonist aerosols. Asthma occurs more frequently in moderate patients. Daily inhaled anti-inflammatory agents and long-acting bronchodilator medications, for control of nighttime symptoms, are recommended. Nocturnal asthma occurs more frequently in these in comparison to mild patients. Severe asthma requires those medications used to control the mild form of the disease as well as systemic steroids. Although nocturnal symptoms constitute a clinical feature of all forms of asthma, especially the moderate and severe ones, the implication from review of the consensus reports is that all sustained bronchodilator medications are comparable in their effectiveness. Except for the specification that tablet steroids be administered in the morning to treat severe asthma, there is no discussion of the effect of treatment time on the pharmacokinetics or pharmacodynamics of sustained-released bronchodilator and other asthma medications. Nor is there any discussion of the possible significance of drug timing in relation to the circadian rhythm-dependencies of the disease. Certain theophyllines are especially formulated to match drug levels to biological requirements by delivering more drug overnight than during the daytime; these formulations are known to be particularly beneficial for treating patients having a medical history of nocturnal asthma. Moreover, clinical studies show the bronchodilator effect of tablet β_2-agonists can be

Table 1. Classification of asthma and recommended pharmacotherapy

Asthma severity	Clinical features before treatment	Lung function levels and stability	Regular medication usually required for control
Mild	• Intermittent, brief symptoms 1–2 times a week • Nocturnal asthma symptoms < 2 times a month • Asymptomatic between exacerbations	• PEF > 80% predicted at baseline • PEF variability < 20% • PEF normal after bronchodilator	• Intermittent inhaled short-acting β_2-agonist (taken as needed) only
Moderate	• Exacerbations > 1–2 times a week • Nocturnal asthma symtpoms > 2 times a month • Symptoms requiring inhaled β_2-agonist almost daily	• PEF 60%–80% predicted at baseline • PEF variability 20%–30% • PEF normal after bronchodilator	• Daily inhaled anti--inflammatory agent • Possibly daily long-acting bronchodilator, especially for nocturnal symptoms
Severe	• Frequent exacerbations • Continuous symptoms • Frequent nocturnal asthma symptoms • Physical activities limited by asthma • Hospitalization for asthma in previous year • Previous life--threatening exacerbation	• PEF < 60% predicted at baseline • PEF variability > 30% • PEF below normal despite optimal therapy	• Daily inhaled anti--inflammatory agent at high doses • Daily long-acting bronchodilator, especially for nocturnal symptoms • Frequent use of systemic corticosteroids

PEF, peak expiratory flow
Adapted from SHEFFER 1992.

enhanced by unequal, morning-evening dosing regimens. New data from studies with tablet and aerosol corticosteroid medications indicate that their efficiency is circadian rhythm-dependent. Thus, although consensus reports recognized the nocturnal occurrence of asthma and/or the need to reduce the circadian variability in airway function by medication, specific treatment strategies for controlling nighttime asthma, which is a chief concern of a significant number of patients, are yet to be recommended.

F. The Chronopharmacology and Chronotherapy of Asthma Medications

The day-night pattern and underlying rhythm-dependencies of asthma form the basis for its chronotherapy. Equal-interval, equal-dose schedules of antiasthma medications oftentime fail to optimally control the disease, especially its more severe forms. This was documented by TURNER-WARWICK (1988); nocturnal asthma and sleep disruptions were found to be all too common problems even though bronchodilator and anti-inflammatory medications were taken as prescribed in equal doses at equal intervals.

G. β_2-Adrenergic Agonist Medications

β_2-Agonist medications may be administered in tablet or aerosol form. Conventional tablet and aerosol bronchodilator preparations have a limited effect duration, generally 4–6 h, while the new generation ones are effective for 12 h or longer. Conventional, short-acting β_2-agonist aerosol medications are relied on to alleviate acute asthma attacks and as a monotherapy to manage very mild and episodic forms of the disease. These β-adrenergic agents bind to cell membranes and activate adenylate cyclase, resulting in relaxation of airway smooth muscle and increase in airway patency and air flow rates. They may also affect airways inflammation as shown by in vitro and in vivo studies (DAHL et al. 1991; JOHNSON 1991; TWENTYMAN et al. 1990; WHELAN and JOHNSON 1992).

Effect Duration of Conventional β_2-Agonist Aerosols. The effect duration of conventional bronchodilator medications is rather short, generally no longer than 4 h (JOHNSON et al. 1992; KEMP et al. 1993). Thus, today these asthma medications are used primarily to control the exacerbation of acute asthma. Nonetheless, some clinicians continue to prescribe them as a maintenance therapy, recommending their administration at equal intervals and in equal doses, 4–5 times daily. Conventional bronchodilators are inappropriate for managing nocturnal asthma, since their effect duration is too short (BUSSE and BUSH 1988; JOAD et al. 1987; ZWILLICH et al. 1989). JOAD et al. (1987) compared conventional twice-daily theophylline tablet treatment alone or in combination with albuterol aerosol versus albuterol alone as maintenance therapy for a three-month period. Data on the clock times of drug administration and asthma symptoms were followed. Albuterol-only treatment was protective for about 4 h; the likelihood of asthma symptoms increased whenever the interval between self-administered doses was longer than this. Moreover, albuterol-only users experienced an excessively high incidence of nocturnal asthma between 04.00 and 08.00 h (Fig. 4). In contrast, the incidence of nighttime symptoms was moderated in twice-daily theophylline users, with or without albuterol.

Circadian Rhythm in Bronchodilator Effect of Conventional β_2-Agonist Therapy. The effect of short-acting β_2-agonist agents on pulmonary function depends on the circadian time of their administration. The peak expiratory flow (PEF), a measure of airway patency, was greatest when epinephrine was infused or inhaled in the afternoon at 16.00 h. It was least when administered during the night at 04.00 h. At all the times of treatment, the airways response to epi-

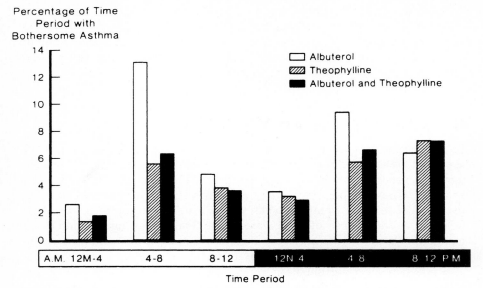

Fig. 4. The occurrence of asthma during each 4-h interval during the 24 h in 18 patients studied during month-long treatment of inhaled albuterol, sustained-release theophylline or both medications. Differences between the albuterol and theophylline-containing regimens were most apparent at night, between 04.00 and 08.00 h; during albuterol-only treatment, asthma was two-times more common than when sustained-release theophylline was used. These findings substantiate the effect-duration of albuterol is too short to be protective against nocturnal asthma. (Reproduced with permission from JOAD et al. 1987)

nephrine was dose dependent; the greater the dose, the greater the PEF level attained. (BARNES et al. 1982; FITZGERALD et al. 1980). However, the percentage increase in PEF, from the respective clock-time baseline level measured under placebo treatment, was greatest when drug administrations were timed at 04.00 h (Fig. 5). The findings were similar in studies with isoproterenol and orciprenaline aerosols (BROWN et al. 1988; GAULTIER et al. 1975; GAULTIER et al. 1988).

Chronotherapy of Sustained-Released (SR) Tablet β_2-Agonist Medications. SR albuterol, salbutamol, and terbutaline have been proven useful in the management of nocturnal asthma when dosed conventionally in equal doses at 12-h intervals (BOGIN and BALLARD 1992; ERIKSSON et al. 1982; ERIKSSON et al. 1988; GROSSMAN et al. 1991; MAESEN and SMEETS 1986). Nonetheless, significant decline in bronchial patency as well as asthma symptoms at night remain problematic.

POSTMA et al. (1985, 1986) and KOETER et al. (1985) were the first to evaluate an unequal morning-evening dosing regimen as a chronotherapy of terbulatine tablets (Brincanyl Depot AB Draco, Sweden). Daily doses were administered in proportion to biological need in terms of the circadian rhythm

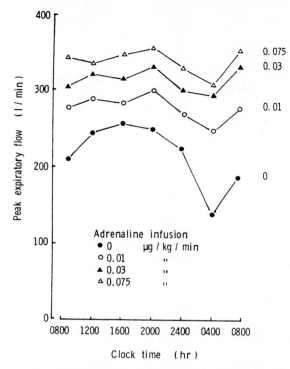

Fig. 5. Circadian rhythm-dependent effect of adrenaline infusion on peak expiratory flow (PEF), a measure of airway caliber, in presumably diurnally active asthmatic subjects. The response to adrenaline is dosage dependent; the greater the dosage, the greater the increase in PEF over the clock-time reference value determined under placebo infusion. Daytime adrenaline infusions resulted in greatest PEFs. However, the percent increase in PEF from baseline was greater when the drug was infused at 04.00 h. Day-night differences in adrenaline effect on pulmonary function result from circadian rhythmicity in β-adrenergic tone. (Reproduced with permission from FITZGERALD et al. 1980)

of pulmonary function; one third (5 mg) of the daily dose was ingested at 08.00 h and the remaining two thirds (10 mg) at 20.00 h. This dosing regimen increased the 24-h mean PEF and FEV_1 and averted the nocturnal decline in airway tone (Fig. 6). The advantage of this type of terbutaline chronotherapy was confirmed by DAHL et al. (1988).

The Proventil Repetabs (Schering, USA) tablet β_2-agonist medication has also been assessed as a chronotherapy for nighttime asthma. This pulse-release formulation essentially works as a tablet within a tablet. An outer color coat of 2 mg albuterol surrounds an inner subcoat. A third "barrier coat" is insoluble in the acid environment of the stomach. It is soluble in the alkaline environment of the small intestine, thus exposing the core of the tablet containing an additional 2 mg albuterol. As a result, one half of the 4 mg tablet dose is

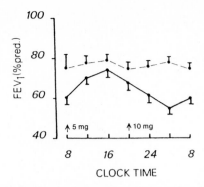

Fig. 6. Chronotherapy of sustained-release terbutaline achieved by an unequal morning (5 mg)–evening (10 mg) dosing regimen in diurnally active patients with reversible airways disease. The *upper curve* shows when one-third the daily dose of terbutaline was timed at 08.00 h and the remaining two-thirds at 20.00 h; the 24-h mean FEV_1 was significantly increased over placebo conditions (*lower curve*) and the nocturnal decline in airway caliber was averted. (Reproduced with permission from KOETER et al. 1985)

released from the outer core during the initial 6 h after ingestion, with the remaining 2 mg released from the barrier-protected inner core during the next 6 h. In this manner, therapeutic levels of albuterol are sustained for a full 12 h (BOGIN and BALLARD 1992; BOLLINGER et al. 1989; HUSSEY et al. 1991; POWELL et al. 1987).

STORMS et al. (1992) studied the chronotherapy of this formulation in a sample of 98 patients who demonstrated a nocturnal fall in PEF of at least 15% and who had a medical history of asthma-induced sleep disruption three nights per week or more. Participants received the study drug in a fixed dose of 4 mg in the morning and varying doses, from 4 to 16 mg, at bedtime. Compared to placebo treatment, the unequal Repetabs regimen resulted in reduction in both the number of nights when PEF declined by more than 15% and the number of nocturnal awakenings due to asthma.

SR albuterol tablet treatment as Volmax (Glaxo, UK; Muro, USA) has also been assessed as a chronotherapy of nighttime asthma. Volmax relies on osmotic-based drug delivery technology to achieve the controlled and steady release of albuterol throughout 12- or even 24-h dosing intervals (GROSSMAN et al. 1991; HIGENBOTTAM et al. 1989; PIERSON et al. 1990; SYKES et al. 1988). Investigations, published mainly in abstract form, document the efficacy and safety of an evening-only dosing regimen of Volmax. In one multicenter investigation, MOORE-GILLON (1988) recruited 34 patients experiencing asthma at least five nights per week. In comparison with placebo treatment, the evening-only 8 mg regimen of Volmax resulted in statistically significantly greater morning PEF plus an increased number of asthma-free mornings. CREEMERS (1988) compared the efficacy and safety of 8 mg evening Volmax dosing to 300 mg evening Theo Dur (Astra Draco, Sweden; Key-Schering, USA) dosing in 55 patients complaining of asthma-induced sleep disruption at least five nights per week. With reference to the run-in period, both medications were equally effective in reducing the number of sleep-disturbed nights. However, evening Volmax treatment better stabilized pulmonary function during the 24 h and better controlled daytime asthma symptoms than did evening-only Theo Dur treatment.

Most published reports pertaining to evening-only and unequal morning-evening SR β_2-agonist tablet chronotherapy document better control and stability of asthma and pulmonary function in comparison to traditional, equal-interval, equal-dose medication schedules. However, compliance by patients to unequal morning-evening dosing strategies is difficult. A new pro-drug of terbutaline, bambuterol (Bambec, Astra Draco, Sweden), exerts a bronchodilator effect for as long as 24 h with lower side-effect potential compared to other tablet terbutaline dosing forms. Bambuterol is the carbamate pro-drug of terbutaline, possessing high affinity for lung tissue and substantial presystemic metabolic stability. Hydrolysis and oxidation of bambuterol are controlled by plasma cholinesterase and cytochrome P-450 enzymes. Because plasma cholinesterase is reversibly inhibited by bambuterol in a dose-dependent manner, the metabolism of the pro-drug takes place in a slow and controlled fashion throughout the 24 h dosing interval (PEDERSEN et al. 1985; PERSSON et al. 1988).

D'ALONZO et al. (1995), using a double-blind, placebo-controlled protocol, assessed the differential effect of morning (07.00 h) versus evening (10.00 h) once-daily 20 mg bambuterol dosing on asthma control and pulmonary function in 25 patients with a medical history of nocturnal symptoms. Weekend around-the-clock studies were conducted to evaluate the kinetics and dynamics of the drug according to treatment timing under rigidly controlled experimental conditions, including meal composition and timing. Airway function and side effects were assessed every 3 h. Outpatient self-assessment diaries were also used to further study drug efficacy and side-effect profiles. Inpatient investigations demonstrated that increase in the 24 h mean FEV_1, relative to the placebo-control baseline, was independent of treatment timing (Fig. 7). In addition, both the morning and evening medication regimen improved pulmonary function overnight to an comparable extent. However, evening dosing was more advantageous since it improved PEF and FEV_1 to a much greater extent at 07.00 h, at the commencement of the daily activity span. Neither one of the bambuterol treatment regimens was associated with significant cardiovascular (tachycardia or hypertension) or nervous system (tremor) side effects.

Sustained-Effect β_2-Agonist Aerosol Medications. Formoterol (Ciba Geigy, Europe) and salmeterol (Glaxo, UK, and USA) are aerosol β-agonist agents; both exert prolonged bronchodilator activity, with reduced adverse-effect profile, making them practical for managing nocturnal and daytime asthma. The two medications possess a similar effect duration of 12 h (KEMPT et al. 1993; MAESEN et al. 1990a, 1990b), but formoterol may have a more rapid onset of effect (ARVIDSSON et al. 1991; KESTEN et al. 1992). To date, salmetrol

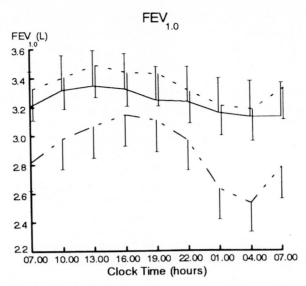

Fig. 7. Mean FEV_1 measured every 3 h throughout the day and night during three steady-state conditions of treatment – morning bambuterol versus evening bambuterol versus placebo – in 22 asthmatic patients. The *lower plot* represents the placebo treatment, and the *solid* and *upper dashed lines* indicate treatment with bambuterol in the morning and evening, respectively. The 24-h mean FEV_1 and, in particular, the 04.00 h FEV_1 were significantly greater for bambuterol versus placebo. Evening bambuterol dosing was superior to the other treatments since it increased the awakening (07.00 h) FEV_1 to the greatest extent. (Reproduced with permission from D'ALONZO et al. 1992)

has been studied more extensively than formoterol; thus, the major focus of this section will feature the former medication.

Salmetrol xinafoate is a chemical analogue of albuterol that is ten times more potent and about 750-fold more selective for β_2- versus β-adrenoceptors than albuterol. The two bronchodilators are similar in chemical structure except that salmetrol possesses an elongated side chain which is thought to bind the molecule firmly to the region of the β_2-adrenoceptor protein, allowing it to repeatedly excite the receptor while anchored to the adjoining exosite (BALL et al. 1991; JOHNSON 1991; KEMP et al. 1993). Salmeterol has a 12 h effect duration in asthma patients and attenuates bronchoconstriction induced by methacholine, histamine, allergens, cold air hyperventilation, and exercise (DEROM et al. 1990; MACONOCHIE et al. 1988; MALO et al. 1992; TWENTYMAN et al. 1990). Although these antiasthma agents are classified as bronchodilator medications, they exhibit substantial effect on airways inflammation. In vitro human lung tissue studies reveal that salmeterol inhibits antigen-induced mast cell mediator release of histamine, leukotriene C_4 and D_4, and prostaglandin D_2 and D_4 and also inhibits plasma protein extravasation (BUTCHERS et al. 1987; JOHNSON 1991; JOHNSON et al. 1991; JOHNSON et al. 1992; PERSSON 1986). Evidence of clinically relevant anti-inflammatory activity of salmeterol in man is inferred from BAL and allergen challenge studies (DAHL et al. 1991; FINNERTY et al. 1991; PEDERSON and DAHL 1990; PEDERSON et al. 1991; TWENTYMANN et al. 1990).

Numerous studies document the efficacy of the extended-duration aerosol formulations over conventional ones in the treatment of asthma. In general, the results of large-scale investigations show that twice-daily, 12-h dosing of 50 μg salmeterol results in better control of overnight and morning flow rates, decreases day and nighttime asthma frequency, decreases dependence on β_2-agonist aerosol rescue medication, and improves nighttime sleep compared to 180–200 μg albuterol inhaled four times daily and certain other types of inhaled or tablet treatment (BRAMBILLA et al. 1994; BRITTON et al. 1992; D'ALONZO et al. 1994; FITZPATRICK et al. 1990; LUNDBACK et al. 1993; MUIR et al. 1992; PEARLMAN et al. 1992; ULLMAN and SVEDMYR 1988; ULLMAN et al. 1990). Although sustained-action aerosols have better effect than conventional aerosols, they do not entirely alleviate the problem of nighttime asthma in many patients when dosed equally in the morning and evening.

To date, only one investigation has explored the possible advantage of timed salmetrol dosing on nocturnal asthma. FAURSCHOU et al. (1994) compared the safety and efficacy of a regimen of 50 μg salmeterol given in equal doses twice daily with a dose of 100 μg given only at night versus placebo in 41 poorly controlled adult asthmatics. The study used a double-blind, randomized, crossover protocol with each treatment period lasting three weeks. Other prescribed asthma medications were continued unaltered in dose and timing during all treatment arms, and albuterol was available for rescue from acute asthma. Morning and evening PEF values were comparable with twice-daily and evening-only salmeterol dosing; however, there was greater diurnal variation in PEF with evening-only than with twice-daily salmeterol dosing. Both β_2-agonist treatment regimens were equally effective in controlling day and nighttime asthma, and there was equal preference for the two salmeterol treatment regimens by patients. Salmeterol had no effect on blood pressure or heart rate. This single investigation infers that once-nightly dosing of 100 μg salmeterol is effective in controlling day and night asthma in poorly controlled patients reliant on oral and inhaled glucocorticoids and/or SR theophylline. However, further investigation is warranted to determine if unequal morning-evening or evening-only salmeterol and formoterol dosing regimens will further enhance drug effectiveness in patients vulnerable to nighttime asthma, as demonstrated previously for SR tablet β_2-agonist medications.

H. Anticholinergic Agents

Cholinergic tone increases during the nighttime, and it has been hypothesized that this contributes to the worsening of asthma nocturnally. Moreover, the amplitude of the circadian rhythm in vagal tone might be amplified in asthmatic patients due to exacerbation of airways inflammation and associated up-regulation of muscarinic receptors at night. Reports on the effect of anticholinergic medications on nocturnal asthma seem to be inconsistent; some investigators have found them to moderate asthma symptoms and attenuate the nocturnal decline of pulmonary function, while others have not (COE and BARNES 1986; CATTERALL et al. 1988; COX et al. 1984; SLY et al. 1987; MORRISON et al. 1988; WOLSTENHOLME and SHETTAN 1989). Since vagal tone differs greatly during the 24 h, failure of conventional anticholinergic dosing regimens to have impact on pulmonary function overnight could be due to too low dosing at bedtime.

Administration-time-dependent differences in the effect of anticholinergic medications have hardly been investigated. GAULTIER et al. (1975) explored the airway response of normal children to 80 or 200 μg of inhaled ipratropium bromide at different clock times. The low dose was most effective in reducing lung resistance when inhaled in the morning at 07.00 h. The 80 μg dose had no effect when inhaled in the evening at 22.00 h, but the higher dose of 200 μg did.

The findings of other investigations are germane to this discussion. COE and BARNES (1986) assessed the effect of low (0.2 mg) and high (0.4 mg) dose oxitropium bromide bedtime treatment of 18 asthma patients for two-week durations. Low-dose bedtime treatment was ineffective in attenuating the nocturnal dip in the group mean PEF. In contrast, high-dose bedtime treatment resulted in a 40% moderation of the decline in PEF overnight compared to placebo. Individual differences were noted between subjects in their response to oxitropium bromide; half showed a dose-response to the drug: the greater the bedtime dose, the greater the attenuation in the PEF dip overnight. The other subjects were nonresponsive to the medication.

CATTERALL et al. (1988) evaluated the effect of 1 mg of nebulized ipratropium bromide in patients with nocturnal bronchoconstriction. Ipratropium bromide and saline were administered by nebulizer at 22.00 h, 02.00 h, and 0.6.00 h in a double-blind, randomized, placebo-controlled protocol. PEF values were increased throughout the night by ipratropium bromide treatment compared to placebo with moderation of the nocturnal decline in airway caliber.

MORRISON et al. (1988) investigated the effect of a large dose of 30 μg/kg atropine, to induce vagal blockade, and of placebo on PEF when dosed intravenously at 16.00 h and 04.00 h. All patients exhibited a 20% or greater dip in PEF overnight prior to the study. The 04.00 h atropine infusion significantly improved the group mean PEF, from 260 L/min with placebo to 390 L/min with drug. The 16.00 h atropine infusion had a lesser, but nonetheless statistically significant, effect on PEF. PEF was increased from 400 L/min with placebo to 440 L/min with atropine (Fig. 8).

GAULTIER et al. (1975), COE and BARNES (1986), CATTERAL et al. (1988), and MORRISON et al. (1988) found that late night and/or early morning dosing of anticholinergic medication attenuates the nocturnal fall in PEF. Taken collectively, the results of these four studies infer that anticholinergic preparations must be administered in high dose at night to be effective in stabilizing the airways of asthmatics overnight. Poor response to these medications may

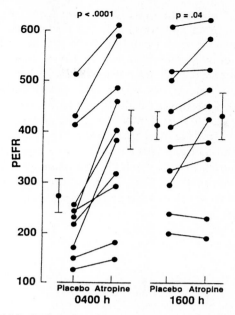

Fig. 8. Effect of vagal blockade induced by intravenous infusion of atropine at 16.00 h and 04.00 h on peak expiratory flow rate (PEFR) of 10 asthmatic subjects. Vagal blockade results in significant bronchodilation at both treatment times; however, the magnitude of improvement in PEFR from the pretreatment baseline level was greatest with the 04.00 h atropine treatment. (Reproduced with permission from MORRISON et al. 1988)

be due to inadequate dose intensity at night, a too short effect duration of agents, poor drug delivery technique of patients, and/or nonspecificity of current agents for muscarinic receptor subtypes. Further study is required to fully assess the time dependency of anticholinergic medications and to determine the relevance of circadian rhythm-adapted dosing strategies.

I. Theophylline

Theophylline is an antiasthma medication that has long been used in the management of nocturnal asthma. It is considered to be a weak bronchodilator compared to β_2-agonist agents, possessing significant anti-inflammatory action (BARNES and PAUWELS 1994; KIDNEY et al. 1994; PAUWELS et al. 1985; SULLIVAN et al. 1994; WARD et al. 1993).

Chronokinetics of SR Theophylline Preparations. Numerous investigations as reviewed elsewhere (Smolensky 1989; Smolensky et al. 1987a; Smolensky et al. 1987b) demonstrate that the kinetics of SR theophylline formulations are different following morning than following evening administration. Most theophyllines exhibit a shorter T_{max} and greater C_{max} when ingested at the commencement of the daily activity span than before bedtime. In general, these effects are more pronounced in children than in adults. Even theophylline bioavailability may differ significantly with administration time (Fig. 9). Time-dependent differences in theophylline kinetics result primarily from circadian rhythms in gastrointestinal processes and functions (Goo et al. 1987; MOORE and ENGLERT 1970; REINBERG et al. 1991; SMOLENSKY 1989; VENER and MOORE 1988; LEMMER and BRUGUEROLLE 1994; LEMMER and NOLD 1991). Day-night differences in drug absorption, rather than drug biotransformation and elimination, seem to be involved (SMOLENSKY 1989). Dosing-time variation in theophylline kinetics is not the result of meal contents and timings, although differences in posture during the 24 h have been reported to contribute to or amplify the described circadian rhythm dependencies (SMOLENSKY 1989; WARREN et al. 1985).

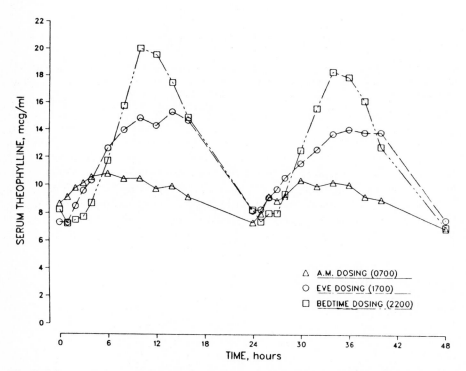

Fig. 9. Mean steady-state serum theophylline concentrations (STC) during 48-h study spans in healthy subjects administered 900 mg Theo-24 (Whitby, USA) once daily at 07.00 h 15.00 h, or 22.00 h. The area under the theophylline-time concentration curve (AUC) differed significantly according to treatment time. AUC was about threefold greater with once-daily 22.00 h than 07.00 h treatment. (Reproduced with permission from SMOLENSKY et al. 1987b)

Chronoeffectiveness of Theophylline. Adverse effects of theophylline may be circadian rhythm-dependent. In rodents, tolerance to theophylline overdose (LD_{50} concentration administered intraperitoneally) is much greater when timed between the middle of the activity span and the beginning of the rest period than at other times during the 24 h. Theophylline-induced mortality in the rodents varied by sixfold according to the circadian timing of the drug overdose (KYLE et al. 1979). However, no studies of day-night differences in theophylline toxicity in humans are known.

Chronotherapy of Theophylline Tablets. The chronotherapy of theophylline entails the delivery of a greater amount of medication during the night when the risk of dyspnea is high and a lesser amount during the day when the risk is lower. The Byk Gulden company in Germany appears to have been the first pharmaceutical company to embrace the concept of theophylline chronotherapy. During the 1980s, it marketed Euphyllin; asthmatics who normally metabolized the drug were prescribed one third the daily dose in the morning at the commencement of the activity span and the remaining two thirds before bedtime (DAROW and STEINIJANS 1987; SCHULZ et al. 1984). This dosing regimen optimizes the effect of theophylline in patients suffering from nocturnal asthma; however, the atypical dosing regimen was not convenient.

Presently, several once-a-day theophylline formulations are marketed for the chronotherapy of nighttime asthma. Two of the most popular ones in Asia, Europe, and the Americas are Euphylong (Byk Gulden, Germany) and Uniphyl/Uniphyllin (Purdue Frederick, USA; Mundipharma, Germany) (D'ALONZO et al. 1990; GOLDENHEIM et al. 1987; MARTIN et al. 1989; NEUENKIRCHEN et al. 1985; STEINIJANS et al. 1986). Figure 10 compares the pharmacokinetics and dynamics of these two preparations to conventional SR ones administered in comparable daily doses. The drug delivery design of the conventional medications is intended to ensure constancy in theophylline levels throughout the 24 h when dosed at equal (12-h) intervals and in equal amounts. The graphs clearly reveal that conventionally formulated theophyllines fail to avert the decline of airway patency overnight. In contrast, the theophylline chronotherapies that deliver a greater amount of medication overnight strongly moderate the nocturnal dip in pulmonary function and lessen the risk of asthma. The increased nighttime protection against asthma derived from theophylline chronotherapy is achieved without compromise of daytime drug effectiveness. Both types of theophylline preparations exert a comparable effect on airway patency during the daytime, even though the once-daily ones give rise to lower serum drug concentrations during the daytime.

The possibility of a day-night difference in the theophylline dose-pulmonary function response was investigated by D'ALONZO et al. (1990). These authors compared the dose-response relationship of theophylline derived from Theo Dur (Schering, USA; Astra Draco, Sweden) as well as Euphylong (Byk Gulden, Germany). Dose-response assessments were explored using the serum theophylline concentrations and FEV_1 values obtained between 02.00 h and

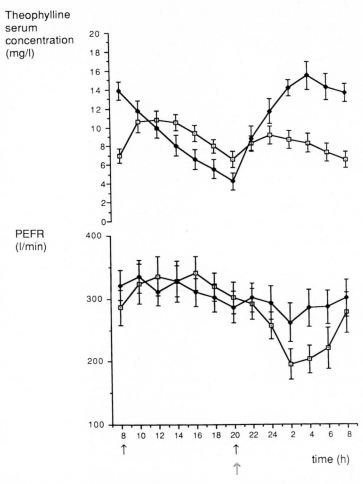

Fig. 10. a Comparison of serum theophylline concentration (STC) and PEFR with 20.00 h once-daily Uniphyllin (*solid line*) chronotherapy versus 08.00/20.00 h twice-daily (*dashed line*) Phyllotemp treatment of nine asthmatic patients while hospitalized. Equal-interval, equal-dose theophylline dosing resulted in little temporal variation in drug level during the 24 h; in contrast, the once-daily treatment resulted in great STC fluctuation with elevated drug levels over night and reduced ones during the daytime. The nocturnal decline in PEFR was great with twice-daily theophylline; it was significantly attenuated with the evening theophylline chronotherapy. Arrows along x-axis indicate times of once- or twice-daily theophylline treatment. **b** Mean FEV_1, PEFR, and STC (Δ, placebo baseline; o, twice-daily, 12h theophylline with Theo-Dur; • once-daily theophylline chronotherapy with Euphylong administered at 20.00 h). Both theophyllines improved airflow over the entire 24 h compared to placebo. However, a significantly higher FEV_1 and PEFR was achieved by Euphylong between 02.00 and 06.00 h, the time of greatest asthma risk, and when STC from this chronotherapy was most elevated; pulmonary function parameters were comparable for the two theophylline medications at all other times of the day and night. (Reproduced with permission from NEUENKIRCHEN et al. 1985 and D'ALONZO et al. 1990)

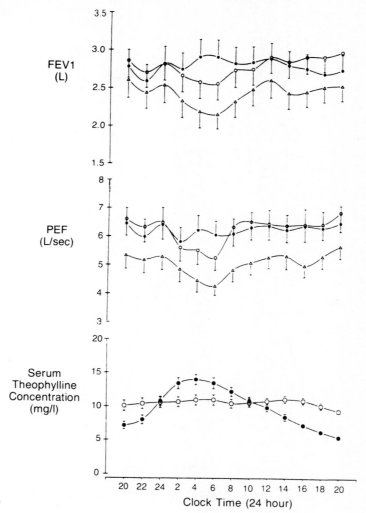

Fig. 10b

06.00 h, a time when asthma risk is great, and also between 14.00 h and 18.00 h when asthma risk is low. No serum drug concentration-airways response relationship was detected with Theo Dur, which produces near constancy in medication levels during the 24 h, neither during the overnight nor afternoon time periods. In contrast, for Euphylong, which delivers more drug during the night than day, a statistically significant relationship was detected for the overnight period only (Fig. 11). Similar results were reported by REINBERG et al. (1987) for another once-a-day theophylline. These findings support the conclusion that the airways response to theophylline during the night is dependent on the nocturnal serum concentration; the greater the nighttime drug level within the therapeutic range, the greater the effect.

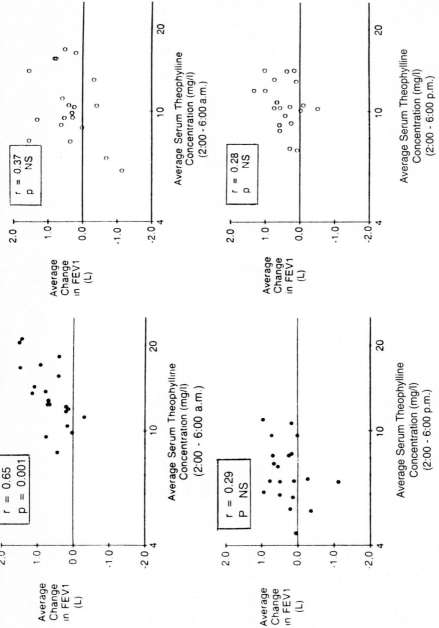

Fig. 11. Average increase from baseline in FEV$_1$ between 02.00 and 06.00 h as well as 14.00 and 18.00 h in relation to average STC during steady-state conditions for evening-only Euphylong (*left*) or twice-daily Theo-Dur (*right*) dosing. Correlation between STC and improvement in FEV$_1$ over placebo values was established only between 02.00 and 06.00 h with evening Euphylong dosing. No correlation was detected during the daytime (14.00 and 18.00 h) with either theophylline preparation or during the nighttime (02.00–06.00 h) with twice-daily Theo-Dur. (Reproduced with permission from D'Alonzo et al. 1990)

The manner in which theophylline exerts its therapeutic effect is currently highly debated. The drug was initially thought to exert bronchodilation through inhibition of one or more of the phosphodiesterase enzyme subtypes. The current consensus is that theophylline possesses significant anti-inflammatory action (BARNES and PAUWELS 1994; SULLIVAN et al. 1994; WARD et al. 1993). In vitro studies demonstrate theophylline decreases the release of: mediators from mast cells; reactive oxygen species from macrophage cells; and basic protein from eosinophils. Furthermore, it inhibits T-lymphocyte cell profileration and cytokine release and decreases the release of reactive oxygen species from neutrophils. In experimental animals, theophylline decreases: the late response to allergen in guinea pigs; airway responsiveness to allergen and platelet activating factor in guinea pigs and sheep; airway inflammation induced by endotoxin and allergen in guinea pigs and rats; and plasma exudation in guinea pigs. In asthmatic persons, theophylline inhibits the late response to allergen, increases $CD8^+$ cells in peripheral blood, and decreases T-lymphocytes in airways.

The exact mechanisms underlying the advantage of nighttime theophylline chronotherapy for nocturnal asthma remain unclear. Evening treatment with Uniphyl in a dose producing an average serum drug concentration at 04.00 h of 15.4 ± 1.4 µg/ml significantly depressed neutrophil content as well as stimulated alveolar macrophage leukotriene B_4 production (LTB_4 is representative of the lipoxygenase pathway) compared to placebo in BAL obtained at 04.00 h from asthma patients (KRAFT et al., 1996; TORVIK et al. 1994; TRUDEAU et al. 1994). Moreover, the drug-induced attenuation of the nocturnal decline in FEV_1 was correlated ($r = 0.53$; $p < 0.01$) with the decrease in neutrophil cell numbers. These studies confirm that theophylline possesses anti-inflammatory action; however, complete understanding of the mechanisms of the dose-response relationship of the medication in patients in terms of bronchodilator and antiphlogistic effects is yet sought.

J. Chronotherapy of Glucocorticoids

Glucocorticoid Tablet and Aerosol Pharmacotherapy. Since airways inflammation constitutes a primary characteristic of asthma, inhaled corticotherapy is recommended at an early stage as a first-line treatment of the disease (BRITISH THORACIC SOCIETY 1993; SHEFFER 1991; SHEFFER 1992; WETTENGEL et al. 1994). Glucocorticoids have direct inhibitory effects on lung macrophages, T-lymphocytes, eosinophils as well as epithelial and other cells involved in airways inflammation (BARNES 1995). These medications decrease eosinophil cell numbers, especially low density ones, by inducing apoptosis, and they reduce mast cell density in lung tissue. Glucocorticoids decrease mucus secretion and inhibit plasma exudation. Aerosol medications are preferred for the long-term management of all forms of asthma in which chronic inflammation is characteristic. Generally, tablet formulations are restricted for use in the treatment of severe exacerbations of dyspnea and severe, unstable asthma.

Circadian Rhythms in Corticosteroid Pharmacokinetics. The chronokinetics of tablet medications have hardly been studied (ENGLISH and MARKS 1981; ENGLISH et al. 1983; MCALLISTER et al. 1981; MORSELLI et al. 1970). ENGLISH et al. (1983) assessed the kinetics of prednisolone tablets (2 mg/kg) in a group of diurnally active healthy subjects. Studies were conducted under carefully controlled conditions at four different times of the day: 06.00 h, 12.00 h, 18.00 h and 00.00 h, each clock-time trial being separated by a 7-day interval. The C_{max} and AUC were greatest and elimination half-life shortest after the 12.00 h drug ingestion; T_{max} and apparent volume of distribution were greatest with the 18.00 h dosing.

Circadian Rhythms in Corticosteriod Pharmacodynamics. Time-dependent differences in the effects of corticosteroid medications have been investigated mainly in regard to adrenal suppression (drug chronotolerance) and to a lesser extent pulmonary function. CERESA et al. (1969) in a series of trials infused methylprednisolone (MP) at a rate of 660 μg/h or placebo at different clock times to diurnally active healthy subjects. When the single MP infusion was timed between 08.00 h and 16.00 h, cortisol secretion remained normal. In contrast, when MP was delivered during the late afternoon and early evening, between 16.00 h and 20.00 h, cortisol suppression was moderate. When the drug infusion was timed between 00.00 h and 04.00 h, cortisol secretion was markedly suppressed (Fig. 12). The day-night difference in MP effect on cortisol secretion is due to the circadian rhythm in drug-induced HPA axis inhibition; ACTH inhibition is more likely when MP is infused late in the day and at night rather than during the morning and early afternoon hours. Additional clinical studies conducted primarily during the 1960s and 1970s extended the findings of CERESA et al. (1969). Morning once-daily ingestion of small to moderate doses of corticosteroid tablet medications results in little or no adrenocortical suppression, while the same daily dose split into four equal administrations to coincide with daily meals and bedtime results in very significant HPA axis suppression (REINBERG 1989; REINBERG et al. 1988).

Investigation of circadian differences in the beneficial effect of MP on airway function began more than 20 years ago by REINBERG et al. (1974). Twelve symptom-free, diurnally active, male asthmatic children between 7 and 12 years of age received single injections of 40 mg of MP at 03.00 h, 07.00 h, 15.00 h, and 19.00 h in random order during different weeks. PEF was self-assessed five times daily for 48 h before and after each injection. With reference to the 48-h baseline PEF, MP was effective in improving airway status at all times of treatment; however, the medication was most effective when administered in the afternoon at 15.00 h (Fig. 13).

Investigations by BEAM et al. (1992) confirm and extend the findings of REINBERG et al. (1974). Seven diurnally active patients with a history of nighttime asthma and documented decline of PEF by at least 15% were treated. Using a double-blind, placebo-controlled crossover protocol, patients received either placebo or a single 50 mg prednisone tablet dose at 08.00 h,

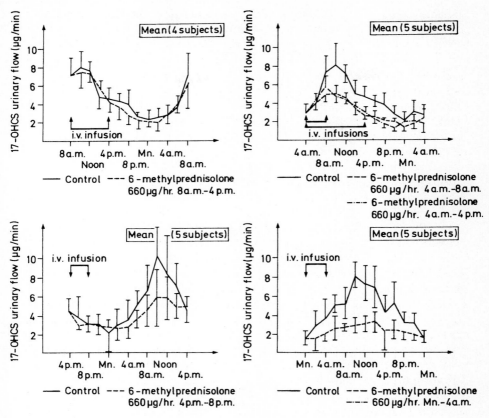

Fig. 12. Infusion of 6-methylprednisolone (660 μg/h) during different 4-, 8-, or 12-h spans of the day or night had a differential effect on adrenal suppression as gauged by changes in urinary concentration of 17-OHCS (the metabolic by-product of cortisol). When the corticosteroid was infused around the commencement of the daily activity span, no adrenal suppression resulted; the 17 OHCS level and circadian pattern following saline and drug infusion were comparable. In contrast, moderate to severe adrenal suppression resulted when methylprednisolone was infused late in the afternoon, evening, and after midnight. (Reproduced with permission from CERESA et al. 1969)

15.00 h, and 20.00 h on different occasions. Prednisone exerted the best effect in attenuating the nocturnal decline in FEV_1 when ingested in the afternoon at 15.00 h. The average percentage decline in the overnight FEV_1 with the 15.00 h placebo treatment amounted to 28.3%; with the 15.00 h glucocorticoid treatment, the decline amounted only to 10.4%. The 08.00 h and 20.00 h prednisone ingestions were ineffective in moderating the overnight dip in FEV_1 (Fig. 14). Moreover, only at the 15.00 h administration was there significant reduction in blood eosinophil count and pancellular reduction in BAL cytology (Fig. 15). This investigation infers that once-a-day afternoon glucocorticoid dosing best attenuates both the overnight decline of pulmonary function and the increase

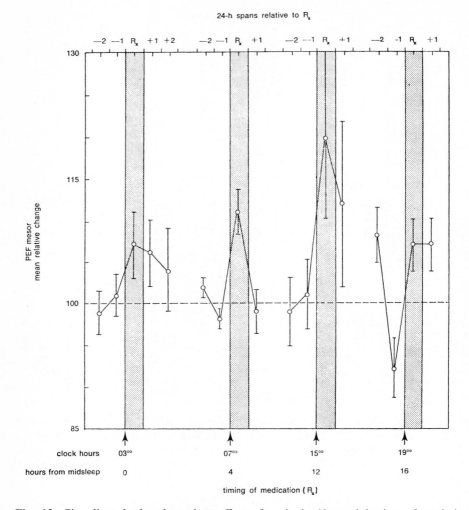

Fig. 13. Circadian rhythm-dependent effect of a single 40 mg injection of methylprednisolone (MP) on the PEF of diurnally active asthmatic boys (7–15 years of age). PEF was assessed every 2 h throughout diurnal activity before, during, and following treatment. The 24-h means for the 2 days prior to MP treatment were averaged and set equal to a normative baseline value of 100% with the MP-induced PEF increase expressed relative to it. MP was effective in increasing PEF at all treatment times; however, the effect was greatest when it was injected in the afternoon at 15.00 h. (Reproduced with permission from REINBERG et al. 1977)

in airways inflammation. However, additional study is yet required to explore whether afternoon-only glucocorticoid chronic dosing of diurnally active asthmatics constitutes the ideal chronotherapy of prednisone tablet medication. Under normal conditions, serum cortisol peaks at the commencement of

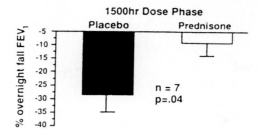
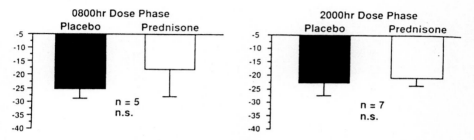

Fig. 14. Effect of a timed single oral dose of 50 mg prednisone versus placebo on the overnight decline in pulmonary function in diurnally active asthmatic persons with a history of nighttime symptoms. Prednisone dosing at 08.00 and 20.00 h was no better than placebo in moderating the nocturnal decline in FEV_1. In contrast, prednisone compared to placebo at 15.00 h resulted in a statistically significant attenuation of the nighttime fall in FEV_1. (Reproduced with permission from BEAM et al. 1992)

the diurnal activity period. Since the circadian rhythm in cortisol clocks numerous metabolic and other critical biological functions, the afternoon once-daily corticotherapy regimen may give rise to a disrupted (desynchronized) circadian time structure as a side effect. To date, this chronobiologic concern has yet to be assessed.

Corticosteroid Tablet Chronotherapy. The daily and alternate-day morning glucocorticoid tablet dosing regimen (Medrol, Upjohn USA) was the first chronotherapy widely adopted into clinical medicine (HARTER et al. 1963). A second chronotherapy of synthetic corticosteroids (Dutimelan, Hoechst, Italy) was introduced during the past decade in selected European countries. Dutimelan is a corticosteroid formulation consisting of a different mix and concentration of synthetic glucocorticoids for morning (08.00 h) and afternoon (15.00 h) dosing. The clock timing, drug composition, and dose strength of the twice-daily administrations are intended to simulate the circadian rhythm of adrenocorticoid secretion of diurnally active persons. The morning strong dose consists of 7 mg prednisolone acetate and 4 mg prednisolone alcohol, and the weaker afternoon dose consists of 3 mg prednisolone alcohol and 15 mg cortisone acetate. Dutimelan is marketed in two different strengths; regular, as

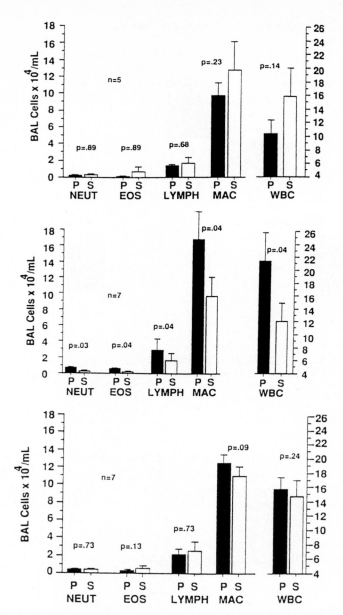

Fig. 15. Comparative effect of a timed single oral dose of placebo versus 50 mg prednisone on 04.00 h bronchoalveolar lavage (BAL) fluid cytology. Compared to placebo, prednisone tablet dosing at 08.00 h (*top*) and 20.00 h (*bottom*) had no effect on 04.00 h BAL cytology. In contrast, prednisone dosing at 15.00 h resulted in a significant pancellular reduction in the 04.00 h BAL cytology (*middle*). The anti-inflammatory effect of the glucocorticoid was realized by afternoon, but not morning or afternoon, dosing. P, placebo; S, steroid; Neut, neutrophils; Eos, eosinophils; Lymph, lymphocytes; Mac, macrophages; WBC, total leukocytes. (Reproduced with permission from BEAM et al. 1992; MARTIN 1993)

described above, and "mite," which contains the same combination of corticosteriods but in one half the dose. Trials by REINBERG et al. (1977) demonstrated the efficacy of this chronotherapy for steroid-dependent asthma. During a one-month course of Dutimelan treatment, PEF improved significantly without induction of adrenal inhibition or other side effects (Fig. 16). Comparable findings were published by CREPALDI and MUGGEO (1974) and SERAFINI and BONINI (1974).

Since asthma is a nocturnal disease, many practitioners believe glucocorticoids should be administered before bedtime to optimize their effects. The findings of Beam et al. (1992) showed a single large tablet dose of prednisone at 20.00 h was ineffective in averting the nighttime deterioration of airway function. REINBERG et al. (1974) also found large-dose evening MP injection to be only marginally effective in improving pulmonary function in asthmatics. In another investigation, REINBERG et al. (1983) compared the effect of the twice-daily Dutimelan medication regimen given as intended – strong dose in the morning at 08.00 h and weak dose at 15.00 h with placebo at 20.00 h (designated as DTM_{8-15}) – versus a regimen in which the strong dose was given at 20.00 h and the weak one at 15.00 h with placebo at 08.00 h (designated as DTM_{15-20}). Eight asthmatic patients were studied using a double-blind, crossover protocol with each DTM treatment regimen administered for eight days duration. DTM_{8-15} in comparison to DTM_{15-20} was more effective in improving and stabilizing pulmonary function throughout the entire 24 h (Fig. 17). These results are consistent with the conclusion that evening-only dosing of synthetic corticosteroids does not result in an optimization of drug effect on nocturnal asthma.

Aerosol Corticosteroid Therapy. Aerosol corticotherapy is advocated as a first-line asthma treatment. Inhaled glucocorticoids are highly lipophilic; glucocorticoid-receptor complexes in the nucleus affect gene transcription activities and inhibit the formation of various cytokines involved in airways inflammation. Inhaled glucocorticoids markedly reduce the number of mast cells, T-lymphocytes, and eosinophils in the bronchial epithelium and submucosa, and they reverse the shedding of epithelial cells and goblet-cell hyperplasia (BARNES 1995). Reduction in airways inflammation results in reduced airways hyperresponsiveness to antigen and nonspecific chemical provocation. (BARNES 1990).

Inhaled glucocorticoid treatment became popular initially as a means of reducing or even eliminating the requirement for tablet corticosteroid doses by severe asthma patients (REED 1980). Inhaled glucocorticoids are now routinely prescribed for the management of airways inflammation in asthma (BRITISH THORACIC SOCIETY 1993; SHEFFER 1991, 1992; WETTENGEL et al. 1994).

Beclomethasone dipropionate, budesonide, flunisolide, fluticasone, and triamcinolone are the most common aerosol glucocorticoid preparations marketed, although not all are available in every country. Desired properties of inhaled glucocorticoids are high binding affinity, high topical potency, low systemic bioavailability, and rapid systemic clearance. The preparations differ in their in vitro topical potency and binding affinity to glucocorticoid receptors (BARNES 1995). The newer fluticasone medication possesses greatest binding affinity and anti-inflammatory potency.

The opportunity for systemic absorption of inhaled glucocorticoids can be great. About 80%–90% of the inhaled dose is deposited in the oropharynx and swallowed; thus, only about 10% of each drug dose reaches the lung (BARNES 1995). Failure to use a spacer for drug delivery or rise one's mouth after dosing may result in significant systemic drug absorption. Flunisolide and budesonide undergo extensive first pass liver metabolism; thus, relatively low amounts of these medications circulate systemically (CHAPLIN et al. 1980; RYRFEDT et al. 1982). Fluticasone possesses low oral availability, thereby minimizing its potential to induce systemic effects (HOLLIDAY et al. 1994). Beclomethasone dipropionate is metabolized by lung tissue to the more active monopropionate form; however, there is no useful human data regarding its absorption and metabolism (WURTHWEIN and ROHDEWALD 1990).

Various types of side effects have been ascribed to the long-term use of beclomethasone dipropionate, budesonide, and triamcinolone, as reviewed elsewhere (BARNES 1995; SMOLENSKY and D'ALONZO 1993). In general, low-dose aerosol therapy (up to 400 ug/day in children and up to 800 ug/day in adults) is well tolerated with minimal systemic effects. However, adverse effects have been reported with long-term use of moderate and high-dose regimens (SMOLENSKY and D'ALONZO 1993). More severe asthmatic patients are occasionally treated with high-dose aerosol as well as tablet glucocorticoids when it is necessary to control severe exacerbations of the disease. Thus, in this type of patient it is not always possible to determine if noted systemic side effects result from long-term use of aerosol and/or episodic tablet corticotherapy (BARNES 1995). Nonetheless, side effects of inhaled corticosteroids have been reported even in patients seldom treated with tablet corticotherapy. Studies of tablet and infused corticosteroid preparations show that circadian timing is an important factor in the risk and extent of HPA axis suppression.

Chronotherapy of Inhaled Glucocorticoids. Daily dosing of inhaled glucocorticoids that possess low first pass effect late in the day, in particular with supper and before bedtime, may be responsible for adrenal suppression and perhaps other reported side effects observed in patients treated with such medications for many months and years. This concern was explored by TOOGOOD et al. (1982) as part of a more comprehensive investigation. TOOGOOD et al. assessed

Fig. 16. Circadian rhythm of urinary 17-OHCS excretion (*top*) 2 days before (no R_x) and at 10-day intervals plus weekly mean 24-h peak expiratory flow and self-rated dyspnea levels (bottom) in nine diurnally active asthmatics undergoing 1-month chronocorticotherapy with low-dose (mite formulation) Dutimelan 8–15. The 1-month chronocorticotherapy increased PEF and decreased dyspnea markedly without induction of adrenal suppression or alteration of the circadian rhythm in 17-OHCS excretion. (Reproduced with permission from REINBERG et al. 1977)

Progress in the Chronotherapy of Nocturnal Asthma

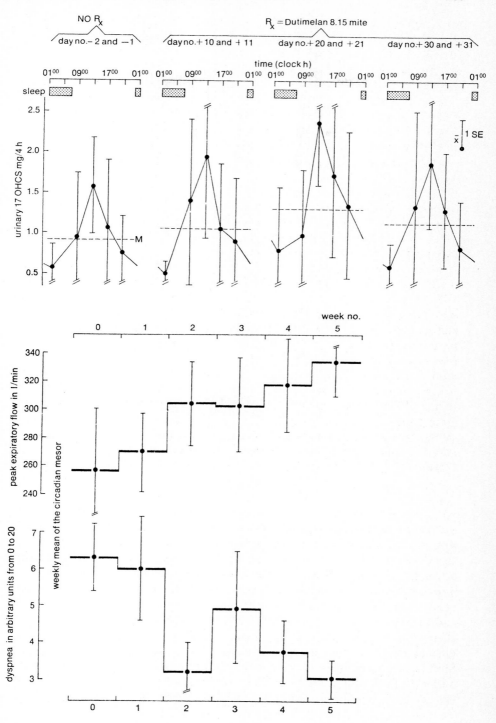

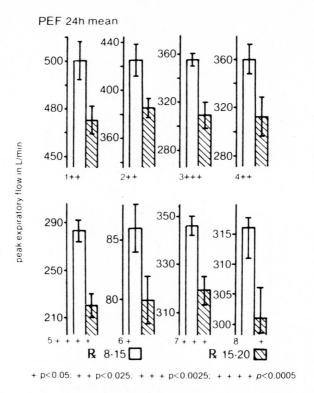

Fig. 17. Effect of 8-day treatment with DTM8-15 versus DTM15-20, in the same total daily dose, on mean 24-h PEF in eight steroid-dependent asthmatics. The effect was best in all asthmatic participants when the largest corticosteroid dose was administered at the start of the activity span at 08.00 h rather than in evening at 20.00 h as part of a twice-daily glucocorticoid treatment regimen with the second dose being fixed at 15.00 h always. (Reproduced with permission from REINBERG et al. 1983)

the factors of dosing frequency (twice- versus four-times-daily drug administration), circadian timing (morning-only versus morning and evening dosing), and dose strength of budesonide on PEF and HPA axis. A significant interaction between daily dose and administration schedule was detected (Fig. 18); morning dosing conserved endogenous cortisol secretion for budesonide doses greater than 800 µg/day. On the other hand, the morning-only (either 08.00 h and 12.00 h or 07.00 h, 09.00 h, 10.30 h, and 12.00 h) dosing schedules were less effective in improving pulmonary function, although not in a statistically significant manner, than the equal-interval ones of comparable dose strength. Shift from a four-times-a-day (08.00 h, 12.00 h, 17.30 h and 22.00 h) to a twice-a-day (08.00 h and 22.00 h) dosing frequency was associated with a loss of efficacy on pulmonary function as indicated by PEF self-assessments.

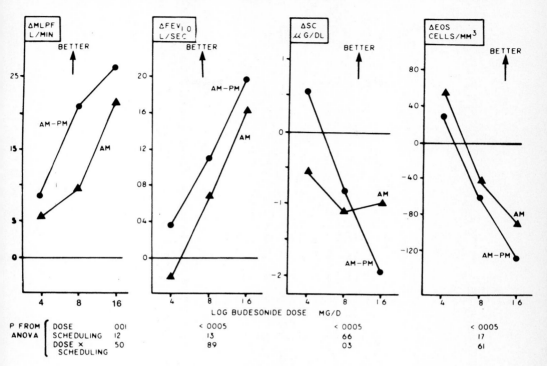

Fig. 18. Effect of budesonide aerosol dose timing, strength, and frequency on change in airways caliber (peak flow: MLPF and FEV_1), serum cortisol ($\triangle SC$), and blood eosinophils ($\triangle EOS$) in asthmatics. The strength of the daily budesonide dose was significantly related to effect on pulmonary function, serum cortisol, and eosinophils. Morning steroid dosing schedules were less adrenal suppressive than multiple daily ones, particularly at the high daily dose of 16 mg. Regimens consisting of both morning and evening dosings were more effective in increasing pulmonary function and reducing blood eosinophils, although not significantly so than those consisting of morning only dosing. (Reproduced with permission from TOOGOOD et al. 1982)

PINCUS et al. (1995) were also concerned with the chronotherapy of inhaled glucocorticoids. They compared a 15.00 h once-daily 800 μg regimen of triamcinolone acetate aerosol with a four-times daily 800 μg one in two comparable groups of asthmatic patients. The 15.00 h treatment time was selected based on the previous findings showing that afternoon-high-dose MP injection and prednisone tablet ingestion produced the best effect on pulmonary function and indicators of airways inflammation (REINBERG et al. 1974; BEAM et al. 1992). Aerosol treatment was administered daily for four weeks during which PEF was self-assessed in the morning and at bedtime, with serum cortisol and urinary 17-OHCS studies carried out before and at the end of treatment. Improvement from the baseline FEV_1 level was comparable between treatments as were the morning and bedtime self-assessed PEFs (Fig. 19). There was a trend ($p < 0.07$) of greater improvement in the bedtime PEF from baseline with the

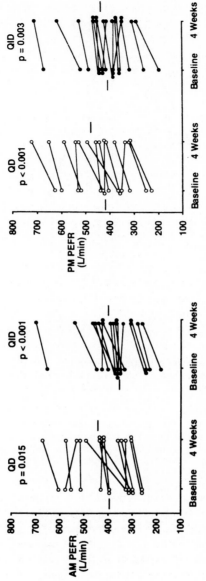

Fig. 19. Morning (*left*) and evening (*right*) PEF at before and after 4-week treatment with 800 μg triamcinolone acetate aerosol dosed all at once at 15.00 h or split into four equal doses throughout the diurnal activity span. Improvement in PEF from baseline was comparable for the two glucocorticoid aerosol treatment regimens. (Reproduced with permission from PINCUS et al. 1995)

once-daily afternoon triamcinolone dosing regimen in comparison to the four-times daily one. The systemic response to the two dosing regimens as assessed by eosinophil count, morning serum cortisol concentration, and 24 h urinary cortisol studies was comparable. Moreover, both groups experienced comparable reduction in the requirement for supplemental β_2-agonist medication.

TOOGOOD et al. (1982), MALO et al. (1989), DAHL and JOHANSSON (1982), among other investigators, found dosing frequency to be critical for achieving an optimal therapeutic effect of aerosol corticosteroid medications. In contrast, PINCUS et al. (1995) demonstrated the circadian timing of once-a-day inhaled triamcinolone to be a factor of at least equal importance as dosing frequency. The studies of BEAM et al. (1992) infer that morning and bedtime dosings of aerosol glucocorticoids are less effective in controlling airways inflammation than is an afternoon one. REINBERG et al. (1974) showed single high-dose afternoon MP injection at 15.00 h to be more effective in improving the 24 h mean PEF than that administered in the morning, at 03.00 h or 07.00 h, or in the evening at 19.00 h. Furthermore, REINBERG et al. (1977) showed one-month treatment of asthma patients with Dutimelan (consisting of high-dose glucocorticoids at 08.00 h and a lesser yet substantial dose at 15.00 h) results in significant improvement in pulmonary function without HPA axis suppression. Collectively, these studies infer that once-daily aerosol and tablet treatment timed in the afternoon, approximately 8–9 h following the commencement of the diurnal activity span, results in a superior effect on pulmonary function and nighttime inflammation. The delivery of glucocorticoid medications at this time most likely interrupts the cascade of events that culminates in the exacerbation of airways inflammation nocturnally (BEAM et al. 1992; LABRECQUE and REINBERG 1989). However, further investigations involving additional times of once-daily corticoste-roid administration are required to determine if this treatment time is ideal.

K. Mast Cell Stabilizers

Cromolyn and Nedocromil Sodium. Cromolyn sodium (Intal, Fisons, UK) and nedocromil sodium (Tilade, Fisons, UK), a pyranoquinoline dicarboxylic acid derivative, exert an anti-inflammatory effect on the airways of asthmatic patients. The exact mechanism of their action is unknown; it is thought that these medications retard release of inflammatory mediators from pulmonary mast cells, thereby decreasing airways hyperresponsiveness (FURUKAWA et al. 1984; KAY 1987). Both medications are effective in reducing the number of nights that patients are disturbed by asthma, although neither seems to advert the nocturnal fall in spirometric indices of pulmonary function (GONZALEZ and BROGDEN 1987; HETZEL et al. 1985; MORGAN et al. 1986; PETTY et al. 1985; RUFFIN et al. 1986; WILLIAMS and STABLEFORTH 1986).

Circadian rhythm-dependencies of these medications have yet to be explored. However, MORGAN et al. (1986) compared the effect of a single high

(160 mg) cromolyn sodium dose versus placebo at bedtime on nocturnal asthma. High-dose cromolyn failed to attenuate the nocturnal dip in pulmonary function; decline in FEV_1 was comparable between the two treatments. However, single-dose studies are not likely to provide an appropriate assessment of these types of medications. WILLIAMS and STABLEFORTH (1986) demonstrated that four-times daily nedocromil treatment (4 mg) by meter dose inhaler increased the 24 h mean and decreased the amplitude of the PEF circadian rhythm in a statistically significant manner during 12 weeks of treatment. Cessation of nedocromil treatment during an ensuing 12-week period resulted in reversal of pulmonary function back to the pretreatment level.

L. Conclusion

The risk and severity of asthma throughout the 24 h is much greater during the night than during the day, reflecting circadian rhythms in physiologic status and perhaps cyclic variation in the strength of and exposure to environmental triggers. Thus, the requirement of patients for bronchodilator and anti-inflammatory medications should not be expected to be the same during the day and night. One way to improve the effect of asthma treatments is to provide them in proportion to biological need, as chronotherapies, during the 24 h.

Very significant progress has been realized in the chronotherapy of theophylline, β_2-agonist, and glucocorticoid tablet medications, in particular. Evening high-dose theophylline and β_2-agonist tablet therapies result in optimization of the effects of these medications in patients who suffer from nocturnal asthma. Morning daily and alternate-day tablet glucocorticoid treatment has been practiced as a chronotherapy of steroid-dependent disease, including asthma, since the 1960s. Recent findings suggest that tablet therapy at 15.00 h, a clock time corresponding to the middle of the daily activity span, exerts the best effect on airways inflammation and function.

Progress in the chronotherapy of inhaled glucocorticoids and sustained-effect β_2-agonist medications is promising. Once-a-day 15.00 h triamcinolone aerosol therapy results in as good an effect on asthma control and pulmonary function as four-times-daily treatment. However, much additional investigation is desired to determine the best circadian time or times for dosing these medications. Only one study has addressed the value of nighttime-only dosing of a sustained-effect β_2-agonist aerosol. Evening once-daily dosing with salmeterol was as good as twice-daily treatment. Yet, it is unknown if the effects of this and other sustained-effect aerosol medications can be further enhanced, for example, by an unequal morning-evening dosing regimen as demonstrated for terbutaline and pulsed-release tablet albuterol preparations. Up to now, there has been inadequate study of circadian rhythm-dependencies of anticholinergic and mast cell stabilizing medications. Nonetheless, the findings of four different investigations on anticholinergic medications illustrate the importance of high bedtime dosing for achieving significant drug effect nocturnally.

Finally, the circadian rhythms of pulmonary function and airways inflammation must be taken into account in the testing of new antiasthma medications, such as leukotriene and phosphodiesterase antagoist agents. The effects of anti-inflammatory and bronchodilator medications may not be the same when administered in the morning and evening. Morever, determination of effect duration of bronchodilator and other medications is dependent on the manner in which they are assessed. Pulmonary function protocols based upon "before" and "after" drug dosing are inappropriate since they assume non-variance in PEF and FEV_1 during the 24 h. High-amplitude circadian rhythms in pulmonary function in asthma patients must be taken into consideration in designing investigative protocols. It is recommended that clock-time placebo-control baseline values be ascertained and used to accurately determine the duration and extent of drug effect when medications are administered in the morning as well as evening since both endpoints may be affected by the circadian time of treatment.

References

Adams F (ed. and trans.) (1856) The extant works of Aretaeus. The Cappadocian. London: Printed for the Syndenham Society

Anonymous (1982) Death from asthma in two regions of England. Br Med J 285:1251–1255

Arvidsson P, Larsson S, Lofdahl CG Melander B, Svedmyr N, Wahlander L (1991) Inhaled formoterol during one year in asthma: a comparison with salbutamol. Eur Respir J 4:1168–1173

Babb RR, Notarangelo J, Smith VM (1970) Wheezing as a clue to gastroesophageal reflux. Am J Gastroenterol 53:230–233

Ball DI, Brittain RT, Coleman RA, Denyer LH, Jack D, Johnson M, Lunts LHC, Nials, AT, Sheldrick KE, Skidmore IF (1991) Salmeterol, a novel, long-acting beta$_2$-adrenergic agonist: characterization of pharmacological activity in vitro and in vivo. Br J Pharmacol 104:665–671

Ballard RD (1993) Effects of sleep on respiratory physiology in nocturnal asthma. In: Martin RJ (ed) Nocturnal asthma: mechanisms and treatment. Futura Publ. Co. Mt Kisco, NY

Ballard RD, Saathoff MC, Patel DK, Kelly PL, Martin RJ (1989) Effects of sleep on nocturnal bronchoconstriction and ventilatory patterns in asthmatics. J Appl Physiol 67:243–249

Barnes PJ (1990) Effect of corticosteroids on airway hyperresponsiveness. Am Rev Respir Dis 141:S70–76

Barnes PJ (1995) Inhaled glucocorticods for asthma. N Engl J Med 332:868-875

Barnes PJ, Pauwels RA (1994) Theophylline in the management of asthma: time for reappraisal? Eur Respir J 7:579–591

Barnes PJ, Fitzgerald G, Brown M, Dollery C (1980) Nocturnal asthma and changes in circulating epinephrine, histamine, and cortisol. N Engl J Med 303:263–267

Barnes PJ, Fitzgerald GA, and Dollery CT (1982) Circadian variation in adrenergic response in asthmatic subjects. Clin Sci 62:349–354

Bateman JRM, Clark SW (1979) Sudden death in asthma. Thorax 34:40–44

Beam WR, Weiner DE, Martin RJ (1992) Timing of prednisone and alteration of airways inflammation in nocturnal asthma. Am Rev Respir Dis 146:1524–1530

Bogin RM, Ballard RD (1992) Treatment of nocturnal asthma with pulse-release albuterol. Chest 102:362–366

Bollinger AM, Young KYL, Gambertoglio JG, Newth CJL, Zureikat G, Powell M, Leung P, Affirme MB, Symchowicz S, Patrick JE (1989) Influence of food on the absorption of albuterol Repatabs. J Allergy Clin Immunol 83:123–126

Bonnet R, Jorres R, Heitmann U, Mangnussen H (1991) Circadian rhythm in airway responsiveness and airway tone in patients with mild asthma. J Appl Physiol 71:1598–1605

Brambilla C, Chastang C, Georges D, Bertin L (1994) Salmeterol compared with slow-release terbutaline in nocturnal asthma. Allergy 49:421–426

British Thoracic Society (1993) Guidelines on management of asthma. Thorax-48(Suppl):1098–1111

Britton MG, Earnshaw JS, Palmer JBD (1992) A twelve month comparison of salmeterol and salbutamol in asthmatic patients. Eur Respir J 5:1062–1067

Brown A, Smolensky M, D'Alonzo G, Frankoff H, Gianotti L, Nilsestuen J (1988) Circadian chronesthesy of the airways of healthy adults to the β-agonist bronchodilator isoproterenol. Ann Rev Chronopharmacol 5:163:166

Busse WW, Bush RK (1988) Comparison of Uniphyl tablets and inhaled albuterol as maintenance therapy in asthmatic adults. Am J Med 85(Suppl 1B):10

Butchers PR, Cousins SA, Vardey CJ (1987) Salmeterol: a potent and long-acting inhibitor of the release of inflammatory and spasmogenic mediators from human lungs. Br J Pharmacol 92(Suppl):745P (abstract)

Calhoun WJ, Bates ME, Schrader L, Sedgwick JB, Busse WW (1992) Characteristics of peripheral blood eosinophils in patients with nocturnal asthma. Am Rev Respir Dis 145:577–581

Callaghan B, Teo NC, Clancy L (1992) Effects of the addition of nedocromil sodium to maintenance bronchodilator therapy in management of chronic asthma. Chest 101:787–792

Catterall JL, Rhind GB, Whyte KF, Shapiro CM, Douglas NJ (1988) Is nocturnal asthma caused by changes in airway cholinergic activity? Thorax 43:720–724

Ceresa F, Angeli A, Boccuzzi A, Molino G (1969) Once-a-day neurally stimulated and basal ACTH secretion phases in man and their responses to corticoid inhibition. J Clin Endocrinol Metab 29:1074–1082

Chan CS, Woolcock AL, Sullivan CE (1988) Nocturnal asthma: role of snoring and obstructive sleep apnea. Am Rev Respir Dis 137:1502–1504

Chaplin MD, Rooks W II, Swenson EW, Cooper WC, Nerenberg C, Chu NI (1980) Flunisolide metabolism and dynamics of a metabolite. Clin Pharmacol Ther 27:402–413

Cochrane GM and Clark TJH (1975) A survey of asthma mortality in patients between 35 and 65 years in the greater London hospitals in 1971. Thorax 30:300–315

Cockcroft DW, Hoeppner VH, Werner GD (1984) Recurrent nocturnal asthma after bronchoprovocation with western red cedar sawdust: association with acute increase in non-allergic bronchial responsiveness. Clin Allergy 14:61–68

Coe CI, Barnes PJ (1986) Reduction of nocturnal asthma by an inhaled anticholinergic drug. Chest 90:485–488

Cox ID, Hughes DTD, McConnell K (1984) Ipratropium bromide in patients with nocturnal asthma. Postgrad Med J 60:526–528

Creemers JD (1988) A multicenter comparative study of salbuterol controlled release (Volmax) and sustained-release theophyllin (Theo-Dur) in the control of nocturnal asthma. Eur Respir J 1(Suppl 2):333s (Abstract)

Crepaldi G, Muggeo M (1974) Plurichronocorticoid treatment of bronchial asthma and chronic bronchitis. Clinical and endocrinometabolic evaluation. Chronobiologia 1(Suppl. 1):407–427

D'Alonzo GE, Smolensky MH, Feldman S, Gianotti LA, Emerson MB, Staudinger H, Steinijans VM (1990) Twenty-four-hour lung function in adult patients with asthma: chronoptimized theophylline therapy once-daily dosing in the evening versus conventional twice-daily dosing. Am Rev Respir Dis 142:84–90

D'Alonzo GE, Nathan RA, Henochowicz S, Morris RJ, Rathner P, Rennard SI (1994) Salmeterol xinafoate as maintenance therapy compared with albuterol in patients with asthma. JAMA 271:1412–1416

D'Alonzo GE, Smolensky MH, Feldman S, Gnosspelius Y, Karlsson K (1995) Bambuterol in the treatment of asthma. A placebo-controlled comparison of once-daily morning vs evening administration. Chest 107:406–412

Dahl R, Johansson S-A (1982) Clinical effect of b.i.d. and q.i.d. administration of inhaled budesonide, a double-blind controlled study. Eur J Respir Dis 63(Suppl 122):268–269

Dahl R, Harving H, Sawedal L, Anchus S. (1988) Terbutaline sustained-release tablets in nocturnal asthma – a placebo-controlled comparison between a high and low evening dose. Br J Dis Chest 82:237–241

Dahl R, Pedersen B, Venge P (1991) Bronchoalveolar lavage studies. Eur Respir Rev 1:272–275

Darow P, Steinijans VW (1987) Therapeutic advantage of unequal dosing of theophylline in patients with nocturnal asthma. Chronobiol Intern'l 4:349–357

Davies KJ, Green M, Schofield NM (1976) Recurrent nocturnal asthma after exposure to grain dust. Am Rev Respir Dis 114:1011–1019

DeMonchy JGR, Kauffman HF, Vernge P, Koeter GH, Jansen HM, Sluiter HJ, DeVries KD (1985) Bronchoalveolar eosinophilia during allergen-induced late asthmatic reaction. Am Rev Respir Dis 131:373–376

Derom E, Pauwels R, Van Der Straeten M (1990) Duration of the protective effect of salmeterol on methacholine challenge of asthmatics. Am Rev Respir Dis 141(Suppl):A469 (Abstract)

Dethlefsen U, Repges R (1985) Ein neues Therapieprinzip bei nächtlichem Asthma. Med Klin 80:44–47

DeVries KJT, Goei H, Booy-Noord H, Orie NGM (1962) Changes during 24 hours in lung function and histamine hyperreactivity of the bronchial tree in asthmatic and bronchitic patients. Arch Allergy Appl Immunol 20:93–101

Douglas NJ (1985) Asthma at night. Clin Chest Med 6:663–674

English J, Marks V (1981) Diurnal variations in methylprednisolone metabolism in the rat. IRCS Med Sci 9:721

English J, Dunne M, Marks V (1983) Diurnal variation in prednisone kinetics. Clin Pharmacol Ther 33:381–385

Eriksson NE, Haglind K, Ljungholm K (1982) A comparison of sustained-release terbutaline and ordinary terbutaline in bronchial asthma. Br J Dis Chest 76:202–204

Eriksson L, Jonson B, Eklundh G, Persson G (1988) Nocturnal asthma: effects of slow-release terbutaline on spirometry and arterial blood gases. Eur Respir J 1:302–305

Faurschou P, Engel A-M, Haanaes OC (1994) Salmeterol in two different doses in the treatment of nocturnal bronchial asthma poorly controlled by other therapies. Allergy 49:827–832

Finnerty JP, Twentyman OP, Holgate ST (1991) The duration of action of salmeterol as a bronchodilator and functional antagonist of histamine-induced bronchoconstriction. Am Rev Respir Dis 143(4pt2):A654 (Abstract)

Fitzgerald GA, Barnes P, Brown, MJ, Dollery CT (1980) The circadian variability of circulating adrenaline and bronchomotor reactivity in asthma. In: Smolensky MH, Reinberg A, Labrecque G (eds) Recent advances in chronobiology of allergy and immunology. Pergamon, Oxford, pp 89–94

Fitzpatrick MF, Mackay T, Driver H, Douglas NJ (1990) Salmeterol in nocturnal asthma: a double-blind, placebo controlled trial of a long acting inhaled β_2 agonist. Br Med J 301:1365–1368

Floyer J (1689) A treatise of the asthma. Wilkin, London

Furukawa CT, Shapiro GG, Bierman CW, Kraemer MJ, Ward DJ, Pierson WE (1984) A double-blind study comparing the effectiveness of cromolyn sodium and sustained-release theophylline in childhood asthma. Pediatrics 74:453–459

Gaultier C, Reinberg A, Girard F (1975) Circadian changes in lung resistance and dynamic compliance in healthy and asthmatic children. Effects of two bronchodilators. Resp Physiol 31:169–182

Gaultier C, Reinberg A, Motohashi Y (1988) Circadian rhythm in total pulmonary resistance of asthmatic children. Effects of a β-agonist agent. Chronobiol Intern'l 5:285–290

Gervais P, Reinberg A, Gervais C, Smolensky MH, Defrance O (1977) Twenty-four-hour rhythm in the bronchial hyperreactivity to house dust in asthmatics. J Allergy Clin Immunol 59:207–213

Goldenheim PD, Conrad EA, Schein LK (1987) Treatment of asthma by a controlled-release theophylline tablet formulation: a review of the North American experience with nocturnal dosing. Chronobiol Intern'l 4:397–408

Goodall RJR, Earis JE, Cooper DN, Bernstein A, Temple JG (1981) Relationship between asthma and gastroesophageal reflux. Thorax 36:116–121

Goo RH, Moore JG, Greenberg E, Alazraki NP (1987) Circadian variation in gastric emptying of meals in man. Gastroenterology 93:515–518

Gonzalez JP, Brogden RH (1987) Nodocromil sodium: a preliminary review of its pharmacodynamic and pharmacokinetic properties, and therapeutic efficacy in the treatment of reversible airways disease. Drugs 34:560–577

Grossman J, Morris RJ, White KD, Cocchetto M (1991) Improved stability in oral delivery of albuterol provides less variability in bronchodilation in adults with asthma. Ann Allergy 66:324–327

Harkness JAL, Richter MB, Pamayi GS, Van DePete K, Unger K, Pownall R, Geddawi M (1982) Circadian variation in disease activity in rheumatoid arthritis. Br Med J 284:551–554

Harter JG, Reddy WIJ, Thorn GW (1963) Studies on an intermittent corticosteroid dosage regimen. N Engl J Med 296:591–595

Hetzel MR, Clark TJH (1980) Comparison of normal and asthmatic circadian rhythms in peak expiratory flow rate. Thorax 35:732–738

Hetzel MR, Clark TJH, Branthwaite MA (1977) Asthma: analysis of sudden deaths and ventilatory arrest in hospital. Br Med J 1:808–811

Hetzel MR, Clarke JH, Gilliam SJ, Isaac P, Perkins M (1985) Is sodium cromoglycate effective in nocturnal asthma? Thorax 40:793:794

Higenbottam MA, Khan A, Williams DO, Mikhail JR, Peake MD, Hughes J (1989) Controlled release salbutamol tablets versus aminophylline in the control of reversible airways obstruction. J Int Med Res 17:435–441

Holliday SM, Faulds D, Sorkin EM (1994) Inhaled fluticasone propionate: a review of its pharmacodynamics and pharmacokinetic properties, and therapeutic use in asthma. Drugs 47:319–331

Hussey EK, Donn KH, Powell JR (1991) Albuterol extended-release products: a comparison of steady-state pharmacokinetics. Pharmacotherapy 11:131–135

Jarjour NN, Calhoun WJ, Busse WW (1992) Enhanced metabolism of oxygen radicals in nocturnal asthma. Am Rev Respir Dis 146:905–911

Joad JP, Ahrens RC, Lindgren SD, Weinberger MM (1987) Relative efficacy of maintenance therapy with theophylline, inhaled albuterol, and the combination for chronic asthma. J Allergy Clin Immunol 79:78–85

Johnson M (1991) The preclinical pharmacology of salmeterol: bronchodilator effects. Eur Respir Rev 1:253–256

Johnson M, Butchers PR, Vardey CJ (1991) Potency and duration of action of salmeterol as an inhibitor of mediator release from human lung. Am Rev Respir Dis 143(4pt2):A655 (abstract)

Johnson M, Butchers PR, Whelan CJ (1992) The therapeutic potential of long-acting $beta_2$-adrenoceptor agonists in allergic inflammation. Clin Exp Allergy 22:177–181

Kay AM (1987) The mode of action of anti-allergic drugs. Clin Allergy 17:154–164

Kesten S, Chapman KR, Broder I, Cartier A, Hyland RH (1992) Sustained improvement in asthma with long-term use of formoterol fumarate. Ann Allergy 69:415–420

Kemp JP, Bierman CW, and Cocchetto DM (1993) Dose-response study of inhaled salmeterol in asthmatic patients with 24-hour spirometry and holter monitoring. Ann Allergy 70:316–322

Kidney JC, Dominguez M, Rose M, Aikman S, Chung K, Barnes PJ (1994) Withdrawing chronic theophylline treatment increases airway lymphocytes in asthma. Thorax 49:396

Koeter GH, Postma DS, Keyzer JJ, Meurs H (1985) Effect of oral slow-release terbutaline on early morning dyspnea. Eur J Clin Pharmacol 28:159–162

Kraft M, Torvik JA, Trudeau J, Wenzel SE, Martin RJ (1996) Theophylline potential antiinflammatory effects in nocturnal asthma. J Allergy Clin Immunol (97: 1242–1246)

Kyle GM, Smolensky, MH, McGovern JP (1979) Circadian variation in the susceptibility of rodents to the toxic effects of theophylline. In: Reinberg A, Halberg F (eds) Chronopharmacology, Pergamon, Oxford, pp 239–244

Labrecque G, Reinberg A (1989) Chronopharmacology of nonsteroidal anti-inflammatory drugs. In: Lemmer B(ed), Chronopharmacology. Cellular and biochemical interactions. Marcel Dekker, New York pp 545–579

Lemmer B, Bruguerolle B (1994) Chronopharmacokinetics: are they clinically relevant. Clin Pharmacokinet 26:419–427

Lemmer B, Nold G (1991) Circadian changes in estimated hepatic blood flow in healthy subjects. Br J Clin Pharmacol 32:627–629

Levi F, LeLouarn C, Reinberg A (1985) Timing optimized sustained indomethacin treatment of osteoarthritis. Clin Pharmacol Ther 37:77-84

Lundback B, Rawlinson DW, Palmer JBD (1993) Twelve month comparison of salmeterol and salbutamol as dry powder formulations in asthmatic patients. Thorax 48:148–153

Maconochie JG, Forster JK, Fowler P, Thomas M (1988) An initial comparison of salmeterol and salbutamol against histamine-induced bronchoconstriction in healthy subjects. Br J Clin Pharm 25:115P (abstract)

Maesen FPV, Smeets JJ (1986) Comparison of a controlled-release tablet of salbutamol given twice daily with a standard tablet given four times daily in the management of chronic obstructive lung disease. Eur Respir J 31:431–436

Maesen FBV, Smeets JJ, Gubbelmans HLL, Zweers PGMA (1990a) Bronchodilator effect of inhaled formoterol vs salbutamol over 12 hours. Chest 97:590–594

Maesen FBV, Smeets JJ, Gubbelmans HLL, Zweers PGMA (1990b) Formoterol in the treatment of nocturnal asthma. Chest 98:866–870

Malo J-L, Cartier A, Merland N, Ghezzo H, Burek A, Morris J, Jennings BH (1989) Four-times-a-day dosing frequency is better than twice-a-day regimen in subjects requiring a high-dose inhaled steroid, budesonide, to control moderate to severe asthma. Am Rev Respir Dis 140:624–628

Malo J-L, Ghezzo H, Trudeau C, L'Archeveque J, Cartier A (1992) Salmeterol, a new inhaled beta$_2$-adrenergic agonist, has longer blocking effect than albuterol on hyperventilation-induced bronchoconstriction. J Allergy Clin Immunol 89:567–574

Martin ME, Grunstein MM, Larsen GL (1982) The relationship of gastroesophageal reflux to nocturnal wheezing in children with asthma. Ann Allergy 49:318–322

Martin RJ (ed) (1993) Nocturnal asthma Mechanisms and treatment. Futura, Mount Kisco, NY

Martin RJ, Cicutto LC, Ballard RD, Goldenheim PD, Cherniack RM (1989) Circadian variations in theophylline concentrations and the treatment of nocturnal asthma. Am Rev Respir Dis 139:475–478

Martin RJ, Cicutto LC, Ballard RD (1990) Factors related to the nocturnal worsening of asthma. Am Rev Respir Dis 141:33–38

Martin RJL, Cicutto LC, Smith HR, Ballard RD, Szefler SJ (1991) Airways inflammation in nocturnal asthma. Am Rev Respir Dis 143:351–357

McAllister WAC, Mitchell DM, Collins JV (1981) Prednisolone pharmacokinetics compared between night and day in asthmatic and normal subjects. Br J Clin Pharmacol 11:303–304

Mohiuddin AA, Martin RJ (1990) Circadian basis of the late asthma response. Am Rev Respir Dis 142:1153–1157
Moore JG, Englert E (1970) Circadian rhythm of gastric acid secretion in man. Nature 226:1261–1262
Moore-Gillon J (1988) Volmax (salbutamol CR 8 mg) in the management of nocturnal asthma: a placebo-controlled study. Eur Respir J 1(Suppl 2):306s (Abstract)
Morgan AD, Connaughton JJ, Caterall JR, Shapiro CM, Douglas NJ, Flenley DC (1986) Sodium cromoglycate in nocturnal asthma. Thorax 41:39–41
Morrison JFJ, Pearson SB, Dean HG (1988) Parasympathetic nervous system in nocturnal asthma. Br Med J 296:1427–1429
Morselli PL, Marc V, Garattini S, Zaccala M (1970) Metabolism of exogenous cortisol in humans-diurnal variations in plasma disappearance rate. Biochem Pharmacol 19:1643–1647
Muir JF, Bertin L, Georges D (1992) Salmeterol versus slow-release theophylline combined with ketotifen in nocturnal asthma: a multicentre trial. Eur Respir J 5:1197–1200
Neuenkirchen H, Wikens JH, Oellerich M, Sybrecht GW. (1985) Nocturnal asthma: effect of a once per evening dose of sustained-release theophylline. Eur J Respir Dis 66:196–204
Newman Taylor AN, Longbottom JL, Pepys J (1977) Respiratory allergy to urine proteins of rats and mice. Lancet 2:847–849
Newman Taylor AN, Davies RJ, Hendrick DJ, Pepys J (1979) Recurrent nocturnal asthmatic reactions to bronchial provocation tests. Clin Allergy 9:213–219
Orie NGM, Sluitter HF, Tammeling GJ, DeVries K, Wal AMVD (1964) Fight against bronchitis. In: Orie NGM and Sluitter HR (eds) Bronchitis: 2nd international bronchitis Symposium. Thomas, Springfield IL, pp 352–369
Pauwels R, Van Renerghem D, Van Der Straeten M, Johannesson N, Persson CGA (1985) The effect of theophylline and emprofylline on allergen induced-bronchoconstriction. J Allergy Clin Immunol 76:583–590
Pearlman DS, Chervinsky P, LaForce C, Seltzer JM, Southern DL, Kemp JP, Dockhorn RJ, Grossman J, Liddle RF, Yancey SW, Cocchetto DM, Alexander WJ, van As A (1992) A comparison of salmeterol with albuterol in the treatment of mild-to-moderate asthma. N Engl J Med 327:1420–1425
Pedersen B, Dahl R (1990) The effect of salmeterol on the early and late reaction to bronchial allergen challenge and hyperreactivity in patients with bronchial asthma. Clin Exp Allergy 20(Suppl 1):94 (abstract)
Pedersen BK, Laursen LC, Gnosspelius Y, Faurschou P, Weeks B (1985) Bambuterol: effects of a new anti-asthmatic drug. Eur J Clin Pharmacol 29:425–427
Pedersen B, Dahl R, Venge P (1991) The effect of salmeterol on the early and late phase reaction to bronchial allergen challenge and postchallenge variation in bronchial hyperreactivity, blood eosinophils and se-eosinophil cationic protein. Am Rev Respir Dis 143(4pt2): A649 (Abstract)
Persson CGA (1986) Role of plasma exudation in asthmatic airways. Lancet 2:1126–1128
Persson G, Gnosspelius Y, Anehus S (1988) Comparison between a new once-daily, bronchodilating drug, bambuterol, and terbutaline sustained-release, twice-daily. Eur Respir J 1:223–226
Petty TL, Rollins DR, Christopher K, Good JT, Oakley R (1989) Cromolyn sodium is effective in adult chronic asthmatics. Am Rev Respir Dis 139:694–701
Pierson WE, LaForce CF, Bell TD, MacCosbe PE, Sykes RS, Tinkelman D (1990). Long-term, double-blind comparison of controlled-release albuterol versus sustained-release theophylline in adolescents and adults with asthma. J Allergy Clin Immunol 85:618–626
Pincus DJ, Szefler SJ, Ackerson LM, Martin RJ (1995) Chronotherapy of asthma with inhaled steroids: the effect of dosage timing on drug efficacy. J Allergy Clin Immunol 95:1172–1178

Postma DS, Koeter GH, v.d. Mark TW, Reig RP, Sluiter HJ (1985) The effects of oral slow-release terbutaline on the circadian variation in spirometry and arterial blood gas levels in patients with chronic airflow obstruction. Chest 87:653–657

Postma DS, Koeter GH, Keyzer JJ, Meurs H (1986) Influence of slow-release terbutaline on the circadian variation of catecholamines, histamine, and lung function in nonallergic patients with partly reversible airflow obstruction. J Allergy Clin Immunol 77:471–477

Powell ML, Weisberger M, Dowdy Y, Gural R, Symchowicz S, Patrick JE (1987) Comparative steady-state bioavailability of conventional and controlled-release formulations of albuterol. Biopharm Drug Dispos 8:461–467

Reed CE (1991) Aerosol steroids as primary treatment of mild asthma. N Engl J Med 325:425–426

Reinberg A (1989) Chronopharmacology of corticosteroids and ACTH. In: Lemmer B (ed) Chronopharmacology. Cellular and biochemical interactions New York, Marcel Dekker, pp 137–178

Reinberg A, Ghata J, Sidi E (1963) Nocturnal asthma attacks: their relationship to the circadian adrenal cycle. J Allergy 34:323–330

Reinberg A, Gervais P, Morin C, Abulker C (1971) Rhythme circadian humain du seuil de la response bronchique a l'acetylcholine. CR Acad Sci 272:1879–1881

Reinberg A, Halberg F, Fallers C (1974) Circadian timing of methylprednisolone effects in asthmatic boys. Chronobiologia 1:333–347

Reinberg A, Guilet P, Gervais P, Ghata J, Vignaud D, Abuker C (1977) One-month chronocorticotherapy (Dutimelan 8–15 mite). Control of the asthmatic condition without adrenal suppression and circadian alteration. Chronobiologia 4:295–312

Reinberg A, Gervais M, Chaussade G, Fraboulet G, Duburque B (1983) Circadian changes in effectiveness of corticosteroids in eight patients with allergic asthma. J Allergy Clin Immunol 71:425–433

Reinberg A, Pauchet F, Ruff F, Gervais A, Smolensky MH, Levi F, Gervais P, Chaout D, Abella ML, Zidani R (1987) Comparison of once-daily evening versus morning sustained-release theophylline dosing for nocturnal asthma. Chronobiol Intern'l 4:409–420

Reinberg A, Smolensky MH, D'Alonzo GE, McGovern JP (1988) Chronobiology and asthma. III. Timing corticotherapy to biological rhythms to optimize treatment goals. J Asthma 25:219–248

Reinberg A, Labrecque G, Smolensky MH (1991) Chronobiologie et chronothérapeutique. Heure optimale d'administration des medicaments. Flammarion, Paris

Ruffin R, Alpers JH, Kromer DK, Rubinfield AR, Pain MCF, Czarny D, and Bowes G (1986) A 4-week Australian multicentre study of nedocromil sodium in asthmatic patients. Eur J Respir Dis 69(Suppl 147):336–339

Ryrfeldt A, Andersson P, Edsbacker S, Tonnesson M, Davies D, Pauwels R (1982) Pharmacokinetics and metabolism of budesonide, a selective glucocorticoid. Eur J Respir Dis 63(Suppl.):86–95

Schulz H-U, Frercks H-J, Hypa F (1984) Vergleichende Theophyllin-Serum-Spiegel Messungen über 24 Stunden nach konventioneller Dosierung einer Theophyllin-retard-Präparation über 4 Tage. TherapieWoche 34:536–543

Serafini U, Bonini S (1974) Corticoid therapy in allergic disease. Clinical evaluation of a chronopharmacological attempt. Chronobiologia 1 (Suppl 1):339–406

Sheffer AL (1991) National Heart, Lung and Blood Institute. National Asthma Education Program. Expert Panel Report: Guidelines for the diagnosis and management of asthma. J Allergy Clin Immunol 88:425–534

Sheffer AL (1992) International consensus report on diagnosis and management of asthma. Eur Respir J 5:601–641

Sly PD, Landau LI, Olinsky A (1987) Failure of ipratropium bromide to modify the diurnal variation of asthma in asthmatic children. Thorax 42:357–60.

Smith E (1860) Recherches experimentales sur la respiration. J Physiol de l' Homme et des Animaux 3:506–21

Smolensky MH (1989) Chronopharmacolgy of theophylline and beta-sympathomimetics. In: Lemmer B (ed) Chronopharmacology. Cellular and biochemical interactions. Marcel Dekker, NY, pp 65–113
Smolensky MH, D'Alonzo GE (1993) Medical chronobiology: concepts and applications. Amer Rev Res Dis 147:511-519
Smolensky MH, Barnes PJ, Reinberg A, McGovern JP (1986) Chronobiology and asthma. I. Day-night differences in bronchial patency and dyspnea and circadian rthythm dependencies. J Asthma 23:321–343
Smolensky MH, McGovern JP, Scott PH, Reinberg A (1987a) Chronobiology and asthma. II. Body-time-dependent differences in the kinetics and effects of bronchodilator medications. J Asthma 24:90–134
Smolensky MH, Scott PH, Harrist RB, Hiatt PH, Wong TK, Baenziger JC, Klank BJ, Marbella A, Meltzer A (1987b) Administration-time-dependency of the pharmacokinetic behavior and therapeutic effect of a once-a-day theophylline in asthmatic children. Chronobiol Intern'l 4:435–448
Smolensky MH, Reinberg A, Labrecque G (1995) Twenty-four hour pattern in symptom intensity of viral and allergic rhinitis: treatment implication. J Allergy Clin Immunol 95:1084–1096
Soutar CA, Costello J, Ijaduola O, Turner-Warwick M (1975) Nocturnal and morning asthma. Relationship to plasma corticosteroids and response to cortisol infusion. Thorax 30:436–440
Soutar CA, Carruthers M, Pickering CA (1977) Nocturnal asthma and urinary adrenaline and nor-adrenaline excretion. Thorax 32:677–683
Steinijans VW, Schulz H-U, Beier W, Radtke HW (1986) Once daily theophylline: multiple-dose comparison of an encapsulated micro-osmotic system (Euphylong) with a tablet (Uniphyllin). Intern'l J Clin Pharmacol 24:43–47
Stewart IC, Rhind GB, Power JT, Flenley DC, Douglas NJ (1987) Effect of sustained release terbutaline on symptoms and sleep quality in patients with nocturnal asthma. Thorax 42:797–800
Storms WW, Nathan RA, Bodman SF, Morris RJ, Morris RJ, Selner JC, Greenstein SM, Zwillich CW (1992) The effect of repeat action albuterol sulfate (Proventil Repetabs) in nocturnal symptoms of asthma. J. Asthma 29:209–216
Storms WW, Bodman SF, Nathan RA, Byer P (1994) Nocturnal asthma symptoms may be more prevalent than we think. J Asthma 31:313–318
Sullivan P, Songul B, Jaffar Z, Page C. Jeffery P, Costello J (1994) Anti-inflammatory effects of low dose oral theophylline in atopic asthma. Lancet 343:1006–1008
Sur S, Mohiuddin AA, Vichyanond P, Nelson HS (1990) A random double-blind trial of the combination of nebulized atropine methylnitrate and albuterol in nocturnal asthma. Ann Allergy 65:384–388
Sykes RS, Reese ME, Meyer MC, Chubb JM (1988) Relative bioavailability of a controlled-release albuterol formulation for twice-daily use. Biopharm Drug Dispos 9:551–556
Tan WC, Martin RJ, Pandey R, Ballard RD (1990) Effects of spontaneous and stimulated gastroesophageal reflux on sleeping asthmatics. Am Rev Respir Dis 141:1394–1399
Toogood JH, Baskerville JC, Jennings B, Lefco NM, Johansson SA (1982) Influence of dosing frequency and schedule on the response of chronic asthmatics to the aerosol steroid, budesonide. J Allergy Clin Immunol 70:288–298
Torvik JA, Borish LC, Beam WR, Kraft M, Wenzel SE, Martin RJ (1994) Does theophylline alter inflammation in nocturnal asthma? Am J Respir Crit Care Med 149:A210 (Abstract)
Trudeau JB, Martin RJ, Kraft M, Torvik JA, Westcott JY, Wenzel SE (1994) Theophylline decreases stimulated alveolar macrophage leukotriene B4 (LTB4) production in nocturnal asthmatics lavaged at 4 a.m. Am J Respir Crit Care Med 149:A941 (Abstract)

Trousseau A (1865) Clinique Medicale de l'Hopital Dieu de Paris, vol. 2. Asthme. Bailliere, Paris, p 373
Turner-Warwick M (1988) Epidemiology of nocturnal asthma. Am J Med 85(Suppl 1B):6–8
Twentyman OP, Finnerty JP, Harris A, Palmer J, Holgate ST (1990) Protection against allergen-induced asthma by salmeterol. Lancet 336:1338–1342
Ullman A, Svedmyr N (1988) Salmeterol, a new long acting inhaled beta$_2$-adrenoreceptor agonist: comparison with salbutamol in adult asthmatic patients. Thorax 43:674–678
Ullman A, Hedner J, Svedmyr N (1990) Inhaled salmeterol and salbutamol in asthmatic patients. Am Rev Resp Dis 142:571–575
Vener KJ, Moore JG (1988) Chronobiologic properties of the alimentary canal affecting xenobiotic absorption. Ann Rev Chronopharmacol 4:259–283
Ward AJM, McKenniff M, Evans JM, Page CP, Costello JF (1993) Theopylline – an immunomodulatory role in asthma? Am Rev Respir Dis 147:518–523
Warren JB, Cuss F, Barnes PJ (1985) Posture and theophylline kinetics. Br J Clin Pharmacol 19:707–709
Wettengel R, Berdel D, Cegla U, Fabel H, Geisler L, Hofmann D, Krause J, Kroidl RF, Lanser K, Leupold W, et al. (1994) Empfehlungen der Deutschen Atemwegsliga zum Asthmamanagement bei Erwachsenen und bei Kindern Med. Klinik. 89:57–67
Whelan CJ, Johnson M (1992) Inhibition by salmeterol of increased vascular permeability and granulocyte accumulation in guinea-pig lung and skin. Br J Pharmacol 101(Suppl):528P (Abstract)
Williams AJ, Stableforth D (1986) The addition of nedocromil sodium to maintenance therapy in the management of patients with bronchial asthma. Eur J Respir Dis 69(Suppl 147):340–343
Wolstenholme RJ, Shettan SP (1989) Comparison of a combination of fenoterol with ipratropium bromide (Duovent) and salbutamol in young adults with nocturnal asthma. Respiration 55:152–157
Wulfsohn NL, Politzer WM (1964) Bronchial asthma during menses and pregnancy. S Afr Med J 38:173
Wurthwein G, Rohdewald P (1990) Activation of beclomethasone dipropionate by hydrolysis to beclomethasone-17-monopropionate. Biopharm Drug Disposition 11:381–394
Zwillich CW, Neagley SR, Cicutto L, White DP, Martin RJ (1989) Nocturnal asthma therapy: inhaled bitolterol versus sustained-release theophylline. Am Rev Respir Dis 139:470–474

CHAPTER 10
Chronopharmacology of Cardiovascular Diseases[1]

B. LEMMER and F. PORTALUPPI

A. Historical Background

Heart rate was one of the earliest physiological functions reported not to be constant throughout 24 h (see P. Lavie, personal communication; ASCHOFF 1992). As early as the beginning of the seventeenth century daily variations, as well as a rapid increase in pulse rate on awakening, were described (STRUTHIUS 1602); later on, in the eighteenth and nineteenth centuries, general observations and even detailed data on daily heart rate variation and pulse quality were published (ZIMMERMANN 1793; REIL 1796; FALCONER 1797; HUFELAND 1797; AUTENRIETH 1801; WILHELM 1806; BARTHEZ 1806; KNOX 1815). Moreover, the pulse of a healthy subject measured in the late afternoon was frequently used by musicians as a readily available "metronome" (QUANTZ 1752). The introduction of the pulse-watch by FLOYER (1707–1710) and, moreover, the introduction of a "third hand" into the clock to measure precisely the seconds allowed determination of changes in pulse rate (ZIMMERMANN 1793). It is of interest to note that the symptoms of "office hypertension or white coat hypertension" – nowadays of increasing medical importance (see PICKERING et al. 1994b; KRAKOFF 1994) – were precisely described more than 250 years ago by HELLWIG (1738), when he reported on pronounced changes in the patient's pulse on the appearance of the doctor ("*Vor allen Dingen stehet aber zu wissen, daß der Pulse sich mercklich verändern ... könne, wozu nicht wenig Anlaß giebet die Ankunft des Medici*"). He, therefore, recommended the doctor to sit down and to talk to the patient for a while before studying the quality of the pulse once or several (!) times, a recommendation reflecting our modern guidelines. Shortly afterwards, DE BORDEU (1756) named the same observation "*le pouls du médicin*" (see LEMMER 1995).

As early as 1631, SANCTORIUS invented the "pulsilogium" in order to record the pulse rate present at different times of the day. With the advent of plethysmographic devices (e.g. VON BASCH 1881; ZADEK 1881; RIVA-ROCCI 1896) it was also observed that the blood pressure in healthy as well as in diseased persons may not be constant throughout 24 h (ZADEK 1881; HOWELL

---

[1] This paper is dedicated to the pioneers in chronopharmacology, Alain Reinberg, and in chronobiology, Jürgen Aschoff.

1897; HILL 1898; JELLINEK 1900; WEISS 1900; HENSEN 1900; BRUSH and FAYERWEATHER 1901; WEYSSE and LUTZ 1915; MÜLLER 1921a,b; KATSCH and PANSDORF 1922). ZADEK (1881) was the first to present detailed data on daily variations (*Tagesschwankungen*) in the blood pressure with an increase in the afternoon and a drop at night. Even different types of hypertension were already verified by their different blood pressure profiles (MÜLLER 1921a,b; KATSCH and PANSDORF 1922).

There are also very early reports describing the nightly occurrence of the symptoms and/or the onset of angina pectoris attacks and of myocardial infarction (ZIMMERMANN 1793; TESTA 1815). In the light of these observations it is not surprising that about 200 years ago REIL (1796) recommended that "the time of day of drug application and the dose must be in harmony with each other".

B. Chronobiological Mechanisms of the Cardiovascular System

I. Physiology and Pathophysiology of Blood Pressure Regulation

Circadian fluctuations in blood pressure (BP) with higher daytime than nighttime values have been established in both normotensive and hypertensive patients (SNYDER et al. 1964; BRISTOW et al. 1969; REINBERG et al. 1970; SMOLENSKY et al. 1976; MILLAR-CRAIG et al. 1978; HALBERG et al. 1984; PORTALUPPI et al. 1989, 1990b; LEMMER et al. 1991c; WITTE and LEMMER 1992; WITTE et al. 1993). Bed-bound subjects, either normal (REINBERG et al. 1970) or hypertensive (TUCK et al. 1985), and patients with fixed heart rate (HR) (DAVIES et al. 1984) still show a significant nocturnal decline in BP, which remains unaltered by antihypertensive therapy (RAFTERY et al. 1981a). The circadian rhythm of uncomplicated essential hypertensive patients is set at higher BP levels, but has the same circadian time pattern as in normal subjects.

The typical circadian pattern of BP exhibits two daytime peaks (around 0900 hours and 1900 hours, respectively), a small afternoon nadir (around 1500 hours) and a profound nocturnal drop (around 0300 hours). The amplitude of the 24-h variation is slightly larger for diastolic than for systolic BP, ranging between 10% and 20% of the daytime mean (DEGAUTE et al. 1991; MIDDEKE and SCHRADER 1994). This blood pressure pattern is modified by age: In elderly patients the amplitudes of systolic BP and heart rate are reduced, a greater ultradian component (12-h period) is present and the secondary afternoon decline is more prominent than in young patients (ATKINS et al. 1994). A decreased nightly drop in BP of elderly male patients was also observed by IMAI et al. (1993). Before awakening, a significant increase in BP was originally found by MILLAR-CRAIG et al. (1978), and questioned by FLORAS et al. (1978) and LITTLER et al. (1979), who believed that it was due to an averaging artefact. Reanalysis of the original work, however, confirmed the initial findings (GOULD and RAFTERY 1991). Over many years, the pre-waking rise was alternately denied (ATHANASSIADIS et al. 1969; MANCIA et al. 1983) and confirmed (DEGAUTE et al. 1991; BROADHURST et al. 1990).

Recent intra-arterial data in a large group of normal subjects, aligned to the time of waking, show that the early morning rise in BP starts hours before awakening, so that its sole attribution to arousal seems highly unlikely (BROADHURST et al. 1990). Moreover, animal data using telemetry clearly gave evidence for a dominant endogenous component of the blood pressure rhythm under free-run conditions (WITTE and LEMMER 1993; LEMMER et al. 1995b), also supported by experiments in which the suprachiasmatic nucleus was lesioned (JANSSEN et al. 1994; WITTE et al. 1995).

Considerable evidence exists for the driving role of the sympathetic system on the circadian rhythm of BP. Intra-arterial studies on the effect of a variety of antihypertensive agents indicate that the morning rise (both before awakening and upon arousal) appears to be due to increased α-adrenoceptor activity (GOULD and RAFTERY 1991). Propranolol reduces the pressor range (difference between basal and maximum BP readings) in most patients and an inverse relationship can be demonstrated between changes in pressor range and noradrenaline (DE LEEUW et al. 1977). The circadian rhythms of BP and sympathoadrenergic activity (catecholamine concentrations in plasma and urine, cAMP concentrations in plasma or in lymphocytes, and β-adrenoceptor density and affinity on lymphocytes) are synchronous both in normotensives and in patients with primary hypertension with normal circadian rhythm, and in abnormal catecholamine secretion with an abnormal circadian BP curve (MIDDEKE 1992). The circadian variations of sympathoadrenal and pressor reactivity to exercise are strongly correlated (HICKEY et al. 1993). Spectral analysis of BP and interbeat interval recordings provides markers of autonomic activity and arterial baroreflex sensitivity. Using this technique, a recent study demonstrated that a clear 24-h variation in sympathetic and vagal tone, but not in arterial baroreflex sensitivity, persists, independent of changes in activity and position (VAN DE BORNE et al. 1994).

In addition to the driving role of the autonomous nervous system, the circadian pattern of BP is likely to be influenced by the circadian rhythmicity of the endogenous opioid system, the hypothalamic-pituitary hormonal axes, the renin-angiotensin-aldosterone system and the vasoactive peptides. Evidence in favour of this view comes from the study of many diverse pathological conditions in which the alterations in the circadian rhythm of the above-mentioned neurohumoral factors, either inherent or secondary to disturbances in autonomic nervous system activity, are reflected by consistent modifications of the 24-h pattern of BP. Reduced or reversed nocturnal decline in BP has been reported with the following conditions: orthostatic autonomic failure (MANN et al. 1983), Shy-Drager syndrome (MARTINELLI et al. 1981), brainstem infarct (STOICA and ENULESCU 1983; SHIMADA et al. 1992; MATSUMURA et al. 1993), neurogenic hypertension (FRANKLIN et al. 1986), fatal familial insomnia (PORTALUPPI et al. 1994), diabetes (HORNUNG et al. 1989; WIEGMANN et al.

1990; LURBE et al. 1993; IKEDA et al. 1993; CHAU et al. 1994), catecholamine-producing tumours (ISSHIKI et al. 1986; OISHI et al. 1988; IMAI et al. 1988a; DABROWSKA et al. 1990; STATIUS VAN EPS et al. 1994), Cushing's syndrome (IMAI et al. 1988b), exogenous glucocorticoid administration (IMAI et al. 1989; VAN DE BORNE et al. 1993a), mineralocorticoid excess syndromes (IMAI et al. 1992; WHITE and MALCHOFF 1992; VEGLIO et al. 1993), hyperaldosteronism (MIDDEKE et al. 1991a), hyperthyroidism (MIDDEKE et al. 1991a), sleep apnoea syndrome (TILKIAN et al. 1976), normotensive and hypertensive asthma (FRANZ et al. 1992), chronic renal failure (HEBER et al. 1989; MIDDEKE et al. 1989, 1991a; BAUMGART et al. 1989, 1991; PORTALUPPI et al. 1990a, 1991, see Fig. 1; HAYASHI et al. 1993; TORFFVIT and AGARDH 1993; TIMIO et al. 1993; DEL ROSSO et al. 1994), severe hypertension (SHAW et al. 1963), pregnancy toxaemia (REDMAN et al. 1976; BEILIN et al. 1983), hypertensive patients with left ventricular hypertrophy (VERDECCHIA et al. 1990; KUWAJIMA et al. 1992), renal (SOERGEL et al. 1992) and cardiac transplantation (REEVES et al. 1986; SEHESTED et al. 1992) in relation to cyclosporine treatment, congestive heart failure (CARUANA et al. 1988; PORTALUPPI et al. 1992; VAN DE BORNE et al.

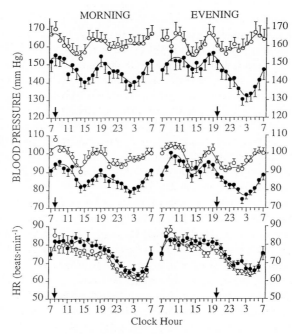

Fig. 1. Effect of the calcium channel blocker isradipine on the 24-h BP and HR profiles of 16 chronic renal diseased patients showing a disturbed BP profile. Isradipine (5 mg/day) was applied in a crossover design for 4 weeks either at 0800 hours or at 2000 hours. Both treatments reduced elevated BP; however, only evening dosing normalized the BP profile. (From PORTALUPPI et al. 1995, with permission. Copyright 1995 American Heart Association)

1992a; SUZUKI et al. 1992), pseudohypoparathyroidism (BRICKMAN et al. 1990) and recombinant human erythropoietin therapy (VAN DE BORNE et al. 1992b).

Alterations in the circadian rhythmicity of the activity of the autonomic nervous system, as well as the various neurohumoral factors involved in central and/or peripheral BP regulation, are clearly involved in the genesis of these alterations of the BP pattern. An imbalance between sympathetic and parasympathetic activity throughout the 24 h is the major determinant of the circadian changes in BP rhythmicity, not only in the various forms of neurogenic dysautonomias but also in diabetes and chronic renal failure, where the alterations in the circadian pattern of BP are minimal (CHAU et al. 1994) or absent (VAN DE BORNE et al. 1993b) before the onset of autonomic neuropathy. In chronic heart failure, a tonic activation of the sympathetic nervous system throughout day and night is present and seems the major determinant of the observed changes in the BP rhythm. Experiments in rats demonstrate that the altered BP pattern of heart failure is independent of changes in locomotor activity (TEERLINK and CLOZEL 1993).

On the other hand, it should be noted that the loss of the nocturnal fall (conventionally defined as a nocturnal decrease of less than 10% of the daytime BP mean) does not necessarily reflect an abolition of the 24-h rhythmicity of BP. This has been demonstrated in chronic renal insufficiency (PORTALUPPI et al. 1990a), where a significant circadian rhythm in BP with reduced amplitude and acrophase shifted toward the daytime hours is frequently found, and in fatal familial insomnia (PORTALUPPI et al. 1994).

Fatal familial insomnia, a prion disease leading to selective degeneration of the anterior ventral and mediodorsal nuclei of the thalamus (LUGARESI et al. 1987; MEDORI et al. 1992), is a rare cause of genetically determined secondary hypertension. In this disease, a dominant 24-h component is detectable by rhythm analysis of both BP and HR for months after total disappearance of any sleep activity (PORTALUPPI et al. 1994). The 24-h component of BP is reduced in amplitude and shifted toward the daytime hours, so that macroscopic analysis of the data reveals only an early disappearance of the nocturnal fall in BP. The alterations in the circadian pattern of BP appear more closely related to a dysfunction of the hypothalamic-pituitary-adrenal axis. In fact, serum cortisol is high and becomes increasingly elevated in fatal familial insomnia, whereas adrenocorticotropic hormone (ACTH) is not suppressed and remains at normal levels, strongly suggesting that some abnormality is present in the feedback suppression of ACTH. In addition, distinct nocturnal peaks are detected in the circadian rhythm of these hormones. This hypersecretion occurs at a time when, in normal conditions, sleep is known to maximally affect ACTH and cortisol secretion, and represents the reversal of the normal sleep-related inhibition. Moreover, the abnormalities in the glucocorticoid and BP patterns precede the development of hypertension and severe dysautonomia. Altogether, these findings in fatal familial insomnia point to a primary role of glucocorticoids in the modulation of the circadian rhythm of BP. Previous studies of Cushing's syndrome (IMAI et al. 1988b) and of subjects under exogenous glucocorticoid administration (IMAI et al. 1989) led to similar findings, although no rhythm analysis of data was performed.

Catecholamines alone seem less important in modulating the 24-h pattern of BP, since cathecolamine-producing tumors obliterate the nocturnal BP fall only when plasma catecholamines reach extremely high values (STATIUS VAN EPS et al. 1994).

The loss of nocturnal BP fall is likely to determine a higher cardiovascular and cerebral risk by prolonging the time over which the increased BP impact is exerted on the target organs. Recent data support this view, showing that the average night-time BP level and the loss of the nocturnal BP fall significantly correlate with increased target organ damage, cardiac (VERDECCHIA et al. 1990; KUWAJIMA et al. 1992; RIZZONI et al. 1992), cerebral (SHIMADA et al. 1992), vascular (TSENG et al. 1994) and renal (TIMIO et al. 1993; DEL ROSSO et al. 1994; BIANCHI et al. 1994; TSENG et al. 1994). This means that restoration of a normal circadian BP pattern is likely to be of prognostic relevance and should be regarded as an additional therapeutic goal in all of the above-mentioned conditions. However, the feasibility and prospective evaluation of such normalization have yet to be addressed in the literature. Drug-induced normalization of the pathological circadian BP profile in patients with haemodynamic brain infarction has only been pursued, and a slow recovery of the blood-brain barrier was obtained (SANDER and KLINGELHOFER 1993). A restoration of a nocturnal fall of BP was first shown to be possible in chronic renal failure after evening administration of isradipine (Fig. 1; PORTALUPPI et al. 1995, see below).

The issue of whether the diurnal pattern of BP reflects a true circadian rhythm or is the result of variations in physical activities and sleep has been debated for years. It is evident that lack of activity cannot be the sole explanation for a 24-h pattern of BP that shows a clear genetic component, is driven by a variety of neurohumoral factors with an established intrinsic circadian rhythmicity, and loses or reverses the nocturnal fall in a variety of pathological conditions in which sleep and activity show only minor changes. Recently, under unmasking conditions, evidence was provided for a clear endogenous rhythm in heart rate in man (KRÄUCHI and WIRZ-JUSTICE 1994).

In addition, data from animals under free-run conditions point to an important endogenous component in the 24-h BP profile (WITTE and LEMMER 1993; LEMMER et al. 1995b). Moreover, telemetric studies in different strains of rats – including transgenic animals exhibiting a dissociation between BP and HR and motility – also indicate the genetic control of the BP profile (Fig. 2; LEMMER et al. 1993a). Moreover, these transgenic rats can serve as an ex-

---

Fig. 2. Circadian rhythms in systolic and diastolic blood pressure and heart rate of two normotensive [Wistar-Kyoto (*WKY*), Sprague-Dawley (*SDR*)] and two hypertensive [spontaneously hypertensive (*SHR*), transgenic-hypertensive (*TGR*(mRen-2)27)] rat strains as evaluated by telemetry. In contrast to WKY, SDR and SHR, the TGR showed a dissociation between BP and HR; in all strains motility peaked during the dark span. (From LEMMER et al. 1993a, with permission. Copyright 1993 American Heart Association)

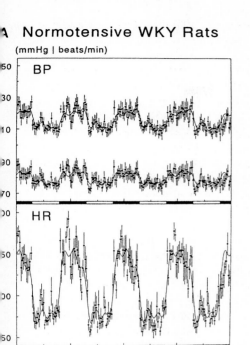
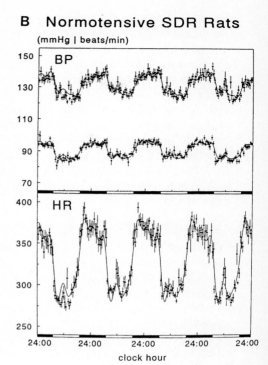
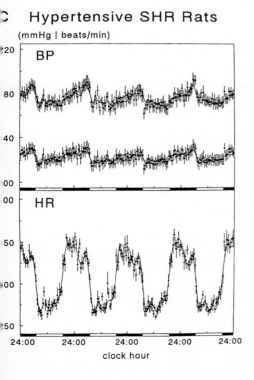
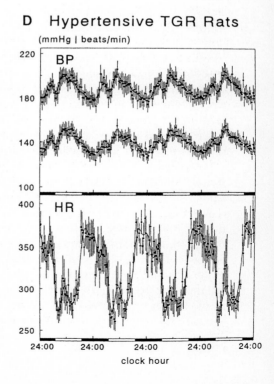

perimental model (LEMMER et al. 1993a) for the study of the mechanisms of regulation leading to a reversed 24-h profile in BP as found in forms of secondary hypertension in man (see Fig. 1). Preliminary data in this transgenic strain show that the 24-h profiles in plasma renin activity are sevenfold (with a peak in the light phase) and in aldosterone concentration are threefold higher than in the normotensive Sprague-Dawley rats (LEMMER et al. 1995a) from which the transgenics were bred. Moreover, corticosterone rhythm in the transgenic animals has a greater amplitude due to higher values exclusively in the rest period (LEMMER et al. 1995a). The findings indicate that these altered rhythms may not only be involved in increasing BP but may also contribute to the disturbed BP profile in this rat strain, and possibly also in some forms of secondary hypertension in man.

II. Cardiovascular Haemodynamics

Our knowledge of circadian variations in cardiovascular haemodynamics is far from complete. Whereas the 24-h variation in BP is a consistent and well-known finding, our understanding of the circadian variations in other parameters of cardiovascular haemodynamics is still based on controversial data.

Cardiac output has been reported to show either a decrease at night parallel to that of BP (KHATRI and FREIS 1967; MILLER and HORVATH 1976; COOTE 1982; TAKAGI 1986; MORI 1990) or no consistent change (BRISTOW et al. 1969). It should be noted that it is extremely difficult to estimate reliably cardiac output at night without interfering with the sleep state of the subject under study. From a haemodynamic point of view, the nocturnal fall in BP can only be related to decreased cardiac output or peripheral vascular resistance or both. Peripheral resistance was reduced by 36% at night in healthy volunteers confined to bed (CASIGLIA et al. 1992). In the same study, total (cutaneous, subcutaneous and muscular) arterial flow in the legs fixed by an orthopaedic device was 28% higher during the night. Another study has detected the same circadian pattern in subcutaneous blood flow of healthy individuals (SINDRUP et al. 1991). As early as 1806, BARTHEZ reported on the dilatation which occurs at the level of the small vessels at night ("*la pléthora relative*") and which is more pronounced in the veins than in the arteries.

A paucity of data exists also on the temporal changes of tissue blood flow. Renal blood flow is markedly reduced at night (DE LEEUW et al. 1985), whereas leg (CASIGLIA et al. 1992), cerebral (TOWNSEND et al. 1973) and gastrointestinal (LEMMER and NOLD 1991) blood flows are increased. This lack of uniformity in flow changes of different body areas suggests that a redistribution of cardiac output should occur with sleep. The only explanation that fits all of these findings would be a generalized reduction in vascular tone, induced by sympathetic withdrawal or parasympathetic activation, with preferential vasodilation in the extremities and relatively less marked dilation in other areas such as the kidneys. In the morning, distribution of cardiac output turns to favour the kidneys at the expense of the extremities.

Arterial vessels seem able to adapt to the nocturnal haemodynamic changes. In fact, nocturnal vasodilation consequent to changes in autonomic activation means increased arterial diameter. Due to the lower BP, this might compromise the buffering function of large arteries, which transform the systolic flow jet from the heart into a more continuous flow for tissue perfusion. However, distensibility of both the elastic common carotid artery and the muscular brachial artery have been shown to decrease at night, leaving the vascular compliance unchanged (KOOL et al. 1991). In conclusion, the circadian patterns of arterial diameter and distensibility are inversely related to each other so that the circadian changes in elastic properties do not affect the buffering function of large arteries throughout day and night.

Changes in pulmonary arterial pressure with wakefulness and sleep have been reported in healthy men: a continuous but moderate (2–4 mmHg) rise in pressure from wakefulness up to non-REM and REM sleep was observed (LUGARESI et al. 1972). A similar diurnal variation in pulmonary artery diastolic pressure was observed in patients with coronary artery disease: values were low during the day and higher at night, with the maximum values between midnight and 0600 hours (LEVY et al. 1987; GIBBS et al. 1989).

The circadian rhythm of autonomic nervous system activity is likely to be the major determinant of circadian changes in circulation. Consistent with this view is the demonstration of a circadian rhythm of α-sympathetic vasoconstrictor activity in the forearm, enhanced in the morning and reduced in the evening (PANZA et al. 1991). Recently, convincing evidence has been reported that plasma noradrenaline variations correlate with peripheral vascular resistance and inversely with cardiac output in resting humans (KENNEDY et al. 1994). Unfortunately, the study was limited to 5 daytime hours, so that we still lack direct evidence on these relationships at night.

A circadian variation in the reactivity of coronary arteries to adrenergic or parasympathetic stimuli, as well as physical exercise, has been demonstrated (for a review see MORRIS 1990; PEPINE 1991). Normal, but not atherosclerotic, coronary arteries maintain basal tone throughout the day. In fact, no circadian differences in coronary blood flow or vasodilating response to a 2-mg intracoronary infusion of nifedipine were seen in normal conditions, whereas a decreased afternoon vasodilation was seen in atherosclerotic coronary arteries, with no change in blood flow (RABY et al. 1993). This reflects increased morning tone in atherosclerotic coronary vessels, due to an endothelial dysfunction which results in a failure to limit the constrictor response associated with the morning increase in sympathetic outflow and circulating catecholamines.

Amplification by serotonin of other mediators, e.g. adrenaline and/or ADP, has a major influence on coronary blood flow (BUSH 1987). Even pineal melatonin production and secretion has been shown to affect directly the circadian rhythmicity of cardiac function, namely the diurnal changes in (Ca^{2+}/Mg^{2+})-dependent ATPase activity in the sarcolemma of heart tissue (CHEN et al. 1993). Among other potential mechanisms that could contribute to the circadian changes in cardiovascular haemodynamics, the endogenous opioid and the plasma renin-angiotensin-aldosterone systems, as well as the hypothalamic-pituitary neurohormonal axes and a number of vasoactive peptides, rank high, as we have seen for the specific case of BP.

Venous haemodynamics have recently been shown to exhibit significant circadian changes (KATZ et al. 1994). Significant changes in venous valvular function (venous-filling index) were detected, indicating progressive insufficiency in the late afternoon as compared to the early morning studies. Calf muscle pump function and ambulatory venous pressures remained constant. The authors concluded that venous haemodynamic changes occur normally as a consequence of daily activity and seem to result from valvular dysfunction. A circadian variation in portal pressure has also been reported, with a progressive decline during the afternoon and evening, followed by a rise during the night and a peak around 0900 hours (GARCIA-PARGAN et al. 1994).

In conclusion, it appears that circadian variations with dominant morning peaks in peripheral vascular resistance, arterial and pulmonary BP, portal pressure, HR, probably cardiac output, myocardial contractility and oxygen demand are all present as a consequence of rising sympathetic outflow and circulating catecholamines. Neurohumoral modulations as well as anatomo-pathological changes may differentially intervene in various disease states. However, the impact of these observations on the expected benefits and timing of drug therapy in the various cardiovascular diseases is still incompletely understood.

III. Electrical Properties of the Heart

Atrial and ventricular rates demonstrate circadian rhythmicity, with peaks during the day and troughs during the night (CHRIST and HOFF 1975; ALBONI et al. 1981) already present in fetal life (VISSER et al. 1982; PATRICK et al. 1982). A circadian rhythm is also demonstrated for sinus node function, AV nodal and myocardial refractoriness, duration of the QT interval and voltage of R and T waves (THORMANN et al. 1983; CINCA et al. 1986, 1990; DE LEONARDIS et al. 1985; BEXTON et al. 1986; GILLIS et al. 1988; MITSUOKA et al. 1990). Diurnal changes of the Q-T interval are more pronounced in normally innervated than in transplanted hearts (BEXTON et al. 1986; ALEXOPOULOS et al. 1988) and are related to variations in autonomic tone and concentrations of circulating catecholamines.

The circadian influence on refractoriness of normal cardiac tissues and accessory pathways exerts a nocturnal protection against electrical inducibility of reciprocating tachycardia but a transient facilitation of arrhythmia in the evening. In fact, diurnal variations occur in the frequency and pattern of ventricular arrhythmias, with significant reduction during sleep and no association with physical activity (LOWN et al. 1975; SENSI et al. 1980; CANADA et al. 1983; LANZA et al. 1986; BIFFI et al. 1987; LUCENTE et al. 1988). In patients with left-sided Kent bundles, a nocturnal protection against electrical induction of reciprocating tachycardia is associated with a prolongation of the atrial, atrioventricular nodal, ventricular and Kent bundle refractoriness (CINCA et al.

1987). More ventricular premature beats are induced by hypoglycaemia during wakefulness than during sleep (SHIMADA et al. 1984). HR (GILLIS et al. 1989) and left ventricular function (GILLIS et al. 1992) are major determinants of the diurnal variation in frequency of ventricular premature depolarization in subsets of patients: only premature beats of the HR-dependent type or in patients with a left ventricular ejection fraction greater than 0.30 show a circadian rhythm. A circadian periodicity is demonstrated also in the occurrence of bradyarrhythmias (OTSUKA et al. 1985) and the ventricular response to atrial fibrillation (RAEDER 1990).

The possible influence of the circadian rhythm of BP has been studied, with unclear results. A strong correlation between the incidence of ventricular extrasystoles and BP and HR values was found in patients with left ventricular hypertrophy, while in the absence of hypertrophy supraventricular extrasystoles had a higher incidence during peaks of BP and HR (Novo et al. 1990). In another study, an independent positive correlation between BP and ectopic beats was found in 8 out of 12 subjects, while the HR was a non-significant negative factor for ectopic beats (SIDERIS et al. 1992).

Altogether, these data suggest a primary role for neural activity in modulating the electrical disturbances underlying ventricular ectopy, even though at least one study has reported no significant circadian variation in any electrophysiological measure of ventricular electrical instability (McCLELLAND et al. 1990). A contribution of other hormonal factors, especially pressure or adrenal stimuli, cannot be excluded.

IV. Pathophysiology in Coronary Heart Disease

There is a body of evidence demonstrating that pathophysiological events within the cardiovascular system do not occur at random. Since many reviews have dealt with this topic (SMOLENSKY et al. 1976; LEMMER 1986, 1989, 1991; WILLICH et al. 1993; WILLICH and MULLER 1996), only some details pertinent to the understanding of time-specified drug treatment are presented here.

The onset of myocardial infarction predominates between 0600 hours and 1200 hours (MASTER 1960; SMOLENSKY et al. 1976; MULLER et al. 1985; MITLER et al. 1987; WILLICH et al. 1989a, 1993; TOFLER et al. 1987; BEHAR et al. 1993; and others). A very similar pattern of onset has been described for sudden cardiac death (MULLER et al. 1987; ARONOW and AHN 1993; ARONOW et al. 1994; GNECCHI-RUSCONE et al. 1994), stroke (MARSHALL 1977; ARGENTINO et al. 1990), ventricular arrhythmias (TWIDALE et al. 1989; LAMPERT et al. 1994) and arterial embolism (DEWAR and WEIGHTMAN 1983). Also the onset of myocardial ischaemia, angina attacks or silent ischaemia is significantly more frequent during the daytime hours than at night (VON ARNIM et al. 1985; ROCCO et al. 1987; HAUSMANN et al. 1987; MULCAHY et al. 1988; EGSTRUP 1991), whereas the onset of angina attacks in variant angina peaks around 0400 hours during the night (WATERS et al. 1984; ARAKI et al. 1983).

C. Chronopharmacology of Hypertension

Non-drug treatment of hypertension includes weight reduction, restriction of sodium intake and activity, aside from psychological treaments. Weight loss in hypertensives was correlated with a decrease in BP, leading to a restoration of a normal circadian rhythm (DasGupta et al. 1991). Though specific studies are lacking, there is some indication that mental and psychological stress may decrease the normal amplitude in the 24-h BP pattern (James et al. 1991; James and Pickering 1993; Pieper et al. 1993).

Drug treament of hypertension includes different types of drugs such as diuretics, β- and α-adrenoceptor-blocking drugs, calcium channel blockers and converting enzyme inhibitors (see below).

There is a great number of studies on the effects of these various antihypertensives on the 24-h BP profile (see Lemmer 1987, 1991, 1996; Stanton and O'Brien 1994). In most of the studies, however, time of day of drug administration was not a specific point of investigation. Since the drug's halflife, its galenic formulation (see below) and the circadian time of drug dosing may influence the degree and the duration of the BP-lowering effect – aside from influencing the pharmacokinetics (see below) – it is difficult to draw final conclusions about the importance of a circadian-time-dependent drug dosing. Thus, only crossover studies (morning vs. evening) will be considered in this review in more detail (see Table 1).

In order to compare adequately the results obtained mainly by ambulatory BP-monitoring (ABPM) devices, the method of analysing the 24-h BP profile consisting of 30–80 data points within 24 h is also of importance. BP profiles were mostly compared by visual inspection, which is not able to detect subtle disturbances in the circadian profile. Less frequently, comparison of mean daytime versus mean night-time values has been carried out using the cosinor method according to Halberg (1969), square-wave-fit method (Idema et al. 1992), cumulative sums method (CUSUMS method, Stanton et al. 1992) and spline functions (Steitberg et al. 1989). Recently analyses by non-linear partial Fourier series (Germano et al. 1984; Mattes et al. 1991; Witte and Lemmer 1992; Lemmer et al. 1993b; Staessen et al. 1993; Zuther et al. 1996) were introduced which take into account the multiphasic and asymmetrical profiles of BP and HR in individual patients as well as in grouped patients. Moreover, one of the methods combines linear and rhythm analyses by partial Fourier series, including calculation of the area under the curve (AUC), slope and statistical evaluations (Lemmer et al. 1993b; Zuther et al. 1996). In general, it is important to note that ABPM is now regarded as the method of choice to evaluate BP profiles, and guidelines have been published (Anlauf et al. 1993; Sheps et al. 1994).

I. β-Adrenoceptor Antagonists

1. Pharmacological Characterization

β-Adrenoceptor antagonists inhibit functions of the sympathetic nervous system in organs supplied with β-adrenoceptors such as the heart, the kidneys and the smooth muscles of the blood vessels and the bronchi. Of great therapeutic importance is the decrease in heart rate, cardiac output and cardiac oxygen consumption and the inhibition of renin release. β-Adrenoceptor blockade, on the other hand, can result in bradycardia, a negative inotropic effect, bronchospasm and an increased peripheral resistance. Two types of β-adrenoceptors are of major importance: $β_1$-adrenoceptors dominate in the heart and the kidneys, whereas within the bronchi and the vascular bed the $β_2$-subtype predominates. β-Adrenoceptor antagonistic drugs can be divided into three groups: non-selective (e.g. propranolol, oxprenolol), $β_1$-selective (e.g. bisoprolol, metoprolol, atenolol) and compounds with intrinsic sympathomimetic/agonist activity (e.g. pindolol, carteolol). Moreover, it is also of importance that these drugs differ greatly in their physicochemical properties such as lipophilicity, which determines the degree of non-specific membrane activity and local anaesthetic property (LEMMER et al. 1972; HELLENBRECHT et al. 1973). In consequence, highly lipophilic β-adrenoceptor antagonists (propranolol, bupranolol, oxprenolol) accumulate in lung and brain tissue, whereas hydrophilic ones (sotalol, atenolol) scarcely pass the blood-brain barrier, exhibiting, however, higher concentrations in heart and muscle tissue (LEMMER and BATHE 1982; LEMMER et al. 1985). Thus, when using β-adrenoceptor antagonists in studies on circadian clocks the different properties of these compounds have to be taken into account in order to avoid false conclusions.

2. Clinical Data

Concerning the circadian phase dependency of the BP-lowering effect of β-adrenoceptor antagonists, final conclusions are difficult to draw, because neither crossover (morning vs. evening) nor equieffective dose studies have been published on hypertensive patients (see LEMMER 1987, 1996; STANTON and O'BRIEN 1994). A resume of about 20 studies (STANTON and O'BRIEN 1994) shows that β-adrenoceptor antagonists – either $β_1$-selective or non-selective, with intrinsic agonist activity – either do not affect, or reduce or even abolish the rhythmic pattern in BP. However, there is a tendency for β-adrenoceptor antagonists not to greatly affect night-time values and to be less effective in reducing the early morning rise in BP (RAFTERY 1983; GOULD and RAFTERY 1991), whereas vasodilators such as α-adrenoceptor antagonists have a pronounced effect (see below). Drugs with partial agonist activity – mainly pindolol – even increase HR at night (QUYYUMI et al. 1984). Decreases in HR by propranolol and oxprenolol are also more pronounced during daytime hours, whereas no circadian phase dependency was found when exercise-induced heart rate was studied under β-adrenoceptor blockade by oxprenolol (Table 2).

There is evidence from a study in healthy volunteers that the concentration-response relationship for propranolol depends on the circadian time (LANGNER and LEMMER 1988). Furthermore, pharmacokinetics of the β-adrenoceptor antagonists propranolol and oxprenolol have also been shown to depend on the circadian dosing time (SEMENOWICZ-SIUDA et al. 1984; LANGNER and LEMMER 1988; KOOPMANS et al. 1993; Table 3) with peak concentrations

Table 1. Effects of calcium channel blockers and ACE inhibitors on the 24-h pattern in blood pressure; only data obtained in crossover studies are considered. Drug dose, duration of treatment, circadian time of drug administration, number of patients, diagnosis as well as effects on **BP** profile are given (for rating see Table 4)

Drug	Dose (mg/day)	Duration Dosing time	Patients (n) Diagnosis	Effect on 24-h blood pressure			References
				Day	Night	24-h profile	
Calcium channel blockers							
Amlodipine	5	4 weeks – a.m. – p.m.	20, EH	++ ++	++ ++	Preserved Preserved	Mengden et al. 1992
Amlodipine	5	3 weeks – 0800 hours – 2000 hours	12, EH	+ +	(+) (+)	Preserved Preserved	Nold et al. 1996
Isradipine	5	4 weeks – 0700 hours – 1900 hours	18, EH	++ ++	++ ++	Preserved Preserved	Fogari et al. 1993
Isradipine	5	4 weeks – 0800 hours – 2000 hours	16, RH[a]	++ +(+)	+ ++(+)	Not normalised Normalised	Portaluppi et al. 1995
Nifedipine	30 GITS	1 or 2 weeks – 1000 hours – 2200 hours	10, EH	++ ++	++ ++	Preserved Preserved	Greminger et al. 1994
Nifedipine	10 i.r.	Single dose – 0800 hours – 1900 hours	12, NT	+ +	+(+) +	Preserved Preserved	Lemmer et al. 1989

Drug	Dose (mg)	Duration/Timing			Profile	Reference
Nitrendipine	20	4 weeks – 0700 hours – 1900 hours	++ ++		Preserved Preserved	Meilhac et al. 1992
Nitrendipine	10	3 days – 0600 hours – 1800 hours	++ ++	++ ++	Preserved Changed	Umeda et al. 1994
ACE inhibitors						
Benazepril	10	Single dose – 0900 hours – 2100 hours	+++ ++	++ ++	Preserved Changed	Palatini et al. 1993
Captopril plus hydrochlorothiazide	25 + 12.5	3 weeks – 0700–0800 hours – 1800–2000 hours	++ +	++ ++	? ?	Middeke et al. 1991b
Enalapril	10	Single dose – 0700 hours – 1900 hours 3 weeks – 0700 hours – 1900 hours	++ ++ ++ +	++ +++ (+) ++	Preserved Changed Preserved Changed	Witte et al. 1993
Quinapril	20	4 weeks – 0800 hours – 2200 hours	++ ++	++ ++	Preserved Preserved	Palatini et al. 1992

Abbreviations: EH, essential (primary) hypertension; RH, renal (secondary) hypertension; NT, normotensives.
[a] Abolished 24-h blood pressure profile.

Table 2. Effects of β-adrenoceptor antagonists on daily variation in heart rate. Drugs were studied after acute dosing at four circadian times. Drug dose, circadian time of drug application (p.o.), number of subjects diagnosed and effects on heart rate are given (for rating see Table 4)

Drug	Dose (mg/day)	Duration Dosing time	Patients (n) Diagnosis	Effect on heart rate Day	Effect on heart rate Night	References
Propranolol[a]	80	Single dose – 0800 hours – 1400 hours – 2000 hours – 0200 hours	6, NT	+ + + 	 + (+) + +	SEMENOWICZ-SIUDA et al. 1984
	80	Single dose – 0800 hours – 1400 hours – 2000 hours – 0200 hours	6, HC	+ + + + 	 + + 	
Propranolol[b]	80	Single dose – 0800 hours – 1400 hours – 2000 hours – 0200 hours	4, NT	+ + + + + + + +	φ + (+) (+) 	LANGER and LEMMER 1988
Oxprenolol[c]	80	Single dose – 0800 hours – 1400 hours – 2000 hours – 0200 hours	6, NT	+ + + + 	 + + + +	KOOPMANS et al. 1993

Abbreviations: NT, normotensives; HC, hepatic cirrhosis.
[a] Effect$_{0-4h}$.
[b] AUC$_{0-10h}$.
[c] AUC$_{0-10h}$ after exercise.

(C_{max}) being higher and time to peak concentrations (t_{max}) being shorter after morning than after evening dosing (for interpretation see Sect. C.II).

In conclusion, the clinical data indicate that β-adrenoceptor-mediated regulation of BP dominates during daytime hours and is of minor importance during the night and early morning hours. This correlates well with the circadian rhythm in sympathetic tone as indicated by the rhythm in plasma noradrenaline and cAMP (see above). Moreover, β-adrenoceptor-mediated signal transduction processes have been shown to vary significantly with circadian time in rodents (see Chap. 6, this volume), giving further evidence for the rhythmicity in sympathetic drive.

II. Calcium Channel Blockers

1. Pharmacological Characterization

Calcium channel blockers inhibit the movement of calcium through the so-called slow channels of the membrane into the cell during the action potential. At the site of the heart this results in a negative inotropic effect; drug-induced heart rate decrease can be bound to an adrenergic counterregulation, mainly by the nifedipine-type calcium channel blockers, resulting in reflexly induced tachycardia. Inhibition of calcium influx relaxes arterial smooth muscle, leading to vasodilatation, mainly of the arteries. The tone of the coronary arteries is reduced and vasospasms are attenuated, which is of importance in variant angina. There are different types of calcium channel blockers, verapamil-, nifedipine- and diltiazem-type compounds. Calcium channel blockers of the verapamil and diltiazem type inhibit the atrioventricular conductance, which is the basis for their use in the treatment of supraventricular tachyarrhythmias. With all calcium channel blockers vasodilatation occurs at lower concentrations than the cardiodepressant effects. However, the difference between vasodilating and cardiodepressant effects is greater with the 1,4-dihydropyridines (e.g. nifedipine, nitrendipine, isradipine, amlodipine) than with the verapamil- and diltiazem-like compounds.

2. Clinical Data

Calcium channel blockers were also analysed mainly by visual inspection of the BP profiles. In primary hypertensives, t.i.d. dosing of non-retarded verapamil did not greatly change the BP profile although it was less effective at night (GOULD et al. 1982a,b). A single morning dosing of sustained-release verapamil showed good 24-h BP control (CARUANA et al. 1987), whereas a sustained-release formulation of diltiazem was less effective at night (LEMMER et al. 1994b). Dihydropyridine derivatives, differing in pharmacokinetics, seem to reduce blood pressure to a varying degree during day and night; drug formulation and dosing interval may play an additional role. However, seven studies using crossover designs have been published (Table 1). In essential hypertensives amlodipine, isradipine and nifedipine, in GITS (gastrointestinal therapeutic system) and in normotensives, immediate-release nifedipine did not

Table 3. Pharmacokinetic parameters of cardiovascular active drugs. Only data obtained in crossover studies are considered. At least two dosing times (around 0600–0800 hours and 1800–2000 hours) were studied; sometimes up to six circadian times were included (see drugs in Table 2). Only the parameters C_{max} = peak drug concentration, and t_{max} = time to C_{max}, are given

Drug	Dose (mg/day, duration)	C_{max} (ng/ml)		t_{max} (h)		References
		Morning	Evening	Morning	Evening	
Cibenzoline	160, sd	446.0	402.0	1.7	2.6	Brazzell et al. 1985
Digoxin	0.5, sd	3.6*	1.8	1.2	3.2	Bruguerolle et al. 1988
Enalapril[a]	10, sd	33.8	41.9	4.4	4.5	Witte et al. 1993
	10, 3w	46.7	53.5	3.5*	5.6	
IS-5-MN i.r.	60, sd	1605.0	1588.0	0.9*	2.1	Scheidel and Lemmer 1991
IS-5-MN s.r.	60, sd	509.0	530.0	5.2	4.9	Lemmer et al. 1991b,c
ISDN i.r.	20, sd	30.7	25.3	0.6	0.6	Lemmer et al. 1989
Nifedipine i.r.[b]	10, sd	82.0*	45.7	0.4*	0.6	Lemmer et al. 1991a
Nifedipine s.r.	2×20, 1w	48.5	50.1	2.3	2.8	Lemmer et al. 1991a
Oxprenolol[c]	80, sd	507.0	375.0	1.0	1.1	Koopmans et al. 1993
Procainamide	500, sd	3.8	3.4	1.7	1.8	Fujimura et al. 1989
Propranolol[d]	80, sd	38.6*	26.2	2.5	3.0	Langer and Lemmer et al. 1988
Propranolol	80, sd	68	60	2.3	2.7	Semenowicz-Siuda et al. 1984
Verapamil s.r.	360, 2w	389.0	386.0	7.2*	10.6	Jespersen et al. 1989
Verapamil	80, sd	59.4*	25.6	1.3	2.0	Hla et al. 1992

Abbreviations: sd, single dose; w, week(s); i.r., immediate-release preparation; s.r., sustained-release preparation.
*Morning vs evening P at least < 0.05.
[a] Pharmacokinetics of enalaprilat.
[b] Bioavailability significantly reduced after evening dosing.
[c] Significant difference in half-life.
[d] Pharmacokinetics of (-)-propranolol

affect the BP profile differently after once morning or once evening dosing, whereas with nitrendipine the profile remained unaffected or changed after evening dosing (Table 1). In primary hypertensive patients twice daily nifedipine was also able to lower the blood pressure throughout a 24-h period (LEMMER et al. 1991b). Most interestingly, the disturbed BP profile in secondary hypertensives due to renal failure was only normalized after evening but not after morning dosing of isradipine (Table 1).

3. Chronopharmacokinetics

The pharmacokinetics of various calcium channel blockers in different galenic formulations were investigated in clinical studies in relation to circadian time of drug dosing (Table 3). In healthy volunteers the bioavailability of an immediate-release formulation of nifedipine was found to be reduced by 40% after evening compared to morning dosing, with C_{max} being higher and t_{max} being shorter after morning dosing (LEMMER et al. 1991a,b). In contrast, pharmacokinetic parameters of a sustained-release formulation of nifedipine studied in primary hypertensives did not depend on the circadian time of dosing (LEMMER et al. 1991b). Also regular (HLA et al. 1992) as well as sustained-release verapamil (JESPERSEN et al. 1989) displayed higher C_{max} and/or shorter t_{max} values after morning dosing (Table 3).

Similar chronokinetics have been reported after oral dosing of other cardiovascular active drugs such as enalapril (WITTE et al. 1993), digoxin (BRUGUEROLLE et al. 1988), β-adrenoceptor antagonists (see above) and antiarrhythmic drugs (see Table 3). Interestingly, intravenously infused nifedipine (1.5 mg infusion for 1 h at 0700 hours or at 1900 hours) did not show daily variations in its pharmacokinetics (LEMMER et al. 1991b), indicating that gastrointestinal mechanisms must be involved in the drugs' chronokinetics. Bearing in mind that drugs are mainly absorbed by passive diffusion, circadian effects on gastric-emptying time and gastrointestinal perfusion may be the main determinants resulting in a circadian phase-dependent absorption. This assumption is supported by the following data: A delayed gastric-emptying time in the evening for solids has been described (GOO et al. 1987). Secondly, of even greater importance may be that hepatic blood flow (representative of gastrointestinal tract perfusion) has been shown to be significantly greater during the late night and early morning than around noon or in the afternoon (LEMMER and NOLD 1991). Both mechanisms could nicely explain higher C_{max} and shorter t_{max} values after oral ingestion of non-retarded drugs in the morning (Table 3; LEMMER and BRUGUEROLLE 1994). Since it is well known that some side effects of drugs used in the treatment of cardiovascular diseases, e.g. heart rate increase or bradycardia, orthostatic dysregulation, are linked to peak drug concentrations (see LEMMER et al. 1991b; BRUGUEROLLE and LEMMER 1993; LEMMER and BRUGUEROLLE 1994), it seems worthwhile to investigate in more detail at different times of day the circadian time-dependent pharmacokinetics simultaneously with the effects in hypertensives patients.

In contrast to β-adrenoceptor-mediated processes of signal transduction (see Chap. 6, this volume), there are no comparable data on circadian variation in calcium-channel-mediated functions either in animals or in man.

III. Converting Enzyme Inhibitors

1. Pharmacological Characterization

These drugs are competitive inhibitors of the converting enzyme by which they reduce the conversion of angiotensin I into angiotensin II in plasma and tissues, angiotensin II being a highly potent vasoconstrictor. The degradation of the vasodilating substance bradykinin is also reduced. Converting enzyme inhibitors reduce peripheral resistance, the formation of aldosterone and the aldosterone-mediated retention of water and sodium. Reduction of angiotensin II formation also diminishes the stimulation of noradrenaline release, which results in an inhibition of sympathetic tone. Captopril is the representative of a short-acting converting enzyme inhibitor, longer-acting ones being enalapril, benazepril, quinapril and ramipril, which are prodrugs and are hydrolysed to the active metabolites in the liver. Converting enzyme inhibitors are not only effective antihypertensive drugs but can also increase life expectancy in congestive heart failure.

2. Clinical Data

Several studies with converting enzyme inhibitors, dosed once in the morning or twice daily, showed that these drugs did not greatly modify the 24-h BP pattern (see STANTON and O'BRIEN 1994). However, intra-arterial studies with enalapril or ramipril have shown that while causing sustained daytime reduction in BP, these drugs had only marginal effects on night-time pressures (GOULD and RAFTERY 1991). Thus, the findings obtained with converting enzyme inhibitors in conventional clinical studies are not equivocal.

Four crossover studies (single morning vs. single evening dosing) with converting enzyme inhibitors in essential hypertensive patients have been published (Table 1). They demonstrated that evening dosing of benazepril and enalapril changed the BP profile, resulting in a more pronounced nightly drop which was even observed after a 3-week single-dose evening treatment with enalapril (Table 1). After chronic treatment with quinapril, evening dosing also resulted in a more pronounced effect than morning dosing; the BP pattern, however, was not greatly modified (Table 1). A fixed combination of captopril and the diuretic hydrochlorothiazide (HCT) only slightly reduced BP after both morning and evening dosing; the effect of the latter treatment, however, was less effective on the next day (Table 3).

Interestingly, the reversed rhythm in BP in hypertensive transgenic rats (LEMMER et al. 1993a) could be normalized by either enalapril (LEMMER et al. 1994a) or the angiotensin II AT_1 receptor antagonist losartan (SCHNECKO et al. 1995), indicating the importance of the renin-angiotensin system for the patholgical regulation of the reversed BP pattern in secondary hypertension.

IV. Other Antihypertensives

Antihypertensives of other classes have rarely been studied in relation to possible circadian variation. Interestingly, once daily morning dosing of the diuretics xipamide (RAFTERY et al. 1981b) and indapamide (OCON and MORA 1990) reduced BP in essential hypertensives without changing the 24-h BP pattern. On twice daily dosing the α-adrenoceptor antagonists indoramin (GOULD et al. 1981) and prazosin (WEBER et al. 1987) also did not change the BP profile. Recently, PICKERING et al. (1994a) reported that a single night-time dose of the α-adrenoceptor antagonist doxazosin reduced both systolic and dystolic BP thoughout day and night, but that the greatest reduction occurred in the morning hours. Since α-adrenoceptor blockade more effectively reduced the peripheral resistance during the early morning hours than at others times of the day (PANZA et al. 1991), these findings point to the importance of α-adrenoceptor-mediated regulation of BP during this time of day. In addition, PICKERING et al. (1994a) reported that the peak treatment effect after night-time dosing of doxazosin was later than predicted from the drug's pharmacokinetics, an observation which nicely supports similar findings on BP described for nifedipine (LEMMER et al. 1991a) and enalapril (WITTE et al. 1993) and on HR for propranolol (LANGNER and LEMMER 1988; see also Table 2). Thus, these observations again support the notion that circadian variations of the regulatory mechanisms of the BP and the circadian phase-dependent responsiveness to drugs greatly determine the treatment profile (see also Sects. D.II,III).

V. Chronopharmacology of Blood Pressure in Congestive Heart Failure

Only very limited data are available to what degree drugs or the timing of drug dosing are able to modify the BP pattern in congestive heart failure. In comparison to control profiles without treatment t.i.d., captopril (0800, 1200, 1700 hours) and once daily dosing of enalapril (0800 hours) equally reduced the BP without significantly affecting amplitude and acrophase of the 24-h pattern (OSTERZIEL et al. 1991, 1992).

VI. Conclusion

The studies on the effects of various antihypertensive drugs belonging to different drug classes indicate that different mechanisms of regulation of the 24-h BP pattern may predominate at certain times of the 24-h day, as already mentioned in Sect. B.I. It would, therefore, be surprising if antihypertensives did not affect the 24-h BP profile differently. In line with this is the observation that a drug's pharmacokinetics may not predict its treatment profile, mainly when considering the correlation between time-to-peak drug concentration (t_{max}) with time-to-peak drug effect (T_{max}) (see above). This again supports the notion that circadian variations in the regulatory mechanisms of the BP have an important impact on the drug effect.

This is also supported by animal data obtained from five strains of normotensive and hypertensive rats in which the long-acting calcium channel blocker amlodipine, the α-adrenoceptor antagonist doxazosin and the converting enzyme inhibitor enalapril differed greatly in their efficacy and their dose and circadian phase dependency of their BP-lowering effects as evaluated in unrestrained rats by telemetry (MATTES and LEMMER 1991; LEMMER et al. 1993a, 1994a).

In general, it can be concluded that in uncomplicated, essential hypertensive patients or those who show a nightly drop in BP (dippers) antihypertensive drugs should be given in the morning. Whether the cardiovascular risk during the morning hours is decreased by evening dosing of an α-adrenoceptor antagonist needs additional confirmation. In secondary hypertension (or in non-dippers) evening dosing may be advantageous in better normalizing the disturbed BP pattern. Possible chronokinetics seem to be of less importance for drug efficacy, with the exception that peak drug effects are often correlated with side effects.

D. Chronopharmacology of Coronary Heart Disease

I. β-Adrenoceptor Antagonists

β-Adrenoceptor antagonists have been shown to reduce the risk of recurrent myocardial infarction (NORWEGIAN MULTICENTER STUDY GROUP 1989). Several studies addressed the question whether or not these drugs could influence the circadian pattern in myocardial infarction (see Tables 4,5). In the MILIS study (MULLER et al. 1985) the morning peak in the onset of myocardial infarction was abolished in patients receiving β-adrenoceptor antagonists. This finding was supported by several other studies (WILLICH et al. 1989a; WOODS et al. 1992). The BHAT study (PETERS et al. 1989) demonstrated a reduction in the number of sudden cardiac deaths by propranolol between 0500 and 1100 hours, whereas the occurrence of deaths at other circadian times was almost equal in untreated and treated patients. Interestingly, non-Q-wave infarction did not show circadian variation, either in untreated patients or in those receiving β-adrenoceptor antagonists (KLEIMAN et al. 1990). Recently, SANDRONE et al. (1994) gave evidence that – in analysing the high- and low-frequency components – heart rate variability after acute myocardial infarction was significantly affected by β-adrenoceptor blockade with either atenolol or metoprolol in reducing sympathetic activation and increasing vagal tone, with the effects being more pronounced during daytime hours.

Concerning the rhythmicity in ischaemic episodes and in the incidence of angina attacks (Tables 4,5), most studies demonstrated an abolition or attenuation of the morning peak by propranolol (COHN et al. 1989) or the more β_1-selective antagonists metoprolol (IMPERI et al. 1987; COY et al. 1990; WILLICH et al. 1989a; EGSTRUP 1991) and atenolol (MULCAHY et al. 1988;

DEEDWANIA et al. 1991). In other studies with atenolol, atenolol/propranolol or bevantolol, no influence on the circadian distribution of angina attacks (KOSTIS 1988; QUYYUMI et al. 1984, 1987; BOWLES et al. 1986) or of ischaemic episodes (BENHORIN et al. 1993) was observed, though their total number was reduced at both times of day. Bisoprolol had a predominant effect during daytime hours (PRAGER et al. 1989). Interestingly, pindolol, a β-adrenoceptor antagonist with intrinsic sympathomimetic activity, even increased heart rate at night and did not reduce angina attacks (QUYYUMI et al. 1984). Despite the preservation of circadian rhythmicity in silent ischaemia in their study, BENHORIN et al. (1993) demonstrated that the time-dependent variation in ischaemic threshold, i.e. heart rate at the onset of ischaemia, was completely abolished by treatment with β-adrenoceptor antagonists.

Whether or not influences of β-adrenoceptor antagonists on circadian rhythms in angina are observed could also depend on the subgroup of patients studied. GILLIS et al. (1992) described significant circadian variation with a morning increase in the frequency of ventricular premature complexes, which was abolished by β-adrenoceptor antagonists. However, the baseline circadian rhythm could be demonstrated only in patients with a left ventricular ejection fraction (LVEF) greater than 0.30, whereas in those with more pronounced ventricular dysfunction (LVEF < 0.30) no rhythmicity was observed. Analysis of all patients without subdivisions would have obscured the circadian time-dependent effects of drug treatment of ventricular arrhythmias. In myocardial infarction, ventricular arrhythmias were also suppressed more effectively during early morning and daytime hours than during the night (LICHTSTEIN et al. 1983). A detailed analysis of data from the ASIS study (ANDREWS et al. 1993) showed the importance of subclassification of ischaemic events with regard to concomitant changes in heart rate. In that study propranolol markedly reduced those ischaemic events which occurred while or directly after heart rate increased, whereas the proportion of episodes not related to heart rate changes was even more pronounced under β-adrenoceptor antagonist treatment. Since in addition ANDREWS et al. (1993) observed that heart-rate-related ischaemia had a different circadian pattern than non-heart-rate-related episodes, anti-ischaemic treatment, affecting only one type of ischaemia, can definitely be expected to show a circadian time dependency.

Unfortunately, prospective chronopharmacological studies comparing the effects of β-adrenoceptor antagonists on ischaemic events after dosing at different times of the day have not been performed so far.

Finally, there is some evidence that responses to defined exercise tests can also be circadian phase dependent in patients with stable angina on or off β-adrenoceptor antagonists (JOY et al. 1982). In variant angina, β-adrenoceptor blockade by propranolol even aggravated the angina attacks (YASUE et al. 1979). Detailed chronopharmacological studies are, therefore, required taking into consideration both chronopharmacokinetics and time-dependent susceptibility of the cardiovascular system to β-adrenoceptor antagonists.

Table 4. Effects of β-adrenoceptor antagonists on the circadian pattern of symptoms in coronary heart diseased patients. In general coronary heart disease was verified by angiography; mostly Holter-ECG was used to assess the ischaemic burden, and myocardial infarction was verified objectively. If available, number of patients (n), drug dosage (after titration), duration of treatment (h, hours; d, days, w, weeks), time of drug dosing (m, morning; e, evening), main diagnosis and reference are given. Though not mentioned in the studies a morning application can be assumed for drugs with once daily dosing

Drug	Patients (n)	Dose (mg/day), duration	Drug effects during			Diagnosis	References
			Morning	Daytime	Night		
Atenolol	23	50/100m, 4w	++	++	++	CHD, stA	Kostis 1988
Atenolol	24	50/100m, 2–4w	+++	+++	+	CHD	Deedwania et al. 1991
Atenolol	41	100, 5d	+++	++	+	CHD, stA	Mulcahy et al. 1988
Atenolol	15	100, 5d		++	++	CHD[e], AA	Quyyumi et al. 1984
Atenolol	9	100, 5d		++	++	CHD[e]	Quyyumi et al. 1987
Atenolol or propranol	18	25/40 or 80, 2w	++	++	++	CHD, stA[c]	Benhorin et al. 1993
Bevantolol	21	200, 4w	++	+	++	CHD, stA	Bowles et al. 1986
Bevantolol	21	2×100, 4w	+	+	++	CHD, stA	Bowles et al. 1986
Bisoprolol	13	10m, 2w	++	++	(+)	CHD	Prager et al. 1989
Metoprolol	9	2×100/200, 1w	+++	+++	++	CHD, stA	Imperi et al. 1987; Willich et al. 1989b
Metoprolol	10	2×200, 1w	++	+++	++	CHD	Coy et al. 1990
Metoprolol	31	2×50/100, 2w	++	+++	(+)	CHD, stA	Egstrup 1991
+ nifedipine	42	+ 3×20/10, 1w	+++	+++	+	CHD, stA	Egstrup 1991
Nadolol	23	40/80m, 4w	++	++	++	CHD, stA	Kostis 1988
Pindolol	15	3×5, 5d		+	φ	CHD[e], AA	Quyyumi et al. 1984

Drug	n	Dose, duration					Diagnosis	Reference
Propranolol	13	60, 2h	−	−			CHD, varA	Yasue et al. 1979
Propranolol	419	?, 6w	++	++		(+)	MI, VA	Lichtstein et al. 1983
Propranolol	9	4×40, 4w		++			CHD, stA	Joy et al. 1982
Propranolol	123	85±28, 116±56w	++	++		φ	MI[f], VA	Aronow et al. 1994
Propranolol LA	50	292, 2w	+++	+++			CHD, stA[a]	Andrews et al. 1993
Propranolol LA	50	292, 2w	++	++			CHD, stA[b]	Andrews et al. 1993
Propranolol LA	24	80, 5d	++	+++		++	CHD,	Cohn et al. 1989
Propranolol	101	180/240, 6–12w	++	(+)		(−)	SCD[d]	Peters et al. 1989
β-Blockers	135	?, ?	φ	φ		φ	MI[d]	Muller et al. 1985
β-Blockers	143	?, ?	++	φ		φ	MI, n-Q[d]	Kleiman et al. 1990
β-Blockers	206	?, ?	++	φ		φ	MI[d]	Willich et al. 1989a
β-Blockers	132	?, ?	+++	+++		φ	MI, VPC[d]	Gillis et al. 1992
β-Blockers	185	?, ?					MI[d]	Woods et al. 1992

Rating of drug effects: +, ++, +++, effective; φ, no effect; − worsening.

Abbreviations: CHD, coronary artery/heart disease; MI, myocardial infarction; stA, stable angina pectoris; varA, variant angina; AA, angina attacks; VA, ventricular arrhythmias; VPC, ventricular premature complex; n-Q, non Q-wave infarction; SCD, sudden cardiac death.

[a] Ischaemia associated with heart rate increase.
[b] Ischaemia not associated with heart rate increase.
[c] Rhythm in ischaemic threshold abolished.
[d] Influence on rhythmic pattern.
[e] Severe angina pectoris.
[f] Rhythm observed with no antiarrhythmics abolished by propranolol.

Table 5. Effects of calcium channel blockers, nitrates and acetylsalicylic acid (aspirin) on the circadian pattern of symptoms in coronary heart diseased patients. For further details see Table 4

Drug	Patients (n)	Dose (mg/day), duration	Drug effects during			Diagnosis	References
			Morning	Daytime	Night		
Calcium channel blockers							
Amlodipine	47	5–10, mean 71d	++	++	(+)	CHD, AA	Taylor 1992
Amlodipine	250	10, 7w	++	++	(+)	CHD, stA	Deanfield et al. 1994
Diltiazem	13	90, 2h	++	++		CHD, varA	Yasue et al. 1979
Diltiazem sr	60	2×180, 2w	++	++	++	CHD, stA	Parmley et al. 1992
Diltiazem sr	50	350, 2w		+		CHD, stA[a]	Andrews et al. 1993
Diltiazem sr	50	350, 2w		++		CHD, stA[b]	Andrews et al. 1993
Nifedipine	50	79, 2w		+		CHD, stA[a]	Andrews et al. 1993
Nifedipine	50	79, 2w		+++		CHD, stA[b]	Andrews et al. 1993
Nifedipine GITS	92	30–180m, 4w	++	++	+	CHD, stA	Parmley et al. 1992
Nifedipine GITS	92	30-180e, 4w	++	++	+	CHD, stA	Parmley et al. 1992
+ β-blockers	115	+ ?,	++	++	+	CHD, stA	Parmley et al. 1992
Nifedipine	33	3×10–20, 5d	φ	(+)	φ	CHD, stA	Mulcahy et al. 1988
Nifedipine	10	4×10–30, 1w	+++	+	+	CHD, stA	Nesto et al. 1991
Nifedipine	16	3×20/30, 2–4w	++	++		CHD	Deedwania et al. 1991
Nifedipine	9	3×10/20, 5d		+	++	CHD[c]	Quyyumi et al. 1987
CC blockers	147	?, ?	φ	φ	φ	MI[d]	Willich et al. 1989a
CC blockers	132	?, ?	++			MI[d]	Woods et al. 1992

Drug	N	Dose				Condition	Reference
Organic nitrates							
ISDN sr	15	120m, 2w	+	+	φ	CHD, stA	Hausmann et al. 1989
ISDN sr	10	120m, 3w	++	++		CHD, stA	Wortmann and Bachmann 1991
IS-5-MN	10	2×20, 3w	++	φ		CHD, stA	Wortmann et al. 1991
IS-5-MN	187	40m, 2w	++	++	++	CHD	Ganzinger 1992
IS-5-MN.	195	2×20, 2w	++	++	++	CHD	Ganzinger 1992
IS-5-MN	9	2×40, 5d		+	++	CHD[c]	Quyyumi et al. 1987
Glyceroltrinitrate	7	0.6, acute	++	φ		CHD, varA[e]	Yause et al. 1979
Nitrates	174	?, ?	φ	φ	φ	MI[d]	Woods et al. 1992
Others							
Aspirin	211	325, ?	+++	++	φ	MI	Tofler et al. 1987

[a] Ischaemia associated with heart rate increase.
[b] Ischaemia not associated with heart rate increase.
[c] Severe angina pectoris.
[d] Influence on rhythmic pattern.
[e] Increase in coronary artery diameter.

At least in myocardial infarction and in stable angina pectoris, the attenuation of the increased sympathetic tone during the early morning hours and during the day seems to be an important therapeutic aspect in the use of the β-adrenoceptor antagonists.

In summary, it appears that treatment of coronary heart diseased patients by β-adrenoceptor antagonists effectively reduces ischaemic events at any time of day. β-Adrenoceptor antagonists seem to be of special therapeutic value in the early morning hours, which are the hours of high risk.

II. Calcium Channel Blockers

In patients receiving calcium channel blockers controversial results have been reported as far as the circadian pattern in ischaemic episodes or in myocardial infarction is concerned (Table 5). In the ISAM study, calcium channel blockers had no effect on the circadian pattern in myocardial infarction (WILLICH et al. 1989a), whereas WOODS et al. (1992) observed a bimodal distribution in treated patients with an attenuated early morning peak and an additional peak around midnight. Unfortunately, in both of the studies the data were not further analysed with regard to the different types of calcium channel blockers used. Such an analysis would be of great importance, since the anti-ischaemic mechanisms of the dihydropyridines differ markedly from that of other calcium channel blockers, which could partly explain the discrepancies in affecting the circadian pattern.

Most studies have focussed on the effects of calcium channel blockers on the circadian pattern in myocardial ischaemia. MULCAHY et al. (1988) reported that, in contrast to atenolol (Table 4), nifedipine did not alter the circadian profile in the episodes and the duration of ischaemia (Table 5). EGSTRUP (1991) compared the effects of metoprolol as monotherapy or in combination with nifedipine, on transient ischaemia: whereas metoprolol had the most prominent effect in the morning when sympathetic tone is highest, combined therapy led to an attenuation of both the morning and evening peaks, indicating that a decrease in the myocardial oxygen supply may be more detrimental in the evening. These results demonstrate that different ischaemic mechanisms could be involved at different circadian times. This hypothesis is supported by the above-mentioned study of ANDREWS et al. (1993), who observed that nifedipine had only minor effects on ischaemic events related to heart rate increases, whereas ischaemia without prior changes in heart rate was markedly reduced by about 58% (Table 5). However, in three studies investigating either immediate-release (DEEDWANIA et al. 1991; NESTO et al. 1991) or controlled-release nifedipine (PARMLEY et al. 1992), reductions in the morning incidence of ischaemic episodes by nifedipine have been observed, although these episodes are regarded as heart rate related (ANDREWS et al. 1993). Interestingly, PARMLEY et al. (1992) additionally studied drug effects after morning versus evening dosing (Table 5) and found that both dosing times resulted in similar reductions in ischaemic events. However, the circadian patterns, which were not analysed in detail by PARMLEY et al. (1992), seem to be slightly different, exhibiting a residual morning peak after evening dosing compared to a steady increase from midnight until the afternoon after nifedipine in the morning.

In a study by JUNEAU et al. (1992), diltiazem appeared to reduce the number of ischaemic episodes more effectively during daytime than at night, although when dividing the 24-h period into 6-h intervals no significant differences were found. Statistical comparisons were done with regard to a relative reduction in ischaemic episodes. If the authors had expressed efficacy in absolute terms, a circadian time-dependent decrease in the anti-ischaemic effects of diltiazem would have become apparent. Accordingly, ANDREWS et al. (1993) found that the anti-ischaemic effects of diltiazem were restricted to episodes with prior or concomitant increases in heart rate, which by itself displayed a pronounced circadian variation.

As already mentioned, in primary hypertensives verapamil has been found to be more effective during daytime (GOULD et al. 1982a,b), whereas the dihydropyridines (either after multiple dosing or after treatment with long-acting preparations) seem to reduce blood pressure during both day and night to a similar extent (RAFTERY 1991; CARR et al. 1992). Whether or not these differences between the classes of calcium channel blockers are of relevance for the treatment of patients with coronary heart disease remains to be elucidated. It is conceivable that nocturnal ischaemic events in treated patients as observed by WOODS et al. (1992) could be related to a nocturnal reduction in coronary flow by dihydropyridines or other antihypertensive drugs that act during day and night (FLORAS 1988). Treatment of nocturnal ischaemia due to vasospasm could be restricted to those drugs which are effective vasodilators at any circadian time. In this context it is interesting to note that, in contrast to smooth coronary arteries, irregular ones have a diminished dilator response to intracoronary nifedipine in the afternoon without an appreciable change in coronary blood flow (RABY et al. 1993). On the other hand, differences in the half-life, the drug formulation and/or the dosing time may play a role. With the exception of one study (ANDREWS et al. 1993), no crossover investigations in coronary heart patients were done with different types of calcium channel blockers. Interestingly, the dihydropyridine derivative amlodipine, which has a long half-life per se, reduced angina attacks and ischaemic events in stable angina pectoris patients on once-a-day dosing throughout 24 h (TAYLOR 1992; DEANFIELD et al. 1994) similar to the sustained-release formulations of diltiazem (JUNEAU et al. 1992) and nifedipine (ANDREWS et al. 1993) (Table 5).

In variant angina pectoris the calcium channel blocker diltiazem (Table 5), in contrast to β-adrenoceptor blockade (Table 4), greatly reduced ischaemic events during early morning and during daytime (YASUE et al. 1979).

In summary, calcium channel blockers – mainly short-acting and non-retarded preparations in contrast to long-acting ones – seem to be less effective than β-adrenoceptor antagonists in reducing ischaemic events during the night and the early morning hours. At least, the early morning peak in the onset of myocardial infarction seems not to be greatly influenced by these drugs.

However, the role of formulation and/or subclasses of the calcium channel blockers as well as their effectiveness in subtypes of the disease remain to be elucidated. This is clearly demonstrated by the 24-h effectiveness of the long-acting calcium channel blocker amlodipine in coronary heart diseased patients (DEANFIELD et al. 1994).

III. Organic Nitrates

1. Pharmacological Characterization

Organic nitrates (glyceroltrinitrate, isosorbide dinitrate, isosorbide mononitrates) are prodrugs from which the active principle, nitric oxide (NO), is formed in vivo. NO in turn stimulates the cytosolic guanylyl cyclase in smooth muscle and other tissues, leading to formation of the second messenger cGMP from guanosine triphosphate (see also Chap. 6, this volume). Organic nitrates are dilators of arterial and venous smooth muscle. Reduction in preload and in afterload reduce the oxygen demand of the heart. Therefore, they are used in the treatment of coronary heart disease and cardiac failure.

2. Clinical Data

Although organic nitrates are widely used in the treatment of coronary heart disease, little is known about their circadian time-dependent effects. The few data available are compiled in Table 5. In 174 patients receiving organic nitrates prior to myocardial infarction, the temporal distribution did not differ from that in untreated patients (WOODS et al. 1992). In contrast, HAUSMANN et al. (1989) demonstrated that in stable angina pectoris once-daily dosing of isosorbide dinitrate (ISDN) reduced the number of ischaemic episodes exclusively during daytime with no effect during the night, resulting in a blunted circadian pattern in myocardial ischaemia. QUYYUMI et al. (1987) compared the anti-anginal efficacy of atenolol, nifedipine and isosorbide-5-mononitrate (IS-5-MN) and observed a similar reduction in nocturnal ischaemia with all drugs, whereas daytime episodes were less affected by nifedipine and the mononitrate compared to atenolol (Tables 4,5). In patients with stable angina pectoris, WORTMANN and BACHMANN (1991) compared the effects of 20 mg IS-5-MN twice daily and 120 mg ISDN once daily on ergometric ST-segment depression at different circadian times (1000, 1400 and 1800 hours). They observed that exercise-induced ST-segment depression depended on the circadian time, the most pronounced depression occurring in the late afternoon. The anti-ischaemic effects of ISDN did not differ between times of exercise testing, whereas those of IS-5-MN achieved statistical significance only during ergometry in the morning (WORTMANN and BACHMANN 1991) (Table 5). The lack of anti-ischaemic effect of the mononitrate at noon and in the late afternoon could be due to declining plasma concentrations several hours after drug intake at 0800 hours, to circadian variation in the sensitivity of the vessels towards nitrovasodilators, or – most likely – to both of these factors. In addition, basal vessel diameter seems to vary with time of day (YASUE et al. 1979) and peripheral resistance is highest in the early morning and declines thereafter (PANZA et al. 1991). Thus, it is understandable that vasodilation by ISDN leads to a more pronounced orthostasis at 0200 hours than at other times of day (LEMMER et al. 1991b). In two other studies after single morning dosing or twice daily dosing of IS-5-MN, significant reductions in angina attacks were described, both treatments being equally effective at different times of the 24-h day, thus not modifying the circadian pattern (GANZINGER 1992).

In Prinzmetal's variant angina, YASUE et al. (1979) observed that morning application of glyceroltrinitrate resulted in coronary artery dilation by about 70%, whereas in the afternoon only minor changes in artery diameter were achieved (Table 5).

Chronopharmacokinetic aspects could also contribute to circadian time-dependent effects of organic nitrates. For both ISDN and IS-5-MN, pharmacokinetics were studied in relation to circadian time of drug dosing in healthy subjects (LEMMER et al. 1989, 1991c; SCHEIDEL and LEMMER 1991) (Table 3). Whereas no significant variations were found with ISDN (GANZINGER 1992) and the slow-release (SR) formulation of IS-5-MN (LEMMER et al. 1991c), t_{max} of the immediate-release (IR) IS-5-MN was significantly shorter after morning dosing (LEMMER et al. 1991a; SCHEIDEL and LEMMER 1991). However, independently from the galenic formulations of the two IS-5-MN preparations, the concentration-response relationships of the two drugs were circadian phase dependent: after morning dosing, peak effects coincided with peak drug concentrations, whereas after evening dosing the time to peak effects occurred significantly earlier than the time to peak concentrations (LEMMER et al. 1991b; SCHEIDEL and LEMMER 1991) (Table 3).

In summary, though the anti-ischaemic property of oral nitrates is well established, its influence on the circadian rhythm in these events needs to be clarified by circadian time-specified studies. Moreover, the important clinical problem of nitrate tolerance has not been addressed in relation to circadian time of drug dosing.

IV. Drugs Affecting Coagulation

After the observation that low-dose aspirin was able to reduce mortality in unstable angina pectoris by about 50% (LEWIS et al. 1983), it was of interest to evaluate whether the morning peak in the incidence of myocardial infarction could also be influenced by inhibition of platelet aggregability, which itself had been reported to be increased in early morning hours (TOFLER et al. 1987). Consequently, RIDKER et al. (1990) studied the preventive effects of alternate-day aspirin intake versus placebo in more than 22000 physicians and analysed the circadian distribution of infarction in both groups (Table 5). In the placebo group a bimodal distribution with a primary peak in the morning hours was observed, whereas in the aspirin group this morning peak was reduced by about 60%, resulting in a loss of circadian variation in the onset of myocardial infarction.

Further support for a chronopharmacological approach to the treatment of coronary heart disease is presented by two studies investigating the circadian-time-dependent effects of heparin (Ogawa et al. 1989) and urokinase (Fujita et al. 1993). In patients with variant angina but not in those with stable angina pectoris, Ogawa et al. (1989) observed a morning peak in the plasma concentration of fibrinopeptide A, which was abolished by repeated administration of heparin in the evening and at night. However, the number of angina attacks and their temporal distribution was not affected by evening dosing of heparin. Fujita et al. (1993) observed in their study on intracoronary thrombolysis a significantly lower success rate, i.e. higher degrees of residual stenosis, in patients with onset of myocardial infarction between 0600 and 1200 hours than at other circadian times.

In summary, these results indicate that rhythmic changes in the coagulation system could be of additional importance in coronary heart disease. Furthermore, it appears that studies on pharmacological modulation of time-dependent coagulatory activity could improve the clinical outcome of these diseased patients.

V. Conclusion

Coronary heart disease covers a family of diseases, and patients exhibit a broad spectrum of symptoms; thus subtype-specific differences in drug efficacy must be expected.

The studies reviewed here clearly demonstrate that biological rhythms in physiological functions as well as in pathophysiological events in coronary heart diseased patients can have an impact on drug treatment.

However, the data presented also make clear that there is still a need for prospective crossover studies (at least morning versus evening dosing) with monitoring of the drug's pharmacokinetics in order to give conclusive recommendations concerning the timing of drug treatment. At present it is still an open question whether or not drug dosing at a specified circadian time may be advantageous in the treatment of subtypes of myocardial infarction and of different forms of coronary heart disease in patients suffering from symptomatic or asymptomatic angina pectoris symptoms. In continuing this line, chronopharmacology is able to make a contribution to improving the drugs' efficacy, to reducing side effects and to increasing patients' compliance.

E. Concluding Remarks

It is now convincingly documented that various functions of the cardiovascular system are well organized in time. Disease is able to disturb, reverse or even destroy a rhythmic pattern. Animal experiments as well as clinical studies give evidence that the pharmacokinetics and the effects of cardiovascular active drugs can greatly depend on the circadian time, leading to a circadian phase-dependent dose response relationship. This can result in a dissociation between a drug's pharmacokinetic and pharmacodynamic profiles. Drug dosing at a certain time of day can improve drug efficacy, can reduce side effects and can even contribute to a normalization of a disturbed 24-h pattern. Finally, using drugs as experimental tools, chronopharmacological studies can also contribute to a better understanding of the mechanisms of regulation of the cardiovascular system in health and disease.

References

Alboni P, Codeca L, Padovan G, Tomaini D, Destro A, Margutti A, Degli Uberti E, Fersini C (1981) Variazioni circadiane della frequenza sinusale in soggetti con nodo del seno normale e patologico. G Ital Cardiol 11:1211–1218

Alexopoulos D, Rynkiewicz A, Yusuf S, Johnston JA, Sleight P, Yacoub MH (1988) Diurnal variations of QT interval after cardiac transplantation. Am J Cardiol 61:482–485

Andrews TC, Fenton T, Toyosaki N, Glasser SP, Young PM, MacCallum G, Gibson RS, Shook TL, Stone PH, for the angina and silent ischemia study group (ASIS) (1993) Subsets of ambulatory myocardial ischemia based on heart rate activity: circadian distribution and response to anti-ischemic medication. Circulation 88:92–100

Anlauf M, Baumgart P, Franz I, Krönig B, Meyer-Sabellek W, Middeke M, Schrader J (1993) Ambulante Blutdruck-Langzeitmessung. Dtsch Med Wochenschr 118:1305–1306

Araki H, Koiwaya Y, Nakagaki O, Nakamura M (1983) Diurnal distribution of ST-segment elevation and related arrhythmias in patients with variant angina: a study by ambulatory ECG monitoring. Circulation 67:995–1000

Argentino C, Toni D, Rasura M, Violi F, Sacchetti M, Allegretta A, Balsano F, Fieschi C (1990) Circadian variation in the frequency of ischemic stroke. Stroke 21:387–389

Aronow WS, Ahn C (1993) Circadian variation of primary cardiac arrest or sudden cardiac death in patients aged 62 to 100 years (mean 82). Am J Cardiol 71:1455–1456

Aronow WS, Ahn C, Mercando AD, Epstein S (1994) Circadian variation of sudden cardiac death or fatal myocardial infarction is abolished by propranolol in patients with heart disease and complex ventricular arrhythmias. Am J Cardiol 74:819–821

Aschoff J (1992) Day-night variations in the cardiovascular system. Historical and other notes by an outsider. In: Schmidt TFH, Engel BT, Blümchen G (eds) Temporal variations of the cardiovascular system. Springer, Berlin Heidelberg New York, pp 3–14

Athanassiadis D, Draper GJ, Honour AJ, Cranston WI (1969) Variability of automatic blood pressure measurements over 24 hour periods. Clin Sci 36:147–156

Atkins G, Witte K, Nold G, Sasse U, Lemmer B (1994) Effects of age on circadian blood pressure and heart rate rhythms in patients with primary hypertension. Chronobiol Int 11:35–44

Autenrieth JHF (1801) Handbuch der empirischen Physiologie, part 1. Heerbrandt, Tübingen, p 209

Barthez PJ (1806) Nouveaux éléments de la science de l'homme, 2nd edn. Goujon et Brunot, Paris, part II, p 147; notes: p 66, 69

Baumgart P, Walger P, Gerke M, Dorst KG, Vetter H, Rahn KH (1989) Nocturnal hypertension in renal failure, haemodialysis and after renal transplantation. J Hypertens 7 [Suppl 6]:S70–71

Baumgart P, Walger P, Gemen S, von Eiff M, Raidt H, Rahn KH (1991) Blood pressure elevation during the night in chronic renal failure, haemodialysis and after renal transplantation. Nephron 57:293–298

Behar S, Halabi M, Reicher-Reiss H, Zion M, Kaplisnky E, Mandelzweig L, Goldbourt U, SPRINT study group (1993) Circadian variation and possible external triggers of onset of myocardial infarction. Am J Med 94:396–400

Beilin LJ, Deacon J, Michael CA, Vandongen R, Lalor CM, Barden AE, Davidson L, Rouse I (1983) Diurnal rhythms of blood pressure, plasma renin activity, angiotensin II and catecholamines in normotensive and hypertensive pregnancies. Clin Exp Hypertens [B] 2:271–293

Benhorin J, Banai S, Moriel M, Gavish A, Keren A, Stern S, Tzivoni D (1993) Circadian variations in ischemic threshold and their relation to the occurrence of ischemic episodes. Circulation 87:808–814

Bexton RS, Vallin HO, Camm AJ (1986) Diurnal variation of the QT interval–influence of the autonomic nervous system. Br Heart J 55:253–258

Bianchi S, Bigazzi R, Baldari G, Sgherri G, Campese VM (1994) Diurnal variations of blood pressure and microalbuminuria in essential hypertension. Am J Hypertens 7:23–29

Biffi A, Cugini P, Pelliccia A, Spataro A, Caselli G, Piovano G (1987) Studio cronobiologico dell'elettrocardiogramma dinamico in atleti sani con battiti ectopici ventricolari frequenti. G Ital Cardiol 17:563–568

Bowles MJ, Khurmi NS, O'Hara MJ, Raftery EB (1986) Usefulness of bevantolol for chronic, stable angina pectoris. Am J Cardiol 58:28–34

Brazzell RK, Khoo KC, Schneck DW (1985) Effect of time of dosing on the disposition of oral cibenzoline. Biopharmaceut Drug Dispos 6:433–440

Brickman AS, Stern N, Sowers JR (1990) Circadian variations of catecholamines and blood pressure in patients with pseudohypoparathyroidism and hypertension. Chronobiologia 17:37–44

Bristow JD, Honour AJ, Pickering TG, Sleight P (1969) Cardiovascular and respiratory changes during sleep in normal and hypertensive subjects. Cardiovasc Res 3:476–85

Broadhurst P, Brigden G, DasatwGupta P, Lahiri A, Raftery EB (1990) Ambulatory intra-arterial blood pressure in normal subjects. Am Heart J 120:160–166

Bruguerolle B, Lemmer B (1993) Recent advances in chronopharmacokinetics: methodological problems. Life Sci 52:1809–1824

Bruguerolle B, Bouvenot G, Bartolin R, Manolis J (1988) Chronopharmacocinétique de la digoxine chez le sujet de plus de soixante-dix ans. Therapie 43:251–253

Brush CE, Fayerweather R (1901) Observations on the changes in blood pressure during normal sleep. Am J Physiol 5:199–210

Bush LR (1987) Effects of the serotonin antagonists, cyproheptadine, ketanserin and mianserin, on cyclic flow reductions in stenosed canine coronary arteries. J Pharmacol Exp Ther 240:674–682

Canada WB, Woodward W, Lee G, De Maria A, Low R, Mason DT, Laddu A, Shapiro W (1983) Circadian rhythm of hourly ventricular arrhythmia frequency in man. Angiology 34:274–282

Carr AA, Bottini PB, Feig P, et al (1992) Effectiveness of once-daily monotherapy with a new sustained release calcium antagonist. Am J Cardiol 69:28E–32E

Caruana M, Heber M, Bridgen G, Raftery EB (1987) Assessment of "once daily" verapamil for the treatment of hypertension using ambulatory, intra-arterial pressure recording. Eur J Clin Pharmacol 32:549–553

Caruana MP, Lahiri A, Cashman PM, Altman DG, Raftery EB (1988) Effects of chronic congestive heart failure secondary to coronary artery disease on the circadian rhythm of blood pressure and heart rate. Am J Cardiol 62:755–759

Casiglia E, Palatini P, Baccillieri MS, Colangeli G, Petucco S, Pessina AC (1992) Circadian rhythm of peripheral resistance: a non-invasive 24-hour study in young normal volunteers confined to bed. High Blood Press 1:249–255

Chau NP, Bauduceau B, Chanudet X, Larroque P, Gautier D. Ambulatory blood pressure in diabetic subjects (1994) Am J Hypertens 7:487–491

Chen LD, Tan DX, Reiter RJ, Yaga K, Poeggeler B, Kumar P, Manchester LC, Chambers JP (1993) In vivo and in vitro effects of the pineal gland and melatonin on $[Ca^{2+} + Mg^{2+}]$-dependent ATPase in cardiac sarcolemma. J Pineal Res 14:178–183

Christ JE, Hoff HE (1975) An analysis of the circadian rhythmicity of atrial and ventricular rates in complete heart block. J Electrocardiol 8:69–72

Cinca J, Moya A, Figueras J, Roma F, Rius J (1986) Circadian variations in the electrical properties of the human heart assessed by sequential bedside electrophysiologic testing. Am Heart J 112:315–321

Cinca J, Moya A, Bardaji A, Figueras J, Rius J (1987) Daily variability of electrically induced reciprocating tachycardia in patients with atrioventricular accessory pathways. Am Heart J 114:327–333

Cinca J, Moya A, Bardaji A, Rius J, Soler-Soler J (1990) Circadian variations of electrical properties of the heart. Ann N Y Acad Sci 601:222–233

Cohn PF, Lawson WE, Gennaro V, Brady D (1989) Effects of long-acting propranolol on a.m. and p.m. peaks in silent myocardial ischemia. Am J Cardiol 63:872–873

Coote JH (1982) Respiratory and circulatory control during sleep. J Exp Biol 100:223–244

Coy KM, Imperi G, Lambert CR, Pepine CJ (1990) Application of time series analysis to circadian rhythms: effect of beta-adrenergic blockade upon heart rate and transient myocardial ischemia. Am J Cardiol 66:22G–24G

Dabrowska B, Feltynowski T, Wocial B, Szpak W, Januszewicz W (1990) Effect of removal of phaeochromocytoma on diurnal variability of blood pressure, heart rhythm and excretion of catecholamines. J Hum Hypertens 4:397–399

DasGupta P, Bridgen G, Ramhamdeny E, Lahiri A, Baird IM, Raftery EB (1991) Circadian variation and blood pressure: response to rapid weight loss by hypocaloric hyponatraemic diet in obesity. J Hypertens 9:441–447

Davies AB, Gould BA, Cashman PM, Raftery EB (1984) Circadian rhythm of blood pressure in patients dependent on ventricular demand pacemakers. Br Heart J 52:93–98

De Bordeu T (1756) Recherches sur le pouls. De Bure l'aîné, Paris, p 471

de Leeuw PW, Falke HE, Kho TL, Vandongen R, Wester A, Birkenhager WH (1977) Effects of beta-adrenergic blockade on diurnal variability of blood pressure and plasma noradrenaline levels. Acta Med Scand 202:389–392

de Leeuw PW, van Leeuwen SJ, Birkenhager WH (1985) Effect of sleep on blood pressure and its correlates. Clin Exp Hypertens [A] 7:179–186

de Leonardis V, de Scalzi M, Fabiano FS, Cinelli P (1985) A chronobiologic study on some cardiovascular parameters. J Electrocardiol 18:385–394

Deanfield JE, Detry L-M, Lichtlen PR, Magnani B, Sellier P, Thaulow E, for the CAPE Study Group (1994) Amlodipine reduces transient myocardial ischemia in patients with coronary artery disease: double-blind circadian anti-ischemia program in Europe (CAPE trial). J Am Coll Cardiol 24:1460–1467

Deedwania PC, Carbajal EV, Nelson JR, Hait H (1991) Anti-ischemic effects of atenolol versus nifedipine in patients with coronary artery disease and ambulatory silent ischemia. J Am Coll Cardiol 17:963–969

Degaute JP, van de Borne P, Linkowski P, Van Cauter E (1991) Quantitative analysis of the 24-hour blood pressure and heart rate patterns in young men [see comments]. Hypertension 18:199–210

Del Rosso G, Amoroso L, Santoferrara A, Fiederling B, Di Liberato L, Albertazzi A (1994) Impaired blood pressure nocturnal decline and target organ damage in chronic renal failure. J Hypertens 12 [Suppl 3]:S15

Dewar HA, Weightman D (1983) A study of embolism in mitral valve disease and atrial fibrillation. Br Heart J 49:133–140

Egstrup K (1991) Attenuation of circadian variation by combined antianginal therapy with suppression of morning and evening increases in transient myocardial ischemia. Am Heart J 122:648–655

Falconer W (1797) Beobachtungen über den Puls. Heinsius, Leipzig, pp 24ff

Floras JS (1988) Antihypertensive treatment, myocardial infarction, and nocturnal myocardial ischemia. Lancet II:994–996

Floras JS, Jones JV, Johnston JA, Brooks DE, Hassan MO, Sleight P (1978) Arousal and the circadian rhythm of blood pressure. Clin Sci Mol Med [Suppl] 4:395s–397s

Floyer J (1707–1710) The physician's pulse watch; or an essay to explain the old art of feeling the pulse and to improve it by the help of a pulse-watch. Smith and Walford, London

Fogari R, Malocco E, Tettamanti F, Tettamanti F, Gnemmi AE, Milani M (1993) Evening vs morning isradipine sustained release in essential hypertension: a double-blind study with 24 h ambulatory monitoring. Br J Clin Pharmacol 35:51–54

Franklin SS, Sowers JR, Batzdorf U (1986) Relationship between arterial blood pressure and plasma norepinephrine levels in a patient with neurogenic hypertension. Am J Med 81:1105–1107

Franz IW, Erb D, Tonnesmann U (1992) Gestörte 24-Stunden-Blutdruckrhythmik bei normotensiven und hypertensiven Asthmatikern. Z Kardiol 81(Suppl 2):13–16

Fujimura A, Kajiyama H, Kumagai Y, Nakashima H, Sugimoto K, Ebihara A (1989) Chronopharmacokinetic studies of pranoprofen and procainamide. J Clin Pharmacol 29:786–790

Fujita M, Araie E, Yamanishi K, Miwa K, Kida M, Nakajima H (1993) Circadian variation in the success rate of intracoronary thrombolysis for acute myocardial infarction. Am J Cardiol 71:1369–1371

Ganzinger U (1992) Influence of isosorbide-5-mononitrate on the circadian rhythm of angina pectoris. Arzneimittelforschung/Drug Res 42:307–310

Garcia-Pagan JC, Feu F, Castells A, Luca A, Hermida RC, Rivera F, Bosch J, Rodes J (1994) Circadian variations of portal pressure and variceal haemorrhage in patients with cirrhosis. Hepatology 19:595–601

Germano G, Damiani S, Civarella M, Appolloni A, Ferucci A (1984) Detection of a diurnal rhythm in arterial blood pressure in the evaluation of 24-hour antihypertensive therapy. Clin Cardiol 7:525–535

Gibbs JS, Cunningham D, Shapiro LM, Park A, Poole-Wilson PA, Fox KM (1989) Diurnal variation of pulmonary artery pressure in chronic heart failure. Br Heart J 62:30–35

Gillis AM, MacLean KE, Guilleminault C (1988) The QT interval during wake and sleep in patients with ventricular arrhythmias. Sleep 11:333–339

Gillis AM, Guilleminault C, Partinen M, Connolly SJ, Winkle RA (1989) The diurnal variability of ventricular premature depolarizations: influence of heart rate, sleep, and wakefulness. Sleep 12:391–399

Gillis AM, Peters RW, Mitchell LB, Duff HJ, McDonald M, Wyse DG (1992) Effects of left ventricular dysfunction on the circadian variation of ventricular premature complexes in healed myocardial infarction. Am J Cardiol 69:1009–1014

Gnecchi-Ruscone T, Piccaluga E, Guzzetti S, Contini M, Montano N, Nicolis E on behalf of GISSI-2 investigators (1994) Morning and Monday: critical periods for the onset of acute myocardial infarction. The GISSI 2 study experience. Eur Heart J 15:882–887

Goo RH, Moore JG, Greenberg E, Alazraki NP (1987) Circadian variation in gastric emptying of meals in humans. Gastroenterology 93:515–518

Gould BA, Raftery EB (1991) Twenty-four-hour blood pressure control: an intraarterial review. Chronobiol Int 8:495–505

Gould BA, Mann S, Davies A, Altman DG, Raftery EB (1981) Indoramin: 24-hour profile of intra-arterial ambulatory blood pressure, a double-blind placebo controlled crossover study. Br J Clin Pharmacol 12 [Suppl]:67s–73s

Gould BA, Mann S, Kieso H, Balasubramanian V, Raftery EB (1982a) The 24-hour ambulatory blood pressure profile with verapamil. Circulation 65:22–27

Gould BA, Hornung RS, Mann S, Balasubramanian V, Raftery EB (1982b) Slow channel inhibitors verapamil and nifedipine in the management of hypertension. J Cardiovasc Pharmacol 4:5369–5373

Greminger P, Suter PM, Holm D, Kobelt R, Vetter W (1994) Morning versus evening administration of nifedipine gastrointestinal therapeutic system in the management of essential hypertension. Clin Invest 72:864–869

Halberg F (1969) Chronobiology. Annu Rev Physiol 31: 675–725

Halberg F, Halberg E, Halberg J, Halberg F (1984) Chronobiologic assessment of human blood pressure variation in health and disease. In: Weber MA, Drayer JIM (eds) Ambulatory blood pressure monitoring. Steinkopf, Darmstadt, p 137–156

Hausmann D, Nikutta P, Hartwig C-A, Daniel WG, Wenzlaff P, Lichtlen PR (1987) ST-Strecken-Analyse im 24-h-Langzeit-EKG bei Patienten mit stabiler Angina pectoris und angiographisch nachgewiesener Koronarsklerose. Z Kardiol 76:554–562

Hausmann D, Nikutta P, Daniel WG, Hatwig CA, Wenzlaff P, Lichtlen PR (1989) Hochdosierte Einmalgabe Isosorbiddinitrat: Einfluß auf die tageszeitliche Verteilung von transistorischen Myokardischämien bei Patienten mit stabiler Angina pectoris. Z Kardiol 78:415–420

Hayashi T, Shoji T, Kitamura E, Okada N, Nakanishi I, Tsubakihara Y (1993) Circadian blood pressure pattern in the patients with chronic glomerulonephritis (in Japanese). Nippon Jinzo Gakkai Shi 35:233–237

Heber ME, Lahiri A, Thompson D, Raftery EB (1989) Baroreceptor, not left ventricular, dysfunction is the cause of hemodialysis hypotension. Clin Nephrol 32:79–86

Hellenbrecht D, Lemmer B, Wiethold G, Grobecker H (1973) Measurement of hydrophobicity, surface activity, local anaesthesia and myocardial conduction velocity as quantitative parameters of the non-specific membrane affinity of nine beta-adrenergic blocking agents. Naunyn Schmiedeberg's Arch Pharmacol 277:211–226

Hellwig C (pseudonym: Kräutermann V) (1738) Curieuser und vernünftiger Urin-Artzt, welcher eines Theils lehret und zeiget, wie man aus dem Urin nicht allein die meisten und vornehmsten Kranckheiten erkennen, Anderen Theils, wie man auch aus dem Pulse den Zustand des Geblüts, die Stärcke und Schwäche der Lebens-Geister, Ab- und Zunahme der Kranckheit ersehen, 3rd edn. JJ Beumelburg, Arnstadt, Leipzig, pp 134, 140

Hensen H (1900) Beiträge zur Physiologie und Pathologie des Blutdruckes. Dtsch Arch Klin Med 67:436–530

Hickey MS, Costill DL, Vukovich MD, Kryzmenski K, Widrick JJ (1993) Time of day effects on sympathoadrenal and pressor reactivity to exercise in healthy men. Eur J Appl Physiol 67:159–63

Hill L (1898) On rest, sleep and work and the concomitant changes in the circulation of the blood. Lancet 1:282–285

Hla KK, Latham AN, Henry JA (1992) Influence of time of administration on verapamil pharmacokinetics. Clin Pharmacol Ther 51:366–370

Hornung RS, Mahler RF, Raftery EB (1989) Ambulatory blood pressure and heart rate in diabetic patients: an assessment of autonomic function. Diabet Med 6:579–585

Howell WHA (1897) Contribution to the physiology of sleep, based on plethysmographic experiments. J Exp Med 2:313

Hufeland CW (1797) Die Kunst das menschliche Leben zu verlängern. Akademische Buchhandlung, Jena, p 552

Idema RN, Gelsema ES, Wenting GJ, Grashuis JL, van den Meiracker AH, Brouwer RML, Tveld A (1992) A new model for diurnal blood pressure profiling, square wave fit compared with conventional methods. Hypertension 19:595–605

Ikeda T, Matsubara T, Sato Y, Sakamoto N (1993) Circadian blood pressure variation in diabetic patients with autonomic neuropathy. J Hypertens 11:581–587

Imai Y, Abe K, Miura Y, Nihei M, Sasaki S, Minami N, Munakata M, Taira N, Sekino H, Yamakoshi K, Yoshinaga K (1988a) Hypertensive episodes and circadian fluctuations of blood pressure in patients with phaeochromocytoma: studies by long-term blood pressure monitoring based on a volume-oscillometric method. J Hypertens 6:9–15

Imai Y, Abe K, Sasaki S, Minami N, Nihei M, Munakata M, Murakami O, Matsue K, Sekino H, Miura Y, Yoshinaga K (1988b) Altered circadian blood pressure rhythm in patients with Cushing's syndrome. Hypertension 12:11–19

Imai Y, Abe K, Sasaki S, Minami N, Munakata M, Nihei M, Sekino H, Yoshinaga K (1989) Exogenous glucocorticoid eliminates or reverses circadian blood pressure variations. J Hypertens 7:113–120

Imai Y, Abe K, Sasaki S, Munakata M, Minami N, Sakuma H, Hashimoto J, Yabe T, Watanabe N, Sakuma M, Yoshinaga K (1992) Circadian blood pressure variation in patients with renovascular hypertension or primary aldosteronism. Clin Exp Hypertens [A] 14:1141–1167

Imai Y, Munakata M, Hashimoto J, Minami N, Sakuma H, Watanabe N, Yabe T, Nishiyama A, Sakuma M, Yamagishi T, Abe K (1993) Age-specific characteristics of nocturnal blood pressure in a general population in a community of northern Japan. Am J Hypertens 6:179S–183S

Imperi GA, Lambert CR, Coy K, Lopez L, Pepine CJ, Shephard C (1987) Effects of titrated beta blockade (metoprolol) on silent myocardial ischemia in ambulatory patients with coronary artery disease. Am J Cardiol 60:519–524

Isshiki T, Akatsuka N, Tsuneyoshi H, Oka H (1986) Periodic fluctuation of blood pressure in a case of norepinephrine secreting extra-adrenal pheochromocytoma. Jpn Heart J 27:437–442

James GD, Pickering TG (1993) The influence of behavioral factors on the daily variation of blood pressure. Am J Hypertens 6:170S–173S

James GD, Moucha OP, Pickering TG (1991) The normal hourly variation of blood pressure in women: average patterns and effect of work stress. J Human Hypertens 5:505–509

Janssen BJ, Tyssen CM, Duindam H, Rietveld WJ (1994) Suprachiasmatic lesions eliminate 24-h blood pressure variability in rats. Physiol Behav 55:307–311

Jellinek S (1900) Über den Blutdruck des gesunden Menschen. Z Klin Med 39:447–472

Jespersen CM, Frederiksen M, Hansen JF, Klitgaard NA, Sorum C (1989) Circadian variation in the pharmacokinetics of verapamil. Eur J Clin Pharmacol 37:613–615

Joy M, Pollard CM, Nunan TO (1982) Diurnal variation in exercise responses in angina pectoris. Br Heart J 48:156–160

Juneau M, Théroux P, Waters D, for the Canadian Multicenter Diltiazem Study Group (1992) Effect of diltiazem slow-release formulation on silent myocardial ischemia in stable coronary artery disease. Am J Cardiol 69:30B–35B

Katsch G, Pansdorf H (1922) Die Schlafbewegung des Blutdrucks. Münch Med Wochenschr 69:1715–1718

Katz ML, Comerota AJ, Kerr RP, Caputo GC (1994) Variability of venous-hemodynamics with daily activity. J Vasc Surg 9:361–365

Kennedy B, Shannahoff-Khalsa D, Ziegler MG (1994) Plasma norepinephrine variations correlate with peripheral vascular resistance in resting humans. Am J Physiol 266:H435–439

Khatri IM, Freis ED (1967) Hemodynamic changes during sleep. J Appl Physiol 22:867–873

Kleiman NS, Schechtman KB, Young PM, Goodman DA, Boden WE, Pratt CM, Roberts R, Diltiazem Reinfarction Study investigators (1990) Lack of diurnal variation in the onset of non-Q wave infarction. Circulation 81:548–555

Knox R (1815) On the relation subsisting between the time of the day, and various functions of the human body; and on the manner in which the pulsation of the heart and arteries are affected by muscular exertion. Edinb Med Surg J 11:52–65

Kool MJ, Wijnen JA, Hoeks AP, Struyker-Boudier HA, Van Bortel LM (1991) Diurnal pattern of vessel-wall properties of large arteries in healthy men. J Hypertens [Suppl] 9:S108–109

Koopmans R, Oosterhuis B, Karemaker JM, Weiner J, van Boxtel CJ (1993) The effect of oxprenolol dosage time on its pharmacokinetics and haemodynamic effects during exercise in man. Eur J Clin Pharmacol 44:171–176

Kostis JB (1988) Comparison of the duration of action of atenolol and nadolol for treatment of angina pectoris. Am J Cardiol 62:1171–1175

Kräuchi K, Wirz-Justice A (1994) Circadian rhythm of heat production, heart rate, and skin and core temperature under unmasking conditions in men. Am J Physiol 267:R819–R829

Krakoff LR (1994) Doxazosin studies provide clearer picture of blood pressure profiles. Editorial. Am J Hypertens 7:853–854

Kuwajima I, Suzuki Y, Shimosawa T, Kanemaru A, Hoshino S, Kuramoto K (1992) Diminished nocturnal decline in blood pressure in elderly hypertensive patients with left ventricular hypertrophy. Am Heart J 123:1307–1311

Lampert R, Rosenfeld L, Batsford W, Lee F MsPherson C (1994) Circadian variation of sustained ventricular tachycardia in patients with coronary artery disease and implantable cardioverter-defibrillators. Circulation 90:241–247

Langner B, Lemmer B (1988) Circadian changes in the pharmacokinetics and cardiovascular effects of oral propranolol in healthy subjects. Eur J Clin Pharmacol 33:619–624

Lanza GA, Lucente M, Rebuzzi AG, Spagnolo A, Dulcimascolo C, Manzoli U (1986) Ventricular parasystole: a chronobiologic study. PACE Pacing Clin Electrophysiol 9:860–867

Lemmer B (1986) The chronopharmacology of cardiovascular medications. Annu Rev Chronopharmacol 2:199–228

Lemmer B (1987) Chronopharmacology of cardiovascular medications. In: Kümmerle H-P, Hitzenberger G, Spitzy KH (eds) Klinische Pharmakologie, 4th edn. Ecomed, Landsberg, II-2.15.3.5, pp 1–14

Lemmer B (1989) Temporal aspects in the effects of cardiovascular active drugs in man. In: Lemmer B (ed) Chronopharmacology – cellular and biochemical interactions. Dekker, New York, pp 525–542

Lemmer B (1991) The cardiovascular system and daily variation in response to antihypertensive and antianginal drugs: recent advances. Pharmacol Ther 51:269–274

Lemmer B (1995) White coat hypertension: described more than 250 years ago. Am J Hypertens 8:437–438

Lemmer B (1996) Differential effects of antihypertensive drugs on circadian rhythm in blood pressure from the chronobiological point of view. Blood Pressure Monitoring 1: 161–169

Lemmer B, Bathe K (1982) Stereospecific and circadian-phase-dependent kinetic behaviour of d,l-, l-, and d-propranolol in plasma, heart, and brain of light-dark-synchronized rats. J Cardiovasc Pharmacol 4:635–644

Lemmer B, Bruguerolle B (1994) Chronopharmacokinetics – are they clinically relevant? Clin Pharmacokinet 26:419–427

Lemmer B, Nold G (1991) Circadian changes in estimated hepatic blood flow in healthy subjects. Br J Clin Pharmacol 32:627–629

Lemmer B, Wiethold G, Hellenbrecht D, Bak IJ, Grobecker H (1972) Human blood platelets as cellular models for investigation of membrane active drugs: beta-adrenergic blocking agents. Naunyn-Schiedeberg's Arch Pharmacol 275:299–313

Lemmer B, Winkler H, Ohm T, Fink M (1985) Chronopharmacokinetics of beta-receptor blocking drugs of different lipophilicity (propranolol, metoprolol, sotalol, atenolol) in plasma and tissues after single and multiple dosing in the rat. Naunyn-Schmiedeberg's Arch Pharmacol 330:42–49

Lemmer B, Scheidel B, Stenzhorn G, Blume H, Lenhard G, Grether D, Renczes J, Becker HJ (1989) Clinical chronopharmacology of oral nitrates. Z Kardiol 87 [Suppl 2]:61–63

Lemmer B, Nold G, Behne S, Kaiser R (1991a) Chronopharmacokinetics and cardiovascular effects of nifedipine. Chronobiol Int 8:485–494

Lemmer B, Scheidel B, Behne S (1991b) Chronopharmacokinetics and chronopharmacodynamics of cardiovascular active drugs: propranolol, organic nitrates, nifedipine. Ann NY Acad Sci 618:166–181

Lemmer B, Scheidel B, Blume H, Becker HJ (1991c) Clinical chronopharmacology of oral sustained-release isosorbide-5-mononitrate in healthy subjects. Eur J Clin Pharmacol 40:71–75

Lemmer B, Mattes A, Böhm M, Ganten D (1993a) Circadian blood pressure variation in transgenic hypertensive rats. Hypertension 22:97–101

Lemmer B, Zuther P, Witte K, Mattes A (1993b) Programm zur automatischen Auswertung von 24-Stunden-Blutdruckprofilen ambulatorischer Blutdruckmeßgeräte: ABPM-FIT. Klin Pharmakol Aktuel 2:38/V5

Lemmer B, Witte K, Makabe T, Ganten D, Mattes A (1994a) Effects of enalaprilat on circadian profiles in blood pressure and heart rate of spontaneously and transgenic hypertensive rats. J Cardiovasc Pharmacol 23:311–314

Lemmer B, Sasse U, Witte K, Hopf R (1994b) Pharmacokinetics and cardiovascular effects of a new sustained-release formulation of diltiazem. Naunyn-Schmiedeberg's Arch Pharmacol 349:R141

Lemmer B, Witte K, Schänzer A, Schnecko A (1995a) Circadian regulation of blood pressure: mechanisms and therapeutic implications. Am J Hypertens 8:23A

Lemmer B, Witte K, Minors D, Waterhouse J (1995b) Circadian rhythms of heart rate and blood pressure in four strains of rat: differences due to, and separate from, locomotor activity. Biol Rhythm Res 26: 493–504

Levy RD, Cunningham D, Shapiro LM, Wright C, Mockus L, Fox KM (1987) Diurnal variation in left ventricular function: a study of patients with myocardial ischaemia, syndrome X, and of normal controls. Br Heart J 57:148–153

Lewis HD, Davis JW, Archibald DG, Steinke WE, Smitherman TC, Doherty JE, Schnaper HW, LeWinter MM, Linares E, Pouget JM, Sabharwal SC, Chesler E, DeMots H (1983) Protective effects of aspirin against acute myocardial infarction and death in men with unstable angina. Results of a Veterans Administration Cooperative Study. N Engl J Med 309:396–403

Lichtstein E, Morganroth J, Harrist R, Hubble E, BHAT Study Group (1983) Effect of propranolol on ventricular arrhythmia. The beta-blocker heart attack trial experience. Circulation 67:5–10

Littler WA (1979) Sleep and blood pressure: further observations. Am Heart J 97:35–37

Lown B, Calvert AF, Armington R, Ryan M (1975) Monitoring for serious arrhythmias and high risk of sudden death. Circulation 52:189–198

Lucente M, Rebuzzi AG, Lanza GA, Tamburi S, Cortellessa MC, Coppola E, Iannarelli M, Manzoli U (1988) Circadian variation of ventricular tachycardia in acute myocardial infarction. Am J Cardiol 62:670–674

Lugaresi A, Baruzzi A, Cacciari E, Cortelli P, Medori R, Montagna P, Tinuper P, Zucconi M, Roiter I, Lugaresi E (1987) Lack of vegetative and endocrine circadian rhythms in fatal familial thalamic degeneration. Clin Endocrinol 26:573–580

Lugaresi E, Coccagna G, Mantovani M, Lebrun R (1972) Some periodic phenomena arising during drowsiness and sleep in man. Electroencephalogr Clin Neurophysiol 32:701–705

Lurbe A, Redon J, Pascual JM, Tacons J, Alvarez V, Batlle DC (1993) Altered blood pressure during sleep in normotensive subjects with type I diabetes. Hypertension 21:227–235

Mancia G, Ferrari A, Gregorini L, Parati G, Pomidossi G, Bertinieri G, Grassi G, di Rienzo M, Pedotti A, Zanchetti A (1983) Blood pressure and heart rate variabilities in normotensive and hypertensive human beings. Circ Res 53:96–104

Mann S, Altman DG, Raftery EB, Bannister R (1983) Circadian variation of blood pressure in autonomic failure. Circulation 68:477–483

Marshall J (1977) Diurnal variation in occurrence of strokes. Stroke 8:230–231
Martinelli P, Coccagna G, Rizzuto N, Lugaresi E (1981) Changes in systemic arterial pressure during sleep in Shy-Drager syndrome. Sleep 4:139–146
Master AM (1960) The role of effort and occupation (including physicians) in coronary occlusion. JAMA 174: 942–948
Matsumura K, Abe I, Fukuhara M, Kobayashi K, Sadoshima S, Hasuo K, Fujishima M (1993) Attenuation of nocturnal BP fall in essential hypertensives with cerebral infarction (letter). J Hum Hypertens 7:309–310
Mattes A, Lemmer B (1991) Effects of amlodipine on circadian rhythms in blood pressure, heart rate and motility – a telemetric study in rats. Chronobiol Int 8:526–538
Mattes A, Witte K, Hohmann W, Lemmer B (1991) PHARMFIT – a non-linear fitting program for pharmacology. Chronobiol Int 8:460–476
McClelland J, Halperin B, Cutler J, Kudenchuk P, Kron J, McAnulty J (1990) Circadian variation in ventricular electrical instability associated with coronary artery disease. Am J Cardiol 65:1351–1357
Medori R, Tritschler HJ, LeBlanc A, Villare F, Manetto V, Chen HY, Xue R, Leal S, Montagna P, Cortelli P, Tinuper P, Avoni P, Mochi M, Baruzzi A, Hauw JJ, Ott J, Lugaresi E, Autilio-Gambetti L, Gambetti P (1992) Fatal familial insomnia, a prion disease with a mutation at codon 178 of the prion protein gene. N Engl J Med 326:444–449
Meilhac B, Mallion JM, Carre A, Chanudet X, Poggi L, Gosse P, Dallochio M (1992) Étude de l'influence de l'horaire de la prise sur l'effet antihypertenseur et la tolérance de la nitrendipine chez des patients hypertendus essentiels légers à modérés. Therapie 47:205–210
Mengden T, Binswanger B, Gruene S (1992) Dynamics of drug compliance and 24-hour blood pressure control of once daily morning vs evening amlodipine. J Hypertens 10 [Suppl 4]:S136
Middeke M (1992) Synchronizität von zirkadianer Blutdruckrhythmik und sympathoadrenerger Aktivität. Z Kardiol 81 [Suppl 2]:55–58
Middeke M, Schrader J (1994) Nocturnal blood pressure in normotensive subjects and those with white coat, primary, and secondary hypertension. Br Med J 308:630–632
Middeke M, Mika E, Schreiber MA, Beck B, Wachter B, Holzgreve H (1989) Ambulante indirekte Blutdrucklangzeitmessung bei primärer und sekundärer Hypertonie. Klin Wochenschr 67:713–716
Middeke M, Klüglich M, Holzgreve H (1991a) Circadian blood pressure rhythm in primary and secondary hypertension. Chronobiol Int 8:451–459
Middeke M, Klüglich M, Holzgreve H (1991b) Chronopharmacology of captopril plus hydrochlorothiazide in hypertension: morning versus evening dosing. Chronobiol Int 8:506–510
Millar-Craig MW, Bishop CN, Raftery EB (1978) Circadian variation of blood-pressure. Lancet 1:795–797
Miller JC, Horvath SM (1976) Cardiac output during human sleep. Aviat Space Environ Med 47:1046–1051
Mitler MM, Hajdukovic RM, Shafor R, Hahn PM, Kripke DF (1987) When people die. Cause of death versus time of death. Am J Med 82:266–274
Mitsuoka T, Ueyama C, Matsumoto Y, Hashiba K (1990) Influences of autonomic changes on the sinus node recovery time in patients with sick sinus syndrome. Jpn Heart J 31:645–660
Mori H (1990) Circadian variation of haemodynamics in patients with essential hypertension. J Hum Hypertens 4:384–389
Morris JJ Jr (1990) Mechanisms of ischemia in coronary artery disease: spontaneous decrease in coronary blood supply. Am Heart J 120:746–756; discussion 769–772
Mulcahy D, Keegan J, Cunningham D, Quyyumi A, Crean P, Park A, Wright C, Fox K (1988) Circadian variation of total ischaemic burden and its alteration with antianginal agents. Lancet II:755–759

Müller C (1921a) Die Messung des Blutdrucks am Schlafenden als klinische Methode – speciell bei der gutartigen (primären) Hypertonie und der Glomerulonephritis I. Acta Med Scand 55:381–442

Müller C (1921b) Die Messung des Blutdrucks am Schlafenden als klinische Methode – speciell bei der gutartigen (primären) Hypertonie und der Glomerulonephritis II. Acta Med Scand 55:443–485

Muller JE, Stone PH, Turin ZG, Rutherford JG, Czeisler CA, Parkers C, Poole WK, Passamani E, Roberts R, Robertson T, Sobel BE, Willerson JT, Braunwald E (1985) The Milis study group: circadian variation in the frequency of onset of acute myocardial infarction. N Engl J Med 313:1315–1322

Muller JE, Ludmer PL, Willich SN, Tofler GH, Aylmer G, Klangos I, Stone PH (1987) Circadian variation in the frequency of sudden cardic death. Circulation 75:131–138

Nesto RW, Phillips RT, Kett KG, McAuliffe LS, Roberts M, Hegarty P (1991) Effect of nifedipine on total ischemic activity and circadian distribution of myocardial ischemic episodes in angina pectoris. Am J Cardiol 67:128–132

Nold G, Strobel G, Thomas-Morr M, Lemmer B (1996) Once-daily amlodipine: antihypertensive efficacy and circadian blood pressure profiles after morning versus evening administration. Naunyn-Schmiedeberg's Arch Pharmacol 353: R160

Norwegian Multicenter Study Group (1989) Timolol-induced reduction in mortality and reinfarction in patients surviving acute myocardial infarction. N Engl J Med 304:801–807

Novo S, Barbagallo M, Abrignani MG, Alaimo G, Nardi E, Corrao S, Papadia C, Strano A (1990) Cardiac arrhythmias as correlated with the circadian rhythm of arterial pressure in hypertensive subjects with and without left ventricular hypertrophy. Eur J Clin Pharmacol 39 [Suppl 1]:S49-51

Ocon J, Mora J (1990) Twenty-four-hour blood pressure monitoring and effects of indapamide. Am J Cardiol 65:58H–61H

Ogawa H, Yasue H, Oshima S, Okumura K, Matsuyama K, Obata K (1989) Circadian variation of plasma fibrinopeptide A level in patients with variant angina. Circulation 80:1617–1626

Oishi S, Sasaki M, Ohno M, Umeda T, Sato T (1988) Periodic fluctuation of blood pressure and its management in a patient with pheochromocytoma. Case report and review of the literature. Jpn Heart J 29:389–399

Osterziel KJ, Dietz R, Lemmer B, Kübler W (1991) Circadian rhythms in blood pressure in congestive heart failure and effects of ACE-inhibitors. Chronobiol Int 8:420–431

Osterziel KJ, Karr M, Lemmer B, Dietz R (1992) Effect of captopril and lisinopril on circadian blood pressure rhythm and renal function in mild-to-moderate heart failure. Am J Cardiol 70:147c–150c

Otsuka K, Sato T, Saito H, Kaba H, Otsuka K, Seto K, Ogura H, Ozawa T (1985) Circadian rhythm of cardiac bradyarrhythmia episodes in rats. Chronobiologia 12:11–28

Palatini P, Racioppa A, Raule G, Zaninotti M, Penzo M, Pessina AC (1992) Effect of timing of administration on the plasma ACE inhibitory activity and the antihypertensive effect of quinapril. Clin Pharmacol Ther 52:378–383

Palatini P, Mos L, Motolese M, Mormino P, DelTorre M, Varotto L, Pavan E, Pessina AE (1993) Effect of evening versus morning benazepril on 24-hour blood pressure: a comparative study with continuous intraarterial monitoring. Int J Clin Pharmacol Ther Toxicol 31:295–300

Panza JA, Epstein SE, Quyyumi AA (1991) Circadian variation in vascular tone and its relation to alpha-sympathetic vasoconstrictor activity. N Engl J Med 325:986–990

Parmley WW, Nesto RW, Singh BN, Deanfield J, Gottlieb SO, and the N-CAP Study Group (1992) Attenuation of the circadian patterns of myocardial ischemia with nifedipine GITS in patients with chronic stable angina. J Am Coll Cardiol 19:1380–1389

Patrick J, Campbell K, Carmichael L, Probert C (1982) Influence of maternal heart rate and gross fetal body movements on the daily pattern of fetal heart rate near term. Am J Obstet Gynecol 144:533–538

Pepine CJ (1991) Therapeutic implications of circadian variations in myocardial ischemia and related physiologic functions. Am J Hypertens 4:442S–448S

Peters RW, Muller JE, Goldstein S, Byington R, Friedman LM, BHAT Study Group (1989) Propranolol and the morning increase in the frequency of sudden cardiac death (BHAT study). Am J Cardiol 63:1518–1520

Pickering TG, Levenstein M, Walmsley P for the Hypertension and Lipid Trial Study Group (1994a) Night-time dosing of doxazosin has peak effect on morning ambulatory blood pressure. Results of the HALT study. Am J Hypertens 7:844–847

Pickering TG, Levenstein M, Walmsley P, Hypertension and Lipid Trial Study Group (1994b) Differential effects of doxazosin on clinic and ambulatory pressure according to age, gender, and presence of white coat hypertension. Results of the HALT study. Am J Hypertens 7:848–852

Pieper C, Warren K, Pickering TG (1993) A comparison of ambulatory blood pressure and heart rate at home and work on work and non-work days. J Hypertens 11:177–183

Portaluppi F, Montanari L, Bagni B, degli Uberti E, Trasforini G, Margutti A (1989) Circadian rhythms of atrial natriuretic peptide, blood pressure and heart rate in normal subjects. Cardiology 76:428–432

Portaluppi F, Montanari L, Ferlini M, Gilli P (1990a) Altered circadian rhythms of blood pressure and heart rate in non-hemodialysis chronic renal failure. Chronobiol Int 7:321–327

Portaluppi F, Bagni B, degli Uberti E, Montanari L, Cavallini R, Trasforini G, Margutti A, Ferlini M, Zanella M, Parti M (1990b) Circadian rhythms of atrial natriuretic peptide, renin, aldosterone, cortisol, blood pressure and heart rate in normal and hypertensive subjects. J Hypertens 8:85–95

Portaluppi F, Montanari L, Massari M, Di Chiara V, Capanna M (1991) Loss of nocturnal decline of blood pressure in hypertension due to chronic renal failure. Am J Hypertens 4:20–26

Portaluppi F, Montanari L, Ferlini M, Vergnani L, Bagni B, Degli Uberti EC (1992) Differences in blood pressure regulation of congestive heart failure, before and after treatment, correlate with changes in the circulating pattern of atrial natriuretic peptide. Eur Heart J 13:990–996

Portaluppi F, Cortelli P, Avoni P, Vergnani L, Contin E, Maltoni P, Pavani A, Sforza E, degli Uberti EC, Gambetti P, Lugaresi E (1994) Diurnal blood pressure variation and hormonal correlates in fatal familial insomnia. Hypertension 23:569–576

Portaluppi F, Vergnani L, Manfredini R, degli Uberti EC, Fersini C (1995) Time-dependent effect of isradipine on the nocturnal hypertension of chronic renal failure. Am J Hypertens 8:719–726

Prager G, Prager W, Hönig B (1989) Effect of β-adrenergic blockade on the circadian rhythm of myocardial ischemia in ambulatory patients with stable angina. J Cardiovasc Pharmacol 13:638–643

Quantz JJ (1752) Versuch einer Anweisung die Flöte traversiere zu spielen. Voß, Berlin, p 261

Quyyumi AA, Wright C, Mockus L, Fox KM (1984) Effect of partial agonist activity in β-blockers in severe angina pectoris: a double blind comparison of pindolol and atenolol. Br Med J 289:951–953

Quyyumi AA, Crake T, Wright CM, Mockus LJ, Fox KM (1987) Medical treatment of patients with severe exertional and rest angina: double blind comparison of β-blocker, calcium antagonist, and nitrate. Br Heart J 57:505–511

Raby KE, Vita JA, Rocco MB, Yeung AC, Ganz P, Fantasia G, Barry J, Selwyn AP (1993) Changing vasomotor responses of coronary arteries to nifedipine. Am Heart J 126:333–338

Raeder EA (1990) Circadian fluctuations in ventricular response to atrial fibrillation. Am J Cardiol 66:1013–1016

Raftery EB (1983) The effects of beta-blocker therapy on diurnal variation of blood pressure. Eur Heart J 4:61–64

Raftery EB (1991) Lacidipine and circadian variation in blood pressure: considerations for therapy. J Cardiovasc Pharmacol 17 [Suppl 4]:S20–S26

Raftery EB, Millar-Craig MW, Mann S, Balasubramanian V (1981a) Effects of treatment on circadian rhythms of blood pressure. Biotelem Patient Monit 8:113–120

Raftery EB, Melville DI, Gould BA, Mann S, Whittington JR (1981b) A study of the antihypertensive action of xipamide using ambulatory intra-arterial monitoring. Br J Clin Pharmacol 12:381–385

Redman CW, Beilin LJ, Bonnar J (1976) Reversed diurnal blood pressure rhythm in hypertensive pregnancies. Clin Sci Mol Med 3:687s-689s

Reeves RA, Shapiro AP, Thompson ME, Johnsen AM (1986) Loss of nocturnal decline in blood pressure after cardiac transplantation. Circulation 73:401–408

Reil JC (1796) Von der Lebenskraft. Arch Physiol 1:8–162

Reinberg A, Ghata J, Halberg F, Gervais P, Abulker C, Dupont J, Gaudeau C (1970) Rythmes circadiens du pouls, de la pression arterielle, des excretions urinaires en 17-hydroxycorticosteroides catecholamines et potassium chez l'homme adulte sain, actif et au repos. Ann Endocrinol (Paris) 31:277–287

Ridker PM, Manson JE, Buring JE, Muller JE, Hennekens CH (1990) Circadian variation of acute myocardial infarction and the effect of low-dose aspirin in a randomized trial of physicians. Circulation 82:897–902

Riva-Rocci S (1896) Un nuovo sfigmomanometro. Gazz Med Dir Torino 47:981–969

Rizzoni D, Muiesan ML, Montani G, Zulli R, Calebich S, Agabiti-Rosei E (1992) Relationship between initial cardiovascular structural changes and daytime and night-time blood pressure monitoring. Am J Hypertens 5:180–186

Rocco MB, Barry J, Campbell S, Nabel E, Cook EF, Goldman L, Selwyn AP (1987) Circadian variation of transient myocardial ischemia in patients with coronary artery disease. Circulation 75:395–400

Sanctorius S (1631) Methodi vitandorum errorum omnium qui in arte medica contingunt. Aubertum, Geneva, p 289

Sander D, Klingelhofer J (1993) Circadian blood pressure patterns in four cases with hemodynamic brain infarction and prolonged blood-brain barrier disturbance. Clin Neurol Neurosurg 95:221–229

Sandrone G, Mortara A, Torzillo D, LaRovere MT, Malliani A, Lombardi E (1994) Effects of beta blockers (atenolol or metoprolol) on heart rate variability after acute myocardial infarction. Am J Cardiol 74:340–345

Scheidel B, Lemmer B (1991) Chronopharmacology of oral nitrates in healthy subjects. Chronobiol Int 8:409–419

Schnecko A, Witte K, Lemmer B (1995) Effects of the angiotensin II receptor antagonist losartan on 24-hour blood pressure profiles of primary and secondary hypertensive rats. J Cardiovasc Pharmacol 26:214–221

Sehested J, Thomas F, Thorn M, Schifter S, Regitz V, Sheikh S, Oelkers W, Palm U, Meyer-Sabellek W, Hetzer R (1992) Level and diurnal variations of hormones of interest to the cardiovascular system in patients with heart transplants. Am J Cardiol 69:397–402

Semenowicz-Siuda K, Markiewicz A, Korczynska-Wardecka J (1984) Circadian bioavailability and some effects of propranolol in healthy subjects and liver cirrhosis. Int J Clin Pharmacol Ther Toxicol 22:653–658

Sensi S, Manzoli U, Capani F, Domenichelli B, Lucente M, Schiavoni G, Coppola E (1980) Circadian rhythm of ventricular ectopy (letter). Chest 77:580

Shaw DB, Knapp MS, Davies DH (1963) Variations in blood pressure in hypertensives during sleep. Lancet 1:797–798

Sheps SG, Pickering TG, White WB, Weber MA, Clement DL, Krakoff LR, Messerli FH, Perloff D (1994) Ambulatory blood pressure monitoring. J Am Coll Cardiol 23:1511–1513

Shimada K, Kawamoto A, Matsubayashi K, Nishinaga M, Kimura S, Ozawa T (1992) Diurnal blood pressure variations and silent cerebrovascular damage in elderly patients with hypertension. J Hypertens 10:875–878

Shimada R, Nakashima T, Nunoi K, Kohno Y, Takeshita A, Omae T (1984) Arrhythmia during insulin-induced hypoglycemia in a diabetic patient. Arch Intern Med 144:1068–1069

Sideris DA, Toumanidis ST, Anastasiou-Nana M, Zakopoulos N, Kitsiou A, Tsagarakis K, Moulopoulos SD (1992) The circadian profile of extrasystolic arrhythmia: its relationship to heart rate and blood pressure. Int J Cardiol 34:21–31

Sindrup JH, Kastrup J, Jorgensen B (1991) Regional variations in nocturnal fluctuations in subcutaneous blood flow rate in the lower leg of man. Clin Physiol 11:491–499

Smolensky MH, Tatar SE, Bergmann SA, Losman JG, Barnard CN, Dacso CC, Kraft IA (1976) Circadian rhythmic aspects of human cardiovascular function:a review by chronobiologic statistical methods. Chronobiologia 3:337–371

Snyder F, Hobson A, Morrison DF, Goldfrank F (1964) Changes in respiration, heart rate and systolic blood pressure in human sleep. J Appl Physiol 19:417–422

Soergel M, Maisin A, Azancot-Benisty A, Loirat C (1992) Ambulante Blutdruckmessung bei nierentransplantierten Kindern und Jugendlichen. Z Kardiol 81 [Suppl 2]:67–70

Staessen A, Fagard R, Thijs L, Amery A (1993) Fourier analysis of blood pressure profiles. Am J Hypertens 6:184S–187S

Stanton A, O'Brien E (1994) Auswirkungen der Therapie auf das zirkadiane Blutdruckprofil. Kardio 3:1–8

Stanton A, Cox J, Atkins N, O'Malley K, O'Brien E (1992) Cumulative sums in quantifying circadian blood pressure patterns. Hypertension 19:93–101

Statius van Eps RG, van den Meiracker AH, Boomsma F, Man in't Veld AJ, Schalekamp MADH (1994) Diurnal variation of blood pressure in patients with catecholamine-producing tumors. Am J Hypertens 7:492–497

Steitberg B, Meyer-Sabellek W, Baumgart P (1989) Statistical analysis of circadian blood pressure recordings in controlled clinical trials. J Hypertens 7 [Suppl 3]:S11–S27

Stoica E, Enulescu O (1983) Inability to deactivate sympathetic nervous system in brainstem infarct patients. J Neurol Sci 58:223–234

Struthius J (1602) Ars sphygmica. Königs, Basel, pp 127ff

Suzuki Y, Kuwajima I, Kanemaru A, Shimosawa T, Hoshino S, Sakai M, Matsushita S, Ueda K, Kuramoto K (1992) The cardiac functional reserve in elderly hypertensive patients with abnormal diurnal change in blood pressure. J Hypertens 10:173–179

Takagi N (1986) Variability of direct arterial blood pressure in essential hypertension–relationships between the fall of blood pressure during sleep and awake resting hemodynamic parameters. Jpn Circ J 50:587–594

Taylor SH (1992) Amlodipine in post-infarction angina. Cardiology 80 [Suppl 1]:26–30

Teerlink JR, Clozel JP (1993) Hemodynamic variability and circadian rhythm in rats with heart failure: role of locomotor activity. Am J Physiol 264:H2111–2118

Testa AJ (1815) Über die Krankheiten des Herzens. Gebauer, Halle, p 323

Thormann J, Schlepper M, Kramer W (1983) Diurnal changes and reproducibility of corrected sinus node recovery time. Cathet Cardiovasc Diagn 9:439–451

Tilkian AG, Guilleminault C, Schroeder JS, Lehrman KL, Simmons FB, Dement WC (1976) Hemodynamics in sleep-induced apnea. Studies during wakefulness and sleep. Ann Intern Med 85:714–719

Timio M, Lolli S, Verdura C, Monarca C, Merante F, Guerrini E (1993) Circadian blood pressure changes in patients with chronic renal insufficiency: a prospective study. Ren Fail 15:231–237

Tofler GH, Brezinski D, Schafer AI, Czeisler CA, Rutherford JD, Willich SN, Gleason RE, Williams GH, Muller JE (1987) Concurrent morning increase in platelet aggregability and the risk of myocardial infarction and sudden cardiac death. N Engl J Med 316:1514–1518

Torffvit O, Agardh CD (1993) Day and night variation in ambulatory blood pressure in type 1 diabetes mellitus with nephropathy and autonomic neuropathy. J Intern Med 233:131–137

Townsend RE, Prinz PN, Obrist WD (1973) Human cerebral blood flow during sleep and waking. J Appl Physiol 35:620–625

Tseng Y-Z, Tseng C-D, Lo H-M, Chang F-T, Hsu K-L (1994) Characteristic abnormal findings of ambulatory blood pressure indicative of hypertensive target organ complications. Eur Heart J 15:1037–1943

Tuck ML, Stern N, Sowers JR (1985) Enhanced 24-hour norepinephrine and renin secretion in young patients with essential hypertension: relation with the circadian pattern of arterial blood pressure. Am J Cardiol 55:112–115

Twidale N, Taylor S, Heddle WF, Ayres BF, Tonkin AM (1989) Morning increase in the time of onset of sustained ventricular tachycardia. Am J Cardiol 12:1204–1206

Umeda T, Naomi S, Iwaoka T, Inoue J, Sasaki M, Ideguchi Y, Sato T (1994) Timing for administration of an antihypertensive drug in the treatment of essential hypertension. Hypertension 23 [Suppl I]:I211–I214

van de Borne P, Abramowicz M, Degre S, Degaute JP (1992a) Effects of chronic congestive heart failure on 24-hour blood pressure and heart rate patterns: a hemodynamic approach. Am Heart J 123:998–1004

van de Borne P, Tielemans C, Vanherweghem JL, Degaute JP (1992b) Effect of recombinant human erythropoietin therapy on ambulatory blood pressure and heart rate in chronic haemodialysis patients. Nephrol Dial Transplant 7:45–49

van de Borne P, Gelin M, Van de Stadt J, Degaute JP (1993a) Circadian rhythms of blood pressure after liver transplantation. Hypertension 21:398–405

van de Borne P, Tielemans C, Collart F, Vanherweghem JL, Degaute JP (1993b) Twenty-four-hour blood pressure and heart rate patterns in chronic hemodialysis patients. Am J Kidney Dis 22:419–425

van de Borne P, Nguyen H, Biston P, Linkowski P, Degaute JP (1994) Effects of wake and sleep stages on the 24-h autonomic control of blood pressure and heart rate in recumbent men. Am J Physiol 266:H548–554

Veglio F, Pinna G, Melchio R, Rabbia F, Molino P, Torchio C, Chiandussi L (1993) Twenty-four-hour power spectral analysis by maximum entropy method of blood pressure in primary hyperaldosteronism. Blood Press 2:189–196

Verdecchia P, Schillaci G, Guerrieri M, Gatteschi C, Benemio G, Boldrini F, Porcellati C (1990) Circadian blood pressure changes and left ventricular hypertrophy in essential hypertension (see comments). Circulation 81:528–536

Visser GH, Goodman JD, Levine DH, Dawes GS (1982) Diurnal and other cyclic variations in human fetal heart rate near term. Am J Obstet Gynecol 142:535–544

von Arnim T, Höfling B, Schreiber M (1985) Characteristics of episodes of ST elevation or ST depression during ambulatory monitoring in patients subsequently undergoing coronary angiography. Br Heart J 54:484–488

von Basch S (1881) Über die Messung des Blutdrucks am Menschen. Z Klin Med 2:79–96

Waters DD, Miller DD, Bouchard A, Bosch X, Theroux P (1984) Circadian variation in variant angina. Am J Cardiol 54:61–64

Weber MA, Tonkon MJ, Klein RC (1987) Effect of antihypertensive therapy on the circadian blood pressure pattern. Am J Med 82 [Suppl 1A]:50–52

Weiss H (1900) Blutdruckmessung mit Gärtler's Tonometer. Münch Med Wochenschr 47:69–71

Weysse AW, Lutz BR (1915) Diurnal variations in arterial blood pressure. Am J Physiol 37:330–347

White WB, Malchoff C (1992) Diurnal blood pressure variability in mineralocorticoid excess syndrome. Am J Hypertens 5:414–418

Wiegmann TB, Herron KG, Chonko AM, MacDougall ML, Moore WV (1990) Recognition of hypertension and abnormal blood pressure burden with ambulatory blood pressure recordings in type I diabetes mellitus. Diabetes 39:1556–1560

Wilhelm GT (1806) Unterhaltungen über den Menschen, Dritter Theil: Von dem Körper und seinen Theilen und Functionen insbesondere. Engelbrechtsche Kunsthandlung, Augsburg, pp 352ff

Willich SN, Muller JE (eds.) (1996) Triggering of acute coronary syndromes – implications for prevention. Kluwer Acad Publ, Dordrecht Boston London

Willich SN, Linderer T, Wegscheider K, Leizorovicz A, Alamercery I, Schröder R, ISAM Study Group (1989a) Increased morning incidence of myocardial infarction in the ISAM study: absence with prior β-adrenergic blockade. Circulation 80:853–858

Willich SN, Pohjola-Sintonen S, Bhatia SJS, Shook TL, Tofler GH, Muller JE, Curtis DG, Williams GH, Stone PH (1989b) Suppression of silent ischemia by metoprolol without alteration of morning increase of platelet aggregability in patients with stable coronary artery disease. Circulation 79:557–565

Willich SN, Maclure M, Mittleman M, Arntz HR, Muller JE (1993) Sudden cardiac death – support for a role of triggering in causation. Circulation 87:1442–1450

Witte K, Lemmer B (1992) Rhythmusanalyse von individuellen 24-Stunden-Blutdruckprofilen essentieller Hypertoniker. Z Kardiol 81 [Suppl 2]:101–104

Witte K, Lemmer B (1993) Free-running rhythms in blood pressure and heart rate in normotensive Sprague-Dawley rats. J Interdiscipl Cycle Res 24:328–334

Witte K, Weisser K, Neubeck M, Mutschler E, Lehmann K, Hopf R, Lemmer B (1993) Cardiovascular effects, pharmacokinetics and converting enzyme inhibition of enalapril after morning versus evening administration. Clin Pharmacol Ther 54:177–186

Witte K, Schnecko A, Buijs R, Lemmer B (1995) Circadian rhythms in blood pressure and heart rate in SCN-Lesioned and unlesioned transgenic hypertensive rats. Biol Rhythm Res 26: 258

Woods KL, Fletcher S, Jagger C (1992) Modification of the circadian rhythm of onset of acute myocardial infarction by long-term antianginal treatment. Br Heart J 68:458–461

Wortmann A, Bachmann K (1991) Chronotherapy in coronary heart disease: comparison of two nitrate treatments. Chronobiol Int 8:399–408

Yasue H, Omote S, Takizawa A, Nagao M, Miwa K, Tanaka S (1979) Circadian variation of exercise capacity in patients with Prinzmetal's variant angina: role of exercise-induced coronary arterial spasm. Circulation 59:938–948

Zadek J (1881) Die Messung des Blutdrucks des Menschen mittels des Basch'schen Apparates. Z Klin Med 2:509–551

Zimmermann JG (1793) Von der Erfahrung in der Arzneykunst. Edlen von Trattnern, Agram, pp 233ff

Zuther P, Witte K, Lemmer B (1996) ABPM-FIT and CV-SORT: an easy-to-use software package for detailed analyses of data from ambulatory blood pressure monitoring. Blood Pressure Monitoring 1

CHAPTER 11
Chronopharmacology of Anticancer Agents

F. LÉVI

A. Introduction

Several strategies are aimed at increasing the selectivity of anticancer agents against cancer cells. Some are based on tumor cell biology: new drug development, resistance revertants, etc. Others are targeted at host cells: development of analogs with less toxicity than the parent drug, combinations of cytostatics without additive toxicities, scheduling and/or supportive care in order to increase chemotherapy tolerability, etc. (DE VITA et al. 1993). A dose-efficacy relationship has been repeatedly established for cancer chemotherapy (HRYNIUK 1988). For this reason, the chronobiology of normal cells has constituted the main basis for attempting to improve the therapeutic index of cytostatic drugs. It was expected that an increase in drug doses, and therefore therapeutic efficacy, would result from an adaptation of drug delivery to circadian rhythms (chronotherapy). We will examine the several steps which have led to the validation of the clinical relevance of chronotherapy in medical oncology that was anticipated more than 20 years ago (HAUS et al. 1972; HALBERG et al. 1973). Only the circadian aspects will be considered.

B. Experimental Chronopharmacology

I. Toxicity Rhythms

Circadian dosing time influences the extent of toxicity of ~30 anticancer drugs, including cytostatics and cytokines, in mice and rats (LÉVI et al. 1988a, 1994a; MORMONT et al. 1989; PERPOINT et al. 1995). The methodology that was mostly used to demonstrate this phenomenon firstly consisted of the administration of the same drug dose to different groups of animals, each group corresponding to a different circadian stage. Six circadian stages, usually located 4 h apart, have commonly been used. Time has usually been expressed in *h*ours *a*fter *l*ight *o*nset (HALO). Nocturnally active mice or rats have mostly been synchronized for 1–3 weeks with an alternation of 12 h light (L) and 12 h darkness (D) (LD 12:12). A 3-week span appears a safe period for biological rhythms to adjust to a synchronization regimen differing by 8 h or more from the previous one. Other photoperiodic schemes have occasionally been used: natural LD, which varies according to the season and latitude; artificial LD 8:16 (so-called winter photoperiod); or LD 16:8 (so-called summer photoperiod). Autonomous chronobiologic facilities have largely improved the feasibility of chronopharmacologic experiments since they allow for any endogenous circadian stage to be explored at any desired local time. For all these drugs, survival rate varies by 50% or more according to circadian dosing time of a potentially lethal dose. Such large differences are observed irrespective of injection route – intravenous or intraperitoneal – or number of injections – single or repeated (Figs. 1,2).

Pirarubicin, an anthracycline compound, mostly exerts myelosuppressive effects, which are lowest following dosing in the second half of the diurnal rest span (~7 HALO) (Lévi et al. 1985). Mitoxantrone, an anthracycline-related compound, displays lowest hematologic toxicity 8 h later (15 HALO) (Lévi et al. 1994b). Platinum complex analogs – cisplatin (cisdichlorodiamineplatinum, CDDP), carboplatin (cyclobutane dicarboxylatoplatinum II, CBDCA) and oxaliplatin [oxalato (2-)-0,0' platinum] (L-OHP) – are also best tolerated near the middle of the nocturnal activity span of mice and rats, despite the differing target tissues of toxicity of these compounds: CDDP is mostly toxic to both kidney and bone marrow, CBDCA to bone marrow and colon mucosa and L-OHP to jejunal mucosa and bone marrow (Hrushesky et al. 1982a; Lévi et al. 1982a; Boughattas et al. 1988, 1989, 1990). Two fluoropyrimidines – 5-fluorouracil (5-FU) and floxuridine (FUDR) – are antimetabolites which dis-

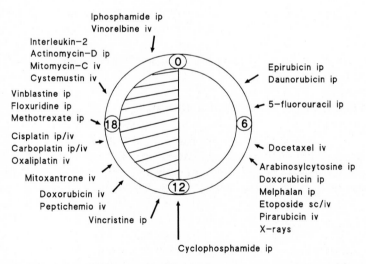

Fig. 1. Circadian change in tolerability of anticancer agents in mice or rats. End point is survival following a potentially lethal dose at one of six dosing times, 4 h apart. Animals are synchronized with an alternation of 12 h light (L) and 12 h darkness (D) (LD 12:12). Time is expressed in *h*ours *a*fter *l*ight *o*nset (HALO), since light onset is the main signal which resets the circadian cycle in these nocturnally active animals. The least toxic time, usually by 50% or more, is indicated with an *arrow* for the corresponding agent(s). (Modified from Lévi et al. 1988)

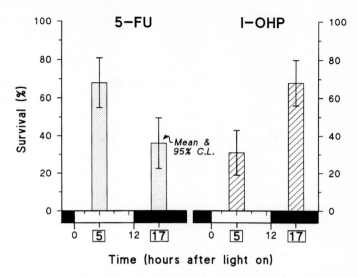

Fig. 2. Relationship between circadian dosing time and lethal toxicity of 5-FU and L-OHP in mice. Time is expressed in *h*ours *a*fter *l*ight *o*nset (HALO) rather than in local clock hours. The circadian times when 5-FU and L-OHP were least toxic are located 12 h apart, being 5 HALO (± 2 h) and 17 HALO (± 2 h), respectively. Tolerability was doubled, by injecting either agent at its "best" rather than at its "worst" dosing time, differences being largely statistically significant. 5-FU (200–600 mg/kg i.p.) in a total of 50 healthy male CD_1 mice at each circadian time (after BURNS and BELAND 1984); L-OHP (17 mg/kg i.v.) in a total of 60 healthy male $B_6D_2F_1$ mice at each circadian time. (After BOUGHATTAS et al. 1989)

play opposite circadian toxicity rhythms (BURNS and BELAND 1984; GONZALEZ et al. 1989).

Knowledge of neither the anticancer drug class nor the main toxicity target organ therefore enables the least toxic time of an anticancer drug to be predicted. Nevertheless, circadian rhythms in the susceptibility of healthy target tissues to the same agent are usually synchronous, supporting circadian-based therapeutic optimization. Chronopharmacologic mechanisms may involve circadian changes in drug pharmacokinetics and/or susceptibility rhythms of target tissues.

II. Chronopharmacokinetics

Despite the technical limitations of working with small rodents, plasma and urinary chronopharmacokinetics have been investigated for several anticancer drugs (Table 1). Times of high toxicity corresponded to longest elimination half-lives for methotrexate, CDDP, CBDCA and mitoxantrone (all between 4 and 8 HALO) (ENGLISH et al. 1982; LÉVI et al. 1982b, 1994b; BOUGHATTAS et al. 1994). However, this was not the case for L-OHP or 5-FU (BOUGHATTAS et al. 1994; CODACCI-PISANELLI et al. 1995). Furthermore, platinum concentration

Table 1. Plasma or urinary chronopharmacokinetics of anticancer drugs in mice or rats

Drug	Dose (mg/kg/day)	Route	Species, strain	Injection time (HALO)	Main results		Reference
Antimetabolites							
Methotrexate	2	i.v.	Rat, Norwegian Hood	0, 6, 12, 18	C_{MAX} $t^{1/2}$ and AUC	0 HALO > 18 HALO	ENGLISH et al. 1982
5-Fluorouracil	23.8	c.s.c.i. (10 days)	Mouse, C57B16	Constant perfusion	24-h rhythm in plasma [5-FU] for 5 days	6 HALO > other HALO (× 10 times)	CODACCI-PISANELLI et al. 1995
6-Mercaptopurine	37.1	???	Rat, Norwegian Hood	0, 6, 12, 18	AUC	6 HALO > 18 HALO (× 3 times)	AHERNE et al. 1987
Platinum complexes							
Cisplatin[a]	5	i.p.	Rat, F344 Fisher	6, 18	Urinary C_{MAX} Elimination rate	6 HALO > 18 HALO 18 HALO > 6 HALO	LÉVI et al. 1982
Carboplatin	72	i.v.	Mouse, B6D2F1	0, 8, 16	$t^{1/2\alpha}$, $t^{1/2\beta}$, DiV, MRT of ultrafiltrated [Pt]	8 HALO > 0 HALO (+ 100%)	BOUGHATTAS et al. 1994
Oxaliplatin	17	i.v.	Mouse, B6D2F1	0, 8, 16	$t^{1/2\alpha}$, $t^{1/2\beta}$, DiV, MRT of ultrafiltrated [Pt]	16 HALO > 0 HALO (+ 100%)	BOUGHATTAS et al. 1994
Anthracyclines (and related)							
Doxorubicin	18	i.p.	Rat, F344 Fisher	2, 6, 10, 14, 18, 22	C_{MAX} [Dox]	9 HALO > 17 HALO	LÉVI et al. (unpublished)
Mitoxantrone	14.5	i.v.	Mouse, B6D2F1	4, 10, 16, 22	$t^{1/2\beta}$, $t^{1/2\gamma}$, AUC DiVβ	16 HALO > 4 HALO (+ 100%) 4 HALO > 16 HALO (× 9 times)	LÉVI et al. (1994b)
Glucocorticoids							
Methylprednisolone	1	i.v.	Rat, Norwegian Hood	0, 6, 12, 18	$t^{1/2}$ and AUC	12 HALO > 6 HALO	ENGLISH et al. 1981

i.v., intravenous; i.p., intraperitoneal injection; c.s.c.i., constant subcutaneous infusion; C_{MAX}, maximal concentration; $t^{1/2}$, half-life; AUC, area under the concentration curve; DiV, distribution volume; MRT, mean residence time; HALO, hours after light onset.
[a] Urinary pharmacokinetics.

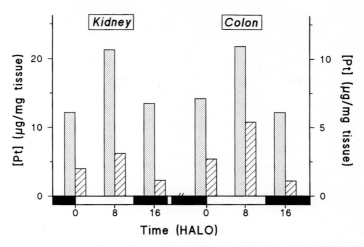

Fig. 3. Renal and intestinal platinum concentration after 6 or 8 weekly injections of cisplatin (5 mg.kg i.v., *dotted bars*) or carboplatin (50 mg/kg, *hatched bars*), respectively, in a total of 288 $B_6D_2F_1$ mice. Dosing-time-related differences were statistically significant with analysis of variance for each compound in each tissue. Highest tissue [*Pt*] corresponded to highest toxicity of either analog. (Modified from BOUGHATTAS et al. 1990)

in kidney or gut mucosa was lowest following CDDP, CBDCA or L-OHP dosing, at 16 HALO, the time of lowest toxicity for these three compounds (BOUGHATTAS et al. 1988, 1989, 1990) (Fig. 3). Thus, it is likely that those mechanisms involved in plasma chronopharmacokinetics adjust the rate of drug exposure of cells, while those involved in cellular chronopharmacokinetics may relate more directly to the extent of healthy tissue damage.

III. Rhythms in Susceptibility of Target Tissues

CDDP-induced renal damage leads to polyuria, serum urea increase and enzymuria. All these toxicity indices were lowest following dosing near the middle of the activity span of rats. The urinary excretion of β-*N*-acetylglucosaminidase (β-NAG) also exhibits a circadian rhythm which reflects normal tubular cell turnover, with high values near the end of the rest span, a circadian time associated with high CDDP kidney toxicity. Other data have further suggested that circadian changes in these enzymes, which contain sulfydryl groups, may contribute to the circadian rhythm in cisplatin toxicity (LÉVI et al. 1982a,b; CAMBAR et al. 1992).

Direct exposure of bone marrow cells to fixed amounts of pirarubicin at different circadian stages allowed a direct assessment of the susceptibility rhythm of this tissue (LÉVI et al. 1988b). Femoral bone marrow from mice was obtained at one of six different circadian times, then incubated in vitro for 1 h with one of several concentrations of pirarubicin, then cultured in agar for 1 week, according to the granulocyte-monocyte colony-forming unit (GM-CFUc) assay, in which the colony count reflects the number of GM progenitors. This variable normally shows a 50% increase near the middle of the active span of mice, when pirarubicin is most toxic in vivo. In vitro, pirarubicin exposure of bone marrow cells decreases GM colony formation in a dose-dependent manner. Furthermore, such an in vitro effect is lowest near the middle of the rest span, when in vivo tolerance is high (Fig. 4). Other experiments support the theory that the circadian rhythm in DNA synthesis of these progenitors may partly contribute to the pirarubicin toxicity being lowest during diurnal rest and highest near the middle of the nocturnal activity span (Lévi et al. 1985, 1988b).

Circadian changes in cellular enzymatic activities of two- to eightfold also account in part for the chronopharmacology of antimetabolites. This is the case for dihydrofolate reductase, a target enzyme for methotrexate cytotoxicity; dehydropyrimidine dehydrogenase (DPDase), the rate-limiting catabolic enzyme of fluoropyrimidines; uridine phosphorylase (URDPase); orotate phophoribosyltransferase (OPRTase); and deoxythymidine kinase (TKase) – all three involved in the anabolism of the cytotoxic forms of fluoropyrimidines, FdUMP and FUMP (MALMARY-NEBOT et al. 1985; HARRIS et al. 1989; EL KOUNI et al. 1990; NAGUIB et al. 1993; ZHANG et al. 1993).

Cellular resistance to many cytostatics involves reduced glutathione, the concentration of which in liver cells doubles along the 24-h time scale; lowest

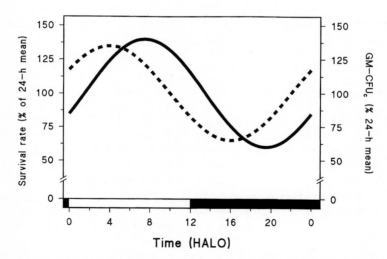

Fig. 4. Circadian rhythm in pirarubicin lethal toxicity in $B_6D_2F_1$ mice (*continuous line*) and its relationship with the rhythm in in vitro susceptibility of granulomonocytic precursors (*GM-CFUc*) for pirarubicin (*dashed line*). A significant circadian rhythm was found for both variables with cosinor analysis. (Data from LÉVI et al. 1985, 1988b)

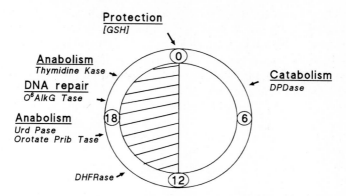

Fig. 5. Cellular mechanisms of anticancer drug chronopharmacology in mice or rats. The 24-h time scale is depicted as in Fig. 1. The time of highest activity or content is indicated with *an arrow*. *DPDase,* dehydropyrimidine dehydrogenase; *DHFRase,* dehydrofolate reductase; *orotate Prib Tase,* orotate phosphoribosyl transferase; *Urd Pase,* uridine phosphorylase; $O^6Alk\ G\ Tase,$ O^6-alkylguanine transferase; *Thymidine Kase,* deoxythymidine kinase; *GSH,* reduced glutathione. (Data from MALMARY-NEBOT et al. (1985); NAGUIB et al. (1993); EL KOUNI et al. (1990); ZHANG et al. (1993); MARTINEAU-PIVOTEAU et al., in press; LI et al., submitted)

and highest values occur near dark onset and near light onset, respectively (BELANGER and LABRECQUE 1992; LI et al., submitted).

Finally, O^6-alkylguanine-DNA alkyltransferase (AGTase), a DNA repair enzyme, was ~12-fold higher at 19 HALO as compared to 7 HALO in mouse liver. These circadian stages corresponded to least and highest toxicity of cystemustin, respectively, a short-half life nitrosourea, which causes the formation of DNA covalent cross-links that are repaired by this enzyme (MARTINEAU et al., submitted). These data are summarized in Fig. 5.

IV. Circadian Rhythms in Antitumor Efficacy

Dosing time not only affects drug tolerability, but may also modify anticancer efficacy. Arabinosylcytosine (Ara-C), an antimetabolite drug, displays high activity against mouse L1210 when it is injected every 3 h for 24 h ("homeostatic" schedule). Several chronomodulated schedules administering the same total dose were compared to such a reference protocol. They consisted in gradually increasing then decreasing 3 hourly dose fractions. The protocols delivering the highest doses at 5 or at 8 HALO, when Ara-C was least toxic, improved median survival threefold and doubled cure rate as compared to the reference schedule or other circadian protocols (HAUS et al. 1972; SCHEVING et al. 1976).

In 13 762 mammary adenocarcinoma-bearing Fisher rats, a doxorubicin–melphalan combination produced twice as many complete tumor regressions following dosing of both drugs near the end of the rest span, when they also exert less toxicity (HALBERG et al. 1980). A threefold greater antitumor efficacy was also obtained when doxorubicin and cisplatin, in combination, were injected at their respective least toxic times, in immunocytoma-bearing LOU rats (SOTHERN et al. 1989). Antitumor efficacy is usually greater with drug combinations than with single agents, and circadian timing appeared even more important for a large improvement in long-term outcome (cure or complete remission rate) with drug combinations rather than with single drug (tumor size decrease, median survival) (Fig. 6). This observation was further sup-

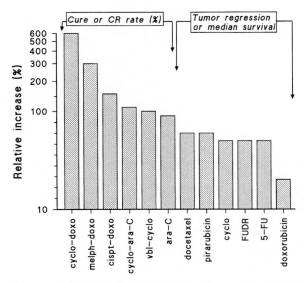

Fig. 6. Relative improvement in antitumor efficacy of anticancer drugs resulting from adequate circadian timing in mouse or rat tumor models. The benefit from circadian timing exceeds 100% for cure or complete remission (CR) rate for drug combinations: concurrent cyclophosphamide-doxorubicin (best time at 13 HALO) or cyclophosphamide-arabinocylcytosine ("best" time at 11–14 HALO) against mouse L1210 leukemia; sequential doxorubicin-melphalan against 13762 mammary adenocarcinoma in Fisher rats ("best" time at 10 HALO) or sequential doxorubicin-cisplatin against LOU rat immunocytoma ("best" respective times at 10 and at 18 HALO) or sequential cyclophosphamide-vinblastine against T_9, T_{10} or MCA sarcoma ("best" time at 18 HALO). (Data from SCHEVING et al. 1977, 1980a,b; HALBERG et al. 1980; FOCAN et al. 1985; SOTHERN et al. 1989). Circadian timing of single drugs usually improves tumor regression rate or median survival by 20%–100% as shown for docetaxel against pancreatic P03 adenocarcinoma in BDF_1 mice ("best" time at 7 HALO), pirarubicin against Lewis lung cancer in $C_{57}Bl_6$ mice ("best" at 15 HALO), cyclophosphamide against mouse mammary carcinoma, T9 or T10 sarcoma, Ehrlich's ascites or L1210 leukemia ("best" time, 2–12 HALO), floxuridine or FUDR against 13762 mammary cancer in Fisher rats ("best" chronomodulated schedule with peak delivery from 22 to 4 HALO), 5-FU against mouse CO26 or CO38 colon tumor ("best" time, 2.5 HALO) or doxorubicin against Lou rat immunocytoma ("best" time, 6–10 HALO). (Data from TAMPELLINI et al. 1995; LÉVI 1995; BADRAN and ECHAVE LLANOS 1965; FOCAN et al. 1985; KODAMA and KODAMA 1982; HAUS et al. 1972; HALBERG et al. 1977; VON ROEMELING and HRUSHESKY 1990; PETERS et al. 1987)

ported by an incremental approach in chronopharmacologic combinations of drugs, which documented that the rate of tumor cures could be brought up from 25% to 80% or more with adequately circadian scheduled chemotherapy involving up to five cytostatics (SCHEVING et al. 1980a,b).

The striking coincidence between times of highest anticancer efficacy and least toxicity suggest common mechanisms may be involved. These may relate to plasma pharmacokinetics: a low C_{max} and an increased AUC could favor both improved tolerability and efficacy. Drug dosing at the least toxic time may also spare the immune system, which in turn could be more effective in clearing up residual tumor cells. Finally, tumor cells may tend to proliferate in a circadian stage different from that of healthy tissues, and thus offer a different "window" of chemotherapy susceptibility along the 24-h time scale.

In any event, chemotherapy dosing at the least toxic time also allowed its tolerable dose to be increased, with improved survival of tumor-bearing mice through this mechanism, as was shown for pirarubicin against Lewis lung carcinoma and for docetaxel against PO3 pancreatic carcinoma (LÉVI 1995; TAMPELLINI et al. 1995).

C. Clinical Chronopharmacology

Several clinical prerequisites have further warranted the assessment of the relevance of adapting chemotherapy to circadian rhythms.

I. Rhythms in Target Tissues

Therefore proliferative activity (DNA synthesis) in hematopoietic or oral mucosa progenitor cells varies by 50% or more along the 24-h time scale in healthy subjects (Fig. 7). This is also the case for ex vivo synthetic activity in human rectal mucosa. For these three tissues, lower mean values in DNA synthesis occurred between midnight and 0400 hours at night, while higher mean values occurred between 0800 hours and 2000 hours (BUCHI et al. 1991; KILLMAN et al. 1962; MAUER 1965; SMAALAND et al. 1991a, 1992a; WARNAKULASURIYA and MACDONALD 1993). Reduced glutathione also varied rhythmically in human bone marrow; however, intersubject variability appeared larger than that of proliferative indices (SMAALAND et al. 1991b).

Mitotic index and/or DNA synthesis have been used to evaluate the proliferative activity of many transplantable tumors and some spontaneously arising ones in laboratory rodents. Data on rapidly and slowly growing experimental models support the hypothesis that well-differentiated slow-growing tumors retain a circadian time structure, whereas poorly differentiated, rapidly growing tumors tend to lose it.

Similar conclusions may be drawn from the few studies performed in humans. In a study on human breast cancer, 12 or more biopsies of subcutaneous tumor nodules were performed in the same patients by TAHTI (1956)

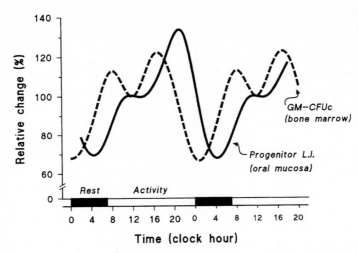

Fig. 7. Twenty-four-hour changes in human normal tissues. The proliferative ability of bone marrow granulomonocytic precursors was assessed from 7 4-hourly bone marrow aspirations in 19 healthy volunteers (after SMAALAND et al. 1992); that of oral mucosa progenitors was documented with a similar sampling scheme in 11 healthy subjects (after WARNAKULASURYIA and MACDONALD 1993). A 4-h span, located between midnight and 0400 hours, corresponds to a low point in the proliferative activity of both of these tissues in man. This also applies to the proliferative activity of rectal mucosa, although its peak occurred near 0800 hours in 24 healthy men (not shown; BUCHI et al. 1991)

and VOUTILAINEN (1953). The mitotic index was determined for each biopsy specimen. Large interindividual differences were observed in the time pattern of tumor cell divisions. Nonetheless, a group circadian rhythm was statistically validated by cosinor analysis, a maximum occurring near 1500 hours and a minimum near 0300 hours (GARCIA-SAINZ and HALBERG 1966). Other authors have attempted to estimate the rhythmicity of tumor cell division by measuring tumor surface temperature. A good correlation was found between this index and ^{32}P incorporation into malignant breast tissue on a 24-h scale (STOLL and BURCH 1968), suggesting that the time variations in malignant breast surface temperature may help to estimate those in tumor cell division and/or metabolism. Surface temperature on the tumor and on the healthy contralateral breast of 26 patients with untreated breast cancer was measured every 20 min for 7–21 days (GAUTHERIE and GROS 1974). Results revealed two kinds of cancer-associated alterations of the physiologic circadian rhythm in breast surface temperature:

1. A shortening of the period itself in patients with rapidly growing tumors, which subsequently proved to be poorly differentiated.
2. The persistence of a 24-h rhythm in the slow-growing, well-differentiated tumors, which however showed a marked decrease in amplitude and a phase advance by 6 h.

Thus, maximal tumor temperature in the malignant breast occurred at 1500 h rather than at 2100 h as observed in the healthy contralateral breast. The coincidence of maximal mitotic index and maximal tumor surface temperature in this disease is striking. This coincidence of maximal mitotic index and maximal tumor surface temperature was also found in four patients with advanced, well-differentiated head and neck tumors (DEKA 1975; FOCAN et al. 1985). Similarly, a circadian rhythm was described for the proportion of ovarian cancer cells in S-phase in women with advanced ovarian cancer and ascites (KLEVECZ et al. 1987). A similar trend was observed in patients with lymphoma (SMAALAND et al. 1993).

In contrast, very large interindividual differences characterized the circadian variation in mitotic index of skin epidermoid carcinomas and melanomas (TAHTI 1956; VOUTILAINEN 1953; GARCIA-SAINZ and HALBERG 1966). In both tumors, no group synchronization of cell proliferation was found. Similarly, the proportion of malignant ovarian cells engaged in DNA synthesis (S-phase) or in G2-M phase of the cell cycle did not exhibit any 24-h synchronized rhythm in patients with ascites from ovarian carcinoma if the tumor was aneuploid, rather than diploid (KLEVECZ and BRALY 1987). Those patients with stage III malignant lymphoma had a significant group circadian rhythm, while those with a more advanced stage IV had not (SMAALAND et al. 1993). Thus, because of intertumor differences in circadian organization, chronopharmacologic effects on malignant tumors need to be specified by tumor chronopathology.

II. Chronopharmacokinetics

Short intravenous infusions of cisplatin, carboplatin, doxorubicin, 5-FU or methotrexate, or oral intake of busulfan or 6-mercaptopurine but not methotrexate, were associated with modifications of plasma and/or urinary pharmacokinetics according to dosing time (Table 2). Physiologic rhythms in urinary excretion or plasma proteins may contribute to drug chronopharmacokinetics, as shown for cisplatin, carboplatin, methotrexate or busulfan (HECQUET et al. 1985; HECQUET and SUCCHE 1986; HRUSHESKY et al. 1982b; KERR et al. 1990; KOREN et al. 1990, 1992; LANGEVIN et al. 1987; ROBINSON et al. 1989; HASSAN et al. 1991; EKSBORG et al. 1989; CANAL et al. 1991; NOWAKOWSKA-DULAWA 1990; VASSAL et al; 1993).

However, the most striking results stemmed from flat intravenous infusion of chemotherapeutic agents. To our knowledge, five such clinical investigations have been reported for 5-FU, in four of which 5-FU was infused at a flat rate for 24 h or 4 or 5 days (PETIT et al. 1988; THIBERVILLE et al. 1992; METZGER et al. 1994; FLEMING et al. 1994). Mean plasma 5-FU nearly doubled from ~ midday to ~ 0100 hours or 0400 hours at night (Fig. 8).

This circadian pattern was similar on all infusional days tested in patients with bladder or gastrointestinal metastatic cancer and in those receiving prior CDDP, concurrent flat infusion of FA alone or associated with L-OHP.

Table 2. Circadian changes in anticancer drug pharmacokinetics in patients

Drug	Dose (units)	Route, schedule	Dosing time (hours) and design	No. of patients (no. of courses) [Age range, years]	Tumor type	Significant main results		Reference
						Variable	Time-related difference (hours)	
Cisplatin	60 mg/m^2	i.v. over 30 min q 3 weeks	0600 or 1800 R and CO	11 (51) [25–73]	Ovary or bladder	Urine C_{MAX} Urine AUC	0600 > 1800 (+50%)[a] 0600 > 1800 (+55%)	Hrushesky et al. 1982b
Carboplatin	400 mg/m^2	i.v. over 30 min q 3 weeks	0600 vs. 1800 CO	7 (12) [47–64]	Ovary	Renal CL	0600 > 1800 (+140%)	Kerr et al. 1990
Doxorubicin	50 mg/m^2	i.v. over 15 min	0900 vs. 2100 R and CO	18 (36) [41–74]	Breast	Dox. CL_B $t^{1/2\gamma}$ AUC Dox. ol AUC	0900 > 2100 (+40%) 2100 > 0900 (+80%) 2100 > 0900 (+50%) 2100 > 0900	Canal et al. 1991
5-Fluorouracil	15 mg/kg	i.v. over 15 min q 4 days for 12 days	0100, 0700, 1300 or 1900 CO	28 [adults]	Gastro-intestinal	$t^{1/2}$ CL_B Vdss AUC	0100 > 1900 (+160%) 1300 > 0100 (+70%) 0100 > 1300 (+35%) 0100 > 1300 (+60%)	Nowakowska-Dulawa et al. 1990
Methotrexate	36–80 mg/m^2	i.v. over 3 min	1000 or 2100 CO	6 (12) [3–11]	High-risk leukemia	CL_B Unbound renal CL Nonglomerular CL	1000 > 2100 (+20%) 1000 > 2100 (+100%) 1000 > 2100 (+150%)	Koren et al. 1992

Drug	Dose	Route	Time of administration	No. of pts (age) [range]	Type of cancer	Parameter	Results	References
Methotrexate	7.5 or 20 mg/m^2	p.o.	0800 or 2000 CO	7 (14) [3–11]	ALL in CR	AUC	0800 = 2000	BALIS et al. 1989
6-Mercaptopurine	75 mg/m^2	p.o.	0800 or 2000 CO	8 (16) [3–11]	ALL in CR	AUC	2000 > 0800 (+20%)	BALIS et al. 1989
6-Mercaptopurine	75 mg/m^2	p.o.	0800 or 1900 R and CO	6 (12) [3–11]	ALL in CR	AUC $t^{1/2}$	1900 > 0700 (+50%) 1900 > 0700 (+50%)	LANGEVIN et al. 1987
Busulfan	1 mg/kg	p.o. q 6 h for 4 days	NR	9 18 [19–50]	AML or ALL		24-h rhythm in plasma [c]; peak at 0200 hours, amplitude decrease with age: ≤5 years (+47%) >5 and ≤16 years (+29%), >16 years (NS)	HASSAN et al. 1991
Busulfan	37.5 mg/m^2	p.o. q 6 h for 4 days	1200, 1800, 0000 or 0600	21 [1–12]	Solid tumors		24-h rhythm in • Plasma trough [c]: peak at 0600 h (+50%) • Urinary excretion: peak at 1200 h (+50%)	VASSAL et al. 1993
5-Fluorouracil	450–970 mg/m^2/day	Flat CVI for 5 days		7 [58–81]	Bladder		24-h rhythm in 5-FU: peak at 0100 h (+80%)	PETIT et al. 1988
5-Fluorouracil	1000–1800 mg/m^2/day (with FA, 500 mg/m^2/day)	Flat CVI for 1 day R and CO (hour of onset 0600 vs. 1800)		28 [42–77]	Gastrointestinal		• 24-h rhythm in 5-FU: peak at 0300 h (+40%) • Amplitude ↓ in pts. with high bilirubin	FLEMING et al. 1994
5-Fluorouracil	250 mg/m^2/day	Flat CVI for 14 days		7 [48–69]	Gastrointestinal		• 24-h rhythm in 5-FU: peak at 1200 h (+280%) • Inverse rhythm in DPDase activity of mononuclear cells, peak at 0000 h (+33%)	HARRIS et al. 1990

Table 2 (*Contd.*)

Drug	Dose (units)	Route, schedule	Dosing time (hours) and design	No. of patients (no. of courses) [Age range, years]	Tumor type	Significant main results — Variable	Significant main results — Time-related difference (hours)	Reference
5-Fluorouracil	600 mg/m²/day (with FA, 300 mg/m²/day and L-OHP, 25 mg/m²/day)	Flat CVI for 5 days		4 [34–47]	Colon		• 24-h rhythm in 5-FU: peak at 0400 h (+ 260%) • Increased amplitude may relate to low toxicity	METZGER et al. 1994
dl-Folinic acid	300 mg/m²/day	Flat CVI for 5 days		4 [34–47]	Colon		• 24-h rhythm in plasma l-FA: peak at 0700 h (+ 20%)	METZGER et al. 1994
Vindesine	2.4–5.85 mg	Flat CVI for 2 days		9 [39–79]	Lung		• 24-h rhythm in plasma [VDS] at steady state: peak at 1200 h (+ 40%)	FOCAN et al. 1989
Doxorubicin	4–5 mg/m²/day	Flat CVI for 4–6 weeks		4 [36–73]	Breast		• No group rhythm • 2–4 fold 24-h variations in plasma dox. and dox. ol, individual peaks between 1200 and 2000 h	LÉVI et al. 1986
	15–20 mg/m²/day	Flat CVI for 4 days		6 [NR]	Adenocarcinoma		As for breast	SQALLI et al. 1989
	30 mg/m²/day	Flat CVI for 2 days		5 (15) [1–13]	ALL, AML		Mean [C]: 0600–1800 > 1800–0600 h (× 2)	SABBAG et al. 1993

i.v., intravenous; p.o., per os; CVI, continuous venous infusion; R, randomized; CO, crossover; CL, clearance All, acute lymphoblastic leukemia; AML, acute myoblastic leukemia.
[a] Percentage increase over lowest mean value.

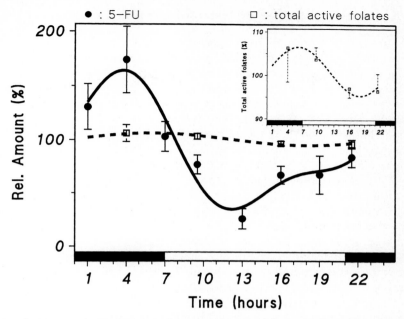

Fig. 8. Circadian variation in relative amounts of 5-fluorouracil (*5-FU, solid circles*) and total active folates (*open squares*). Curves are calculated with the parameters from cosinor analysis with periods of 24 h and 12 h for 5-FU (*solid line*) and 24 h only for total active folates (*broken line*). *Inset,* folate variation with expansion of *y*-axis. (After METZGER et al. 1994)

However, after a 14-day venous infusion of 5-FU, the plasma 5-FU rhythm peaked near noon in nine patients with gastrointestinal malignancy; furthermore, DPDase activity in circulating mononuclear cells displayed a circadian rhythm, with a peak occurring near 0100 hours at night, e.g., ~12 h out of phase of the 5-FU rhythm (HARRIS et al. 1990). We hypothesize that high-dose flat infusion or bolus injection of 5-FU saturates DPDase activity, which no longer represents a relevant mechanism in such experimental conditions. In both of these cases, 5-FU clearance is lowest near 0100 hours at night, while this parameter is highest near 1300 hours when 5-FU is infused for 14 days at a lower dose, and its 24-h mean plasma level is ~20 times as low as that observed during 4- or 5-day infusion (16 vs. 340 ng/ml) (LÉVI et al. 1993a). Thus, either infusion duration or daily dose of 5-FU may influence the peak time location of the circadian rhythm in drug plasma level. Despite flat infusion, circadian changes in drug plasma level were also observed at equilibrium state for doxorubicin (2 days, 4 days or 4–6 weeks infusion) or vindesine (4 days infusion) (LÉVI et al. 1986; FOCAN et al. 1989; SQUALLI et al. 1989; SABBAG et al. 1993).

D. Clinical Validation of Chronotherapy in Oncology

Most of the recent work has been based upon toxicity circadian rhythms. Early clinical trials had suggested significant clinical benefits from specific circadian timing of chemotherapy (combined vinblastine, cyclophosphamide and methotrexate or 5-FU) or radiotherapy (FOCAN 1979; DEKA 1975). Furthermore, the survival rate of children with acute lymphoblastic leukemia (ALL) differed markedly depending on the time of maintenance chemotherapy (RIVARD et al. 1985). Thus 80% of the patients dosed with 6-mercaptopurine and methotrexate in the evening were alive and disease free 5 years after disease onset, as compared with 40% of children receiving the same drugs in the morning ($P < 0.001$). Despite the lack of randomization in this study, the magnitude of the dosing-time-related difference is impressive. These findings suggest that residual malignant lymphoblasts might be more susceptible to antimetabolites in the evening than in the morning. Most of the recent clinical trials, however, have been based upon circadian toxicity rhythms.

As a working hypothesis, expected times of least toxicity in human patients were extrapolated from those experimentally demonstrated in mice or rats, by referring them to the respective rest/activity cycle of each species, e.g., with a ~12 h time lag. For instance, least toxicity of 5-FU occurred near 5 HALO in mice and was predicted to correspond to 0400 hours in human subjects resting from 2300 hours to 0700 hours. Similarly, L-OHP, which was best tolerated near 17 HALO in mice, was expected to be least toxic in cancer patients near 1600 hours (Fig. 2).

Circadian rhythms in most biological functions are ~12 h out of phase between mice or rats and human beings, and therefore display a similar phase relationship with the rest-activity cycle. Thus, bone marrow proliferation was highest near 19 HALO in mice and near 1600 hours in healthy human subjects (LÉVI et al. 1988b; SMAALAND et al. 1991a, 1992a); DPDase activity was highest near 4 HALO in mice or rats, and near midnight in human beings (HARRIS et al. 1989, 1990; NAGUIB et al. 1993; ZHANG et al. 1993); even flat infusional 5-FU resulted in a circadian rhythm in its plasma level, with the maximal value in the first half of the rest span, both in mice and in cancer patients (CODACCI-PISANELLI et al. 1995; PETIT et al. 1988; METZGER et al. 1994; FLEMING et al. 1994).

The first clinical trials involved a randomized comparison of toxic effects according to dosing time of a short venous infusion of an anthracycline (doxorubicin or pirarubicin) and cisplatin in women with advanced ovarian cancer. Both trials indicated that bone marrow suppression and renal damage were less by 20% or more in those patients receiving the anthracycline at 0600 hours and cisplatin in the late afternoon as compared to anthracycline injection at 1800 hours and cisplatin administration between 0400 hours and 0800 hours. Such an increase in chemotherapy tolerability was statistically significant, despite it being possible to increase dose intensity (a concept which takes into account both dose per course and interval duration between courses) by ~15% in the least toxic schedules (HRUSHESKY 1985; LÉVI et al. 1990).

Since one of the simplest mathematical models of a circadian rhythm is a sine wave with a 24-h period, this function was used for modulating chemotherapy delivery, as soon as programmable-in-time drug delivery systems appeared. Their use has required the assessment of drug pharmacokinetics during chronomodulated infusion and the establishment of maximum tolerated dose and toxicities (phase I trials), to assess the antitumor activity of this chronomodulated schedule (phase II trial) and to evaluate the clinical relevance of the schedule in a randomized comparison versus flat infusion (phase III trial). Reviews of these topics have recently appeared in the oncologic literature (HRUSHESKY and BJARNASON 1993a,b; BJARNASON 1995; LÉVI et al. 1995a).

I. Phase I Trials of Chronomodulated Chemotherapy

The potential clinical usefulness of chronotherapy was assessed against flat infusion either in randomized trials (FUDR, doxorubicin, L-OHP, 5-FU–FA–L-OHP combination) or through a comparison with well-known data from the current practice of flat delivery (5-FU) (Table 3).

Doxorubicin was delivered either constantly or with peak delivery in the second half of the night (from 0300 to 0700 hours) with an implanted programmable pump (Medtronic, Synchromed, USA) for several weeks until toxicity. Thirteen patients with previously treated metastatic breast cancer (one unevaluable) participated in this randomized crossover study, which suggested a better tolerability of the chronomodulated schedule than for flat infusion. Furthermore, such protracted low-dose infusional therapy (3–7 mg/m^2 per day) displayed good antitumor activity (BAILLEUL et al. 1987). Subsequent investigations were discontinued because of the high cost of these implanted systems and of the risk of severe local damage in the case of doxorubicin extravasation. Synchronized 24-h rhythms in plasma concentrations of doxorubicin and doxorubicinol (the main liver metabolite) were found in two patients receiving the chronomodulated regimen, with peak levels near 0700 hours and trough levels near 1900 hours, suggesting that plasma levels varied in parallel with pump output despite the last half-life of doxorubicin exceeding 24 h and the equilibrium state long having been reached.

FUDR was infused for 14 consecutive days, every 28 days at a flat rate or according to a circadian-shaped four-step delivery schedule (15% of daily dose from 0900 to 1500 hours, 68% from 1500 to 2100 hours, 15% from 2100 to 0300 hours and 2% from 0300 to 0900 hours, with an implanted pump (Medtronic, Synchromed, United States). In a first stage, 30 patients were randomized to receive either treatment modality with an FUDR dose of 0.15 mg/kg per day. Diarrhea, the dose-limiting toxicity, occurred in half of the patients receiving flat infusion as compared to none of those treated with the chronomodulated schedule. In a second stage, involving 39 additional patients, the

Table 3. Summary of phase I and II clinical trials of chronomodulated venous infusions of cancer chemotherapy

Phase	Trial design	Drug	Dose, schedule	Tumor type No. of patients (no. pretreated)	Main conclusions	Reference
I–II	Intrapatient, dose escalation if toxicity < grade 2	5-FU	800–1900 mg/m²/day (peak at 0400 h) × 5 days q 21 days	Colorectal 35 (15)	• Dose-limiting toxicities: mucositis, diarrhea, hand-foot syndrome • Recommended dose: 1400 mg/m²/day × 5 days q 21 day • Objective responses: 3/15 previous tt. (20%) 7/20 naive (35%)	Lévi et al. 1995c
I	Inter- and intrapatient, dose escalation if toxicity < grade 2	5-FU and *l*-FA	600–1100 mg/m²/day 150 mg/m²/day (peak at 0400 h) × 5 days q 21 days	Colorectal 34 (17)	• Dose-limiting toxicities: mucositis, diarrhea • Recommended dose of 5-FU: 900 mg/m²/day × 5 days q 21 day • Objective responses: 3/17 pretreated (18%) 6/16 naive (38%)	Garufi et al., submitted
I	Interpatient, dose escalation	5-FU and *dl*-FA	200–300 mg/m²/day 5–20 mg/m²/day (peak at 0400 h) × 14 days q 28 days	Solid tumors 14 (7)	• Dose-limiting toxicities: mucositis, hand-foot syndrome • MTD: 250 mg/m²/day 5-FU and 20 mg/m²/day FA	Bjarnason et al. 1993
I	Randomized, intrapatient, dose escalation	l-OHP	25–40 mg/m²/day flat vs. chrono (peak at 1600 h) × 5 days q 21 days	Breast or liver 23 (16)	• Dose-limiting toxicities: neutropenia, vomiting, peripheral sensitive neuropathy (cumulative) • Recommended dose: chronomodulated: 35 mg/m²/day × 5 days flat: 25 mg/m²/day × 5 days	Caussanel et al. 1990

Phase	Study design	Drug	Dose/schedule	Cancer (n)	Results	Reference
I	Randomized, intrapatient, dose escalation	FUDR	0.15 mg/kg/day flats vs. chrono. (peak from 1500 to 2100 h) × 14 days q 28 days	Solid tumors 36	• Dose-limiting toxicity: diarrhea: ~0 with chrono. • Dose intensity: + 45% with chrono. • Recommended dose: chronomodulated: 0.23 mg/kg/day × 14 days flat: 0.15 mg/kg/day × 14 days	von Roemeling et al. 1989
I	Randomized, intrapatient, dose escalation	Doxo.	3–7 mg/m²/day flat vs. chrono. (peak from 0300 to 0700 hours) × 6–12 weeks	Breast cancer 13 (13)	• Dose-limiting toxicities: neutropenia, thrombopenia, stomatitis, more frequent in flat than chrono. • Recommended dose: chronomodulated: 4–6 mg/m²/day × 6 weeks flat: 3–4 mg/m² day × 6 weeks	Bailleul et al. 1987
I	Intrapatient, dose escalation of mitox.	Mitox. 5-FU dl-FA	2–2.75 mg/m²/day (peak at 1600 h) 600 mg/m²/day 300 mg/m²/day (peak at 0400 h) × 5 days q 21 days	Breast cancer 18 (17)	• Dose-limiting toxicity: neutropenia • Recommended dose of mitox: 2.75 mg/m² day × 5 days	Deprés-Brummer et al. 1995
I–II	Intrapatient, dose escalation of CBDCA	CBDCA 5-FU dl-FA	40–55 mg/m²/day (peak at 1600 h) 700 mg/m²/day 300 mg/m²/day × 5 days q 21 days	NSC lung cancer 32 (0)	Good quality of life Dose can be further increased Objective responses: 31%	Focan et al. 1995
I	Intrapatient, dose escalation	Interferon-α	15–20 MU/m²/day (peak from 1800 to 2200 h) × 21 days q 28 days	Kidney cancer or melanoma 10 (6)	• Dose-limiting toxicity: fatigue • Recommended dose: 15 MU/m² day × 21 days	Deprés-Brummer et al. 1991
II	Multicenter (France 2, Belgium 1, Italy 1)	5-FU and dl-FA	600–800 mg/m²/day 300 mg/m²/day (peaks at 0400 h) × 5 days q 21 days	Colorectal 36 (17)	• No toxicity > grade 2 • Objective responses: 1/17 pretreated 6/19 naive (35%)	Chollet et al. 1994

Table 3 (*Contd.*)

Phase	Trial design	Drug	Dose, schedule	Tumor type No. of patients (no. pretreated)	Main conclusions	Reference
II	Multicenter (France 3, Belgium 1, Italy 1)	L-OHP	30–40 mg/m^2/days (peak at 1600 h) × 5 days q 21 days	Colorectal 29 (26)	• Median dose: 35 mg/m^2/day × 5 days • Objective responses: 3 (10%)	Lévi et al. 1993b
II	Single institution	5-FU, *dl*-FA, L-OHP	700 mg/m^2/day 300 mg/m^2/day (peaks at 0400 h) 25 mg/m^2/day (peak at 1600 h) × 5 days q 21 days	Colorectal 93 (46)	• Dose-limiting toxicities: diarrhea, vomiting, peripheral sensitive neuropathy (cumulative) • Objective responses: 54 (58%)	Lévi et al. 1992
II	single institution, intrapatient, dose escalation	FUDR	0.15 mg/kg/day (peak from 1500 to 2100 h) × 14 days q 28 days	Kidney cancer 68 (25)	Objective responses: 11/56 assessable (20%)	Hrushesky et al. 1990
II	Single institution	FUDR	0.15 mg/kg/day (peak from 1500 to 2100 h) × 14 days q 28 days	Kidney cancer 42 (0)	Objective responses: 14%	Damascelli et al. 1990
II	Single institution	FUDR	0.15 mg/kg/day (peak from 1500 to 2100 h) × 14 days q 28 days	Kidney cancer 42	Objective responses: 4/40 assessable (10%)	Dexeus et al. 1991
I–II	Bicentric	5-FU i.v. FUDR i.a.	1000–1400 mg/m^2/day (peak at 0400 h) 80–120 mg/m^2/day (peak at 1600 h) × 5 days q 21 days	Colon cancer (liver metastases) 56 (16)	• Chrono. less toxic than flat • Response rate similar: 48% vs. 38% • Marked center-effect for long-term outcome	Focan et al. 1994
I–II	Single institution	5-FU	1200–1400 mg/m^2/day × 5 days q 21 days	Pancreatic cancer 16 (7)	• Good tolerability • Response rate: 21% • Five pts. alive ≥ 2 years	Bertheault-Cvitkovic et al. 1993

maximum tolerated dose of chronomodulated FUDR was defined as being ~45% greater than that with flat infusion (VON ROEMELING and HRUSHESKY 1989).

L-OHP was delivered either at a flat rate or according to 24-h sinusoidal modulation, using an external programmable pump (IntelliJect, Aguettant, Lyon, France) with 25 cancer patients. Neutropenia, nausea or vomiting and peripheral sensitive neuropathy occurred 2–10 times more often in patients receiving flat infusion, despite the dose needing to be reduced by ~30% as compared to chronomodulated delivery (CAUSSANEL et al. 1990). These results were the first which suggested that peripheral sensitive neuropathy might not necessarily prohibit the use of high-dose L-OHP, as was felt then.

5-FU infusion rate was chronomodulated along the 24-h time scale for five consecutive days (every 3 weeks), with peak delivery at 0400 hours and no infusion from 1600 to 2200 hours. 5-FU was administered with a single-reservoir external programmable pump (Autosyringe, Baxter Travenol, Hooksett, United States). Thirty-five patients with metastatic colorectal cancer participated in this phase I-II trial with intrapatient dose escalation according to defined toxicity criteria. As a result of good tolerability ($<8\%$ courses with severe – WHO grade 3 – toxic symptoms), the recommended dose could be escalated up to 1400 mg/m^2 per day or more for 5 days in 80% of assessable patients. This represents a 40%–100% increase in dose or dose intensity as compared to flat infusion, for which the recommended dose ranges from 800 to 1000 mg/m^2 per day for 5 days every 3–4 weeks (LÉVI et al. 1995b).

Three phase I trials have evaluated the tolerability and recommended dose of chronomodulated infusions of the 5-FU and FA combination. In a German study, 75% of the 5-FU was given by flat infusion from 2400 to 0700 hours combined with concurrent low-dose FA infusion for 5 consecutive days. The maximum tolerated dose (MTD) of 5-FU was 600 mg/m^2 per day (2500 mg/m^2 per course). That of FA was 60 mg/m^2 per day (300 mg/m^2 per course). As a result, 5-FU dose intensity was 625 mg/m^2 per week (with low-dose FA). The authors suggested that a sinusoidal shape of 5-FU–FA infusion might be more tolerable and effective (ADLER et al. 1994). In a Canadian study, 5-FU was infused for 14 days, together with low-dose FA. Both drugs were admixed and infused according to a quasi-sinusoidal 24-h rhythmic chronomodulated pattern, with peak flow rate near 0400 hours at night. Each course was followed by 2 weeks treatment-free interval. MTD was 250 mg/m^2 per day of 5-FU (3500 mg/m^2 per course) together with 20 mg/m^2 per day of FA. Theoretical dose intensity over three courses was 875 mg/m^2 per week (with low-dose FA) (BJARNASON et al. 1993).

GARUFI et al. (submitted) conducted a study in order to establish the MTD of 5-FU given as a 5-day chronomodulated infusion in combination with L-folinic acid (L-FA) in ambulatory metastatic colorectal cancer patients in consecutive cohorts of six patients. Both 5-FU and L-FA were infused from 2200 to 1000 hours with peak delivery at 0400 hours, using a multichannel programmable pump. L-FA dose level remained fixed (150 mg/m^2 per day). 5-

FU dose level was escalated by 500 mg/m^2 per course, starting at 3000 mg/m^2. In 34 patients with metastatic colorectal cancer (17 previously treated), MTD of 5-FU was reached at the 4th level (first course at 900 mg/m^2 per day – 4500 mg/m^2 per course). Nine of 33 assessable patients displayed a partial response, with an apparent dose-response relationship. Therefore the latter 5-FU–FA regimen is the chronomodulated schedule which allowed the highest 5-FU dose intensity to be delivered together with high-dose FA.

Interferon-α was administered by chronomodulated venous infusion consisting of six 4-hourly steps with highest flow rate from 1800 to 2200 hours and null rate from 0600 to 1000 hours for 21 days every 4 weeks. Eight of ten patients with metastatic kidney cancer or malignant melanoma tolerated daily doses ranging between 15 and 20 MU/m^2, e.g., ~two to three times those usually administered (DEPRÉS-BRUMMER et al. 1991).

5-FU–FA–L-OHP. A three-drug combination with 5-FU (600 mg/m^2 per day), FA (300 mg/m^2 per day) and L-OHP (20 mg/m^2 per day) was infused for five consecutive days to nine patients with previously untreated metastatic colorectal cancer either at a flat rate or according to a chronomodulated delivery regimen (peak 5-FU–FA at 0400 hours and peak L-OHP at 1600 hours) (Figs. 9, 10). This randomized trial documented that severe mucositis (grade 3 or 4) occurred in three of the four patients receiving the flat regimen as compared to none of the five patients treated with the chronomodulated regimen.

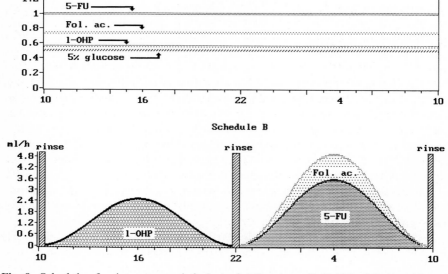

Fig. 9. Schedules for intravenous infusion of 5-fluorouracil (*5-FU*), folinic acid and oxaliplatin (*L-OHP*). Drug delivery was constant (schedule A) or chronomodulated (schedule B) over 24 h. This cycle was repeated automatically for five consecutive days and followed by a 16-day interval

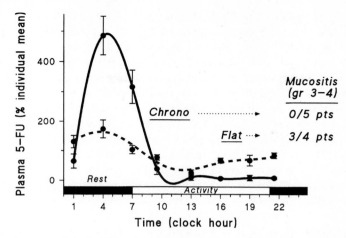

Fig. 10. Imposed amplification of circadian rhythmicity in 5-FU plasma level, through chronomodulation of drug delivery, abolished its oral mucosa severe toxicity. Figure shows mean relative variation of plasma 5-FU along the 24-h time scale during flat or chronomodulated three-drug 5-day infusion in nine patients with metastatic colorectal cancer (5-FU, 600 mg/m^2 per day; folinic acid, 300 mg/m^2 per day; oxaliplatin, 25 mg/m^2 per day). In the chronomodulated schedule, peak delivery was at 0400 hours for 5-FU and FA and at 1600 hours for L-OHP; 24-h mean concentration of plasma 5-FU (designated here as "100%") was 470 ng/ml. An apparent relationship exists between extent of 24 h change in 5-fluorouracil (*5-FU*) plasma concentration and toxicity to the oral mucosa. (After METZGER et al. 1994 and LÉVI et al. 1995b)

In the latter schedule, plasma 5-FU and FA levels varied in parallel with pump output, and mean 5-FU AUCs were similar in both treatment groups (METZGER et al. 1994).

5-FU–FA–Mitoxantrone. This 5-day three-drug regimen administered chronomodulated 5-FU–FA (peak at 0400 hours; respective doses: 600 and 300 mg/m^2 per day) and mitoxantrone (2–2.75 mg/m^2 per day) to women with metastatic breast cancer. In these heavily pretreated patients, tolerability was rather good since only a single sepsis was encountered. Recommended dose of mitoxantrone for phase II assessment was 2.75 mg/m^2 per day (13.85 mg/m^2 per course). This dose is ~30% higher than that used with standard schedules of mitoxantrone–5-FU–FA (DEPRES-BRUMMER et al. 1995).

In summary, phase I trials of chronomodulated delivery have usually involved small groups of patients and showed a potential improvement in drug tolerability, allowing for 30% or greater increase in safe doses.

II. Phase II Trials

Response rate, the proportion of patients in whom tumor size decreased by 50% or more (according to WHO criteria), was assessed in patients receiving chronomodulated regimens against several tumor types, at an advanced or metastatic stage (Table 3). These chronomodulated regimens were administered to patients with metastases from colorectal, pancreatic, lung, breast or kidney cancer. Results confirmed the improved tolerability of chronotherapy that was shown in phase I trials, and often produced an increase in antitumor efficacy by 20% or more as compared to standard chemotherapy schedules. In no instance that we know of was chronotherapy apparently less active than standard schedules of administration of the same drug(s). In addition, the activity of L-OHP against metastatic colorectal cancer was first demonstrated using a chronomodulated delivery scheme (LÉVI et al. 1993a). This was also the case for the activity of FUDR against metastatic renal cell cancer (HRUSHESKY et al. 1990; DAMASCELLI et al. 1990; DEXEUS et al. 1991).

The clinical benefits of these chronomodulated schedules need further phase III multicenter clinical validation before being recommended for clinical practice. This stage of evaluation has been undertaken for chronomodulated FUDR against metastatic renal cell and chronomodulated 5-FU–FA–L-OHP against metastatic colorectal cancer. Whereas the American kidney cancer trial is still ongoing, results of the European colorectal study have just been reported and will be briefly summarized.

III. Chronotherapy of Metastatic Colorectal Cancer with 5-Fluorouracil, Folinic Acid and Oxaliplatin (Chrono-FFL)

5-FU–FA combination chemotherapy constitutes the reference treatment against metastatic colorectal cancer. According to a meta-analysis, 5-FU–FA yields 23% objective responses and a ~12-month median survival in previously untreated patients. These figures appear to be independent of the 5-FU–FA ratio and scheduling, among the several regimens in current clinical use (ADVANCED COLORECTAL CANCER META-ANALYSIS PROJECT 1992). Similar results stem from a seven-arm multicenter trial performed by the Intergroup in the United States, who further suggest that neither 5-FU–FA, nor continuous venous infusion of 5-FU, are greatly superior to standard bolus injection of 5-FU and have called for further ongoing large-scale phase III trials (LEICHMANN et al. 1995).

In a phase II trial of chrono-FFL (Fig. 9, schedule B), a 58% objective response rate was obtained in 93 patients with metastatic colorectal cancer, 46 of whom had received previous chemotherapy. Moreover, all treatments were administered on an outpatient basis and less than 10% of the 784 courses given were associated with severe toxicity (LÉVI et al. 1992). These results compared very favorably with those achieved by standard regimens. Two factors probably accounted for this high antitumor efficacy: the new active drug L-OHP, and chronomodulation, which allowed the safe delivery of high drug doses.

A randomized, multi-institutional trial was then undertaken in patients with previously untreated metastatic colorectal cancer in order to assess the role of chronomodulation. From May 1990 to May 1991, 92 consecutive patients with metastatic colorectal cancer were registered in the first stage of this phase III evaluation. Eight centers participated in this trial: four in France, three in Italy and one in Belgium (LÉVI et al. 1994c). Treatment consisted of a 5-day course of continuous intravenous infusion of 5-FU (600 mg/m^2 per day), FA (300 mg/m^2 per day) and L-OHP (20 mg/m^2 per day), which was repeated every 21 days (after a 16-day interval). Two schedules of drug delivery were compared: 5-FU, FA and L-OHP automatically delivered to outpatients either at a flat rate or at a chronomodulated rate (Fig. 9). Either complex delivery schedule of these three drugs was administered to outpatients with a programmable in-time multichannel ambulatory pump. Per course, stomatitis was by far the most frequent cause of acute dose-limiting toxicity of this protocol. Grade 3 or 4 stomatitis occurred 8.7 times as often on schedule A as on schedule B ($\chi^2=82$; $P<0.0001$). The incidence of severe toxic symptoms (grade 3 or 4) was less than 5% for diarrhea, for nausea or vomiting or for skin toxicity. No hematologic suppression greater than grade 2 was observed.

The proportion of patients experiencing grade 3 or 4 toxicity was fivefold higher in schedule A than in schedule B for stomatitis (89% vs. 18%; $P<0.0001$) and 2.5-fold higher for hand-foot syndrome (11% vs. 4%; NS). It was similar for diarrhea (24% vs. 20). The median dose of 5-FU which was effectively delivered was 700 mg/m^2 per day on schedule B and 500 mg/m^2 per day on schedule A. The median dose intensity of 5-FU was 22% higher in schedule B than in schedule A ($P<0.0001$). Thus response rates of all registered patients were 32% (95% confidence limits: 18%, 46%) for schedule A and 53% (38%, 68%) for schedule B ($\chi^2=4.3$; $P=0.038$).

A risk of partial chemical inactivation of L-OHP with the basic pH of 5-FU in the flat schedule warranted performing a second stage of this multicenter evaluation, where any risk of chemical drug interaction was avoided. One hundred and eighty-six patients were registered in this second stage. The results support the main findings and conclusions above (LÉVI et al. 1994c,d, submitted).

In both of these trials, involving a total of 278 patients with metastatic colorectal cancer, 5-day chronomodulated infusion of 5-FU, FA and L-OHP produced about twice as many objective responses as current chemotherapeutic schedules, or flat three-drug infusion. Furthermore, in this European multicenter randomized setting, the most active chronomodulated schedule was also the least toxic one (LÉVI et al. 1994c,d, 1995c).

Because of the high activity and good tolerability of chrono-FFL, two areas of investigation have been actively explored: (1) can this regimen be further intensified and does this further improve efficacy? (2) Can patients with previously unresectable metastases undergo surgical resection of their metastases after effective chronotherapy, and does this combined approach impact on survival?

Early results and retrospective analyses suggest positive answers to these questions. In two phase II clinical trials, one involving nine European centers, biweekly intensified chrono-FFL produced ~65% objective responses in previously untreated patients with metastatic colorectal cancer (BERTHEAULT-

CVITKOVIC et al., in press; LÉVI et al. 1995d). A retrospective analysis of outcome was performed in 252 patients with previously unresectable metastatic colorectal cancer receiving chrono-FFL at our institute between 1987 and 1993 (55% had received previous chemotherapy). An attempt to resect metastases after effective chrono-FFL was performed in 25% patients. An impact of this combined strategy on long-term outcome was suggested since median projected survival of the whole population was 18 months. This figure is approximately twice that usually obtained in similar patient populations. Combining chrono-FFL with surgery may thus permit some of these patients to be cured (GIACCHETTI et al. 1995).

E. Conclusions and Perspectives

The strategy of extrapolating the least toxic times of chemotherapy from mice to cancer patients has been validated in a clinical phase III trial. More such trials will be needed to firmly establish the role of chronotherapy in medical oncology. This "group chronotherapy," where all patients receive the same chronomodulated chemotherapy regimen, relies on the fact that groups of cancer patients who enter clinical trials do exhibit significant circadian rhythms in almost every variable investigated. Nevertheless, investigations of rhythms in individual patients and/or in subgroups of cancer patients with very advanced disease and/or in poor general condition (with a so-called poor performance status, according to the World Health Organization) have shown marked alterations in circadian rhythms (BAILLEUL et al. 1986; HRUSHESKY et al. 1983; BENAVIDES 1991; SMAALAND et al. 1992; TOUITOU et al. 1995). These alterations may consist in amplitude decrease and/or acrophase shift and/or circadian period suppression. The incidence of these rhythm alterations, as well as their significance and relevance for the outcome of patients receiving chronotherapy, is totally unknown at present, but is an area under active investigation.

Thus, ongoing or future developments of cancer chronotherapy involve (1) the examination of the respective roles of chronomodulation vs. circadian peak time of drug delivery in groups of patients, (2) the further assessment of the possible role of intervals and sequences between drugs in chronomodulated regimens, (3) the assessment of the relevance of a "normal" circadian system for a favorable outcome after chronotherapy, (4) the further evaluation of the relevance of this approach for improving survival, both in metastatic disease and in adjuvant situations, and (5) the testing of the relevance of this approach for cytokines, hormones or other new drugs. Finally other time periods for chronomodulated drug delivery will deserve to be explored.

References

Adler S, Lang S, Langenmayer I et al (1994) Chronotherapy with 5-fluoruracil and folinic acid in advanced colorectal carcinoma. Results of a chronopharmacologic phase I trial. Cancer 73:2905–2912

Advanced Colorectal Cancer Meta-Analysis Project (Piedbois P et al) (1992) Modulation of fluorouracil by leucovorin in patients with advanced colorectal cancer: evidence in terms of response rate. J Clin Oncol 10:896–903

Aherne GW, English J, Burton N et al (1987) Chronopharmacokinetics and their relationship to toxicity and effect with reference to methotrexate, 6-mercaptopurine and morphine. Proceedings of the satellite symposia of the 4th European Conference of Clinical Oncology (ECCO-4), Madrid 1–4 Nov 1987, p 40

Badran AF, Echave Llanos JM (1965) Persistence of mitotic circadian rhythm of a transplantable mammary carcinoma after 35 generations: its bearing on the success of treatment with endoxan. J Natl Cancer Inst 35:285–290

Bailleul F, Lévi F, Reinberg A, Mathe G (1986) Interindividual differences in the circadian hematologic time structure of cancer patients. Chronobiol Int 3:47–54

Bailleul F, Lévi F, Metzger G et al (1987) Chronotherapy of advanced breast cancer with continuous doxorubicin infusion (C DOX I) via an implantable programmable device. Proc Am Assoc Cancer Res 28:194 (abstract 771)

Balis FM, Jeffries SL, Lange B et al (1989) Chronopharmacokinetics of oral methotrexate and 6-mercaptopurine: is there diurnal variation in the disposition of antileukemic therapy? Am J Pediatr Hematol Oncol 11:324–326

Belanger PM, Labrecque G (1992) Biological rhythms in hepatic drug metabolism and biliary systems. In: Touitou Y, Haus E (eds) Biologic rhythms in clinical and laboratory medicine. Springer, Berlin Heidelberg New York, pp 403–409

Benavides M (1991) Cancer avancé de l'ovaire: approche chronobiologique comme nouvelle stratégie du traitement et de la surveillance clinique et biologique. Thèse de doctorat en cancérologie, Université Paris XI

Bertheault-Cvitkovic F, Lévi F, Soussan S et al (1993) Circadian-rhythm modulated chemotherapy with high-dose 5-fluorouracil: a pilot study in patients with pancreatic adenocarcinoma. Eur J Cancer 29A (13):1851–1854

Bertheault-Cvitkovic F, Jami A, Ithzaki M et al (1996) Biweekly dose intensification of circadian chronotherapy with 5-fluorouracil, folinic acid and oxaliplatin in patients with metastatic colorectal cancer. J Clin Oncol 14 (11) (in press)

Bjarnason GA (1995) Clinical cancer chronotherapy trials: a review. J Infus Chemother 5:29–37

Bjarnason GA, Kerr IG, Doyle N et al (1993) Phase I study of 5-fluorouracil and leucovorin by a 14-day circadian infusion in metastatic adenocarcinoma patients. Cancer Chemother Pharmacol 33:221–228

Boughattas NA, Lévi F, Hecquet B et al (1988) Circadian time dependence of murine tolerance for carboplatin. Toxicol Appl Pharmacol 96:233–247

Boughattas N, Lévi F, Fournier C et al (1989) Circadian rhythm in toxicities and tissue uptake of 1,2-diaminocyclohexane (trans-1) oxalatoplatinum (II) in mice. Cancer Res 49:3362–3368

Boughattas N, Lévi F, Fournier C et al (1990) Stable circadian mechanisms of toxicity of two platinum analogs (cisplatin and carboplatin) despite repeated dosages in mice. J Pharmacol Exp Ther 255:672–679

Boughattas NA, Hecquet H, Fournier C et al (1994) Comparative pharmacokinetics of oxaliplatin (L-OHP) and carboplatin (CBDCA) in mice with reference to circadian dosing time. Biopharm Drug Dispos 15 (864):1–13

Buchi KN, Moore JG, Hrushesky WJM et al (1991) Circadian rhythm of cellular proliferation in the human rectal mucosa. Gastroenterology 101:410–415

Burns ER, Beland SS (1984) Effect of biological time on the determination of the LD_{50} of 5-fluorouracil in mice. Pharmacology 28:296–300

Cambar J, L'Azou B, Cal JC (1992) Chronotoxicology. In: Touitou Y, Haus E (eds) Biologic rhythms in clinical and laboratory medicine. Springer, Berlin Heidelberg New York, pp 138–150

Canal P, Sqalli A, De Forni M et al (1991) Chronopharmacokinetics of doxorubicin in patients with breast cancer. Eur J Clin Pharmacol 40:287–291

Caussanel JP, Lévi F, Brienza S et al (1990) Phase I trial of 5-day continuous infusion of oxaliplatinum at circadian-modulated vs constant rate. J Natl Cancer Inst 82:1046–1050

Chollet P, Cure H, Garufi C et al Phase II trial with chronomodulated 5-fluorouracil (5-FU) and folinic acid (FA) in metastatic colorectal cancer. Proceedings of the 6th international conference on chronopharmacology and chronotherapy, Amelia Island, 5–9 July 1994, abstract VIIIb-4

Codacci-Pisanelli G, van der Wilt C, Pinedo H et al (1995) Antitumor activity, toxicity and inhibition of thymidylate synthase of prolonged administration of 5-fluorouracil in mice. Eur J Cancer 31A:1517—1525

Damascelli B, Marchiano A, Spreafico C et al (1990) Circadian continuous chemotherapy of renal cell carcinoma with an implantable, programmable infusion pump. Cancer 66:237–241

Deka AC (1975) Application of chronobiology to radiotherapy to tumours of the oral cavity. MD thesis, Postgraduate Institute of Medical Education and Research, Chandigarh, India

Deprés-Brummer P, Lévi F, Di Palma M et al (1991) A phase I trial of 21-day continuous venous infusion of alpha-interferon at circadian rhythm modulated rate in cancer patients. J Immunother 10:440–447

Deprés-Brummer P, Bertheault-Cvitkovic F, Lévi F et al (1995) Circadian rhythm modulated (CRM) chemotherapy of metastatic breast cancer with mitoxantrone, 5-fluorouracil and folinic acid: preliminary results of a phase I trial. J Infus Chemother. 5 (3, suppl 1): 144–147

De Vita VT, Hellman S, Rosenberg SA (1993) Cancer. Principles and practice of oncology. Lippincott, Philadelphia

Dexeus FH, Logothetis CJ, Sella A et al (1991) Circadian infusion of floxuridine in patients with metastatic renal cell carcinoma. J Urol 146:709–713

Eksborg S, Stendahl U, Antila K (1989) Pharmacokinetics of 4'epiadriamycin after morning and afternoon intravenous administration. Med Oncol Tumor Pharmacother 6:195–197

El Kouni MH, Naguib FNM, Park KS et al (1990) Circadian rhythm of hepatic uridine phosphorylase activity and plasma concentration of uridine in mice. Biochem Pharmacol 40:2479–2485

English J, Marks V (1981) Diurnal variation in methylprednisolone metabolism in the rat. IRSC Med Sci 9:721

English J, Aherne GW, Marks V (1982) The effect of timing of a single injection on the toxicity of methotrexate in the rat. Cancer Chemother Pharmacol 9:114–117

Fleming GF, Schilsky RL, Mick R et al (1994) Circadian variation of 5-fluorouracil (5-FU) and cortisol plasma levels during continuous-infusion 5-FU and leucovorin (LV) in patients with hepatic or renal dysfunction. Proc Am Soc Clin Oncol 13:352 (abstract)

Focan C (1979) Sequential chemotherapy and circadian rhythm in human solid tumors. Cancer Chemother Pharmacol 3:197–202

Focan C (1985) Le rythme nycthéméral de la prolifération tumorale. Aspects expérimentaux et cliniques. Implication pour la chimiothérapie oncolytique. Thèse de doctorat en sciences biomédicales expérimentales, Université de Liège

Focan C, Doalto L, Mazy V et al (1989) Vindesine en perfusion continue de 48 heures (suivie de cisplatine) dans le cancer pulmonaire avancé. Données chronopharmacocinétiques et efficacité clinique. Bull Cancer (Paris) 76:909–912

Focan C, Denis B, Kreutz F et al (1994) Ambulatory chronotherapy with 5-fluorouracil (5-FU), folinic acid (F) and carboplatin for advanced non small cell lung cancer (NSCLC). Proc Eur Soc Med Oncol Ann Oncol 1994 5 [Suppl 8]:P760 (abstract)

Focan C, Denis B, Kreutz F et al (1995) Ambulatory chronotherapy with 5-fluorouracil, folinic acid and carboplatin for advanced non small cell lung cancer. A phase II feasibility trial. J Infus Chemother 5(Suppl 1):148–152

Garcia-Sainz M, Halberg F (1966) Mitotic rhythms in human cancer reevaluated by electronic computer programs. Evidence for chronopathology. J Natl Cancer Inst 37:279–292

Garufi C, Lévi F, Aschelter AM et al A Phase I trial of five day chronomodulated infusion of 5-fluorouracil and l-folinic acid in patients with metastatic colorectal cancer (submitted for publication)

Gautherie M, Gros C (1974) Circadian rhythm alteration of skin temperature in breast cancer. Chronobiologia 4:1–17

Giacchetti S, Gruia G, Itzhaki M et al (1995) Surgery after chronomodulated chemotherapy with 5-fluorouracil (5-FU), folinic acid (FA) and oxaliplatin (L-OHP) (chrono) allows long term survival of patients (pts) with unresectable colorectal liver metastases. Proc Am Assoc Clin Oncol 492 (abstract)

Gonzalez RB, Sothern RB, Thatcher G et al (1989) Substantial difference in timing of murine circadian susceptibility to 5-fluorouracil and FUDR. Proc Am Assoc Cancer Res 30:A2452 (abstract)

Halberg F, Haus E, Cardoso SS et al (1973) Toward a chronotherapy of neoplasia: tolerance of treatment depends upon host rhythms. Experientia 29:909–934

Halberg F, Gupta BD, Haus E, Halberg E, Deka AC, Nelson W, Sothern RB, Cornelissen G, Klee J, Lakatua DJ, Scheving LE, Burns ER (1977) Steps toward a cancer chronopolytherapy. Proc. XIVè Congrès International de Thérapeutique, 8–10 Septembre 1977, Montpellier, France, pp 151–196

Halberg F, Nelson W, Lévi F et al (1980) Chronotherapy of rat mammary cancer. Int J Chronobiol 7:85–90

Harris B, Song R, Soong S et al (1989) Circadian variation of 5-fluorouracil catabolism in isolated perfused rat. Cancer Res 49:6610–6614

Harris B, Song R, Soong S et al (1990) Relationship between dihydropyrimidine dehydrogenase activity and plasma 5-fluorouracil levels:evidence for circadian variation of plasma drug levels in cancer patients receiving 5-fluorouracil by protracted continuous infusion. Cancer Res 50:197–201

Hassan M, Oberg G, Bekassy AN et al (1991) Pharmacokinetics of high-dose busulphan in relation to age and chronopharmacology. Cancer Chemother Pharmacol 28:130–134

Haus E, Halberg F, Scheving L et al (1972) Increased tolerance of leukemic mice to arabinosylcytosine with schedule-adjusted to circadian system. Science 177:80–82

Hecquet B, Sucche M (1986) Theoretical study of the influence of the circadian rhythm of plasma protein binding in cisplatin area under the curve. J Pharmacokinet Biopharmacol 14:79–93

Hecquet B, Meynadier J, Bonneterre J et al (1985) Time dependency in plasmatic protein binding of cisplatin. Cancer Treat Rep 69:79–83

Hrushesky W (1985) Circadian timing of cancer chemotherapy. Science 228:73–75

Hrushesky WJ, Lévi F, Halberg F et al (1982a) Circadian stage dependence of cis-diamminedichloroplatinum lethal toxicity in rats. Cancer Res 42:945–949

Hrushesky WJM, Borch R, Lévi F (1982b) Circadian time dependence of cisplatin urinary kinetics. Clin Pharmacol Ther 32:330–339

Hrushesky WJ, Haus E, Lakatua F et al (1983) Marker rhythms for cancer chronochemotherapy. In: Haus E, Kabat HF (eds) Chronobiology 1982–1983. Karger, Basel, pp 493–499

Hrushesky W, von Roemeling R, Lanning RM et al (1990) Circadian-shaped infusions of floxuridine for progressive metastatic renal cell carcinoma. J Clin Oncol 8:1504–1513

Hrushesky WJM, Bjarnason GA (1993a) The application of circadian chronobiology to cancer chemotherapy. In: DeVita VT, Hellman S, Rosenberg SA (eds) Cancer. Principles and practice on oncology. Lippincott, Philadelphia, pp 2666–2686

Hrushesky WJM, Bjarnason GA (1993b) Circadian cancer therapy. J Clin Oncol 11:1403–1417

Hryniuk WM (1988) The importance of dose intensity in the outcome of chemotherapy. In: De Vita VT, Holman S, Rosenberg SA (eds) Important advances in oncology. Lippincott, Philadelphia, pp 121–142

Kerr DJ, Lewis C, O'Neil B et al (1990) The myelotoxicity of carboplatin is influenced by the time of its administration. Hematol Oncol 8:59–63

Killman SA, Cronkite ZEP, Fliedner TM et al (1962) Mitotic indices of human bone marrow cells. 1. Number and cytologic distribution of mitoses. Blood 19:743–750

Klevecz R, Braly P (1987) Circadian and ultradian rhythms of proliferation in human ovarian cancer. Chronobiol Int 4 (4):513–523

Klevecz R, Shymko R, Braly P (1987) Circadian gating of S phase in human ovarian cancer. Cancer Res 47:6267–6271

Kodama M, Kodama T (1982) Influence of corticosteroid hormones on the therapeutic efficacy of cyclosphosphamide. Gann 73:661–666

Koren G, Langevin AM, Olivieri N et al (1990) Diurnal variation in the pharmacokinetics and myelotoxicity of mercaptopurine in children with acute lymphocytic leukemia. Am J Dis Child 144:1135–1137

Koren G, Ferrazini G, Sohl H et al (1992) Chronopharmacology of methotrexate pharmacokinetics in childhood leukemia. Chronobiol Int 9:434–438

Langevin AM, Koren G, Soldin SJ et al (1987) Pharmacokinetic case for giving 6-mercaptopurine maintenance doses at night. Lancet 2:505–506

Leichman CG, Fleming TR, Muggia FM et al (1995) Phase II study of fluorouracil and its modulation in advanced colorectal cancer: a Southwest Oncology Group Study. J Clin Oncol 13:1303–1311

Lévi F (1995) Chronochimiothérapie et intensité de dose. Bull Cancer 82 [Suppl 1]:29s–35s

Lévi F, Hrushesky W, Blomquist CH et al (1982a) Reduction of cis-diaminedichloroplatinum nephrotoxicity in rats by optimal circadian drug timing. Cancer Res 42:950–955

Lévi F, Hrushesky WJM, Borch RF et al (1982b) Cisplatin urinary pharmacokinetics and nephrotoxicity: a common circadian mechanism. Cancer Treat Rep 66:1933–1938

Lévi F, Mechkouri M, Roulon A et al (1985) Circadian rhythm in tolerance of mice for the new anthracycline analog 4'-0-tetrahydropyranyl-adriamycin (THP). Eur J Cancer Clin Oncol 21:1245–1251

Lévi F, Metzger G, Bailleul F et al (1986) Circadian-varying plasma pharmacokinetics of doxorubicin (DOX) despite continuous infusion at constant rate. Proc Am Assoc Cancer Res 27:693

Lévi F, Boughattas NA, Blazsek I (1988a) Comparative chronotoxicity of anticancer agents and related mechanisms. Annu Rev Chronopharmacol 4:283–331

Lévi F, Blazsek I, Ferle-Vidovic A (1988b) Circadian and seasonal rhythms in murine bone marrow colony-forming cells affect tolerance for the anticancer agent 4'-0-tetrahydropyranyladriamycin (THP). Exp Hematol 16:696–701

Lévi F, Benavides M, Chevelle C et al (1990) Chemotherapy of advanced ovarian cancer with 4'-0-tetrahydropyranyl adriamycin (THP) and cisplatin: a phase II trial with an evaluation of circadian timing and dose intensity. J Clin Oncol 8:705–714

Lévi F, Misset JL, Brienza S et al (1992) A chronopharmacologic phase II clinical trial with 5-fluorouracil, folinic acid and oxaliplatin using an ambulatory multichannel programmable pump. High antitumor effectiveness against metastatic colorectal cancer. Cancer 69:893–900

Lévi F, Brienza S, Metzger G et al (1993a) Implications of chronobiology for 5-fluorouracil (5-FU) efficacy. In: Rustum YM (ed) Novel approaches to selective treat-

ments of human solid tumors: laboratory and clinical correlation. Plenum, New York, pp 169–186
Lévi F, Perpoint B, Garufi C et al (1993b) Oxaliplatin activity against metastatic colorectal cancer. A phase II study of 5-day continuous venous infusion at circadian rhythm modulated rate. Eur J Cancer 29 (9):1280–1284
Lévi F, Bourin P, Deprés-Brummer P et al (1994a) Chronobiology of the immune system: implication for the delivery of therapeutic agents. Clin Immunother 2 (1):53–64
Lévi F, Tampellini M, Metzger G et al (1994b) Circadian changes in mitoxantrone toxicity in mice:relationship with plasma pharmacokinetics. Int J Cancer 59:543–547
Lévi F, Zidani R, Vannetzel JM et al (1994c) Chronomodulated versus fixed infusion rate delivery of ambulatory chemotherapy with oxaliplatin, 5-fluorouracil and folinic acid in patients with colorectal cancer metastases. A randomized multi-institutional trial. J Natl Cancer Inst 86:1608–1617
Lévi F, Zidani R, Dipalma M, International Organization for Cancer Chronotherapy (IOCC) (1994d) Improved therapeutic index through ambulatory circadian rhythmic delivery (CRD) of high dose 3-drug chemotherapy in a randomized phase III multicenter trial. Proc Am Soc Clin Oncol 13: 574 (abstract)
Lévi F, Giacchetti S, Adam R, International Organization for Cancer Chronotherapy (1995a) Chronomodulation of chemotherapy against metastatic colorectal cancer. Eur J Cancer 31A:1264–1270
Lévi F, Soussan A, Adam R et al (1995b) A Phase I-II trial of five-day continuous intravenous infusion of 5-fluorouracil delivered at circadian rhythm modulated rate in patients with metastatic colorectal cancer. J Infus Chemother 5(Suppl 1):153–158
Lévi F, Zidani R, Vannetzel JM et al (1995c) Chronomodulated versus flat infusion of 5-fluorouracil (5-FU), folinic acid (FA) and oxaliplatin (L-OHP) against metastatic colorectal cancer (MCC) in 2 consecutive European randomized multicenter trials (T). Proc Am Assoc Clin Oncol 493 (abstract)
Lévi F, Dogliotti L, Perpoint B et al (1995d) Ambulatory intensified 4-day (d) every 2 weeks chronotherapy with oxaliplatin (L-OHP), 5-fluorouracil (5-FU) and folinic acid (FA) in patients (pts) with metastatic colorectal cancer (MCC). Proc Am Soc Clin Oncol 568 (abstract)
Li XM, Metzger G, Filipski E et al Pharmacologic modulation of reduced glutathione (GSH) circadian rhythms by buthionine sulfoximine (BSO): relationship with cisplatin (CDDP) toxicity in mice (submitted for publication)
Malmary-Nebot M, Labat C, Casanovas A et al (1985) Aspect chronobiologique de l'action du methotréxate sur la dihydrofolate réductase. Ann Pharm Fr 43:337–343
Martineau-Pivoteau N, Lévi F, Cussac C (in press) Circadian rhythm in toxic effects of cystemustine and its relationship with alkytransferase activity. Int J Cancer
Mauer AM (1965) Diurnal variation of proliferative activity in the human bone marrow. Blood 26:1–7
Metzger G, Massari C, Etienne MC et al (1994) Spontaneous or imposed circadian changes in plasma concentrations of 5-fluorouracil coadministered with folinic acid and oxaliplatin: relationship with mucosal toxicity in cancer patients. Clin Pharmacol Ther 56:190–201
Mormont C, Boughattas N, Lévi F et al (1989) Mechanisms of circadian rhythms in the toxicity and the efficacy of anticancer drugs: relevance for the development of new analogs. In: Lemmer B (ed) Chronopharmacology: cellular and biochemical interactions. Dekker, New York, pp 395–437
Naguib FNM, Soong SJ, El Kouni MH (1993) Circadian rhythm of orotate phosphoribosyltransferase, pyrimidine nucleoside phosphorylases and dihydrouracil dehydrogenase in mouse liver. Biochem Pharmacol 45:667–673
Nowakowska-Dulawa E (1990) Circadian rhythm in 5-fluorouracil (FU) pharmacokinetics and tolerance. Chronobiologia 17:27–35

Perpoint B, Le Bousse-Kerdiles C, Clay D et al (1995) In vitro chronopharmacology of recombinant mouse IL-3, mouse GM-CSF and human G-CSF on murine myeloid progenitor cells. Exp Hematol 23:362–368

Peters GJ, Van Dijk J, Nadal JC, Van Groeningen CS, Lankelma J, Pinedo HM (1987) Diurnal variation in the therapeutic efficacy of 5-fluorouracil against murine colon cancer. In Vivo 1:113–118

Petit E, Milano G, Lévi F et al (1988) Circadian varying plasma concentration of 5-FU during 5-day continuous venous infusion at constant rate in cancer patients. Cancer Res 48:1676–1679

Rivard G, Infante-Rivard C, Hoyeux C et al (1985) Maintenance chemotherapy for childhood acute lymphoblastic leukemia: better in the evening. Lancet i:1264–1266

Robinson BA, Begg EJ, Colls BM et al (1989) Circadian pharmacokinetics of methotrexate. Cancer Chemother Pharmacol 24:397–399

Sabbag R, Sunderland M, Leclerc JM et al (1993) Chronopharmacokinetic of a 48-hour continuous infusion of doxorubicin to children with acute lymphoblastic leukemia. Proceedings of the XXIst conference of the International Society of Chronobiology, abstract XI-5

Scheving LE, Haus E, Kuhl JFW et al (1976) Different laboratories closely reproduce characteristics of circadian rhythm in tolerance of mice for arabinofuranosylcytosine. Cancer Res 36:1133–1137

Scheving LE, Burns ER, Pauly JE et al (1977) Survival and cure of leukemic mice after circadian optimization of treatment with cyclophosphamide and 1-β-arabinofuranosylcytosine. Cancer Res 37:3648–3655

Scheving LE, Burns ER, Halberg F et al (1980a) Combined chronochemotherapy of L1210 leukemic mice using β-D-arabinofuranosylcytosine, cyclophosphamide, vincristine, methylprednisolone and cis-diaminedichloroplatinum. Chronobiologia 17 (1):33–40

Scheving LE, Burns ER, Pauly JE, Halberg F (1980b) Circadian bioperiodic response of mice bearing advanced L1210 leukemia to combination therapy with adriamycin and cyclophosphamide. Cancer Res 40:1511–1515

Smaaland R, Laerum OD, Lote K et al (1991a) DNA synthesis in human bone marrow is circadian stage dependent. Blood 77:2603–2611

Smaaland R, Svardal AM, Lote K et al (1991b) Glutathione content in human bone marrow and circadian stage relation to DNA synthesis. J Natl Cancer Inst 83:1092–1098

Smaaland R, Laerum OD, Sothern RB et al (1992a) Colony-forming unit-granulocytemacrophage and DNA synthesis of human bone marrow are circadian stage-dependent and show covariation. Blood 79:2281–2287

Smaaland R, Abrahamsen JF, Svardal AM (1992b) DNA cell cycle distribution and glutathione (GSH) content according to circadian stage in bone marrow of cancer patients. Br J Cancer 66:39–45

Smaaland R, Lote K, Sothern RB et al (1993) DNA synthesis and ploidy in non-Hodgkin's lymphomas demonstrate variation depending on circadian stage of cell sampling. Cancer Res 53:3129–3138

Sothern R, Lévi F, Haus E et al (1989) Control of a murine plasmacytoma with doxorubicin-cisplatin: dependence on circadian stage of treatment. J Natl Cancer Inst 81:135–145

Squalli A, Oustrin J, Houin G et al (1989) Clinical chronopharmacokinetics of doxorubicin (DXR). Annu Rev Chronopharmacol 5:393–396

Stoll BA, Burch W (1968) Surface detection of circadian rhythm in 32p content of cancer of the breast. Cancer 21:193–196

Tahti E (1956) Studies of the effect of X-irradiation on 24 hour variations in the mitotic activity in human malignant tumors. Acta Pathol Microbiol Scand 117:1–61

Tampellini M, Filipski E, Vrignaud P et al (1995) Circadian dosing time dependency of docetaxel toxicity in mice: implication for antitumor efficacy. Proc Am Assoc Cancer Res 36:2713 (abstract)

Thiberville L, Compagnon P, Moore N et al (1992) Accumulation plasmatique du 5-fluorouracile chez l'insuffisant respiratoire. Rev Mal Respir 111 (abstract)

Touitou Y, Lévi F, Bogdan A et al (1995) Circadian desynchronization of blood variables in patients with metastatic breast cancer. Role of prognostic factors. J Cancer Res Clin Oncol 121:181–188

Vassal G, Challine D, Koscielny S et al (1993) Chronopharmacology of high-dose busulfan in children and its relationship with liver toxicity. Cancer Res 53:1534–1537

Von Roemeling R, Hrushesky W (1989) Circadian patterning of continuous floxouridine infusion reduces toxicity and allows higher dose intensity in patients with widespread cancer. J. Clin. Oncol., 7:1710–1719

Von Roemeling R, Hrushesky WJM (1990) Determination of the therapeutic index of floxuridine by its circadian infusion pattern. J Natl Cancer Inst 82:386–393

Voutilainen A (1953) Über die 24-Stunden-Rhythmik der Mitozfrequenz in malignen Tumoren. Acta Pathol Microbiol Scand [Suppl] 99:1–104

Warnakulasuriya KAAS, MacDonald DG (1993) Diurnal variation in labelling index in human buccal epithelium. Arch Oral Biol 38 (12):1107–1111

Zhang R, Lu Z, Liu T et al (1993) Relationship between circadian-dependent toxicity of 5-fluorodeoxyuridine and circadian rhythms of pyrimidine enzymes: possible relevance to fluoropyrimidine chemotherapy. Cancer Res 53:2816–2822

CHAPTER 12
The Endocrine System and Diabetes

L. MEJEAN, A. STRICKER-KRONGRAD, and A. LLUCH

A. Introduction

Diabetes mellitus therapy, in which the treatment must maintain a moment-to-moment glycaemic balance, is a major domain in which the principles of chronopharmacology must be included in the therapeutic regime. The therapeutic action, via insulin treatment in the case of insulin-dependent diabetes, or oral sulfamides or biguanides in the case of non-insulin-dependent diabetes, must take into account the physiological variations in induced blood glucose levels in order to avoid hyper or hypoglycaemia.

The bases of this chronopharmacology are the following : firstly blood glucose and plasma insulin levels show their own physiological periodicity, which must be described as precisely as possible so that the periodicity can be reproduced by the therapeutic regime. Secondly, the chronobiology of blood glucose and plasma insulin levels are dependent on other factors that also show a periodicity, such as eating behaviour, which superimpose their own circadian variations. Thirdly, they are dependent on external phenomena, such as responses to stress, social pressure or sleep, which are under the control of neuroendocrinological factors which themselves show a well-known periodicity.

One major physiopathological fact to keep in mind is that the absence or reduction in insulin control observed in diabetes produces pathological modifications related to the three factors described above. These modifications may themselves counteract the periodicity of blood glucose. For example, retinal degenerative processes observed in association with diabetes are related to growth hormone secretion and may counteract the chronopharmacological aspects of the treatment. Furthermore, blood glucose levels are under the control of several hormones, i.e. insulin and the counterregulatory hormones. The circulating levels and metabolic action of these hormones undergo different circadian variations (APARICIO et al. 1974; HAUTECOUVERTURE et al. 1975; HOLST et al. 1983; LEFEBVRE et al. 1987; WALDHAUST 1989). These circadian oscillations may influence the hormonal action on glucose uptake by peripheral tissues, glycogenolysis and gluconeogenesis. The exogenous supply of glucose depends on food intake, which is known to be regulated by pre- and peri-ingestive factors. These factors inhibit or stimulate motivational states such as hunger, satiety, satiation and appetite (NICOLAIDIS and BURLET 1988).

These states are under the control of central systems among which the hypothalamus is strongly implicated (LEIBOWITZ 1988). Among the numerous neurotransmitters that play a major role within the hypothalamus, research, first focused on classical aminergic transmission, has more recently widened to include neuropeptidergic transmission. These peptides are located within hypothalamic sites known to be implicated in the genesis of biological rhythms, in the suprachiasmatic nucleus, the lateral hypothalamus and the ventromedial nucleus (BLOCK and PAGE 1978) and in the hypothalamo–hypophyseal axis (LEIBOWITZ 1988). Among them, orexigenic peptides, such as neuropeptide Y (LEVINE and MORLEY 1984), or anorectic peptides such as corticotropin releasing hormone (CRF), neurotensin, vasopressin and oxytocin (BECK 1992), or endogenous opioids (FANTINO 1986), have been most widely investigated.

This interrelationship between the chronobiology of food intake and the treatment of diabetes mellitus was suspected for a long time (MOLLERSTROM 1928, 1953). Previous studies on chronobiological aspects of nutrition (DEBRY et al. 1977) and on the chronobiology of glucose tolerance (HAUTECOUVERTURE et al. 1975) have demonstrated that it shows a high degree of complexity.

The present review will look briefly at recent developments in the chronobiology of glucose levels, plasma insulin levels, and food intake and will explore their potential consequences for the chronopharmacological approach in the treatment of insulin-dependent (type I) diabetes mellitus.

B. Chronobiology of Blood Glucose Levels

Blood glucose levels are a common marker of carbohydrate metabolism and they are influenced by multiple metabolic phenomena. Circadian modulation of the human endocrine function is well established for systems under hypothalamopituitary control. Human studies have suggested that the set point of glucose regulation is under circadian control, but this is still a matter of controversy. Circadian variations were found by FREINKEL et al. (1968), FLOYD et al. (1974) and MEJEAN et al. (1988), while FAIMAN and MOORHOUSE (1967), SCHLIERF and RAETZER (1972), DESCHAMPS et al. (1969), HAUTECOUVERTURE et al. (1974) and MALHERBE et al. (1969) failed to demonstrate consistent circadian modulations.

Although the use of different experimental designs may explain these discrepancies, it is more likely that the disagreements originate in both detection and quantification of the blood glucose level.

Low amplitudes in the circadian rhythm of blood glucose levels were observed in heathy adults of both sexes by SWOYER et al. (1984): amplitude and acrophase were found to be 4.1 ± 1.6% of the mesor and 6.3 ± 1.5 h, respectively. This circadian rhythm could not be detected by MEJEAN et al. (1988), probably because of the small number of subjects. However, both cosinor (according to NELSON et al. 1979) and spectral (according to DE PRINS and MALBECQ 1983) analysis revealed an ultradian rhythm of blood glucose with a 6-h period (amplitude of 13% and acrophase at 2.6 h). Analysis of individual circadian and ultradian rhythms in blood glucose levels indicated that the multiplicity of peaks over the 24-h time scale contributed to the obliteration of a common circadian rhythm. In this study, the four peaks detected over 24 h coincided with the three main meals (08:45 h, 12:30 h and 19:30 h, with a duration of 15, 30 and 30 min, respectively) and with the so-called dawn phenomenon (around 05:00 h). Interestingly, the only two subjects with a prominent circadian rhythm of blood glucose presented this typical spontaneous rise in the early morning.

C. Chronobiology of Glucose Tolerance

The ultradian rhythmicity observed in healthy volunteers suggests a circadian modulation of glucose tolerance in relation to carbohydrate intake. Changes in blood glucose and insulin in response to an oral glucose load are reduced in the afternoon when compared with the morning (JARRETT et al. 1972). This phenomenon is not influenced by the age or sex of the subject (JARRETT and KEEN 1969, 1970; ZIMMER et al. 1974), nor by the duration of fasting before the test (BOWEN and REEVES 1967; JARRETT and KEEN 1969), the last meal composition (CAROLL and NESTEL 1973), nor the administered glucose load (BEN DYKE 1971; JARRETT et al. 1972).

Moreover, digestive absorption does not influence the circadian modulation of glucose tolerance since the same results are found in intravenous glucose tolerance tests (ABRAMS et al. 1974). A circadian modulation of glucose tolerance is also observed in the hypoglycaemic effect of intravenous tolbutamide, this effect being lower during the afternoon (BAKER and JARRETT 1972).

However, SIMON et al. (1988) did not find comparable blood glucose variations in their subjects when studying the so-called dawn phenomenon in normal subjects.

D. Chronobiology of Plasma Insulin

There is a general agreement about the time-dependent variations of plasma insulin levels. Studies by FREINKEL et al. (1968), FLOYD et al. (1974), MALHERBE et al. (1969), HAUTECOUVERTURE et al. (1974), AHLERSOVA et al. (1992), BERMAN et al. (1993), LEFEBVRE et al. (1987) and OPARA and GO (1991) all agree in regard to a circadian variation of plasma insulin both in man and animals. SWOYER et al. (1984) detected an insulin circadian rhythmicity in 33 healthy males and 25 females with an acrophase at 18.04 ± 01.32 h in males and 16.24 ± 02.16 h in females. Exploration of plasma insulin level variations in five healthy subjects by MEJEAN et al. (1988) revealed not only a circadian rhythmicity (mesor of 33.1 ± 3.3 µU/ml, amplitude of 13.4 µU/ml and acrophase at 14.22 h), but also, by spectral analysis the presence of an ultradian rhythm in two out of five subjects (Fig. 1). Furthermore, the magnitude of the meal-related insulin response was found to be higher after lunch than after breakfast or dinner (MEJEAN et al. 1988) or later in the evening (VAN CAUTER et al. 1992), these discrepancies being related to the morning-to-evening decline in cortisol levels and the meal-related increases in cortisol levels (VAN CAUTER et al. 1989, 1991). The same mechanisms are observed when insulin secretion is measured after various stimuli such as the oral or intravenous glucose tolerance

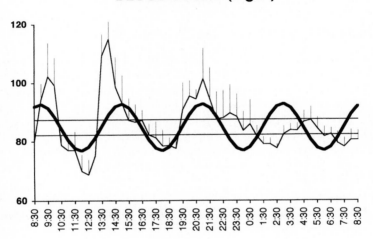

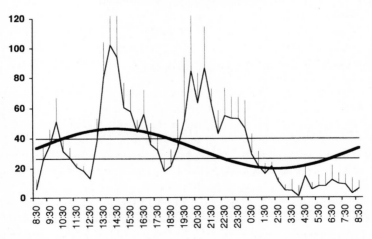

Fig. 1. Blood glucose and plasma insulin rhythms determined in five healthy volunteers

test (Jarrett et al. 1972, 1974; Zimmer et al. 1974) or the intravenous tolbutamide test (Baker and Jarrett 1972).

The chronobiological variation of plasma insulin levels may be linked to either an endogenous rhythm of β-cell function or a rhythmicity of insulin receptors and insulin binding.

The first hypothesis is supported by several lines of evidence, namely the demonstration of a circadian rhythm of the volume of the β cell (Hellman and Hellestrom 1959) and circadian variation of the in vivo pulsatility of pancreatic islet peptides (Jaspan et al. 1986). Other lines of evidence support this hypothesis: the normal pulsatility of insulin release (Lefebvre et al. 1987; Ahlersova et al. 1992; Berman et al. 1993) is abolished in patients bearing an islet-cell tumour (Villaume et al., 1984) and the presence of a circadian rhythm of C-peptide in healthy subjects (Nicolau et al. 1983). Stagner et al. (1987) demonstrated that the oscillating pacemaker driving insulin secretion is located within the pancreas on the basis of their observation that this pulsatility persists in the isolated organ. The site of this pacemaker was further narrowed down to the pancreatic islets, which also secrete insulin in an oscillatory fashion (Bergstrom et al. 1990; Chou and Ipp 1990; Opara et al. 1989). The β cell is most likely the origin of the primary oscillator rhythm (Chou et al. 1991).

The second hypothesis is also supported by several investigations indicating a circadian variation of insulin binding to the receptors of erythrocytes or monocytes (Beck-Nielsen and Pedersen 1978; Hung et al. 1986; Pedersen et al. 1982; Schultz et al. 1983a,b; Wu et al. 1986) and to epididymal fat (Feuers et al. 1990).

There are also reports of circannual rhythmicities of glucose tolerance and insulin secretion. A significant increase in the glucose:insulin ratio was observed by Campbell et al. (1975a,b) in autumn compared to spring in 12 healthy volunteers living in the Antarctic. Similarly, Mejean et al. (1977) reported that the insulin response to an oral glucose tolerance test was lower and delayed in spring compared to autumn (Fig. 2). The same pattern of response was observed in elderly subjects after a test meal (Nicolau et al. 1983), in young women after a test meal (Haus et al. 1983) and in healthy subjects after a tolbutamide test (Del Ponte et al. 1984).

E. Chronobiology of the Corticotropic Axis

The circadian periodicity of the hypothalamic-pituitary-adrenal (HPA) system in mammals is well established. Cortisol secretion by the human adrenals is episodic and clustered as intense episodes early in the morning (Krieger et al. 1971; Weitzman 1976; Weitzman et al. 1971). Cosinor analysis revealed a strict period of 24 h in the time series of cortisol pulses (Minors et al. 1989). Besides this circadian surge of cortisol, another peak follows the midday meal (Follenius et al. 1982). This rhythm was found to be resistant to manipula-

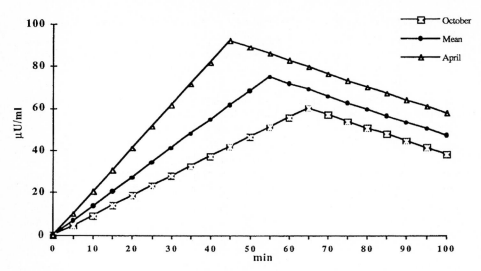

Fig. 2. Seasonal variations of post-stimulative insulin secretion during oral glucose tolerance test in healthy subjects (unpublished observations)

tions of the sleep–wake cycle (KRIEGER 1979) when these were not associated with very bright light exposure (CZEISLER et al. 1986). The circadian variations of plasma cortisol reflect those of the adrenocorticotrophic hormone (ACTH) and, to a lesser extent, those of pro-opiomelanocortin (POMC)-related hormones (FOLLENIUS et al. 1987; GENNAZZANI et al. 1983; KRIEGER et al. 1971; NICOLAU et al. 1986). In turn, the circadian variations in ACTH result from changes in corticotropin-releasing factor (CRF), which has been accepted as the major factor driving the release of ACTH (VALE et al. 1981). External synchronizers, such as stressors on light–dark variations, are relayed to specialized secretory cells of the paraventricular hypothalamic nuclei (PVN). These neurons project to the median eminence where CRF is released into the primary portal plexus (GIBBS 1985; PLOTSKY 1987) and reaches the ACTH secreting cells of the anterior pituitary. Other ACTH secretagogues are arginine vasopressin (AVP), oxytocin and angiotensin II.

F. Chronobiology of Feeding Behaviour

The existence of a circadian rhythm of food intake was shown by COLLIER et al. (1972, 1973) in rats and by MIGRAINE and REINBERG (1974) in humans. A circadian rhythm of spontaneous food intake in 4-year-old children was also obtained by DEBRY et al. (1975). These authors also found a circaseptan rhythm with an acrophase on Saturdays and Sundays. The analysis of the weekly variations in feeding behaviour performed by MEJEAN et al. (1992) in 51

Table 1. Seasonal variations of nutritions status measured in 57 healthy young female volunteers

	Spring		Summer		Autumn		Winter		
	Mean	SEM	Mean	SEM	Mean	SEM	Mean	SEM	
Energy	1816	18	1840	19	1893	18	1875	18	$p<0.05$
Carbohydrates	209	2	207	2	218	2	215	2	$p<0.01$
Lipids	78	1	81	1	82	1	81	1	ns
Protids	69	1	69	1	72	1	72	1	$p<0.05$
Alcohol	11.3	1.2	14.5	1.4	12.0	1.3	14.0	1.4	$p<0.001$
Sugars	74.0	1.2	72.0	1.2	80.0	1.4	79.0	1.2	ns
Breakfast	329	6	312	6	348	6	330	6	$p<0.001$
Lunch	688	10	667	10	723	10	694	4	$p<0.001$
Dinner	568	9	615	10	606	10	604	4	$p<0.05$
Nibbling	231	9	246	9	216	8	239	4	ns

young women showed similar results. Saturdays and Sundays were linked to a high intake of protein and fats, whereas Wednesdays and Thursdays were linked to a high intake of carbohydrates and sugars. The periodicity of food intake does not only show daily and weekly, but also annual rhythms. With regard to the circannual variations of food intake, the group of women investigated by MEJEAN et al. (1992), showed a progressive increase in caloric intake from spring to winter. Protein intake increased in parallel, but the highest fat intake was observed during summer, whereas carbohydrate consumption increased during autumn and winter. The results of this study are summarized in Table 1. They confirmed previously published studies, such as those of SARGENT (1954) and DEBRY et al. (1975), regardless of the different methodologies used.

G. Chronopharmacological Approach to Diabetes

The implications of these various results must be taken into account in the treatment of diabetes mellitus. The question:"At what clock hours must insulin be administered to insulin-dependent diabetics?" was raised by HOPPMANN (1940), and later by MOLLERSTROM (1953), and is still awaiting an appropriate chronotherapeutic answer.

Obviously, the problem of insulin-dosing time or times has been associated with two other pertinent questions: "What type of insulin(s) and what dose(s) are required?". However, investigators have paid more attention to these latter two questions than to the question: "When to treat?". This is presumably due to a lack of appropriate methodology in both data gathering and analysis, leading to a poor chronophysiologic background of the insulin-mediated blood glucose control. Methods and instruments allowing a diabetic to self-determine blood glucose four times a day over a period of years provided a typical autorhythmometric procedure.

Using this chronophysiological background, a chronotherapeutic strategy can be proposed: appropriate doses and types of insulin should be injected at specific times over the 24-h period so as to mimic its physiologic circadian rhythm, which in turn leads to the control of glucose levels not only in terms of the 24-h mean but also in terms of small-amplitude circadian rhythms.

The work of our group (KOLOPP et al 1986) is pertinent here: the aim of our study was to test this strategy on an individual basis by retrospective analyses of data gathered daily over 12–27 months from six insulin-dependent adult diabetics. In addition, such time periods were appropriate to test the hypothesis of annual changes in insulin needs of such patients, again with reference to data suggesting the existence of such rhythms in healthy subjects as well as in animal models.

The detection of circadian and ultradian rhythms of blood glucose from cosinor analyses, using trial period τ of 24 h and 6 h as well as prominent periods from spectral analyses, were performed on a monthly basis. The number of months (out of 12) for which a 24-h period was validated varied from 1 to 12 months, while the number of months for which a 6-h rhythm was validated was more consistent: 8 months for two patients and 12 out of 12 months for the four other patients. According to these first results, it seems that an ultradian rhythm with $t = 6$ h was almost invariably detected in all patients and months, while a circadian rhythm with $\tau = 24$ h was not consistently validated.

For all patients and all months, with few exceptions, the acrophases were located between 22:00 h and 03:00 h. Thus, inter- and intra individual differences in the circadian peak time of blood glucose were small. However, some patients exhibited a great stability in blood glucose peak time acrophase. For instance, the acrophase was located between 01:00 h and 03:00 h for two patients, and around 01:00 h for the patient, while some erratic acrophase estimates were seen for three patients. This result has to be related to the so-called dawn phenomenon described by several authors over subsequent years (SCHMIDT et al. 1981; KERNER et al. 1984; BOLLI 1988; CAMPBELL et al. 1985, 1986; DE FEO et al. 1986; GALE 1985; HOLL and HEINZE 1992).

Mesor values were within the limits of 1–2 g/l. For three patients all monthly circadian values were lower than 1.65 g/l. However, rather large intraindividual differences can be observed from month to month in mesor values. This is also the case of circadian amplitude values.

Can a normalization of mesor blood sugar level be expected to be associated with a nil or small circadian amplitude as is the case in healthy subjects? To test this hypothesis, the correlation coefficient r between mesor and amplitude values was calculated. A positive correlation was found for three patients, with respectively $r = 0.60$ ($p < 0.05$), $r = 0.86$ ($p < 0.01$) and $r = 0.85$ ($p < 0.01$), i.e. leading us to conclude that the smaller the mesor, the smaller the amplitude. Data from a further patient with rather small mesors and frequent nil or small amplitudes also fits with this hypothesis. A correlation between mesor and amplitude was not found for two patients.

Since insulin was injected at least three times a day, with the evening dose usually being a mixture of retard forms, it is not surprising that a circadian rhythm was detected in almost all monthly means (with a few exceptions in two patients). Great stability in the circadian acrophase location can be observed in all cases. The acrophases are located at various times. Computed circadian acrophases of both blood glucose and injected insulin coincided roughly in phase in four patients, while the circadian acrophase of injected insulin preceeded in phase the blood glucose acrophase by 12 h and 15 h in two patients (Fig. 3). The number of hypoglycaemic events were few, and even absent in one patient. They occurred mostly in the morning hours (between 07:00 h and 13:00 h) in two patients, but were randomly distributed for the other three patients. Such findings may have practical implications with regard to the timing of either insulin injections and/or blood sugar determinations.

In type II diabetes, another chronophysiologic event is the pulsatility of plasma insulin concentrations. This phenomenon was previously described by LANG et al. (1979, 1982) and MATHEWS et al. (1978, 1983) and implies: (1) that basal plasma insulin concentrations oscillate in a healthy male with a periodicity of 14 min; (2) that the oscillations persist in response to stimuli; and (3) that they are not found in non-insulin-dependent diabetes. Studies performed

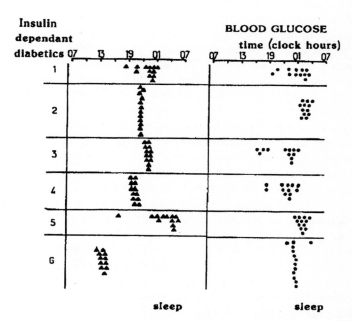

Fig. 3. Circadian acrophase (peak time) location of injected insulin (*left*) and blood glucose (*right*)

by POLONSKY et al. (1988) and O'RAHILLY et al. (1988) have concluded that profound alterations in the amount and temporal organization of stimulated insulin secretion may be important in the physiopathology of β-cell dysfunction and in the development of non-insulin-dependent diabetes. According to O'RAHILLY et al. (1988) "absent or abnormal oscillatory insulin secretion may be an early phenomenon in the development of non-insulin-dependent diabetes". The chronopharmacological application of this phenomenon may be found in the study of STURIS et al. (1995) which compared the effects of infusions of exogenous insulin administered during a constant infusion of glucose in nine normal young men: the endogenous insulin secretion was suppressed by somatostatin and the insulin infusions were either constant or in a sinusoidal pattern. The hypoglycaemic effect of insulin was greater when insulin infusions showed a sinusoidal pattern then when they were constant. This study shows that the ultradian insulin oscillations appear to promote more efficient glucose utilization.

The most interesting aspect of this study was observed when the annual rhythms were analyzed. Data were gathered over a 12-to 27-month period (12 months for one patient, more than 2 years for the others). Monthly means have been combined into chronograms of 12 months from January to December. An annual change of blood glucose data pooled on a seasonal basis with statistically significant differences between peaks and troughs can be seen in each patient's curve. Peak time never occurred in the autumn. However, apart from this, no clear group pattern can be described with regard to time distribution of peaks and troughs (Fig. 4). Peak time occurred in the winter for one patient, in the spring for three patients and in the summer for two. The trough time occurred in the autumn and/or winter for four patients, in the spring for one patient and in the summer for the last patient.

As for blood glucose, a statistically significant annual change in injected insulin can be demonstrated. The injected dose varied from 40 to 80 U in 24h. Contrasting with the annual changes in blood glucose, a group pattern in annual changes of insulin requirements can be clearly seen. Peak time occurred in the autumn or in the summer. However, differences between autumn and summer values were small and not statistically significant. Therefore, only reference to an autumn peak has been made. In addition, trough time occurred in the winter and/or spring for four patients and in the summer for two patients.

Considered on this seasonal basis, annual changes in blood glucose and injected self-administered insulin follow neither a correlated trend nor a roughly complementary pattern. Such facts can be verified for any of the patients studied. It seems therefore that annual changes in injected insulin correspond to a long-term rhythm, reflecting annual changes in requirements of insulin independent, to some extent, of blood sugar levels. To test this possibility, data were analysed in another way.

Are annual changes in requirement of insulin independent of annual changes in blood glucose? To test this hypothesis, data were reanalysed on a

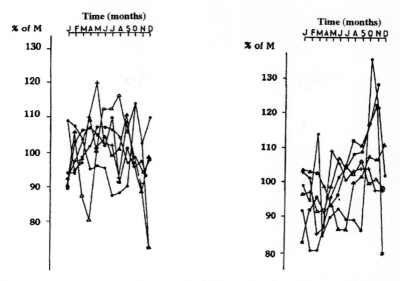

Fig. 4. Annual changes (monthly means) of blood glucose (*left*) and injected insulin (*right*)

monthly basis and expressed as a percentage of the individual annual mean for both variables. From these non-smoothed curves it can again be seen that larger doses of insulin were injected in the summer and autumn months than at other times of the year, while blood glucose changes were almost randomly distributed in terms the location of peaks and troughs.

However, the point of major interest deals with relative amplitude of both variables. Expressed as a percent of annual mean, the magnitude (double amplitude or difference between peak and trough) of insulin changes were greater than the magnitude of glucose changes for four patients. The magnitude of insulin changes were 51%, 40%, 26% and 54%, respectively, while that of glucose changes were 14%, 17%, 21% and 27%, respectively. For the other two patients, relative glucose changes were larger than those of injected insulin. In other words, with regard to four patients it seems that a rather small amplitude of annual change of blood glucose has been obtained by a rather large amplitude of change of injected insulin with a peak time in November (two patients), December or October.

These results once again suggest that annual changes of insulin correspond to a long-term and stable process reflecting requirements of insulin rather than overall annual blood glucose changes.

Insulin administration to insulin-dependent diabetics corresponds to the therapeutic replacement of a hormone which, in a healthy subject, has both a circadian and a circannual rhythmicity, as shown previously. In addition, hypoglycaemia and other metabolic effects also have a circadian rhythm (GIBSON et al. 1975; Sensi et al. 1970). Therefore, the chronotherapeutic approach to insulin-dependent diabetics must take into account these periodic phenomena in order to mimic rhythmic physiological patterns of both insulin secretion and blood glucose.

Healthy subjects usually have large amplitude circadian rhythms of insulin associated with small (or nil) amplitude in circadian rhythms of blood glucose. The present study shows that:

(a) Ultradian rhythms of blood glucose are more readily detectable than circadian ones
(b) In four out of six patients, circadian amplitude and mesor had a clear relationship: the smaller the mesor, the smaller the amplitude

In addition, healthy subjects usually have a circannual rhythm of basal insulin secretion, presumably reflecting insulin needs. For diabetic patients, peak time of insulin requirement occurred in the autumn. As demonstrated (Mejean et al. 1977). IRI response of healthy young adults to the standardized blood glucose tolerance test also exhibited a circannual rhythm with the strongest (highest C_{max} of IRI) and fastest (shortest t_{max} response occurring in September with regard to other months (e.g. the trough of both C_{max} and t_{max} of IRI occurred in April).

Emphasis has to be put on the fact that in long-term regulation, blood glucose rhythms are relatively independent of rhythmic changes in insulin. Despite large inter-individual differences in injected-insulin circadian acrophases, blood glucose acrophases clustered between 22:00 h and 03:00 h for all subjects. With regard to circannual rhythms, acrophases of insulin requirements clustered in the autumn or summer for all patients, while annual peaks and troughs of blood glucose were almost randomly distributed.

The cross comparison between circadian acrophases of immunoreactive insulin determined in healthy subjects and injected insulin in diabetic patients is not validated. However, results reported by Kolopp et al. (1986) encourage such a chronotherapeutic study, while the long-term regulation of glucose appears to be an important feature to monitor in insulin-dependent diabetes: Since the location of the circadian acrophase of this metabolic parameter is both stable and relatively independent of insulin-dosing times, it should be of interest to request self-controlled diabetics to perform one of the daily determinations at around 23:00 h, i.e., before retiring to bed. Recent papers published by several research groups (Garrel et al. 1992; Parillo et al. 1992; Hirsch et al. 1990; Sherwin et al. 1987; Bolli et al. 1993; Shapiro et al. 1991; Walker and Viberti 1990) show the importance, and the difficulties, of controlling these nocturnal glycaemic excursions.

References

Abrams RL, Cherchio GM, Graver AI (1968) Circadian variation of intravenous glucose tolerance in man. Diabetes 17:314

Ahlersova E, Ahlers I, Smajda B, Kassayova M (1992) The effect of various photoperiods on daily oscillations of serum corticosterone and insulin in rats. Physiol Res 41:315–321

Aparicio NJ, Puchulu FE, Gagliardino JJ, Ruiz R, Llorens JM, Ruiz J, Lamas A, Miguel R (1974) Circadian variation of the blood glucose plasma insulin and human growth hormone levels in response to an oral glucose load in normal subjects. Diabetes 23:132–137

Baker G, Jarrett RJ (1972) Diurnal variation in the blood sugar and plasma insulin response to tolbutamide. Lancet 2:945–947

Beck B (1992) Cholescystokinine, neurotensine et corticotropin releasing factor: trois importants peptides anorexigènes. Ann Endocrinol, 53:44–56

Beck-Nielsen H, Pedersen O (1978) Diurnal variation in insulin binding to human monocytes. J Clin Endocrinol Metab 47:385–390

Ben Dyke R (1971) Diurnal variation of oral glucose tolerance in volunteers and laboratory animals. Diabetologia 7:156–159

Bergstrom RW, Fujimoto WY, Teller DC, De Haen C (1990) Oscillatory insulin secretion in perifused isolated rats islets. Am J Physiol 257:E479–E485

Berman N, Chou HF, Berman A, Ipp E (1993) A mathematical model of oscillatory insulin secretion. Am J Physiol 264:839–851

Block GD, Page TL (1978) Circadian pacemakers in the nervous system. Ann Rev Neuroscience 1:19–34 Bolli GB, Gerich JE (1984) The "dawn phenomenon", a common occurrence in both non-insulin-dependent and insulin-dependent diabetes mellitus. N Engl J Med 310:746–750

Bolli GB (1988) The dawn phenomenon: its origin and contribution to early morning hyperglycaemia in diabetes mellitus. Diabet Metab 14:675–686

Bolli GB, Perriello G, Fanelli CG, De Feo P (1993) Nocturnal blood glucose control in type I diabetes mellitus. Diabetes Care 16:71–89

Bowen AJ, Reeves RL (1967) Diurnal variation in glucose tolerance. Arch Intern Med 119:261–264

Bruns W, Jutzi E, Fisher U, Bombor H, Woltanski KP, Jahr D, Wodrig W, Albrecht G (1981) Circadian varations in insulin concentrations and blood glucose in non-diabetics as well as insulin requirements in insulin-dependent diabetics. Z Gesamte Inn Med 36:258–260

Bruns W, Steinborn F, Menzel R, Staritz B, Bibergeil H (1990) Nocturnal continuous subcutaneous insulin infusion – a therapeutic possibility in labile type I diabetes under exceptional conditions. Z Gesamte Inn Med 45:154–158

Campbell IT, Jarrett RJ, Keen H (1975a) Diurnal and seasonal variation in oral glucose tolerance: studies in the Antartic. Diabetologia 11:139–146

Campbell IT, Jarrett RJ, Rutland P, Stimmler L (1975b) The plasma insulin and growth hormone response to oral glucose: diurnal and seasonal observations in the Antartic. Diabetologia 11:147–150

Campbell PJ, Bolli GB, Cryer PE, Gerich JE (1985) Pathogenesis of the dawn phenomenon in patients with insulin-dependent diabetes mellitus. N Engl J Med 312:1473–1479

Campbell PJ, Cryer PE, Gerich JE (1986) Occurrence of the dawn phenomenon without a change in insulin clearance in patients with insulin-dependent diabetes mellitus. Diabetes 35:749–752

Caroll KF, Nestel PJ (1973) Diurnal variation in glucose tolerance and in insulin secretion in man. Diabetes 22:333–348

Castillo M, Nemery A, Verdin E, Lefebvre PJ, Luyckx AS (1983) Circadian profiles of blood glucose and plasma free insulin during treatment with semisynthetic and

biosynthetic human insulin and comparison with conventional monocomponent preparations. Eur J Clin Pharmacol 25:767–771
Chou HF, Ipp E (1990) Pulsatile insulin secretion in isolated rat islets. Diabetes 39:112–117
Chou HF, Ipp E, Bowsher RR, Berman N, Ezrin C, Griffiths S (1991) Sustained pulsatile insulin secretion from adenomatous human beta cells: synchronous cycling of insulin, C-peptide and proinsulin. Diabetes 40:1453–1458
Collier G, Hirsch E, Hamkin P (1972) The ecological determinants of reinforcement in the rat. Physiol Behav 9:705–726
Collier G, Hirsch E, Kanarek R, Marwin A (1973) Environmental determinants of feeding patterns. Int J Chronobiol 1:322 (abstract)
Czeisler CA, Allan JS, Strogatz SH, Ronda JM, Sanchez R, Rios CD, Freitag WO, Richardson GS, Kronauer RE (1986) Bright light resets in the human circadian pacemaker independent of the timing of the sleep-wake cycle. Science 233:667–671
De Feo P, Perriello G, Ventura MM, Calcinaro F, Basta G, Lolli C, Cruciano C, Dell'olio A, Santeusanio F, Brunetti P (1986) Studies on overnight insulin requirements and metabolic clearance rate of insulin in normal and diabetic man: relevance to the pathogenis of the dawn phenomenon. Diabetologia 29:475–480
De Prins J, Malbecq W (1983) Analyse spectrale pour données non équidistantes. Bull Classe Sci Acad Royale de Belgique 64:287–294
Debry G, Bleyer R, Reinberg A (1975) Circadian, circannual and other rhythms in spontaneous nutrient and caloric intake of healthy four year olds. Diabetes Metab 1:91–99
Debry G, Méjean L, Villaume C, Drouin M, Martin JM, Pointel JP, Gay G (1977) Chronobiologie et nutrition humaine. XIVéme Congrés International de Thérapeutique. L'Expansion Scientifique Française, Paris 225–245
Del Ponte A, Guagnano MT, Sensi S (1984) Circannual rhythm of insulin release and hypoglycemic effect of tolbutamide stimulus in healthy man. Chronobiol Int 1:225–228
Deschamps I, Heilbronner J, Canivet J, Lestradet H (1969) Les variations spontanées de l'insuline au cours des vingt-quatre heures chez des sujets normaux Presse Méd 77:1815–1817
Faiman C, Moorhouse JA (1967) Diurnal variation in the levels of glucose and related substances in healthy and diabetic subjects during starvation. Clin Sci 31:111–126
Fantino M (1986) Opiacés endogènes et prise alimentaire. Cah Nutr Diet 10:347–358
Feuers RJ, Hunter JD, Tsai TH, Cardoso SS, Scheving LE (1990) Circadian stage dependent 125I-insulin binding in the liver and epididymal fat of the mouse. Prog Clin Biol Res 341A:529–534
Floyd JC, Pek S, Schteingart DE, Fajans SS (1974) Diurnal changes of plasma glucose (G), insulin (IRI) and massively obese subjects during fasting. Diabetes 33:371
Follenius M, Brandenberger G, Simeoni M, Reinhardt (1982) Diurnal cortisol peaks and their relationships to meals. J Clin Endocrinol Metab 55:757–761
Follenius M, Simon C, Brandenberger G, Lenzi P (1987) Ultradian plasma corticotropin and cortisol rhythms time-series analyses. J Endocrinol Invest 10:261–266
Freinkel M, Mager M, Vinnick L (1968) Cyclicity in the interrelationships between plasma insulin and glucose during starvation in normal young men. J Lab Clin Med 71:171–178
Gagliardino JJ, Hernandez RE, Rebolledo OR (1984) Chronobiological aspects of blood glucose regulation: a new scope for the study of diabetes mellitus. Chronobiologia 11:357–379
Gale EA (1985) The dawn phenomenon – fact or artefact? Neth J Med 28:50–52
Garrel DR, Bajard L, Harfouche M, Tourniaire J (1992) Decreased hypoglycemic effect of insulin at night in insulin-dependent diabetes mellitus and healthy subjects. J Clin Endocrinol Metab 75:106–109
Gennazzani AR, Petraglia F, Nappi C, Martignoni E, de Leo M, Facchinetti F (1983) Endorphins in peripheral plasma: origin and influencing factors. In: Muller EE,

Gennazzani AR (eds) Central and peripheral endorphins: basic and clinical aspects. Raven, New York, pp 89-97

Ghata J, Reinberg A (1979) Circadian rhythms of blood glucose and urinary potassium in normal adults males, and in similarly aged insulin dependent diabetic subjects. In: Reinberg A, Halberg F (eds) Chronopharmacology. Pergamon, Oxford, 315-322

Gibbs DM (1985) Measurement of hypothalamic corticotropin releasing factors in hypophyseal portal blood. Fed Proc 44: 203-209

Gibson T, Stimmler L, Jarrett RJ, Rutland P, Shiu M (1975) Diurnal variation in the effects of insulin in blood glucose, plasma non-estirified fatty acids and growth hormone. Diabetologia 11:83-88

Haus E, Nicolau G, Halberg F, Lakatua D, Sackett-Lunden L (1983) Circannual variations in plasma insulin and C-peptide in clinically healthy subjects. Chronobiologia 10:132

Hautecouverture M, Slama G, Assan R, Tchobroustky G (1974) Sex diurnal variations in venous blood and plasma insulin levels. Effect of estrogens in men. Diabetologia 10:725-730

Hautecouverture M, Slama G, Tchobroustky G (1975) Cycle nycthéméral de la glycémie et de l'insulino sécrétion. In: Journées annuelles de la diabétologie de l'Hotel Dieu. Flammarion, Paris, 79-83

Hellman B, Hellestrom C (1959) Diurnal changes in the function of the pancréatic islets of rats as indicated by nuclear size in the islet cells. Acta Endocrinol 31:267-281

Hirsch IB, Smith LJ, Havlin CE, Shah SD, Clutter WE, Cryer PE (1990) Failure of nocturnal hypoglycemia to cause daytime hyperglycemia in patients with IDDM. Diabetes Care 13:133-142

Holl RW, Heinze E (1992) The dawn or Somogyi phenomenon? High morning fasting blood sugar values in young type I diabetics. Dtsch Med Wochenschr 117:1503-1507

Holst JJ, Schwartz TW, Lovgren NA, Pedersen O, Beck-Nielsen H (1983) Diurnal profile of pancreatic polypeptide, pancreatic glucagon, gut glucagon and insulin in human morbid obesity. Int J Obes 7:529-538

Hopmann R (1940) The 24-h variation into account when injecting insulin. Acta Med Scand 108:165

Hung CT, Beyer J, Schulz G (1986) Fasting and feeding variations of insulin requirements and insulin binding to erythrocytes at different times of the day in insulin-dependent diabetics assessed under the condition of glucose-controlled insulin infusion. Horm Metab Res 18:466-469

Jarrett RJ, Keen H (1969) Diurnal variation of oral glucose tolerance; a possible pointer to the evolution of diabetes mellitus. Brit Med J 2:341-344

Jarrett RJ, Keen H (1970) Further observations on the diurnal variation in oral glucose tolerance. Brit Med J 4:334-337

Jarrett RJ, Baker IA, Keen H, Oakley NW (1972) Diurnal variation in oral glucose tolerance blood sugar and plasma insulin levels morning, afternoon and evening. Brit Med J 1:199-201

Jarrett RJ (1974) Diurnal variation in glucose tolérance: associated changes in plasma insulin, growth hormone and non-esterified fatty acids and insulin sensitivy. In: Aschoff J, Ceresa F, Halberg F (eds) Chronobiological aspects of endocrinology. Schattauer Stuttgart, Symposi Medica Hoechst 9:229-238

Jaspan JB, Lever E, Polonsky KS, Van Cauter E (1986) In vivo pulsatility of pancreatic islet peptides. Am J Physiol 251:E215-E226

Kerner W, Navascues I, Torres AA, Pfeiffer EF (1984) Studies on the pathogenesis of the dawn phenomenon in insulin-dependent diabetic patients. Metabolism 33:458-464

Kolopp M, Bicakova-Rocher A, Reinberg A, Drouin P, Mejean L, Levi F, Debry G (1986) Ultradian, circadian and circannual rhythms of blood glucose and injected insulins documented in six self-controlled adult diabetics. Chronobiology Intern 3: 265–280

Krieger DT, Allen W, Rizzo F (1971) Characterization of the normal pattern of plasma corticosteroid levels. J Clin Endocrinol Metab 32:266–284

Krieger DT (1979) Rhythms in CRF, ACTH and corticoteroids. In: Krieger DT (Ed) Endocrine rhythms. Raven: New York, pp 123–142

Lang DA, Matthews DR, Peto J, Turner RC (1979) Cyclic oscillations of basal plasma glucose and insulin concentrations in human beings. N Engl J Med 301:1023–1027

Lang DA, Matthews DR, Ward GM, Burnett M, Turner RC (1982) Pulsatile synchronous basal insulin and glucagon secretion in man. Diabetes 31:22–26

Lefebvre PJ, Paolisso G, Scheen AJ, Henquin JC (1987) Pulsatility of insulin and glucagon release: physiological significance and pharmalogical implications. Diabetologia 30:443–452

Leibowitz SF (1988) Neurotransmetteurs centraux et contrôle des appétits spécifiques pour les macronutriments. Ann Endocrinol 49:133–140

Levine AS, Morley JE (1984) Neuropeptide Y: a potent inducer of consummatory behavior in rats. Peptides 5:1025–1029

Malherbe C, De Gasparo M, De Hertogh R, Hoett JJ (1969) Circadian variations of blood sugar and plasma insulin levels in man. Diabetologia 5:397–404

Matthews R, Lang DR, Peto J, Turner RC (1978) Cyclical variation (Hunting) of basal plasma glucose and insulin concentrations in normal and diabetic man. Diabetologia 15:254

Matthews R, Naylor BA, Jones RG, Ward GM, Turner RC (1983) Pulsatile insulin has a greater hypoglycemic effect than continuous delivery. Diabetes 32:617–621

Méjean L, Reinberg A, Gay G, Debry G (1977) Circannual changes of the plasma insulin response to glucose tolerance test of healthy young human males. Proceedings of the XXVII th international congress physiol sciences 498

Méjean L, Bicakova-Rocher A, Kolopp M, Villaume C, Levi F, Debry G, Reinberg A, Drouin P (1988). Circadian and ultradian rhythms in blood glucose and plasma insulin of healthy adults. Chronobiol Int 5:227–236

Mejean L, Kolopp M, Drouin P (1992) Chronobiology, nutrition and Diabetes mellitus. in: Touitou Y, Haus E, eds, Biologic Rhythms in Clinical and Laboratory Medicine. Springer-Verlag: New York, pp 375–385

Migraine C, Reinberg A (1974) Persistance des rhytmes circadiens de l'alternance veille sommeil et du comportement alimentaire d'un homme de 20 ans pendant son isolement souterrain et sans montre. C R Acad Sci 279:331–334

Minors DS, Rabbit PM, Worthington H, Waterhouse JM (1989) Variation in meals and sleep-activity patterns in aged subjects; its relevance to circadian rhythm studies. Chronobiol Int 6:139–146

Mollerstrom J (1928) Om dygnsvariationer, blod-och urinsoker-kurvan hos diabetiker. Hygieia (Stockolm) 91:23

Mollerstrom J (1953) Rhythmus, Diabetes und Behandlung. Verhandlungen der dritten Konferenz der Internationalen Gesellschaft für biologische Rhythmusforschung. Acta Med Scand (Suppl) 278:1949

Nelson W, Tong YK, Juen -Kueng L, Halberg F (1979) Methods for cosinor-rhythmometry. Chronobiologia 6:305–323

Nemeth S, Vigas M, Macho L, Stukovsky R (1970) Diurnal variations of the disappearance rate glucose from the blood of healthy subjects is revealed by iv GTT. Diabetologia 6:641

Nicolaidis S, Burlet C (1988) Les régulations de la prise alimentaire. Ann Endocrinol 49:87–88

Nicolau GY, Haus E, Lakatua DJ, Bogdan C, Popescu M, Petrescu E, Sackett-Lundeen L, Stelea P, Stelea S (1983). Circadian and circannual variations in plasma immunoreactive insulin (IRI) and C-peptide concentrations in elderly subjects. Endocrinologie 21:243–255

Nicolau GY, Haus E, Lakatua DJ, Bodgan C, Sackett-Lundeen L, Petrescu E, Reilly C (1986) Circannual rhythms of parameters in serum of elderly subjects. Evaluation by cosinor analysis. Rev Roum Med Endocrinol 24:281–292

Opara EC, Atwater ECI, Go VLW (1989) Characterization and control of pulsatile secretion of insulin and glucagon. Pancreas 3:484–487

Opara EC, Go VL (1991) Effect of nerve blockade on pulsatile insulin and glucagon secretion in vitro. Pancreas 6:653–658

O'Rahilly S, Turner RC, Matthews DR (1988) Impaired pulsatile secretion of insulin in relatives of patients with non-insulin-dependent diabetes. N Engl J Med 318:1225–1230

Parillo M, Mura A, Iovine C, Rivellese AA, Lavicoli M, Riccardi G (1992) Prevention of early-morning hyperglycemia in IDDM patients with long-acting zinc insulin. Diabetes Care 15:173–177

Pedersen O, Hjollund E, Lindsky HO, Beck-Nielson H, Jensen J (1982) Circadian profiles of insulin receptors in insulin-dependent diabetics in usual and poor metabolic control. Am J Physiol 242:E127–E136

Plotsky PM (1987) Regulation of hypophysiotropic factors mediating ACTH secretion. Ann NY Acad Sci 512:205–217

Polonsky KS, Given BD, Hirsch LJ, Tillh H, Shapiro ET, Beebe C, Frank BH, Galloway JA, Van Cauter E (1988) Abnormal patterns of insulin secretion in non-insulin-dependent diabetes mellitus. N Engl J Med 318:1231–1239

Sargent F (1954) Season and the metabolism of fat and carbohydrate, a study of vestigial physiology. Meteorol Monogr (USA) 2:68–80

Schlierf G, Raetzer H (1972) Diurnal patterns of blood sugar, plasma insulin free fatty acid and triglyceride levels in normal subjects and patients with type IV hyperlipoproteinemia and the effect of meal frequency. Nutr Metab 17: 123–126

Schmidt MI, Hadji-Georgopoulos A, Rendell M, Kowarsky A, Margolis S, Kowarsky D (1981) The dawn phenomenon, an early morning glucose rise implications for diabetic intraday blood glucose variation. Diabetes Care 4:579–585

Schulz B, Ratzmann KP, Albrecht G, Bibergeil H (1983a) Diurnal rhythm of insulin sensitivity in subjects with normal and impaired glucose tolerance. Exp Clin Endocrinol 81:263–272

Schulz B, Greenfied M, Reaven GM (1983b) Diurnal variation in specific insulin binding to erythrocytes. Exp Clin Endocrinol 81:273–279

Sensi S, Capani F, Caradonna P, Policiccho D, Carotenuto M (1970) Diurnal variation of insulin response to glycemic stimulus. Biochim Biol Sper 9: 153–156

Shapiro ET, Polonsky KS, Copinschi G, Bosson D, Tillil H, Blackman J, Lewis G, Van Cauter E (1991). Nocturnal elevation of glucose levels during fasting in non insulin-dependent diabetes. J Clin Endocrinol Metab 72:444–454

Sherwin RS, Tamborlane WV, Ahern J (1987) Lessons from glucose monitoring at night. Diabetes Care 10:249–251

Simon C, Brandenberg G, Follenius M (1988) Absence of the dawn phenomenon in normal subjects. J Clin Endocrinol Metab 67:203–205

Stagner JI, Samols E, Weir GC (1987) Sustained oscillations of insulin, glucagon and somatostatin from the isolated canine pancreas during exposure to a constant glucose concentration. J Clin Invest 65:939–942

Sturis J, Scheen AJ, Leproult R, Polonsky KS, Van Cauter E (1995) 24-hour glucose profiles during continuous or oscillatory insulin infusion. Demonstration of the functional significance of ultradian insulin oscillations. J Clin Invest 95:1464–1467

Swoyer J, Haus E, Lakatua D, Sackett-Lundeen L, Thompson M (1984) Chronobiology in the clinical laboratory. In: Haus E, Kabat J (eds) Proceedings of the XV th International Conference of the International Society of Chronobiology. Basel, Karger 533–543

Vale W, Spiess J, Rivier C, Rivier J (1981) Characterization of a 41-residue ovine hypothalamic peptide that stimulates secretion of corticotropin an β-endorphin. Science 213:1394–1397

Van Cauter E, Desir D, Decoster C, Fery F, Balasse EO (1989) Nocturnal decrease in glucose tolerance during constant glucose infusion. J Clin Endocrinol Metab 6:604–611

Van Cauter E, Blackman JD, Roland D, Spire JP, Refetoff S, Polonsky KS (1991) Modulation of glucose regulation and insulin secretion by circadian rhythmicity and sleep. J Clin Invest 88:934–942

Van Cauter E, Shapiro ET, Tillil H, Polonsky KS (1992) Circadian modulation of glucose and insulin responses to meals: relationship to cortisol rhythm. Am J Physiol 262:467–475

Villaume C, Beck B, Dollet JM, Pointel JP, Drouin P, Debry G (1984) 28-hour profiles of blood glucose (BG), plasma immunoreactive insulin (IRI) and IRI/BG ratio in four insulinomas. Ann Endocrinol 45:155–160

Waldhaus W (1989) Circadian rhythms of insulin needs and actions. Diabetes Res Clin Pract 6:17–24

Walker JD, Viberti G (1990) Recurrent nocturnal hypoglycemia in an insulin-dependent diabetic patient receiving a small daily dose of insulin. Am J Med 88:537–539

Weitzman ED, Fukushima D, Nogeire C (1971) Twenty-four hour pattern of episodic secretion of cortisol in normal subjects. J Endocrinol Metab 33:14–22

Weitzman ED (1976) Circadian rhythms and episodic hormone secretion in man. Ann Rev Med 27:225–243

Whichelow MJ, Sturge RA, Keen H, Jarrett RJ, Stimmler L, Grainger S (1974). Diurnal variation in response to intravenous glucose. Brit Med J 1:488–491

Wu MS, Ho LT, Jap TS, Chen JJ, Kwok CF (1986) Diurnal variation of insulin clearance and sensitivity in normal man. Proc Natl Sci Counc Repub China 10:64–69

Zimmer PZ, Wall JR, Rome R, Stimmler L, Jarrett RJ (1974) Diurnal variation in glucose tolerance: associated changes insulin growth hormone and non esterified fatty acids. Brit Med J 1:485–491

CHAPTER 13
Gastrointestinal Tract

J.G. MOORE and H. MERKI

A. Introduction

This chapter will focus on biological rhythms of gastrointestinal (GI) function insofar as they influence the pharmacokinetics, pharmacodynamic and toxicological behavior of oral and, in some instances, parenterally administered medications. Rhythms around the circadian (~24 h) and ultradian (< 22 h) time domains will be emphasized; infradian (> 28 h, to include monthly and yearly rhythms) rhythms will not be discussed simply because there is no published evidence of their influence, or even existence, in human gastrointestinal function. Human data will be highlighted; animal data will be included only if supportive of a human observation or in support of a particular theoretical view proposed by the authors. In addition, rhythms of motor, secretory or absorptive function of the hollow-viscus GI organs will receive greater attention; those of the liver – important as this organ is in drug metabolism – and pancreas are beyond the intended scope of this chapter and are dealt with in more detail elsewhere (BELANGER and LABRECQUE 1992; MEJEAN et al. 1992).

B. Gastrointestinal Motility Rhythms

Rhythmic gastrointestinal motor patterns influence the absorption and disposition of oral medications in several ways. Orally administered drugs must exit the stomach before absorption begins. The rate of transfer of drugs into the small bowel is thus dependent on the rate of gastric emptying that, in turn, is influenced by the presence or absence of food, meal physical composition (liquid or solid), meal nutrient content (lipid, carbohydrate or protein) and a host of other physiological factors (e.g., posture, exercise and gender) (MOORE et al. 1990). In fed subjects, gastric motor activity is dominated by caudally directed peristaltic contractions that, in man, occur at the frequency of three per minute. These muscular contractions are responsible for the grinding of digestible solid food particles to a size (< ~1.5 mm) that permits passage from the gastric antrum into the duodenum. The antrum, or distal stomach, is the major anatomical site for this function. Liquid meal emptying is also dependent on normal distal stomach peristalsis and, to a lesser extent, on a gastric-

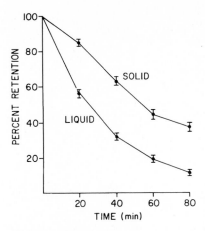

Fig. 1. Gastric emptying of liquid and solid foods. Note the near linear and slower emptying of solids when compared to liquids. A 300-g combined solid (beef stew) and liquid (orange juice) meal, labeled with technetium-99m sulfur colloid (solid-phase marker) and indium-111 (liquid-phase marker), was ingested at 0 min. The percentage retention refers to the percentage of the radiolabeled marker remaining in the stomach over time (minutes). *Each point* represents mean ± SEM values for 32 studies in 8 male healthy subjects (4 studies each). (Brophy et al. 1986)

duodenal pressure gradient generated by muscular forces in the proximal stomach. As observed in Fig. 1, liquid and solid phases of a mixed liquid-solid meal empty at different rates with different patterns (Brophy 1986). Liquids typically empty more rapidly than digestible solids and, if ingested alone without added nutrients, obey first-order emptying kinetics. Digestible solid food emptying obeys zero-order kinetics. Indigestible solids, represented by food particles greater than ~1.5 mm after grinding and trituration, and therefore unable to pass through the pylorus, are emptied by another type of ultradian motor wave termed the migrating myoelectric complex (MMC), a particularly powerful "housekeeping" muscular contraction that serves to sweep indigestible solids from the stomach into the small bowel. The MMC originates in the stomach and travels aborally to the ileocecal valve. These waves occur at the frequency of 90–120 min in the fasted human and are inhibited by feeding as observed in Fig. 2, which illustrates gastric emptying rates for inert markers – representing indigestible solids – under fasted and fed conditions (Smith and Feldman 1986). Feeding markedly inhibited emptying of the markers and in proportion to the frequency of feeding.

Circadian rhythmic motor patterns superimposed on ultradian patterns are also characteristic of the healthy human GI tract. Figure 3 demonstrates circadian rhythmicity in the speed of MMC propagation along the small bowel in groups of healthy and functional bowel syndrome patients intubated with pressure-sensor-equipped small bowel intestinal tubes (Kumar et al. 1986). The daytime velocity (centimeters per minute) of MMC propagation was more than

Gastrointestinal Tract

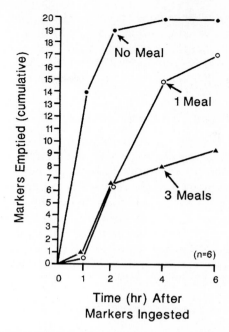

Fig. 2. Cumulative mean gastric emptying of indigestible solids (2-mm and 10-mm inert markers) in six healthy subjects. Gastric emptying was most rapid under fasted conditions and delayed in proportion to the frequency of feeding. (SMITH and FELDMAN 1986)

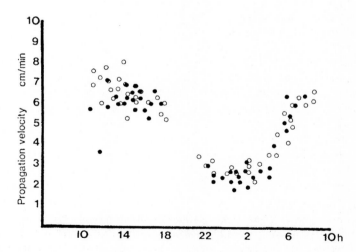

Fig. 3. Propagation velocity of human small bowel MMC motor activity over 24 h. *Filled circles* show plot of ten patients with irritable bowel syndrome; *open circles* show plot of ten healthy subjects. The gap at 1800 hours demonstrates the inhibiting influence of a single meal on MMC initiation. In a companion meal-controlled porcine study, feeding did not alter the circadian patterns. (KUMAR et al. 1986)

double the nocturnal value in both groups. Gastric emptying rates for meals also display circadian variation (Goo et al. 1987). In a two-time-point study depicted in Fig. 4, gastric emptying rates for meals administered at 2000 hours were significantly slower than emptying rates for the same meal administered to the same subjects at 0800 hours.

I. Pharmacological and Therapeutic Implications

The rate of gastric emptying, gastrointestinal transit and ultimate absorption of an oral medication is thus dependent on whether the medication is ingested as a liquid or digestible or undigestible solid that, in turn, is strongly influenced by ultradian and circadian GI motor events. For example, circadian alterations in gastric emptying result in delayed absorption of most oral medications when administered during the evening (REINBERG and SMOLENSKY 1982; LEMMER and BRUGUEROLLE 1994). The delay is reflected by lower maximum plasma concentrations (C_{max}) and longer times-to-peak (T_{max}) drug plasma concentrations. With few exceptions, however, the extent of absorption (bioavailability) for most drugs, reflected by the area under the plasma concentration-time

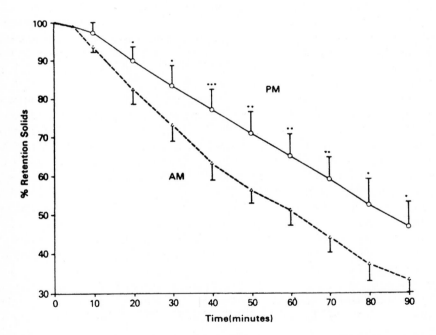

Fig. 4. Gastric emptying rates for the solid portion of a meal eaten in the morning (*AM*) and evening (*PM*) by 16 healthy male volunteers. Identical 300-g 208-kcal radiolabeled beef stew meals were eaten at 0800 and 2000 hours on separate days. *Points* represent mean ± SEM values. *Asterisks* represent significant differences ($P < 0.05$) at indicated time points. (Goo et al. 1987)

curve (AUC), does not differ with day or night administration. However, slower evening-time absorption has therapeutic ramifications for several drugs. Slower evening-time gastric emptying and intestinal absorption rates may result in an increase in hepatic first-pass effect for high-extraction compounds such as propranolol and lead to lower bioavailability for proton pump-inhibiting antiulcer drugs that are subject to gastric acid degradation (LANGNER and LEMMER 1988; PRICHARD et al. 1985). Slower evening-time compared to morning absorption of theophylline may also be to the benefit of asthmatic patients, with increased nocturnal patterns of airway resistance (REINBERG et al. 1987). Evening dosage may also be more desirable for drugs with lower toxicity thresholds because acute side effects of drugs are correlated to initially high drug plasma concentrations as, for example, with isorbide-5-mononitrate administration. Orthostatic hypotension was more pronounced and T_{max} for the drug significantly shorter after morning compared to evening dosage (LEMMER et al. (1991). Conversely, morning dosage may be more desirable for some drugs because of more rapid onset-of-action.

Drug absorption rates are also influenced by the degree to which drugs disintegrate and dissolve within the gastric lumen. Drugs that completely dissolve are, in the main, emptied as liquids although with large combined liquid-solid meals, liquid and solid phase-emptying becomes indistinguishable, particularly during the late meal-emptying phase (MOORE et al. 1981). Drugs that completely dissolve and are entirely emptied in a non-nutrient liquid phase do not display the marked circadian variation in absorption rates when compared to drugs ingested with solids (Goo et al. 1987). Solid meal ingestion always delays drug absorption because of the inhibiting effect of nutrient solids on gastric emptying. This delay is exaggerated with drugs administered following or with evening meals because gastric emptying is slower in the evening compared to morning hours (Goo et al. 1987). Gastric emptying and absorption of enteric-coated or matrix-type sustained-release medications are even further delayed by concurrent meal ingestion. As discussed above, food inhibits the gastric-originating MMC wave that sweeps indigestible solids, such as enteric-coated capsules or tablets, out of the stomach and into the small bowel where dissolution and absorption begins. In the absence of MMCs gastric residence time for these drug formulations is prolonged, as can be seen for enteric-coated aspirin in Fig. 5 (BOGANTOFT et al. 1978). It is thus possible that a morning ingested enteric-coated or matrix-type sustained-release preparation will neither be emptied nor absorbed for the entire daytime hours in a frequently snacking patient. In addition, because the propagation velocity of MMCs is slower at night, when compared to morning (Fig. 3), these drug formulations taken without meals may be expected to empty and absorb at slower rates at night.

II. Other Circadian Rhythms Influencing Drug Bioavailability

Ultradian and circadian motor rhythmic events thus alter the pharmacokinetic behavior of many drug formulations. However, a number of other facts con-

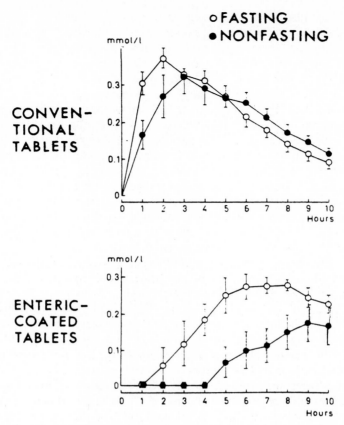

Fig. 5. Mean ± SEM plasma concentrations of salicylic acid in eight subjects after administration of 1.0 g acetylsalicylic acid (ASA) as conventional or enteric-coated tablets under fasting and nonfasting conditions. (BOGENTOFT et al. 1978)

found and make predictions concerning the effects of such rhythms on drug disposition uncertain. In the rodent model, for example, hepatic microsomal enzyme activity and flood flow – crucial in the metabolism and disposition of many drugs – have their own circadian rhythms (BELANGER and LABRECQUE 1992). In man, a circadian rhythm in hepatic blood flow has been described (LEMMER and NOLD 1991). In this study in supine healthy subjects, shown in Fig. 6, estimated hepatic blood flow as assessed by indocyanin green dye clearance was significantly higher at 0800 hours when compared to other times of the 24-h period. Assuming that hepatic and intestinal blood flow patterns run parallel, this finding could explain the higher C_{max} and/or shorter T_{max} values observed after morning compared to evening ingestion with some drugs just as readily as the difference in gastric emptying rates between the two time frames. The liver is the major site of drug metabolism in man, and for drugs with high extraction ratios, extraction and biotransformation rates are closely

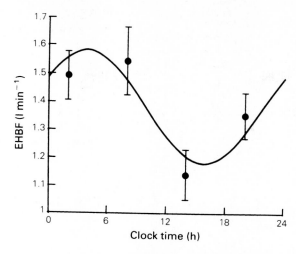

Fig. 6. Circadian rhythm in estimated hepatic blood flow as measured by indocyanine green clearance in ten healthy male subjects. Tested subjects were studied in the supine position with measurements at 0200, 0800, 1400 and 2000 hours. Mean ± SEM values are shown. A significant ($P < 0.025$) circadian rhythm was determined by cosinor analysis with a rhythm-estimated peak time of 0400 hours. (LEMMER and NOLD 1991)

correlated with hepatic blood flow. For drugs with low extraction ratios, in contrast, metabolism and biotransformation depend principally on the microsomal-oxidizing and cytosolic-conjugating enzymatic activities within the hepatocyte (BELANGER and LABRECQUE 1992) and are, therefore, less well correlated to hepatic blood flow. Moreover, circadian alterations in circulating plasma protein levels also affect the binding, and therefore availability, of many drugs (NAKANO et al. 1984). However, notwithstanding these confounding multiple influences, most orally administered drugs will be absorbed more rapidly after morning compared to evening administration and for a few of these drugs, as discussed above, this may lead to clinical consequences.

C. Rhythms in Gastric Acid Secretion

Hydrochloric acid is secreted by oxyntic glands of the human stomach. Acid secretion may be stimulated by histamine, acetylcholine and gastrin, resulting in parietal cell secretion of hydrogen ions into the gastric lumen against a concentration gradient of greater than 2 million (Fig. 7). Histamine is released by ECL cells, mast cells and histaminergic neurons after meal stimulation and/ or by other endocrine and paracrine stimulants; acetylcholine is released from parasympathetic neurons after cephalic-vagal or gastric-vagal stimulation; and gastrin is secreted from G cells of the antral and duodenal mucosa into the blood stream in response to cephalic-vagal stimulation, chemical reactions to

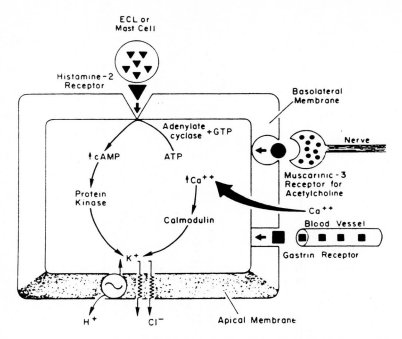

Fig. 7. Schema of parietal cell with cytosolic pathways and membrane acceptor sites pertinent to the stimulation and secretion of hydrochloric acid. Histamine from mast cells or enterochromatin-like (*ECL*) cells activates the proton pump on the apical membrane of the cell by increasing intracellular cyclic AMP. Acetylcholine from nerves and gastrin from the blood acts by increasing intracellular calcium. (GOLDSCHMIEDT and FELDMAN 1995)

food (amino acids and peptides) and gastric distention. The fact that acid secretion may be independently blocked at each of these receptor sites forms the basis for modern day peptic ulcer therapy. In addition, the H^+-K^+ ATPase receptor site at the cell apex is blocked by proton pump inhibitors. The proton pump is believed to be the final common pathway for hydrogen ion secretion into the gastric lumen.

In the absence of any exogenous stimulation, including meals, acid is secreted in relatively low amounts to maintain a gastric pH between 0.8 and 2.5 over a 24-h period. This is termed basal acid secretion (BAO). A 24-h rhythm in BAO secretion has been described in healthy individuals and in patients with active duodenal ulcer disease (MOORE and HALBERG 1986). In the absence of food, acid output is highest in the evening (from 1600 hours to midnight) and lowest during the morning hours (Fig. 8); higher rates of mean acid secretion were observed in duodenal ulcer patients (approximately 30% above controls) when compared to normals, but individual overlap between the two groups was considerable.

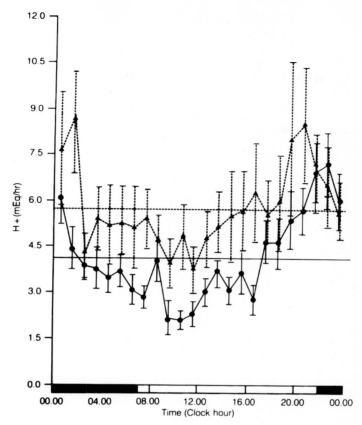

Fig. 8. Twenty-four-hour gastric acid (H^+) chronogram from 14 healthy volunteers (*circles*) and 21 patients with active peptic ulcer (*triangles*). *Points* represent hourly mean (±SEM) acid secretory rates. Note low morning and high evening secretion in both groups. *Dashed line* shows the mean 24-h rate (5.76±0.98 mEq/h) for the ulcer group; *solid line* shows mean rate (4.12 ± 0.40 mEq/h) for the healthy group. (Moore and Halberg 1986)

Using the method of monitoring gastric pH with glass electrodes intragastrically placed, increased acidity has been found in patients with active duodenal ulceration as compared with normals, especially in the evening hours (Merki et al. 1988a) (Fig. 9). In agreement with the study of Moore et al. and others, however, a considerable overlap of median 24-h acidity was observed between duodenal ulcer (DU) patients and controls.

The mechanisms underlying gastric acid circadian rhythmicity are not fully understood. No correlation or even a negative correlation was described when basal acid secretion was compared with serum gastrin concentrations (Moore and Wolfe 1973; Gedde-Dahl 1974; Trudeau and McGuigan 1971). Vagal stimulation may be responsible for the rhythmicity in the fasting state of acid

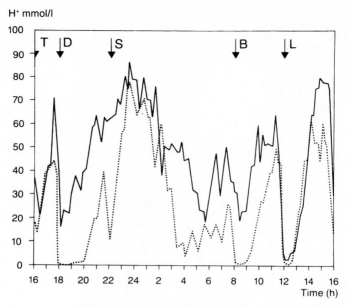

Fig. 9. Median intragastric acidity profiles of duodenal ulcer patients (*solid line*) and healthy controls (*broken line*) over 24 h. Meals are shown above as tea (*T*), dinner (*D*), snack (*S*), breakfast (*B*) and lunch (*L*). Acidity is expressed as hydrogen ion activity in millimoles per liter. (MERKI et al. 1988)

secretion, as anticholinergics effectively inhibit basal acid secretion. In addition, rhythmicity in acid secretion is lost in postvagotomy patients who continue to secrete significant amounts of basal gastric acid, suggesting that vagal nerve innervation is necessary to maintain the circadian pattern of secretion (MOORE 1973).

Under physiological conditions, the rhythmicity of gastric secretion is greatly influenced by the intake of meals and liquids. After the ingestion of a meal, intragastric hydrogen ion concentration falls immediately due to the buffering effect of food despite rapidly increasing acid output (Fig. 10). One hundred and twenty to 150 min after a meal, pH values drop to low levels as food buffers become ineffective due to saturation and because of gastric emptying. The powerful influence of meals on acid secretion, overwhelming basal acid and the circadian rhythm, are mediated via cephalic-vagal-mediated stimulation, gastric distention and chemical reactions to food involving cholinergic, histaminergic and gastrinergic mechanisms. During interdigestive periods, especially at night time, intragastric pH levels remain low. It is believed that this nocturnal period is the time span during which gastric mucosa is most vulnerable to damage and also most susceptible to acid-inhibiting treatment strategies (SOLL 1989).

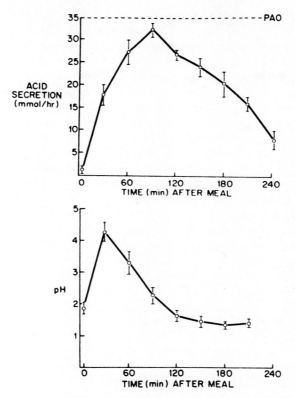

Fig. 10. Mean (±SEM) acid secretion (*top*) and intragastric pH (*bottom*) following a sirloin steak in healthy subjects. On day 1 (*top*), after the meal was eaten, acid secretion was measured by in vivo intragastric titration to pH 5.5 in six subjects. On day 2 (*bottom*), intragastric pH in ten subjects was allowed to seek its natural level after the meal was eaten. The mean basal secretion rate (*top*) and basal pH (*bottom*) prior to the meal are shown at 0 min. Peak acid output (*PAO*) is also indicated. (GOLDSCHMIEDT and FELDMAN 1993)

I. Pharmacological Implications

Reduction of 24-h acid output or acidity results in improvement or healing of acid-related disease (peptic ulcer, gastro-esophageal reflux, stress erosive gastritis) and is achieved by different pharmacological approaches. Antacids buffer acidity for approximately 60–90 min after their intake, but have little influence on night-time acid secretion and, therefore, play a minor role in modern peptic ulcer therapy.

Histamine H_2-receptor antagonists (H_2RAs) are the most widely used antisecretory agents. A wide array of dosages and recommendations exist for the oral or intravenous administration of this class of drugs (including cimetidine, ranitidine, famotidine, nizatidine, roxatidine and others). Simultaneous collection of plasma H_2RA concentrations and intragastric pH data reveal a sigmoidal dose-response relationship consistent with competitive inhibition of histamine action (ECHIZEN et al. 1988; SANDERS et al. 1989).

The influence of the circadian rhythm on basal acid secretion is reflected in the varying pH response to constant rate intravenous infusions of histamine H_2-receptor antagonists; intragastric pH decreases during the late evening (SANDERS et al. 1988; BALLESTEROS et al. 1990) (Fig. 11), matching the time period of increased basal acid secretion and highest acidity. This rhythmic alteration in acid secretion results in a changing dose-response relationship for H_2-receptor antagonists over a 24-h period, with higher doses needed to inhibit acid output at times of peak acid secretion (SANDERS et al. 1988, 1991; MERKI et al. 1988b). This was clearly reflected in a study using individually adapted H_2RA infusion rates to achieve target pH levels over a 24-h period (MERKI et al. 1991a,b). A novel computerized infusion pump was used to adjust the individual need for antisecretory drugs; infusion rates of H_2RAs were maximal in the late afternoon and evening and decreased gradually during the night, with the lowest infusion rates during the morning (HANNAN et al. 1990) (Fig. 12) clearly matching the rhythmic ebb and flow of basal acid secretion. Meals significantly alter the rhythmicity of 24-h gastric secretion (FELDMAN and RICHARDSON 1986) (Fig. 13) and intragastric pH levels (MERKI et al. 1988a) (Fig. 9). A dramatic loss of antisecretory effects was seen after the intake of a meal, both after oral (MERKI et al. 1990; JOHNSTON and WORMSLEY 1988; FRISLID and BERSTAD 1985; POUNDER et al. 1977) and intravenous administration of an H_2-receptor (MERKI et al. 1991b). A full morning dose of ranitidine increased gastric pH significantly less than the same dose given at night time (Fig. 14), and a significant interaction between ad lib snacks and

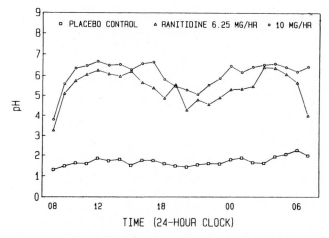

Fig. 11. Mean gastric pH measured in 15 fasted duodenal ulcer patients during three 24-h studies. Note the decrease in pH between 1800 and 0000 hours despite continuous intravenous infusion of ranitidine at 6.25 mg/h and 10 mg/h. (SANDERS et al. 1988)

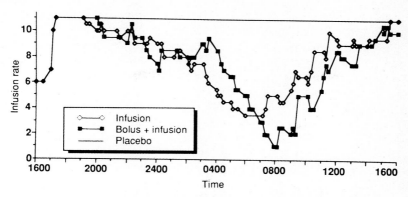

Fig. 12. Median pump rate changes during placebo and famotidine infusion using an intragastric pH-famotidine infusion feedback system. The computer was programmed to increase or decrease famotidine infusion rates below or above an intragastric pH of 6.0, respectively. The graph shows 15-min median pump rates; a clear trend for rates to decrease in the morning hours is seen. (HANNAN et al. 1990)

antisecretory effects was seen when food was taken after a full evening dose of this drug (Fig. 15). Even high doses of intravenous famotidine were unable to suppress meal-stimulated acid secretion in a group of duodenal ulcer patients (MERKI et al. (1988b) (Fig. 16). When meals were administered via nasogastric tubes, however, the interaction between meal stimulation and antisecretory effects was significantly less (BRATER et al. 1982). These differences may reflect the importance of the cephalic-vagal component of the meal response; modified sham-feeding, indeed, dramatically reduced the effect of high intravenous doses

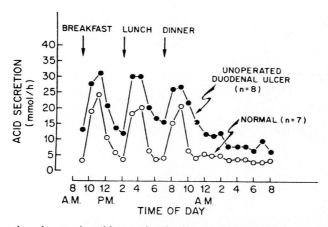

Fig. 13. Mean hourly gastric acid secretion in duodenal ulcer patients and normal men over 24 h. Breakfast, lunch and dinner were infused into the stomach at 0900, 1400 and 1700 hours, respectively. Differences in acid secretion between groups are greatest at night. (FELDMAN and RICHARDSON 1986)

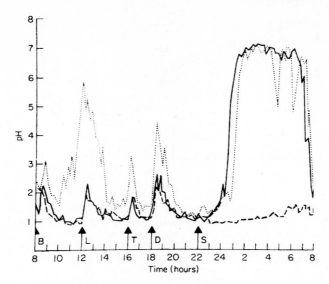

Fig. 14. Median 24-h intragastric acidity in 12 duodenal ulcer patients receiving placebo (*dashed line*), ranitidine 300 mg nocte (*continuous line*) or ranitidine 300 mg b.d. (*dotted line*). Meals are shown *at the bottom*. Ranitidine was administered at 0830 hours and 2215 hours. (MERKI et al. 1987)

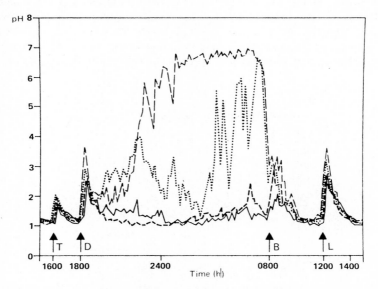

Fig. 15. Median 24-h profiles in 20 duodenal ulcer patients during placebo at 1830 hours followed by no food (*dashed line, bottom*), placebo at 1830 hours followed by ad lib snacks (*continuous line*), ranitidine 300 mg at 1830 hours with no additional food during the evening (*dashed line, top*) and ranitidine 300 mg at 1830 hours followed by ad lib snacks before bedtime (*dotted line*). Meals are shown *at bottom by the arrows*; *T*, tea; *S*, snack; *D*, late dinner; *B*, breakfast; *L*, lunch. (MERKI et al. 1990)

Gastrointestinal Tract

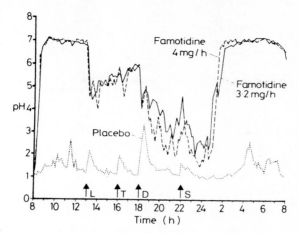

Fig. 16. Median 24-h intragastric pH profiles during placebo (*dotted line*), famotidine 3.2 mg/h (*solid line*) and famotidine 4 mg/h (*broken line*). Meals are shown *at bottom by the arrows*; L, lunch; D, dinner; S, snack. (MERKI et al. 1988)

of H_2RAs for 2–3 h (MERKI et al. 1991b) (Figs. 17, 18). The observation that the addition of anticholinergic drugs only partially attenuated the secretory effect of sham-feeding and feeding supports the view that complete blockade of food-stimulated acid secretion requires simultaneous receptor-antagonist drug

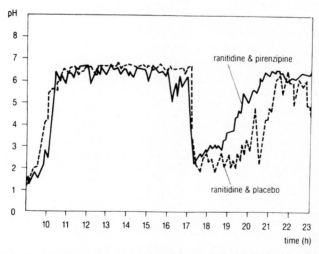

Fig. 17. Median intragastric pH time-profiles of ten healthy volunteers sham-feeding at 1700 hours. Ranitidine was administered intravenously by a pH feedback-ranitidine infusion system. If a pH > 6.0 was not maintained after 1700 hours, pirenzipine or placebo were given additionally. Note that the addition of an anticholinergic only partially attenuated the effect of sham-feeding. (MERKI et al. 1991)

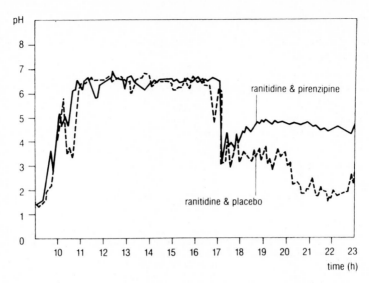

Fig. 18. Median intragastric pH-time profiles of ten healthy volunteers fed a standardized meal at 1700 hours. Ranitidine was administered intravenously by a pH feedback-ranitidine infusion system. If a pH > 6.0 was not maintained after 1700 hours, pirenzipine or placebo were given additionally. Note that the addition of an anticholinergic only partially attenuated the acid-stimulating effect of a meal. (MERKI et al. 1991)

inhibition at several sites. The cephalic phase of gastric acid secretion is mediated by vagal mechanisms, including direct stimulation of the parietal cell, promotion of gastrin release and release of inhibitory hormones (RICHARDSON et al. 1977; STENQUIST 1979; LAM et al. 1980). These pathways may be implicated in the inability of H_2-blockers to substantially counteract the postprandial secretory drive.

H^+K^+-ATPase inhibitors such as omeprazole, lansoprazole and pantoprazole block the final step of acid secretion. These drugs are characterized by a long duration of action and little interindividual variability in response when given in high doses. Only the H^+K^+-ATPase in actively secreting membranes is inactivated with these drugs. A 40-mg dose of omeprazole administered before breakfast inhibits daytime H^+ concentration by over 90% and nighttime acidity by 99% (BURGET et al. 1990). Doses lower than 20 mg are inconsistent in their suppression of acidity and secretion (HOWDEN et al. 1986); this may well be explained by the low bioavailability of the drug and high interindividual variability in drug absorption. Morning dosing of either 30 or 40 mg omeprazole was shown to be superior to night-time dosing in suppressing gastric secretion (CHIVERTON et al. 1992; PRICHARD et al. 1985), despite the lack of a difference between the two time points in the AUC of serum drug concentration.

II. Therapeutic Implications

Ulcer healing directly correlates with the degree and duration of the inhibition of acidity as measured by intragastric pH techniques; a much smaller correlation was found between DU healing and reduction in total acid output (JONES et al. 1987). In order to achieve optimal suppression of acidity with H_2-receptor antagonists, the influence of the rhythm in BAO and the effect of meals on acid secretion need to be taken into account.

The optimal effects of oral H_2RAs on intragastric pH have been found when the full daily dose was administered after the last meal in the evening, therefore suppressing acidity during the longest interdigestive period of the day (MERKI et al. 1987). If intravenous administration of this class of drugs is required, patients should remain fasted throughout the treatment period, as meals interact markedly with the antisecretory effects of H_2RAs.

In contrast, H^+-K^+-ATPase inhibitors should be taken shortly before or with a meal, which will activate the pumps in the secreting membranes, a requirement for proton-pump blockade effect. The intake of omeprazole and lansoprazole is compromised if taken during interdigestive periods or administered intravenously in fasting subjects. Morning administration of H^+-K^+-ATPase inhibitors has shown to be superior to night-time dosing in suppressing 24-h acidity and gastric acid secretion.

D. Rhythm in Gastric Mucosal Defense

The major therapeutic aim for most peptic ulcer regimes is to reduce gastric acidity. However, it is known that more than half of all peptic ulcer patients exhibit normal or less than normal gastric acid secreting rates and for this reason experimental efforts in recent years have been directed to possible alterations in gastric mucosal defense factors as a pathogenetic explanation. In this regard, an association with antral gastritis, peptic ulcer disease and *Helicobacter pyloris* – a bacterium that resides on the surface of antral mucosa and is now known to reduce some factors of mucosal defense – has been described and antibiotic treatment of this organism has led to a dramatic reduction in ulcer recurrence rates (MARSHALL et al. 1988). However, it is estimated that 70% of the world's population harbors this organism and, moreover, not all peptic ulcer patients harbor *H. pylori*. Several of these defense factors display circadian rhythmicity as illustrated in Fig. 19 (MOORE et al. 1994). This rat study demonstrated that gastric mucosal defense against the aggressive action of gastric acid is afforded by different mechanisms over the circadian time period. During the dark phase, when this species is active and gastric acid secretion is highest, protection against the potentially damaging effect of high acidity is provided by increased gastric mucosal blood flow. During the light phase, when this species is inactive and acidity is lowest, protection against acid damage is provided by a relative increase in mucous, bicarbonate and tissue prostaglandin tissue concentrations. In addition, a circadian rhythm in mu-

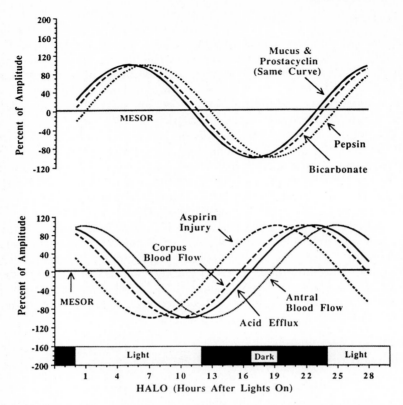

Fig. 19. Normalized cosinor curves of mucus effluent, tissue prostacyclin activity, pepsin secretion, corpus and antral blood flow, bicarbonate and acid secretion and aspirin injury in the rat stomach over 24 h. In each plot, the MESOR (~ mean value) was set at zero and the amplitude at ±100%. All plots demonstrated significant ($P < 0.05$) circadian rhythmicity by cosinor analysis. (MOORE et al. 1994)

cosal damage produced by topically applied acetylsalicylic acid (ASA) was identified, with peak damage occurring during the dark phase. The amount of damage was correlated to the relative difference in acid and bicarbonate secretion and could not be compensated for by the relative increase in dark-phase mucosal blood flow. The relevance of this rodent study to human peptic ulcer disease is that susceptibility to gastric mucosal damage can be explained by a *relative* change in the ratio of gastric aggressive and defensive factors without the requirement for either an *absolute* increase or decrease in either factor(s). Similar circadian-based investigations of factors in mucosal defense have not been reported in humans but some data exist to suggest that important day-night differences in mucosal defense do occur in man. In a two-time point endoscopic study performed on healthy males, an orally administered 1300-mg dose of ASA produced significantly more damage to the gastric mucosa when given at 0800 hours than when given at 2000 hours (Fig. 20) (MOORE and GOO

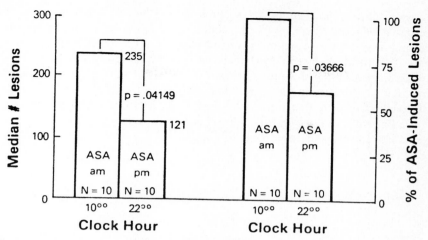

Fig. 20. Median number of aspirin (*ASA*)-induced gastric lesions following single 1300-mg ASA doses in the morning and evening in ten healthy fasted male volunteers. Studies were performed on separate days. Evening ASA administration produced 37% fewer lesions. (Moore and Goo 1987)

1987). However, a recent crossover study after orally administered low- (75 mg) and high-dose (1000 mg) ASA, in which the drug-induced lesions were rated by video-endoscopy in a blinded fashion, did not support the circadian phase-dependency in ASA-induced gastric lesions (Nold et al. 1995).

E. Circadian Influence in Cancer Chemotherapeutic Treatment Regimens

Cancer chemotherapeutic agents produce predictable and often treatment-limiting bone marrow and gastrointestinal toxic side effects. However, side effects can be reduced by circadian-staged chemotherapeutic infusion schedules, in which the largest administered doses are confined to a particular segment of the 24-h period (Roemeling and Hrushesky 1989; Lévi et al. 1992; Bjarnason et al. 1993). The basis for this salutary response to circadian-staged infusion schedules, as opposed to traditional flat-rate continuous infusion schedules, may rest on the timing of the delivery of the chemotherapeutic agent to the proliferating tissue and, in particular, with those inhibiting agents active in the synthesis phase (S phase) of cell replication. There is now ample evidence in randomized controlled clinical trials in a variety of human cancers (colorectal, ovarian, renal, others) that chronomodulated treatment schedules are less toxic and, in a few studies, at least suggestive of greater efficacy (i.e., prolonged survival) (Roemeling and Hrushesky 1989). In metastatic adenocarcinoma of the colon, for example, less toxicity and higher maximum tolerated doses are allowed if the largest doses of 5-fluoropyrimidine (5-FU) are given during the evening (BJARNASON et al. 1993). It is hypothesized that the basis for reduced toxicity with evening drug administration is that this is the time frame when bone marrow and gastrointestinal epithelial replication rates for healthy cells are least active and thus relatively spared from the damaging effects of anti-S phase drugs. The study of BUCHI et al., documenting circadian rhythmicity in rectal mucosal cell replication in the healthy human, with peak rates of replication occurring at 0700 hours, would lend support to an evening time anticancer drug delivery schedule, at least for the anti-S phase agents employed in the treatment of colorectal cancer (Fig. 21) (BUCHI et al. 1991).

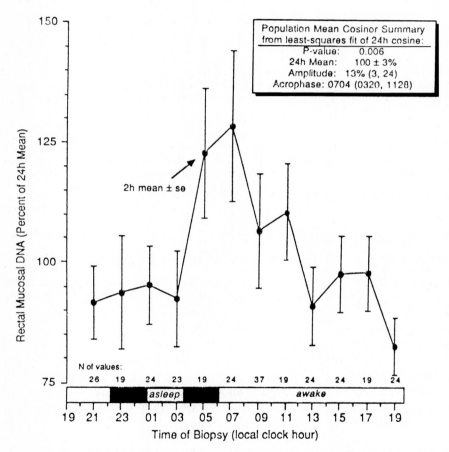

Fig. 21. Circadian rhythm in incorporation of tritiated thymidine into rectal cell DNA (dpm/µg DNA) in 16 men. Values expressed as percentages of the mean value of incorporation over the entire 24-h study period normalized to 100%. (BUCHI et al. 1991)

References

Ballesteros MA, Hogan DF, Koss MA, Isenberg JI (1990) Bolus or intravenous infusion of ranitidine: effects on gastric pH and acid secretion. Ann Intern Med 112:334–339

Bélanger PM, Labrecque G (1992) Biological rhythms in the hepatic drug metabolism. In: Touitou Y, Haus E (eds) Biological rhythms in clinical and laboratory medicine. Springer, Berlin Heidelberg New York, pp 403–409

Bjarnason G, Kerr I, Doyle N et al (1993) Phase I study of 5-fluoruracil (5-FU) and leucovorine (LV) by a 14 day circadian infusion in patients with metastatic adenocarcinoma. Cancer Chemother Pharmacol 33:221–228

Bogentoft C, Carlson I, Ekenved G, Magnusson A (1978) Influence of food on the absorption of acetylsalicylic acid from enteric-coated dosage forms. Eur J Clin Pharmacol 15: 351–355

Brater, DC, Peters MN, Eshelman FN, Richardson CT (1982) Clinical comparison of cimetidine and ranitidine. Clin Pharmacol Ther 32:484–489

Brophy CM, Moore JG, Christian PE, Taylor AT (1986) Individual variability of gastric emptying measurement employing standardized radiolabeled meals. Dig Dis Sci 21(8):799–806

Buchi KN, Moore JG, Hrushesky WJM et al (1991) Circadian rhythm of cellular proliferation in the human rectal mucosa. Gastroenterology 101:410–416

Burget DW, Chiverton SG, Hunt RH (1990) Is there an optimal degree of acid suppression for healing of duodenal ulcers? A model of the relationship between ulcer healing and acid suppression. Gastroenterology 99:345

Chiverton SG, Howden CW, Burget DW, Hunt RH (1992) Omeprazole (20 mg) daily given in the morning or evening: a comparison of effects on gastric acidity, and plasma gastrin and omeprazole concentration. Aliment Pharmacol Ther 6:103–111

Echizen H, Shoda R, Umeda N, Ishizaki T (1988) Plasma famotidine concentration versus intragastric pH in patients with upper gastrointestinal bleeding and in healthy subjects. Clin Pharmacol Ther 44:690–698

Feldman M, Richardson CT (1986) Total 24-hour gastric acid secretion in patients with duodenal ulcer. Comparison with normal subjects and effects of cimetidine and parietal cell vagotomy. Gastroenterology 90(3):540–544

Frislid K, Berstad A (1985) Prolonged influence of a meal on the effect of ranitidine. Scand J Gastroenterol 20:711–714

Gedde-dahl D (1974) Radioimmunoassay of gastrin. Fasting serum levels in humans with normal and high gastric acid secretion. Scand J Gastroenterol 9:41–47

Goldschmiedt M, Feldman M (1995) Gastric secretion in health and disease. In: Scharschmidt MHC, Feldman JSF (eds) Gastrointestinal disease/pathophysiology/diagnosis/management, 5th edn. Saunders, Philadelphia, pp 524–544

Goo RH, Moore JG, Greenberg E, Alazraki NP (1987) Circadian variation in gastric emptying of meals in humans. Gastroenterology 93:515–518

Hannan A, Chesner I, Merki HS, Mann S, Walt RP (1990) Use of automatic computerized pump to maintain constant intragastric pH. Gut 31:1246–1249

Howden CW, Derodra JK, Burget DW, Hunt RH (1986) Effects of low dose omeprazole on gastric secretion and plasma gastrin in patients with healed duodenal ulcer. Hepatogastroenterology 33:267–270

Johnston DA, Wormsley KG (1988) The effect of food on ranitidine-induced inhibition of nocturnal gastric secretion. Aliment Pharmacol Ther 2:507–511

Jones DB, Howden CW, Burget DW, Kerr GD, Hunt RH (1987) Acid suppression in duodenal ulcer: a meta-analysis to define optimal dosing with antisecretory drugs. Gut 28:1120–1127

Kumar D, Wingate D, Ruckebusch Y (1986) Circadian variation in the propagating velocity of the migrating motor complex. Gastroenterology 91:926–930

Lam SK, Isenberg JS, Grossman MI, Lane WH, Walsh JH (1980) Gastric acid secretion is abnormally sensitive to endogenous gastrin released after peptone test meals in duodenal ulcer patients. J Clin Invest 65:555–562

Langner B, Lemmer B (1988) Circadian changes in the pharmacokinetics and cardiovascular effects of oral propranolol in healthy subjects. Eur J Clin Pharmacol 33:619–624

Lemmer B, Bruguerolle B (1994) Chronopharmacokinetics – are they clinically relevant? Clin Pharmacokinet 26:419–427

Lemmer B, Nold G (1991) Circadian changes in estimated hepatic blood flow in healthy subjects. Br J Clin Pharmacol 32:627–629

Lemmer B, Scheidel B, Behne S (1991) Chronopharmacokinetics and chronopharmacodynamics of cardiovascular active drugs: propranolol, organic nitrates, nifedipine. Ann NY Acad Sci 618:166–181

Levi F, Misset JL, Brienza S et al (1992) A chronopharmacologic phase II clinical trial with 5-fluorouracil, folinic acid and oxaliplatin using an ambulatory multichannel programmable pump. Cancer 69:893–900

Marshall BJ, Goodwin CS, Warren JR et al (1988) Prospective double-blind trial of duodenal ulcer relapse after eradication of Campylobacter pylori. Lancet 2:1437–1441

Mejean L, Kolopp M, Provin P (1992) Chronobiology, nutrition and diabetes mellitus. In: Touitou Y, Haus E (eds) Biological rhythms in clinical and laboratory medicine. Springer, Berlin Heidelberg New York, pp 375–385

Merki HS, Witzel L, Harre K, Scheurle E, Neumann J, Rohmel J (1987) Single dose treatment with H_2-receptor antagonists: is bedtime administration too late? Gut 28:451–454

Merki HS, Fimmel CJ, Walt RP, Harre K, Rohmel J, Witzel L (1988a) Pattern of 24 hour intragastric acidity in active duodenal ulcer disease and in healthy controls. Gut 29:1583–1587

Merki HS, Witzel L, Kaufman D, Kempf M, Neumann J, Rohmel J, Walt RP (1988b) Continuous intravenous infusions of famotidine maintain high intragastric pH in duodenal ulcer. Gut 29:453–457

Merki HS, Halter F, Wilder-Smith CW, Allemann P, Witzel L, Kempf M, Rohmel J, Walt RP (1990) Effect of food on H_2-receptor blockade in normal subjects and duodenal ulcer patients. Gut 31:148–150

Merki HS, Hunt RH, Walt RP, Wilder-Smith CH, Gennoni M, Ernst T, Zeyen B, Rohmel J, Halter F (1991a) A new programmable infusion pump for individual control of intragastric pH: validation and effect of ranitidine. Eur J Gastroenterol Hepatol 3:9–13

Merki HS, Wilder-Smith CW, Walt RP, Halter F (1991b) The cephalic and gastric phases of gastric secretion during H_2-antagonist treatment. Gastroenterology 101:599–606

Moore JG (1973) High gastric acid secretion after vagotomy and pyloroplasty in man – evidence for non-vagal mediation. Am J Dig Dis 18:661–669

Moore JG, Goo RH (1987) Day and night aspirin-induced gastric mucosal damage and protection by ranitidine in man. Chronobiol Int 4:111–116

Moore JG, Halberg F (1986) Circadian rhythm of gastric acid secretion in men with active duodenal ulcer. Dig Dis Sci 31:1185–1191

Moore JG, Wolfe HG (1973) The relation of plasma gastrin to the circadian rhythm of gastric acid section in man. Digestion 9:97–105

Moore JG, Christian PE, Coleman RE (1981) Gastric emptying of varying meal weight and composition in man. Dig Dis Sci 26:16–22

Moore JG, Datz FL, Christian PE (1990) Exercise increases solid food gastric emptying rates in men. Dig Dis Sci 35(4):448–452

Moore JG, Larsen KR, Barattini P, Dayton MT (1994) Asynchrony in circadian rhythms of gastric function in the rat. A model for mucosal injury. Dig Dis Sci 39(8):1619–1624

Nakano S, Watanabe H, Nagai K, Ogawa N (1984) Circadian stage-dependent changes in diazepam kinetics. Clin Pharmacol Ther 36:271–277

Nold G, Drossard W, Lehmann K, Lemmer B 1995) Gastric mucosal lesions after morning versus evening application of 75 mg or 1000 mg acetylsalicylic acid (ASA). Naunyn Schmiedebergs Arch Pharmacol 351:R17

Pounder RE, Williams JG, Hunt RH, Vincent SH, Milto-Thompson CJ, Misiewicz JJ (1977) The effects of oral cimetidine on food-stimulated gastric acid secretion and 24-hour intragastric acidity. In: Burland W, Simkins M (eds) Proceedings of the second international symposium on histamine H_2-receptor antagonists. Excerpta Medica, Amsterdam, p 189

Prichard PJ, Yeomans ND, Mihaly GW, Jones DG, Buckle PJ, Smallwood RA, Louis WJ (1985) Omeprazole: a study of its inhibition of gastric pH and oral pharmacokinetics after morning and evening dosage. Gastroenterol 88:64–69

Reinberg A, Smolensky MH (1982) Circadian changes of drug disposition in man. Clin Pharmacokinet 7:401–402

Reinberg A, Pauche F, Ruff F, Gervais A, Smolensky MH, Levi F, Gervais P, Chaouat D, Abella L, Zindai R (1987) Comparison of once daily evening versus morning sustained-release theophylline dosing for nocturnal asthma. Chronobiol Int 4(3):409–420

Richardson CT, Walsh JH, Cooper KA, Feldman M, Fordtran JS (1977) Studies on the role of cephalic-vagal stimulation in the acid secretory response to eating in normal subjects. J Clin Invest 60:435–441

Roemeling R, Hrushesky WJM (1989) Circadian patterning of continuous FUDR intrusion reduces toxicity and allows higher dose intensity. J Clin Oncol 7:1710–1717

Sanders SW, Moore JG, Buchi KN, Bishop AL (1988) Circadian variation in the pharmacodynamic effect of intravenous ranitidine. Annu Rev Chronopharmacol 5:335–338

Sanders SW, Buchi KN, Moore JG, Bishop AL (1989) Pharmacodynamics of intravenous ranitidine after bolus and continuous infusion in patients with healed duodenal ulcers. Clin Pharmacol Ther 46:545–551

Sanders SW, Ballesteros MA, Hogan DK, Koss MA, Isenberg JI (1991) Effect of basal gastric acid secretion on the pharmacodynamics of ranitidine. Chronobiol Int 8:186–193

Smith HJ, Feldman M (1986) Influence of food and marker length on gastric emptying of indigestible radiopaque markers in healthy humans. Gastroenterology 91:1452–1455

Soll AH (1989) Duodenal ulcer and drug therapy. In: Sleisenger MH, Fordtran JS (eds) Gastrointestinal disease pathophysiology, diagnosis, management, 4th edn. Saunders, Philadelphia, pp 814–879

Stenquist B (1979) Studies on vagal activation of gastric acid secretion in man. Acta Physiol Scand [Suppl] 465:1–31

Trudeau WL, McGuigan JE (1971) Relations between serum gastrin levels and rates of gastric hydrochloric acid secretion. N Engl J Med 284:408–412

CHAPTER 14
The Pineal Gland, Circadian Rhythms and Photoperiodism

J. ARENDT

A. Introduction

The pineal gland serves the same function in all species studied to date. Essentially it conveys information concerning light-dark cycles to body physiology for the organisation of seasonal and circadian rhythms. The pattern of secretion of the pineal hormone melatonin (N-acetyl-5-methoxytryptamine) forms the basis for this message, and a considerable amount of information has accumulated regarding the coding of photoperiodic information by melatonin. How the message is read at a cellular and molecular level remains largely unclear; however recently great strides have been made in identifying, localising and characterising melatonin receptors. The current development of melatonin agonists and antagonists will no doubt help to clarify the pharmacology of melatonin.

The relationship of the pineal gland to the circadian system appears to be hierarchically more important in lower vertebrates than in mammals. For example pinealectomy in the lizard *Anolis carolinensis* leads to a splitting of circadian activity rhythms into several components and the pineal can be regarded as a coupling device for the circadian system (UNDERWOOD 1977). In some birds the pineal appears to act as a central circadian rhythm generator. Pinealectomy of *Passer domesticus* leads to arrythmicity which can be restored by transplanting a pineal from another bird. Moreover the phase of the donor bird is conveyed to the host with the transplant (MENAKER et al. 1981). It is possible to culture pineal explants or indeed dispersed pineal cells from lizards and birds and these preparations retain their circadian melatonin production in vitro (UNDERWOOD and GOLDMAN 1987). More is known of the pharmacology and physiology of the pineal in mammals than in lower vertebrates and birds, and the latter will not be treated here. The interested reader is referred to articles by UNDERWOOD and GOLDMAN (1987), GERN and KARN (1983), COLLIN (1971), GWINNER (1989) and CASSONE (1990).

In mammals pinealectomy or denervation of the gland clearly abolishes the ability to respond to changing artificial daylength in terms of variations in seasonal functions (HERBERT 1981; TAMARKIN et al. 1985; ARENDT 1986). Moreover in sheep and ferrets pinealectomy leads to desynchronisation of seasonal rhythms in reproductive function from the annual periodicity. This latter observation leads to the suggestion that in species with endogenous annual cycles the pineal essentially synchronises the endogenous cycle to a yearly periodicity (WOODFILL et al. 1991, 1994).

Pinealectomy in mammals has rather more subtle effects on the circadian system. As long ago as 1970, QUAY reported that following pinealectomy the rate of resynchronisation of rats to a phase shift of the light-dark cycle was faster than in intact animals (QUAY 1970) and this was confirmed by ARMSTRONG and REDMAN (1985). More recently CASSONE (1992) has found that pinealectomy leads to disrupted circadian rhythms in rats kept in constant light. Some mice do not secrete melatonin but yet appear to have normal behavioural rhythms (MENAKER et al. 1981). Some few humans have undetectable endogenous melatonin production without overt manifestation of circadian rhythm disturbance, and suppression of melatonin secretion by adrenoceptor antagonists (ARENDT 1985; ARENDT et al. 1985b) does not appear to disturb circadian rhythms in the short term, although this has not been carefully assessed. Thus the pineal and indeed melatonin secretion cannot be regarded as essential in the mammalian circadian system in the sense that, even if they are important under normal conditions, there must be back-up systems which replace pineal circadian input in circumstances where it is absent.

In general more information concerning the relationship of the pineal to the circadian system in mammals has been obtained by administration of melatonin than by pinealectomy. Unusually most of this information is derived from human work. This is due to the fact that very early in the investigation of pineal function, melatonin was perceived to have potential therapeutic uses in human rhythm disturbance (ARENDT 1983).

The present review will consider the important features of the production and effects of melatonin which impinge upon its circadian and seasonal functions and which have proved of use in the general study of biological rhythms.

B. Melatonin Production

Melatonin is synthesised within the pineal gland itself (AXELROD 1974) and in the retina (PANG et al. 1977), and there are reports of its synthesis in the gastrointestinal (GI) tract (RAIKHLIN et al. 1975), in platelets (LAUNAY et al. 1982) and in the Harderian gland (PANG et al. 1977). In mammals pinealectomy leads to a great reduction, in most cases to undetectable levels, of circulating melatonin (ARENDT et al. 1980; NEUWELT and LEWY 1983). Thus most if not all of the hormone reaching peripheral sites is pineal derived. There is some controversy as to whether pineal-synthesised melatonin can reach the brain without passing via the circulation. In sheep there is evidence that it is primarily secreted into the circulation (ROLLAG et al. 1978) but in some species very high levels in the CSF (KANEMATSU et al. 1989) suggest that it may reach the brain by a more direct route, perhaps via the third ventricle.

Melatonin is synthesised from tryptophan via 5-hydroxylation by tryptophan-5-hydroxylase to 5-hydroxytryptophan, decarboxylation by aromatic amino-acid decarboxylase to 5-hydroxytryptamine (serotonin), N-acetylation by N-acetyl transferase to N-acetylserotonin (NAT) and O-methylation by hydroxyindole-O-methyltransferase (HIOMT) to melatonin (N-acetyl-5-methoxytryptamine) (Fig. 1). All these steps take place within the gland (KLEIN 1993). The cardinal feature of this synthetic pathway is its rhythmicity. Within the pineal there is a substantial day-night variation in serotonin content with high levels during the day (QUAY 1963). Most important, however, is the variation in the activity of the enzyme NAT, which can increase from 30- to 70-fold at night (KLEIN and WELLER 1970). This, followed by a small increase in HIOMT at night, leads to the nocturnal rise in melatonin observed in nearly all species, whether nocturnal or diurnal, under synchronised conditions. In most circumstances NAT is considered to be rate-limiting in melatonin synthesis. It is, however, possible to increase synthesis by augmenting the supply of precursor molecules by administration of either tryptophan (HUETHER et al. 1992) or 5-hydroxytryptophan (NAMBOODIRI et al. 1983). Serotonin reuptake inhibitors also increase circulating melatonin by unknown mechanisms, which may be neural and/or metabolic (DEMISCH et al. 1986; SKENE et al. 1994). The large change in activity of NAT is peculiar to the enzyme found in the pineal and probably the retina. Other isoforms in, for example, the liver do not share this property. The rhythm of melatonin production is a true circadian rhythm persisting in the absence of a light-dark cycle and other 24-h time cues (e.g. ARENDT et al. 1985a; WEVER 1989).

The neural mechanisms controlling NAT activity are of the utmost importance in understanding the control of pineal function. The gland is innervated primarily from a peripheral sympathetic tract arising in the superior cervical ganglion (KAPPERS 1960). There is good evidence for direct central innervation (KORF and MOLLER 1984; MOLLER et al. 1991) but the role of this input remains to be determined. Sectioning the sympathetic nerve supply abolishes melatonin rhythmicity and essentially mimics the effects of pinealectomy (LINCOLN 1979; BITTMAN 1985; REITER 1980). Observations in rodents (MOORE and KLEIN 1974) and more recently sheep (TESSONEAUD 1994) have shown that the endogenous circadian rhythm of melatonin like most other circadian rhythms is generated in the suprachiasmatic nuclei (SCN) and entrained principally by the light-dark cycle acting via the retinohypothalamic tract, probably with a contribution from the lateral geniculate nucleus. The pathway from the SCN to the superior cervical ganglion (SCG) is not fully elucidated but certainly passes via the paraventricular nucleus and the median forebrain bundle (KLEIN 1993) (Fig. 1).

Sympathetic pineal input terminates in adrenoceptors, characterised as β_1 and α_1 in rodents and humans, with a recent report of α_{2D}-adrenoceptors being present in rodents and bovines (KLEIN 1993). In rats it is clear that β-adrenoceptor stimulation of pineal activity is potentiated by concomitant α-adrenoceptor stimulation and there is some evidence for this in humans (SUGDEN

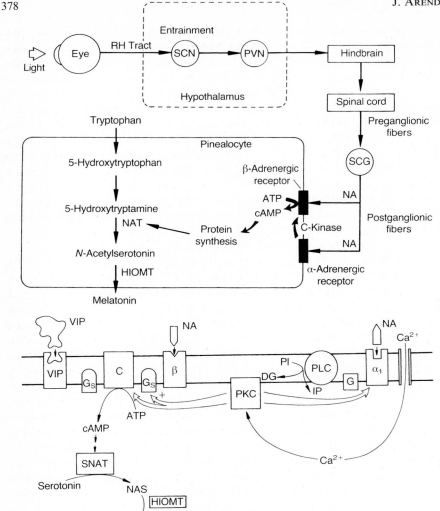

Fig. 1. a Major control mechanisms in melatonin synthesis. A summary diagram of the noradrenergic control of melatonin production. The suprachiasmatic nucleus (*SCN*) generates rhythmic signals, entrained to 24 h by light entering by the retina, relayed via the paraventricular nucleus (*PVN*), the hindbrain, spinal cord and superior cervical ganglion (*SCG*) to β- and α-adrenoceptors within the pineal. *RH*, retinohypothalamic; *NAT*, *N*-acetyltransferase; *HIOMT*, hydroxyindole-*O*-methyltransferase. In mammals noradrenaline (*NA*) is the primary transmitter. **b** Transduction mechanisms involved in the biosynthesis of melatonin. α1, α-adrenergic receptor, β, β-adrenergic receptor, *C*, adenylate cyclase; *ATP*, adenosine triphosphate; *cAMP*, cyclic adenosine monophosphate; *SNAT*, serotonin-*N*-acetyltransferase; *HIOMT*, hydroxyindole-*O*-methyltransferase; *PI*, phosphatidylinositol; *IP*, inositol phosphate; *DG*, diacylglycerol; *PKC*, protein kinase C; *Gs* stimulatory guanine-nucleotide-binding protein; *PLC*, phospholipase C; *VIP*, vasoactive intestinal peptide. (Redrawn from SUGDEN 1989, by permission)

et al. 1984; PALAZIDOU et al. 1989). The transmitter is noradrenaline. The events surrounding pineal adrenoceptor stimulation have been used extensively to characterise β- and α-adrenoceptor function (KLEIN 1993) (Fig. 1). Subsequent to stimulation the cascade of postreceptor events leads to an increase in cAMP potentiated by the α-adrenoceptor-mediated increase in phospholipid-dependent protein kinase C and subsequent increase in NAT activity involving transcription, translation and activation. There is close coupling between receptor numbers and noradrenaline variations in the rat pineal (ZATZ 1978). Effectively noradrenaline increases in the pineal towards the end of the light phase of the day and is accompanied by an increase in β-adrenoceptor numbers. During the phase of night time melatonin secretion β-adrenoceptor numbers decrease and the system becomes subsensitive although it is possible to override the physiological control mechanisms by administration of β-adrenoceptor agonists during the latter half of the night in rats (ILLNEROVA and VANECEK 1982; VAUGHAN and REITER 1987).

Whilst it is possible to stimulate melatonin production in vivo by administration of adrenoceptor agonists in rodents, this has proved difficult in humans where cardiovascular effects preclude the use of large doses of drug. The use of noradrenaline uptake inhibitors such as desmethylimipramine and oxaprotiline and antagonists such as atenolol and propanolol shows clearly that human melatonin production is adrenergically mediated (HANSSEN et al. 1977; COWEN et al. 1983; ARENDT et al. 1985b; FRANEY et al. 1986; CHECKLEY and PALAZIDOU 1988; PALAZIDOU et al. 1992). Other evidence indicates that the control of melatonin secretion in humans is similar to that of rats. For example tryptophan loading, monoamine (MAO) A inhibitors and some serotonin uptake inhibitors (BIECK et al. 1988; SKENE et al. 1994; DEMISCH et al. 1986) all stimulate production (Fig. 2). The peptide vasoactive intestinal polypeptide (VIP) has clear stimulatory effects on rodent melatonin production and it may well have a physiological function in the control of synthesis (YUWILER 1983). The pineal contains many other neuroreceptors and potential neuromediators but their physiological significance is unclear (EBADI and GOVITRAPONG 1986). To date there is no specific inhibitor of melatonin synthesis although there are reports that some benzodiazepines, non-steroidal anti-inflammatory drugs and decarboxylase inhibitors are capable of suppressing secretion. Chlorpromazine increases circulating melatonin presumably by inhibition of metabolism (BEEDHAM et al. 1987).

Recently STEHLE et al. (1993) have discovered a product of the CREM (cyclic AMP-responsive element modulator) gene, highly expressed in the pineal (and other neuroendocrine tissues), which encodes a potent repressor of cAMP-induced gene transcription. This inducible cAMP early repressor (ICER) shows a marked circadian rhythm in the pineal. ICER mRNA is induced at night under adrenergic control via cAMP. The peak of ICER expression corresponds to the declining phase of NAT and thus it is possible that it is the major component of a negative feedback loop regulating circadian rhythmic gene expression in the pineal.

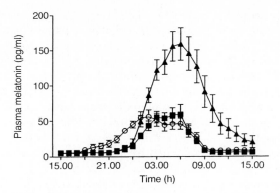

Fig. 2. Response of human plasma melatonin to a serotonin uptake inhibitor (fluvoxamine, 100 mg, *triangles*), a primarily noradrenaline uptake inhibitor (desmethylimipramine, 100 mg, *circles*) or placebo (*squares*) at 1600 h. Fluvoxamine increases the amplitude whereas desmethylimipramine advances the onset and increases the duration of the secretory rhythm. (From SKENE et al. 1994, by permission)

C. Light Control of Melatonin Secretion

One of the most important observations on control of melatonin production is that light of sufficient duration and intensity suppresses night-time production. As little as 1 min exposure to light of 150 lux will rapidly suppress rat melatonin production at night (ILLNEROVA and VANECEK 1979). The amount of light required is dependent on both species and photoperiodic environment. For example in some laboratory-raised animals less light is needed than in the same species raised in the wild (REITER 1985). Humans require 2500 lux for complete suppression (LEWY et al. 1980) although partial suppression can be observed with as little as 200–300 lux (McINTYRE et al. 1989; BOJKOWSKI et al. 1987a). There are very large individual variations. This human requirement for bright light has proved to be of fundamental importance in our knowledge of human circadian and photoperiodic physiology.

For the determination of melatonin phase response curves to light, the phase reference points that have been used are either the onset of the evening rise in plasma melatonin, the calculated peak time or the morning and evening onset and offset of *N*-acetyl transferase activity. The phase of the free-running rhythm can be reset by a single light pulse, the magnitude and direction of shift being dependent on the circadian time at which light is applied. Phase delays are induced by light in the late subjective day and early subjective night whereas advances follow light in the middle to late subjective night and early subjective day. A 'dead zone' is present during the subjective day when no shifts can be induced. The direction and magnitude of shift in response to light is summarised as the phase response curve (PRC).

This approach has been used extensively in the rat by ILLNEROVA (ILLNEROVA and VANECEK 1982) to investigate the circadian control of melatonin production, but using animals pulsed with 1 min of light at night under ambient light-dark cycles and then maintained in constant darkness (DD) to assess the phase of the melatonin rhythm. In these circumstances the onset of evening activity and offset of morning activity of N-acetyl transferase were used as phase reference points. Quite different shaped phase response curves for the evening rise and the morning decline were obtained and Illnerova has interpreted these observations as indicating separate control of evening onset and morning offset of melatonin by two distinct oscillators termed E and M.

Until the human requirement for bright light to suppress melatonin was demonstrated, it was generally assumed that light was not of major importance for the synchronisation of human circadian rhythms. However, it is now clear that sufficiently bright light is a strong circadian zeitgeber in humans (WEVER et al. 1983; ARENDT and BROADWAY 1986; BROADWAY et al. 1987; CZEISLER et al. 1986; MINORS et al. 1991). Human free-running tau is usually greater than 24 h and on average is reported as 24.9 h (WEVER 1979). Thus for synchronisation to 24 h the clock must be reset daily by an advance of about 0.9 h. Appropriately timed single light pulses of 5000 lux are capable of inducing phase shifts according to a PRC of a magnitude such as to provide this resetting mechanism (MINORS et al. 1991) and if even greater intensity light and 3–5 daily pulses are administered the phase shifts induced can be very large (JEWETT et al. 1991). At present this phenomenon is being exploited for aiding adaptation to shift work, to time zone change and in pathology (CZEISLER et al. 1990; TERMAN et al. 1995). Moreover less intense light (1200 lux) given over a period of 8–9 h daily and combined with imposed darkness can also induce large phase shifts over 5 days (DEACON and ARENDT 1994b, 1995b, c). It is possible that the light-induced phase shifts depend on suppression of melatonin production for their efficiency but this remains to be proven.

The control mechanisms of melatonin synthesis: rhythm generation in the SCN, synchronisation by light-dark cycles and suppression by light, mean that for a given length of darkness melatonin is produced during the dark phase (Fig. 3) and the duration of its secretion is dictated by the length of darkness up to a defined length of secretion, which is species dependent. In sheep the duration of melatonin secretion extends to 16 h in 8:16 LD (but does not expand further in DD) and retracts to 8 h in 16:8 LD (ROLLAG and NISWENEDER 1976; ARENDT 1986; LINCOLN et al. 1985), whereas in hamsters whilst the duration of secretion positively reflects the length of darkness it occupies rather less of the dark phase than in sheep (TAMARKIN et al. 1985; HOFFMAN 1981).

It is the duration of secretion which is the critical parameter signalling daylength for the organisation of seasonal rhythms. This change in duration as a function of night length is found in the vast majority of species studied with the possible exception of domestic pigs. It has, however, proved difficult to demonstrate duration changes in humans. Even in polar regions only small increases in duration are (sometimes) found in winter (BECK-FRIIS et al. 1984) compared to summer and usually the only seasonal change in human mela-

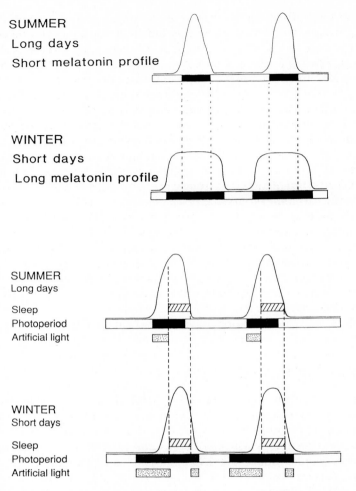

Fig. 3. Diagrammatic representation of melatonin secretion in long days and short days in for example the sheep (*upper panel*). Humans (*lower panel*) living in urban environments are normally not fully exposed to the variations in natural photoperiod, resulting in little change in duration of melatonin but a small phase delay in winter. Duration changes are seen if complete darkness is imposed for short or long periods. (WEHR 1991)

tonin is in phase with delayed phase being characteristic of winter in normal healthy individuals (ILLNEROVA et al. 1985; BROADWAY et al. 1987). Two daily light pulses of 2500 lux given as a skeleton spring photoperiod in the dim light conditions of winter in the Antarctic will phase advance the delayed melatonin to a summer phase position (ARENDT and BROADWAY 1986; BROADWAY et al. 1987). Thus the elements of a photoperiodic response remain in human physiology. Even more striking are the effects of keeping humans in an artificial light-dark cycle akin to that used for experiments in sheep. If humans are kept for 2 months in 8:16 LD where the 16 h of darkness are completely dark, it is possible to show an increase in the duration of melatonin secretion compared to the same individuals kept for 2 months in 14:10 LD (WEHR 1991). Evidently these are very artificial conditions. The normal light-dark cycle characteristic of urban communities would include a period of absolute darkness for around 8 h at night, preceded by 5–6 h of domestic intensity light and succeeded by 1–2 h of domestic intensity light with small changes from winter to summer. Given therefore that duration changes are not normally seen, this suggests that evening domestic intensity light is capable of suppressing melatonin secretion in spite of its low intensity (< 500 lux). It is rare for urban humans to experience more than 1000 lux during a working day (OKUDAIRA et al. 1983) and thus we are not exposed as a general rule to a strong light-dark zeitgeber. This suggests that other factors such as social cues and imposed behaviour, combined with a weak light-dark cycle, are important for synchronisation of human rhythms. This is illustrated very nicely by synchronisation of melatonin, cortisol and activity rhythms on a British Antarctic base where a strict regime is imposed (BROADWAY et al. 1987; O'CONOR 1992; GRIFFITHS et al. 1986) and free-run on a Greenpeace base where there were no behavioural restraints and where the light intensity was if anything higher than that on the British base (KENNAWAY and VAN DORP 1991).

D. Control of Seasonal Cycles

It is possible to mimic the duration change in melatonin in various ways. This approach has been used to define the function of melatonin in the control of seasonal cycles. By infusion of appropriate duration profiles of physiological levels of melatonin to pinealectomised hamsters and sheep, GOLDMAN and coworkers (CARTER and GOLDMAN 1983; GOLDMAN 1983; BARTNESS et al. 1993) and KARSCH and coworkers (KARSCH et al. 1984, 1988; BITTMAN 1985; WOODFILL et al. 1991) have shown the critical role of melatonin duration in signalling daylength in hamsters and sheep. In the short-day breeding sheep, long-duration melatonin is inductive to reproductive activity whereas in the long-day breeding hamster it is inhibitory and short-duration melatonin is inductive. The frequency of melatonin infusions does not appear to be critical over periodicities ranging from 20 to 26 h (MAYWOOD et al. 1992) but in order for an infusion to be read efficiently it must be perceived as a single block. For example two daily infusions of 4 h and 4 h separated by a 2-h gap in hamsters are read as 4 h not 8 h, whereas if the gap is sufficiently small the two periods are perceived as 8 h (GOLDMAN 1983). Photoperiodic history is important in the interpretation of any melatonin duration signal. For example transferring animals from 8:16 LD to 12:12 LD leads to perception of 12:12 LD as a long day whereas transferring from 16:8 LD to 12:12 LD leads to perception of 12:12 LD as a short day (HASTINGS et al. 1989).

KARSCH'S group have provided evidence in pinealectomised sheep that the seasonal breeding cycle is an endogenous annual rhythm which desynchronises from 365 days in the absence of the pineal and can be resynchronised by a single block of 70 consecutive days of long day melatonin infusions (WOODFILL et al. 1991, 1994). Thus the long days of spring are presumably both necessary and sufficient to cue the entire annual cycle in the absence of other seasonal time cues. Long days are similarly essential for the appropriate timing of pubertal development in sheep (EBLING and FOSTER 1989).

In ruminants it is possible to create a winter duration melatonin in natural or artificial summer photoperiod by feeding 3–5 mg adsorbed onto a food pellet 5–6 h before onset of darkness (KENNAWAY et al. 1982; ARENDT et al. 1983). The rumen essentially acts to give a slow-release profile of exogenous melatonin and combined with the endogenous production generates a winter long-duration profile. If animals are fed daily from midsummer in this way, they respond as if exposed to winter photoperiod by early onset of seasonal reproductive function, winter coat growth and the suppression of prolactin secretion found in short winter days (ARENDT 1986). Melatonin mimics the winter photoperiod. The use of continuous release implants or oral bolus preparations leads to elevated melatonin levels throughout the light-dark cycle. This, in summer, is read by body physiology as a 'super short' day and seasonal functions are appropriately modified (KENNAWAY et al. 1987; ENGLISH et al. 1986). Thus for agricultural purposes this is a practical approach to induction of early breeding – for early lamb production or manipulation of other commercially important animal products such as winter pelage and milk production. A bonus in some reports is an increased fertility and better synchronisation of the breeding season in a herd (KENNAWAY et al. 1987). It is necessary for the animal to experience a period of long days before exogenous melatonin will induce a short-day effect. This phenomenon is known as refractoriness and is common to photoperiodic responses in many species (FOLLETT 1982). It may represent the 'dead' zone of a seasonal PRC. Refractoriness is broken either by the natural long days of spring or by treatment with artificially long days early in the year (CHEMINEAU et al. 1986). For example in the short-day breeding goat, it is possible to advance oestrous by 178 days by keeping the animals in 20:4 LD for 2 months during January to March and follow this with either a melatonin implant or daily oral administration in the afternoon. After one successful oestrus advance in this way, a second suitably timed application of long days followed by melatonin leads to a complete reversal of the normal breeding cycle (DEVESON et al. 1991, 1992a).

For animals kept out of doors the disadvantage of this manoeuvre is that winter coat growth occurs in summer and vice versa leading to insufficient winter protection. However, by judicious timing of treatment it is possible to advance oestrous substantially but leave sufficient time for winter coat growth to resynchronise with the appropriate season (GEBBIE 1993).

Young sheep born out of season in autumn attain puberty at the normal time of year, i.e. in the following autumn (EBLING and FOSTER 1989). Young goats born in autumn rather than spring attain puberty the same winter 10–12 weeks earlier than if they had been born in the spring and may in principle be used for breeding earlier than animals born at a normal time of year (DEVESON et al. 1992b).

Pubertal development in photoperiodic species depends on exposure to an appropriate sequence of long and short days, indicated by changing melatonin secretion (EBLING and FOSTER 1989). Moreover prenatal photoperiod dictates the rate and timing of puberty presumably by foetal perception of maternal melatonin patterns (WEAVER and REPPERT 1986; DEVESON et al. 1992b).

Commercial preparations of melatonin for use in agriculture for the manipulation of seasonal breeding in sheep have been developed and at least one is registered and available in a number of countries including Australia and the United Kingdom. The long-term results of its use have yet to be fully evaluated.

E. Mechanism of Action of Melatonin in Control of Seasonal Rhythms

The target sites of melatonin are currently under intense investigation. Lesions of the SCN and the anterior hypothalamic area can block photoperiodic and/or circadian effects of melatonin in some rodents – but with a degree of disparity between laboratories (HASTINGS et al. 1991; EBLING et al. 1992; BARTNESS et al. 1993; BITTMAN 1993).

Implants of melatonin in the hypothalamus suppress luteinising hormone (LH) release in rats (FRASCHINI et al. 1968). Implants or infusion of melatonin in the hypothalamus mimics or blocks photoperiodic responses in several species (GLASS and LYNCH 1982; HASTINGS et al. 1991; LINCOLN and MAEDA 1992; LINCOLN 1992). In prepubertal rats melatonin inhibits LH-releasing hormone (LHRH)-induced LH release in pituitary cultures at concentrations comparable to those circulating in the blood (MARTIN and KLEIN 1975). There is evidence that melatonin influences LHRH secretion from the hypothalamus in co-cultures of median eminence and pars tuberalis (NAKAZAWA et al. 1991).

Early work using low specific activity radiolabelled melatonin served to focus attention on the brain as a site of uptake and binding of melatonin (CARDINALI et al. 1979). The development of $2\text{-}^{125}\text{I}$-iodomelatonin as a high specific activity ligand (VAKKURI et al. 1985a) has permitted the identification of high-affinity (K_d 25–175 pM), saturable, specific and reversible melatonin binding to cell membranes in the central nervous system initially in the SCN (VANECEK et al. 1987) and the pars tuberalis of the pituitary (MORGAN et al. 1989a; DE REVIERS et al. 1989). Both peripheral and central high-affinity binding sites for melatonin are found using 2-iodomelatonin as ligand. In ungulates binding is found in the pars tuberalis, rarely in the SCN, in septal and preoptic areas, in the inner and outer molecular layers of the hippocampus and the stratum lacunosum, entorhinal cortex, subiculum and apical interpeduncular nucleus (BITTMAN 1993). Evidently some of these sites (anterior hypothalamus, preoptic area, SCN) represent potential targets for the effects of melatonin on seasonal and circadian rhythms. The most consistent binding site is within the pars tuberalis of the pituitary gland, but with frequent binding in the retina, the SCN and many other brain sites. Various authors have reported melatonin-binding sites in anterior hypothalamic areas concerned with gonadotrophin release (see STANKOV and FRASCHINI 1994). More dense binding is found in lower vertebrates compared with mammals especially in retino-recipient areas (BITTMAN 1993; STANKOV and FRASCHINI 1994), underlying perhaps the more important role of the pineal and melatonin in these species. The evidence for sites of action in the brain and the pituitary gland in the control of seasonal rhythms is compelling. Melatonin implants within the brain act on seasonal functions in sheep when placed in the medial basal hypothalamus (LINCOLN and MAEDA 1992; LINCOLN 1992). This area together with adjoining brain and pituitary areas such as the pars tuberalis are obvious candidates. In one photoperiodic species, the ferret, binding is only found in the pars tuberalis and pars distalis; the pars tuberalis, of as yet unknown function, is a prime candidate for seasonal control mechanisms (WEAVER and REPPERT 1990) Whilst much is known of the pharmacology of the pars tuberalis binding site (see later) evidence for its physiological importance has been lacking.

There are changes in detectable binding with age, for example in foetal rats the first appearance of binding is in the pars distalis and pars tuberalis of the pituitary with SCN labelling appearing in later gestation (WILLIAMS et al. 1991) and with season (SKENE et al. 1993). Pars distalis binding is absent in adult rats but persists after birth in the neonate. This suggests that binding may indeed underlie function as melatonin will inhibit LHRH-induced pituitary LH release in prepuberty but not in adults.

Recently, however, LINCOLN and CLARKE (1994) have investigated seasonal rhythms in hypothalamic-pituitary-disconnected Soay rams, animals in which the pars tuberalis and pituitary are intact but have no input from hypothalamic neurohormones and transmitters. In such conditions tonic inhibitory, presumably dopaminergic, control of prolactin secretion is abolished as is the circadian rhythm, but nevertheless the well-characterised seasonal variations with high values in long days and low in short days persist and respond to both changing artificial daylength and melatonin administration. The inference is therefore that melatonin acts at the pituitary level for this particular seasonal variation via pars tuberalis receptors. The gonadotrophic hormones do not respond, however, and clearly there is scope for seasonal effects at hypothalamic level as well as at the pituitary. It has been known for some time that gonadal and prolactin responses can be differentiated. Notably both short-and long-day breeders show high prolactin in long days and low prolactin in short days. This is consistent with a role for prolactin in the control of winter coat growth (DUNCAN and GOLDMAN 1984), which of course must occur during the cold season irrespective of the breeding pattern of the animal. The role of the SCN and its melatonin-binding sites in seasonality is unclear and somewhat controversial. In sheep, a highly photoperiodic species like ferrets, most authors report a lack of SCN-binding sites.

F. Human Seasonality

As stated above humans show a melatonin response to changing photoperiod and thus in principle have the machinery necessary for a photoperiodic response. Human seasonality is evident in many areas. For example there is a seasonal variation in conception rate which has recently been carefully evaluated over different latitudes and over very large number of births (ROENNE-BERG and ASCHOFF 1990). This study considers that both photoperiod and ambient temperature may play a role in human reproduction. But there are major inconsistencies. PARKES (1976) described the inverse relationship of human seasonal conception rates in the United States and in Europe at similar latitudes during the 1950–1960s and clearly social and possibly fiscal considerations must play a major role. Many hormones, incidence of diseases, duration of sleep and aspects of the immune system show seasonal variations in humans (TOUITOU and HAUS 1992). Whether or not these are endogenous or exogenously derived is a matter for conjecture and little is known of their control mechanisms. Most attention has been paid to seasonal differences in psychiatric disorder whereby incidence of suicide (major depressive disorder) is highest in the spring with a secondary peak in autumn (EASTWOOD and PEACOCK 1976). Seasonal affective disorder or recurrent winter depression is best characterised (ROSENTHAL et al. 1984). Initially it was thought that this disorder was triggered by changing melatonin profiles in winter and summer and that the initial treatment of an artificial summer photoperiod with bright (2500 lux) light acted via a shortening of melatonin duration. Although this attractive hypothesis appears to be accepted publicly, i.e. in the national and international media, there is in fact very little evidence that changes in melatonin secretion are causal in the disease or indeed that light acts by its effects on melatonin secretion (WEHR et al. 1986; Wirz-Justice, personal communication).

G. Melatonin as a Circadian Marker Rhythm

Before considering the role of the pineal and melatonin in circadian rhythms it is of interest to describe the uses to which this molecule has been put in the evaluation of rhythm physiology and pathology. The melatonin rhythm in either plasma or saliva is arguably the best marker of the phase of the endogenous 'biological clock' principally because it is demonstrably generated in the SCN (MOORE and KLEIN 1974) and because very few 'masking' factors influence its production (WEVER 1989). This is in contrast to other marker rhythms such as core body temperature, which is affected by many things especially activity, and cortisol, which is modified inter alia by stress and meals

(MINORS and WATERHOUSE 1992). The major masking factor on melatonin is bright light, which can readily be controlled. A secondary factor is posture: passing from a recumbent to a standing posture leads to a small (ca 25%) increase in plasma levels (DEACON and ARENDT 1994a). Given control of these factors and in the absence of drug treatment the melatonin rhythm is essentially 'pure'. Moreover in the same individual it is highly reproducible in amplitude, in details of profile and in timing in a synchronised environment, almost like a hormonal fingerprint (ARENDT 1988). This means that small differences can be highly significant. Between individuals there are large differences in amplitude (e.g. ARENDT 1985, 1988) and the relationship of melatonin amplitude to clock function is ill defined.

Various characteristics of the rhythm have been used to define phase. The dim light melatonin onset (DLMO) is the onset of the evening rise in light, sufficiently dim as not to mask the secretion by suppression (usually < 50 lux, LEWY et al. 1984). Depending on methodology either a specific plasma or saliva concentration may be used as the cutoff point; alternatively the point where values exceed the limit of assay detection by a factor (usually of 2), provided that this point is succeeded by further increases, or the first point 3 standard deviations above baseline (light phase), provided that this point is succeeded by further increases, have all been used. The problem with using absolute plasma concentration is that individuals with high amplitude will appear to have earlier onsets than individuals with low amplitudes. The calculated acrophase (= time of peak value usually by cosinor techniques), the time of offset of secretion, usually defined as the reverse of the onset definition, and the duration of secretion (onset to offset) are all important and in general all these factors should be used to define the circadian profile.

The disadvantage of plasma melatonin is that only 'snapshots' can be taken of evolving circadian status in order to avoid unacceptable blood loss. The use of saliva avoids this, but brings with it the problem that it can only be collected during the night by waking the subject (VAKKURI et al. 1985b; NOWAK et al. 1987; ENGLISH et al. 1993).

The major urinary melatonin metabolite 6-sulphatoxy melatonin (aMT6s) has been extensively used in circadian studies. It is especially useful for long-term and field studies together with clinical evaluations in view of the non-invasive nature of sampling. It is present in much higher concentrations than melatonin in urine, is fairly robust, does not degrade in urine kept at 0–4 °C for 24 h and accurately reflects both the quantitative and qualitative aspects of melatonin secretion albeit with some loss of detail (ARENDT et al. 1985b; BOJKOWSKI et al. 1987; DEACON and ARENDT 1994b; ROSS et al. 1995) (Fig. 4). Acrophase, duration and amplitude can all be derived from sufficient numbers of sequential urine samples. There is, however, a need for a better descriptor of the melatonin profile than can be produced by cosinor techniques, even using harmonics.

aMT6s has been used extensively in humans to describe free-running rhythms, response to fractional desynchronisation, adaptation to shift work

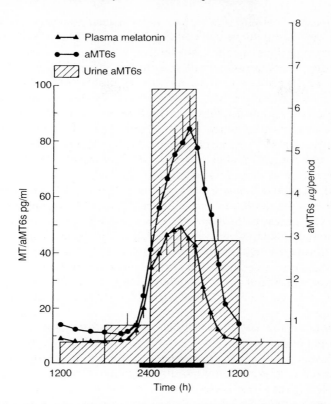

Fig. 4. Melatonin and 6-sulphatoxymelatonin (*aMT6s*) in hourly plasma samples and aMT6s in 6-h urine samples in normal human subjects. Note that the timing of the evening rise in both compounds is very comparable in all three fluids but that there is a delay in aMT6s clearance from plasma and urine in the morning. (From ARENDT 1988, by permission)

and jet lag and to characterise circadian rhythm status in blind subjects, the aged and in delayed sleep phase insomnia (see ARENDT 1994 for references).

In an entrained environment healthy subjects show a peak of melatonin between 0200 and 0500 hours with an onset (depending on definition) between 1900 and 2300 hours and an offset (depending on definition) between 0600 and 1000 hours. aMT6s in plasma and urine closely parallels the evening rise in melatonin, peaks about 2 h later than melatonin and declines to baseline values around 0900–1100 hours (ARENDT et al. 1985b; BOJKOWSKI et al. 1987).

Whilst the melatonin rhythm has been extensively used to characterise photoperiodic responses in large animals, there is much less information on detailed circadian profiles in rodents. This would appear to be due to the problem of sampling sufficient blood given assay sensitivities although very sensitive assays under development may eliminate this problem. Most authors have used measures of pineal melatonin in individual animals which have been shown to relate closely to plasma melatonin in rats (WILKINSON et al. 1977). In rodents and mink there is good evidence that urinary aMT6s can be measured non-invasively over lengthy periods and this may well provide another useful circadian tool (e.g. MAUREL et al. 1992; BROWN et al. 1991).

In humans a careful study (KENNAWAY et al. 1992) suggests that the onset of a circadian melatonin rhythm with peak values at night is established by 9 months of age. Secretion reaches a lifetime peak between 3 and 5 years, subsequently declining to adult levels by 15–18 years (WALDHAUSER and STEGER 1986). Amplitude remains relatively stable until old age when a drastic decline is reported in most studies (IGUCHI et al. 1982). Low amplitude in old age may be related to general lack of robustness of the circadian system. Whether or not the declining plasma concentrations during puberty have any causal role in pubertal development remains to be proven. In precocious puberty amplitude is high for age and in delayed puberty amplitude is low for age (DAS GUPTA et al. 1983; WALDHAUSER et al. 1991). Moreover hypothalamic amenorrhoea is associated with high melatonin levels (BERGA et al. 1988). High daytime melatonin is associated with anovulatory cycles in Finland (KAUPPILA et al. 1987). Thus evidence is accumulating for a role of melatonin in human reproduction.

Abnormal timing of melatonin has rarely been reported in clinical studies. There is some evidence that seasonal affective disorder (SAD) patients show slightly delayed melatonin rhythms compared to normal subjects in winter but this is by no means a consistent finding (LEWY et al. 1987; WINTON et al. 1989). Likewise a decline in amplitude in major depression is seen in some patient groups but not others and there is one report of increased amplitude in mania (see ARENDT 1989 for references). The substantial decline in melatonin amplitude in old age appears to be particularly related to insomnia and may be causal (HAIMOV et al. 1994).

One problem with many clinical observations is the lack of control of environmental and postural variables and much work needs reassessment with this in mind. Many reports of melatonin abnormalities in relation to different pathologies exist. They are summarised in other reviews (e.g. VAUGHAN 1984; WEBB and PUIG-DOMINGO 1995) and those that are not specifically concerned with rhythm abnormality will not be treated here.

The most obvious abnormalities in melatonin secretion are seen in blind subjects with no light perception (NLP) (SMITH et al. 1981; LEWY and NEWSOME 1983; ARENDT et al. 1988; ALDHOUS and ARENDT 1991; TZISCHINSKY et al. 1991) and in circadian dysrhythmia during adaptation to phase shift (FEVRE-MONTANGE et al. 1981; ARENDT et al. 1987). The number of blind subjects studied worldwide is small to date. The first reports of circadian problems in blind people described free-running rhythms of cortisol, core body temperature and sleep-wake in NLP patients in spite of strong 24-h zeitgebers and a 'normal' lifestyle living or attempting to live on a 24-h day (MILES et al. 1977). In a study using few sampling points SMITH et al. (1981) described an 'abnormal' melatonin rhythm. Only when sampling is frequent throughout 24 h and the patient is followed continuously or at intervals over several weeks does an accurate picture emerge. Several groups have described both free-running rhythms and entrained but abnormally phased rhythms of melatonin and other variables in the blind (ALDHOUS and ARENDT 1991; LEWY and NEWSOME 1983). Recently CZEISLER et al. (1995) have shown that in some patients with no light perception circadian entrainment is possible and that this is linked to the ability of light to suppress melatonin in these individuals. This phenomenon of 'hypothalamic light perception' is known in animals where the retinohypothalamic projection is intact but the primary and accessory optic tracts have been sectioned (MOORE-EDE et al. 1983). It is possible that blind people who have retained the ability to synchronise to the 24-h day have intact 'circadian' photoreceptors and retinohypothalamic tract. The nature of the photoreceptors concerned with circadian responses is as yet unknown. The use of retinally degenerate mice who show a PRC to light in terms of activity suggests that they may be specialised cones (FOSTER et al. 1991). An interesting feature of free-run in blind people attempting to live on a 24-h day is that they suffer periodic sleep disturbance and this appears to occur mostly when the melatonin rhythm and indeed other endogenous circadian rhythms are 180° out of phase with the normal cycle (FOLKARD et al. 1990; ARENDT et al. 1988).

H. Effects of Melatonin on Circadian Rhythms

The relationship of the pineal to the circadian system in mammals has already been referred to as modulatory. It is difficult to make a case for a major role of the pineal or melatonin in mammalian circadian control. There are nevertheless very interesting effects of melatonin on mammalian circadian rhythms which are proving of considerable therapeutic use.

In rats free running in DD, daily melatonin injections (5 μg to 1 mg) can resynchronise activity rest cycles to 24 h when the injection time approaches the time of free-running activity onset (REDMAN et al. 1983). This observation, initially controversial, has been amply confirmed. In rats exhibiting a persistent phase delay of activity rest cycles in entrained conditions, melatonin will phase advance the rhythm and, following forced 8-h advance phase shift, when given at the original dark onset, it can dictate the direction of re-entrainment as an advance (ARMSTRONG and REDMAN 1985; ARMSTRONG 1989). Curiously the PRC of activity-rest to melatonin injections in rats shows only phase advances. Whether or not these phase-shifting and entraining properties apply to all mammals is at present doubtful. Although some authors report similar effects in hamsters, others consider that arousal is the primary effect of the injection, as vehicle injections have similar effects (HASTINGS et al. 1992). Hamsters are very sensitive to arousal of various sorts: merely the presence of a new activity wheel in the animals' cage is sufficient to immediately re-entrain hamsters after a forced phase shift (MROSOVSKY and SALMON 1987). Hastings and coworkers have provided evidence that arousal-induced phase shifts in hamsters are mediated through serotoninergic mechanisms. However, there is clear evidence for entrainment of hamster pups by prenatal melatonin treatment (DAVIS and MANNION 1988).

In humans both pharmacological and physiological doses of melatonin can induce phase shifts. Early work suggested that 2 mg melatonin given daily in the late afternoon (1700 h) for a period of 1 month was sufficient to induce a phase advance of the endogenous melatonin rhythm of 1–3 h (when the endogenous and exogenous components could be distinguished) (ARENDT et al. 1985a; WRIGHT et al. 1986) (Fig. 5). This was accompanied by an earlier onset of evening fatigue or sleep (ARENDT et al. 1984), suggesting an effect on the timing mechanisms of sleep, and also an earlier timing of the prolactin rhythm. This study clearly suggested that melatonin had 'chronobiotic' effects on the human circadian system. An acute sleep-inducing effect of pharmacological doses of melatonin has been known since 1974 (CRAMER et al. 1974; see VOLLRATH et al. 1981). There is evidence that melatonin shifts sleep time in delayed sleep phase insomnia (DAHLITZ et al. 1991). Lavie and coworkers (TZICHINSKY et al. 1992a) have recently reported that low-dose (5 mg) mela-

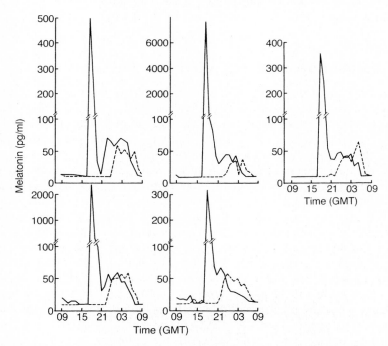

Fig. 5. Melatonin (2 mg) in the late afternoon (1700 hours, see Fig. 11) phase advances the onset of its own endogenous secretion (*solid line*) compared to placebo (*dotted line*) in those subjects where the endogenous and exogenous plasma melatonin can be distinguished. Note the supraphysiological peak of exogenous melatonin followed by the endogenous night time secretion. (From ARENDT et al. 1985a, by permission)

tonin increases sleep propensity in a time-dependent manner, DOLLINS et al. (1994) describe an increase in fatigue after acute administration of physiological amounts of melatonin and DEACON and ARENDT (1995a) report acute, dose-dependent advances in sleep timing over the dose range 0.05–5.0 mg.

A number of other hormones have been assessed after chronic low doses of melatonin. No significant effects were seen on gonadotrophins, testosterone, GH, T4 or cortisol (WRIGHT et al. 1986). Whilst subsequent work shows similar phase shifts in cortisol to those of prolactin (J. Arendt et al. 1995, unpublished), and stimulation of prolactin secretion by acute melatonin treatment (WALDHAUSER et al. 1987), the fact that no changes were seen in the concentration of reproductive hormones after chronic low-dose treatment indicated that it was unlikely that such amounts of melatonin would affect human reproductive function in contrast to the potent effects in photoperiodic seasonal breeders. Very large doses of melatonin (80–300 mg) have been shown variously to suppress LH and in combination with norethistrone (an oral contraceptive minipill) to suppress ovulation when given at night (VOORDOW et al. 1992), to increase the amplitude of LH pulses when given in the morning (CAGNACCI et al. 1991) and to potentiate testosterone-induced suppression of LH when given in the late afternoon (ANDERSON et al. 1993). However, the therapeutic potential of low doses of melatonin with regard to circadian rhythm disorders is not compromised by these pharmacological effects.

Melatonin has subsequently been shown to induce phase delays in humans when given in circadian early morning (LEWY et al. 1992) and two full human PRCs using endogenous melatonin as the marker rhythm have been described (LEWY et al. 1992; ZAIDAN et al. 1994) (Fig. 6). ZAIDAN and coworkers have demonstrated a PRC using melatonin infused at physiological concentrations and LEWY and coworkers used oral melatonin sufficient to achieve or exceed night-time levels. The melatonin PRC approximately mirrors the light PRC (MINORS et al. 1991), lending further credence to the concept of melatonin as a darkness hormone with respect to the circadian system as well as in photoperiodism. Given the PRCs to light and melatonin, and the general tendency of human circadian rhythms to delay, it is possible that early morning light resets the endogenous clock daily by phase advance and the evening melatonin rise may reinforce this effect. In rare individuals with a periodicity of less than 24 h, evening light may reset the system by delay and the melatonin still present in the early morning may reinforce this. Thus it is possible to propose a physiological role for the pineal and melatonin in circadian organisation in mammals. Nevertheless it is noteworthy that the maximum phase changes induced by both light and melatonin are found at times when under normal environmental conditions these two zeitgebers are not experienced, i.e. in the second half of the light phase for melatonin and in the second half of the dark phase for light, and that the magnitude of the induced dawn and dusk phase changes is very small according to the PRCs.

For melatonin to be a principal zeitgeber in humans it would be necessary to show that it can extend the range of entrainment analogous to bright light

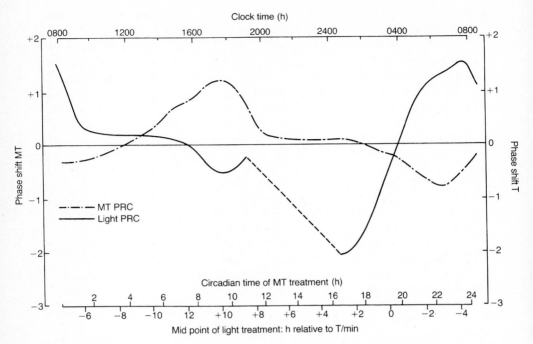

Fig. 6. A comparison of the phase response curves of core body temperature to bright white light redrawn from MINORS and WATERHOUSE (1992), and of the endogenous melatonin rhythm to exogenous melatonin (LEWY et al. 1992). Note that the responses mirror each other to some extent, e.g. morning bright and evening melatonin phase advance, evening bright light and morning melatonin phase delay. *MT*, melatonin; *T*, core body temperature; *PRC*, phase response curve. Approximate clock time is indicated

and to re-entrain free-running individuals in the absence of other time cues. WEVER (1989) attempted to extend the range of entrainment during fractional desynchronisation with melatonin without success, but it is likely that the timing of administration was inappropriate. Effectively he administered it in a phase-advancing position when a phase delay was required. Experiments in the blind have addressed the free-running problem but in circumstances where other 24-h zeitgebers were present. In the case of one blind (NLP) man melatonin was able to entrain his sleep wake cycle but without entrainment of core body temperature, cortisol or endogenous melatonin (ARENDT et al. 1988; FOLKARD et al. 1990). In a number of other reports melatonin (SARRAFZADEH et al. 1990; PALM et al. 1991; TZISCHINSKY et al. 1992b) has improved sleep in some blind subjects and, where assessed, without entrainment of the endogenous melatonin rhythm (ALDHOUS and ARENDT 1991). SACK et al. (1991) have reported the (rather wobbly) entrainment of the endogenous melatonin rhythm in another blind subject. JAN et al. (1994) have reported stabilisation of sleep in children with disabilities and apparently desynchronised and disrupted sleep wake cycles.

However, when melatonin is used in concert with other time cues to hasten adaptation to phase shift it clearly does affect endogenous variables such as core temperature, cortisol and endogenous melatonin as well as the sleep wake cycle (Arendt et al. 1987; Samel et al. 1991).

I. Melatonin and Core Body Temperature

Some of the most important correlative associations of melatonin observed as early as 1979 (Akerstedt et al. 1979, 1982) are that in humans the nadir of core temperature occurs within 1 h of the peak of melatonin together with the peak of fatigue and the trough in ability to perform certain tasks (reflected in high night-time accident rates in night shift workers, Smith et al. 1994). In the case of core temperature this association may in part be causal. Melatonin (0.5–5 mg) acutely suppresses core body temperature and Cagnacci et al. (1992) have calculated that possibly 40% of the amplitude of the rhythm in core body temperature may be due to night time melatonin secretion. The acute suppression induced by a single dose of exogenous melatonin is dose dependent, accompanied by decreased alertness, and correlates closely to the acute phase advance induced when given at 1700 h (Deacon et al. 1994; Deacon and Arendt 1995a) (Fig. 7). Importantly bright light increases body temperature and alertness when given at night and this can be reversed by melatonin (Strassman et al. 1991). Whether or not the changes in core body temperature are integral to phase-shifting mechanisms remains to be proven but this hypothesis may prove fruitful in the future.

J. Therapeutic Uses of Melatonin

The above discussion underlies the current appreciation of melatonin as a treatment for circadian rhythm disorder. This condition affects very large numbers of people. Encompassed within this category are shift work, jet lag, blindness, insomnia and other problems of old age, some psychiatric disorders and in conditions where natural zeitgebers are very weak (e.g. dim light in high-latitude winters). Circadian rhythm disturbance may also accompany many other conditions as a secondary effect (Touitou and Haus 1992).

The problems of shift work, where individuals are required to work during the low point of performance and high point of fatigue and to sleep at an inappropriate circadian time, are of real economic importance (Smith et al. 1994; Akerstedt 1991). Moreover the health problems of shift workers include sleep disorder, gastrointestinal disturbance and increased susceptibility to cardiovascular disease (Minors and Waterhouse 1981; Knutsson 1989). All of these may be related to rhythm disruption. Jet lag can be thought of as just a

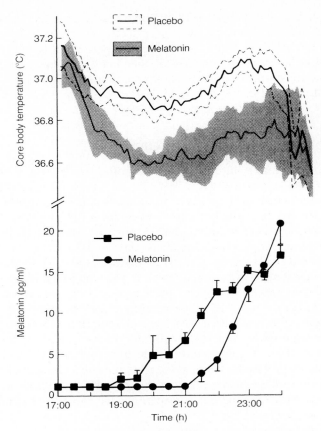

Fig. 7. Phase shift of melatonin and suppression of core body temperature using 5 mg melatonin compared to placebo. A single treatment of melatonin (5 mg) in the late afternoon (1700 h) acutely suppresses temperature and phase advances the endogenous melatonin rhythm (in saliva) the following day when subjects are kept seated and in dim (<200 lux) light from 1700 h until bedtime (2300 h). (From data in DEACON et al. 1994, published by Elsevier, figure from ARENDT 1994, by permission)

nuisance and some people appear to be immune to its effects, but it can seriously compromise the businessman's ability to function and spoil the 1st week of a holiday. In some respects shift work and jet lag are similar, in others they are not. For example night-shift workers operate throughout their night shift counter to the natural zeitgebers whereas time zone travellers adapt with the help of the environment. The use of increased light intensity at night (>1000 lux) and with specified phase-shifting light regimes is proving to be successful as an aid to adaptation and improved efficiency (ROSA et al. 1990; MOORE-EDE 1993; CZEISLER et al. 1990). However, not all circumstances allow the use of bright light, it is expensive to install and maintain and its potential deleterious effects on the eyes remain to be fully assessed. The use of melatonin as a phase-shifting mechanism offers a convenient alternative. The combined use of timed melatonin and bright light is likely to provide optimum phase-shifting conditions.

Early work indicated, in a placebo-controlled study, that over an advance phase shift of eight time zones suitably timed melatonin was able significantly to improve night sleep latency and quality, daytime alertness and to hasten the resynchronisation of endogenous melatonin and cortisol rhythms (ARENDT et al. 1986, 1987). In this study melatonin (5 mg) was given for three consecutive days at 1800 h to initiate a phase advance before the flight and was subsequently taken at bedtime postflight (phase advance position) for 4 days. Subsequent work has confirmed the behavioural effects in field studies (ARENDT and ALDHOUS 1988; PETRIE et al. 1989; CLAUSTRAT et al. 1992) and the circadian re-entrainment in simulation studies (SAMEL et al. 1991). We have further shown that a single preflight administration with 4 days postflight treatment is effective (unpublished) and it is possible that preflight treatment is unnecessary. In a study of aircrew taking melatonin, preflight and postflight treatment proved ineffective (PETRIE et al. 1993). Aircrew, however, are very difficult subjects as their circadian status is variable and it is possible that treatment was inappropriately timed. The timing of melatonin treatment for westward time zone transitions poses problems over less than eight time zones as according to the PRC it should be taken in the middle of the night. Moreover preflight early morning treatment to initiate a phase delay can lead to loss of alertness at a very inappropriate time of day. In the authors' experiments (Fig. 8) melatonin is not given preflight but is taken at bedtime after westward flight for 4 days. To date, with 474 people taking melatonin and 112 taking placebo, the overall reduction in perceived jet lag (visual analogue scale, 100 = very bad jet lag, 0 = negligible jet lag) over all time zones and in both directions is 50%. Included in these figures are both placebo controlled and uncontrolled studies with no difference between these. Side effects reported more than once are (melatonin%-placebo%): sleepiness (8.3%-1.8%), headache (1.7%-2.7%), nausea, (0.8%-0.9%), giddiness (0.6%-0%), light-headedness (0.8%-0%).

Melatonin has been used in two small controlled studies of night-shift workers, compared to both placebo and no treatment, and timed to phase delay by administration after the night shift and prior to daytime sleep. It was successful in improving night-shift alertness, sleep duration and quality in one study and improved circadian adaptation assessed by endogenous melatonin production in the other: clearly many more data are required and especially its effects on work related performance must be assessed (FOLKARD et al. 1993; SACK et al. 1994).

Recently we have developed a means of simulating time-zone transition or shift work without environmental isolation using a combination of timed moderately bright light (1200 lux, 9 h followed by 8 h imposed darkness or sleep), delaying or advancing by 3 h each day followed by 2 days stable treatment at the new phase. In this way it is possible to induce 9 h phase shifts

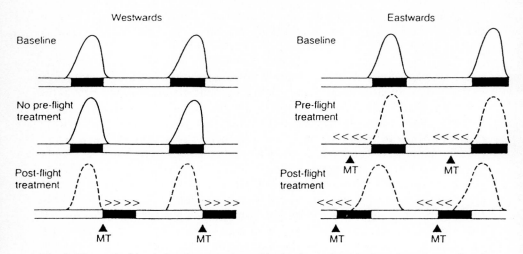

Fig. 8. Theoretical basis for the treatment of jet lag by melatonin, based on the predicted phase position of the endogenous melatonin rhythm as a function of the local light dark cycle (*dark bar*, darkness). Preflight melatonin in *solid lines*, postflight-predicted melatonin in *dotted lines*. *Arrowheads* indicate the time of melatonin (*MT*) treatment. Westwards postflight melatonin (4 days) at bedtime should initiate phase delays. Eastwards pre (1 day) and post (4 days) flight melatonin as indicated should initiate and maintain phase advances. Using this approach more than 500 subjects have participated in jet lag studies during the last 9 years. Melatonin reduces perceived jet lag overall by 50%. (ARENDT, ENGLISH and WRIGHT, unpublished)

whilst maintaining internal synchronisation. Subsequently subjects are abruptly returned to the local time cues thus simulating time zone transition or rotating shift systems. After an abrupt phase advance of 9 h, melatonin (5 mg), timed to phase advance and with no preshift treatment, immediately improves sleep quality and duration together with alertness and the ability to perform low and high memory load cancellation tests (DEACON and ARENDT 1995b, c) (Fig. 9). This improvement is evident before any major phase shift has occurred in endogenous markers such as core body temperature. Although resynchronisation of endogenous variables is hastened, the acute effects of melatonin appear to be as important as any induced phase shift. In the author's opinion melatonin acutely and chronically reinforces physiological phenomena connected with darkness, in particular sleep.

K. Mechanism of Action of Melatonin in the Control of Circadian Rhythms

The most obvious target tissue for the actions of melatonin on the circadian system is the SCN and, in view of its effects on the photomechanics of light transduction (STEINLECHNER 1991), receptors in the retina may also be of

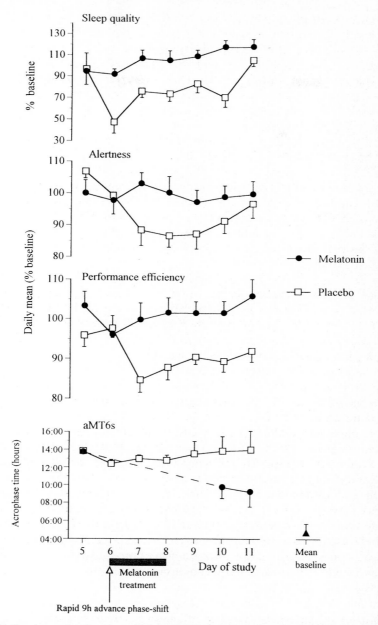

Fig. 9. Following a simulated, rapid 9-h advance phase shift, melatonin (5 mg), timed to phase advance, improves sleep, alertness and performance efficiency, and hastens the resynchronisation of the endogenous melatonin rhythm in seven healthy volunteers in their normal environment. (From data in DEACON and ARENDT 1995, unpublished figure)

importance in circadian light perception. SCN-lesioned rats do not respond to melatonin by restoration of activity rest cycles, although of course if the SCN itself is essential for generation and expression of activity-rest rhythms this experiment is susceptible to different interpretations (CASSONE et al. 1986). Given that the phase-shifting effects of melatonin (and possibly light) are so closely related to changes in body temperature, it is possible that the systems concerned with minute to minute body temperature control will prove to be of equal if not greater importance.

The SCN shows clear melatonin binding in human postmortem tissue (REPPERT et al. 1988; WEAVER et al. 1993). In parallel with the brain studies, Dubocovich and coworkers have demonstrated a functional melatonin receptor initially in rabbit and chicken retina. Activation of the receptor is associated with inhibition of calcium-dependent dopamine release. The receptors are localised in the inner plexiform layer containing dopamine amacrine cells in rabbits, also in the outer and inner segments in mice and possibly the pigmented layer in some mammals (DUBOCOVICH 1985, 1988; DUBOCOVICH and TAKAHASHI 1987).

By far the most convincing evidence that the SCN is a target site for circadian effects of melatonin is that of Gillette and coworkers (MCARTHUR et al. 1991). Using SCN-containing hypothalamic slice preparations in vitro, it is possible to show that the 24-h rhythm of electrical activity persists in vitro for several cycles and is phased in relation to the donor's previous light-dark cycle. Physiological concentrations of melatonin added to the cultures at different circadian times induce substantial phase advances according to a phase response curve which resembles that seen in the intact animal using activity rest as a circadian marker rhythm or indeed to that in humans using endogenous melatonin as a marker rhythm.

A range of low nanomolar to low picomolar affinity binding to membranes has been reported by some authors using ^{125}I-melatonin (LAUDON and ZISAPEL 1986; ZISAPEL et al. 1991; DUBOCOVICH 1988) although more recently the existence of different classes of sites has been disputed (SUGDEN and CHONG 1991). Until the endogenous concentration of melatonin at its target sites is known with certainty the physiological relevance of low- or high-affinity binding will remain controversial. In most mammals melatonin is thought to be secreted primarily into the blood, subsequently entering the brain, and blood concentrations are in the pico- to femtomolar range. These concentrations are physiologically relevant, as shown by infusion studies in hamsters and sheep, and receptor affinity should be in this range. Melatonin, however, has many effects if administered in quantities sufficient to generate nanomolar concentrations or above. Such effects may not be 'physiological' but they are nevertheless of interest and some are exploitable therapeutically. Both low- and high-affinity receptors have been shown to vary systematically with physiological status (ZISAPEL et al. 1991; SKENE et al. 1993; MASSON-PEVET and GAUER 1994).

L. Melatonin Receptors

ROLLAG and coworkers (WHITE et al. 1987) demonstrated that melatonin-induced pigment aggregation in amphibian melanophores is a pertussis-toxin-sensitive system and that melatonin inhibits forskolin-activated cAMP formation. Intensive investigation of the properties of the pars tuberalis binding site has revealed that physiological doses of melatonin inhibit forskolin-activated cAMP production in vitro in a time- and dose-related manner (MORGAN et al. 1989a,b, 1991; MORGAN and WILLIAMS 1989). There is other good evidence that the binding sites are coupled to G-proteins. GTP analogues which interfere with the regeneration of G_1-coupled receptors decrease the affinity and sometimes the capacity of ^{125}I-melatonin binding in reptiles, birds and mammals (MORGAN and WILLIAMS 1989). However, the receptors in the pars tuberalis may act via both pertussis-sensitive and -insensitive G-proteins (MORGAN et al. 1990) and there is preliminary evidence that some neuronal-binding sites are GTPγs insensitive and have a different molecular weight to the pars tuberalis site (MORGAN et al. 1995).

Interestingly the inhibition of forskolin-activated cAMP formation by melatonin is sensitised by pretreatment with melatonin, with a maximum effect at 16 h. This suggests a mechanism for the perception of duration of melatonin secretion for the control of seasonal functions (HAZELRIGG et al. 1993).

The amphibian melatonin receptor was cloned in 1994 (EBISAWA et al. 1994) and cloning of the sheep and human receptors was reported shortly thereafter (REPPERT et al. 1994) with high structural homolgy (80%) between sheep and human clones. In situ hybridisation studies in several mammals have shown signals in both the pars tuberalis and the SCN and the pharmacology of the recombinant receptors is identical to the endogenous G-protein-linked receptor (REPPERT et al. 1994). These receptors are members of a new receptor group that is distinct from other G-protein-linked groups. This work opens large new perspectives and approaches not only for the study of the mechanism of action of melatonin but also for the development of new molecules for therapeutic use. A model for the melatonin-binding site has been proposed using computer-modelling techniques (LEWIS et al. 1990; SUGDEN et al. 1995). With the recent publication of the receptor sequence (Fig. 10) the importance of this model has been reinforced by the substantial sequence agreements revealed between prediction and actuality.

M. Melatonin Antagonists and Agonists

In view of the very large potential market for melatonin-like effects in both occupational health and pathology a number of pharmaceutical companies have initiated development of specific formulations of melatonin and of novel analogues. The 6- and 2-substituted halogenated melatonins have been known for many years and are good agonists both in vivo and in vitro (FLAUGH et al.

```
                                                                                                        I
Xenopus     MMEVNSTCLDCRTPGTI RTEQDAQDS ASQG.....LTSALAVVLIFTI VVDVLGNI LVILSVLRNKKLQNA GN  68
Sheep       MAGRLWGSPGGTPKGNGSSALL NVSQAAP GAGDGVRPRPSWLAATLASI LIFTI VVDI VGNLLVVLSVYRNKKLRNA GN  79
Human                   MQGNGSALPNASQPVLR GDGA...RPSWLASALACVLIFTI VVDI LGNLLVILSVYRNKKLRNA GN  63

Consensus                                              L....LIFTI VVD..GN..LV.LSV.RNKKL..NA GN

                                                                              III
Xenopus     LFVVSLSI ADLVVAVYPYPVILI AI FQNGWTLGNI HCQI SG FLMGLSVI GSVFNI TAI AI NRYCYI CHSLRYDKLYN QR 147
Sheep       VFVVSLAVADLLVAVYPYPLAL ASI VNNGWSLSSLHCQLSG FLMGLSVI GSVFS I TGI AI NRYCCI CHSLRYGKLYS GT 158
Human       IFVVSLAVADLVVAI YPYPLVL MSI FNNGWNLGYLHCQVSG FLMGLSVI GSI FNI TGI AI NRYCYI CHSLKYDKLYS SK 142
Hamster                                                                CYI CHSLKYDRLYS NK
Rat                                                                    CYI CHSLKYDRI YS NK

Consensus   .FVVSL..ADL..VA.YPYP..L...I..NGW.L...HCQ.SG FLMGLSVI GS.F..I T.I AI NRYC.I CHSL.Y.K.Y...

                                                                              V
Xenopus     STWCYLGLTWI LTI AI VPNFFVGSLQYDP RI FSCTFAQTV SSSYTI TVVVHFI VPLSVVTFCYLRI WVLVI QVKH RV 226
Sheep       NSLCYVFLI WTLTLVAI VPNLCVGTLQYDP RI YSCTFTQSV SSAYTI AVVVFHF L VPML VVVFCYLRI WALVLQVRWKV 237
Human       NSLCYVLLI WLLTLAAVLPNLRAGTLQYDP RI YSCTFAQSV SSAYTI AVVVFHF L VPMI I VI FCYLRI WI LVLQVRQ RV 221
Hamster     NSLCYVFLI WVLTLVAI MPNLQTGTLQYDP RI TSCTFTQSV SSAYTI AVVVFHF L VPMI I VPMI I VTFCYLRI WI LVLQVRR RV
Rat         NSLCYVFLI WTLTLI AI MPNLQTGTLQYDP RI YSCTFTQSV SSAYTI ALVVFHF V VPMI I VTFCYLRI WI LVLQVRR RV

Consensus   ..CY...L..W.LT..A..PN....G.LQYDP RI .SCTF.Q.VSS.YTI...VV.HF.VP....V.FCYLRI W.LV.QV...V

                                                                 VII
Xenopus     RQDF FQKL TQTDLRNFLT MFVVFVLFAVCWAPLNFI GLAVA I NPFHVAPKI PEWL FVLSYFMAYFNSCLNAVI YGVL NQ 305
Sheep       KPDNKPKLKPQDFRNFVTMFVVFVLFAI CWAPLNFI GLVVA SDPASMAPRI PEWL FVASYMAYFNSCLNAI I YGLL NQ 316
Human       KPDRKPKLKPQDFRNFVTMFVVFVLFAI CWAPLNFI GLAVA SDPASMAPRI PEWL FVASYYMAYFNSCLNAI I YGLL NQ 300
Hamster     KPDSKPRLKPQDFRNFVTMFVVFVLFAI CWAPLNFI GLI VA SDPATMAPRI PEWL FVASYY
Rat         KPDSKPKLKPQDFRNFVTMFVVFVLFALCWAPLNFI GLI VA SDPATMAPRI PEWL FVASYY

Consensus   ..D...K...L.P.D.RNF.TMFVVFVLFA.CWAPLNFI GL.VA....?....I PEWL FV.SY.MAYFNSCLNA..I YG.LNQ

Xenopus     NFRKEYKRI L MSLLTPRLL FLDTSRGGTEG LKSKPSPAVTN NNQADML GEARSL WLSRRNGAKMVI I I RPRKAQI AI I H 384
Sheep       NFRQEYRKI I VSLCTTKMFFVDSSNHVADRI KRKPSPLI AN HNLI KVDSV                              366
Human       NFRKEYRRI I VSLCTARVFFVDSSNDVADR VKWKPSPLMTN NNVVKVDSV                               350

Consensus   NFR.EY...I...SL.T.....F.D.S......KPSP....N.....

Xenopus     QI FWPQSSWATCRQDTKI TGEEDGCRELCKDGI SQR  426
```

1979). Presumably in both cases but particularly the 6-substituted molecule there is protection against metabolic degradation and hence a lower dose requirement for the induction of a given effect.

A series of agonists and antagonists have been developed from napthalene derivatives recently (Yous et al. 1992), and also using benzofuran, benzothiophene or benzimidazole as the bicyclic ring (Caignard 1994). They show a range of affinity for the pars tuberalis melatonin receptor, some being of much higher affinity than melatonin. The most interesting napthalenic agonists have similar effects to melatonin on SCN electrical activity, on in vivo rhythm physiology in rodents (Guardiola-Lemaitre 1994) and more recently on human circadian rhythms (Arendt, English and Defrance, unpublished; Wirz-Justice and Defrance, unpublished).

Sugden and coworkers have reported that a number of N-acyl-4-aminomethyl-5-methoxy-9-methyl-1,2,3,4-tetrahydrocarbazoles have high affinity for the melatonin receptor in chick brain and are agonists in dermal melanophore migration assays for melatonin (Davies et al. 1995). This group have also described the high affinity of 2-phenyltryptamines for the chick brain receptor with agonist or antagonist activity depending on the structure of the side chain, and consider that N-cyclopropanecarbonyl-2-phenyltryptamide is the most potent agonist so far reported (Jones et al. 1995).

Besides these analogues many other potentially useful molecules are under investigation. Two early antagonists were reported but neither has proved to be consistently useful in vivo (Laudon et al. 1988; Krause and Dubocovich 1990). However, the new antagonists may prove more interesting and their detailed pharmacology is eagerly awaited.

Melatonin has provoked the development of a whole new pharmacology of so-called chronobiotic drugs. It would be reasonable to identify clearly this class of melatonin-like molecules by a suitable generic name applied to molecules with defined phase-shifting and resynchronising properties. The important acute effects of melatonin on core body temperature and on alertness should also be included as a formal property of the class. 'Melatoninomimetic' drugs is the logical choice; however, this clumsy appelation might well be shortened to 'melatomimetic'.

Fig. 10. Deduced amino acid sequence of mammalian melatonin receptors and their comparison with the *Xenopus* high-affinity melatonin receptor. The full-length coding regions of the *Xenopus*, sheep and human receptor are shown. Reverse transcriptase polymerase chain reaction (RT-PCR)-generated fragments of the coding regions of the Siberian hamster and rat receptors are also depicted. To maximise homologues, gaps (represented by *dots*) have been introduced into the sequences. The seven presumed transmembrane domains (I–VII) are highlighted by *solid bars*. Consensus sites for N-linked glycosylation are *underlined*. The sequences have been deposited in GenBank under accession numbers U14108 (sheep), U14109 (human), U14110 (Siberian hamster and U14409 (rat). (From Reppert et al. 1994, by permission)

References

Akerstedt T (1991) Sleepiness at work: effects of irregular work hours.In: Monk T (ed) Sleep, sleepiness and performance. Wiley, New York, chap 5

Akerstedt T, Froberg JE, Friberg W, Wetterberg L (1979) Melatonin excretion, body temperature and subjective arousal during 64 hours of sleep deprivation. Psychoneuroendocrinology 4:219

Akerstedt T, Gillberg M, Wetterberg I (1982) The circadian covariation of fatigue and urinary melatonin. Biol Psychiatry 17:547–554

Aldhous ME, Arendt J (1991) Assessment of melatonin rhythms and the sleep wake cycle in blind subjects. In: Arendt J, Pevet P (eds) Proceedings of the European Pineal Society, Guildford, 1990. Adv Pineal Res 5:307–311

Anderson RA, Lincoln GA, Wu FC (1993) Melatonin potentiates testosterone-induced suppression of luteinising hormone secretion in normal men. Hum Reprod 8:1819–1822

Arendt J (1983) Biological rhythms (review). Int Med 3:6–9

Arendt J (1985) Mammalian pineal rhythms. Pineal Res Rev 3P:161–213

Arendt J (1986) Role of the pineal gland and melatonin in seasonal reproductive function in mammals. Oxford Rev Reprod Biol 8:266–320

Arendt J (1988) Melatonin. Clin Endocrinol (Oxf) 29:205–229

Arendt J (1989) Melatonin – a new probe in psychiatric investigation. Br J Psychiatry 155:585–590

Arendt J (1994) Melatonin and the mammalian pineal gland. Chapman Hall, London

Arendt J, Aldhous M (1988) Further evaluation of the treatment of jet-lag by melatonin: a double blind crossover study. Annu Rev Chronopharmacol 5:53–55

Arendt J, Broadway J (1986) Phase response of human melatonin rhythms to bright light in Antarctica. J Physiol (Lond) 377:68

Arendt J, Brown WB, Forbes JM, Marston A (1980) Effect of pinealectomy on immunoassayable melatonin in sheep. J Endocrinol 85:1–2P

Arendt J, Symons AM, Laud CA, Pryde SJ (1983) Melatonin can induce early onset of the breeding season in ewes. J Endocrinol 97:395–400

Arendt J, Borbely AA, Franey C, Wright J (1984) The effect of chronic, small doses of melatonin given in the late afternoon on fatigue in man: a preliminary study. Neurosci Lett 45:317–321

Arendt J, Bojkowski C, Folkard S, Franey C, Minors DS, Waterhouse JM, Wever RA, Wildgruber C, Wright J (1985a) Some effects of melatonin and the control of its secretion in man. In: Evered D, Clark S (eds) Photoperiodism, melatonin and the pineal. Pitman, London, pp 266–283 (Ciba Foundation symposium 117)

Arendt J, Bojkowski C, Franey C, Wright J, Marks V (1985b) Immunoassay of 6-hydroxymelatonin sulfate in human plasma and urine: abolition of the urinary 24-hour rhythm with atenolol. J Clin Endocrinol Metab 60:1166–1173

Arendt J, Aldhous M, Marks V (1986) Alleviation of jet-lag by melatonin: preliminary results of controlled double-blind trial. Br Med J 292:1170

Arendt J, Aldhous M, English J, Marks V, Arendt JH, Marks M, Folkard S (1987) Some effects of jet lag and their alleviation by melatonin. Ergonomics 30:1379–1393

Arendt J, Aldhous M, Wright J (1988) Synchronisation of a disturbed sleep-wake cycle in a blind man by melatonin treatment. Lancet i:772–773

Armstrong SM (1989) Melatonin and circadian control in mammals. Experientia 45:932–939

Armstrong SM, Redman J (1985) Melatonin administration: effects on rodent circadian rhythms. In: Evered D, Clark S (eds) Photoperiodism, melatonin and the pineal. Pitman, London, pp 188–207 (Ciba Foundation symposium 117)

Axelrod J (1974) The pineal gland: a neurochemical transducer. Science 184:1341–1348

Bartness TJ, Powers B, Hastings MH, Bittman EL, Goldman BD (1993) The timed infusion paradigm for melatonin delivery: what has it taught us about the mela-

tonin signal, its reception, and the photoperiodic control of seasonal responses. J Pineal Res 15:161–190
Beck-Friis J, von Rosen D, Kjellman BF, Ljungen JG, Wetterberg L (1984) Melatonin in relation to body measures, sex, age, season and the use of drugs in patients with major affective disorders and healthy subjects. Psychoneuroendocrinology 9:261–277
Beedham C, Smith J, Steele D, Wright J (1987) Chlorpromazine inhibition of melatonin metabolism by normal and induced rat liver microsomes. Eur J Drug Metab Pharmokinet 12 (4):299–302
Berga SL, Mortola JF, Yen SSC (1988) Amplification of nocturnal melatonin secretion in women with functional hypothalamic amenorrhea. J Clin Endocrinol Metab, 66:242–244
Bieck PR, Antonin K, Balon R, Oxenkrug G (1988) Effect of brofaromine and pargyline on human plasma melatonin concentrations. Proc Neuropsychopharmacol Biol Psychiat 12:93–101
Bittman EL (1985) The role of rhythms in the response to melatonin. In: Evered D, Clark S (eds) Photoperiodism, melatonin and the pineal. Pitman, London, pp 149–169 (Ciba Foundation symposium 117)
Bittman EL (1993) The sites and consequences of melatonin binding in mammals. Am Zool 33:200–211
Bojkowski C, Aldhous M, English J et al (1987a) Suppression of nocturnal plasma melatonin and 6-sulphatoxymelatonin by bright and dim light in man. Horm Metab Res 19:437–440
Bojkowski C, Arendt J, Shih M, Markey SP (1987b) Assessment of melatonin secretion in man by measurement of its metabolite: 6-sulphatoxymelatonin. Clin Chem 19:437–440
Broadway J, Arendt J, Folkard S (1987) Bright light phase shifts the human melatonin rhythm during the Antarctic winter. Neurosci Lett 79:185–189
Brown GM, Bar-Or A, Grossi D, Kashur S, Johannson E, Yie SM (1991) Urinary 6-sulphatoxymelatonin, an index of pineal function in the rat. J Pineal Res 10:141–147
Cagnacci A, Elliot JA, Yen SSC (1991) Amplification of pulsatile LH secretion by exogenous melatonin in women. J Clinical Endocrinol Metab 73: 210–212
Cagnacci A, Elliott JA, Yen SSC (1992) Melatonin: a major regulator of the circadian rhythm of core temperature in humans. J Clin Endocrinol Metab 75 (2):447–452
Caignard D-H (1994) Structural activity relationships of melatonin agonists/antagonists. Adv Pineal Res 8:349–356
Cardinali DP, Vacas MI, Boyer EE (1979) Specific binding of melatonin in bovine brain. Endocrinology 105:437–441
Carter DS, Goldman BD (1983) Antigonadal effects of timed melatonin infusion in pinealectomised male Djungarian hamsters (Phodopus sungorus sungorus); duration is the critical parameter. Endocrinology 113:1261–1267
Cassone V (1990) Effects of melatonin on vertebrate circadian systems. Trends Neurosci 13:457–463
Cassone V (1992) The pineal gland influences rat circadian activity rhythms in constant light. J Biol Rhythms 7:27–40
Cassone VM, Chesworth MJ, Armstrong SM (1986) Entrainment of rat circadian rhythms by daily injections of melatonin depend upon the hypothalamic suprachiasmatic nuclei. Physiol Behav 36:1111
Checkley SA, Palazidou E (1988) Melatonin and anti-depressant drugs: clinical pharmacology. In: Miles A, Philbrick DRS, Thompson C (eds) Melatonin: clinical perspectives. Oxford University Press, Oxford, pp 190–204
Chemineau PE, Normandt JP, Ravault J, Thimonier (1986) Induction and persistence of pituitary and ovarian activity in the out-of-season lactating dairy goat after a treatment combining a skeleton photoperiod, melatonin and the male effect. J Reprod Fertil 78:497–594

Claustrat B, Brun J, David M, Sassolas G, Chazot G (1992) Melatonin and jet-lag: confirmatory result using a simplified protocol. Biol Psychiat 32:8:705–711

Collin JP (1971) Differentiation and regression of the cells of the sensory line in the epihysis cerebri. In: Wolstenholme GEW, Knight J (eds) The pineal gland. Churchill Livingstone, Edinburgh, pp 79–125

Collin JP, Arendt J, Gern NA (1988) Le 'troisieme oeil'. Recherche 203:1154–1165

Cowen PJ, Fraser S, Sammons R, Green AR (1983) Atenolol reduces plasma melatonin concentrations in man. Br J Clin Pharmacol 15:579–581

Cramer H, Rudolph J, Consbruch V (1974) On the effects of melatonin on sleep and behaviour in man. In: Costa E, Gessa GL, Sandler M (eds) Serotonin – new vistas: biochemistry and behavioural and clinical studies. Raven, New York, pp 187–191 (Advances in biochemical psychopharmacology, vol 11)

Czeisler CA, Allan JS, Strogatz JS, Ronda JM, Sandrez R, Rios CD, Freitag WO, Richardson GS, Kronauer RE (1986) Bright light resets the human circadian pace-maker independent of the timing of the sleep-wake cycle. Science 233:667–671

Czeisler CA, Johnson PJ, Duffy JF et al (1990) Exposure to bright light and darkness to treat physiologic maladaptation to night-work. N Engl J Med 322:1253–1259

Czeisler CA, Shanaghan TL, Klerman E, Martens, H, Brotman DJ, Emens JS, Klein T, Rizzo J (1995) Suppression of melatonin secretion in some blind patients by exposure to bright light. N Engl J Med 332:6–11

Dahlitz M, Alvarez B,Vignau J,English J, Arendt J, Parkes JD (1991) Delayed sleep phase syndrome response to melatonin. Lancet 337:1121–1124

Das Gupta D, Riedel L, Frick JH, Attanasio A, Ranke MB (1983) Circulating melatonin in children: in relation to puberty, endocrine disorders, functional tests and racial origin. Neuroendocrinol Lett 5:63–78

Davies D, Vonhoff S, Garratt P, Sugden D (1995) Chicken brain and Xenopus laevis melatonin receptors discriminate between enantiomers of terahydrocarbazoles. Abstracts of the British Chronobiology Meeting, Jan 1995

Davis FC, Mannion J (1988) Entrainment of hamster pup circadian rhythms by prenatal melatonin injections. Am J Physiol 255:R439–R448

De Reviers MM, Ravault JP, Tillet,Y Pelletier J (1989) Melatonin binding sites in the sheep pars tuberalis. Neurosci Lett 100:89–93

Deacon S, Arendt J (1994a) Posture influences melatonin concentrations in plasma and saliva in humans. Neurosci Lett 167:191–194

Deacon SJ, Arendt J (1994b) Phase shifts in melatonin, 6-sulphatoxymelatonin and alertness rhythms after treatment with moderately bright light at night. Clin Endocrinol (Oxf) 40:413–420

Deacon S, Arendt J (1995a) Melatonin-induced temperature suppression and its acute phase-shifting effects correlate in a dose dependent manner in humans. Brain Res 688:77-85

Deacon S, Arendt J (1995b) Adapting to phase-shifts, I. An experimental model for jet lag and shift work. Physiol Behav (in press)

Deacon S Arendt J (1995c) Adapting to phase-shifts, II. Effects of melatonin and conflicting light treatment. Physiol Behav (in press)

Deacon S, English J, Arendt J (1994) Acute phase-shifting effects of melatonin associated with suppression of core body temperature. Neurosci Lett 1 (78):32–34

Demisch K, Demisch L, Bodinik HJ, Nickelson,T, Althoff PH, Schoffling K, Rieth R (1986) Melatonin and cortisol increase after fluvoxamine. Br J Clin Pharmacol 22:620–622

Deveson S, Arendt J, Forsyth I (1992a) Role of the pineal and melatonin rhythms on reproductive performance in female domestic ungulates. Rev Anim Reprod Sci 30:113–134

Deveson S, Forsyth IA, Arendt J (1992b) Retardation of pubertal development by prenatal long days in goat kids born in autumn. J Reprod Fertil 95:629–637

Deveson SL, Arendt J, Forsyth IA (1992) Induced out-of-season breeding in British Saanen dairy goats: use of artificial photoperiods and/or melatonin administration. Anim Reprod Sci 29:1-15

Dollins AB, Zhdanova IV, Wurtman RJ, Lynch HJ, Deng MH (1994) Effect of inducing nocturnal serum melatonin concentrations in daytime on sleep, mood, body temperature, and performance. Proc Natl Acad Sci USA 91:1824–1828

Dubocovich M (1988) Pharmacology and function of melatonin receptors. J FASEB 2:2765–2773

Dubocovich M (1985) Characterisation of a retinal melatonin receptor. J Pharmacol Exp Ther 234:395–401

Dubovich ML, Takahashi JS (1987) Use of 2-[^{125}I]-iodomelatonin to characterize melatonin binding sites in chicken retina. Proc Natl Acad Sci USA 84:3916–3920

Duncan MJ, Goldman BD (1984) Hormonal regulation of the annual pelage colour cycle in the Djungarian hamster, Phodopus sungorus. II. Role of prolactin. J Exp Zool 230:97–103

Eastwood MR, Peacock J (1976) Seasonal patterns of suicide, depression and electroconvulsive therapy. Br J Psychiatry 129:472–475

Ebadi M, Govitrapong P (1986) Orphan transmitters and their receptor sites in the pineal gland. Pineal Res Rev 4:1–54

Ebisawa T, Karne S, Lerner M, Reppert SM (1994) Expression cloning of a high affinity melatonin receptor from Xenopus melanophores. Proc Natl Acad Sci USA 91:6133–6137

Ebling FJP, Foster DL (1989) Pineal melatonin rhythms and the timing of puberty in mammals. Experientia 45:946–955

Ebling F, Maywood E, Humby T, Hastings M (1992) Circadian and photoperiodic time measurement in male Syrian hamsters following lesions of the melatonin-binding sites of the paraventricular thalamus. J Biol Rhyth 73:241–254

English J, Poulton, AL, Arendt J, Symons, AM (1986) A comparison of the efficiency of melatonin treatments in advancing oestrus in ewes. J Reprod Fertil 77:321–327

English J, Middleton B, Arendt J, Wirz-Justice (1993) A rapid, direct measurement of melatonin in saliva using an iodinated tracer and solid phase second antibody. Ann Clin Biochem 30:415-416

Fevre-Montange M, van Cauter E, Refetoff S, Desir D, Tourniaire J, Copinschi G (1981) Effects of "Jet Lag" on hormonal patterns. II. Adaptation of melatonin circadian periodicity. J Clin Endocrinol Metab 52:642–649

Flaugh ME, Crowell TA, Clemens JA, Sawyer BD (1979) Synthesis and evaluation of the anti-ovulatory activity of a variety of melatonin analogues. J Med Chem 22:63–69

Folkard S, Arendt J, Aldhous M, Kennett H (1990) Melatonin stabilises sleep onset time in a blind man without entrainment of cortisol or temperature rhythms. Neurosci Lett 113:193–198

Folkard S, Arendt J, Clarke M (1993) Can melatonin improve shiftworkers tolerance of the night shift? Some preliminary findings. Chronobiol Internat 10:315–320

Follett BK (1982) Physiology of photoperiodic time measurement. In: Aschoff J, Daan S, Groos G (eds) Vertebrate circadian systems. Springer, Berlin Heidelberg New York, pp 268–275

Foster RG, Provencio I, Hudson D, Fiske S, De Grip W, Menake M (1991) Circadian photoreception in the retinally degenerate mouse (rd/rd). J Comp Physiol 169:39–50

Franey C, Aldhous A, Burton S, Checkley S, Arendt J (1986) Acute treatment with desipramine stimulates melatonin and 6-sulphatoxymelatonin in man. Br J Clin Pharmacol 22:73–79

Fraschini FO, Mess B, Martini L (1968) Pineal gland, melatonin and the control of luteinizing hormone secretion. Endocrinology 82:919–924

Gebbie FE (1993) Control of seasonal breeding and coat development in the goat. Thesis, University of Surrey, Guildford, UK

Gern WA, Karn CM (1983) Evolution of melatonin's functions and effects. Pineal Res Rev 1:49–91
Glass JD, Lynch GR (1982) Evidence for a brain site of melatonin action in the white-footed mouse, Peromyscus leucopus. Neuroendocrinology 34:1–6
Goldman BD (1983) The physiology of melatonin in mammals. Pineal Res Rev 1 (145):182
Griffiths PA, Folkard S, Bojkowski C, English J, Arendt J (1986) Persistent 24th variations of urinary 6-hydroxymelatonin sulphate and cortisol in Antarctica. Experientia 42:430–432
Guardiola-Lemaitre B (1994) Melatonin agonist/antagonist: from the receptor to therapeutic applications. In: Moller M, Pevet P (eds) Adv Pineal Res 8:333–348
Gwinner E (1989) Melatonin in the circadian system of birds: model of internal resonance. In: Hiroshige T, Honma K (eds) Circadian clocks and ecology. Hokkaido University Press, Sapporo, pp 127–153
Haimov I, Laudon M, Zisapel N, Souroujon M, N of D, Shlitner A, Herer P, Tzischinsky O, Lavie P (1994) Sleep disorders and melatonin rhythms in elderly people. Br Med J 309:167
Hanssen T, Heyden T, Sundberg I, Wetterberg L (1977) Effect of propanolol on serum-melatonin. Lancet ii:309
Hastings MH, Walker AP, Powers JB et al (1989) Differential effects of photoperiodic history on the responses of gonadotrophins and prolactin to intermediate daylengths in the male Syrian hamster. J Biol Rhythms 4:335–350
Hastings MH, Maywood ES, Ebling FJP (1991) Sites and mechanism of action of melatonin in the photoperiodic control of reproduction. Adv Pineal Res 5:147–157
Hastings MH, Mead SM, Vindlacheruvu RR, Ebling FJP, Maywoold ES, Grosse J (1992) Non-photic phase shifting of the circadian activity rhythm of Syrian hamsters: the relative potency of arousal and melatonin. Brain Res 591 (1):20–26
Hazelrigg DG, Gonzalez-Brito A, Lawson W et al (1993) Prolonged exposure to melatonin leads to time dependent sensitization of adenylate cyclase and down regulates melatonin receptors in pars tuberalis cells from ovine pituitary. Endocrinology 132:285–292
Herbert J (1981) The pineal gland and photoperiodic control of the ferret's reproductive cycle. In: Follett BK, Follett DE (eds) Biological clocks in seasonal reproductive cycles. Wright, Bristol, pp 261–276
Hoffman K (1981) The role of the pineal gland in the photoperiodic control of seasonal cycles in hamsters. In: Follett BK, Follett DE (eds) Biological clocks in seasonal reproductive cycles. Bristol Scientechnica Pitman Press, Bath, UK, pp 237–250
Huether G, Poeggeler B, Reimer A, George A (1992) Effect of tryptophan administration on circulating melatonin levels in chicks and rats: evidence for stimulation of melatonin synthesis and release in the gastrointestinal tract. Life Sci 51 (2):945–943
Iguchi H, Kato K, Ibayashi H (1982) Age dependent reduction in serum melatonin concentration in healthy human subjects. J Clin Endocrinol Metab 55:27–29
Illnerova H, Vanecek J (1979) Response of rat pineal serotonin N-acetyltransferase to one minute light pulse at different night times. Brain Res 167:431–434
Illnerova H, Vanecek J (1982) Complex control of the circadian rhythm in N-acetyltransferase activity in the rat pineal gland. In: Aschoff J, Daan S, Groos G (eds) Vertebrate circadian systems. Springer, Berlin Heidelberg New York, pp 285–296
Illnerova H, Zvolsky P, Vanecek J (1985) The circadian rhythm in plasma melatonin concentration of the urbanised man – the effect of summer and winter time. Brain Res 328:186–189
Jan JE, Espezel H, Appleton RE (1994) The treatment of sleep disorders with melatonin. Dev Med Child Neurol 36:97–107

Jewett M, Kronauer R, Czeisler C (1991) Light-induced suppression of endogenous circadian amplitude in humans. Nature 350:59–62

Jones R, Garratt P, Sugden D (1995) Design and synthesis of melatonin agonists and antagonists derived from 2-phenyltryptamines. Abstracts of the British Chronobiology Meeting

Kanematsu N, Mori Y, Hayashi S, Hoshino K (1989) Presence of a distinct 24-hour melatonin rhythm in the ventricular cerebrospinal fluid of the goat. J Pineal Res 7:143–152

Kappers JA (1960) Innervation of the epiphysis cerebri in the albino rat. Anat Rec 136:220–221

Karsch FJ, Bittman EL, Foster DL et al (1984) Neuroendocrine basis of seasonal reproduction. Rec Prog Horm Res 40:185–232

Karsch FJ, Malpaux B, Wayne NL, Robinson JE (1988) Characteristics of the melatonin signal that provide the photoperiodic code from timing seasonal reproduction in the ewe. Reprod Nutr Dev 28:459–472

Kauppila A, Kivela A, Pakarinen A, Vakkuri O (1987) Inverse seasonal relationship between melatonin and ovarian activity in humans in a region with a strong seasonal contrast in luminosity. J Clin Endocrinol Metab 65:823–828

Kennaway DJ, Van Dorp CF (1991) Free running rhythms of melatonin, cortisol, electrolytes and sleep in humans in Antarctica. Am J Physiol 260:R1137–1144

Kennaway DJ, Gilmore TA, Seamark RF (1982) Effect of melatonin feeding on serum prolactin and gonadotrophin levels and the onset of seasonal oestrous cyclicity in sheep. Endocrinology 110:1766–1722

Kennaway DJ, Dunstan EA, Staples LD (1987) Photoperiodic control of the onset of breeding activity and fecundity in ewes. J Reprod Fertil [Suppl] 34:187–199

Kennaway DJ, Stamp GE, Goble FC (1992) Development of melatonin production in infants and the impact of prematurity. J Clin Endocrinol Metab 75:367–369

Klein DC (1974) Circadian rhythms in indole metabolism in the rat pineal gland. In: Schmitt FO, Worden FG (eds) The neurosciences, third study program. MIT Press, Cambridge, Massachusetts, pp 509–516

Klein DC (1974) Circadian rhythms in the pineal gland. In: Krieger DT (ed) Endocrine rhythms. Raven Press, New York

Klein DC (1993) The mammalian melatonin rhythm-generating system. In: Wetterberg L (ed) Light and biological rhythms in man. Pergamon, Oxford, pp 55–72 (Wenner-Gren international series, vol 63)

Klein DC, Weller JL (1970) Indole metabolism in the pineal gland a circadian rhythm in N-acetyltransferase. Science 169:1093–1095

Knutsson A (1989) Shift work and coronary heart disease. Scand J Soc Med 44:1–36

Korf H-W, Moller M (1984) The innervation of the mammalian pineal gland with special reference to central pinealopetal projections. Pineal Res Rev 2:41–86

Krause DN, Dubocovich ML 1990) Regulatory sites in the melatonin system of mammals. Trends Neurosci 13:464–470

Laudon M, Zisapel N (1986) Characterization of central melatonin receptors using I^{125}-melatonin. FEBS Lett 197:9–12

Laudon M, Yaron Z, Zisapel N (1988) N-(3,5-dinitrophenyl)-5-methoxytryptamine, a novel melatonin antagonist: effects on sexual maturation of the male and female rat. J Endocrinol 116:43–53

Launay JM, Lemaitre B, Husson HP (1982) Melatonin synthesis by rabbit platelets. Life Sci 31:1487–1494

Lewis DFV, Arendt J, English J (1990) Quantitative structure-activity relationships within a series of melatonin analogues and related indolealkylamines. J Pharmacol Exp Ther 252:370–373

Lewy AJ, Newsome DA (1983) Different types of melatonin circadian secretory rhythms in some blind subjects. J Clin Endocrinol Metab 56:1103–1107

Lewy AJ, Wehr TA, Goodwin FK et al (1980) Light suppresses melatonin secretion in humans. Science 210:1267–1269

Lewy AJ, Sack RL, Singer CM (1984) Assessment and treatment of chronobiologic disorders using plasma melatonin levels and bright light exposure, the clock-gate model and the phase response curve. Psychopharmacol Bull 20:561–565

Lewy AJ, Sack RL, Miller LS, Hoban TM (1987) Anti-depressant and circadian phase-shifting effects of light. Science 235:352–354

Lewy AJ, Saeeduddin A, Latham Jackson J, Sack R (1992) Melatonin shifts human circadian rhythms according to a phase-response curve. Chronobiol Int 9 (5):380–392

Lincoln G (1979) Photoperiodic control of seasonal breeding in the ram: participation of the cranial sympathetic nervous system. J Endocrinol 82:135–147

Lincoln G, Maeda K (1992) Effects of placing micro-implants of melatonin in the mediobasal hypothalamus and preoptic area on the secretion of prolactin and β-endorphin in rams. J Endocrinol 134 (3):437–448

Lincoln GA (1992) Administration of melatonin into the mediobasal hypothalamus as a continuous or intermittent signal affects the secretion of follicle-stimulating hormone and prolactin in the ram. J Pineal Res 12:135–144

Lincoln GA, Clarke IJ (1994) Photoperiodically-induced cycles in the secretion of prolactin in hypothalamo-pituitary disconnected rams: evidence for translation of the melatonin signal in the pituitary gland. J Neuroendocrinol 6:251–260

Lincoln GA, Ebling FJP, Almeida OFX (1985) Generation of melatonin rhythms. In: Evered D, Clark S (eds) Photoperiodism, melatonin and the pineal. Pitman, London, pp 129–141 (CIBA Foundation symposium 117)

Martin JE, Klein DC (1975) Melatonin inhibition of neonatal pituitary response to luteinising hormone-releasing factor. Science 191:301–302

Masson-Pevet M, Gauer F (1994) Differential daily and seasonal regulations of melatonin receptors in the supra chiasmatic nuclei and the pars tuberalis of mammals. Adv Pineal Res 8:321–332

Maurel D, Mas N, Roch G, Boissin J, Arendt J (1992) Diurnal variations of urinary 6-sulphatoxymelatonin in male intact or ganglionectomised mink. J Pineal Res 13:117–123

Maywood ES, Grosse J, Lindsay JO, Karp JD, Hastings M (1992) The effect of signal frequency on the gonadal response of male Syrian hamsters to programmed melatonin infusions. J Neuroendocrinol 4:3–6

McArthur AJ, Gillette MU, Prosser RA (1991) Melatonin directly resets the rat suprachiasmatic circadian clock in vitro. Brain Res 565:158–161

McIntyre IM, Norman TR, Burrows GD, Armstrong SM (1989) Human melatonin suppression by light is intensity dependent. J Pineal Res 6:149

Menaker M, Hudson DJ, Takahashi JS (1981) Neural and endocrine components of circadian clocks in birds. In: Follett BK, Follett DE (eds) Biological clocks in seasonal reproductive cycles. Wright, Bristol, pp 171–183

Miles LE, Raynal DM, Wilson MA (1977) Blind man living in normal society has circadian rhythms of 24.9 hours. Science 198:421–423

Minors DS, Waterhouse JM (1981) Circadian rhythms and the human. Wright, Bristol, pp 143–148

Minors DS, Waterhouse J (1992) Investigating the endogenous component of human circadian rhythms: a review of some simple alternatives to constant routines. Chronobiol Int 9 (1):55–78

Minors DS, Waterhouse JM, Wirz-Justice A (1991) A human phase response curve to light. Neurosci Lett 133:36–40

Moller M, Ravault JP, Cozzi B, Zang ET, Phansuwam-Pujito P, Larsen PJ, Mikkelsen JD (1991) The multineuronal input to the mammalian pineal gland. In: Foldes A, Reiter RJ (eds) Advances in pineal research, vol 6. Libbey, London, pp 3–12

Moore RY, Klein DC (1974) Visual pathways and the central neural control of a circadian rhythm in pineal serotonin N-acetyltransferase. Brain Res 71:17–33

Moore-Ede M (1993) The twenty four hour society. Addison Wesley, Massachusetts

Moore-Ede MC, Czeisler CA, Richardson GS (1983) Circadian timekeeping in health and disease, part 1. Basic properties of circadian pacemakers. N Engl J Med 309:469-476

Morgan PJ, Williams LM (1989) Central melatonin receptors; implications for a mode of action. Experientia 45:955-965

Morgan PJ, Williams LM, Davidson G et al (1989a) Melatonin receptors on ovine pars tuberalis: characterisation and autoradiographical localisation. J Neuroendocrinol 1:1-4

Morgan PJ, Lawson W, Davidson G et al (1989b) Guanine nucleotides regulate the affinity of melatonin receptors on the ovine Pars tuberalis. Neuroendocrinology 50:359-362

Morgan PJ, Davidson G, Lawson W, Barrett P (1990) Both pertussis toxin-sensitive and insensitive G-proteins link melatonin receptor to inhibition of adenylyl cyclase in the ovione pars tuberalis. J Neuroendocrinol 2:773-776

Morgan PJ, Lawson W, Davidson G (1991) Interaction of forskolin and melatonin on cyclic AMP generation in pars tuberalis from ovine pituitary. J Neuroendocrinol 3:497-501

Morgan PJ, Williams LM, Maclean A, Hazelrigg D (1995) Non G-protein coupled high affinity melatonin receptors are present in the sheep. Abstracts of the British Chronobiology Meeting, Jan 1995

Mrosovsky N, Salmon PA (1987) A behavioural method for accelerating re-entrainment of rhythms to new light-dark cycles. Nature 330:372-373

Nakazawa K, Marubayashi U, McCann SM (1991) Mediation of the short-loop feedback of luteinising hormone (LH) on LH-releasing hormone release by melatonin-induced inhibition of LH release from the pars tuberalis. Proc Natl Acad Sci USA 88:7576-7579

Namboodiri MA, Sugden D, Klein CC, Mefford IN (1983) 5-Hydroxytryptophan elevates serum melatonin. Science 221:659-661

Neuwelt EA, Lewy AJ (1983) Disappearance of plasma melatonin after removal of a neoplastic pineal gland. N Engl J Med 19:1132-1135

Nowak R, McMillen C, Redman J, Short RV (1987) The correlation between serum and salivary melatonin concentrations and urinary 6-hydroxymelatonin sulphate: two non invasive techniques for monitoring human circadian rhythmicity. Clin Endocrinol (Oxf) 27:445-452

O'Conor RAA (1992) A year in Antarctica: sleep and mood studies on Halley Base, 75S. In: Proceedings of the 8th Meeting of the European Society for Chronobiology, Leiden, The Netherlands, May 1992. J Interdisciplinary Cycle Res 23:158-159

Okudaira N, Kripke DF, Webster JB (1983) Naturalistic studies of human light exposure. Am J Physiol 245:R613-615

Palazidou E, Franey C, Arendt J, Stahl S, Checkley S (1989) Evidence for a functional role of alpha-1-adrenoreceptors in the regulation of melatonin secretion in man. Psychoneuroendocrinology 14:131-135

Palazidou E, Skene DJ, Arendt J, Everitt B, Checkley SA (1992) The acute and chronic effects of (+) and (-) oxaprotiline upon melatonin secretion in normal subjects. Psychol Med 22:61-67

Palm L, Blennow G, Wetterberg L (1991) Correction of a non-24-hour sleep wake cycle by melatonin in a blind retarded boy. Ann Neurol 29:336-339

Pang SF, Brown GM, Grota LJ et al (1977) Determination of N-acetylserotonin and melatonin activities in the pineal gland, retina, and Harderian gland, brain and serum of rats and chickens. Neuroendocrinology 23:1-13

Parkes AS (1976) Patterns of sexuality and reproduction. Oxford University Press, Oxford

Petrie K, Dawson AG, Thompson L, Brook R (1993) A double blind trial of melatonin as a treatment for jet lag in international cabin crew. Biol Psychiatry 33:526-530

Petrie K, Conaglen JV, Thompson L et al (1989) Effect of melatonin on jet-lag after long haul flights. Br Med J 298:705–707

Quay WB (1963) Circadian rhythm in rat pineal serotonin and its modifications by estrous cycle and photoperiod. Gen Comp Endocrinol 3:473–479

Quay WB (1970) Precocious entrainment and associated characteristics of activity patterns following pinealectomy and reversal of photoperiod. Physiol Behav 5:1281–1290

Raikhlin NT, Kvetnoy IM, Tokachev VN (1975) Melatonin may be synthesized in enterochromaffin cells. Nature 255:344

Redman J, Armstrong S, Ng KT (1983) Free-running activity rhythms in the rat: entrainment by melatonin. Science 219:1089–1091

Reiter RJ (1980) The pineal and its hormones in the control of reproduction in mammals. Endocr Rev 1:109–131

Reiter RJ (1985) Action spectra, dose-response relationships and temporal aspects of light's effects on the pineal gland. In: Wurtman RJ, Baum MJ, Potts JR Jr (eds) The medical and biological effects of light. Ann NY Acad Sci 453:215–230

Reppert SM, Weaver DR, Rivkees SA et al (1988) Putative melatonin receptors in a human biological clock. Science 242:78–81

Reppert SM, Weaver DR, Ebisawa T (1994) Cloning and characterisation of a mammalian melatonin receptor that mediates reproductive and circadian responses. Neuron 13:1177–1185

Roenneberg T, Aschoff J (1990) Annual rhythms in human reproduction: I. Biology, sociology or both? J Biol Rhythms 5:195–216

Rollag MD, Niswender GD (1976) Radioimmunoassay of serum concentrations of melatonin in sheep exposed to different lighting regimens. Endocrinology 98:482–489

Rollag MD, Morgan RJ, Niswender GD (1978) Route of melatonin secretion in sheep. Endocrinology 102:1–8

Rosa RR, Bonnet MH, Bootzin RR, Eastman CI, Monk T, Penn PE, Tepas DI, Walsh JK (1990) Intervention factors for promoting adjustment to nightwork and shiftwork. Occup Med State Art Rev 5:391–415

Rosenthal NE, Sack DA, Gillin JC et al (1984) Seasonal affective disorder. A description of the syndrome and preliminary findings with light therapy. Arch Gen Psychiatry 41:72–79

Ross JK, Arendt J, Horne J, Haston W (1995) Night-shift work during Antarctic winter: sleep characteristics and adaptation with bright light treatment. Physiol Behav 57:1169-1174

Sack RL, Lewy AJ, Blood ML, Stevenson J, Keith LD (1991) Melatonin administration to blind people: phase advances and entrainment. J Biol Rhythms 6:249–261

Sack RL, Blood ML, Lewy AJ (1994) Melatonin administration promotes circadian adaptation to shift work. Sleep Res 23:509

Samel A, Wegman HM, Vejvoda M, Maas H (1991) Influence of melatonin treatment on human circadian rhythmicity before and after a simulated 9 hour time shift. J Biol Rhythms 6:235–248

Sarrafzadeh A, Wirz-Justice A, Arendt J, English J (1990). Melatonin stabilises sleep onset in a blind man. In: Horne JA (ed) Sleep '90. Pontenagel, Dortmund, pp 51–54

Skene DJ, Masson-Pevet M, Pevet P (1993) Seasonal changes in melatonin binding sites in the pars tuberalis of male European hamsters and the effect of testosterone manipulation. Endocrinology 132:1682–1686

Skene DJ, Bojkowski CJ, Arendt J (1994) Comparison of the effects of acute fluvoxamine and desipramine administration on melatonin and cortisol production in humans. Br J Clin Pharmacol 37:181–186

Smith JA, O'Hara J, Schiff AA (1981) Altered diurnal serum melatonin rhythm in a blind man. Lancet ii:933

Smith L, Folkard S, Poole CJM (1994) Increased injuries on night-shift. Lancet 344:1137–1139

Stankov B, Fraschini F (1994) Distribution of the high affinity melatonin-binding sites in the vertebrate brain. Adv Pineal Res 8:295–307

Stehle JH, Foulkes NS, Molina CA, Simonneaux V, Pevet P, Sassone-Corsi P (1993) Adrenergic signals direct rhythmic expression of transcriptional repressor CREM in the pineal gland. Nature 365:314–320

Steinlechner S (1991) In search of a physiological role for retinal melatonin. Adv Pineal Res [Suppl] 5:123–128

Strassman RJ, Qualls CR, Lisansky EJ, Peake GT (1991) Elevated rectal temperature produced by all night bright light is reversed by melatonin infusion in man. J Appl Physiol 71:2178–2181

Sugden D, Chong NWS (1991) Pharmacological identity of 2-(^{125}I)iodomelatonin binding sites in chicken brain and sheep pars tuberalis. Brain Res 539:151–154

Sugden D, Weller JL, Klein DC, Kirk KL, Creveling CR (1984) α-Adrenergic potentiation of β-adrenergic stimulation of rat pineal N-acetyltransferase: studies using citazoline and fluorine analogs of norepinephrine. Biochem Pharmacol 33:3947–3950

Sugden D, Chong N, Lewis D (1995) Structural requirements at the melatonin receptor. Br J Pharmacol 114:618–623

Tamarkin K, Baird CJ, Almeida OFX (1985) Melatonin: a coordinating signal for mammalian reproduction. Science 227:714–720

Terman M, Boulos Z, Campbell SS, Djik D-J, Eastman C, Lewy AJ (1995) Light treatment for sleep disorders. Joint Task Force Report. J Biol Rhythms 10:105–112

Tessonneaud A (1994) Thesis, INRA, Centre de Recherche de Tours, France

Touitou Y, Haus E (eds) (1992) Biologic rhythms in clinical and laboratory medicine. Springer, Berlin Heidelberg New York

Tzischinsky O, Skene D, Epstein R, Lavie P (1991) Circadian rhythms in 6-sulphatoxymelatonin and nocturnal sleep in blind children. Chronobiol Int 8:168–175

Tzischinsky O, Lavie P, Pal I (1992a) Time-dependent effects of 5 mg melatonin on the sleep propensity function. 11th European congress on sleep research, July, Helsinki, Finland. J Sleep Res 1 [Suppl 1]:234

Tzischinsky O, Pal I, Epstein R, Dagan Y, Lavie P (1992b) The importance of timing in melatonin administration in a blind man. J Pineal Res 12:105–108

Underwood H (1977) Circadian organisation in lizards: the role of the pineal organ. Science 195:587–589

Underwood H, Goldman BD (1987) Vertebrate circadian and photoperiodic systems: role of the pineal gland and melatonin. J Biol Rhythms 2:279–315

Vakkuri O, Leppaluoto J, Vuolteenaho O (1985a) Development and validation of a melatonin radioimmunoassay using radioiodinated melatonin as tracer. Acta Endocrinol (Copenh) 106:152–157

Vakkuri O, Leppaluoto J, Kauppila A (1985b) Oral administration and distribution of melatonin in human serum, saliva and urine. Life Sci 37:489–495

Vanecek J, Pavlik A, Illnerova H (1987) Hypothalamic melatonin receptor sites revealed by autoradiography. Brain Res 453:359–362

Vaughan GM (1984) Melatonin in humans. Pineal Res Rev 2:141–201

Vaughan GM, Reiter RJ (1987) The Syrian hamster pineal gland responds to isoproterenol in vivo at night. Endocrinology 120:1682–1684

Vollrath L, Semm P, Gammel G (1981) Sleep induction by intranasal application of melatonin. Adv Biosci 29:327–329

Voordow BCG, Euser R, Verdonk RER, Alberda BT, DeJong FH, Drogendijk AC, Fauser BCJM, Cohen M (1992) Melatonin and melatonin-progestin combinations alter pituitary-ovarian function in women and can inhibit ovulation. J Clin Endocrinol Metab 74:108–117

Waldhauser F, Steger H (1986) Changes in melatonin secretion with age and pubescence. J Neural Transm [Suppl] 21:183–198

Waldhauser F, Steger H, Vorkapic P (1987) Melatonin secretion in man and the influence of exogenous melatonin on some physiological and behavioural variables. Adv Pineal Res 2:207–223

Waldhauser F, Boepple P, Schemper M, Crowley WF (1991) Serum melatonin in central precocious puberty is lower than in age matched pre-pubertal children. J Clin Endocrinol Metab 73:793–796

Weaver DR, Reppert SM (1986) Maternal melatonin communicates daylength to the fetus in Djungarian hamsters. Endocrinology 119:2861–2863

Weaver DR, Reppert SM (1990) Melatonin receptors are present in the ferret pars tuberalis and pars distalis, but not the brain. Endocrinology 127:2607–2609

Weaver DR, Stehle JH, Stopa EG et al (1993) Melatonin receptors in human hypothalamus and pituitary: implications for circadian and reproductive responses to melatonin. J Clin Endocrinol Metab 76:295–230

Webb S, Puig-Domingo M (1995) Melatonin in health and disease. Clin Endocrinol (Oxf) 42:221-234

Wehr TA (1991) The durations of human melatonin secretion and sleep respond to changes in daylength (photoperiod). J Clin Endocrinol Metab 73:1276–1280

Wehr TA, Jacobsen FM, Sack DA, Arendt J, Tamarkin L, Rosenthal NE (1986) The efficacy of phototherapy in seasonal affective disorder appears not to depend on its timing or its effect on melatonin secretion. Arch Gen Psychiatry 43:870–875

Wever RA (1979) The circadian system of man: results of experiments under temporal isolation. Springer, Berlin Heidelberg New York

Wever RA (1989) Light effects on human circadian rhythms: a review of recent Andechs experiments. J Biol Rhthms 4:161–186

Wever RA, Polasek J, Wildgruber CM (1983) Bright light affects human circadian rhythms. Pflugers Arch 396:85–87

White BH, Sekura RD, Rollag MD (1987) Pertussis toxin blocks melatonin-induced aggregation in Xenopus dermal melanophores. J Comp Physiol [B] 157:153–159

Wilkinson M, Arendt J, Bradtke J, de Ziegler D (1977) Determinatin of dark- induced elevation of pineal N-acetyl-transferase with simultaneous radioimmunoassay of melatonin in pineal, serum and pituitary of the male rat. J Endocrinol 72:243–244

Williams LM, Martinoli MG, Titchener LT, Pelletier G (1991) The ontogeny of central melatonin binding sites in the rat. Endocrinology 128:2083–2090

Winton F, Checkley SA, Corn T, Huson LW, Franey C, Arendt J (1989) Effects of light treatment upon mood and melatonin in seasonal affective disorder. Psychol Med 15:19–25

Woodfill CJI, Robinson JE, Malpaux BM, Karsch FJ (1991) Synchronisation of the circannual reproductive rhythm of the ewe by discrete photoperiodic signals. Biol Reprod 45:110–121

Woodfill CJI, Wayne N, Moenter SM, Karsch F (1994) Photoperiodic synchronisation of a circannual reproductive rhythm in sheep: identification of season-specific time cues. Biol Reprod 50:965–976

Wright J, Aldhous M, Franey C, English J, Arendt J (1986) The effects of exogenous melatonin on endocrine function in man. Clin Endocrinol (Oxf) 24:375–382

Yous S, Antrieux J, Howell HE et al (1992) Novel napthalenic ligands with a high affinity for the melatonin receptor. J Med Chem 35:1484–1486

Yuwiler A (1983) Vasoactive intestinal peptide stimulation of pineal serotonin-N-acetyltransferase activity: general characteristics. J Neurochem 41:141–156

Zaidan R, Geoffriau M, Brun J, Taillard J, Bureau C, Chazot G, Claustrat B (1994) Melatonin is able to influence its secretion in humans: description of a phase-response curve. Neuroendocrinology 60:105–112

Zatz M (1978) Sensitivity and cyclic nucleotides in the rat pineal gland. J Neural Transm [Suppl 13]:97–114

Zisapel N, Oaknin S, Anis Y (1991) Melatonin receptors in discrete areas of the rat and Syrian hamster brain: modulation by melatonin, pinealectomy, testosterone and the photoperiod. Adv Pineal Res 5:175–181

CHAPTER 15
Problems in Interpreting the Effects of Drugs on Circadian Rhythms

N. MROSOVSKY

A. Multiplicity of Sites for Chronotypic Action

The aim of this chapter is to provide a guide to some of the questions that might be raised when a drug is found to have effects on the temporal profile of variables. It discusses possible explanations, especially those involving the newly discovered major influence that behaviour can have on rhythms. Some people, of course, will be more concerned with learning whether or not a drug produces a desired temporal effect; discussion about whether the effect represents masking or entrainment, or whether an agent is a *true* zeitgeber or not, may seem academic. It is no more so than knowing whether an agent is an agonist or a partial agonist. In the long run medicine can be improved if the complexities of the circadian system are taken into account, and if the mode and site of the actions of chronobiotic drugs are understood.

There have been some penetrating analyses of possible modes of action of particular drugs that affect circadian rhythms (e.g. for lithium, by KLEMFUSS 1992). Nevertheless, as recently as 1988 (TUREK), it appeared necessary to ask the following basic question: Do circadian biologists and chronopharmacologists talk the same "language"? This question is still pertinent. When a drug or hormone is found to alter the phase or period of a circadian rhythm, there remains a tendency to suppose that the drug acts on the oscillator. If one thinks of the circadian system as comprising various components, then it becomes apparent that overt rhythmicity might be changed by drugs in a variety of ways other than by direct action on the pacemakers themselves. Figure 1 is a schematic representation of the mammalian circadian system. In fact, it is much more complicated anatomically than shown (MOORE and CARD 1994; MIKKELSEN and VRANG 1994; MIKKELSEN et al. 1994), but even a simplified version is enough to illustrate the point that there are a number of potential sites of action for chronobiotic drugs.

In light-dark (LD) cycles, drugs could affect the potency of entrainment and phase angle at a number of points along the pathway from the eye to the suprachiasmatic nucleus (SCN). In addition to the well-known effects of light on biological rhythms, non-photic behavioural events can also exert powerful influences on phase and period. Inducing a hamster to become active by confining it to a novel running wheel can produce large phase shifts, even when

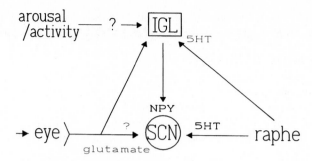

Fig. 1. Schematic and simplified diagram of the circadian system in mammals. *IGL*, intergeniculate leaflet; *SCN*, suprachiasmatic nucleus. Non-photic events associated with arousal are thought to activate the IGL but the pathways are not known (MROSOVSKY 1995)

the manipulation occurs in complete darkness (DD). Drugs could therefore influence circadian rhythms not by acting directly on pacemakers but by creating states similar to those in an excited and active hamster.

Even a drug that has no direct action on timing mechanisms could result in phase shifts. REDFERN et al. (1994) have pointed out that if a substance *directly* induced sleep (i.e. had a *masking* as opposed to a *phase-shifting* effect on a sleep rhythm), then if that substance were given regularly it would be likely to alter the timing of meals and social engagements. Both periodic feeding and social interactions are known entraining agents. Altered schedules of these activities during drug treatment might in turn lead to altered synchronization. Thus, masking effects could alter the "zeitgeber package".

There follow some examples of drug effects that initially were considered likely to depend on actions on the pacemaker, but after further research are more probably explained by indirect influences.

B. Alternative Interpretations for Mode of Drug Action
I. Gonadal Steroids and Changes in Period

In a variety of rodents, the presence of gonadal steroids is associated with a speeding up of free-running period (τ). It has been suggested that this depends on some action of the steroids on the pacemaker, the SCN in mammals. For instance, DAAN et al. (1975) speculated that testosterone in mice acted on the pacemaker, or on components of the pacemaker such as dawn and dusk oscillators, to shorten τ. MORIN et al. (1977) were able to show that the ability of oestrogens to shorten period was not dependent on alterations of the perceived intensity of light, or on some mechanism within the eye, because steroid state affected τ even in enucleated hamsters. They thought that "estradiol may change τ by directly influencing the SCN". However, because the SCN does not selectively concentrate oestradiol, it was proposed that there might be some non-classical binding system in the SCN involving membrane potentials, or that a part of the brain with direct inputs to the SCN, such as the medial preoptic area, might be the site of action.

A very different interpretation is that the chronobiological effects of steroids are indirectly mediated by changes in locomotor activity. In intact untreated mice, individuals with relatively long free-running periods tend to be most active toward the end of their subjective night (EDGAR et al. 1991). It is also known that the circadian rhythms of rats and mice lacking access to running wheels have longer periods than those with wheels (YAMADA et al. 1990; EDGAR et al. 1991). So decreased activity and a peak of activity late in the night are in themselves associated with long periods. It has been pointed out (EDGAR et al. 1991) that a combination of decreased activity and a shift of activity toward the end of the subjective night were just the changes noted by DAAN et al. (1975) in their castrated mice.

Support for the view that steroid-dependent τ changes are mediated by associated changes in activity comes from comparisons between ovariectomized rats with and without access to running wheels (DE ELVIRA et al. 1992). In this work it was possible to show, by the use of implanted devices sensitive to movement, that rats without a wheel were generally less active than those with a wheel. If the rats had no wheel, it made little difference to their periodicity whether or not they were ovariectomized; in both cases they were relatively quiescent. However, if the rats had access to a wheel, then ovariectomy resulted in decreased activity, and in τs that were longer than those of intact animals. The ovariectomized rats did not use their wheels very much; their τs were similar to those of rats without wheels (Fig. 2).

Additional work is desirable, especially with hamsters, given the report that in this species the τ of animals without wheels is shorter rather than longer than that of animals with wheels (ASCHOFF et al. 1973; REFINETTI et al. 1994). This is puzzling (and needs confirming) because if relatively sedentary hamsters

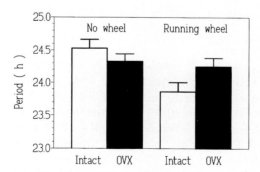

Fig. 2. Free-running periods (means ± SEMs) of intact and ovariectomized (*OVX*) rats kept in constant dim red light. Periods are derived from average periods of body temperature and activity rhythms. (Data from DE ELVIRA et al. 1992)

(those without access to wheels) have shorter τs, and if, as is evident, ovariectomy decreases activity, then ovariectomy should to some extent mimic removal of running wheels: according to the view that period changes are mediated by activity changes, ovariectomy and low oestrogen in hamsters should lead to shorter τs. Yet the opposite is true: low levels of oestrogens are associated with longer, slower rhythms in hamsters, as in rats (MORIN et al. 1977; DE ELVIRA et al. 1992).

II. Benzodiazepines and Phase Shifts

In 1986 TUREK and LOSEE-OLSON showed that peripheral injections of triazolam produced phase shifts in hamsters. Triazolam is a short-acting agonist at the benzodiazepine receptor which forms an integral part of the $GABA_A$ receptor-chloride ion channel complex. The actions of triazolam are believed to arise from augmentation of GABA-mediated inhibition. The presence of GABA-containing neurons in the SCN (VAN DEN POL 1986) raised the possibility that phase-shifting by triazolam depended on its acting on the SCN. TUREK and LOSEE-OLSON also considered the possibility that triazolam worked on some input pathway to the SCN. If so, on what pathway? Triazolam induces phase advances of about 90 min in the middle of the subjective day, and smaller delays at the end of the subjective night. So the phase response curve (PRC) for this drug is different in shape from that for light (Fig. 3). This makes it unlikely that triazolam acts on the photic input pathway to the SCN. In later work it was shown that triazolam produces phase shifts in enucleated hamsters, ruling out mediation by GABA-containing neurons present in the eyes (VAN REETH et al. 1987).

From the actograms presented in the original report of phase-shifting by triazolam, it was evident that hamsters sometimes ran in their wheels after receiving this drug. Since wheel-running by an untreated hamster placed in a novel wheel is in itself associated with phase shifts, the suggestion was made that the phase-shifting effects of triazolam in hamsters are mediated by the activity it induces (MROSOVSKY and SALMON 1987). The attraction of this view is that the PRC for triazolam is similar in shape to that for novelty-induced activity (REEBS and MROSOVSKY 1989a; MROSOVSKY et al. 1992). Both have prominent phase advances during the subjective day (Fig. 3).

In support of an activity-mediated effect for triazolam, preventing hamsters from being active, by confinement to a small area, blocks or negates the ability of this benzodiazepine to produce phase shifts; this blocking action is evident both at circadian time (CT) 21, in the delay portion of the triazolam PRC, and at CT 6, in the advance portion (VAN REETH and TUREK 1989; MROSOVSKY and SALMON 1990). In addition, the enhancement by triazolam of the rate of re-entrainment to an 8-h advance in an LD 14:10-h schedule correlates significantly with the incidence of wheel-running in the few hours after the drug is injected (Fig. 4). Evidently the pharmacological state does not result in a phase shift unless the hamster is in an appropriate environment.

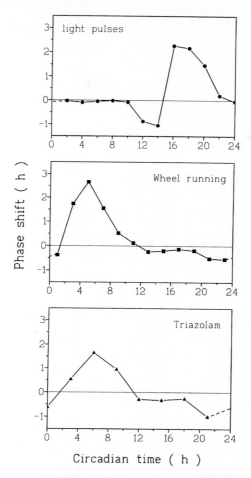

Fig. 3. Phase response curves (PRCs) of hamsters for 1-h light pulses (TAKAHASHI et al. 1984), 3-h pulses of novelty-induced wheel-running (MROSOVSKY et al. 1992) and injections of 2.5 mg triazolam/animal (TUREK and LOSEE-OLSON 1986). The PRCs were all obtained from hamsters kept in the dark (DD). Circadian times (CT) on the x-axis refer to the start of manipulations; CT 12 by definition is the onset of activity for a nocturnal animal

When the first report of phase-shifting by triazolam was published, commentators said that the data provided "the strongest hint yet that the elusive jet-lag pill is within reach" (WINFREE 1986). If, however, triazolam is no more than a fancy way of inducing a state that can also be produced by appropriate environments and behaviour without any drug administration, then the attractiveness of taking a pharmacological approach and of jet-lag pills may be diminished.

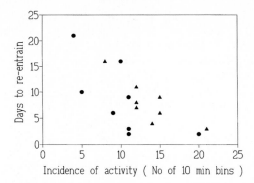

Fig. 4. Rate of re-entrainment after an 8-h advance in an LD cycle as a function of incidence of wheel-running in the 36 10-min time bins over the 6-h postinjection of triazolam. *Circles*, 1.5 mg/hamster; *triangles*, 2.5 mg/hamster. Injections were made at approximately CT 5 (data replotted from Mrosovsky and Salmon 1990). A small number of days to re-entrain implies that the circadian system has adjusted rapidly to the shifted LD cycle. Median values for re-entrainment of vehicle-injected animals were close to 10 days in these experiments

III. Periodicity After a Single Treatment

Single doses of triazolam produce phase-dependent changes in free-running periodicity as well as in phase (Joy et al. 1989). A single 3-h pulse of novelty-induced wheel-running also produces similar changes in τ (Mrosovsky 1993a). Again then, the period changes following triazolam could be mediated by activity (or a correlate) rather than by an action of the benzodiazepine on GABA-containing neurons in the SCN.

A change in τ after a single administration of a drug, or after a single bout of running in a novel wheel, may be distinguished, at least operationally, from the period changes that are related to continued access to a running wheel, and to the amount and distribution of running in the wheel (Yamada et al. 1990; Edgar et al. 1991; Sect. B.I). Whether the effects of single manipulations depend on them having produced longer-lasting effects on activity level and distribution has not yet been investigated. Mediation by activity possibly applies to the period-shortening action of quipazine in rats, since following a single injection of this drug the distribution of activity moves toward the latter part of the subjective night (Edgar et al. 1993).

IV. Melatonin Injections and Infusions

Daily injections of high doses (1 mg/kg) of melatonin are capable of entraining the rhythms of rats (Redman et al. 1983). Because a small percentage of saline-injected rats showed signs of entrainment, it was suggested that non-specific effects should be more closely investigated (Moore-Ede 1985; Mrosovsky 1988). Subsequent research showed that melatonin injections maintain entrainment even when activity is reduced by locking running wheels (REDMAN and ROBERTS 1991). In addition, entrainment of rats has now been demonstrated with lower doses of melatonin, 5.5 µg/kg (CASSONE et al. 1986). Thus it seems that rats can be entrained by melatonin.

In hamsters, however, the situation appears to be different. Subcutaneous melatonin injections given at CT 10 do indeed cause phase advances, as they do in rats, but so do injections of saline (Fig. 5; HASTINGS et al. 1992). When melatonin or its saline vehicle is infused through indwelling cannulae, neither of these substances produces significant shifts. HASTINGS et al. (1992) argued that arousal associated with picking the animal up and injecting it was acting as a zeitgeber, and that using infusions eliminates these disturbances, thereby eliminating phase shifts.

The data from hamsters on differences in route of administration of melatonin are quite clear, but so are those from rats showing a superiority of melatonin over saline injections for producing shifts. Does this mean that in rats melatonin produces phase shifts by a specific pharmacological mechanism but that in hamsters the observed effect is something non-specific associated with the injection procedure? HASTINGS et al. (1992) suggest that the species difference in outcome depends on a quantitative rather than a qualitative difference in mechanism: in both rats and hamsters, melatonin increases the sensitivity to the phase-shifting effects of arousal by other stimuli, rather than acting as a true zeitgeber. Phase-shifting in rats is obtained because melatonin amplifies the response to the arousing stimuli associated with injection. Hamsters, it was suggested, are more arousable – sufficiently so that they do not need this amplifying effect to show a phase shift in response to being picked up and injected; therefore they respond even to administration of saline.

Recently, GOLOMBEK and CARDINALI (1993) have shown that re-entrainment to an advanced LD cycle can be speeded up by melatonin, apparently in

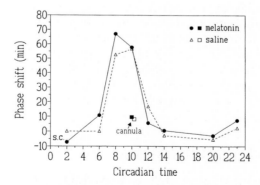

Fig. 5. Phase shifts (means) produced by administration of 1 mg/kg melatonin (*solid symbols*) and saline (*open symbols*) to hamsters kept in dim red light. *Points joined by lines* show PRCs when subcutaneous injection was the route of administration. *Squares* show the lack of phase shifts when these substances were infused at CT 10 by remote control through indwelling cannulae. (Data replotted from HASTINGS et al. 1992)

the absence of enhanced wheel-running. This is the first demonstration in hamsters of an effect of melatonin on circadian phasing that is greater than that of saline. However, this does not disprove the notion that melatonin acts to amplify arousal because the test procedure involved a change in lighting as well as giving melatonin; it is possible that melatonin acted synergistically with the new LD cycle in this situation. Chronic elevation of melatonin facilitates entrainment of sparrows to dim light-dark cycles that are otherwise only marginally effective zeitgebers (HAU and GWINNER 1994).

Further work and confirmatory experiments are needed before these matters can be resolved. Whether the suggestion of HASTINGS et al. (1992) that melatonin acts to amplify arousing stimuli is correct or not, it represents a theoretical way in which a substance could result in a phase shift without in itself having an entraining action on the circadian system.

C. Tactics for Research on Behavioural Mediation

The question arises as to what methods are most appropriate for discovering whether or not phase-shifting by a drug depends on the induction of arousal or activity. Four approaches to the problem will now be outlined.

I. Measuring Activity

One tactic is to measure activity after the drug is given (e.g. MROSOVSKY and SALMON 1990; EDGAR et al. 1993; TOMINAGA et al. 1992) and to look for increases and for correlations between post-drug activity and subsequent phase shifts. If found, these can be compared to correlations between novelty-induced activity and shifts (e.g. Fig. 6). Finding correlations does not, of course, prove a causal link, but it does suggest that activity mediation should be considered. But if no increases of activity are found, this does not disprove some mediating role. It depends on the correct choice of methods of measuring activity. Selection of wheel-running assumes that increased activity outside the wheel is of no consequence (e.g. EDGAR et al. 1993; TOMINAGA et al. 1992). To gain an impression of a greater variety of behaviours, some investigators have preferred direct observation (BIELLO and MROSOVSKY 1993), but this involves some subjectivity in scoring and runs the risk of disturbing the animals under observation.

II. Altering Opportunities for Activity

Another tactic has been to prevent running wheels from rotating for a while after the drug treatment, or to include control groups housed in cages without wheels (EDGAR et al. 1993). This again neglects behaviour outside the wheel. Such behaviour has been implicated in at least one experiment. Phase-shifting effects of dark pulses in hamsters appear to be mediated by activity increases. If

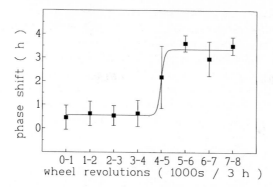

Fig. 6. Phase shift (means ± SDs) as a function of wheel-running (bins of 1000 wheel turns) during confinement to a novel wheel for 3 h, starting at zeitgeber time (ZT) 4 for the Aschoff (1965) type II procedure (data from Janik and Mrosovsky 1993). Although relationships between wheel-running and shifts have been found in other experiments (Reebs and Mrosovsky 1989b; Mrosovsky 1991), it should not be assumed that the same shape curve will be obtained at other cycle phases – or with other ages and strains of hamster

hamsters are confined to a small area during dark pulses, then phase-shifting is virtually abolished (Reebs et al. 1989; van Reeth and Turek 1989). If, however, the wheels are blocked but the hamsters are left free to move around the rest of their cages, then phase shifts are merely attenuated, not abolished (Reebs et al. 1989); evidently in the case of dark pulses, activity other than running in a wheel can affect the circadian system.

III. Confinement and Restraint

Given the findings with dark pulses just mentioned, the method of choice would seem to be to prevent drug-treated animals from moving at all by confining them to a small area, but this too has some potential drawbacks. Physical restraint such as putting the animal in a small tube (e.g. van Reeth and Turek 1989) is obviously stressful. In the case of triazolam this is probably not critical for blocking phase shifts because in other experiments (Mrosovsky and Salmon 1990) phase shifting by triazolam was also abolished when hamsters were merely confined to a small nest box with which they were thoroughly familiar and in which, given a choice, they voluntarily spent much of their time. All the same, confinement may have effects other than preventing activity. If motivation to be active is reduced, or if frustration is produced, it could be these that were responsible for blocking the ability of a drug to produce phase shifts, rather than the reduction of activity.

There is a further interpretational problem with any manipulation involving activity restriction. Suppose such restriction in itself produced phase shifts: then restriction, rather than *preventing* a drug from inducing a phase shift, might *cancel* out a drug-induced shift by providing a counteracting shift.

This can be illustrated with a specific example. VAN REETH et al. (1991) have reported that a PRC exists for immobilizing a hamster in a small tube for 3 h: if this occurs in the first half of the subjective night, that is CT 12–18, then delays of about 50 min are produced. Now suppose that a drug given a few hours before this time, say at CT 10, were found to cause a phase advance. To test for activity mediation an experimenter confines animals to a small area after injecting the drug. If no shifts are found, one conclusion would be that shifts in unconfined animals are mediated by activity. An alternative interpretation would be that the drug produced an activity-independent shift that *rapidly* advanced the clock to CT 12, and that restraint then produced a phase delay, thus cancelling out or reducing the advance. Therefore, experiments with restraint need to provide information not just on the effects (or lack of effects) of restraint alone at the same time as the drug is given, but also information on the effects of restraint at times to which the drug might have reset the clock. For further discussions of this problem see REEBS et al. (1989) and BIELLO and MROSOVSKY (1993).

IV. Testing Drugs In Vitro

One way to *cut* clear of these tangles is to work on the SCN slice. The SCN slice does not (one hopes!) locomote around the perfusion dish. If substances applied directly to the isolated SCN produce phase shifts, with PRCs corresponding in shape to those in the whole animal, then it is implausible to raise the possibility of activity mediation (e.g. PROSSER et al. 1993; SHIBATA et al. 1992 with serotonin agonists).

However, the large-amplitude PRCs that are characteristic of drug action at the slice level suggest that some modulating influences from outside the SCN may be lost. Also slices are not usually maintained long enough to assess steady state shifts. Some caution is in order in extrapolating to the circadian system as it normally functions.

D. Problems in Assessing Behavioural Mediation

Underlying many of the difficulties in determining whether behavioural mediation is involved in the chronobiotic actions of drugs is a gap in basic knowledge: although it is evident that non-photic manipulations such as novelty-induced wheel-running are capable of exerting major influences on the circadian system, the critical variable and its input pathway remains unknown. In some circumstances the amount of running during confinement to a novel wheel correlates with subsequent phase-shifting (Fig. 6), and the plasma cortisol level correlates with phase-shifting after saline injections (SUMOVA et al. 1994). But running and cortisol level may well simply be indices of arousal. Is it arousal, a particular motivational state, the accompanying running, or the interruption of rest that results in the phase shift? It is no accident that the term "non-photic" defines by exclusion. At present, "non-photic" is a widely used and useful descriptor precisely because it does not imply more than is known.

Because of this state of affairs, recommendations as to the most appropriate tests for behavioural mediation cannot be made today with confidence. If there is a concern with this question, probably a variety of the tactics listed above should be tried. Nevertheless, it is encouraging that the possibility of non-photic behavioural mediation of drug effects on rhythms is becoming recognized to the extent that it is considered a problem either worth thinking about, even if only to reject it, or investigating with additional experiments. The matter has been taken seriously in papers on the chronobiotic effects of a number of drugs: melatonin (REDMAN and ROBERTS 1991), vitamin B_{12} (TSUJIMARU et al. 1992), chlordiazepoxide (BIELLO and MROSOVSKY 1993), lithium (HAFEN and WOLLNIK 1994), quipazine (EDGAR et al. 1993), ($\pm$)-8-hydroxy-2-(di-*n*-propylamino) tetralin (8-OH-DPAT) (TOMINAGA et al. 1992), neuropeptide Y (NPY) (BIELLO et al. 1994) and clorgyline (DUNCAN and SCHULL 1994). In most cases not surprisingly, given the methodological difficulties described, the issue of activity mediation is not fully resolved.

Readers of a handbook may expect something more definitive. The answers are not yet available because systematic investigation of non-photic behavioural effects is a relatively new development in rhythms research. But there is one concrete fact to hold onto: simple alterations of the environment abolish the phase-shifting effects of some drugs but not of others. Figure 7 illustrates this point.

E. Mimicking and Modulation of Photic Effects

An extensive review on the circadian visual system has been recently published (MORIN 1994). Only a few additional points about drugs and photic effects will be covered here.

I. The Retinohypothalamic Tract Transmitter

For photic entrainment, the input pathway to the SCN, the retinohypothalamic tract (RHT), has been known for some time. The neurotransmitter for the RHT remains unknown (see discussion in MORIN 1994). If this were discovered, numerous opportunities for pharmacological manipulation of rhythms would arise, so it is not surprising that much research has been directed at this question. One method is to inject substances close to the SCN; if any produce shifts with a PRC similar to that for light pulses (Fig. 3), then they become candidates for the RHT transmitter. One difficulty with this strategy is deciding whether the PRC is indeed similar to that for light pulses. For example, EARNEST and TUREK (1985) considered their PRC for intraventricular injection of carbachol to be "strikingly similar in amplitude and waveform to existing phase-response curves generated with 15- to 60-min light

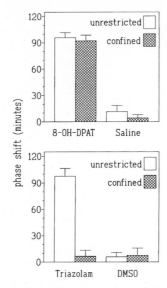

Fig. 7. Phase shifts of male hamsters (9 ± 2 months old) injected at ZT 8 with 5 mg/kg 8-OH-DPAT or 5 mg/kg triazolam (data from BOBRZYNSKA). Shifts were calculated as in JANIK and MROSOVSKY (1993) using an ASCHOFF (1965) type II procedure with darkness prevailing from shortly after the injections. The shifts seen in vehicle-injected animals are attributable to the change in the lighting. The hamsters were either free to move about their cages, including entering their wheels, or were confined to small nest boxes at the sides of their cages (diagram in MROSOVSKY and SALMON 1990)

pulses". They did, however, note some differences in the extent of the advance portion of the PRC. Contemplating the same data, RUSAK and BINA (1990) said that "carbachol effects do not closely resemble those of light pulses". Objective criteria of similarity are needed, together with PRCs with many time points, ideally with a variety of doses (MILLER 1993); constructing such PRCs is laborious and seldom undertaken.

Even if a drug produces a pattern of phase shifts that closely resembles the photic PRC, that does not prove the substance is an RHT transmitter. Suppose some general process such as membrane depolarization is an essential step in photic phase-shifting, as seems likely for certain molluscs (BLOCK et al. 1993). Then it should be possible to find a variety of substances that would mimic photic phase-shifting but not normally be involved in this process.

Some other cautions are that there might be more than one RHT neurotransmitter; also substances mimicking photic shifts might be acting on interneurons within the SCN rather than on cells postsynaptic to RHT neurons (ZATZ and HERKENHAM 1981). It is possible of course that some RHT terminals synapse directly on pacemaker cells. One way to discover if drugs act on clock elements as opposed to afferents to such cells is to block action potentials by tetrodotoxin and then see if shifts are still obtained when drugs are applied to the SCN (PROSSER et al. 1992). Given all these cautions, probably a combination of experimental approaches is needed to implicate a substance as an RHT transmitter (see MORIN 1994, for discussion of carbachol and other possibilities).

II. Period and Aschoff's Rule

In continuous light (LL), pharmacological effects on free-running period could depend on actions in the eye: if perceived brightness were affected, then, according to Aschoff's rule relating the level of illumination to periodicity (ASCHOFF 1960), the drug might mimic the effects of changes in illumination. This can be investigated by conducting the tests in continuous darkness (DD). For instance, that period lengthening by lithium occurs in DD and in blind animals suggests that its effect is not dependent on simple changes in sensitivity to light (KLEMFUSS 1992).

III. Sensitivity to Light and Phase Shifts

A substance might attenuate or enhance photic phase-shifting by altering the sensitivity of the eye to light. One way to investigate this would be to produce curves relating the size of shift to intensity of light pulses (e.g. TAKAHASHI et al. 1984; MEIJER et al. 1992), and determine in what way the shape was changed by the drug. It is important that phase-shifting and not some other response to light be used to assess effects on the input side because the circadian system appears to depend on different photoreceptors than those essential for spatiotemporal discriminations and normal vision: mice with virtual total loss of rods and cones are as sensitive to phase-shifting stimuli as mice with normal retinas (FOSTER 1993). Thus, reductions produced by lithium in the ability to report the presence of light in threshold tests (SEGGIE 1988), although of interest, are not necessarily relevant to circadian rhythms. It is, nevertheless, important to learn more about effects of lithium on the eye because therapeutic procedures might be influenced. The possibility that lithium reduces retinal sensitivity, thereby correcting abnormal phase relationships in bipolar affective disorder – presumed to arise from supersensitivity to light – has led to the suggestion that lithium be given in eye drops (SCHREIBER and AVISSAR 1991; for other opinions see KLEMFUSS 1992).

IV. Behavioural Inhibition of Photic Shifts

The phase advances produced by light pulses in the late subjective night can be attenuated in hamsters by inducing the animals to be active just before and during the pulse (RALPH and MROSOVSKY 1992; MROSOVSKY 1993b). The attenuation is probably not a result of the active animals receiving less light. If anything, aroused active animals would have more widely dilated pupils, and their posture would expose their eyes to more light than that received by curled up resting animals. The phenomenon of behavioural inhibition of photic shifting is not well researched, and probably occurs only with low levels of light. Nevertheless, its existence, in some circumstances at least, is enough to raise questions about the interpretation of the effects of certain drugs on circadian rhythms.

For instance, RALPH and MENAKER (1989) found that triazolam reduced the phase advances occurring in response to a pulse of light in the late subjective night in hamsters. The authors' hypothesis was that GABA reduces synaptic transmission at the termination of the RHT in the SCN; triazolam was thought to be a way of potentiating the activity of GABA at GABA-benzodiazepine receptors. However, since triazolam also makes hamsters hyperactive, the attenuation of phase shifts might be mediated by increased activity (or correlates) rather than by direct action on SCN synapses.

The same question can be raised about the ability of MK 801 [(+)-5-methyl-10, 11-dihydro-5-H-dibenzo[a,d] cyclohepten-5,10-imine maleate] to block or attenuate photic phase shifts (COLWELL et al., 1990). This has been interpreted as evidence that excitatory amino acids are neurotransmitters for the RHT, because MK 801 is a glutamate NMDA-receptor antagonist. However, at some doses at least, MK 801 also makes rats hyperactive (TRICKLEBANK et al. 1989; DANYSZ et al. 1994).

The point of raising these matters is not to debate the conclusions made from the experiments on triazolam and MK 801. An assessment of what are the neurotransmitters for the RHT should include additional information, for instance that excitatory amino acid receptor antagonists can block field potentials in the SCN elicited by stimulation of the cut end of the optic nerve (CAHILL and MENAKER 1989). An assessment of the role of behavioural arousal in phase-shifting by benzodiazepines should include information on those benzodiazepines that do not appear to increase locomotor activity (MISTLBERGER et al. 1991; BOULOS and HOUPT 1994; BIELLO and MROSOVSKY 1993). Activation by triazolam and by MK 801 is mentioned here to introduce a theoretical possibility: some substances that reduce phase advances might do so indirectly through producing behavioural arousal. More generally, if there are both photic and non-photic (behavioural) ways of resetting clocks, then the possibility arises for interactions between these two entrainment systems. Drugs might affect these interactions.

F. Saline Injections

I. Peripheral Injections

Perhaps the most important point about the work of HASTINGS et al. (1992) is not whether melatonin acts as a true zeitgeber or an amplifier, but the demonstration that saline injections alone can reset circadian rhythms, including those of induction of c-*fos* expression to pulses of light (MEAD et al. 1992).

In our laboratory, using a different strain of hamsters, we were not able to detect significant shifts in hamster rhythms following injections of saline at zeitgeber time (ZT) 8 and circadian time (CT) 8 (Bobrzynska and Mrosovsky, unpublished). Possibly some apparently trivial details are important. About 20% of the hamsters studied in HASTINGS' laboratory do not shift in response to saline (SUMOVA et al. 1994). Different strains of hamsters might have much higher proportions of non-shifters.

In any case, it is evident (HASTINGS et al. 1992; SUMOVA et al. 1994) that in some circumstances simply injecting saline can produce phase shifts. Therefore, if one is trying to detect a drug effect, one needs to use an injection procedure that is the least likely to cause shifts through arousal and non-specific means. To find such a procedure for a given strain of animals, a particular type of test, and even a particular experimenter (people handle animals in different ways) may require some initial comparisons between the rhythms of vehicle-injected animals and those of animals left undisturbed in their home cages.

If injection of a drug produces phase shifts that are significantly different from those after saline, then one is on fairly firm ground. Even then, however, if saline is having some effect on its own, it may be advisable to include an uninjected home-cage control group; this will provide some assessment of the amount of shift that is attributable to an injection procedure (though this could be complicated by synergistic effects). Another approach is to avoid manipulations at circadian times at which saline effects have been demonstrated (Fig. 5).

II. Central Injections

Phase shifts, sometimes as great as several hours, have been noted by a number of authors following microinjections of vehicles (saline, or artificial cerebrospinal fluid) to the SCN (DE VRIES and MEIJER 1991; ALBERS and FERRIS 1984; BIELLO et al. 1994). The presence of such shifts can complicate assessment of what a drug does. One possibility is that shifts are caused by the handling and the injection procedure, rather than by the vehicle itself, and are thus another example of clock resetting by arousal. In this case one would expect saline-induced shifts to occur at circadian times when behavioural arousal has been found to be effective; for hamsters this would be in the middle of the subjective day. Another possibility is that saline causes shifts by some non-specific (mechanical or chemical) stimulation of the SCN, perhaps resulting in release of transmitters. In experiments by SMITH et al. (1989), the two saline injections producing shifts were among the closest to the SCN (see also DE VRIES and MEIJER 1991).

Investigation of saline effects, although important methodologically, is not something that appeals to many researchers. Data from a small unpublished experiment are reported here as one approach to the problem. Hamsters had guide cannulae implanted with the tip aimed to be 2.7 mm above the SCN. For the first injection of saline, the injector was of a length designed to deliver saline

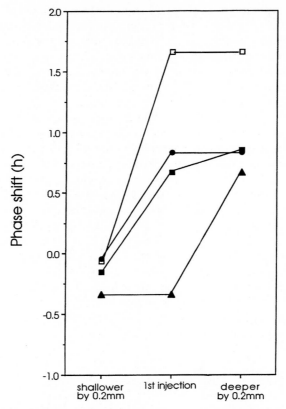

Fig. 8. Phase shifts produced by injections of 200 μl physiological saline into the region of the SCN. Male hamsters, approximately 3–4 months old at the start of the tests, were kept in an LD 14:10-h cycle. Injections took place at ZT 5. The lights were then turned off, and phase shifts assessed by an Aschoff type II procedure. The animals were re-entrained to the LD cycle before each additional test. *Different symbols* indicate different individuals. (Data from BIELLO and MROSOVSKY, unpublished)

at the dorsal margin of the SCN. For the second and third injections, 0.2 mm longer or 0.2 mm shorter injectors were used in the same animals, so that the saline was delivered either lower down, presumably nearer or into the SCN, or higher up, presumably further away from the SCN. The data (Fig. 8) suggest that distance above the SCN may be important in avoiding saline-induced shifts. This is in line with the idea that such shifts are produced by non-specific disturbance of the SCN. However, further work is needed, not only because of the small sample but also because none of the shifts to saline were especially large.

G. Concluding Remarks

This paper does not assert that any of the interpretations of drug actions discussed above are indeed correct. These examples – mostly limited to hamsters – are given to embody some of the general questions that might be entertained by chronopharmacologists. The difficulties in arriving at unambiguous interpretations of drug actions on rhythms do indeed seem numerous, but – in the long run – there may be compensations. The very complexity of the circadian system and the multiplicity of ways in which drugs might affect overt rhythms provide opportunities for a variety of different but specific interventions, tailored to particular research or therapeutic needs.

Acknowledgements. Support came from the Medical Research Council of Canada. I thank M.R. Ralph for helpful comments.

References

Albers HE, Ferris CF (1984) Neuropeptide Y: role in light-dark cycle entrainment of hamster circadian rhythms. Neurosci Lett 50:163–168

Aschoff J (1960) Exogenous and endogenous components in circadian rhythms. Cold Spring Harbor Symp Quant Biol 25:11–28

Aschoff J (1965) Response curves in circadian periodicity. In: Aschoff J (ed) Circadian clocks. North-Holland, Amsterdam, pp 95–111

Aschoff J, Figala J, Pöppel E (1973) Circadian rhythms of locomotor activity in the golden hamster (*Mesocricetus auratus*) measured with two different techniques. J Comp Physiol Psychol 85:20–28

Biello SM, Mrosovsky N (1993) Circadian phase-shifts induced by chlordiazepoxide without increased locomotor activity. Brain Res 622:58–62

Biello SM, Janik D, Mrosovsky N (1994) Neuropeptide Y and behaviorally induced phase shifts. Neuroscience 62:273–279

Block GD, Khalsa SBS, McMahon DG, Michel S, Guesz M (1993) Biological clocks in the retina: cellular mechanisms of biological timekeeping. Int Rev Cytol 146:83–144

Bobrzynska KJ, Godfrey MH, Mrosovsky N (1996) Serotonergic stimulation and nonphotic phase-shifting in hamsters. Physiol Behav 59:221–230

Boulos Z, Houpt TA (1994) Failure of triazolam to alter circadian reentrainment rates in squirrel monkeys. Pharmacol Biochem Behav 47:471–476

Cahill GM, Menaker M (1989) Effects of excitatory amino acid receptor antagonists and agonists on suprachiasmatic nucleus responses to retinohypothalamic tract volleys. Brain Res 479:76–82

Cassone VM, Chesworth MJ, Armstrong SM (1986) Dose-dependent entrainment of rat circadian rhythms by daily injection of melatonin. J Biol Rhythms 1:219–229

Colwell CS, Ralph MR, Menaker M (1990) Do NMDA receptors mediate the effects of light on circadian behavior? Brain Res 523:117–120

Daan S, Damassa D, Pittendrigh CS, Smith ER (1975) An effect of castration and testosterone replacement on a circadian pacemaker in mice (*Mus musculus*). Proc Natl Acad Sci USA 72:3744–3747

Danysz W, Essmann U, Bresink I, Wilke R (1994) Glutamate antagonists have different effects on spontaneous locomotor activity in rats. Pharmacol Biochem Behav 48:111–118

de Elvira Ruiz MC, Persaud R, Coen CW (1992) Use of running wheels regulates the effects of the ovaries on circadian rhythms. Physiol Behav 52:277–284

De Vries MJ, Meijer JH (1991) Aspartate injections into the suprachiasmatic region of the Syrian hamster do not mimic the effects of light on the circadian activity rhythm. Neurosci Lett 127:215–218

Duncan WC Jr, Schull J (1994) The interaction of thyroid state, MAOI drug treatment, and light on the level and circadian pattern of wheel-running in rats. Biol Psychiatry 35:324–334

Earnest DJ, Turek FW (1985) Neurochemical basis for the photic control of circadian rhythms and seasonal reproductive cycles: role for acetylcholine. Proc Natl Acad Sci USA 82:4277–4281

Edgar DM, Martin CE, Dement WC (1991) Activity feedback to the mammalian circadian pacemaker: influence on observed measures of rhythm period length. J Biol Rhythms 6:185–199

Edgar DM, Miller JD, Prosser RA, Dean RR, Dement WC (1993) Serotonin and the mammalian circadian system: II. Phase-shifting rat behavioral rhythms with serotonergic agonists. J Biol Rhythms 8:17–31

Foster RG (1993) Photoreceptors and circadian systems. Curr Direct Psychol Sci 2:34–39

Golombek DA, Cardinali DP (1993) Melatonin accelerates reentrainment after phase advance of the light-dark cycle in Syrian hamsters: antagonism by flumazenil. Chronobiol Int 10:435–441

Hafen T, Wollnik F (1994) Effect of lithium carbonate on activity level and circadian period in different strains of rats. Pharmacol Biochem Behav 49:975–983

Hastings MH, Mead SM, Vindlacheruvu RR, Ebling FJP, Maywood ES, Grosse J (1992) Non-photic phase shifting of the circadian activity rhythm of Syrian hamsters: the relative potency of arousal and melatonin. Brain Res 591:20–26

Hau M, Gwinner E (1994) Melatonin facilitates synchronization of sparrow circadian rhythms to light. J Comp Physiol [A] 175:343–347

Janik D, Mrosovsky N (1993) Nonphotically induced phase shifts of circadian rhythms in the golden hamster: activity-response curves at different ambient temperatures. Physiol Behav 53:431–436

Joy JE, Losee-Olson S, Turek FW (1989) Single injections of triazolam, a short-acting benzodiazepine, lengthen the period of the circadian activity rhythm in golden hamsters. Experientia 45:152–154

Klemfuss H (1992) Rhythms and the pharmacology of lithium. Pharmacol Ther 56:53–78

Mead S, Ebling FJP, Maywood ES, Humby T, Herbert J, Hastings MH (1992) A nonphotic stimulus causes instantaneous phase advances of the light-entrainable circadian oscillator of the Syrian hamster but does not induce the expression of c-fos in the suprachiasmatic nuclei. J Neurosci 12:2516–2522

Meijer JH, Rusak B, Gänshirt G (1992) The relation between light-induced discharge in the suprachiasmatic nucleus and phase shifts of hamster circadian rhythms. Brain Res 598:257–263

Mikkelsen JD, Vrang N (1994) A direct pretectosuprachiasmatic projection in the rat. Neuroscience 62:497–505

Mikkelsen JD, Vrang N, Larsen PJ (1994) Neuropeptides in the mammalian suprachiasmatic nucleus. Adv Pineal Res 8:57–67

Miller JD (1993) On the nature of the circadian clock in mammals. Am J Physiol 264:R821–R832

Mistlberger RE, Houpt TA, Moore-Ede MC (1991) The benzodiazepine triazolam phase-shifts circadian activity rhythms in a diurnal primate, the squirrel monkey (*Saimiri sciureus*). Neurosci Lett 124:27–30

Moore RY, Card JP (1994) Intergeniculate leaflet: an anatomically and functionally distinct subdivision of the lateral geniculate complex. J Comp Neurol 344:403–430

Moore-Ede MC (1985) Discussion. Photoperiodism melatonin and the pineal. In: Evered D, Clark S (eds) Ciba Foundation symposium, vol 117. Pitman, London, p 203

Morin LP (1994) The circadian visual system. Brain Res Rev 67:102–127

Morin LP, Fitzgerald KM, Zucker I (1977) Estradiol shortens the period of hamster circadian rhythms. Science 196:305–307

Mrosovsky N (1988) Phase response curves for social entrainment. J Comp Physiol [A] 162:35–46

Mrosovsky N (1991) Double pulse experiments with nonphotic and photic phase-shifting stimuli. J Biol Rhythms 6:167–179

Mrosovsky N (1993a) τ changes after single nonphotic events. Chronobiol Int 10:271–276

Mrosovsky N (1993b) Photic phase shifting in hamsters: more than meets the eye. Light Treatment Biol Rhythms 5:34–36

Mrosovsky N (1995) A nonphotic gateway to the circadian clock of hamsters. Ciba Found Symp 183:154–174

Mrosovsky N, Salmon PA (1987) A behavioural method for accelerating re-entrainment of rhythms to new light-dark cycles. Nature 330:372–373

Mrosovsky N, Salmon PA (1990) Triazolam and phase-shifting acceleration re-evaluated. Chronobiol Int 7:35–41

Mrosovsky N, Salmon PA, Menaker M, Ralph MR (1992) Nonphotic phase shifting in hamster clock mutants. J Biol Rhythms 7:41–49

Prosser RA, Heller HC, Miller JD (1992) Serotonergic phase shifts of the mammalian circadian clock: effects of tetrodotoxin and high Mg^{2+}. Brain Res 573:336–340

Prosser RA, Dean RR, Edgar DM, Heller HC, Miller JD (1993) Serotonin and the mammalian circadian system: I. *In vitro* phase shifts by serotonergic agonists and antagonists. J Biol Rhythms 8:1–16

Ralph MR, Menaker M (1989) GABA regulation of circadian responses to light. I. Involvement of $GABA_A$-benzodiazepine and $GABA_B$-receptors. J Neurosci 9:2858–2865

Ralph MR, Mrosovsky N (1992) Behavioral inhibition of circadian responses to light. J Biol Rhythms 7:353–359

Redfern P, Minors D, Waterhouse J (1994) Circadian rhythms, jet lag, and chronobiotics: an overview. Chronobiol Int 11:253–265

Redman J, Armstrong S, Ng KT (1983) Free-running activity rhythms in the rat: entrainment by melatonin. Science 219:1089–1091

Redman JR, Roberts CM (1991) Entrainment of rat activity rhythms by melatonin does not depend on wheel-running activity. Soc Neurosci Abstr 17:673

Reebs SG, Mrosovsky N (1989a) Effects of induced wheel running on the circadian activity rhythms of Syrian hamsters: entrainment and phase response curve. J Biol Rhythms 4:39–48

Reebs SG, Mrosovsky N (1989b) Large phase-shifts of circadian rhythms caused by induced running in a re-entrainment paradigm: the role of pulse duration and light. J Comp Physiol [A] 165:819–825

Reebs SG, Lavery RJ, Mrosovsky N (1989) Running activity mediates the phase-advancing effects of dark pulses on hamster circadian rhythms. J Comp Physiol [A] 165:811–818

Refinetti R, Kaufman CM, Menaker M (1994) Complete suprachiasmatic lesions eliminate circadian rhythmicity of body temperature and locomotor activity in golden hamsters. J Comp Physiol [A] 175:223–232

Rusak B, Bina KG (1990) Neurotransmitters in the mammalian circadian system. Annu Rev Neurosci 13:387–401

Schreiber G, Avissar S (1991) Lithium administered by eye drops: a better treatment for bipolar affective disorder? Prog Neuropsychopharmacol Biol Psychiatry 15:315–321

Seggie J (1988) Lithium and the retina. Prog Neuropsychopharmacol Biol Psychiatry 12:241–253

Shibata S, Tsuneyoshi A, Hamada T, Tominaga K, Watanabe S (1992) Phase-resetting effect of 8-OH-DPAT, a serotonin$_{1A}$ receptor agonist, on the circadian rhythm of firing rate in the rat suprachiasmatic nuclei in vitro. Brain Res 582:353–356

Smith RD, Inouye S-IT, Turek FW (1989) Central administration of muscimol phase-shifts the mammalian circadian clock. J Comp Physiol [A] 164:805–814

Sumova A, Ebling FJP, Maywood ES, Herbert J, Hastings MH (1994) Non-photic circadian entrainment in the Syrian hamster is not associated with phosphorylation of the transcriptional regulator CREB within the suprachiasmatic nucleus, but is associated with adrenocortical activation. Neuroendocrinology 59:579–589

Takahashi JS, DeCoursey PJ, Bauman L, Menaker M (1984) Spectral sensitivity of a novel photoreceptive system mediating entrainment of mammalian circadian rhythms. Nature 308:186–188

Tominaga K, Shibata S, Ueki S, Watanabe S (1992) Effects of 5-HT_{1A} receptor agonists on the circadian rhythm of wheel-running activity in hamsters. Eur J Pharmacol 214:79–84

Tricklebank MD, Singh L, Oles RJ, Preston C, Iversen SD (1989) The behavioural effects of MK-801: a comparison with antagonists acting non-competitively and competitively at the NMDA receptor. Eur J Pharmacol 167:127–135

Tsujimaru S, Ida Y, Satoh H, Egami H, Shirao I, Mukasa H, Nakazawa Y (1992) Vitamin B_{12} accelerates re-entrainment of activity rhythms in rats. Life Sci 50:1843–1850

Turek FW (1988) Do circadian biologists and chronopharmacologists talk the same 'language'? Annu Rev Chronopharmacol 4:205–208

Turek FW, Losee-Olson S (1986) A benzodiazepine used in the treatment of insomnia phase-shifts the mammalian circadian clock. Nature 321:167–168

van den Pol AN (1986) Gamma-aminobutyrate, gastrin releasing peptide, serotonin, somatostatin, and vasopressin: ultrastructural immunocytochemical localization in presynaptic axons in the suprachiasmatic nucleus. Neuroscience 17:643–659

van Reeth O, Turek FW (1989) Stimulated activity mediates phase shifts in the hamster circadian clock induced by dark pulses or benzodiazepines. Nature 339:49–51

van Reeth O, Losee-Olson S, Turek FW (1987) Phase shifts in the circadian activity rhythm induced by triazolam are not mediated by the eyes or the pineal gland in the hamster. Neurosci Lett 80:185–190

van Reeth O, Hinch D, Tecco JM, Turek FW (1991) The effects of short periods of immobilization on the hamster circadian clock. Brain Res 545:208–214

Winfree AT (1986) Benzodiazepines set the clock. Nature 321:114–115

Yamada N, Shimoda K, Takahashi K, Takahashi S (1990) Relationship between free-running period and motor activity in blinded rats. Brain Res Bull 25:115–119

Zatz M, Herkenham MA (1981) Intraventricular carbachol mimics the phase-shifting effect of light on the circadian rhythm of wheel-running activity. Brain Res 212:234–238

CHAPTER 16
Rhythms in Retinal Mechanisms

G.D. BLOCK and S. MICHEL

A. Introduction

Historically the search for circadian pacemakers has most often focused on the visual pathways, the rationale being that to perform useful work for the organism, pacemakers must be synchronized to local time and thus, perforce, the biological clock should be intimately associated with the visual pathways. In some instances this strategy has resulted in identifying pacemaker structures outside but in close contact with the retina such as the suprachiasmatic nucleus of the rodent (MOORE and EICHLER 1972; STEPHAN and ZUCKER 1972) or the optic lobes of the cockroach (SOKOLOVE 1975). In a few instances the search for pacemaker structures has stopped almost immediately as it became clear that the retina contained a biological clock.

It is now well documented that retinal circadian pacemakers exist in some species of birds, amphibians, and mollusks and it is likely that endogenous retinal rhythmicity extends to additional phyla. In most cases it has been possible to study these retinal clocks in vitro and this has opened up experimental opportunities to evaluate the cellular and molecular organization of the pacemaking system. Investigation of retinal pacemakers in vitro has often relied on pharmacological approaches which have provided significant insights into the circadian pacemaker and about the mechanisms by which these clocks are synchronized by environmental light cycles. The review which follows is a summary of some of these efforts.

B. Retinal Rhythmicity in *Xenopus*

Daily rhythms in retinal melatonin release or content have been reported in a number of species (for review see CAHILL et al. 1991); in a few cases these changes have been shown to persist under constant conditions (i.e., exhibit circadian rhythmicity). Circadian rhythmicity has been demonstrated in the chicken (HAMM and MENAKER 1980), quail (UNDERWOOD and SIOPES 1985; UNDERWOOD et al. 1988), and *Xenopus* (CAHILL and BESHARSE 1989). Recently Tosini and Menaker (1996) have demonstrated a circadian rhythm in melatonin release in the rodent retina. Thus, there is now unambiguous evidence for a circadian rhythm in the mammalian retina.

While retinal melatonin rhythmicity has been known for some time, the recent development of in vitro preparations has improved greatly our understanding of the cellular and molecular mechanisms underlying this rhythmicity. The most intensively studied of these preparations is the eyecup of the African clawed toad, *Xenopus laevis*. BESHARSE and IUVONE (1983) first demonstrated that a circadian oscillator controlling *N*-acetyltransferase (NAT) resides within the eye of *Xenopus*. Circadian rhythms in retinal NAT activity were measured in eyecup preparations that were maintained in constant darkness in vitro.

I. Pacemaker Localization

The retinal circadian pacemaker in *Xenopus* is located among the photoreceptors. Cultured photoreceptor layers in which approximately 94% of the viable cells were rod or cone photoreceptors continued to exhibit circadian periodicities in melatonin synthesis (CAHILL and BESHARSE 1993).

Shown in Fig. 1 are examples of melatonin rhythms from whole neural retinas and from fragments containing only the photoreceptor layers. As indicated, the retinas as well as isolated photoreceptor layers exhibit a precise circadian rhythm, which can be phase shifted by light pulses. Isolated photoreceptor layers release melatonin rhythmically when cultured in continuous darkness at a constant temperature and can be phase shifted by light pulses (CAHILL and BESHARSE 1993).

II. Rhythm Generation

It is not yet clear whether the rods, cones, or both are responsible for rhythmicity and whether interactions between cell types are necessary for rhythmicity. Nothing is presently known about how the circadian rhythm is generated within or among these cells.

III. Pacemaker Entrainment

The isolated eyecup preparation contains critical photoreceptors for entrainment (Fig. 2). This was first demonstrated by reversing the light cycle in vitro and finding that after 2 days the NAT rhythm was reversed (BESHARSE and IUVONE 1983).

A "best guess" of the entrainment pathway (and expression pathway) for the *Xenopus* retinal pacemaker is shown in Fig. 3. The analysis of the light entrainment pathway in *Xenopus* is made somewhat complicated by virtue of the fact that light has two classes of effects on the circadian rhythm of NAT activity and melatonin production. First, light causes acute suppression of retinal NAT content and melatonin release (BINKLEY et al. 1979; HAMM and MENAKER 1980; IUVONE and BESHARSE 1983). Second, it exerts effects on the circadian oscillator driving NAT activity and melatonin production (BESHARSE and IUVONE 1983; CAHILL et al. 1991).

Evidence in hand suggests that light normally exerts its effects on the pacemaker through a non-dopaminergic pathway and, perhaps, through a

Rhythms in Retinal Mechanisms 437

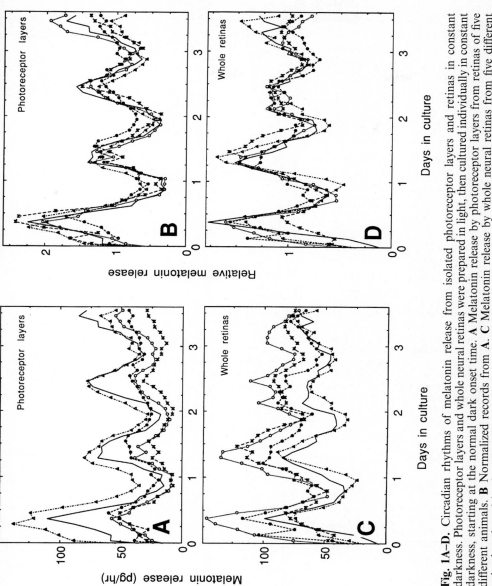

Fig. 1A–D. Circadian rhythms of melatonin release from isolated photoreceptor layers and retinas in constant darkness. Photoreceptor layers and whole neural retinas were prepared in light, then cultured individually in constant darkness, starting at the normal dark onset time. **A** Melatonin release by photoreceptor layers from retinas of five different animals. **B** Normalized records from **A**. **C** Melatonin release by whole neural retinas from five different animals cultured in the same experiment as the photoreceptor layers in **A** and **B**. **D** Normalized records from **C**. (Figure reproduced from CAHILL and BESHARSE 1993)

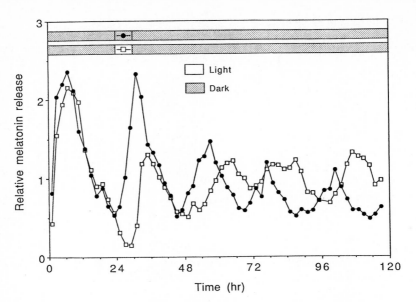

Fig. 2. Light-induced phase shift of melatonin rhythm expressed in isolated *Xenopus* eyecups. A pulse of light during the early subjective night suppresses melatonin and delays the phase of the circadian clock in cultured *Xenopus* eyecups. Records of melatonin release by single eyecups (not paired) are representative of five eyecups per condition. Both eyecups were maintained in constant darkness, except that the eyecup represented by *open squares* was exposed to a 6-h pulse of fluorescent white light starting 24 h after dark onset at the beginning of the experiment. Both eyecups were exposed to medium containing 0.5% ethanol at the time of the light pulse, as a control for a pharmacological experiment that is not shown. Eyecups were cultured in medium containing 100 µM 5-hydroxytryptophan, a serotonin precursor. (Figure reproduced from CAHILL et al. 1991)

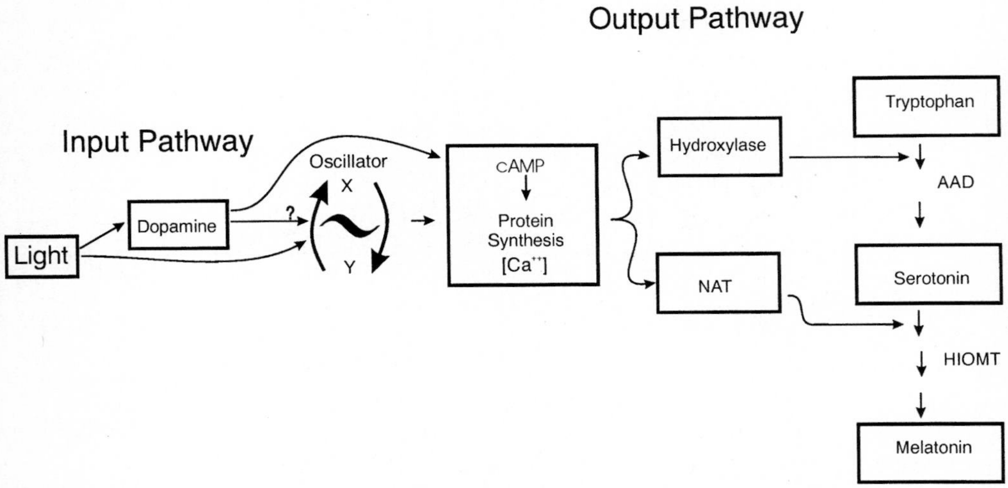

Fig. 3. *Xenopus* entrainment and rhythm expression pathways. *NAT*, *N*-acetyltransferase; *AAD*, amino acid decarboxylase; *HIOMT*, hydroxyindole-*O*-methyltransferase. See text for details

dopaminergic pathway as well (CAHILL and BESHARSE 1991; CAHILL et al. 1991). The observations leading to these conclusions are as follows: CAHILL and BESHARSE (1991) report that the eticlopride (50 μM), a D_2 dopamine receptor antagonist, while completely suppressing melatonin release and blocking the phase delay induced by 500 nM dopamine, did not block the suppression of melatonin release or the phase shift induced by light. These results indicate that dopamine-independent pathways exist for light input to the pacemaker. While dopaminergic pathways are not necessary for light-induced phase shifts, dopamine receptor agonists can mimic light in suppressing melatonin release and resetting the phase of the circadian rhythm. Phase shifts induced by quinpirole, a D_2 dopamine receptor agonist, are similar to phase shifts by light. In addition, pharmacological analysis with selective catecholamine receptor agonists and antagonists indicates that there are pathways to the melatonin-generating system and the circadian oscillator that include D_2 dopamine receptors. Thus, it seems likely that there are two potential entrainment pathways that may converge at the pacemaker or at some step prior to the pacemaker (CAHILL et al. 1991).

IV. Rhythm Expression

A good deal is known about the regulation of circadian rhythms within the amphibian retina (Fig. 3). Primary circadian control over melatonin synthesis is mediated by the enzyme N-acetyltransferase (NAT). cAMP, Ca^{2+}, and protein synthesis have all been implicated in the activation of retinal melatonin synthesis (IUVONE and BESHARSE 1983, 1986a, b; IUVONE 1986; BESHARSE et al. 1988). When eyecup preparations are maintained in darkness, there is an increase in NAT during the early subjective night. Exposure to light suppresses this rise (IUVONE and BESHARSE 1983). The action of light can be blocked by treatments that increase the intracellular levels of cAMP (IUVONE and BESHARSE 1983). The cAMP-induced rise in NAT is blocked by protein synthesis inhibitors (IUVONE and BESHARSE 1983, 1986a). The cAMP-induced rise in NAT is not blocked by the removal of extracellular calcium (IUVONE and BESHARSE 1986a). However, the dark-induced rise does require extracellular Ca^{2+} (IUVONE and BESHARSE 1986b). This suggests that the nocturnal rise in melatonin synthesis involves synaptic transmission or some other calcium-dependent mechanism.

While circadian modulations in melatonin production are driven by a rhythm in NAT, the rate of melatonin synthesis is significantly limited by the availability of the precursor, serotonin (5-hydroxytryptamine, 5-HT) (CAHILL and BESHARSE 1990). Evidence in hand suggests that the rate-limiting enzyme for serotonin synthesis, tryptophan hydroxylase, also shows rhythmicity in release and content (see review by CAHILL et al. 1991) and in its mRNA expression (GREEN and BESHARSE 1994). If eyecups are maintained in superfusion culture, 5-hydroxytryptophan increases melatonin production at all times of the circadian cycle (CAHILL and BESHARSE 1990), indicating that none of the enzymes that follow tryptophan hydroxylase in the melatonin synthetic pathway is normally saturated. Therefore, there may be a coordinated regulation scheme for NAT and tryptophan hydroxylase that may have the effect of increasing the amplitude of retinal melatonin rhythms (CAHILL et al. 1991).

C. Retinal Rhythmicity in *Aplysia californica*

The *Aplysia* eye preparation has provided important insights into the mechanisms of pacemaker regulation and entrainment (for review see; JACKLET 1989; KOUMENIS and ESKIN 1992; COLWELL et al. 1992a). The *Aplysia* eye exhibits a robust circadian rhythm in optic nerve impulse frequency both in the animal (BLOCK 1981) and when removed from the animal and placed into continual darkness at constant temperature (JACKLET 1969a; Fig. 4).

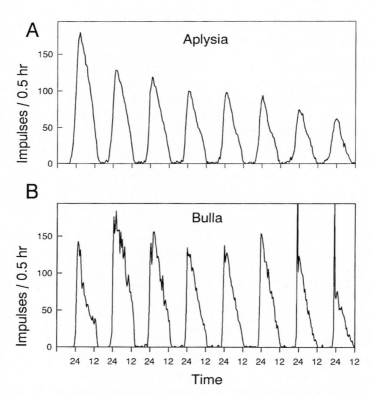

Fig. 4A,B. Ocular rhythm of *Bulla* and *Aplysia*. Spontaneous impulse frequency from eyes recorded in darkness at 15 °C. A suction electrode is placed around the optic nerve. Impulses occur as compound action potentials. (Figure reproduced from BLOCK et al. 1993)

I. Pacemaker Localization

The *Aplysia* retina consists of a number of cell types (JACKLET 1969b; JACKLET et al. 1972; LUBORSKY-MOORE and JACKLET 1977; HERMAN and STRUMWASSER 1984). The compound optic nerve impulses that express the circadian rhythm appear to be generated by a population of neurosecretory-like neurons (D-type cells) that are outside the receptor layer. Upon impalement with microelectrodes, these cells fire in synchrony with the large compound impulses in the optic nerve (JACKLET et al. 1982). While the precise retinal circuitry is uncertain, it appears that the D-type cells are electrically coupled to one another and fire in synchrony to generate the compound action potentials that express the circadian rhythm (JACKLET et al. 1982). The relationship among these cells and others within the retina is uncertain. Illumination of the receptor layer with a small light fiber results in low-amplitude desynchronized activity in the optic nerve with no compound action potentials. Illumination near the base of the retina elicits primarily compound action potentials (BLOCK and MCMAHON 1983). These results suggest that the D-type cells or other neurons in the outer layer of the retina are responsible for the light response in compound potentials and that the photoreceptor layer constitutes a separate visual pathway. Pairwise electrophysiological recording between D-type and photoreceptor cells that might reveal functional connections has yet to be attempted in the *Aplysia* retina.

The precise location of the circadian pacemaker(s) remains uncertain. In an early study, JACKLET and GERONIMO (1971) suggested that a circadian rhythm emerged from the interactions of cells within the retina. However, subsequent experiments failed to support this view. STRUMWASSER (1973) reported that eyes reduced to less than 20% of their original volume still exhibited a circadian cycle following surgery. In addition, WOOLUM and STRUMWASSER (1980) found that X-irradiation, which can leave the eye aperiodic, was most effective when directed towards the posterior portion of the retina, the location of the D-type cells. These results suggest that rhythm generation is most likely due to a subpopulation of neurons outside the receptor layer. A recent preliminary report (JACKLET and BARNES 1995) indicates that retinal neurons in dispersed cell culture continue to exhibit a circadian rhythm in membrane conductance. Thus, it seems quite likely that circadian rhythm generation occurs within individual retinal neurons.

II. Rhythm Generation

Relatively little is known about the cellular and molecular mechanisms that give rise to circadian rhythmicity within the *Aplysia* retina. There is some evidence that transcription and protein synthesis are involved in the "timing loop." The reversible transcription inhibitor 5,6-dichlorobenzimidazole riboside (DRB) lengthens the free-running period in a dose-dependent manner, and at higher concentrations of inhibitor no periodicity is observed (RAJU et al.

1991). Similar effects are observed with translation inhibitors (JACKLET 1980; LOTSHAW and JACKLET 1986). A more complete analysis of the behavior of circadian pacemakers at high doses of transcription and translation inhibitors in *Bulla* (see below) suggests that daily transcription and translation are required for the pacemaker to complete a circadian cycle.

III. Pacemaker Entrainment

The *Aplysia* ocular rhythm can be entrained by light pulses both in vivo and in vitro (ESKIN 1971). Single pulses of light shift the pacemaker in a phase-dependent manner, with the eye displaying a conventional low-amplitude phase response curve (JACKLET 1974; CORRENT et al. 1982). In addition to phase shifts by light, the neuromodulator serotonin is also effective in generating phase shifts (CORRENT et al. 1978; ESKIN 1979; COLWELL 1990). In this case the phase response curve is shifted approximately 180° on the time axis when compared to phase shifts obtained to light (CORRENT et al. 1978). A summary of the identified and suspected processes in the entrainment (and expression pathway) is shown in Fig. 5.

1. Light-Induced Phase Shifts

The following is the most likely sequence of events in the entrainment cascade. Light pulses, acting on photopigments within light-sensitive pacemaker neurons, lead to an increase in cyclic GMP, which generates membrane depolarization, and an increased transmembrane calcium flux. Calcium entry may exert its action via protein synthesis.

a) Cyclic GMP

Light appears to act through an increase in cGMP. Six-hour pulses of light increase the levels of cGMP without increasing cAMP levels (ESKIN et al. 1984a). Furthermore, an analog of cGMP, 8-bromoguanosine-3'5'-cyclic monophosphate, can shift the phase of the ocular pacemaker, and the resultant phase response curve appears identical to that of light pulses. Since application of the cGMP analog leads to impulse production similar to that observed during light pulses, it seems likely that cGMP exerts its effects through a change in membrane potential.

b) Membrane Depolarization

Several studies in *Aplysia* implicate membrane depolarization in generating phase shifts of the ocular pacemaker. ESKIN (1972) and JACKLET and LOTSHAW (1981) report that elevated K^+ seawater solutions phase shift the ocular rhythm and the resulting phase response curve appears similar, although not identical, to that obtained from light pulses.

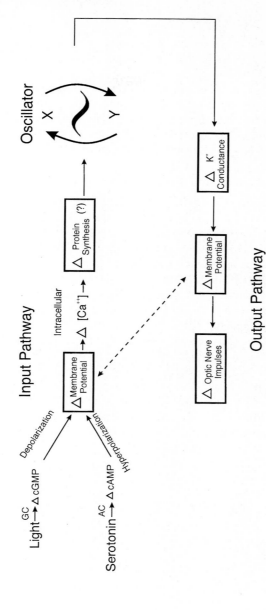

Fig. 5. *Aplysia* entrainment and rhythm expression pathways. *AC*, adenylate cyclase; *GC*, guanylate cyclase. See text for details

c) Calcium Flux

In *Aplysia* there is some evidence that a transmembrane Ca^{2+} flux is involved in light-induced phase shifts. Eskin (1977) reported that light-induced phase shifts are blocked in ethyleneglycoltetraacetic acid (EGTA)-containing seawater solutions with a Ca^{2+} concentration calculated at 0.13 μM but are not blocked with a calcium concentration of 450 μM. More recently, Colwell et al. (1994) have reported that light-induced phase advances and delays are blocked in either low-Ca^{2+} solutions (calculated at 0.17 μM free Ca^{2+}) or in the presence of 1 mM Ni^{2+}, a calcium channel blocker.

d) Protein Synthesis

The role of protein synthesis in the light phase-shifting pathway has been addressed and what has emerged is a complex set of results. In *Aplysia*, the translation inhibitors anisomycin and cycloheximide block light-induced phase advances (Raju et al. 1990). The action of cycloheximide on light-induced phase delays, however, is less clear. Although it appears that phase delays are not blocked, interpretation of the data is complicated by the observation that the cycloheximide treatment alone generates a phase shift at this phase. The light-induced and the cycloheximide-induced delays do not seem to be strictly additive as would be expected for concurrent treatments acting through independent mechanisms (Raju et al. 1990). Therefore, phase delays may be somewhat dependent upon protein synthesis. The interpretation of these results is complicated further by recent results in *Bulla* indicating that the pacemaker appears stopped by protein synthesis inhibitors applied near dawn (Khalsa et al. 1992). This result raises the issue of whether the effect of protein synthesis inhibitors on preventing phase advances is due to a direct action on the pacemaker rather than on elements along the light entrainment pathway.

Whereas the exact role of protein synthesis in phase shifting by light is uncertain, there is evidence in *Aplysia* that several proteins are influenced by light. In examining proteins separated by two-dimensional gel electrophoresis, Raju and coworkers (1990) found that incorporation of amino acids into 11 proteins was changed during a 6-h light pulse. Nine of these proteins were affected by light in a phase-dependent manner. It is not yet certain whether these protein changes occur in the pacemaker neurons or within other cells within the retina. This is a serious concern in the *Aplysia* retina since a recent report indicates that at least some photoreceptors appear to have circadian modulations impressed on them by pacemaker cells (Jacklet 1991).

2. Serotonin-Induced Phase Shifts

Serotonin (5-HT) pulses can phase shift the ocular pacemaker of *Aplysia* (Corrent et al. 1978; Eskin 1979; Colwell 1990). The phase-response curve for 5-HT pulses is shifted approximately 12 h on the time axis with respect to light pulses (Corrent et al. 1978). Six-hour 5-HT pulses lead to phase delays in the early subjective day, phase advances in the late subjective day, and no shifts during the early subjective night. The most likely sequence of events is that serotonin leads to an increase in cAMP, leading to an increase in K^+ conductance and membrane hyperpolarization, a reduction in transmembrane calcium flux, and, finally, synthesis of specific proteins involved in the phase-shifting mechanism.

a) cAMP

5-Hydroxytryptamine appears to exert its effects on the pacemaker through a cAMP-dependent process. Forskolin, which increases cAMP through activation of adenylate cyclase, and cAMP analogs generate phase response curves that are indistinguishable from that of 5-HT (ESKIN and TAKAHASHI 1983; ESKIN et al. 1982).

b) Membrane Hyperpolarization

There is fairly convincing evidence that membrane hyperpolarization is involved in the 5-HT phase-shifting pathway. Treatments such as K^+ free seawater, which should produce membrane hyperpolarization, generate a phase response curve identical to that of serotonin (ESKIN 1982). In addition, pulses of elevated K^+ seawater block the phase shift of a simultaneously applied 5-HT pulse (ESKIN 1982). Furthermore, experiments with bathing solutions containing barium, which is known to block K^+ channels, prevented the phase-shifting action of 5-HT (COLWELL et al. 1992b).

c) Calcium Flux

It is most likely that membrane hyperpolarization affects the pacemaker through a reduction in a transmembrane Ca^{2+} flux. Pacemaker neurons generate spontaneous impulses during the subjective day, the time when 5-HT is effective in producing phase shifts. Hyperpolarization, induced by serotonin application, would be expected to reduce Ca^{2+} entry that would typically occur due to membrane depolarization and impulse production. Low-Ca^{2+} treatments generate phase shifts which resemble those of serotonin (COLWELL et al. 1994). In addition, unlike 5-HT, low Ca^{2+} does not increase ocular cAMP levels nor can the low Ca^{2+} phase shifts be prevented by increasing extracellular K^+ concentration (COLWELL et al. 1994). These results would be expected if a reduction in a diurnal transmembrane Ca^{2+} flux were downstream of the increase in cAMP and hyperpolarization.

d) Protein Synthesis

Protein synthesis appears to play a critical role in the serotonin phase-shifting pathway. Simultaneous application of the translation inhibitor, anisomycin, blocks serotonin-induced phase advances (ESKIN et al. 1984b). ESKIN and coworkers have identified proteins whose synthesis is altered by serotonin (YEUNG and ESKIN 1987). The best characterized of these proteins is a 34-kDa protein. The synthesis of the protein is increased by 5-HT at phases where phase advances and delays occur, but not increased at phases where 5-HT has no effect. In addition, the effects of 5-HT on synthesis of this protein are more pronounced in surgically reduced retinas containing the proximal portion of the retina, the retinal region that is required for pacemaker function.

The most abundant protein, a 40-kDa protein, has been sequenced (RAJU et al. 1993). A peptide derived from this protein yielded a sequence of 38 amino acids. A significant sequence similarity exists between this peptide and a family of proteins called lipocortins or annexins. Members of this family are all Ca^{2+}/phospholipid-binding proteins.

IV. Rhythm Expression

Relatively little is known about the mechanisms by which the retinal pacemaker is coupled to optic nerve activity. Recent evidence from patch clamp recordings from isolated pacemaker neurons suggests that the pacemaker modulates a K^+ current, possibly a delayed rectifier (JACKLET and BARNES 1995). It is not known whether this current plays a role in controlling the circadian rhythm in impulse activity.

D. Retinal Rhythmicity in *Bulla gouldiana*

While the *Aplysia* eye represented a major advance in the development of new in vitro preparations for circadian study, it became apparent that there were some experimental shortcomings. A significant problem was the difficulty the preparation presented for electrophysiological study. The small size of many of the neurons in the retina prevented long-duration electrophysiological recordings, although some short-duration studies were successfully completed (JACKLET 1969b; JACKLET et al. 1982). The observation that several opisthobranch species contained ocular circadian pacemakers (for review, see BLOCK et al. 1993) raised the possibility that some of these other retinas might be better suited for electrophysiological study.

One system, discovered during such a systematic search of different opisthobranchs, was *Bulla gouldiana*, the cloudy bubble snail (BLOCK and WALLACE 1982). The *Bulla* retina exhibits a robust circadian rhythm, but contains fewer and larger cells than its *Aplysia* counterpart (BLOCK and WALLACE 1982; JACKLET and COLQUHOUN 1983). When the *Bulla* eye is removed from the animal and recorded in isolation, a circadian rhythm can be measured in the frequency of spontaneous optic nerve potentials (Fig. 4). If the animal has been previously placed on a light cycle consisting of 12 h of light and 12 h of dark, peak impulse activity occurs just after projected dawn. The eye will exhibit a robust free-running rhythm for at least nine cycles and the average free-running period is 23.7 h (BLOCK et al. 1993). Two eyes from the same animal have comparable phase and period, making one eye an ideal control in experiments in which the phase or period of the eye is experimentally perturbed.

I. Pacemaker Localization

The *Bulla* retina is a typical stratified gastropod retina consisting of a solid lens surrounded by a photoreceptor layer containing approximately 1000 receptors (JACKLET and COLQUHOUN 1983; BLOCK et al. 1984). Distinct from the photoreceptor layer and located near the retinal base is a group of approximately 100 neurons referred to as basal retinal neurons or BRNs. These cells are electrically coupled to one another (BLOCK et al. 1984; GEUSZ and BLOCK 1992b) and produce the compound action potentials that express the circadian rhythm recorded in the optic nerve. The large photoreceptors surrounding the lens do not appear to make contact with the BRNs; however, other cells in both the receptor layer and at the base of the retina do make functional connections with the BRN population (BLOCK et al. 1984; GEUSZ and BLOCK 1992a).

The BRNs are the cells responsible for generating the circadian rhythm. The evidence for this conclusion is as follows: Electrophysiological recordings from the *Bulla* retina reveal that the large photoreceptor cells that surround the lens do not express a circadian rhythm in membrane potential (BLOCK and WALLACE 1982). On the other hand, there is a circadian rhythm in the membrane potential of basal retinal neurons (MCMAHON et al. 1984). Surgical tissue reduction of the retina reveals that only the base of the retina containing the basal retinal neurons is required to support a circadian periodicity. Eyes in which only a small number of basal retinal neurons remained were capable of expressing circadian periodicities (BLOCK and MCMAHON 1984). Finally, fully isolated neurons are capable of expressing a circadian periodicity. MICHEL and coworkers (1993a) disassociated neurons from the base of the retina. These cells were dispersed or, in some experiments, isolated individually in microwell dishes. Membrane conductance measurements made before and after dawn revealed a circadian rhythm in membrane conductance similar to that observed in the intact retina (Fig. 6).

II. Rhythm Generation

The *Bulla* retina has also been useful in addressing the issue of rhythm generation. Although the entrainment pathway involves a transmembrane Ca^{2+} flux, and rhythm expression appears to require an outward K^+ current (see below), there is little evidence that transmembrane fluxes are involved in generating circadian periodicities. The observations supporting this conclusion are as follows: removal of Cl^- (KHALSA et al. 1990; MICHEL et al. 1992), Na^+ (Khalsa, unpublished experiments), or Ca^{2+} (KHALSA et al. 1993a) from the bathing medium does not block circadian rhythmicity, although rhythm expression is often obscured in low-Ca^{2+} seawater. Large changes in the extracellular K^+ concentration likewise do not profoundly affect the period or phase of the pacemaker (MCMAHON and BLOCK 1987b; Khalsa, unpublished experiments). These experiments suggest that transmembrane fluxes are not involved in rhythm generation.

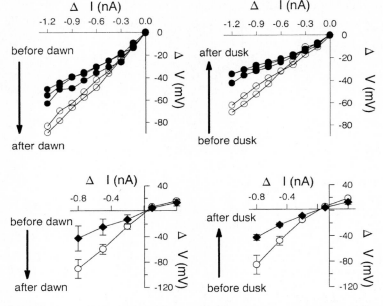

Fig. 6. Circadian rhythm in conductance in semi-intact retina and in dispersed *Bulla* pacemaker neurons. *Upper panels* show change in membrane conductance of two basal retinal neurons of semi-intact eye preparations during projected dawn (*left*) and dusk (*right*). Current pulses of 6 s duration were given from −1.2 to 0.6 nA while in the discontinuous current clamp mode with membrane potential brought to −65 mV. Measurements were made within 2 h before and after projected dawn and dusk. *Left panel*, change in membrane potential of pacemaker neuron in response to current pulses delivered at times before (*filled circles*) or after (*open circles*) projected dawn. *Right panel*, change in membrane potential of pacemaker neuron before (*open circles*) or after (*filled circles*) projected dusk.

Lower panels indicate membrane conductance of dispersed basal retinal neurons during the first circadian cycle in culture. Two basal retinae were treated with 0.01% protease and then triturated with fire-polished pipettes onto a glass cover slip coated with poly-D-lysine. Cells were maintained in dim fluorescent room light at 22 °C. *Left panel*, change in membrane potential to current pulses before (ZT 19–22, $n=4$, *filled circles*) or after (ZT 2–5, $n=7$, *open circles*) projected dawn. *Right panel*, membrane conductance of basal retinal neurons before (ZT 10–12, $n=5$, *open circles*) and after (ZT 14–16, $n=4$, *closed circles*) projected dusk. Membrane conductance of basal retinal neurons decreases at projected dawn and increases at projected dusk. (Redrawn from MICHEL et al. 1993)

Unlike transmembrane fluxes, the molecular events of transcription and translation appear to be intimately associated with rhythm generation. Increasing concentrations of the translation inhibitors cycloheximide or anisomycin led to a systematic lengthening of the free-running period (KHALSA et al. 1993a). At higher concentrations, spontaneous activity is lost. Nevertheless, when the higher concentration of inhibitor is washed out, the phase of the restored rhythm suggests that the pacemaker has been held motionless at the higher concentration at a phase near ZT 0.

A similar role has been proposed for transcription (KHALSA et al. 1993b). Low concentrations of the reversible transcription inhibitor DRB lead to period lengthening and higher concentrations to a cessation of pacemaker motion if the inhibitor is present near ZT 0. These data suggest that both translation and transcription may play a role in timing the *Bulla* ocular rhythm. A similar role for transcription and translation has been proposed in *Aplysia* (LOTSHAW and JACKLET 1986; RAJU et al. 1991). In addition, both processes have been implicated in timing the circadian cycle in *Drosophila* (EDERY et al. 1994; HARDIN et al. 1990) and in the bread mold *Neurospora* (ARANSON et al. 1994a,b).

III. Pacemaker Entrainment

The *Bulla* eye is entrained by light cycles in vivo (BLOCK and WALLACE 1982). The isolated eye can likewise be phase shifted by light pulses (BLOCK et al. 1984). The phase response curve to light is typical of known light phase response curves, with delays in the early subjective night followed by advances in the late subjective night. A summary of processes involved or most likely involved in pacemaker entrainment (and expression) is shown in Fig. 7.

The entrainment cascade appears overall similar to that reported in *Aplysia*. Light appears to act directly on photopigments located within the BRNs (GEUSZ and PAGE 1991). The transduction process results in membrane depolarization, calcium influx, and an increase in intracellular calcium, possibly protein synthesis, and ultimately a shift in the phase of the rhythm.

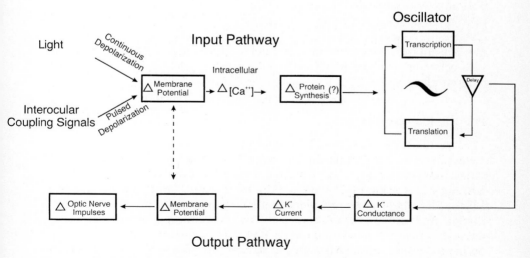

Fig. 7. *Bulla* entrainment and rhythm expression cascade. See text for details

a) Membrane Depolarization

A number of experiments implicate membrane potential changes in light-induced phase shifts. Light pulses lead to membrane depolarization of BRN pacemaker neurons, and depolarization of these cells, either by intracellular current injection (McMahon and Block 1987a) or by depolarization induced through ionic substitution in the bath (McMahon and Block 1987a; Khalsa and Block 1988a), generates phase shifts that are identical to those produced by light. In addition, light-induced phase shifts can be blocked by injecting sufficient hyperpolarizing current to cancel out the membrane potential change (McMahon and Block 1987a).

b) Transmembrane Calcium Flux

Membrane depolarization leads to a transmembrane Ca^{2+} flux. Blocking Ca^{2+} influx, either by lowering extracellular Ca^{2+} (McMahon and Block 1987a; Khalsa and Block 1988a) or by blocking Ca^{2+} channels with Ni^{2+} (Khalsa and Block 1988a), prevents light-induced or depolarization-induced phase shifts. These data suggest that light leads to membrane depolarization and Ca^{2+} influx through Ca^{2+} channels. High-threshold voltage-activated Ca^{2+} channels of basal retinal neurons have been described in a preliminary report (Michel et al. 1994).

c) Sustained Increase in Intracellular Calcium

Recently it has been possible to directly measure the change in intracellular calcium concentration $[Ca^{2+}]_i$ associated with membrane depolarization. Using the intracellular probe fura-2, Geusz and coworkers (1994) were able to demonstrate that $[Ca^{2+}]_i$ remains elevated for the duration of the depolarization – for durations measured up to 1 h. Upon depolarization isolated neurons exhibited a large transient increase in $[Ca^{2+}]_i$ followed by a sustained plateau at an elevated level.

It is not clear what is the next step in the entrainment cascade. Experiments with calmodulin inhibitors (Khalsa and Block 1988b) failed to block light-induced phase shifts.

IV. Rhythm Expression

The circadian rhythm in impulse frequency is driven by a circadian rhythm in membrane potential. Intracellular recordings from BRNs reveal that the membrane is relatively hyperpolarized during the subjective night and depolarizes by approximately 13 mV during the subjective day (McMahon et al. 1984). This rhythm in membrane potential appears to be controlled by a rhythm in membrane conductance. Conductance measurements from *Bulla* retinas reveal that membrane conductance decreases by approximately 50% during the subjective day (Ralph and Block 1990; Michel et al. 1993a). This change in conductance appears to be due to a K^+ conductance. Blocking K^+

conductance with 100 mM tetraethylammonium (TEA) prior to dawn leads to an approximately 50% reduction in membrane conductance. Following dawn TEA does not have a measurable effect. This suggests that a TEA-sensitive conductance is under clock control.

Recent experiments employing perforated patch recording techniques have studied three K^+ currents as possible sources of circadian modulation. A rapidly inactivated current and two sustained outward currents, one Ca^{2+} modulated, were evaluated (MICHEL et al. 1993b). Only the sustained non-Ca^{2+}-regulated current was found to be unambiguously modulated by a circadian rhythm. The data, while preliminary, suggest that a delayed rectifier K^+ current is responsible for at least part of the conductance change measured in basal retinal neurons.

E. Conclusions

The cellular and molecular mechanisms underlying circadian rhythmicity present a great challenge to both molecular and cell sciences. Circadian pacemakers are ubiquitous and influence myriad cellular processes. Whatever the precise molecular events that comprise the central "timing loop," it will be necessary to understand how the pacemaker is synchronized by light and how it regulates circadian processes and behaviors. These questions will require model systems that are amenable to physiological analysis. While elegant molecular genetics applied to *Drosophila* will reveal important insights into circadian rhythm genesis, ultimately we may have to turn to other preparations with more experimentally tractable nervous systems to complete our understanding of biological timing.

The three extant in vitro retinal preparations offer many opportunities for continued experimental investigation. Most importantly, the ability to record circadian rhythmicity in isolated neurons should allow for an increasingly detailed understanding of pacemaker synchronization and the mechanisms involved in expression of the rhythm. Retinal pacemaker models have much to offer the chronobiology community.

Acknowledgement. We thank H. Noakes for assistance with graphics. Support was provided by NIH NS15264 and the NSF Center for Biological Timing.

References

Aronson BD, Johnson KA, Dunlap JC (1994a) The circadian clock locus frequency: a single ORF defines period length and temperature compensation. Proc Natl Acad Sci USA 91:7683–7687

Aronson BD, Johnson KA, Loros JJ, Dunlap JC (1994b) Negative feedback defining a circadian clock: autoregulation in the clock gene frequency. Science 263:1578–1584

Besharse JC, Iuvone PM (1983) Circadian clock in Xenopus eye controlling retinal serotonin N-acetyltransferase. Nature 305:133–135

Besharse JC, Iuvone PM, Pierce ME (1988) Regulation of rhythmic photoreceptor metabolism: a role for post-receptoral neurons. Prog Retinal Res 7:21–61

Binkley S, Hryshchyshyn M, Reilly K (1979) N-Acetyltransferase activity responds to environmental lighting in the eye as well as in the pineal gland. Nature 281:479–481

Block GD (1981) In vivo recording from the Aplysia eye. Brain Res 222:138–143

Block GD, McMahon D (1983) Localized illumination of retinal layers in the Aplysia and Bulla eye reveal new relationships between retinal cell types. Brain Res 265:134–137

Block GD, McMahon D (1984) Cellular analysis of the Bulla ocular circadian pacemaker system: III. Localization of the circadian pacemaker. J Comp Physiol A 155: 387–395

Block GD, Wallace S (1982) Localization of a circadian pacemaker in the eye of a mollusc, Bulla. Science 217:155–157

Block GD, McMahon D, Wallace S, Friesen W (1984) Cellular analysis of the Bulla ocular circadian pacemaker system: I. A model for retinal organization. J Comp Physiol [A] 155:365–378

Block GD, Khalsa S, McMahon D, Michel S, Geusz M (1993) Biological clocks in the retina: cellular mechanisms of biological timekeeping. Int Rev Cytol 146:83–143

Cahill GM, Besharse JC (1989) A D-2 dopamine receptor agonist resets the phase of a circadian clock in cultured eyecups from Xenopus laevis. Soc Neurosci Abstr 15:24

Cahill GM, Besharse JC (1990) Circadian regulation of melatonin in the retina of Xenopus laevis: limitation of serotonin availability. J Neurochem 54:716–719

Cahill GM, Besharse JC (1991) Resetting the circadian clock in cultured Xenopus eyecups: regulation of retinal melatonin rhythms by light and D_2 dopamine receptors. J Neurosci 11:2959–2971

Cahill GM, Besharse JC (1993) Circadian clock functions localized in Xenopus retinal photoreceptors. Neuron 10:573–577

Cahill GM, Grace MS, Besharse JC (1991) Rhythmic regulation of retinal melatonin: metabolic pathways, neurochemical mechanisms, and the ocular circadian clock. Cell Mol Neurobiol 11:529–560

Colwell CS (1990) Light and serotonin interact in affecting the circadian system in Aplysia. J Comp Physiol [A] 167:841–845

Colwell CS, Khalsa S, Block G (1992a) Cellular basis of entrainment. Chronobiol Int 9:163–179

Colwell CS, Michel S, Block G (1992b) Evidence that potassium channels mediate the effect of serotonin on the ocular circadian pacemaker of Aplysia. J Comp Physiol [A] 171:651–656

Colwell CS, Michel S, Block GD (1994) Calcium plays a central role in phase shifting the ocular pacemaker in Aplysia. J Comp Physiol [A] 175:415–423

Corrent G, Eskin A, Kay I (1982) Entrainment of the circadian rhythm from the eye of Aplysia: role of serotonin. Am J Physiol 242:R326–R332

Corrent G, McAdoo DJ, Eskin A (1978) Serotonin shifts the phase of the circadian rhythm from the Aplysia eye. Science 202:977–979

Edery J, Ruitila JE, Rosbash M (1994) Phase shifting of the circadian clock by induction of the Drosophila period protein. Science 263:237–240

Eskin A (1971) Properties of the Aplysia visual system: In vitro entrainment of the circadian rhythm and centrifugal regulation of the eye. Z Vgl Physiol 74:353–371

Eskin A (1972) Phase shifting a circadian rhythm in the eye of Aplysia by high potassium pulses. J Comp Physiol [A] 80:353–376

Eskin A (1977) Neurophysiological mechanisms involved in photo-entrainment of the circadian rhythm from the Aplysia eye. J Neurobiol 8:273–299

Eskin A (1979) Circadian system of the Aplysia eye: properties of the pacemaker and mechanisms of its entrainment. Fed Proc 38:2573–2579

Eskin A (1982) Increasing external K^+ blocks phase shifts in a circadian rhythm produced by serotonin or 8-benzylthio-cAMP. J Neurobiol 13:241–249

Eskin A, Takahashi JS (1983) Adenylate cyclase activation shifts the phase of a circadian pacemaker. Science 220:82–84

Eskin A, Corrent G, Lin C-Y, McAdoo DJ (1982) Mechanism of shifting the phase of a circadian oscillator by serotonin: involvement of cAMP. Proc Natl Acad Sci USA 79:660–664

Eskin A, Takahashi JS, Zatz M, Block GD (1984a) Cyclic guanosine 3': 5'-monophosphate mimics the effects of light on a circadian pacemaker in the eye of Aplysia. J Neurosci 10:2466–2471

Eskin A, Yeung SJ, Klass MR (1984b) Requirement for protein synthesis in the regulation of a circadian oscillator by serotonin. Proc Natl Acad Sci USA 1:7637–7641

Geusz ME, Block G (1992a) The retinal cells generating the circadian small impulses in the Bulla optic nerve. J Biol Rhythms 7:255–268

Geusz ME, Block G (1992b) Measurements of electrical coupling between circadian pacemaker cells of the Bulla eye. Neurosci Abstr 18:1

Geusz ME, Page T (1991) An opsin-based photopigment mediates phase shifts of the Bulla circadian pacemaker. J Comp Physiol [A] 168:565–570

Geusz ME, Michel S, Block G (1994) Intracellular calcium responses of circadian pacemaker neurons measured with fura-2. Brain Res 638:109–116

Green CB, Besharse J (1994) Tryptophan hydroxylase expression is regulated by a circadian clock in Xenopus laevis. J Neurochem 62:2420–2428

Hamm HE, Menaker M (1980) Retinal rhythms in chicks: circadian variation in melatonin and serotonin N-acetylferase activity. Proc Natl Acad Sci USA 77:4998–5002

Hardin PE, Hall JC, Rosbash M (1990) Feedback of the Drosophila period gene product on circadian cycling of its messenger RNA levels. Nature 343:536–540

Herman K, Strumwasser F (1984) Regional specializations in the eye of Aplysia, a neuronal circadian oscillator. J Comp Neurol 230:593–613

Iuvone PM (1986) Rhythms of melatonin biosynthesis in retina: Involvement of calcium, cyclic AMP accumulation and serotonin N-acetyltransferase activity. Life Sci 38:331–342

Iuvone PM, Besharse JC (1983) Regulation of indoleamine N-acetyltransferase activity in the retina: effects of light and dark, protein synthesis inhibitors and cyclic nucleotide analogues. Brain Res 273: 111–119

Iuvone PM, Besharse JC (1986a) Cyclic AMP stimulates serotonin N-acetyltransferase activity in Xenopus retina in vitro. J Neurochem 46:33–39

Iuvone PM, Besharse JC (1886b) Involvement of calcium in the regulation of serotonin N-acetyltransferase in retina. J Neurochem 46:82–88

Jacket JW (1969a) Circadian rhythm of optic nerve impulses recorded in darkness from isolated eye of Aplysia. Science 164:562–563

Jacklet JW (1969b) Electrophysiological organization of the eye of Aplysia. J Gen Physiol 53:21–42

Jacklet JW (1974) The effects of constant light and light pulses on the circadian rhythm in the eye of Aplysia. J Comp Physiol [A] 90:33–45

Jacklet JW (1980) Protein synthesis requirement of the Aplysia circadian clock, tested by active and inactive derivatives of the inhibitor anisomycin. J Exp Biol 85: 33–42

Jacklet JW (1989) Circadian neuronal oscillators. In: Jacklet JW (ed) Neuronal and cellular oscillators. Decker, New York, pp 483–527

Jacklet JW (1991) Photoresponsiveness of Aplysia eye is modulated by the ocular pacemaker and serotonin. Biol Bull 180:284–294

Jacklet JW, Barnes S (1995) Circadian phase differences in a retinal pacemaker neuron delayed rectifier K^+ current and increased current induced by a phase-shifting neurotransmitter, serotonin. Physiologist 38:A-22

Jacklet JW, Colquhoun W (1983) Ultrastructure of photoreceptors and circadian pacemaker neurons in the eye of a gastropod, Bulla. J Neurocytol 12:673–696

Jacklet JW, Geronimo J (1971) Circadian rhythm: population of interacting neurons. Science 174: 299–302

Jacklet JW, Lotshaw DP (1981) Light and high potassium cause similar phase shifts of the Aplysia eye circadian rhythm. J Exp Biol 94:345–349

Jacklet JW, Alvarez R, Bernstein B (1972) Ultrastructure of the eye of Aplysia. J Ultrastruct Res 38:246–261
Jacklet JW, Schuster L, Rolerson C (1982) Electrical activity and structure of retinal cells of the Aplysia eye. I. Secondary neurons. J Exp Biol 99:369–380
Khalsa SBS, Block G (1988a) Calcium channels mediate phase shifts of the Bulla circadian pacemaker. J Comp Physiol [A] 164:195–206
Khalsa SBS, Block G (1988b) Phase shifts of Bulla pacemaker are not blocked by calmodulin antagonists. Life Sci 43:1151–1156
Khalsa SBS, Ralph M, Block G (1990) Chloride conductance contributes to period determination of a neuronal circadian pacemaker. Brain Res 520:166–169
Khalsa SBS, Whitmore D, Block G (1992) Stopping the biological clock with inhibitors of protein synthesis. Proc Natl Acad Sci USA 89:10862–10866
Khalsa SBS, Ralph M, Block G (1993a) The role of extracellular calcium in generating and in phase-shifting the Bulla ocular circadian rhythm. J Biol Rhythm 8:125–139
Khalsa SBS, Whitmore D, Bogart B, Block G (1993b) Evidence for the direct involvement of transcription in the timing mechanism of the circadian pacemaker. Soc Neurosci Abstr 19:1703
Koumenis C, Eskin A (1992) The hunt for mechanisms of circadian timing in the eye of Aplysia. Chronobiol Int 9:201–221
Lotshaw D, Jacklet JW (1986) Involvement of protein synthesis in the circadian clock of the Aplysia eye. Am J Physiol 250:R5–R17
Luborsky-Moore J, Jacklet JW (1977) Ultrastructure of the secondary cells in the Aplysia eye. J Ultrastruct Res 60(2):235–245
McMahon DG, Block G (1987a) The Bulla circadian pacemaker: I. Pacemaker neuron membrane potential controls phase through a calcium-dependent mechanism. J Comp Physiol [A] 161:335–346
McMahon DG, Block G (1987b) The Bulla circadian pacemaker: II. Chronic changes in membrane potential lengthen free running period. J Comp Physiol [A] 161:347–354
McMahon DG, Wallace S, Block G (1984) Cellular analysis of the Bulla ocular circadian pacemaker system: II. Neurophysiological basis of circadian rhythmicity. J Comp Physiol [A] 155:379–385
Michel S, Khalsa S, Block G (1992) Phase shifting the circadian rhythm in the eye of Bulla by inhibition of chloride conductance. Neurosci Lett 146:219–222
Michel S, Geusz M, Zaritsky J, Block G (1993a) Circadian rhythm in membrane conductance expressed in dissociated molluscan neurons. Science 259:239–241
Michel S, Manivanna K, Zaritsky J, Block G (1993b) Whole cell currents in cultured circadian pacemaker neurons of Bulla. Soc Neurosci Abstr 19:1616
Michel S, Geusz M, Block G (1994) Calcium current in circadian pacemaker neurons of the marine snail Bulla. Eur J Neurosci 7 [Suppl]:145
Moore RY, Eichler VB (1972) Loss of a circadian adrenal corticosterone rhythm following suprachiasmatic lesions in the rat. Brain Res 42:210–216
Raju U, Yeung SJ, Eskin A (1990) Involvement of proteins in light resetting ocular circadian oscillators in Aplysia. Am J Physiol 258:R256–R262
Raju U, Koumenis C, Nunez-Regueiro M, Eskin A (1991) Alteration of the phase and period of a circadian oscillator by a reversible transcription inhibitor. Science 253:673–675
Raju U, Nunez-Regueiro M, Cook R, Kaetzel M, Yeung S, Eskin A (1993) Identification of an annexin-like protein and its possible role in the Aplysia eye circadian system. J Neurochem 61:1236–1245
Ralph MR, Block G (1990) Circadian and light-induced conductance changes in putative pacemaker cells of Bulla gouldiana. J Comp Physiol [A] 166:589–595
Sokolove PG (1975) Localization of the cockroach optic lobe circadian pacemaker with microlesions. Brain Res 87:13–21
Stephan FK, Zucker I (1972) Circadian rhythms in drinking behavior and locomotor activity of rats are eliminated by hypothalamic lesions. Proc Natl Acad Sci USA 69:1583–1586

Strumwasser F (1973) Neural and humoral factors in the temporal organization of behavior. Physiologist 16:9–42
Tosini G, Menaker M (1996) Circadian rhythms in cultured mammalian retina. Science 272:419–421
Underwood H, Sciopes T (1985) Melatonin rhythms in quail: regulation by photoperiod and circadian pacemakers. J Pineal Res 2:133–143
Underwood H, Siopes T, Barrett RK (1988) Does a biological clock reside in the eye of the quail? J Biol Rhythms 3:323–331
Woolum JC, Strumwasser F (1980) The differential effects of ionizing radiation on the circadian oscillator and other functions in the eye of Aplysia. Proc Natl Acad Sci USA 77:5542–5546
Yeung SJ, Eskin A (1987) Involvement of a specific protein in the regulation of a circadian rhythm in the Aplysia eye. Proc Natl Acad Sci USA 84:279–283

CHAPTER 17
Circadian Rhythms and Depression: Clinical and Experimental Models

A.M. ROSENWASSER and A. WIRZ-JUSTICE

A. Introduction and Scope

In this chapter we review clinical and experimental evidence linking biological rhythms and affective state, and explore possible mechanisms underlying these relationships. Alterations in circadian rhythmicity have been observed in association with mood disorders (ANDERSON and WIRZ-JUSTICE 1991; WIRZ-JUSTICE 1995) as well as in putative animal models of depression and/or altered affective state (ROSENWASSER 1992). However, much of the evidence for covariation of chronobiological and affective parameters is correlative, and the causal bases for such observations have not been fully elucidated.

As illustrated in Fig. 1, altered rhythmicity could be either a cause or an effect of affective state, or there could be reciprocal causal interactions. Further, covariation of rhythmicity and affective state could be due to common influences of other variables, including genetic, experiential, neurochemical, and neuroendocrine factors. For example, monoaminergic neurotransmitter systems appear to influence both affective and circadian processes, such that alterations in these transmitter systems could lead to parallel alterations in affect and rhythmicity. Of course, identifying a possible causal linkage is not the same as showing that this linkage normally contributes to the regulation of the system, and systematic study of these various interactions is required for a full understanding of the role of circadian rhythms in mood disorders.

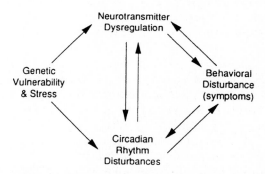

Fig. 1. Potential causal relationships underlying associations among circadian rhythms, behavioral state, neurotransmitters, and predisposing factors. (From GOODWIN and JAMISON 1990)

B. Measurement of Circadian Rhythms

Circadian rhythmicity is expressed in a wide variety of overt, measurable functions, including locomotor activity, psychomotor performance, sleep timing and architecture, core body temperature, and neuroendocrine function. Rhythmic expression of these functions is thought to reflect, at least in part, the behavior of an underlying circadian pacemaker. However, the measured variable may also be influenced by other, non-pacemaker-dependent factors, such as ambient light and temperature, stress, fatigue, sleep, exercise, and food intake; such effects are referred to as "masking."

The underlying circadian pacemaker may be characterized by its fundamental properties: (1) the free-running period as measured in the absence of environmental synchronizers or zeitgebers and (2) the phase-response curve(s) (PRCs) describing the pacemaker's response to pertubation by such zeitgebers. Direct measurement of these properties is easiest during long-term maintenance under controlled constant conditions, in the absence of all potential time-cues. Such conditions are easiest to achieve in experimental animals, more difficult in human subjects, and all the more difficult in depressed patients. Therefore, it is not surprising that free-running circadian rhythms have been assessed in very few patients with major affective disorder.

Instead, much of the human literature, and nearly all of the clinical research, depends upon the measurement of circadian phase under entrained conditions, either in the "natural" environment or under imposed light-dark cycles and/or scheduled social interactions. Unfortunately, assessment of circadian phase is problematic for several reasons: (1) entrainment phase is a complex parameter, influenced by both free-running period and the photic PRC, so that changes in entrainment phase could reflect changes in either or both of these more fundamental pacemaker properties. (2) Masking effects may directly alter overt temporal order without affecting the phase of the underlying pacemaker. This problem has led to development of both experimental designs and analytical techniques that attempt to "unmask" the phase of the underlying pacemaker, including the use of "constant routine" and "forced desynchrony" protocols. (3) Overt circadian rhythms may display nonsinusoidal waveforms that require special analytical techniques to determine the phase of a best-fitting function composed of multiple sinusoidal harmonics. In other cases, phase is estimated by some empirical phase-markers, such as the time of the actual body temperature minimum, or the time of sleep onset.

Historically, much less attention has been paid to variations in circadian amplitude than to phase and period. Recently, however, there has been increasing recognition that circadian amplitude may be altered in depression (e.g., SOUETRE et al. 1989). Since low-amplitude oscillators would be expected to display different entrainment dynamics than "strong" oscillators, an underlying difference in pacemaker amplitude could account for differences in overt entrainment phase. However, the same caveats apply to assessment of circadian amplitude, as well as other aspects of the circadian waveform, since masking effects could alter the amplitude of an expressed rhythm without affecting the underlying pacemaker.

C. Circadian Rhythms and Depression: Clinical Research

I. Phase-Advanced Circadian Rhythms

As summarized in several earlier reviews (WEHR and GOODWIN 1983; WEHR et al. 1983), there is considerable evidence for phase-advanced circadian rhythms – for example, in core body temperature and hormone secretion – in both unipolar and bipolar depressed patients, even though not all studies have found phase-advanced rhythms in depression. Some of the variability in research findings appears to be due to identified methodological differences. For example, frequent reports of nonstatistically significant phase alterations in depression suggest that many clinical studies have been lacking in statistical power. Another important factor may be study design, since longitudinal studies appear much more sensitive to phase alterations than cross-sectional studies. Indeed, probably the most impressive data correlating circadian phase with clinical state come from long-term studies of individual rapid-cycling bipolar patients (WEHR and GOODWIN 1983). In these patients, the phase of circadian temperature and sleep rhythms was found to systematically covary with affective state (i.e., depressed vs. manic), with the most advanced phase occurring just before the switch out of mania into depression. Similarly, circadian phase abnormalities are often reversed during clinical remission (LINKOWSKI et al. 1987), suggesting that circadian phase may be state rather than trait dependent.

Differences in statistical methods are undoubtedly another significant source of variant findings. For example, one of the most consistently detected abnormalities is early timing of the nadir of the plasma cortisol rhythm, while the peak of the cortisol waveform is usually not altered, or may even be delayed. Cortisol secretion is characterized by a complex circadian waveform with superimposed ultradian variations due to pulsatile secretion, and both circadian and ultradian aspects of the waveform are altered in depression. In view of this complexity, the single cosinor method of fitting a simple sinusoidal function to the data is inadequate for measurement of depression-related phase-shifts. Recent studies using very frequent sampling and a multiharmonic curve-fitting approach have confirmed that depressed patients show a clear phase-advance in the timing of the cortisol nadir but not in the peak of the fitted function (LINKOWSKI et al. 1987). This result is consistent with the findings of twin studies utilizing the same analytical approach, which showed that the timing of the cortisol nadir is under significant genetic control, but that the timing of the acrophase is not (LINKOWSKI et al. 1993).

II. Depression, Sleep, and Rhythmicity

Sleep disturbances are inextricably linked with depressive illness (WEHR et al. 1983; BENCA et al. 1992; BERGER and RIEMANN 1993). Two of the best-established sleep-related biological markers in depressive illness, early-morning awakening and shortened REM (rapid eye movement) latency, may both reflect a phase advance of the underlying circadian pacemaker, either with respect to objective environmental time or with respect to the timing of sleep onset. Similarly, core body temperature and cortisol rhythms may be phase-advanced with respect to the sleep-wake cycle, rather than with respect to objective time, in depressed patients.

Experimental manipulation of the amount and timing of sleep has provided evidence for a causal relationship between sleep and depression. Rapid and dramatic improvement in depressive symptoms is seen following one night of total sleep deprivation in about 60% of depressed patients (SCHULTE 1971; WU and BUNNEY 1990), while symptoms usually return rapidly after subsequent recovery sleep. A propensity for diurnal variation of mood with amelioration in the evening predicts sleep deprivation response, further implicating a circadian correlate.

Antidepressive effects are also seen following less dramatic sleep manipulations, including (1) partial sleep deprivation restricted to the second half of the night; (2) selective deprivation of REM sleep (which tends to increase in the second half of the night); and (3) phase-advancing sleep times by several hours without reducing the amount of sleep per se (WEHR and WIRZ-JUSTICE 1982; WU and BUNNEY 1990). According to the now-classic "phase-advance hypothesis" of depression, a sleep-dependent depressogenic process occurs during a critical circadian phase; this critical phase occurs sometime after awakening in normal individuals, but occurs earlier – during sleep – in depressive individuals (WEHR and WIRZ- JUSTICE 1982). Thus, the therapeutic effects of various sleep manipulations may all be due to preventing sleep at the sleep-sensitive phase. This hypothesis is also consistent with the spontaneous switching out of depression and into hypomania or mania in rapid cycling patients, which is often accompanied by a night of total or partial sleep deprivation, and which is by far most likely to occur during the second half of the night (WEHR and GOODWIN 1983).

According to the phase-advance hypothesis, one might predict that experimentally imposed phase-delays of sleep timing should induce depression-like symptoms in healthy normal subjects, and, indeed, modest but reliable mood decrements have been reported. Certain individuals, however, may become noticeably depressed (e.g., two out of ten in the study by SURRIDGE-DAVID et al. 1987), and in an extreme case a presumptively healthy subject committed suicide after a phase-shift experiment (ROCKWELL et al. 1978). This individual was found to have had abnormal and instable circadian rhythms during the baseline period. These observations suggest that sudden circadian phase-shifts can induce depressive symptoms in predisposed individuals, as documented for subjects with a history of affective illness becoming depressed after a westbound flight, or manic after an eastbound flight across several time zones (JAUHER and WELLER 1982).

The most parsimonious model to explain the timing of human sleep and wakefulness is the two-process model (BORBÉLY and WIRZ-JUSTICE 1982; DAAN et al. 1984). According to this model, the sleep-wake cycle is dependent on a homeostatic depletion-repletion process or relaxation oscillator ("process S"). This process controls a state variable representing slow-wave sleep "drive" whose level increases during wakefulness and decreases exponentially during sleep; this process has been experimentally modeled by slow-wave power or intensity in electroencephalographic recordings. Process S triggers sleep at an upper threshold and waking at a lower threshold, while both thresholds are modulated by a circadian pacemaker ("process C"), presumed to be the same pacemaker that underlies other behavioral and physiological rhythms. In this way, the two-process model can account for the complex differential behavior of sleep-wake and other circadian rhythms, such as occurs during "internal desynchronization" (WEVER 1979), with only a single underlying circadian pacemaker.

If the depressogenic process involves an alteration in the timing of sleep relative to other circadian rhythms, it is possible that the underlying mechanism involves process S rather than process C. Specifically, it has been suggested that depression is characterized by a deficient process S, while the antidepressant effects of sleep deprivation are attributed to augmentation of process S (BORBÉLY and WIRZ-JUSTICE 1982). A deficient S process could also plausibly account for a delay in sleep timing relative to other rhythms, leading to the apparent circadian phase advance in depression. To date, the few studies that have attempted to explicitly test this hypothesis by assessment of slow-wave intensity in depressed patients have yielded conflicting results (see VAN DEN HOOFDAKKER 1994), and the hypothesis thus remains in need of stringent testing.

III. Seasonal Depression: A Circadian Rhythm Disorder?

Seasonal affective disorder (SAD) is characterized by recurrent episodes of depression, hypersomnia, augmented appetite with carbohydrate craving, and weight gain in autumn and winter (cf. ROSENTHAL and BLEHAR 1989). In contrast to the substantial evidence for phase-advanced circadian rhythms in depressed patients desribed above, patients with SAD have been reported to show phase-delayed circadian temperature and melatonin rhythms (LEWY et al. 1987; DAHL et al. 1993; WIRZ-JUSTICE et al. 1993b, 1995). SAD patients show a therapeutic response to bright light treatment, leading to a variety of circadian and noncircadian hypotheses that attempt to account for SAD pathophysiology and for the therapeutic efficacy of light exposure. These hypotheses include: (1) SAD is triggered by the decrease in light availability during autumn and winter, while light therapy increases total daily light exposure; (2) SAD is triggered by the shortening of daylengths in autumn and winter, while light therapy lengthens the effective photoperiod; and (3) SAD is triggered by a seasonally dependent abnormally delayed circadian phase position, while light therapy acts to phase advance the circadian pacemaker and thus correct the phase abnormality. The first of these hypotheses essentially treats light as a drug, and assumes that bright light exposure corrects some neurovegetative dysfunction, while the second and third hypotheses are based on mechanisms known to underlie seasonality in animals.

Seasonal variation in photoperiod serves as the primary proximate signal triggering seasonally dependent changes in animals; so it is possible that seasonal photoperiod variation also triggers SAD symptomatology. In many species, the circadian system is involved in mediating photoperiod-dependent changes, suggesting that a circadian abnormality could be responsible for triggering seasonal depressions. If light therapy acts by increasing the apparent photoperiod duration, then one might predict that light therapy should be effective when presented either in the morning or in the evening; but if light therapy acts by altering circadian phase, then this treatment should be effective only when presented in the morning, at a time when light is known to phase-advance circadian rhythms. In contrast to these hypotheses, the light deprivation hypothesis leads to the expectation that light therapy should be equally effective regardless of time of presentation (except, of course, that circadian variation in visual responsiveness could render light more or less effective at different times of day).

Despite considerable attention, the possible circadian phase-dependent efficacy of light therapy is still controversial. Therapeutic effects have been reported for morning, evening, and midday light exposure, although a slight superiority for morning light seems likely (TERMAN et al. 1989). Most studies comparing the therapeutic response to different light treatment regimens have not assessed circadian phase; of those that have measured both circadian phase and clinical status, both morning light and combined morning plus evening light have been shown to normalize mood and phase-advance the melatonin and/or temperature rhythms, suggesting that evening light does not fully antagonize morning light (LEWY et al. 1987; TERMAN et al. 1988; SACK et al. 1990; DAHL et al. 1993).

Wirz-Justice et al. (1995) recently examined circadian core temperature and melatonin rhythms in SAD patients and controls, both before and after midday (1000–1400 hours) light treatment, using the unmasking conditions of the constant routine protocol. While the melatonin rhythm was not influenced by either clinical status or light treatment, SAD patients as a group showed a significant delay in the core temperature rhythm as well as phase advances after light treatment (Fig. 2). Since midday light might not generally be expected to induce circadian phase-shifts, these results highlight the importance of measuring both mood and circadian phase in such studies. Further, it must be emphasized that light exposure may independently induce a phase shift and a therapeutic response, even if phase-shifting is not part of the mechanism un-

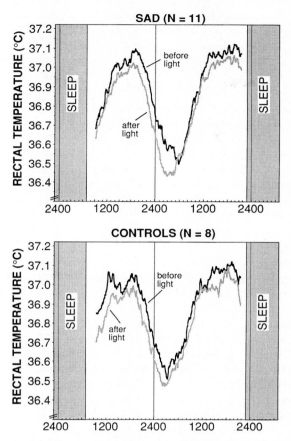

Fig. 2. Circadian core body temperature rhythm in female SAD patients ($n=11$) and controls ($n=8$), measured in winter, before and after midday light treatment (5 days of 6000 lux from 1000 to 1400 hours), under constant routine conditions. (From WIRZ-JUSTICE et al. 1995)

derlying remission. In accord with this caveat, WIRZ-JUSTICE et al. (1993a) have shown that, although many SAD patients showed delayed rhythms, phase position was correlated with neither depth of depression nor with a preferential response to morning or evening light.

If light therapy does act by normalizing circadian phase, one might expect that appropriately timed light exposure would also have a therapeutic effect on nonseasonal depressions, since these patients tend toward phase-advanced rhythms. Studies of light therapy in nonseasonal depression have been equivocal, including both negative and positive outcomes. In perhaps the most systematic study of this kind, positive therapeutic effects approaching the magnitude generally seen in SAD patients were observed for evening light and for combined evening plus morning light in a group of depressed inpatients

(KRIPKE et al. 1992). However, much more work will be required to determine whether time of treatment is a factor in light therapy for nonseasonal depression, or whether circadian phase-shifting is involved.

IV. Circadian Amplitude and Waveform

Several authors have suggested that blunted amplitude may be the main chronobiological abnormality associated with affective disorder (VON ZERSSEN et al. 1985; SOUETRE et al. 1989). However, the majority of relevant studies have been carried out under normal entrainment conditions, so diminished circadian amplitude could reflect alterations in external (e.g., by the light-dark cycle) or internal (e.g., by the sleep-wake or rest-activity cycle) masking, rather than disturbed pacemaker function. Indeed, two recent studies using constant routine protocols have failed to reveal consistent amplitude differences between SAD patients and controls, or as a consequence of light treatment (DAHL et al. 1993; WIRZ-JUSTICE et al. 1995).

A relationship between the relative durations of daily rest and active phases and depression was first suggested by KRIPKE (1981), based on models of seasonality in animals. In species expressing seasonal variation, changes in photoperiod are often accompanied by seasonal changes in the relative durations of the daily activity and rest phases of the circadian cycle, thought to reflect alterations in the phase relationship between putative dawn- and dusk-controlled oscillators. Although humans do manifest a variety of seasonal rhythms (LACOSTE and WIRZ-JUSTICE 1989), little is known concerning possible photoperiod-dependent changes in circadian patterning or, indeed, even whether changes in photoperiod underlie seasonal rhythms in this species.

One remarkable recent study has demonstrated an increased duration of the nocturnal circadian phase in several parameters, including sleep time, melatonin, and prolactin secretion, when healthy subjects remained in long or short nights of absolute darkness, a condition that is never experienced under usual typical conditions with artificial light readily available (WEHR et al. 1993). These data suggest that humans may indeed retain photoperiodic responsiveness similar to that seen in animals. In addition, one subject in this study became severely suicidal under simulated long nights, even though all other subjects felt remarkably well under this protocol (WEHR et al. 1993).

This last observation suggests that changes in the relative timing of dawn- and dusk-related phase markers may influence mood. Similar evidence has been found in a study of women with premenstrual depression, who showed an earlier morning decrease in melatonin levels, thereby reducing the duration of nocturnal secretion (PARRY et al. 1990). Cortisol rhythms in major affective disorder show an earlier evening nadir but no change in the timing of the morning peak (LINKOWSKI et al. 1987), which may be interpreted as an increased duration of nocturnal secretion. In contrast to these results linking changes in the rest-activity ratio to mood changes, constant-routine studies in SAD patients have thus far failed to show any consistent differences in activity-rest ratios between patients and controls, or before and after light treatment (WIRZ-JUSTICE et al. 1995), suggesting that changes in activity-rest ratios, like changes in circadian amplitude, may be largely due to alterations in masking.

V. Instability and Zeitgeber Coupling

It has been suggested that the pathophysiology of depression is primarily characterized by intra- (SIEVER and DAVIS 1985) or intersubject (PILETZ et al. 1994) instability of rhythms, rather than by a particular rhythm abnormality, analogous to neurotransmitter dysregulation hypotheses of depression. Phase instability could arise through any of several mechanisms, including (1) an endogenous circadian period either very close to or very far from 24 h; (2) a low-amplitude pacemaker; or (3) weak coupling between the pacemaker and its photic and nonphotic entraining stimuli.

Depressed patients have been reported to display increased variability of circadian phase under entrainment conditions (TSUJIMOTO et al. 1990; DAIMON et al. 1992; PILETZ et al. 1994; RAOUX et al. 1994), as well as greater mood variability and higher amplitude ultradian fluctuations than controls (HALL et al. 1991). In a group of SAD patients, winter mood ratings showed greater between-day and within-day variability than in controls, while light treatment reduced or eliminated group differences in both mean level and variability (KRAUSS et al. 1992).

The alterations in overt phase, amplitude, waveform, and variability discussed above could plausibly all result from weakening of normal coupling between the circadian pacemaker and its zeitgebers. While light is traditionally considered the primary synchronizing agent, both human and animal systems are responsive to nonphotic cues including social stimulation, food availability, and behaviorally derived feedback. In humans, photic and social zeitgebers can be quite difficult to disentangle, since social behavior and cognitive factors influence light exposure, and since people may respond to imposed changes in lighting as informational signals, rather than as photic cues (WEVER 1979, 1989).

Depressed patients have been reported to show increased light sensitivity (LEWY et al. 1981) and blunting of light sensitivity induced by antidepressants (SEGGIE 1989). These observations are somewhat surprising, since they imply the possibility of stronger rather than weaker zeitgeber coupling in patients, and seem difficult to resolve with the generally lower amplitude and increased variability seen in depression. On the other hand, the direct relevance of these studies to circadian entrainment may be questioned, since they did not employ photic phase-shifting as the index of sensitivity.

Social zeitgebers – defined as personal relationships, jobs, or interpersonal demands that serve to entrain biological rhythms – have played an increasingly important role in circadian hypotheses of affective disorders (e.g., EHLERS et al. 1988). According to this approach, psychosocial precipitants such as stressful life events or lack of appropriate social support systems can destabilize social zeitgebers, leading to disrupted circadian rhythms. This phenomenon had been described in qualitative terms by SCHULTE (1971) more than 20 years ago, who proposed that sudden changes in habits or duties could precipitate depressive episodes.

Recent studies on social zeitgebers in depression have been carried out using scales developed to quantify daily social rhythms. For example, in recently widowed subjects, individuals with most highly disrupted social rhythms had the highest depression scores (EHLERS et al. 1988). Similarly, low social activity scores were associated with high depression ratings in hospitalized depressives (SZUBA et al. 1992) and in the elderly (PRIGERSON et al. 1994). While social activity scores in remitted depressives did not differ from controls, they did show greater variability (MONK et al. 1991), suggesting that low social zeitgeber strength may be linked with rhythm instability.

VI. Free-Running Rhythms

One of the first hypotheses linking circadian rhythms and depression proposed that manic-depressive cycling arose as a beat phenomenon between an underlying free-running circadian rhythm and an entrained sleep-wake cycle (HALBERG 1968). In agreement with this hypothesis, KRIPKE et al. (1978) reported that five of seven bipolar patients showed evidence for a short-period free-running component in their temperature rhythm as well as in several psychomotor task variables, even while other aspects of rhythmicity appeared to be entrained (but see PFLUG 1983). Such findings are consistent with the phase-advance hypothesis, in that a short endogenous pacemaker period would be associated with an advanced entrainment phase. Possibly, in some patients with very short underlying periods – perhaps in combination with altered zeitgeber coupling – entrainment fails, leading to expression of a free-running rhythm. The rare investigation of depressed subjects in isolation from time cues has not shown any consistent abnormally short period (DIRLICH et al. 1981; WEHR et al. 1985). However, interpretation of these already difficult studies is complicated by the fact that clinical state itself changed throughout the time in temporal isolation.

VII. Activity Feedback

Psychomotor disturbance is an important hallmark of depression and mania (cf. Wolff et al. 1985). Recent animal research described more fully below has revealed that alterations in activity level can phase-shift, entrain, or alter the period of free-running circadian rhythms, and preliminary studies suggest that exercise may also shift human rhythms (SCHMIDT et al. 1990; VAN REETH et al. 1994). Such findings raise the possibility that depression-related psychomotor disturbances may be viewed as a cause of, rather than an effect of, altered rhythmicity.

TEICHER et al. (1988) have described phase-delayed locomotor activity rhythms and locomotor hyperactivity in a group of depressed geriatric inpatients. In contrasting these findings with the phase advances that typically characterize anergic depression, TEICHER et al. (1988, 1989) suggested that circadian phase-type (i.e., advanced vs. delayed) may be related to psychomotor type (retardation vs. agitation) in different subgroups of patients. In a related manipulative study in SAD patients, 2 h of regular early morning exercise decreased depressive symptoms, and also phase-advanced the temperature minimum (KÖHLER et al. 1993). These studies suggest that manipulation of activity level may be a promising approach to untangling relationships between arousal, mood, and circadian rhythmicity.

VIII. Summary

Considerable evidence indicates that circadian rhythms are altered in depression. While alterations in circadian phase position – both delayed and advanced – have been documented, such effects may be attributed to underlying differences in the free-running period, amplitude, or entrainment dynamics of the circadian pacemaker, or to a complex combination of these. In addition, depression-related psychomotor or sleep disturbances could alter pacemaker function through feedback mechanisms. Serious limitations of interpretation are imposed by methodological constraints, including (1) the lack of long-term free-running studies in depressed patients and (2) the relative dominance of correlative versus manipulative research in this area.

To an extent, the increasing use of "unmasking" protocols and sophisticated statistical analyses may partially overcome these limitations. However, the few available studies using such unmasking paradigms have not revealed any consistent, specific rhythm abnormality in depressed patients. For example, Fig. 3 shows results of a constant routine protocol conducted in a single SAD patient using two well-validated markers of the circadian pacemaker: core body temperature and melatonin secretion (unpublished data, study reported in WIRZ-JUSTICE et al. 1992). The protocol was repeated three times: (1) while the patient was severely depressed in winter, (2) after successful light treatment, and (3) after successful pharmacotherapy with citalopram. Prior to treatment, the core body temperature rhythm was phase delayed but showed normal amplitude, while very little melatonin was secreted. Light treatment induced remission, phase-advanced both rhythms, and slightly augmented melatonin secretion without altering the amplitude of the core temperature rhythm. Citalopram also induced remission but produced a different spectrum of effects on the rhythms: this treatment reduced the amplitude of the core temperature rhythm without causing a phase change, and greatly augmented and extended the duration of melatonin secretion. The complexity of these results suggests that inconsistencies in the existing literature are not entirely due to uncontrolled masking effects. Instead, these observations imply the absence of any simple relationship between specific circadian parameters and clinical state.

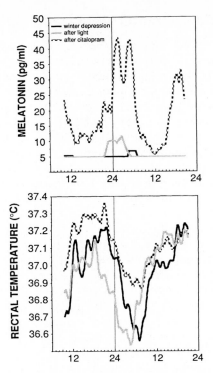

Fig. 3. Circadian melatonin (*top*) and core temperature (*bottom*) rhythms measured in winter in a single SAD patient under constant routine conditions. The subject was tested on three separate occasions; once while depressed, once after successful light treatment (5 days of 6000 lux from 1000 to 1400 hours), and once after successful pharmacotherapy with citalopram (20 mg/day for 6 weeks). (Wirz-Justice et al., unpublished data; study described in Wirz-Justice et al. 1992)

D. Circadian Rhythms and Depression: Animal Models

In this section, we review the evidence for circadian alterations in animal depression models, as well as related observations that were not originally presented in the explicit context of animal depression models. As noted above, the relative ease of studying free-running rhythms in animal systems has led to an emphasis on mechanisms controlling free-running period, and this emphasis will be reflected in the following discussion.

I. Psychopharmacology of Circadian Rhythms

1. Phase-Shifting and Free-Running Period

Monoaminergic and cholinergic neurotransmitter systems have been implicated in both human depression and animal depression models, and pharmacological manipulation of these transmitter systems can alter circadian rhythms in animals. Typically, studies of drug effects on rhythms have examined either the effects of chronic drug administration on the period or coherence of free-running rhythms, or the ability of acute pharmacological pertubation to phase-shift the free-running rhythm.

Acute treatment with cholinergic and monoaminergic agents can induce phase-dependent circadian phase-shifts. Light-pulse type phase-shifting has been reported for carbachol, a cholinergic agonist (ZATZ and HERKENHAM 1981; EARNEST and TUREK 1985; but see MEIJER et al. 1988), and for clonidine, a selective α_2-adrenergic autoreceptor agonist (ROSENWASSER et al., 1995; see Fig. 4), while cholinergic antagonists have been reported to block the phase-shifting effects of light (KEEFE et al. 1987). In contrast, the PRC for serotonergic agonists is dissimilar to that seen with light pulses, and instead appears to resemble the PRC seen with dark pulses, induced activity, and benzodiazepines (EDGAR et al. 1993; cf. Chaps. 1, 4, 15, this volume).

Free-running period is shortened during chronic administration of either carbachol (administered intracerebrally via an implanted pellet; FURUKAWA et al. 1987) or clonidine (administered systemically via the drinking water; ROSENWASSER 1989, 1990a; see Fig. 5), and lengthened by neurotoxic lesions of the serotonergic system (MORIN and BLANCHARD 1991) and following mono-

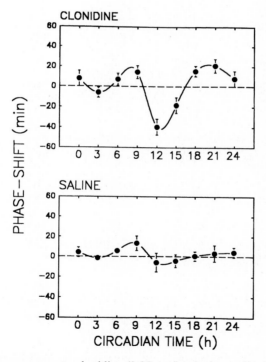

Fig. 4. Phase-response curves to clonidine (0.25 mg/kg, i.p.) or saline injections in male Syrian hamsters. Animals were maintained in running wheel cages in either constant darkness or in dim red light, and injected under dim red light. (Adapted from ROSENWASSER et al., 1995)

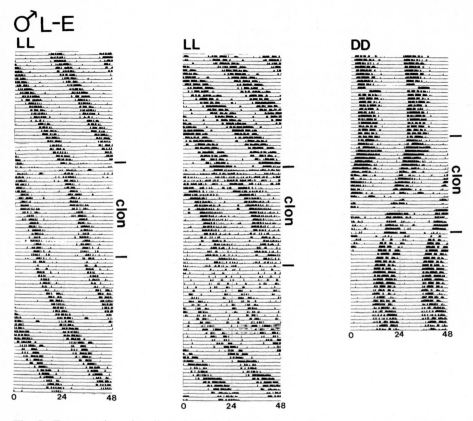

Fig. 5. Free-running circadian rhythms in male Long-Evans rats measured before, during, and after chronic clonidine treatment (5.0 µg/ml via the drinking water). The animals were maintained in either constant light (*left, middle*) or constant darkness (*right*) throughout the experiment. (Rosenwasser, previously unpublished data)

amine depletion by reserpine (PENEV et al. 1993). Monoaminergic effects on rhythms may interact with lighting conditions, since both 5,7-DHT lesions (MORIN and BLANCHARD 1991) and chronic clonidine treatment (ROSENWASSER 1989) consistently alter free-running period in constant light but not in constant darkness, and since monoamine depletion may alter photic entrainment and phase-shifting (SMALE et al. 1990; PENEV et al. 1993). These effects are likely related to the effects of serotonergic and noradrenergic agents on spontaneous activity and photic responsiveness in neurons of the suprachiasmatic nucleus (SCN), site of the primary mammalian pacemaker (see Chaps. 1, 4, this volume).

2. Antidepressants

A number of antidepressants that target the monoamine neurotransmitter systems, including clorgyline, imipramine, desipramine, fluoxetine, and lithium, have been reported to alter free-circadian running period and/or photic entrainment (Wirz-Justice and Campbell 1982; Wirz-Justice et al. 1982; Tamarkin et al. 1983; Kripke et al. 1987; Duncan et al. 1988; Possidente et al. 1992; Wollnik 1992; Klemfuss and Kripke 1994). Most of these agents produce a complex spectrum of effects on monoaminergic systems, and even relatively selective agents like the serotonin reuptake inhibitor fluoxetine would be expected to indirectly affect other systems via normal transmitter interactions. The reported effects of these agents on circadian rhythms are also complex and, indeed, rather inconsistent, both across and within studies. Both lengthening and shortening of free-running period have been reported during antidepressant treatment, and there is no obvious way to account for these variations by differential pharmacological activity. Thus, period shortening has been reported during chronic treatment with the selective serotonin reuptake inhibitor fluoxetine (Possidente et al. 1992), and with the preferential norepinephrine reuptake inhibitor desipramine (Wollnik 1992). Although the data are not as extensive as for circadian period, the reported effects of chronic antidepressant treatment on circadian amplitude, rhythm coherence, and activity level appear to be similarly inconsistent.

Delayed entrainment phase, dramatic lengthening of free-running period, and dissociation of activity rhythms into multiple components have all been reported during chronic treatment with methamphetamine, another agent producing a broad spectrum of monoaminergic agonist effects (e.g., Honma et al. 1986, 1991). Similar effects have also been observed during long-term maintenance on a serotonin-depleting, tryptophan-free diet, especially the emergence of a long-period activity component dissociating from the main daily activity bout (Kawai et al. 1994). One particularly intriguing finding with methamphetamine administration is the induction of a long-period, light-insensitive rhythm during drug treatment in otherwise-arrhythmic SCN-lesioned rats, suggesting that the long-period component is normally coupled to, but potentially independent of, the SCN-based pacemaker (Honma et al. 1987). Like other antihypertensive agents, clinical use of both reserpine and clonidine has been associated with increased risk of depression, presumably as a consequence of reduced monoaminergic function. Pharmacological antagonism of clonidine and reserpine treatments has been used as a research tool for screening antidepressant drugs, and depression-like behavioral changes have been reported in animals during treatment with these agents. Despite these common effects, and despite the fact that both agents generally cause reduced activity and circadian coherence, clonidine and reserpine appear to produce opposite effects on free-running period.

3. A Pharmacogenic-Developmental Model

Several laboratories have reported that neonatal antidepressant treatment produces in rats a constellation of behavioral and neurochemical changes in adulthood that may model clinical depression (MIRMIRAN et al. 1981; HILAKIVI and HILAKIVI 1987; VOGEL et al. 1990). These changes include decreases in sexual, aggressive, and other motivated behaviors, increased REM sleep, and alterations in monoaminergic systems. Interestingly, a similar depression-like syndrome is seen following neonatal treatment with both serotonin-selective and norepinephrine-selective reuptake inhibitors, suggesting that some common effect of the two drug classes, such as inhibition of juvenile REM sleep, is responsible.

In a recent study (ROSENWASSER and HAYES 1994), neonatal desipramine treatment was shown to lengthen free-running period in constant darkness and to blunt the period-lengthening effect of constant light. Surprisingly, circadian amplitudes were increased in all lighting conditions relative to controls. In addition, neonatal desipramine treatment significantly potentiated the period-altering effects of adult desipramine treatment. This last observation raises the largely unexplored possibility that more consistent effects of monoaminergic drugs on circadian rhythms might be uncovered by studying drug effects in "depressed," rather than in "normal," rats.

II. Stress-Induced Depression Models

Behavioral change evoked by chronic stress is considered to provide one of the most valid animal models of depression (WILLNER 1984), despite continuing controversy concerning the role of stress in the etiology of human depression. Stress-induced behavioral "pathology" includes alterations in activity level, affective behaviors, attention, motivation, and learning. Further, these behavioral symptoms are associated with alterations in monoaminergic systems and are reversible by antidepressant drugs with reasonable selectivity.

An early study demonstrated that some stress-exposed monkeys showed either 48-h sleep-wake cycles or failure to entrain in the presence of an otherwise effective light-dark cycle (STROEBEL 1969); both of these effects are reminiscent of circadian disturbances previously reported in some bipolar affective patients. In more recent studies (STEWART et al. 1990a, b), some rats exposed to repeated sessions of either handling or shock stress showed clear lengthening of free-running period following stress exposure. Concurrent assessment of escape performance (i.e., learned helplessness) and activity level, two variables previously suggested to serve as markers for behavioral depression, revealed that stress-induced period-lengthening was associated with persistent activity deficits, but, surprisingly, with intact escape performance. In other words, alterations in circadian activity rhythms were generally associated with resistance to helplessness. These results should be considered in the context of other studies also reporting a lack of correspondence among different indices of behavioral depression in stress-depression models.

III. Behaviorally Characterized Inbred Strains

Inbred strains provide useful experimental models in the study of affective behavior and, in some cases, particular strains have been explicitly suggested to provide animal models of depression, or at least of altered affective regulation. Such models may be especially useful for understanding the role of genetic vulnerability in the etiology of depression.

1. Selection for Hypertension

Numerous studies have compared the behavioral and neurochemical profiles of the spontaneously hypertensive rat (SHR) and the normotensive Wistar-Kyoto (WKY) progenitor strain. SHRs show reduced fear and anxiety in stressful situations, but exaggerated physiological (e.g., cardiovascular) responsiveness to stress or novelty. One limitation of this research is that SHRs and WKYs have only rarely been compared to other typical laboratory rat strains. Indeed, it has been suggested that WKYs, not SHRs, typically deviate from standard laboratory strains in behavioral tests, and that the WKY strain may provide a model for genetic susceptibility to stress-induced depression (PARE 1989).

Recent studies of circadian rhythms in SHRs and WKYs indicate that SHRs show shorter free-running periods, reduced phase-delays, and increased phase-advances in response to brief light pulses, advanced entrainment phase, and increased activity levels, relative to WKYs (ROSENWASSER and PLANTE 1993; ROSENWASSER 1993; PETERS et al. 1994). Although SHRs consistently showed longer free-running periods in constant light, PETERS et al. (1994) also showed strain-differentiated free-running period in constant darkness, while ROSENWASSER (1993) did not find strain differences in free-running period in constant darkness or in very dim light. However, even in the study by PETERS et al. the strain difference in constant darkness was somewhat smaller than in constant light; therefore, the results of both studies taken together suggest a strain by light interaction influencing free-running period. Since these studies did not directly compare SHRs and WKYs to other strains, it is impossible to determine which strain shows atypical rhythms. Nevertheless, informal comparison to other published results seems to suggest that SHRs may show unusually advanced entrainment phase, while WKYs may show unusually long free-running periods.

HENDLEY et al. (1991) have produced two new strains from SHR and WKY progenitors, one of which (WKHT) displays hypertension without hyperactivity, and the other of which (WKHA) displays hyperactivity without hypertension. ROSENWASSER et al. (1994) have recently found that free-running period and photic entrainment parameters cosegregate differently among the strains. Thus, markedly advanced entrainment phase and a reduced ratio of light-induced delay to advance phase-shifts are seen in the two hyperactive strains (SHRs and WKHAs), while unusually short free-running periods are seen in the two hypertensive strains (SHRs and WKHTs). These results suggest that entrainment dynamics may genetically cosegregate with responsiveness to stress or novelty, while free-running period may cosegregate with some as yet unidentified neurochemical correlate of hypertension.

2. Selection for Drug Sensitivity

Although the Flinders Sensitive (FSL) and Flinders Resistant (FRL) inbred rat lines were originally selected for differences in sensitivity to a specific anticholinesterase, FSL rats also show marked hypersensitivity to various cholinergic agonists as well as a generalized hypercholinergic tone (OVERSTREET 1986). Since depression has also been associated with increased cholinergic tone, an intensive research program has been conducted to behaviorally and neurochemically characterize the FSL rat. The results of this research indicate that FSL rats display hypoactivity, increased REM sleep, impaired avoidance performance, exaggerated stress responses, and alterations in monoaminergic and other neurotransmitter systems, all reminiscent of human depression (OVERSTREET 1986). Studies of circadian rhythms indicate that FSL rats show advanced entrainment phase of the core temperature rhythm and shortened free-running period of the drinking rhythm in constant darkness, relative to FRL rats (SHIROMANI et al. 1991; SHIROMANI and OVERSTREET 1994).

3. Inbred Mice Differing in Affective Behavior

Inbred strains of mice differ in affective behavior, neurochemical profile, and various aspects of circadian rhythmicity (POSSIDENTE and STEPHAN 1988; ROSENWASSER 1990b; SCHWARTZ and ZIMMERMAN 1990). For example, the BALB/c mouse has classically been considered to be hyperemotional and aggressive, while recent studies indicate that this strain may be particularly susceptible to stress-induced behavioral and neurochemical deficits (SHANKS and ANISMAN 1988). The circadian rhythms of BALB/c mice also appear to be exceptional when compared to other inbred strains. For example, BALB/c mice show unusually short free-running periods in constant darkness (POSSIDENTE and STEPHAN 1988; SCHWARTZ and ZIMMERMAN 1990), and unusual lability of free-running period, activity level, and rhythm coherence, when maintained under dim constant light (ROSENWASSER 1990b). As a further indication of the lability of their activity rhythms, BALB/c mice, but not C57s, showed significant alterations in free-running period as a consequence of running wheel availability (SCHWARTZ and ZIMMERMAN 1990).

Free-running activity rhythms have also been compared in strains of mice explicitly selected for either high or low intraspecific aggression (i.e., long or short attack latency; BENUS et al. 1988). In apparent contrast to the results for BALB/c mice discussed above, the less aggressive strain showed shorter free-running periods and greater responsiveness to the period-altering effects of wheel availability. With the exception of this one report, affective and circadian parameters have not been directly compared in the same study; this would appear to be a promising avenue for further investigation.

IV. Ablation-Based Depression Models

1. Olfactory Bulbectomy

Behavioral and neurochemical studies of olfactory bulbectomized (OBX) rats suggest that these animals may provide a model for agitated or anxious depression (RICHARDSON 1991). OBX leads to increased activity, emotionality, and aggressiveness, decreased sexual behavior, and learning deficits, as well as generally reduced levels of monoamine neurotransmitters throughout the limbic brain; while these effects are reversed by chronic antidepressant treatment with rather good selectivity. While the OBX depression model has been developed mainly in rats, OBX consistently lengthens circadian period and delays entrainment phase in rats, mice, and hamsters (POSSIDENTE et al. 1990; PIEPER and LOBOCKI 1991; LUMIA et al. 1992). OBX has also been reported to decrease (in running wheels; LUMIA et al. 1992) or increase (in ambulatory activity monitors; GIARDINA and RADEK 1991) the amplitude of the circadian activity rhythm, and both changes were reversible by antidepressant treatment. OBX also dramatically increases cAMP content in the SCN (VAGELL et al. 1991), providing a possible mechanism for these effects.

2. Thyroidectomy

Another possible ablation-based depression model is the thyroidectomized rat. RICHTER (1965) showed many years ago that thyroidectomy leads to the emergence of unusual long-term cycles in activity level, and proposed that this preparation could serve as a model for manic-depressive illness. This suggestion is consistent with the high incidence of thyroid abnormalities in affective disorder patients, as well as the high incidence of affective disturbances in patients with primary thyroid disease. Thyroidectomized and thyroparathyroidectomized (TPX) rats have been reported to display shortening of free-running period and increased activity levels (DUNCAN and SCHULL 1994; SCHULL et al. 1988, 1989), while rats made hypothyroid by pharmacological treatment showed blunting of the period-lengthening effects of increasing light intensity. These effects appear to be specifically related to alterations in thyroid hormone levels, since thyroxine administration lengthened free-running period in TPX rats but not in controls (MCEACHRON et al. 1993). In contrast, while antithyroid drug treatments have been reported to lengthen free-running period in hamsters, surgical thyroidectomy failed to alter circadian period in this species (MORIN 1988). It has been suggested that relationships between thyroid status and affect are mediated by the effects of thyroid hormones on monoaminergic transmitter systems (WHYBROW and PRANGE 1981), and this may also be true of thyroid effects on the circadian system, since TPX reduced and thyroxine increased the levels of adrenergic receptors in the SCN (VESSOTSKIE et al. 1993).

V. Animal Models of SAD?

Both people and animals display seasonal changes in behavioral and physiological state, and it has been suggested that SAD reflects an exaggerated form of normal human seasonal variation. Indeed, the original impetus to attempt light therapy for SAD was based on animal studies showing that photoperiod duration – acting through changes in the duration of nocturnal melatonin release – could trigger seasonal changes in animals. Since neither photoperiod duration nor melatonin secretion have definitively been shown to play a role in SAD, the utility of animal photoperiodic responses as a model for SAD may be quite limited.

Another neurobiological model for SAD attempts to integrate the atypical vegetative symptoms characteristic of SAD patients – increased carbohydrate appetite, weight gain, and hypersomnia – within a framework of known neurobiological mechanisms related to medial hypothalamic function (KRÄUCHI and WIRZ-JUSTICE 1992). In mammals, carbohydrate selection is regulated by serotonergic and α_2-adrenergic mechanisms in the paraventricular hypothalamus (PVN), while neural input from the SCN to the PVN transduces circadian and seasonal information. Since serotonergic and noradrenergic mechanisms also appear to be involved in affective regulation, changes in these systems may link food selection, mood, time of day, and time of year. This hypothesis is consistent with results showing that both light therapy and serotonergic antidepressant drugs selectively suppress carbohydrate intake and alleviate depressed mood in SAD patients. Additionally, the high proportion of women SAD patients, and the known sex difference in hypothalamic serotonergic mechanisms, adds a hormonal factor that may interact with putative genetic or experiential factors, possibly acting through alterations in circadian function, to predispose individuals to exaggerated seasonality.

VI. Behavioral Feedback Effects

Until recently, it was thought that human circadian rhythms were primarily influenced by social and cognitive factors, rather than by environmental lighting, and that the opposite was true of animals. However, it is now clear that photic stimuli can phase shift and entrain the human circadian pacemaker (CZEISLER et al. 1989; WEVER 1989; MINORS et al. 1991), and that behaviorally derived stimuli are of considerable importance in the control of rhythmicity in animals (MROSOVSKY et al. 1989; TUREK 1989). Further, the effects of so-called social stimuli in both humans and other animals may be at least partially dependent on unidentified correlates related to arousal, activity, and/or exercise.

As described more fully elsewhere in this volume (see Chaps. 1, 4), social stimulation, cage-cleaning, inert saline injections, and novelty-induced voluntary activity have all been shown capable of phase-shifting free-running activity rhythms in hamsters (MROSOVSKY et al. 1989; TUREK 1989; HASTINGS et al.

1992). Furthermore, the phase-shifting effects of at least some pharmacological agents appear to be mediated by drug effects on activity level (VAN REETH and TUREK 1989). In addition, chronically elevated activity induced by access to running wheels is associated with shortening of free-running period in rats and mice (EDGAR et al. 1991; SHIOIRI et al. 1991), while the degree of period-shortening during wheel access is related to the level of activity (SHIORI et al. 1991). The effects of chronic methamphetamine administration on circadian rhythmicity also differ between wheel-housed and non-wheel-housed rats, in that wheel-access seems to promote rhythm coherence (HONMA et al. 1991).

The mechanisms underlying the effects of behavioral activity on the circadian system are not fully known, but increasing evidence suggests that such effects are mediated by ascending monoaminergic projections to the SCN and/or the intergeniculate leaflet (IGL) of the thalamus, which in turn projects to the SCN (see Chaps. 1, 4, this volume). The PRC for serotonergic agonists is similar to that for induced activity (EDGAR et al. 1993), while triazolam-induced phase-shifts – thought to be mediated by induced locomotor activity – are blocked by IGL lesions (JOHNSON et al. 1988), by the selective serotonergic neurotoxin 5,7-DHT (PENEV et al. 1994a), and by nonselective monoamine depletion with reserpine (PENEV et al. 1994b). Serotonergic input to the SCN is also implicated by a report that SCN serotonin levels correlate positively with activity levels and negatively with free-running period in rats (SHIORI et al. 1991).

Although these behavioral feedback phenomena have not generally been presented in the context of animal depression models, they do demonstrate that alterations in affective state may be associated with alterations in circadian period and phase. These relationships are at least reminiscent of longitudinal studies in bipolar patients showing covariation of activity level, circadian phase (period?), and mood. Importantly, such observations suggest that the altered rhythmicity seen in affective disorders could largely be a consequence of psychomotor alterations associated with depression and mania.

VII. Summary

Studies of drug effects on circadian rhythms clearly support the hypothesis that the monoaminergic transmitter systems are involved in modulating circadian rhythmicity, either by acting directly on the pacemaker, on the light-entrainment pathway, or on other pathways afferent to the pacemaker. Since these same transmitter systems have long been implicated in both abnormal and normal variations in affective state, such findings raise the possibility that alterations in monoaminergic systems may serve as an "intervening variable" mediating observed relationships between mood and rhythmicity. On the other hand, the diversity and variability of monoaminergic drug effects on circadian period, amplitude, rhythm coherence, and activity level do not lead to any simple models of these interactions.

Alterations in affective behavior in animals are clearly associated with alterations in circadian rhythms. Indeed, altered rhythmicity has been detected in every animal depression model examined, whether induced by pharmacological, genetic, surgical, or environmental means. In addition, it seems likely that covariation of affective and circadian parameters is mediated by monoaminergic neurotransmitter systems, since these systems are generally altered in animal models of depression and since drugs that target these systems can alter the expression of circadian rhythms.

Despite this apparent wealth of positive findings, we are not yet able to ascertain specific, generalizable linkages between circadian rhythms, affective state, and neurotransmitter function. While lengthening of free-running period has been observed in several putative depression models (e.g., reserpine treatment, chronic stress, WKY strain, olfactory bulbectomy, neonatal desipramine), others show shortening of period (e.g., chronic clonidine, FSL strain, thyroidectomy). Similarly, both lengthening and shortening of free-running period have been observed during antidepressant and depressogenic drug treatments.

Can short- and long-period "subtypes" of animal depression model be identified based on associated behavioral or neurochemical characteristics? TEICHER and coworkers (1988, 1989) have hypothesized that circadian subtypes might be differentiated by psychomotor changes, in both patients and animals. Specifically, they suggested that phase-delayed rhythms may be associated with psychomotor agitation and unipolar depressions, while phase-advanced rhythms may be associated with the anergic depressions of bipolar patients. However, changes in locomotor activity level are not clearly predictive of circadian changes in animal studies. For example, clonidine treatment leads to reduced activity and period shortening, while desipramine treatment leads to increased activity and period shortening. Similarly, olfactory bulbectomy leads to increased activity and period lengthening, while chronic stress decreases activity and lengthens period. Of course, the spontaneous locomotor activity of animals may not be a reasonable analogue to psychomotor changes seen in patients.

Another possibility is that changes in free-running period may be related to other behavioral predictors, such as anxiety, aggression, or emotionality, rather than to activity levels per se. Unfortunately, these traits are assessed by a multitude of behavioral tests which do not always covary systematically, and the existing data are not sufficient to allow comparison across all the models discussed in this review. Nevertheless, as a tentative hypothesis, it may be noted that period lengthening is seen in several models generally associated with exaggerated anxiety, including the chronic stress model, the WKY rat model, the short-attack latency mouse model, and the olfactory bulbectomy model. In contrast, the short-period FSL model appears to test negatively on anxiety measures (SCHILLER et al. 1991), and the period-shortening agent, clonidine, also has clear anxiolytic properties. While this hypothesis appears to present a worthwhile avenue for future investigations, untangling the behavioral features of anxiety and depression is likely to be as difficult in animal models as in clinical situations.

E. Conclusions

Both depressed patients and depression-model rats show evidence for alterations in circadian rhythms, including changes in free-running period, amplitude and coherence, entrainment, and photic responsiveness. However, the direction of such changes is not consistent in either patients or animals. For example, while advanced entrainment phase is commonly described in depressed patients (WEHR et al. 1983), phase-delayed circadian rhythms have been seen in some SAD patients (LEWY et al. 1987) and in a group of geriatric depressed inpatients (TEICHER et al. 1988). Similarly, both lengthening and shortening of free-running period have been observed in putative animal models of depression. This variability should probably not be surprising, in light of the behavioral heterogeneity in both patients and animal models. Indeed, either increases or decreases in other so-called neurovegetative markers, including sleep, appetite, and libido, may be diagnostic for depression.

While the critical variables that distinguish phase-advanced/short-period and phase-delayed/long-period subtypes have not been elucidated in either patients or animals, one tentative hypothesis, based on the animal research discussed above, is that period-lengthening might be associated with anxiety and/or psychomotor agitation. Such a hypothesis may also be consistent with clinical observations, since delayed phase has been reported in patients with SAD, a syndrome characterized by associated anxiety in nearly all patients, and in a population of agitated and depressed psychiatric inpatients. Much more work will be required to refine and test this hypothesis further.

Finally, what can be said about the nature and direction of the causal mechanisms underlying these relationships? Certainly, rhythm disturbances could play a causal role in depression, as suggested by the phase-advance hypothesis (WEHR and WIRZ-JUSTICE 1982; WEHR et al. 1983) and incorporated into more recent integrative theories (HEALY 1987; HEALY and WILLIAMS 1988; EHLERS et al. 1988). Conversely, studies of activity feedback in animals suggest that depression-associated alterations in arousal, activity, and affect may lead to altered rhythmicity. It is also possible that no causal mechanisms link affective state and circadian rhythmicity, but that covariation between these systems may instead reflect common controlling variables, including genetic predisposition, environmental stress, and neurotransmitter dysregulation. Indeed, these hypotheses are clearly not mutually exclusive. Possibly, changes in mood, behavior, and circadian rhythmicity are linked by a network of interacting causal loops, such that stimuli interacting with any of the nodes in this network can potentially alter the overall behavior of the system.

References

Anderson J, Wirz-Justice A (1991) Biological rhythms in the pathophysiology and treatment of affective disorders. In: Horton R, Katona C (eds) Biological aspects of affective disorders. Academic, London, pp 223–269

Benca RM, Obermeyer WH, Thisted RA, Gillin JC (1992) Sleep and psychiatric disorders: a meta-analysis. Arch Gen Psychiatry 49:651–668

Benus RF, Koolhaas JM, van Oortmerssen GA (1988) Aggression and adaptation to the light-dark cycle: role of intrinsic and extrinsic control. Physiol Behav 43:131–137

Berger M, Riemann D (1993) REM sleep in depression – an overview. J Sleep Res 2:211–223

Borbély AA, Wirz-Justice A (1982) Sleep, sleep deprivation and depression. A hypothesis derived from a model of sleep regulation. Hum Neurobiol 1:205–210

Czeisler CA, Kronauer RE, Allan JS, Duffy JF, Jewett ME, Brown EN, Ronda JM (1989) Bright light induction of strong (type 0) resetting of the human circadian pacemaker. Science 244:1328–1333

Daan S, Beersma DGM, Borbely AA (1984) Timing of human sleep: recovery process gated by a circadian pacemaker. Am J Physiol 246:R161–R178

Dahl K, Avery DH, Lewy AJ et al (1993) Dim light melatonin onset and circadian temperature during a constant routine in hypersomnic winter depression. Acta Psychiatr Scand 88:60–66

Daimon K, Yamada N, Tsujimoto T, Takahashi S (1992) Circadian rhythm abnormalities of deep body temperature in depressive disorders. J Affect Disord 26:191–198

Dirlich G, Kammerloher A, Schulz H, Lund R, Doerr P, von Zerssen D (1981) Temporal coordination of rest-activity cycle, body temperature, urinary free cortisol, and mood in a patient with 48-hour unipolar-depressive cycles in clinical and time-cue-free environments. Biol Psychiatry 16:163–179

Duncan WC, Schull J (1994) The interaction of thyroid state, MAOI drug treatment, and light on the level and circadian pattern of wheel-running in rats. Biol Psychiatry 35:324–334

Duncan WC, Tamarkin L, Sokolove PG, Wehr TA (1988) Chronic clorgyline treatment of Syrian hamsters: an analysis of effects on the circadian pacemaker. J Biol Rhythms 3:305–322

Earnest DJ, Turek F (1985) Neurochemical basis for the photic control of circadian rhythms and seasonal reproductive cycles: role for acetylcholine. Proc Natl Acad Sci USA 82:4277–4281

Edgar DM, Martin CE, Dement WC (1991) Activity feedback to the mammalian circadian pacemaker: influence on observed measures of period length. J Biol Rhythms 6:185–199

Edgar DM, Miller JD, Prosser RA, Dean RR, Dement WC (1993) Serotonin and the mammalian circadian system: II. Phase-shifting rat behavioral rhythms with serotonergic agonists. J Biol Rhythms 8:17–31

Ehlers CL, Frank E, Kupfer DJ (1988) Social zeitgebers and biological rhythms: a unified approach to understanding the etiology of depression. Arch Gen Psychiatry 45:948–952

Furukawa T, Murakami N, Takahashi K, Etoh T (1987) Effect of implantation of carbachol pellet near the suprachiasmatic nucleus on the free-running period of rat locomotor activity rhythm. Jpn J Physiol 37:321–326

Giardina WJ, Radek RJ (1991) Effects of imipramine on the nocturnal behavior of bilateral olfactory bulbectomized rats. Biol Psychiatry 29:1200–1208

Goodwin FK, Jamison KR (1990) Manic-depressive illness. Oxford University Press, New York

Halberg F (1968) Physiologic considerations underlying rhythmometry with special reference to emotional illness. In: de Ajuriaguerra J (ed) Cycles biologiques et psychiatrie. Masson and Cie, Paris, pp 73–126

Hall DP Jr, Sing HC, Romanoski AJ (1991) Identification and characterization of greater mood variance in depression. Am J Psychiatry 148:418–419

Hastings MH, Mead SM, Vidlacheruvu RR, Ebling FJP, Maywood ES, Grosse J (1992) Non-photic phase-shifting of the circadian activity rhythm of Syrian hamsters: the relative potency of arousal and melatonin. Brain Res 591:20–26

Healy D (1987) Rhythm and blues. Neurochemical, neuropharmacological and neuropsychological implications of a hypothesis of circadian rhythm dysfunction in the affective disorders. Psychopharmacology 93:271–285

Healy D, Williams JMG (1988) Dysrhythmia, dysphoria, and depression: the interaction of learned helplessness and circadian dysrhythmica in the pathogenesis of depression. Psychol Bull 103:163–178

Hendley ED, Ohlsson WG (1991) Two new inbred rat strains derived from SHR: WKHA, hyperactive, and WKHT, hypertensive rats. Am J Physiol 261:H584–H590

Hilakivi LA, Hilakivi I (1987) Increased adult behavioral "despair" in rats neonatally exposed to desipramine or zimeldine: an animal model of depression? Pharmacol Biochem Behav 28:367–369

Honma K, Honma S, Hiroshige T (1986) Disorganization of the rat activity rhythm by chronic treatment with methamphetamine. Physiol Behav 38:687–695

Honma K, Honma S, Hiroshige T (1987) Activity rhythms in the circadian domain appear in suprachiasmatic nuclei lesioned rats given methamphetamine. Physiol Behav 40:767–774

Honma S, Honma K, Hiroshige T (1991) Methamphetamine effects on rat circadian clock depend on actograph. Physiol Behav 49:787–795

Jauhar P, Weller MPI (1982) Psychiatric morbidity and time zone changes. Br J Psychiatry 140:231–235

Johnson RF, Smale L, Moore RY, Morin LP (1988) Lateral geniculate lesions block circadian phase-shift responses to a benzodiazepine. Proc NatL Acad Sci USA 85:5301–5304

Kawai K, Yokota N, Yamawaki S (1994) Effect of chronic tryptophan depletion on the circadian rhythm of wheel-running activity in rats. Physiol Behav 55:1005–1013

Keefe DL, Earnest DJ, Nelson D, Takahashi JS, Turek FW (1987) A cholinergic antagonist, mecamylamine, blocks the phase-shifting effects of light on the circadian rhythm of locomotor activity in the golden hamster. Brain Res 403:308–312

Klemfuss H, Kripke DF (1994) Antidepressant and depressogenic drugs lack consistent effects on hamster circadian rhythms. Psychiatry Res 53:173–184

Köhler WK, Fey P, Schmidt KP, Pflug B (1993) Zeitgeber und das circadiane System bei depressiven Patienten. In: Baumann P (ed) Biologische Psychiatrie der Gegenwart. Springer, Vienna New York, pp 233–237

Kräuchi K, Wirz-Justice A (1992) Seasonal patterns of nutrient intake in relation to mood. In: Anderson GH, Kennedy SH (eds) The biology of feast and famine: relevance to eating disorders. Academic, Orlando, pp 157–182

Krauss SS, Depue RA, Arbisi PA, Spoont M (1992) Behavioral engagement level, variability, and diurnal rhythm as a function of bright light in bipolar II seasonal affective disorder: an exploratory study. Psychiatry Res 43:147–160

Kripke DF (1981) Photoperiodic mechanisms for depression and its treatment. In: Perris C, Struwe G, Jansson B (eds) Biological psychiatry 1981. Elsevier/North-Holland Biomedical, Amsterdam, pp 1249–1252

Kripke DF, Mullaney DJ, Atkinson M, Wolf SR (1978) Circadian rhythm disorders in manic-depressives. Biol Psychiatry 13:335–351

Kripke DF, Mullaney DJ, Gabriel S (1987) The chronopharmacology of antidepressant drugs. Annu Rev Chronopharmacol 2:275–289

Kripke DF, Mullaney DJ, Melville RK, Risch SC, Gillin JC (1992) Controlled trial of bright light for nonseasonal major depressive disorders. Biol Psychiatry 31:119–134

Lacoste V, Wirz-Justice A (1989) Seasonal variation in normal subjects: an update of variables current in depression research. In: Rosenthal NE, Blehar MC (eds) Seasonal affective disorders and phototherapy. Guilford, New York, pp 167–229

Lewy AJ, Wehr TA, Goodwin FK, Newsome DA, Rosenthal NE (1981) Manic-depressive patients may be supersensitive to light. Lancet 1:383–384

Lewy AJ, Sack RL, Miller S, Hoban TM (1987) Antidepressant and circadian phase-shifting effects of light. Science 235:352–354

Linkowski P, Mendlewicz J, Kerkhofs M, Leclercq R, Golstein J, Brasseur M, Copinschi G, van Cauter E (1987) 24-hour profiles of adrenocorticotropin, cortisol, and growth hormone in major depressive illness: effect of antidepressant treatment. J Clin Endocrinol Metab 65:141–152

Linkowski P, Van Onderbergen A, Kerkhofs M, Bosson D, Mendlewicz J, Van Cauter E (1993) A twin study of the 24-hour cortisol profile: evidence for genetic control of the human circadian clock. Am J Physiol 264:E173–E181

Lumia AR, Teicher MH, Salchli F, Ayers E, Possidente B (1992) Olfactory bulbectomy as a model for agitated hyposerotonergic depression. Brain Res 587:181–185

McEachron DL, Lauchlan CL, Midgley DE (1993) Effects of thyroxine and thyroparathyroidectomy on circadian wheel running in rats. Pharmacol Biochem Behav 45:243–249

Meijer JH, van der Zee E, Dietz M (1988) The effects of intraventricular carbachol injections on the free-running activity rhythm of the hamster. J Biol Rhythms 3:333–348

Minors D, Waterhouse JM, Wirz-Justice A (1991) A human phase-response curve to light. Neurosci Lett 133:36–40

Mirmiran M, Brenner E, Van der Gugten J, Swaab DF (1985) Neurochemical and electrophysiological disturbances mediate developmental behavioral alterations produced by medicines. Neurobehav Toxicol Teratol 7:677–683

Monk TH, Kupfer DJ, Frank E, Ritenour AM (1991) The Social Rhythm Metric (SRM): measuring daily social rhythms over 12 weeks. Psychiatry Res 36:195–207

Morin LP (1988) Propylthiouracil, but not other antithyroid treatments, lengthens hamster circadian period. Am J Physiol 255:R1–R5

Morin LP, Blanchard J (1991) Depletion of brain serotonin by 5,7-DHT modifies hamster circadian rhythm response to light. Brain Res 566:173–185

Mrosovsky N, Reebs SG, Honrado GI, Salmon PA (1989) Behavioural entrainment of circadian rhythms. Experientia 45:696–702

Overstreet DH (1986) Selective breeding for increased cholinergic function: development of a new animal model for depression. Biol Psychiatry 21:49–58

Pare WP (1989) Stress ulcer susceptibility and depression in Wistar Kyoto (WKY) rats. Physiol Behav 46:993–998

Parry BL, Berga SL, Kripke DF et al (1990) Altered waveform of plasma nocturnal melatonin secretion in premenstrual depression. Arch Gen Psychiatry 47:1139–1146

Penev PD, Turek FW, Zee PC (1993) Monoamine depletion alters the entrainment and the response to light of the circadian activity rhythm in hamsters. Brain Res 612:156–164

Penev PD, Turek FW, Zee PC (1994a) A serotonin neurotoxin blocks the phase shifting effects of an activity-inducing stimulus on the circadian clock in hamsters. Abstr Soc Res Biol Rhythms 4:74

Penev PD, Zee PC, Turek FW (1994b) Monoamine depletion blocks triazolam-induced phase advances of the circadian clock in hamsters. Brain Res 637:255–261

Peters RV, Zoeller RT, Hennessey AC, Stopa EG, Anderson G, Albers HE (1994) The control of circadian rhythms and the levels of vasoactive intestinal peptide mRNA in the suprachiasmatic nucleus are altered in spontaneous hypertensive rats. Brain Res 639:217–227

Pflug B, Johnsson A, Martin W (1983) Alterations in the circadian temperature rhythms in depressed patients. In: Wehr TA, Goodwin FK (eds) Circadian rhythms in psychiatry. Boxwood, Pacific Grove, pp 71–76

Pieper DR, Lobocki CA (1991) Olfactory bulbectomy lengthens circadian period of locomotor activity in golden hamsters. Am J Physiol 261:R973–R978

Piletz JE, DeMet E, Gwirtsman HE, Harris A (1994) Disruption of circadian MHPG rhythmicity in major depression. Biol Psychiatry 35:830–842

Possidente B, Stephan FK (1988) Circadian period in mice: analysis of genetic and maternal contributions to inbred strain differences. Behav Genet 18:109–117

Possidente B, Lumia AR, McGinnis MY, Teicher MH, deLemos E, Sterner L, Deros L (1990) Olfactory bulb control of circadian activity rhythm in mice. Brain Res 513:325–328

Possidente B, Lumia AR, McEldowney S, Rapp M (1992) Fluoxetine shortens circadian period for wheel running activity in mice. Brain Res Bull 28:629–631

Prigerson HG, Reynolds III CF, Frank E, Kupfer DJ, George CJ, Houck PR (1994) Stressful life events, social rhythms, and depressive symptoms among the elderly: an examination of hypothesized causal linkages. Psychiatry Res 51:33–49

Raoux N, Benoit O, Dantchev N, Denise P, Franc B, Allilaire J-F, Widlocher D (1994) Circadian pattern of motor activity in major depressed patients undergoing antidepressant therapy: relationship between actigraphic measures and clinical course. Psychiatry Res 52:85–98

Richardson JS (1991) The olfactory bulbectomized rat as a model of major depressive disorder. In: Boulton AA, Baker GB, Martin-Iverson MT (eds) Animal models in psychiatry II. Humana, Clifton, p 61

Richter CP (1965) Biological clocks in medicine and psychiatry. Thomas, Springfield

Rockwell DA, Winget CM, Rosenblatt LS, Higgins EA, Hetherington NW (1978) Biological aspects of suicide: circadian disorganisation. J Nerv Ment Dis 166:851–858

Rosenthal NE, Blehar MC (eds) Seasonal affective disorder and phototherapy. Guilford, New York

Rosenwasser AM (1989) Effects of chronic clonidine administration and withdrawal on free-running circadian activity rhythms. Pharmacol Biochem Behav 33:291–297

Rosenwasser AM (1990a) Free-running circadian activity rhythms during long-term clonidine administration in rats. Pharmacol Biochem Behav 35:35–39

Rosenwasser AM (1990b) Circadian activity rhythms in BALB/c mice: a weakly-coupled circadian system? J Interdiscipl Cycle Res 21:91–96

Rosenwasser AM (1992) Circadian rhythms and depression: animal models? LTBR Bull 4:35–39

Rosenwasser AM (1993) Circadian drinking rhythms in SHR and WKY rats: effects of increasing light intensity. Physiol Behav 53:1035–1041

Rosenwasser AM, Plante L (1993) Circadian activity rhythms in SHR and WKY rats: strain differences and effects of clonidine. Physiol Behav 53:23–29

Rosenwasser AM, Hayes MJ (1994) Neonatal desipramine treatment alters free-running circadian drinking rhythms in rats. Psychopharmacology 115:237–244

Rosenwasser AM, Pellowski MP, Hendley ED (1996) Circadian time keeping in hyperactive and hypertensive inbred rat strains. Am J Physiol 40:R 787–R 790

Rosenwasser AM, Vogt LJ, Pellowski MP (1995) Circadian phase shifting induced by clonidine injections in Syrian hamsters. Biol Rhythms Res 26:553–572

Sack RL, Lewy AJ, White DM, Singer CM, Fireman MJ, Vandiver R (1990) Morning vs evening light treatment for winter depression. Arch Gen Psychiatry 47:343–351

Schiller GD, Daws LC, Overstreet DH, Orbach J (1991) Lack of anxiety in an animal model of depression with cholinergic supersensitivity. Brain Res Bull 26:433–435

Schmidt KP, Kîhler WK, Fleissner G, Pflug B (1990) Locomotor activity accelerates the adjustment of the temperature rhythm in shiftwork. J Interdiscipl Cycle Res 21:243–245

Schull J, McEachron DL, Adler NT, Fiedler L, Horvitz J, Noyes A, Olson M, Shack J (1988) Effects of thyroidectomy, parathyroidectomy and lithium on circadian wheelrunning in rats. Physiol Behav 42:33–39

Schull J, Walker J, Fitzgerald K, Hiilivirta L, Ruckdeschel J, Schumacher D, Stanger D, McEachron DL (1989) Effects of sex, thyro-parathyroidectomy, and light regime on levels and circadian rhythms of wheel-running in rats. Physiol Behav 46:341–346

Schulte W (1971) Zum Problem der Provokation und Kupierung von melancholischen Phasen. Arch Neurol Neurochir Psychiatry 109:427–435

Schwartz WJ, Zimmerman P (1990) Circadian timekeeping in BALB/c and C57BL/6 inbred mouse strains. J Neurosci 10:3685-3694

Seggie J, Canny C, Mai F, McCrank E, Waring E (1989) Antidepressant medication reverses increased sensitivity to light in depression: preliminary report. Prog Neuropsychopharmacol Biol Psychiatry 13:537-541

Shanks N, Anisman H (1988) Stressor-provoked behavioral changes in six strains of mice. Behav Neurosci 102:894-905

Shioiri T, Takahashi K, Yamada N, Takahashi S (1991) Motor activity correlates negatively with free-running period, while positively with serotonin contents in SCN in free-running rats. Physiol Behav 49:779-786

Shiromani PJ, Overstreet D (1994) Free-running period of circadian rhythms is shorter in rats with a genetically upregulated central cholinergic system. Biol Psychiatry 36:622-626

Shiromani PJ, Klemfuss H, Lucero S, Overstreet DH (1991) Diurnal rhythm of core body temperature is phase advanced in a rodent model of depression. Biol Psychiatry 29:923-930

Siever LJ, Davis KL (1985) Overview: toward a dysregulation hypothesis of depression. Am J Psychiatry 142:1017-1031

Smale L, Michels KM, Moore RY, Morin LP (1990) Destruction of the hamster serotonergic system by 5,7-DHT: effects on circadian rhythm phase, entrainment and response to triazolam. Brain Res 515:9-19

Souêtre E, Salvati E, Belugou J-L et al (1989) Circadian rhythms in depression and recovery: evidence for blunted amplitude as the main chronobiological abnormality. Psychiatry Res 28:263-278

Stewart KT, Rosenwasser AM, Hauser H, Volpicelli JR, Adler NT (1990a) Circadian rhythmicity and behavioral depression I: effects of stress. Physiol Behav 48:149-155

Stewart KT, Rosenwasser AM, Levine JD, McEachron DL, Volpicelli JR, Adler NT (1990b) Circadian rhythmicity and behavioral depression II: effects of lighting schedules. Physiol Behav 48:157-164

Stroebel C (1969) Biological rhythm correlates of disturbed behavior in the rhesus monkey. In: Rohles F (ed) Circadian rhythms in nonhuman primates. Karger, Basel, p 91

Surridge-David M, MacLean AW, Coulter ME, Knowles JB (1987) Mood change following an acute delay of sleep. Psychiatry Res 22:149-158

Szuba MP, Yager A, Guze BH, Allen EM, Baxter LR (1992) Disruption of social circadian rhythms in major depression: a preliminary report. Psychiatry Res 42:221-230

Tamarkin L, Craig C, Garrick N, Wehr T (1983) Effect of clorgyline (a MAO type A inhibitor) on locomotor activity in the Syrian hamster. Am J Physiol 245:R215-R221

Teicher MH, Lawrence JM, Barber NI, Finklestein SP, Lieberman HR, Baldessarini RJ (1988) Increased activity and phase-delay in circadian motility rhythms in geriatric depression. Arch Gen Psychiatry 45:913-917

Teicher MH, Barber MI, Lawrence JM, Baldessarini RJ (1989) Motor activity and antidepressant drugs: a proposed approach to categorizing depression syndromes and their animal models. In: Koob GF, Ehlers CL, Kupfer DJ (eds) Animal models of depression. Birkhauser, Boston, p 135

Terman M, Terman JS, Quitkin FM, Cooper TB, Lo ES, Gorman JM, Stewart JW, McGrath PJ (1988) Response of the melatonin cycle to phototherapy for seasonal affective disorder. J Neural Transm 72:147-165

Terman M, Terman JS, Quitkin FM, McGrath PJ, Stewart JW, Rafferty B (1989) Light therapy for seasonal affective disorder: a review of efficacy. Neuropsychopharmacology 2:1-22

Turek FW (1989) Effects of stimulated physical activity on the circadian pacemaker of vertebrates. J Biol Rhythms 4:135-147

Tsujimoto T, Yamada N, Shimoda K, Hanada K, Takahashi S (1990) Circadian rhythms in depression. Part II: circadian rhythms in inpatients with various mental disorders. J Affect Disord 18:199–210

Vagell ME, McGinnis MY, Possidente BP, Narasimhan VN, Lumia AR (1991) Olfactory bulbectomy increases basal suprachiasmatic cyclic AMP levels in male rats. Brain Res Bull 27:839–842

Van den Hoofdakker RH (1994) Chronobiological theories of nonseasonal affective disorders and their implications for treatment. J Biol Rhythms 9:157–183

Van Reeth O, Turek FW (1989) Stimulated activity mediates phase shifts in the hamster circadian clock induced by dark pulses or benzodiazepines. Nature 339:49–51

Van Reeth O, Sturis J, Byrne MM, Blackman JD, L'Hermite-Baleriaux M, Leproult R, Oliner C, Refetoff S, Turek FW, Van Cauter E (1994) Nocturnal exercise phase delays circadian rhythms of melatonin and thyrotropin secretion in normal men. Am J Physiol 266:E964–E974

Vessotskie JM, McGonigle P, Molthen RC, McEachron DL (1993) Thyroid and thyroxine effects on adrenoceptors in relation to circadian activity. Pharmacol Biochem Behav 46:251–257

Vogel G, Neill D, Hagler M, Kors D (1990) A new animal model of endogenous depression: a summary of present findings. Neurosci Biobehav Rev 14:85–91

von Zerssen D, Barthelms H, Dirlich G et al (1985) Circadian rhythms in endogenous depression. Psychiatry Res 16:51–63

Wehr TA, Goodwin FK (1983) Biological rhythms in manic-depressive illness. In: Wehr TA, Goodwin FK (eds) Circadian rhythms in psychiatry. Boxwood, Pacific Grove, pp 129–184

Wehr TA, Wirz-Justice A (1982) Circadian rhythm mechanisms in affective illness and in antidepressant drug action. Pharmacopsychiatry 15:31–39

Wehr TA, Gillin JC, Goodwin FK (1983) Sleep and circadian rhythms in depression. In: Chase M, Weitzman ED (eds) Sleep disorders: basic and clinical research. Spectrum, New York, p 195

Wehr TA, Sack DA, Duncan WC et al (1985) Sleep and circadian rhythms in affective patients isolated from external time cues. Psychiatry Res 15:327–339

Wehr TA, Moul DE, Barbato G et al (1993) Conservation of photoperiod-responsive mechanisms in humans. Am J Physiol 265:R846–R857

Wever R (1979) The circadian system of man: results of experiments under temporal isolation. Springer, Berlin Heidelberg New York

Wever RA (1989) Light effects on human circadian rhythms: a review of recent Andechs experiments. J Biol Rhythms 4:161–185

Whybrow P, Prange AA (1981) A hypothesis of thyroid-catecholamine receptor interactions: its relevance to affective illness. Arch Gen Psychiatry 38:106–113

Willner P (1984) The validity of animal models of depression. Psychopharmacology 83:1–16

Wirz-Justice A (1995) Biological rhythms in mood disorders. In: Bloom FE, Kupfer, DJ (eds) Psychopharmacology: the fourth generation of progress. Raven, New York, pp 999–1017

Wirz-Justice A, Campbell IC (1982) Antidepressant drugs can slow or dissociate circadian rhythms. Experientia 38:1301–1309

Wirz-Justice A, Groos GA, Wehr TA (1982) The neuropharmacology of circadian timekeeping in mammals. In: Aschoff J, Daan S, Groos G (eds) Vertebrate circadian systems: structure and physiology. Springer, Berlin Heidelberg New York, pp 183–193

Wirz-Justice A, van der Velde P, Bucher A, Nil R (1992) Comparison of light treatment with citalopram in winter depression: a longitudinal single case study. Int Clin Psychopharmacol 7:109–116

Wirz-Justice A, Graw P, Kräuchi K et al (1993a) Light therapy in seasonal affective disorder is independent of time of day or circadian phase. Arch Gen Psychiatry 50:929–937

Wirz-Justice A, Graw P, Kräuchi K, Haug H-J, Leonhardt G, Brunner DP (1993b) Effect of light on unmasked circadian rhythms in winter depression. In: Wetterberg L (ed) Light and biological rhythms in man. Pergamon, Oxford, 385–393

Wirz-Justice A, Kräuchi K, Brunner DP, Graw P, Haug H-J, Leonhardt G, Sarrafzadeh A, English J, Arendt J (1995) Circadian rhythms and sleep regulation in seasonal affective disorder. Acta Neuropsychiatr 7:41–43

Wolff EA III, Putnam FW, Post RM (1985) Motor activity and affective illness. Arch Gen Psychiatry 42:288–294

Wollnik F (1992) Effects of chronic administration and withdrawal of antidepressant agents on circadian activity rhythms in rats. Pharmacol Biochem Behav 43:549–561

Wu JC, Bunney WE Jr (1990) The biological bias of an antidepressant response to sleep deprivation and relapse: review and hypothesis. Am J Psychiatry 147:14–21

Zatz M, Herkenham MA (1981) Intraventricular carbachol mimics the phase-shifting effect of light on the circadian rhythm of wheel-running activity. Brain Res 212:234–238

CHAPTER 18
Chronobiology and Chronopharmacology of the Haemopoietic System

R. SMAALAND

A. Outline of Haemopoiesis

The bone marrow, an extremely complex tissue comprising approximately 4.5% of an adult's body weight (a mass comparable to the liver) (NATHAN 1988), is found in the ends of flat bones (sternum, ribs, skull, vertebrae and innominates) and contains the haemopoietic stem cells, which give rise to the many developing functional blood cell lineages within the marrow spaces. After birth, the bone marrow is the production site for all types of blood cells, which are released through vascular channels into the peripheral blood according to the needs of the body, mediated through different feedback mechanisms. Haemopoiesis is the multi-phase process of cell proliferation and gradual maturation, until the end stage is reached with a population of mature cells that can exert their specialized functions, but are no longer capable of cell proliferation (LAERUM et al. 1989).

The continuous, extremely high proliferative capacity of the bone marrow is rivalled only by the skin and the intestinal mucosa, both of which have been shown to exhibit circadian rhythms in humans (SCHEVING 1959; FISHER 1968; BUCHI et al. 1991). It has been estimated that, every second, up to 2 million red cells (ERSLEV 1983; SPIVAK 1984), 2 million platelets (SPIVAK 1984) and 700 000–800 000 granulocytes are produced in the human bone marrow (DANCEY et al. 1976; Spivak 1984). Thus, being a labile, rapidly proliferating organ system, it is no wonder that haemopoiesis changes with time, both quantitatively and qualitatively.

Until recently little recognition has been paid to possible temporal aspects in either experimental or clinical conditions. Most haematologists have assumed that a bone marrow sample taken at one time point of the day represents a constant organ function. Neither in laboratory animals, nor in man, is this true. On the contrary, both circadian and seasonal variations occur in haemopoiesis. Although this phenomenon is not generally recognized, a large, well-documented literature on this topic is now available (for recent reviews, see LAERUM and SMAALAND 1989; LAERUM et al. 1989; SLETVOLD et al. 1991; HAUS 1992; SMAALAND and SOTHERN 1994).

Accordingly, peripheral blood, while thought of as the mixture of fluid and formed elements which circulate within the cardiovascular system, is also more appropriately considered as a unique organ, with its own anatomy, physiology and developmental history (SPIVAK 1984).

In order to understand the biology of the bone marrow, with its temporal organization, as well as the mechanisms and effects of perturbation of the haematopoietic system, a brief overview of the developmental stages of haemopoiesis is necessary. A scheme for haemopoiesis is given in Fig. 1.

Of the total haemopoietic cells, less than 0.5% are pluripotent stem cell progenitors (1 in 250 to 1 in 1000 bone marrow cells) (SPIVAK 1984). This nondescript primitive cell population is maintained by self-renewal and their multilineage differentiation potential is the fundamental base from which all the major blood lines are derived. A single pluripotent stem cell is capable of giving rise in a seemingly random manner to increasingly committed progenitor cells which are destined to form differentiated, recognizable precursors of the specific types of blood cells: erythroid (red cells), granulocytic-macrophagic (white cells), megakaryocytic (platelets) and lymphoid cell lines (Fig. 1). These "invisible" haemopoietic progenitor cells have no distinguishing morphologic features and can only be identified by functional clonal assays in vivo and in vitro. This definition regards the progenitor cell as a colony-forming unit (CFU), followed by a suffix designating the colony type, e.g. CFU-G (granulocyte), CFU-GM (granulocyte-macrophage). These committed progenitor cells constitute about 3% of the total haematopoietic cells and are transit cells on a suicide maturation pathway, destined to die by differentiation. Although they have some residual self-replicative capacity, this serves only to amplify the population prior to its entering the functional maturation stages. Committed progenitor cells are capable of responding to external humoral influences, such as growth factors, which convey essential regulatory information.

Under normal circumstances it takes approximately 14 days from immature stem cell proliferation onset and 7 days from the myeloblast or pronormoblast stage until mature cells are released into the circulating blood (GORDON et al. 1985). Proliferation of stem cells, i.e. CFU-S, appears to be under the control of competing glycoprotein inhibitory factors (interferons, tumour necrosis factor-α (TNF-α), transforming growth factor-β (TGF-β) and haemoregulatory peptides

---

Fig. 1. Scheme for haemopoiesis: the development of lymphoid, myeloid and erythroid blood cells in the bone marrow. The maturation sequence (1) begins with pluripotent stem cells (with multilineage differentiation potential), giving rise to committed progenitor cells (intermediate classes of cells committed to one or, at most, two lines of cellular development), (2) which leads to morphologically recognizable immature cells belonging to one of several lines of cellular development, (3) which are boosted by rapid proliferation to over 95% of the total cells in the bone marrow and (4) are ultimately released into the circulation. The proliferation and differentiation of the haemopoietic progenitor cells and the function of mature blood cells are regulated by a complex network of hormones, growth factors and cytokines, some of which affect multiple steps in both the early and late stages of maturation. *CFU,* colony-forming units; *CSF,* colony-stimulating factor; *IL,* interleukin. (From SMAALAND and SOTHERN 1994)

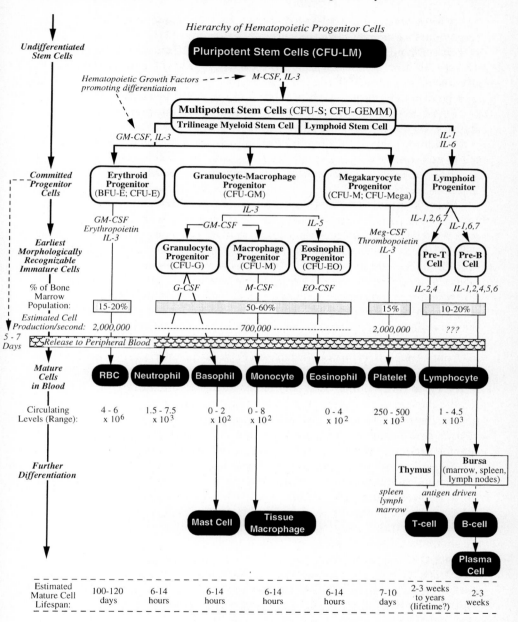

(PAUKOVITS et al. 1990a,b), as well as stimulatory factors [colony-stimulating factors (CSFs), interleukins (IL-1 to IL-13)]; the inhibitory factors block entry to DNA synthesis and accumulate cells in the G1-phase, while the stimulatory factors trigger cells from G1 rapidly into DNA synthesis.

After a series of amplifying divisions, the committed precursor cells undergo a further change when the cells take on the morphological characteristics of their cell type. This may be the result of a further differentiation step, such as the switching on of haemoglobin synthesis by the hormone erythropoietin (a glycoprotein produced in the kidney and to a lesser extent in the liver), or it may be a prolongation of the proliferation sequence throughout the maturation phase by the appropriate growth factor. Proliferation boosts these populations to over 95% of the total cells, and it is on the basis of their number and conditions that, traditionally, the physician has diagnosed disease. These cell populations are morphologically recognizable as belonging to one of several lines of cellular development, fully capable of exerting their specialized functions in the body.

Both the proliferation and differentiation of haemopoietic progenitor cells and the function of mature blood cells are dependent on a large family of glycoprotein hormonal growth factors, known as colony-stimulating factors (CSFs). It has been found that practically all healthy tissues in the body, and some tumour cells, contain measurable, if minute, amounts of one or more types of CSF (METCALF 1986). Cellular sources of all CSFs except erythropoietin are monocytes or macrophages, endothelial and fibroblastic cells, and stimulated T lymphocytes. Indeed, the growth of bone marrow progenitor cells in clonal cultures requires the continuous presence of these growth factors in order to dispose the marrow stem cells to form clones of granulocytes, macrophages and erythroid clusters. CSFs differ in their level of action, with some acting on multipotential cells, as well as on progenitor cells that are restricted to differentiate along any of the lineages. The overlap of target cells for different growth factors is a striking feature of the regulation of haemopoiesis. However, a given CSF may not only be acting directly on the progenitor cells stimulated to proliferate, but also indirectly through regulatory cellular networks, which will result in the production of other growth factors.

Several classes of proteins are involved in haemopoietic cell proliferation, differentiation, programmed cell death (apoptosis), as well as specialized functions. Extracellular growth factors (e.g. G-CSF, GM-CSF, CSF-1) bind to their receptors, which initiates a cascade of intracellular signalling responses, including activation of tyrosine kinases and phosphatases, enhancement of phosphoinositol metabolism, activation of GTP-binding proteins and their cognate GTPase-activating proteins (GAPs) and guanine nucleotide exchange factors (GNEFs), and activation of serine/threonine kinases. This leads to modification (phosphorylation, protein complex formation) of nuclear transcription factors and cell cycle components, producing the altered gene expression and growth regulation required for the appropriate phenotypic response, i.e. proliferation, differentiation, cell death and specialized functions (VARMUS and LOWELL 1994). Thus, although it is not known how a specific response occurs to a specific message received at the cell surface, a strict and dynamic balance must exist between inhibitory and stimulatory effects to keep the haemopoietic system functioning in a controlled manner. All of these intermediate signalling reactions will potentially be a target for endogenous as well as exogenous factors, and will be susceptible to temporal variation in these factors.

It has been proposed that haemopoietic stem cells are in G0 and begin active cell cycling randomly (LAJTHA 1963, 1979). It has also been reasoned that the G0 state provides stem cells with time to repair DNA damage, thus allowing maintenance of the genetic integrity of the stem cell populations. Early- and intermediate-acting lineage-non-specific factors, e.g. cytokines, appear to support the proliferation of multipotential progenitors, but only after they exit from G0 (OGAWA 1993; SUDA et al. 1985). Triggering of cycling by dormant primitive progenitors appears to require interactions of early-acting cytokines, including interleukin-6 (IL-6), G-CSF, IL-11, IL-12, leucocyte inhibition factor (LIF) and stem cell factor (SF).

Available evidence indicates that qualitative changes in haemopoietic stem cells and progenitors, such as the decision of stem cells to self-renew or differentiate, or selection of lineage potentials by the multipotential progenitors during differentiation (commitment), are intrinsic properties of the progenitors and are stochastic in nature. In contrast, proliferative kinetics of the progenitors, namely survival and expansion of the progenitors, appear to be controlled by a number of interacting cytokines.

B. The Haemopoietic System and the Clinician

From the foregoing outline of the haemopoietic process it is obvious that the haemopoietic system is regulated in a very complex manner so that its vital functions can be preserved: (1) production of red blood cells, which are the transport vehicles for oxygen to all tissues; (2) production of white blood cells, which are essential for fighting infections (viral, bacterial, protozoan, etc.); and (3) production of platelets for prevention of haemorrhagic diathesis.

By far the most frequent cause of drug-induced bone marrow hypoplasia is cytotoxic anticancer drugs. However, other drugs or groups of drugs have occasionally been reported to induce bone marrow failure. While most drugs linked to bone marrow failure appear only very occasionally to precede a haematological catastrophe, careful observation may disclose milder, more regular haemopoietic effects, such as mild leucopenia in chronic users of phenothiazines (PISCIOTTA 1969) and the alterations in erythropoiesis due to chloramphenicol (VOLINI et al. 1950). However, aplastic anaemia is a rare complication of chloramphenicol. Benzene is the chemical most convincingly linked to bone marrow failure (FISHBEIN 1984), probably because benzene and its metabolites are potent antimetabolites. Non-steroidal anti-inflammatory drugs have also been reported to induce aplastic anaemia (MIESCHER and POLA 1986); however, the complication occurs extremely rarely. Finally, viruses may interact with drugs (YOUNG and MORTIMER 1984).

Animal experiments provide ample evidence that acute exposure to cytotoxic drugs causes changes in the size of the pluripotent stem cell compartment, with some cytotoxic drugs depleting this compartment considerably, whereas others affect it only slightly (LOHRMANN and SCHREML 1982a). A number of variables are known to influence the pattern of depletion and replenishment of the stem cell compartment, including dose, schedule of administration, and proliferative state of the stem cells at the time of drug administration. Furthermore, the type of cytotoxic drug and its mode of action is also very important. Following exposure to most cytotoxic agents, the size of the stem cell compartment is reduced, but it recovers rapidly to pretreatment size. This is achieved by a transiently increased proliferative activity of those stem cells that survived exposure to the drug. Some cytotoxic drugs, however, such as busulphan and the nitrosoureas, deplete the stem cell compartment for a prolonged period of time. It is conceivable that premature readministration of these drugs before numerical normalization will lead to cumulative toxicity, i.e. to progressive loss of haemopoietic stem cells. However, return of the haemopoietic stem cell compartment to pretreatment size following exposure to a cytotoxic agent does not necessarily indicate a complete restoration, since changes in the age structure of the stem cell population may persist even after normalization of the number of cells. There is, in fact, experimental evidence for the existence of qualitative changes within the pluripotent stem cell compartment after cytotoxic exposure in vivo with a reduced capacity of pluripotent stem cells (CFU-S) for self-replication (BOTNICK et al. 1976; FRIED and BARONE 1980; HELLMAN et al. 1978; SCHOFIELD 1978). This loss of proliferative potential is irreversible: stem cells can lose, but not regain, the capacity for self-replication.

All chemotherapeutic agents affect both normal cells and malignant cells. The ratio of these two effects, or the therapeutic index, determines the toxicity of the drug at effective doses. Scheduling the administration of chemotherapy to maximize tumour reduction while minimizing toxicity to organs has been the focus of research in recent years, including investigation of prolonged drug exposure through infusional therapy. In addition, there is growing recognition that chronobiological phenomena may be important in the effective administration of chemotherapy and that the rhythm of biological activity induced by regular light and dark periods (circadian), by shorter periods (ultradian) and by longer seasonal periods (infradian) may be used to increase the difference in cytotoxicity between normal and malignant cells.

Susceptibility of organs to the effects of anticancer chemotherapy is partially determined by the rate of cell division, because the activity of many of these drugs depends on this activity. Peripheral blood cell depression is the chief measurable effect and reflects the suppression of bone marrow cells by chemotherapeutic drugs. The effect of various agents on particular cell lines varies, but is to a large extent predictable. The degree of suppression of the major cell lines, erythrocytes, white blood cells and platelets, is determined by the different effects on the precursor cells (stem cells) and the kinetics of the cell line in the peripheral blood compartment. Anaemia occurs as a late effect because of the half-life of red blood cells (120 days), thrombocytopenia appears in an intermediate time frame (platelet half-life 5–7 days) and granulocytopenia appears earliest (granulocyte half-life of 4–6 h).

Two major approaches to overcoming myelotoxicity have been developed recently. First, changes in administration rate and duration may change the therapeutic index of chemotherapeutic agents. Second, the use of growth factors to shorten the duration of myelosuppression has offered a technological solution to the toxicity of chemotherapy through recombinant DNA techniques.

Similar doses of chemotherapeutic agents may suppress the bone marrow to different extents, depending on the mode of action, the cycle in which the cells are affected or the area under the concentration curve of the chemotherapeutic agents, which depends on the metabolism of the drug. The dose-limiting sensitivity of bone marrow to cytotoxic anticancer drugs represents a general and daily problem to oncologists deciding on dose intensity in the treatment of cancer patients. Doses of the drugs may be altered or cycles withheld on the basis of the degree of myelosuppression from the previous cycle and the current organ function. Prediction of toxicity is imperfect because of imprecise knowledge of the complex mechanism of action of many drugs.

There is limited evidence for the existence of persistent cytotoxic drug-induced damage to the haemopoietic system in humans, especially to the myelopoietic cells, comprising 50%–60% of the maturing bone marrow population (erythropoiesis comprises 15%–20%, while lymphopoiesis may account for 10%–20% of the cells) (WINTROBE 1981). Normally, the haemopoietic system recovers rather promptly and predictably following cytotoxic drug therapy against haematological malignancies and solid tumours. However, it is well known to oncologists that the haemopoietic system of heavily pretreated patients is often unusually sensitive to further administration of cytotoxic agents.

The potentially long lasting cytotoxic effects on the bone marrow are of particular concern in the clinical setting of adjuvant chemotherapy. Cytotoxic agents are administered more and more in routine treatment following surgery in an attempt to eradicate clinically undetectable micrometastatic disease, and many of these patients so treated become long-term survivors. It is known that these patients are particularly at risk of developing long-term damage, i.e. persistent chromosomal aberrations, irreversible damage to various organs and increased risk of secondary acute leukaemias, following exposure to cytotoxic drugs. However, in addition to these well-characterized sequelae of cytotoxic drug treatment, the possibility of long-term damage to self-renewal tissues has only recently gained attention. This is a highly relevant concern as a significant proportion of these patients undergoing adjuvant therapy will ultimately have a relapse of their malignant disease. A damaged bone marrow with a reduced proliferative capacity will pose a serious problem confronting additional cytotoxic therapy in an attempt to eradicate gross metastatic cancer.

Over the past 35 years a critical mass of data has emerged indicating that the timing within the day of anticancer cytotoxic agents and more recently protein and peptide cytokines is highly relevant to their toxic-therapeutic ratios. This chapter will concentrate on how chronobiology and chronopharmacology can help to reduce side effects to the bone marrow and thereby possibly increase the therapeutic effect.

C. Circadian Aspects of the Haemopoietic System

Since myelosuppression is a frequent and clinically important consequence of antineoplastic drugs, there is a need to reduce bone marrow toxicity. In seeking to achieve this goal, time-dependent aspects of toxicity have generally been neglected when approaching the problem of increasing the therapeutic index.

Until recently, little has been known about biological rhythms of different aspects of the human bone marrow. It may not therefore be surprising to see the systematic neglect of consideration of biological rhythms in the haematological literature. The homeostatic notion that production rates precisely equal destruction rates is so pervasive that several large surveys on haemopoiesis have not even considered time-dependent variations (GORDON et al. 1985; LOHRMANN and SCHREML 1982b). Since the production and migration of mature granulocytes into peripheral blood are dependent both on the actual needs of the body and on several hormonal and regulatory factors (reactive homeostasis) (ARENDT et al. 1989), the proliferative pool in the marrow may vary considerably from time to time (BUTCHER 1990). For example, physical exercise and cortisone/cortisol are strong mobilizers of granulocytes, as are acute bacterial infections. In addition, however, there are strong endogenous rhythmic variations in these proliferative and mobilizing processes, which further complicate the picture, but are also a part of the organism's homeostasis (predictive homeostasis) (ARENDT et al. 1989).

Therefore, as new information related to temporal aspects of proliferation has emerged, it has become increasingly clear that haemopoiesis is not a temporally constant phenomenon. This can be taken advantage of in several ways, both in relation to cytotoxic side effects of cancer therapy and in relation to primary and other secondary diseases of the bone marrow, such as aplastic anaemia, myelodysplastic conditions, AIDS and other immunodeficiency diseases.

Furthermore, the existence of interindividual differences in circadian time structure implies that monitoring physiological marker variables could be of benefit in allowing individualized chronotherapy.

I. Circadian Aspects of Proliferative Activity in Murine Bone Marrow

Synchronous circadian gating of DNA synthesis and cell proliferation in rodent model systems are described extensively in the literature (SCHEVING 1984; SCHEVING et al. 1991). It began with measurements of mitotic index and later radiolabelling, and has continued to the present using flow cytometry (CLAUSEN et al. 1979).

In the murine system, several laboratories have independently reported circadian variations of different aspects of haemopoiesis as well as in peripheral blood (AARDAL and LAERUM 1983; LAERUM et al. 1988). In addition, seasonal variations are documented, even in mice that have been living under complete isolation during their whole lifetime (LAERUM et al. 1988).

Several studies in mice have demonstrated circadian variation in the proliferation of total bone marrow cells or specific subpopulations by measurement of DNA synthesis and/or mitotic index/activity or duration of mitosis, by using [^{3}H]TdR labelling, percentage of labelled mitoses or flow cytometry (PIZZARELLO and WITCOFSKI 1970; SHARKIS et al. 1971, 1974; MOSKALIK 1976; SCHEVING et al. 1978; BURNS 1981). For example, studying erythropoiesis, DÖRMER et al. (1970) reported that DNA synthesis in mice underwent circadian variation, with the acrophase during the dark (activity) period. In addition, with the introduction of assay methods for the various classes of stem and progenitor cells, data have become available about stem cell proliferation of animals. Several authors have demonstrated circadian and seasonal variations of multipotent and committed stem cells in mice (STONEY et al. 1975; LAERUM and AARDAL 1981; AARDAL et al. 1982; AARDAL 1984; BARTLETT et al. 1984). For instance, both CFU-S and CFU-GM have been reported to undergo circadian variation (STONEY et al. 1975; AARDAL and LAERUM 1983; BARTLETT et al. 1984; LÉVI et al. 1988b; SLETVOLD and LAERUM 1988; SLETVOLD et al. 1988a).

However, although large differences in proliferation according to circadian time have been found, the results have not been consistent. An explanation for this variation may be related to investigation of different species, interstrain differences, animals of different age and sex, different lighting schedules, and possibly infradian variations in rhythm characteristics [mesor (rhythm-adjusted average), amplitude (extent of predictable change above or below the mesor) and/or acrophase (peak of a fitted curve indicating location of high values)] at different times of the year.

We traced 40 reports on circadian variation of bone-marrow-related proliferative parameters published between 1969 and 1992 which have previously been reported by us (SMAALAND and SOTHERN 1994) and which include reports on DNA synthesis (S-phase), mitoses, mitotic index and CFU (various sites and all cell types).

It was found that in six out of six studies the DNA synthesis (S-phase) was highest during the activity (dark) span in mice. On the other hand, the maximal number of mitoses and highest mitotic index were generally found during the second half of the resting phase, i.e. almost 12 h removed from the location of peak DNA synthesis. Furthermore, most of the 14 studies relating to circadian variation of colony-forming units had their acrophase late in the activity span or during the rest span, although the acrophase was found to be in the activity phase in several of the studies, especially for the myeloid lineage.

Overall it seems that the proliferative activity in haemopoiesis estimated as DNA synthesis has a peak during the animals' active period and a trough during the period of rest. On the basis of these data it may be inferred that S-phase-specific drugs or drugs with their major effect on cells in S-phase would be least toxic to the bone marrow of rats and mice when administered during the resting (light) part of the 24-h period. Accordingly, least toxicity may also be achieved for M-phase-specific drugs if they are administered during the daily activity (dark) span. Reports on murine bone marrow circadian proliferative activity are summarized in Fig. 2. However, extrapolations in timing from mice to men cannot be based on the assumption of a 12-h time difference between diurnally active humans and nocturnal rodents.

It should be kept in mind that age-related changes in physiological variations, including haemopoiesis, may be a modifying factor influencing the circadian stage-dependent effect of cytotoxic therapy. Reports on haemopoiesis in aging mice are inconsistent, however, with regard to both multipotent stem cells (CFU-S) (HARRISON 1979; WILLIAMS et al. 1986; SLETVOLD and LAERUM 1988) and myelopoietic progenitor cells (CFU-GM) (AKAGAWA et al. 1984; METCALF and STEVENS 1972). SLETVOLD et al. (1988a) demonstrated that both the 24-h mean values and the amplitudes of the circadian rhythm declined in aging mice. In addition, a phase shift of the CFU-GM peak was observed, indicating that equal timing of chemotherapy in young and old mice might have different effects. Phase differences between young and aged animals have also been observed by others (STONEY et al. 1975; BARTLETT et al. 1984; SLETVOLD 1987; LÉVI et al. 1988b; SLETVOLD and LAERUM 1988; SLETVOLD et al. 1988a,b, 1991). Therefore, direct comparison between old and young animals at only one time point may lead to erroneous conclusions.

Is there synchrony or are there phase differences between different bone marrow cell lines? In the mouse, significant differences were found in the timing of the circadian rhythms in the proliferation of erythroid and granulocyte-macrophage colonies in soft agar cultures (BARTLETT et al. 1984; HAUS et al. 1983). AARDAL and LAERUM (1983) have, however, reported partial synchronization of circadian rhythms of CFU-S and CFU-GM.

SCHEVING et al. (1991; SCHEVING 1984), summarizing the work of many years in rodents, have simply stated that the peak in DNA synthesis for most tissues occurs somewhere between the middle and the end of the dark (activity) phase, i.e. activity phase, while mitosis occurs a few hours later. This is to a large extent in accordance with our literature review of the bone marrow. Scheving has also pointed to the large literature showing that the duration of

Fig. 2. Peak times of circadian variation in murine bone marrow DNA synthesis (S-phase) and colony-forming units (*CFU*). The 95% confidence interval (CI) for the peak of the best-fitting 24-h cosine (*closed dots,* acrophase) can be compared with the actual peak in each time series (*open squares,* macrophase). Following overall summary of acrophases by population mean cosinor, highest values for DNA occur during the animal's activity (dark) span (95% CI = 12:40–22:36 HALO), while highest values for CFU occur late in activity or during the daily rest (light) span (95% CI = 21:48–06:40 HALO). (From SMAALAND and SOTHERN 1994)

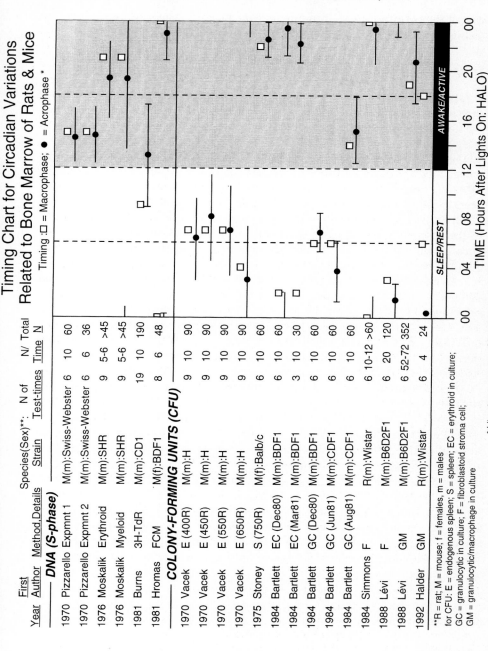

the cell cycle and substages determined from fraction-of-labelled-mitosis curves also varies depending upon when during the day the label was added.

II. Circadian Stage-Dependent Cytotoxicity of Murine Bone Marrow

Temporal sensitivity-resistance cycles of haemopoietic cells to agents used in the treatment of human malignancies, including leukaemias, have been demonstrated in a large number of experimental studies. The results suggest that an adaption of the treatment schedule to the circadian variation in susceptibility of the host to the drug can minimize the undesirable side effects of the treatment to a large degree. In such experimental models, the importance of timing of antileukaemic treatment for survival of the animals and curative effect has been widely documented (HAUS et al. 1972; SCHEVING et al. 1980; HAUS et al. 1988; SCHEVING et al. 1989). Chronotherapeutic treatment schedules adjusted to the susceptibility-resistance cycles of the host increased the survival of leukaemic animals, as compared to the effect of treatment schedules established without regard to the organism's time structure.

As an example, BOUGHATTAS et al. (1989) have investigated the toxic effects of a new platinum analogue, oxaliplatin, on different tissues, including the bone marrow, determining the concentrations of this drug in 18 different tissues. It was found that despite the fact that the highest platinum concentrations in tissues usually corresponded to drug dosing at 8 HALO (hours after light onset), no correlation was documented between such variables and tissue toxicity. Thus, this study demonstrates that tissue pharmacokinetics may only in part contribute to the circadian rhythm in haematological and jejunal toxicity of a drug.

LÉVI et al. have demonstrated that in vivo administration of the anticancer drug 4′-0-tetrahydropyranyl (THP) Adriamycin was best tolerated when given at 7–10 HALO. This corresponded to the time of lowest proliferative activity of the bone marrow, as measured by DNA synthesis of myeloid progenitor cells and by the number of myeloid progenitor cell colonies (CFU-GM) (LÉVI et al. 1988b).

Also chronopharmacological aspects are important to the circadian variation in bone marrow toxicity. In the laboratory it has been possible to relate directly circadian-stage-dependent changes in the pharmacokinetics of methotrexate with corresponding changes in bone marrow toxicity measured by the fall in peripheral white cell count (WBC) (AHERNE 1989).

Interferons (α, β and γ) are biological response modifiers that have antiviral and antitumour activity. Interferon-α (IFN-α) is employed in a number of countries for the treatment of malignant and viral diseases. In addition to their antitumour and antiviral activity, interferons cause undesirable side effects even when employed at low dosage levels. These adverse effects include an influenza-like syndrome, neurotoxicty and haematological toxicity with a clinically significant reduction of peripheral WBC counts (FENT and ZBINDEN 1987). Interferon-induced peripheral WBC suppression has been studied as a function of the time of recombinant human IFN-αA/D administration in a mouse model (KOREN and FLEISCHMANN 1993a). Administration of rhIFN-αA/D at the end of the activity phase caused the greatest suppressive effect, while administration of IFN-αA/D during the middle of the rest phase caused the least suppressive effect. In fact, a tenfold difference in toxicity to rhIFN-αA/D between these two circadian time points was found.

The same authors found that recombinant murine INF-γ (rMuIFN-γ) induced a WBC-suppressive effect that varied in its intensity in a cyclical manner. This differential sensitivity to the WBC-suppressive effects of rMuIFN-γ was found to be a general effect, occurring throughout the circadian cycle. Mice treated with rMuIFN-γ at 14 HALO were found to be about 20-fold less sensitive to the peripheral WBC-suppressive effects of rMuIFN-γ than mice treated at 4 HALO (KOREN and FLEISCHMANN 1993b).

III. Circadian and Circannual Proliferative Activity in Healthy Human Bone Marrow

Until recently few data have been available concerning temporal variations of parameters in the human bone marrow, even though as early as 1948 GOLDECK showed a marked circadian variation in reticulocytes sampled in the bone marrow and blood at three times of the day (GOLDECK 1948; GOLDECK and SIEGEL 1948). Stem cells and progenitor cells are present in the circulation in small numbers, and therefore they have been subject to investigation by sampling peripheral blood. LASKY et al. (1983) measured circulating uncommitted pluripotential precursors (CFU-GEMM) and committed precursors (CFU-GM and BFU-E) at two times of the day (1600 hours and 0800 hours). The greatest number of colonies was found late in the day (as compared to 0800 hours); values at 1600 hours were 83% higher for CFU-GEMM, 84% higher for CFU-GM and 130% higher for BFU-E. In a subsequent study of 45 subjects sampled at 0800 hours and 1500 hours these authors reported that CFU-GM was significantly higher in the afternoon (MORRA et al. 1984), corroborating the earlier findings. In studies by Ross et al. (1980) and VERMA et al. (1980), a significant circadian variation was shown in CFU-C (myeloid progenitor cells) in peripheral blood sampled at several different times of the day. However, only the study by Ross et al. had samples through the night as well, with highest values early in the day (acrophase = 0904 hours).

In 1962 and 1965, two limited studies were published in which DNA synthesis (by use of the [^{3}H]TdR technique) and mitotic index of the human bone marrow were measured at different circadian stages (KILLMANN et al. 1962; MAUER 1965). By sampling bone marrow at four different times (0600 hours and 1800 hours, noon and midnight) in four different individuals during a 42-h period, MAUER (1965) found that the [^{3}H]TdR-labelled cells of the myeloid lineage were clearly higher during the day (at 0600 hours and noon) as compared to midnight in three of four individuals, and with a trend towards lower DNA synthesis at midnight in the fourth individual. However,

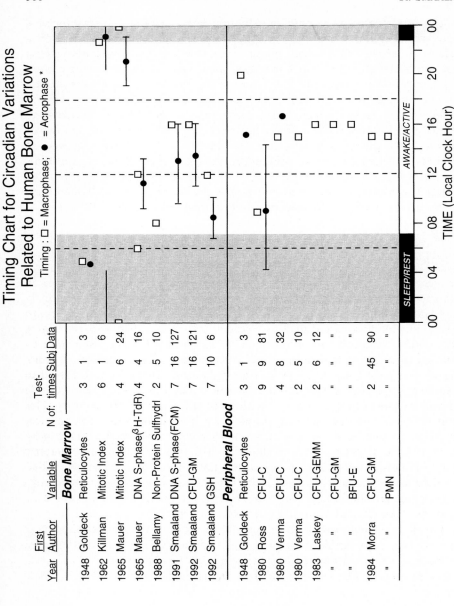

he was unable to find any circadian variation of the [^{3}H]TdR-labelling index of erythropoiesis. The mitotic index was found to be highest at 1800 hours or at midnight and lowest at 0600 hours in five of six individuals. Our single cosinor analysis of these data computed the acrophase for DNA synthesis at 1121 hours, and at 2109 hours for the mitotic index (SMAALAND and SOTHERN 1994). Thus, the time of the greatest percentage of cells incorporating the labelled thymidine was computed to precede the time of the greatest number of mitotic figures by 10 h. KILLMAN et al. made corresponding observations in one human volunteer in an earlier study with regard to mitotic indices, demonstrating an increase in this proliferative parameter during the day until late evening (KILLMANN et al. 1962).

Our review of the literature on circadian variations related to human bone marrow is summarized as a timing chart of macrophases and acrophases in Fig. 3.

In our own laboratory, we have for several years been performing studies on temporal variations of the human bone marrow in order to obtain more extensive data on time-dependent variations in haemopoiesis in the human. We have carried out a temporal mapping of the different parts of haemopoiesis in about 30 men, of which data from 21 have so far been published.

Serial sampling of bone marrow from human volunteers was carried out to investigate circadian variations of cell cycle distribution of nucleated cells in bone marrow by using flow cytometry (FCM). In addition, we have cultured myeloid progenitor cells in a CFU-GM assay to measure a possible circadian stage dependence in proliferative activity of these essential cells with regard to defence against infection. The cell cycle distribution of bone marrow cells from 16 healthy male volunteers (mean age 33.7 years, range 19–47 years) was investigated, 5 of them undergoing the sampling procedure twice, making altogether twenty-one 24-h periods. Bone marrow was sampled seven times by puncturing the sternum and anterior iliac crests in a randomized sequence every 4 h during a single 24-h period (SMAALAND et al. 1991a). Venous blood was obtained from the same subjects at the same time as bone marrow sampling to determine peripheral blood parameters, including total and differential blood cell counts, in addition to cortisol measurements.

The mean S-phase value of the 24-h sampling period varied from 10.9% to 16.6% for the different individuals, i.e. a difference of 52.3%. However, more importantly, the percentage of bone marrow cells synthesizing DNA at each time point showed a large variation along the circadian axis for all twenty-one

Fig. 3. Circadian phase chart showing peak times for temporal variations of human bone marrow derived cells. Single cosinor-derived acrophase (*closed dots*) and a 95% confidence interval (CI) for the peak of the best-fitting 24-h cosine are shown if three or more time points were available for analysis. The actual peak in each time series (macrophase) is represented by an *open square*. DNA and CFU are maximal during the daily activity span, while mitoses are greatest later in the evening. (From SMAALAND and SOTHERN 1994)

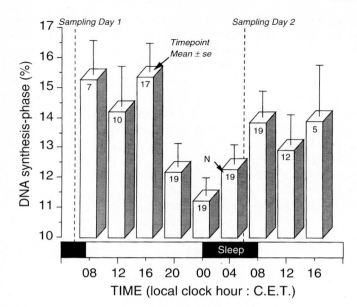

Fig. 4. Circadian variation in human bone marrow DNA synthesis seen in pooled data from 16 diurnally active, clinically healthy men. Mean values vs. original times of collection (total $n = 127$, five to seven samples/profile). Timepoint means and standard errors derived from nineteen 24-h series. Due to different sampling start times of 0800, 1200 and 1600 hours, the pooled data cover two consecutive day-periods, corroborating the pattern of DNA synthesis values measured during daytime. Lowest values are found at midnight, with the difference between lowest and highest values smaller than found for individual subjects due to different phasing between subjects. (From SMAALAND et al. 1991a)

24-h periods. The range of change from lowest to highest value during the 24-h period for each subject varied between 29% and 339%, with a mean difference of 118%. The mean values of the periods with lowest and highest S-phase were $8.9\% \pm 0.5\%$ (SE) and $17.6\% \pm 0.6\%$, respectively, i.e. a difference of nearly 100%.

Lowest values were always seen between 2000 hours and 0400 hours (late evening to midsleep). When pooling the data for all subjects for the mean S-phase values, a consistent pattern was observed, with a lower DNA synthesis around midnight as compared to during the day (Fig. 4). Due to different phasing between the subjects, the difference between the lowest and highest average value is smaller than that for the individual subjects. However, the circadian stage-dependent variation for the pooled data is statistically significant, analysed both by analysis of variance (ANOVA) and cosinor (NELSON et al. 1979) methods; $P = 0.018$ and $P = 0.016$, respectively. The time of highest DNA synthesis (acrophase) estimated by cosinor analysis was 1316 hours.

DNA synthesis may potentially vary as a function of the location in the bone marrow from which the sample is taken. However, conceptually one would regard the total red bone marrow as one organ or tissue, being affected by the same endogenous physiological and hormonal factors, making site-dependent variations of less importance. This is supported by previously reported data by Dosik et al. (1980), who demonstrated a very close correlation between DNA synthesis in bone marrow samples obtained simultaneously by biopsies from right and left iliac crests. A good reproducibility, although with larger individual variations, was also demonstrated by simultaneous bilateral aspirations. In our study, sampling was also done from sternum, in addition to the left and right iliac crests. In agreement with the results of Dosik et al. (1980), we found no statistical difference in the S-phase between the left and right iliac crests, being $12.1\% \pm 0.5\%$ and $11.6\% \pm 0.5\%$, respectively ($P=0.47$). A significantly higher S-phase for samples taken from the sternum ($n=52$), as compared to the iliac crests ($n=89$), was found: $14.6\% \pm 0.5\%$ vs. $12.6\% \pm 0.3\%$, respectively ($P=0.001$). However, DNA synthesis in cells from samples of both sternum and iliac crests demonstrated the same circadian pattern. This finding rules out the possibility of different sampling sites being the reason for the observed circadian stage dependence of DNA synthesis and contradicts the possibility that the overall circadian rhythm detected could be attributed to a difference in level of S-phase dependent on sampling site. Also, the finding of no significant difference in mean value of DNA synthesis in samples of the left and right iliac crests strongly indicates that the total red bone marrow must be looked upon as a functional entity. The demonstration of the same circadian variation in the bone marrow of the sternum and in the iliac crests further corroborates this functional homogeneity.

When we compared the level of the stress-related hormone cortisol at the start and at the end of the sampling procedure, i.e. 24 h apart, no statistically significant difference was observed ($P=0.98$). Neither did we find any statistically significant difference in level of DNA synthesis 24 h apart ($P=0.16$). These findings negate the possibility of a stress-induced circadian rhythm of DNA synthesis.

Thus, our studies demonstrating a circadian stage-dependent variation in the proliferative activity of total bone marrow cells, i.e. DNA synthesis, corroborate the earlier findings of Mauer (1965) in myeloid cells. A crucial observation is that the circadian variation of the myeloid progenitor cells (CFU-GM) demonstrated the same circadian phasing as that for DNA synthesis in the total bone marrow nucleated population (Smaaland et al. 1992b) (Fig. 5).

At the present time we are studying subpopulations in the human bone marrow by use of multiparameter flow cytometric analysis (Lund-Johansen et al. 1990; Abrahamsen et al. 1995) in order to see if this pattern is reflected in all lineages as well as in different stages of maturation. Preliminary data have verified that the various subpopulations of myelo- and erythropoiesis undergo similar variations, although their phasing may be slightly different (Abrahamsen et al. 1994) (Fig. 6).

In addition, by analysing the DNA synthetic activity according to time of year, we found a highly significant circannual rhythm in S-phase with an ac-

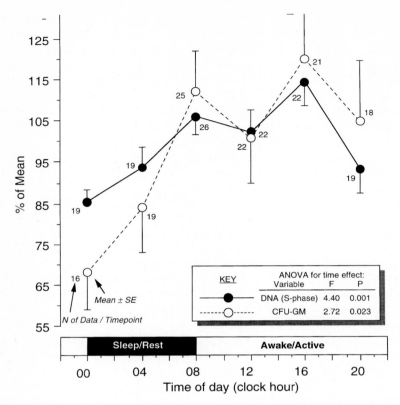

Fig. 5. Circadian covariation in human bone marrow DNA synthesis and colony-forming units–granulocytes/macrophages (CFU-GM) in healthy men suggests a common circadian phasing of both myeloid progenitor cells and more mature proliferating cells. Both indices of bone marrow proliferation are lowest at midnight and show peak values during the day (acrophase = 1225 and 1333 hours for DNA and CFU-GM, respectively). (From Smaaland et al. 1992b)

rophase in late summer, and an amplitude similar to that of the circadian rhythm (Sothern 1994).

IV. Circadian Variation in Bone Marrow DNA Synthesis in Cancer Patients

A major question is to what extent are disease-related alterations of a patient's temporal structure compatible with the chronotherapeutic model provided by studies on healthy humans? In order to find out whether these results in healthy subjects were also valid in cancer patients, in whom the circadian rhythmicity might be disturbed due to the malignant disease, we performed a study of 15 patients (6 women, 9 men; mean age 49.4 years, range 27–70 years) with various malignancies. All patients had a regular diurnal rest-activity schedule

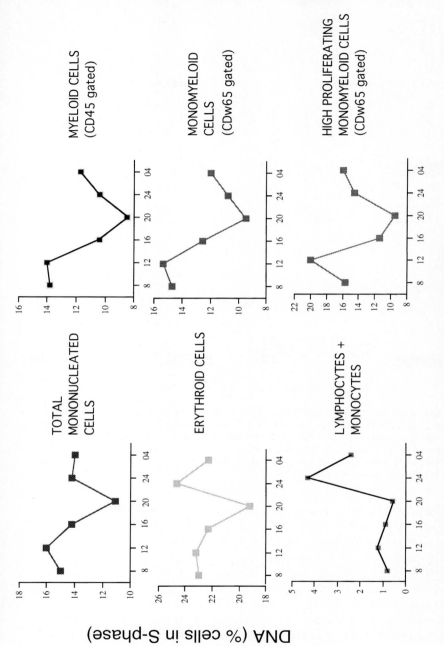

Fig. 6. Different subpopulations of the bone marrow in a healthy person demonstrating some phase difference between cell subpopulations. A drop in the S-phase of the myeloid cells occurs before the drop in the erythroid (and lymphoid) cells

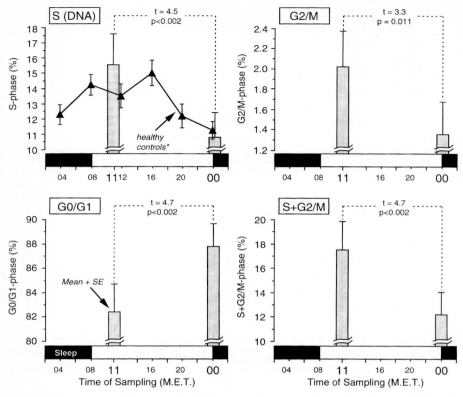

Fig. 7. Cell cycle distribution (G0/G1, S, G2/M and S+G2/M phases) according to circadian stage (1100 hours vs. midnight) for cancer patients with normal cortisol pattern and no bone marrow infiltration (six men and three women with various malignancies). The circadian variation in S-phase of bone marrow cells in these cancer patients suggests a circadian pattern similar to that of healthy male subjects sampled every 4 h and shown for reference. (From SMAALAND et al. 1992a)

for at least 3 weeks prior to bone marrow sampling at two different times of the day, i.e. at 1100 hours and midnight. A significantly higher fraction of cells in S-phase and G2/M-phase was found during the day than at midnight when excluding patients with an abnormal circadian variation in cortisol (SMAALAND et al. 1992a) (Fig. 7). Thus, these data indicate that patients that are not too afflicted with disease do have a circadian variation in DNA synthesis which corresponds to that in healthy subjects.

Recently KLEVECZ and BRALY (1994) have performed repeated sampling of the bone marrow of ovarian cancer patients undergoing resection for their disease. The halogenated pyrimidine bromodeoxyuridine (BUDR) was injected before the start of the operative procedure, and thus BUDR incorporation into proliferative cells of the bone marrow of these same patients could be followed in serial bone marrow aspirates during the day. Preliminary results obtained by combining a number of samples from six patients who underwent surgery at different times of day indicated that the fraction of cells incorporating BUDR declined rapidly through the morning hours to a minimum in the late morning, and with low values at 2000 hours and 2200 hours. BUDR infused at 2000 hours gave a very low labelled fraction. This study using an agent that is incorporated when the cell is actively synthesizing DNA corroborates the pattern of circadian variation in DNA synthesis in our studies of healthy volunteers and cancer patients measured by conventional flow cytometry, and thus contradicts the notion that cells found to be in S-phase as measured by one-dimensional flow cytometry may not be actively synthesizing DNA.

D. Regulation of Haemopoietic Circadian Rhythms

Generally, studies on rhythmic variations in haemopoiesis have been descriptive. Very little is known about regulatory aspects underlying these phenomena, although the regulation of erythropoiesis seems to follow a rhythmic pattern through the circadian variation of erythropoietin (WIDE et al. 1989; WOOD et al. 1990). A temporal relation of DNA synthesis in marrow to cortisol rhythm in peripheral blood with a certain phase difference must be considered as a descriptive pattern and no proof of a causal relation (SMAALAND et al. 1991b). Accumulating data indicate, however, that growth factors may be involved in the temporal regulation of proliferative activity for different types of tissues, including haemopoietic tissue. For instance, the naturally occurring opioid peptide enkephalin (opioid growth factor) inhibits DNA synthesis in mouse epithelium in a circadian-rhythm-dependent manner (ZAGON et al. 1994).

Evidence has been presented in the literature indicating that the pineal gland has a physiological role in the control of proliferation of granulocyte/macrophage colony-forming unit (CFU-GM). By pinealectomy and administration of melatonin to pinealectomized rats, the rhythm of CFU-GM was obliterated or changed. Thus, the pineal gland or its main hormone melatonin seems to have a regulatory role in the proliferation of CFU-GM in rat bone marrow cultures (HALDAR et al. 1992).

In accordance with this, PERPOINT et al. (1994) have reported that the responsiveness of mouse CFU-GM to different CSFs varies with a circadian pattern. They found that the femoral myeloid progenitors had a peak of responsiveness to recombinant mouse IL-3, mouse GM-CSF and human G-CSF around 3 HALO, corresponding to their early rest span. This was independent of both CSF type and dose. Also, in the absence of exogenously added CSF, the numbers of clusters varied with a circadian pattern, and with a similar peak at 3 HALO. Their conclusion was that proliferation, circulation and functions of haemopoietic cells are under circadian control. These findings also indicate that the circadian variations of haemopoiesis are not a passive, adaptive process, but actively regulated at the cellular level.

A phase adaption of haematological rhythms has been reported in human subjects after changes in the sleep/wakefulness pattern (SHARP 1960) and the activity/rest cycle as encountered in shift workers or in subjects exposed to transmeridian flights over several time zones. It is noteworthy that by changing the activity/rest cycle to the opposite circadian pattern, it may take more than 3 weeks for important physiological factors (e.g. cortisol, lymphocytes) to adapt to the new pattern (HAUS 1992).

E. Chronopharmacological Aspects of the Haemopoietic System

Dosing-time dependency in the effects of drugs is the basis for chronotherapeutics. However, the determination of dosing-time dependencies is usually based on results obtained as group phenomena involving populations of laboratory animals and/or human subjects (Table 1). Thus, it is assumed that in a given species all subject and even all the biological circadian rhythms of a given subject are synchronized and that, correspondingly, the chron-

Table 1 Chronopharmacological studies of bone marrow cytotoxic anticancer agents – experimental (E) and clinical (C) studies

Alkylating agents	Cyclophosphamide (E)
	Ifosfamide (E)
	Melphalan (E)
	Peptichemio (E)
	Threosulfan (E)
	Mitomycin C (E)
Anthracyclines and other intercalating agents	Doxorubicin (E, C)
	Daunorubicin (E)
	4'-Epidoxorubicin (E)
	Tetrahydropyranyl doxorubicin (THP-Dox) (E, C)
	Actinomycin D
Vinca alkaloids and epipodophyllotoxins	Vincristine (E)
	Vinblastine (E)
	VP-16 (E, C)
Fluoropyrimidines and DNA synthesis inhibitors	5-Fluorouracil (5-FU) (E, C)
	Fluodeoxyuridine (E, C)
	Methotrexate (E, C)
	Arabinocylcytosine (Ara-C)
Platinum analogs	Carboplatin (E, C)
	Cisplatin (E, C)
	Oxaliplatin (E, C)
Biological response modifiers	Interferon-α (E, C)
	Interferon-γ (E)
	Tumour necrosis factor (TNF) (E)

opharmacological effects of a given drug will be highly predictable. It is, however, obvious that subjects are exposed to different synchronizing regimens and that there will be interindividual differences in the phasing of pharmacokinetics and pharmacodynamics.

In this context it is important to recognize that nocturnal rodents have an almost inverted chronotoxicity curve as compared to humans. All of the drugs tested and found to be not-overly-toxic and efficacious to rodents would be those that were non-toxic in the light phase of the light-dark cycle (the rodent's sleep phase) (KLEVECZ and BRALY 1994). This implies that many potentially effective drugs may be discarded in phase I trials due to extensive toxicity, e.g. bone marrow toxicity, because they have been tested at a circadian time when normal human tissue proliferative activity was at its peak!

In experimental studies strain-related differences have been reported in chronotoxicity studies for several anticancer agents. Therefore, the genetic aspect of chronotolerance must be taken into account when selecting an animal model as a guide for devising clinical chronotherapeutics. From a practical point of view, one cannot assume an identical circadian pattern of healthy persons and patients, even if synchronizers are present. As a consequence, chronopharmacological effects of medications may differ in a given individual from those expected with reference to group phenomena. Gender-related differences represent an additional aspect of genetic-related differences, which may be very striking. Finally, age-related differences should be taken into account.

Significant circadian variations in the pharmacokinetics and pharmacodynamics of several classes of drugs have now been documented and it is clear that elimination and metabolic processes are not equally effective over 24 h. It should not be assumed therefore that a particular dose of an anticancer drug will be equally effective throughout the day. The magnitude of the circadian differences in drug handling will not be the same for all anticancer drugs and is determined by the amplitude of the circadian variation of the main mode of elimination, e.g. liver metabolism and renal excretion, for that drug. For example, circadian rhythms have been observed in a number of aspects of hepatic microsomal drug metabolism in the golden hamster (LAKE et al. 1976). Also, significant circadian variations in the activity of the cytochrome P-450 system (HAEN and GOLLY 1986) and in dihydropyrimidine dehydrogenase (HARRIS et al. 1988) have been reported. The concentration of the anticancer drug in plasma will thus be determined by these factors, as well as by transport of drug into cells (LESZCZYNSKA-BISSWANGER and PFULF 1985) and by variations in target cell susceptibility, i.e. number and/or quality of receptors, membrane function, etc.

Glutathione and its related enzymes play an important role in the detoxification of xenobiotics. Concentrations of glutathione display significant circadian fluctuations in several tissues including the liver (FAROOQI and AHMED 1984). We have also demonstrated a small but significant circadian variation in glutathione content in human bone marrow (SMAALAND et al. 1991c). In addition, the low level of glutathione content in the human bone marrow may in part contribute to the low tolerance for cytotoxic drugs of this tissue (SMAALAND et al. 1991c).

The extent to which drugs are bound by plasma proteins also determines to some extent the pharmacodynamic effects, since effective blood concentrations and the availability of drug at the end organ receptor are altered. One should therefore be aware that plasma protein concentrations display a significant rhythm (RENBOURN 1947; ANGELI et al. 1978) reflected in circadian rhythms of plasma binding of, for example, cisplatin (HECQUET et al. 1985) and prednisolone (ENGLISH et al. 1983) in humans.

Optimization of cancer chemotherapy by chronopharmacological principles must, however, ultimately include circadian differences in cell proliferation of normal versus neoplastic tissues, since these circadian variations in DNA synthesis are important determinants of therapeutic outcome (BURNS and BELAND 1983).

If optimum timing is clinically inconvenient, it may be possible to manipulate or modulate circadian rhythms to improve the therapeutic index of drugs. It has been shown that the toxicity of methotrexate in the rat can be modulated by both exogenous corticosteroids and melatonin (ENGLISH et al. 1987). In the rat it was initially found that greatest toxicity following methotrexate administration occurred at the time of greatest drug exposure, which corresponded to the nadir in endogenous corticosterone (i.e. late activity span). In a subsequent study the corticosterone level was kept high by exogenous administration of this steroid and the circadian variation of methotrexate toxicity was found to be trivial, even when it was administered at the time normally associated with greatest toxicity and lethality. In contrast, in rats pretreated with dexamethasone and in which corticosterone levels were suppressed throughout the day, methotrexate toxicity was so great that all animals died from toxic effects within 5 days regardless of the time of drug administration.

This effect on the circadian system of corticoids may partly be due to a phase shift of biological rhythms, for example, in liver enzymes. As a rule, adrenal suppression results from evening and nocturnal administration of corticoids to diurnally active humans; therefore dramatic alteration of some drug effects may result from the evening dosing of corticoids. Thus, either directly or indirectly imposed alteration of the cortisol circadian rhythm may have consequences on the metabolism of cytotoxic drugs.

The circadian rhythm of melatonin production may also influence that of corticosterone. ENGLISH et al. (1987) demonstrated that the toxicity of methotrexate in the rat could also be modulated by dosing the animals in the light/rest span each day for several weeks with melatonin. In the melatonin-treated animals, toxicity was not only observed as expected following a dose of methotrexate but also at other time points. Such experiments illustrate that it may be possible to modulate circadian rhythms in pharmacology in clinical practice by the appropriately timed administration or inhibition of endogenous hormones. Modulation or elimination of the melatonin rhythm may also be possible by the appropriate use of bright lights (AHERNE 1989).

It should be recognized that the magnitude of time-dependent variations of pharmacodynamic/pharmacokinetic parameters may be very large. A threefold variation in disappearance half-life has been demonstrated depending on time of administration of the bone marrow cytotoxic drug cytosine arabinoside. Similar variations in pharmacokinetics have been demonstrated for other anticancer agents such as methotrexate, vincristine and bleomycin (AHERNE 1989).

Chronopharmacokinetic effects may be important determinants for successful 6-mercaptopurine therapy. In one clinical study (LANGEVIN et al. 1987), in children with acute lymphoblastic leukaemia, night-time administration of 6-mercaptopurine resulted in greater drug exposure than morning administration. These results may partly explain the results of a retrospective study (RIVARD et al. 1985, 1993) in which children with acute lymphoblastic leukaemia who had their maintenance dose of 6-mercaptopurine and methotrexate in the evening had a lower risk of disease relapse than children receiving their medication in the morning.

However, it is not necessarily the case that the peak of therapeutic response needs to coincide with the peak of blood/plasma concentrations (AHERNE 1989), because the ultimate effect of a circadian drug administration schedule will depend on other factors as well.

In conclusion, several factors and processes involved in pharmacodynamics and pharmacokinetics are susceptible to temporal variations, and the variation observed in the overall outcome of drug therapy will depend on how these factors interact with each other throughout the day and night.

F. Disease-Dependent Chronotherapeutic Effects and Consequences

The strategy that has been adopted in the chronotherapy of cancer by potent antitumour medications so far is to take advantage of the circadian rhythm of the host tolerance relative to the effect of the anticancer drug. The highest concentration of the daily dose is given at the time of greatest tolerance and the least amount 12 h later and/or earlier. By this approach, both daily dose and frequencies of drug courses can be increased compared to the conventional homeostatic approach (HAUS et al. 1972; SCHEVING et al. 1994). It is thus likely that chronotherapy, by increasing the dose intensity of cancer medications, intensifies the destruction of tumour cells and reduces the probability of developing clones of chemoresistant cells (EVANS 1988; GALE 1988). However, this treatment strategy is based on the assumption that the temporal structure of the patient with cancer is not altered. From a clinical point of view, critical circadian rhythms (with regard to the role they play in drug metabolism) may be altered in patients with malignant disease. This is the case for plasma proteins in patients with advanced cancer (BRUGUEROLLE et al. 1986) as well as plasma cortisol in patients with metastatic breast cancer (TOUITOU et al. 1990). In addition, alteration of performance status (i.e. level of daily activity) is strongly correlated with the alteration of at least certain circadian rhythms (BAILLEUL et al. 1986). Observations in patients with advanced breast or ovarian cancer suggest that the number of altered haematological rhythms is associated with decreased performance status (BENAVIDES-ORGAZ 1991). As a consequence, patients with good performance status and presumably normal circadian time structure most likely will benefit to a greater extent from chronotherapy than do patients with poor performance status (LÉVI et al. 1988a). Accordingly, if physical activity is critical to the maintenance of temporal structure, it may well be that outpatients will benefit better from chronotherapy than will in-hospital patients.

There is as yet no study showing that light therapy (exposure to bright light, i.e. ≥ 2500 lux, for several hours at a specified time of day) can resynchronize the entire circadian time structure of a patient for subsequent chronotherapy; however, artificial light therapy or natural light may reinforce the synchronization (EDGAR et al. 1991), and thereby also increase the effect of ambulatory cancer chemotherapy.

In experiments on rodents, the presence of a transplantable tumour has been shown to alter cell proliferation rhythms in host tissues (HAUS et al. 1979). In cancer patients, BLANK (1987) found that by serially independent sampling of bone marrow cells (each patient was sampled only once) a circadian acrophase in the number of mitoses per 1000 cells was observed at night (0316 hours), which differs only slightly from the data presented by Mauer (see earlier). In our studies on cancer patients with a normal cortisol pattern, the same circadian pattern of DNA synthesis was found as for the healthy individuals. However, a lower G2/M-phase was found at midnight than in late morning.

Because of differences in the populations studied and in the methods used, the question of a phase difference in the circadian rhythms in the bone marrow of cancer patients as compared to healthy individuals awaits further investigation.

Thus, if a circadian rhythm of bone marrow proliferative activity can also be determined on a group basis in cancer patients, and possibly optimized by use of marker rhythms, timing the administration of cell-cycle-specific (and probably also non-cycle-specific) chemotherapeutic agents may lead to a relative protection of the marrow from the undesirable side effects of the agent. This may increase the dose intensity, which may be critical for its success. On the other hand, optimal circadian timing of the administration of growth factors may allow increased specific stimulation of one or another bone marrow function.

G. Marker Rhythms for Proliferative Activity and Cytotoxicity of the Bone Marrow

How can we define the "correct" relevant circadian rhythm of the bone marrow in an individual patient? In view of the non-sinusoidal shape of the data from individual patients, and because there may be a rapid alteration of the rate of DNA synthesis, one cannot necessarily extrapolate from the group to the individual except in a very broad sense. Also, the acrophase, as given by cosinor analysis, does not necessarily represent the peak of the data; neither can we assume that the trough is 12 h displaced from the acrophase. Although cosinor analysis gives an indication of when the high and low values are to be found with a known degree of statistical probability on a group basis, this analysis may not be satisfactory in the individual case.

Which marker rhythms could provide information on an individual patient's bone marrow cell proliferation rhythms is a pertinent question, as the invasive nature of the direct measurement of such rhythms is difficult. Such marker rhythms should show a fixed phase relation to the rhythm in bone marrow cell proliferation and should be measurable by non-invasive or minimally invasive methods. Therefore, it would be of interest to examine a possible coherence in the bone marrow data between the timing of the circadian rhythm in bone marrow cell proliferation and those of plasma cortisol and of the circulating formed elements in the peripheral blood. However, based on experience from the mouse, no fixed-phase relation between the circadian rhythm in the number of GM-CFUs in soft agar culture and the circulating leucocytes and corticosterone has been found (BARTLETT et al. 1984). On the other hand, the observed small but significant circadian rhythm found in circulating reticulocytes may indicate a circadian periodic release of these cells from the bone marrow and thus conceivably may serve as a marker rhythm for bone marrow rhythmicity (HAUS 1992).

Experimentally, SOTHERN et al. (1977) have demonstrated that the time of the circadian trough for body temperature corresponded to the best tolerance time for Adriamycin and therefore temperature would seem to be a marker rhythm for the dosing time of Adriamycin.

For cancer patients, both body temperature and haematological rhythms may be used as a marker rhythm for the same patients since their alteration correlates with the WHO performance status (BAILLEUL et al. 1986; BENAVIDES-ORGAZ 1991). The persistence of detectable circadian rhythms in these two parameters as well as good performance status may be predictive of a positive response to cancer chronotherapy (LÉVI et al. 1988a).

H. Chronobiology of Peripheral Blood in Health and Disease

In addition to the bone marrow, rhythmic events of different frequencies have been found at several levels of organization of the haemopoietic system, i.e. in the lymphoid elements in lymph nodes, thymus, spleen and peripheral blood (Scheving 1981; Haus et al. 1984b).

In the peripheral blood the great variability in the number of circulating formed elements has been observed since techniques for counting these structures became available during the second half of the last century. It was soon recognized that some of these variations do not occur at random, but are the expression of regularly recurring rhythmic events (Japha 1900; Sabin et al. 1927). It later became evident that some of these periodic variations, especially in the circadian range, are highly reproducible and predictable in their timing and, in some instances, are large enough to be clinically relevant.

In the study of haematological parameters in peripheral blood, rhythmic events have been described in the frequency range of a few hours (ultradian) (Sabin et al. 1927), and more frequently in the circadian range (Halberg and Visscher 1950; Halberg et al. 1953; Brown and Dougherty 1956; Haus 1959; Bartter et al. 1962; Malek et al. 1962), and in the frequency range of a week (Derer 1960; Haus et al. 1981, 1983, 1984a; Lévi and Halberg 1982). These rhythmic variations may be superimposed upon rhythms with periods between 15 and 30 days (Morley 1966), including the menstrual frequency range in women, and upon seasonal changes or circannual variations (Kusnetsova et al. 1977; Berger 1980a,b; Bratescu and Teodorescu 1981; Reinberg et al. 1980; Rocker et al. 1980).

In the peripheral blood, the periodic changes in the number of circulating cells may be the results of rhythmic influx and distribution of some young formed elements, the distribution between the circulating and the marginal cell compartments, and the distribution between different tissues or organs of the body.

The number of circulating red blood cells, haemoglobin and haematocrit show a highly reproducible and regular but low-amplitude circadian rhythm, both in clinically healthy young adults and in elderly subjects (Touitou et al. 1979, 1986; Haus et al. 1983, 1988; Swoyer et al. 1989). It is of interest that in HIV-infected patients circadian periodicity was found for haemoglobin, haematocrit and a number of circulating red cells, but with significantly lower amplitudes than for healthy individuals (Swoyer et al. 1990).

In comparison with young adults the elderly subjects show differences in timing (phase advance) of the circadian rhythms in circulating neutrophil leucocytes and lymphocytes, a decrease in circadian amplitude of circulating platelets, a decrease in circadian-rhythm-adjusted mean (mesor) in the red cell count, and a decrease in neutrophil band counts (Swoyer et al. 1989).

I. Total White Blood Cells

The circadian rhythm in total white blood cell count is highly reproducible, as has been shown in many studies by numerous investigators in different geographical locations, on populations of different ethnic background with different living habits, and during different seasons (Roitman et al. 1975;

HALBERG et al. 1977; REINBERG et al. 1977). The acrophase is found during the evening hours (between 2100 hours and midnight, with a change in total white cell count over the 24-h time span in clinically healthy subjects from about 0.9–2.0 × 10^9/l (HAUS et al. 1983, 1988). However, the circadian rhythm in total white blood cells is the composite of the circadian rhythms of the different types of leucocytes (neutrophils, lymphocytes, monocytes, etc.), some of which have different circadian phasing.

II. Neutrophil Leucocytes

The circulating neutrophil leucocytes show a circadian rhythm with acrophase during the late afternoon and evening hours (around 1900 hours with a 95% CI from 1730 to 2130 hours) (HAUS 1992). The changes in the number of circulating neutrophils observed during the 24-h time span are, in clinically healthy subjects, around 0.6–1.0 × 10^9/l.

III. Lymphocytes

The circadian rhythm in the number of circulating lymphocytes is a very regular and highly reproducible phenomenon, although these blood cells are not a homogeneous population but consist of functionally very different subtypes. The acrophase of the circadian rhythm in circulating lymphocytes occurs during the night, with highest values found between midnight and 0400 hours. A difference between morning and night samples of 1.0 × 10^9/l has been reported.

A circadian rhythm has been demonstrated for several lymphocyte subsets. For example, circadian variations in circulating T cells and/or T-cell subsets have been described by several authors (ABO et al. 1981; BERTOUCH et al. 1983; HAUS et al. 1983; RITCHIE et al. 1983; CANON et al. 1985; MIYAWAKI et al. 1984; SIGNORE et al. 1985; LÉVI et al. 1988a, b). Generally, the most consistent variations have been found in the CD3+ and CD4+ cells, while the CD8+ cells have been found by some to remain rhythmically more stable over the 24-h time span (RITCHIE et al. 1983) or to exhibit a 12-h rhythm only (LÉVI et al. 1985, 1988a).

Not only the number of circulating lymphocytes, but also numerous aspects of lymphocyte function show circadian rhythms. The T-cell response to phytohaemaglutinin (PHA) has been found to be circadian periodic by several investigators studying this parameter over the entire 24-h time span (HAUS et al. 1974, 1983; TAVADIA et al. 1975). There appears to be some variability in circadian timing, possibly due to seasonal or other factors, which may lead to negative results, if for example only two time points along the 24-h scale are examined (FELDER et al. 1985).

Alterations of circadian periodicity of human lymphocytes have been reported in patients with lymphoid tumours (SWOYER et al. 1975) and in patients infected with the lymphotropic human immunodeficiency virus (HIV). In fact, alterations in the circadian rhythm of circulating lymphocytes are an early event in the course of the development of the AIDS-related syndrome (ARS) and AIDS (MARTINI et al. 1988a,b; BOURIN et al. 1989; SWOYER et al. 1990). A circadian rhythm of circulating neutrophils, monocytes, total lymphocytes, CD3+, CD8+ and B1+ lymphocytes, and in platelets, could not be demonstrated in a group of HIV-infected patients by cosinor analysis (SWOYER et al. 1990).

IV. Eosinophil Leucocytes

The circadian rhythm in the number of circulating eosinophils demonstrates a large amplitude, with the acrophase during the night (RUD 1947; HALBERG and VISSCHER 1950). However, a large variation in the number of these cells in different individuals will limit the diagnostic value of a single time-specified eosinophil count. Of importance is that the physician is aware of the high-amplitude circadian periodicity of the eosinophil leucocyte count in the evaluation of consecutive samples in the same subject.

In patients with nocturnal asthma a greater circadian variation with peak eosinophilia at 0400 hours was found (CALHOUN et al. 1992). Also, a significant circadian increase in low-density eosinophils was found at 0400 hours, but only in patients with nocturnal asthma. These observations suggest that a circadian variation in low-density eosinophils may contribute to nocturnal exacerbations of asthma (CALHOUN et al. 1992).

V. Natural Killer Cells

There have been conflicting reports on circadian rhythms in natural killer (NK) cell numbers and activity. A considerable number of regulatory factors, both soluble and cellular, have been shown to influence NK cell functions, including some of the cytokines, i.e. interferon (INF)-γ and interleukin (IL)-2 (HERBERMAN and CALLEWAERT 1985). Several authors have, however, demonstrated a reproducible circadian rhythm in the NK cell activity in peripheral blood of clinically healthy adult subjects, with activity being high in the morning and then declining to a minimum during the night hours (WILLIAMS et al. 1979; ABO et al. 1981; GATTI et al. 1986, 1988; MOLDOFSKY et al. 1986, 1989), and thus being out of phase with the numbers of circulating lymphocytes of most other subtypes.

The immune defences are organized along both 24-h and yearly time scales. Two circadian systems have been isolated in man: (a) the circulation of T, B or NK lymphocyte subsets in peripheral blood and (b) the density of epitope molecules (CD3, CD4, etc.) at their surface, which may relate to cell reactivity to antigen exposure. These two systems may be desynchronized.

A recent study has demonstrated seasonal variations in a number of peripheral blood immune cells, such as leucocytes, monocytes, neutrophils, lymphocytes, CD3(+) T, CD8(+) T, CD25(+) T, CD20(+) B, serum interleukin-6 (IL-6), soluble IL-6 receptor (sIL-6R) and sIL-2R levels in normal volunteers. Altogether 26 individuals had monthly blood samples taken during one calendar year. It was found that most of the immune variables changed rhythmically during the seasons as a group phenomenon. It was concluded that the immune system is characterized by a multifrequency time-structure with *significant high-amplitude* yearly variations in the number of some peripheral blood leucocyte subsets (MAES et al. 1994).

VI. Platelets

The number of platelets circulating in the peripheral blood shows a statistically highly significant circadian variation as a group phenomenon. However, the variance between different individuals in the population is large and the extent of the circadian variation in platelet numbers relatively small.

More important are the functional changes in the aggregability to stimulation, for example with adenosine diphosphate (ADP) or epinephrine (PETRALITO et al. 1982; TOFLER et al. 1987; MEHTA et al. 1989; HAUS et al. 1990) or in platelet adhesiveness (HAUS et al. 1990). The circadian rhythm in platelet function may contribute to the circadian variations in the incidence of sudden cardiac death (RABKIN et al. 1980; MULLER et al. 1987), myocardial infarction (REINBERG et al. 1973; MULLER et al. 1985) and cerebral infarction (REINBERG et al. 1973; MARSHALL 1977; MARLER et al. 1989), which have been reported to occur most frequently during the early and mid-morning hours.

A very significant circadian variation of dihydropyrimidine dehydrogenase (DPD) in human peripheral blood mononuclear cells has been demonstrated by HARRIS et al. (1990). This enzyme had an activity that was inverse to the level of 5-fluorouracil (5-FU) administered by continuous infusion. Thus, information about the DPD activity in a specific cell type of peripheral blood could be used as a chronopharmacological marker in planning continuous infusion schedules of fluoropyrimidines so that optimal plasma drug concentration may be maintained over a 24-h cycle.

I. Cyclic Haemopoiesis

Are frequencies other than the circadian rhythm of importance for bone marrow cell proliferation? In reanalysing the data of BLANK (1987) and BLANK et al. (1992), a circaseptan rhythm in bone marrow cell proliferaton was found in the bone marrow of cancer patients. Infradian rhythms (with periods > 28 h) also characterize bone marrow regeneration after melphalan treatment and are found widely in parameters of humoral and cellular immunity, including regeneration after immune suppression (LÉVI and HALBERG 1982; HAUS et al. 1983). Lower-frequency infradian rhythms have been noted in healthy human subjects and find their extreme expression in cyclic neutropenia and thrombocytopenia (HAUS et al. 1983). Infradian rhythms of low frequencies have also been found in the circulating leukaemic cells of some patients with chronic granulocytic leukaemia (MORLEY et al. 1967; KENNEDY 1970; GATTI et al. 1973).

Cyclic neutropenia and cyclic thrombocytopenia are characterized by approximate 21-day fluctuations of circulating blood neutrophils, monocytes, eosinophils, lymphocytes, reticulocytes and platelets (GUERRY et al. 1973; DALE and HAMMOND 1988). The numbers of monocytes, platelets and reticulocytes and the concentration of colony-stimulating factor (CSF) frequently oscillate in cyclic neutropenia with the same period as, but out of phase with, the neutrophil count. The recurrent severe neutropenia causes patients to experience periodic symptoms of fever, malaise, mucosal ulcers and, in rare instances, life-threatening infections. Studies of the pathophysiology of cyclic haemopoiesis demonstrate that the abnormality lies in the regulation of cell production and not in peripheral destruction (GUERRY et al. 1973; DALE and HAMMOND 1988).

It has been shown that the haemopoietic progenitor cell numbers in these patients fluctuate cyclically (BRANDT et al. 1975; GREENBERG et al. 1976; VERMA et al. 1982). It is of interest that bone marrow transplantation from a person to her sibling with leukaemia resulted in the transfer of cyclic neutropenia, suggesting that the basic defect represents a stem cell disorder (KRANCE et al. 1982). These patients have been treated with G-CSF, although their progenitor cells are less responsive to this haemopoietic growth factor than are normal progenitors. In the small number of patients thus far treated over a prolonged time span, no untoward side effects upon the haemopoietic elements have been reported (MIGLIACCIO et al. 1990).

Several investigators have reported marked cyclic oscillation in the white blood cell count in a population of 10%–20% of patients with Phl+ and/or Phl− chronic myelocytic leukaemia (CML) (MORLEY et al. 1967; KENNEDY 1970; SHADDUCK et al. 1972; VODOPICK et al. 1972; GATTI et al. 1973; CHIKKAPPA et al. 1976; IUBAL et al. 1983; MEHTA and AGARWAL 1980; UMEMURA et al. 1986). The periods reported varied between 30 and 120 days (median 70 days). Although the period varies between different subjects, it recurs with considerable regularity in each individual. Platelet numbers and haemoglobin concentrations may vary in synchrony with the leucocytes (UMEMURA et al. 1986). It has been found that the cycling of the numbers of granulocytes and their precursors in patients with CML is due to a periodic contraction and expansion of the total blood granulocyte pool (VODOPICK et al. 1972) and the periodic influx of progenitor cells from the haemopoietic stem cell pool (UMEMURA et al. 1986). It appears to be the same phenomenon as in cyclic haemopoiesis, except that the cycle period is longer, possibly due to a longer maturation time of some blood cell precursors in CML (WHELDON et al. 1974). Any mathematical models trying to explain the phenomenon of cyclic haemopoiesis and cyclic CML will have to incorporate the concept of non-stationary, rhythmically changing cell functions and cell sensitivity to appropriate and abnormal time-dependent stimulation or inhibition (KLEIN and VALLERON 1977; GUIGUET et al. 1978; VON SCHULTHESS and MAZER 1982; HAUS 1992). In a clinical context, cycling of the leukaemic elements in CML can pose problems in the diagnosis and monitoring of therapy, in which the spontaneous cyclic variations may be confounded by treatment effects or lack thereof. On the other hand, the cyclicity of many CMLs may provide an opportunity to learn more about the nature of this disorder.

J. Scientific and Clinical Implications

The recognition of a multifrequency time structure in the numbers and functions of haemopoietic cells is essential for the scientific and clinical exploration of haemopoietic parameters. Circadian acrophase maps of haematological variables and extent of their circadian variations have become available (HAUS et al. 1983; HAUS et al. 1988). The high-amplitude variations of some parameters, e.g. the number of circulating neutrophils and lymphocytes, may have diagnostic implications in clinical medicine. Therefore, time-qualified reference ranges are of importance in functions with high-amplitude rhythms. In some of these rhythms, for instance those of the circulating eosinophils, monocytes and blood platelets, the wide spread of values found in the clinically healthy population makes the time-qualified normal ranges very broad, sometimes to the point that they do not add much diagnostic value to the presently used non-time-qualified reference ranges. Nevertheless, even for these parameters, the circadian rhythms have to be kept in mind if consecutive samples are obtained in the same subject (HAUS 1992).

Rhythm alterations may be of pathobiological and clinical interest, e.g. in viral infections with HIV, and in neoplasia. Low-amplitude rhythms usually do not represent diagnostic problems, but their timing may indicate differences in responsiveness of one or other component of the haemopoietic or the immune system.

The recognition of time-dependent changes in the effects of haematological regulatory and growth factors and in the susceptibility and resistance to environmental agents, including drugs used in clinical medicine, environmental toxins and chemical carcinogens, is expected to lead to the development of chronopharmacological treatment schedules. Especially in the case of drugs which have unavoidable side effects, such as those used in cancer chemotherapy, improvement of host tolerance by timing of treatment appears to be promising.

In order to alleviate the bone marrow toxicity of many anticancer drugs, clinical trials, based on experimental results (LÉVI et al. 1988b; PERPOINT et al. 1994) and human studies (SMAALAND et al. 1991a,b,c, 1992a,b), should be implemented in order potentially to improve the clinical efficacy of CSF treatment.

Although timing of chemotherapy and radiotherapy in chronobiologically determined schedules have been shown to improve host tolerance and therapeutic effects in animal models, there has been relatively little application to human oncology. However, in the treatment of acute lymphoblastic leukaemia in children, RIVARD et al. (1985, 1993) reported that the disease-free survival was better and the risk of relapsing substantially lower when 6-mercaptopurine (6-MP) and methotrexate (MTX) were given in the evening than in the morning hours. These findings with potentially great clinically implications should be followed up in a controlled multicentre prospective study.

Although this time-dependent difference in survival may be due to chronopharmacokinetic and chronopharmacodynamic factors, these results may also be explained on the basis of previously discussed cyclic and especially circadian variations in number and proliferative activity of leukaemic cells, i.e. the tumour itself may also show a time structure, with circadian rhythms or rhythms in other frequency ranges. Thus, identification of such rhythmic cycles in sensitivity may imply improved therapeutic response and/or less toxicity. Periodicities in haematological malignancies have indeed been detected in some patients with multiple myeloma (KACHERGENE et al. 1972; ZINNEMAN et al. 1972, 1974; RAMOT et al. 1976). It is noteworthy that GUERCI et al. (1990) found that DNA synthesis as well as the G2/M-phase was statistically related to the time of day of sampling, with highest proliferative activity at the end of the day. Circadian variations of the blast cell count were also observed (GUERCI et al. 1990). Furthermore, such temporal variations are in accordance with our findings of a circadian variation in DNA synthetic activity in non-Hodgkin's lymphomas (SMAALAND et al. 1993) (Fig. 8). Alternatively, the difference in clinical outcome may be due to circadian variations in proliferative activity of normal bone marrow cells.

This leads us to a pertinent question. What is the relationship between the rhythms in bone marrow cell proliferation and the rhythms in tumours? If tumour cell proliferation is in phase with that of the bone marrow, little advantage is gained by trying to protect the bone marrow by timed treatment. Unfortunately, the timing of rhythms in human tumours can be determined only under very special circumstances. Whenever this is possible, the rhythms of both the host cells and the tumour can be used to design an optimal chronotherapeutic regimen. However, if the tumour is not directly accessible for study, tumour markers may be explored. Their rhythms, if found, may indicate a metabolic rhythm in the tumour, although the time of maximal metabolic activity in a tissue usually is out of phase with the peak in mitotic activity. Such metabolic tumour markers may be noted, e.g. in multiple myeloma in the form of excreted light chains or in some tumours in the form of polyamine excretion (ZINNEMAN et al. 1974).

Some human tumours have been described as arrhythmic; others have shown ultradian (< 20 h) periodicity and some circadian periodicity, often with a phase difference between host and tumour cells (KLEVECZ and BRALY 1991). It has to be realized that in the course of treatment both tumour and host can change periodicity; change may range from synchronization to rhythm induction (circaseptan) to free running. Such a change in periodic behaviour during treatment may be critical for the timing of chronotherapy, and a "favourable" time may in the course of treatment become "unfavourable" if the pertinent rhythms in host and tumour are not appropriately monitored.

As much as "correct" timing of treament can lead to a substantial improvement in therapeutic effect, with tolerance of a higher dose intensity and

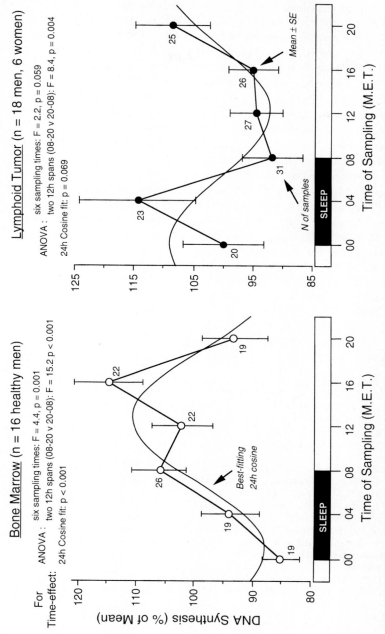

Fig. 8. Circadian variation in DNA synthesis in non-Hodgkin's lymphoma (NHL) versus healthy human bone marrow (BM). The prognostic power of S-phase in classifying NHL may be increased by proper circadian sampling (i.e. at night). In addition, the observation that peaks in NHL and BM appear to be out of phase suggests that delivery of a peak dose of chemotherapy near midnight may induce maximal damage to the tumour and minimal toxicity to the bone marrow. (From SMAALAND et al. 1993)

improved long-term remission or cure, and to decreased toxic side effects, a wrong timing of cytotoxic anticancer drug administration may lead to reduced therapeutic effect and serious side effects (HRUSHESKY 1985; ROEMELING 1991).

Finally, as has been discussed earlier, alterations in the organism's time structure may be of importance for the early recognition of abnormal function, often before structural disease can be identified.

In conclusion, what has not until recently been appreciated is that there is in the human a 24-h circadian coordination of entry into active proliferation of bone marrow cells (KILLMANN et al. 1962; MAUER 1965; SMAALAND et al. 1989, 1991a, b; KLEVECZ and BRALY 1994; ABRAHAMSEN et al. 1994, 1995). Available information related to proliferation makes it increasingly clear that haemopoiesis is not a temporary fixed phenomenon but rhythmic during each day and year. This rhythmicity can be exploited in several ways, to diminish cytotoxic side effects of cancer therapy, as well as in relation to primary diseases of the bone marrow. Chronotherapy protocols have to be properly designed, taking into account the chronobiology of both the haemopoietic system and the tumour, as well as the chronopharmacodynamic and chronopharmacokinetic properties of the drugs.

Acknowledgements. This study was supported by the Norwegian Cancer Society. I am indebted to Robert B. Sothern, PhD, for help with statistical analyses and compilation of animal data

References

Aardal NP (1984) Circannual variations of circadian periodicity in murine colony-forming cells (CFU-C). Exp Hematol 12:61–67
Aardal NP, Laerum OD (1983) Circadian variations in mouse bone marrow. Exp Hematol 11:792–801
Aardal NP, Laerum OD, Paukovits WR (1982) Biological properties of partially purified granulocyte extract (chalone) assayed in soft agar culture. Virchows Arch [B] Cell Pathol 38:253–261
Abo T, Kawate T, Itoh K, Kumagai K (1981) Studies on the bioperiodicity of the immune response. I. Circadian rhythms of human T, B, and K cell traffic in the peripheral blood. J Immunol 126:1360–1363
Abrahamsen JF, Smaaland R, Laerum OD (1994) Circadian stage dependent variations in the DNA synthesis phase of human bone marrow subpopulations. 6th international conference on chronopharmacology and chronotherapeutics, abstract VIIIa-2
Abrahamsen JF, Lund-Johansen F, Laerum OD, Schem BC, Sletvold O, Smaaland R (1995) Flow cytometric assessment of peripheral blood contamination and proliferative activity of human bone marrow cell populations. Cytometry 19:77–85
Aherne GW (1989) An introduction to chronopharmacology. In: Arendt J, Minors S, Waterhouse JM (eds) Biological rhythms in clinical practice. Wright, London, pp 8–19
Akagawa T, Onari K, Peterson WJ, Makinodan T (1984) Differential effect on mitotically active and inactive bone marrow stem cells and splenic stem T cells in mice. Cell Immunol 86:53–63
Angeli A, Frajria R, Depaoli R, Fonzo D, Ceresa F (1978) Diurnal variation of prednisolone binding to serum corticosteroid binding globulin in man. Clin Pharmacol Ther 23:47–53

Arendt J, Minors DS, Waterhouse JM (1989) Basic concepts and implications. In: Arendt J, Minors S, Waterhouse JM (eds) Biological rhythms in clinical practice. Wright, London, pp 3–7

Bailleul F, Lévi F, Reinberg A (1986) Interindividual differences in the circadian hematologic time structure of cancer patients. Chronobiol Int 3:47–54

Bartlett P, Haus E, Tuason T, Sackett-Lundeen L, Lakatua D (1984) Circadian rhythm in number of erythroid and granulocytic colony forming units in culture (ECFU-C and GSFU-C) in bone marrow of BDF1 male mice. Chronobiology 1982-1983. Karger, Basel

Bartter FC, Delea CS, Halberg F (1962) A map of blood and urinary changes related to circadian variations in adrenal cortical function in normal subjects. Ann NY Acad Sci 98:969–983

Bellamy WT, Alberts DS, Dorr RT (1988) Daily variation in non-protein sulfhydryl levels of human bone marrow. Eur J Cancer Clin Oncol 11:1759–1762

Benavides-Orgaz M (1991) Cancer avancé de l'ovaire: approche chronobiologique comme nouvelle stratégie du traitement et de la surveillance clinique et biologique. Thesis, University of Paris

Berger J (1980a) Circannual rhythms in the blood picture of laboratory rats. Folia Haematol (Leipz) 107:54–60

Berger J (1980b) Seasonal influences on circadian rhythms in the blood picture of laboratory mice. Z Versuchstierkd 22:122–134

Bertouch JV, Roberts-Thompson P, Bradley J (1983) Diurnal variation of lymphocyte subsets identified by monoclonal antibodies. Br Med J 286:1171–1172

Blank MA (1987) Characteristics of the distribution of mitotic activity indices in human malignant neoplasms. Dokl Akad Nauk SSSR 297:979–981

Blank MA, Cornélissen G, Neishtadt EL, Kochrev VA, Yakovlev GY, Haus E, Halberg E, Halberg F (1992) Circadian-circaseptan-circannual mitotic aspects of the bone marrow chronome of patients with malignancy. Workshop on computer methods on chronobiology and chronomedicine. Medical Review, Tokyo, pp 245–262

Botnick LE, Hannon EC, Hellman S (1976) Limited proliferation of stem cells surviving alkylating agents. Nature 162:68–70

Boughattas AN, Lévi F, Fournier C et al (1989) Circadian rhythm in toxicities and tissue uptake of 1,2-diaminecyclohexane (trans-1) oxalatoplatinum (II) in mice. Cancer Res 49:3362–3368

Bourin P, Mansour I, Lévi F, Vilette JM, Roué R, Fiet J, Rouger P, Doinel C (1989) Perturbations précoces des rythmes circadiens des lymphocytes T et B au cours de l'infection par le virus de l'immunodéficience humaine (VIH). C R Acad Sci (Paris) 308:431–436

Brandt L, Forssman O, Mitelman F, Odeberg H, Olofsson T, Olsson I, Svensson B (1975) Cell production and cell function in human cyclic neutropenia. Scand J Haematol 15:228–240

Bratescu A, Teodorescu M (1981) Circannual variations in the B cell/T cell ratio in normal human peripheral blood. J Allergy Clin Immunol 68:273–280

Brown HE, Dougherty TF (1956) The diurnal variation of blood leukocytes in normal and adrenalectomized mice. Endocrinology 58:365–375

Bruguerolle B, Lévi F, Arnaud C, Bouvenot G, Mechkouri M, Vannetzel J, Touitou Y (1986) Alteration of physiologic circadian time structure of six plasma proteins in patients with advanced cancer. Annu Rev Chronopharmacol 3:207–210

Buchi KN, Hrushesky WJM, Sothern RB, Rubin N, Moore JG (1991) Circadian rhythm of cellular proliferation in the human rectal mucosa. Gastroenterology 101:410–415

Burns ER (1981) Circadian rhythmicity in DNA synthesis in untreated and saline-treated mice as a basis for improved chronochemotherapy. Cancer Res 41:2795–2802

Burns ER, Beland SS (1983) Induction by 5-fluorouracil of a major phase difference in the circadian profiles of DNA synthesis between the Ehrlich ascites carcinoma and five normal organs. Cancer Lett 20:235–239

Butcher EC (1990) Cellular and molecular mechanisms that direct leukocyte traffic. Am J Pathol 136:3–11

Calhoun WJ, Bates ME, Schrader L, Sedgwick JB, Busse WW (1992) Characteristics of peripheral blood eosinophils in patients with nocturnal asthma. Am Rev Respir Dis 145:577–581

Canon C, Lévi F, Reinberg A, Mathé G (1985) Circulating calla-positive lymphocytes exhibit circadian rhythms in man. Leuk Res 9:1539–1546

Chikkappa G, Borner G, Burlington H, Chanana AD, Cronkite EP, Ohl S, Pavelec M, Robertson JS (1976) Periodic oscillations of blood leukocytes, platelets and reticulocytes in a patient with chronic myelocytic leukemia. Blood 47:1023–1030

Clausen OPF, Thorud E, Bjerknes R, Elgjo K (1979) Circadian rhythms in mouse epidermal basal cell proliferation. Cell Tissue Kinet 12:319

Dale DC, Hammond WP (1988) Cyclic neutropenia: a clinical review. Blood Rev 2:178–185

Dancey JT, Deubelbeiss KA, Harker LA, Finch CA (1976) Neutrophil kinetics in man. J Clin Invest 58:705–715

Derer L (1960) Rhythm and proliferation with special reference to the 6-day rhythms of blood leukocyte count. Neoplasma 7:117–133

Dörmer P, Schmolke W, Muschalik P, Brinkman W (1970) Die DNS-Synthesegeschwindigkeit im Verlaufe der DNS-Synthesephase von Erythroblasten der Maus in vivo. Beitr Pathol 141:174–186

Dosik GM, Barlogie B, Göhde W, Johnston D, Tekell J L, Drewinko B (1980) Flow cytometry of DNA content in human bone marrow: a critical reappraisal. Blood 55:734–740

Edgar DM, Martin CE, Dement WC (1991) Activity feedback to the mammalian circadian pacemaker. Influence on observed measures of rhythm period length. J Biol Rhythms 6:185–199

English J, Dunne M, Marks W (1983) Diurnal variation in prednisolone kinetics. Clin Pharmacol Ther 33:381–385

English J, Aherne GW, Marks V (1987) The effect of abolition of the endogenous corticosteroid rhythm on the circadian variation in methotrexate toxicity in the rat. Cancer Chemother Pharmacol 19:287–290

Erslev AJ (1983) Production of erythrocytes. McGraw-Hill, New York

Evans WE (1988) Clinical pharmacodynamics of anticancer drugs: a basis for extending the concept of dose-intensity. Blut 56:241–248

Farooqi NYH, Ahmed AE (1984) Circadian periodicity of tissue glutathione and its relationship with lipid peroxidation in rats. Life Sci 34:2413–2418

Felder M, Doré CJ, Knight SC, Ansell BM (1985) In vitro stimulation of lymphocytes from patients with rheumatoid arthritis. Clin Immunol Immunopathol 37:253–261

Fent K, Zbinden G (1987) Toxicity of interferon and interleukin. Trends Pharmacol Sci 8:100

Fishbein L (1984) An overview of environmental and toxicological aspects of aromatic hydrocarbons. I. Benzene. Sci Total Environ 40:189–218

Fisher LB (1968) The diurnal mitotic rhythm in the human epidermis. Br J Dermatol 60:75

Fried W, Barone J (1980) Residual marrow damage following therapy with cyclophosphamide. Exp Haematol 8:610–614

Gale RP (1988) Myelosuppressive effects of antineoplastic chemotherapy. In: Testa NG, Gale RP (eds) Hematopoiesis. Long-term effects of chemotherapy and radiation. Dekker, New York, pp 63–71

Gatti G, Cavallo R, Sartori ML, Marinone C, Angeli A (1986) Cortisol at physiological concentrations and prostaglandin E2 are additive inhibitors of human natural killer cell activity. Immunopharmacology 11:119–128

Gatti G, Masera R, Cavallo R, Delponte D, Sartory ML, Salvadori A, Carignola R, Angeli A (1988) Circadian variation of interferon-induced enhancement of human natural killer (NK) cell activity. Cancer Detect Prevent 12:431–438

Gatti RA, Robinson WA, Deinard AS, Nesbit M, McCullough JJ, Ballow M, Good RA (1973) Cyclic leukocytosis in chronic myelogenous leukemia: new perspectives on pathogenesis and therapy. Blood 41:771–782

Goldeck H (1948) Der 24-stunden-rhytmus der erythropoese. Arztl Forsch 2:22–27

Goldeck H, Siegel P (1948) Die 24-Stunden-Periodik der Blutreticulocyten unter vegetativen Pharmaka. Arztl Forsch 2:245–248

Gordon MY, Barrett AJ, Gordon-Smith EC (1985) Bone marrow disorders. The biological basis of clinical problems. Blackwell Scientific, Oxford

Greenberg PL, Bax I, Levin J, Andrews TM (1976) Alteration of colony-stimulating factor output, endotoxemia, and granulopoiesis in cyclic neutropenia. Am J Haematol 1:375–385

Guerci A, Scheid P, Feugier P, Pierrez J, Frenkiel N, Guerci O (1990) Time-variations of pretreatment peripheral blood S+G2/M-phase size determined by flow cytometry in adult acute myeloid leukaemia. Eur J Hematol 45:5–10

Guerry D, Dale DC, Omine M, Perry S, Wolff SM (1973) Periodic hematopoiesis in human cyclic neutropenia. J Clin Invest 52:3220–3230

Guiguet M, Klein B, Valleron AJ (1978) Diurnal variation and the analysis of percent labelled mitosis curves. Biomathematics and cell kinetics. Elsevier Biomedical, North Holland

Haen E, Golly I (1986) Circadian variation in the cytochrome P-450 system of rat liver. Annu Rev Chronopharmacol 3:357–361

Halberg F, Visscher MB (1950) Regular diurnal physiological variation in eosinophil levels in five stocks of mice. Proc Soc Exp Biol Med 75:846–847

Halberg F, Visscher MB, Bittner JJ (1953) Eosinophil rhythm in mice: range of occurrence; effects of illumination, feeding and adrenalectomy. Am J Physiol 174:109–122

Halberg F, Sothern RB, Roitman B, Halberg E, Benson E, von Mayersbach H, Haus E, Scheving LE, Kanabrocki EL, Bartter FC, Delea C, Simpson HW, Tavadia HB, Fleming K, Hume P, Wilson C (1977) Agreement of circadian characteristics for total leukocyte counts in different geographic locations. XIIth international conference of the International Society of Chronobiology, pp 3–17

Haldar C, Haussler D, Gupta D (1992) Effect on the pineal gland on circadian rhythmicity of colony forming units for granulocytes and macrophages (CFU-GM) from rat bone marrow cell cultures. J Pineal Res 12:79–83

Harris B, Song R, Soong S, Diasio RB (1990) Relationship between dihydropyrimidine dehydrogenase activity and plasma 5-fluorouracil levels: evidence for circadian variation of plasma drug levels in cancer patients receiving 5-fluorouracil by protracted continuous infusion. Cancer Res 50:197–201

Harris BE, Song R, He Y-J, Soong S-J, Diasio RB (1988) Circadian rhythm of rat liver dihydropyrimidine dehydrogenase, possible relevance to fluoropyrimidine chemotherapy. Biochem Pharmacol 37:4759

Harrison DE (1979) Proliferative capacity of erythropoietic stem cell lines and aging: an overview. Mech Ageing Dev 9:409–426

Haus E (1959) Endokrines System und Blut. Urban und Schwarzenberg, Munich (Handbuch der gesamten Hämatologie, vol 2)

Haus E (1992) Chronobiology of circulating blood cells and platelets. In: Touitou Y, Haus E (eds) Biologic rhythms in clinical and laboratory medicine. Springer, Berlin Heidelberg New York, pp 504–526

Haus E, Halberg F, Scheving LE, Pauly JE, Cardoso S, Kuhl JFW, Sothern RB, Shiotsuka RN, Hwang DS (1972) Increased tolerance of leukemic mice to arabinosyl cytosine with schedule adjusted to circadian system. Science 177:80–82

Haus E, Halberg F, Kuhl JFW, Lakatua DJ (1974) Chronopharmacology in animals. Chronobiologia 1 [Suppl 1]:122–156

Haus E, Halberg F, Scheving LE, Simpson H (1979) Chronotherapy of cancer – a critical evaluation. Int J Chronobiol 6:67–107

Haus E, Sackett LL, Haus M, Babb WK, Bixby EK (1981) Cardiovascular and temperature adaption to phase shift by intercontinental flights − longitudinal observations. Adv Biosci 30:375–390

Haus E, Lakatua DJ, Swoyer J, Sackett-Lundeen L (1983) Chronobiology in hematology and immunology. Am J Anat 168:467–517

Haus E, Taddeini L, Larson K, Bartlett P, Sackett-Lundeen L (1984a) Circadian rhythm in spontaneous ^{3}H-thymidine uptake and in PHA response of splenic cells of BDF1 male mice in vitro. Phase relations to hematologic rhythms in vivo. Chronobiology 1982–1983. Karger, Basel

Haus M, Sacket-Lundeen L, Lakatua D, Haus E (1984b) Circadian variation of ^{3}H-thymidine uptake in DNA of lymphatic organs irrespective of relative length of light and dark span. J Minn Acad Sci 49:19

Haus E, Nicolau GY, Lakatua D, Sackett-Lundeen L (1988) Reference values for chronopharmacology. Annu Rev Chronopharmacol 4:333–424

Haus E, Cusulos M, Sackett-Lundeen L, Swoyer J (1990) Circadian variations in blood coagulation parameters, alpha-antitrypsin antigen and platelet aggregation and retention in clinically healthy subjects. Chronobiol Int 7:203–216

Hecquet B, Meynadier J, Bonneterre J, Adenis L, Demaille A (1985) Time dependency in plasmatic protein binding of cisplatin. Cancer Treatment Rep 69:79–83

Hellman S, Botnick LE, Hannon EC, Vigneulle RM (1978) Proliferative capacity of murine hematopoietic stem cells. Proc Natl Acad Sci USA 75:490–494

Herberman RB, Callewaert DH (1985) Mechanism of cytotoxicity by NK cells. Academic, Orlando

Hrushesky WJM (1985) Circadian timing of cancer chemotherapy. Science 228:73–75

Hromas RA, Hutchins JT, Marke DE, Scholes VE (1981) Flow cytometric analysis of the effect of ara-C on the chronobiology of the bone marrow synthesis. Chronobiologia 8:369–373

Iubal A, Aktein E, Barak I, Meytes D, Many A (1983) Cyclic leukocytosis and long survival in chronic myeloid leukemia. Acta Haematol 69:353–357

Japha A (1900) Die Leukozyten beim gesunden und kranken Säugling. Jahrb Kinderheilkd 52:242–270

Kachergene NB, Koshel IV, Nartsissov RP (1972) Circadian rhythm of dehydrogenase activity in blood cells during acute leukemia in childhood. Pediatria 51:81–85

Kennedy BJ (1970) Cyclic leukocyte oscillations in chronic myelogenous leukemia during hydroxy-urea therapy. Blood 35:751–760

Killmann S-Å, Cronkite EP, Fliedner TM, Bond VP (1962) Mitotic indices of human bone marrow cells. I. Number and cytologic distribution of mitosis. Blood 19:743–750

Klein B, Valleron AJ (1977) A compartmental model for the study of diurnal rhythms in cell proliferation. J Theor Biol 64:27–42

Klevecz RR, Braly PB (1991) Circadian and ultradian cytokinetic rhythms of spontaneous human cancer. Ann NY Acad Sci 618:257–276

Klevecz RB, Braly PS (1994) Circadian and ultradian cytokinetics of human cancers. In: Hrushesky WJM (ed) Circadian cancer therapy. CRC Press, Boca Raton, pp 165–183

Koren S, Fleischmann R Jr (1993a) Circadian variations in myelosuppressive activity of interferon-α in mice: identification of an optimal treatment time associated with reduced myelosuppressive activity. Exp Haematol 21:552–559

Koren S, Fleischmann R Jr (1993b) Optimal circadian timing reduces the myelosuppressive activity of recombinant murine interferon-gamma administered to mice. J Interferon Res 13:187–195

Krance RA, Spruce WE, Forman SJ, Rosen RB, Hecht T, Hammond WP, Blume KG (1982) Human cyclic neutropenia transferred by allogeneic bone marrow grafting. Blood 60:1263–1266

Kusnetsova SS, Parvdina GM, Yezhova VM (1977) Seasonal variations of some parameters of peripheral blood and haematogenetic organs in mice. Zh Obsliteh Biol 38:133–140

Laerum OD, Aardal NP (1981) Chronobiological aspects of bone marrow and blood cells. In: von Mayersbach H, Scheving LE, Pauly JE (eds) 11th international congress of anatomy, part C, biological rhythms in structure and function. Liss, New York, pp 87–97

Laerum OD, Smaaland R (1989) Circadian and infradian aspects of the cell cycle: from past to future. Chronobiologia 16:441–453

Laerum OD, Sletvold O, Riise T (1988) Circadian and circannual variation of cell cycle distribution in the mouse bone marrow. Chronobiol Int 5:19–35

Laerum OD, Smaaland R, Sletvold O (1989) Rhythms in blood and bone marrow: potential therapeutic implications. In: Lemmer B (ed) Chronopharmacology. Cellular and biochemical interactions. Dekker, New York, pp 371–393

Lajtha LG (1963) On the concept of the cell cycle. J Cell Comp Physiol 62:143

Lajtha LG (1979) Stem cell concepts. Differentiation 14:23

Lake BG, Tredger JM, Burke MD, Chakraborty J, Bridges JW (1976) The circadian variation of hepatic microsomal drug and steroid metabolism in the golden hamster. Chem Biol Interact 12:81–90

Langevin AM, Koren G, Soldin S, Greenberg M (1987) Pharmacokinetic case for giving 6-mercaptopurine maintenance doses at night (letter to the editor). Lancet ii:505–506

Lasky LC, Ascencao J, McCullough J, Zanjian ED (1983) Steroid modulation of naturally occurring diurnal variations in circulating pluripotential haemotopoietic cells. Br J Haemotol 55:615–622

Leszczynska-Bisswanger A, Pfulf E (1985) Diurnal variation of methotrexate transport and accumulation in hepatocytes – a consequence of variations in cellular glutathione. Biochem Pharmacol 34:1635–1638

Lévi F, Halberg F (1982) Circaseptan (about 7-day) bioperiodicity – spontaneous and reactive – and the search for pacemaker. La Ricerca 12:323–370

Lévi F, Canon C, Blum JP, Mechkouri M, Reinberg A, Mathé G (1985a) Circadian and/or circahemidian rhythms in nine lymphocyte-related variables from peripheral blood of healthy subjects. J Immunol 134:217–222

Lévi F, Mechkouri M, Roulon A, Bailleul F, Horvath C, Reinberg A, Mathé G (1985b) Circadian rhythm in tolerance of mice for etoposide. Cancer Treat Rep 69:1443–1445

Lévi F, Adam R, Soussan A (1988a) Ambulatory 5-day chronotherapy of colorectal or pancreatic cancer with continuous venous infusion of 5-fluorouracil at circadian-modulated rate. Annu Rev Chronopharmacol 5:419–422

Lévi F, Blazcek I, Ferlé-Vidovic A (1988b) Circadian and seasonal changes in murine bone marrow colony forming cells affect tolerance for 4-tetrahydropyranyladriamycin. Exp Hematol 16:696–701

Lévi F, Canon C, Touitou Y, Reinberg A, Mathé G (1988c) Seasonal modulation of the circadian time structure of circulating T and natural killer lymphocyte subsets from healthy subjects. J Clin Invest 81:407–413

Lévi F, Canon C, Touitou Y, Sulon J, Mechkouri M, Ponsart ED, Touboul JP, Vannetzel JM, Mowzowicz I, Reinberg A, Mathé G (1988d) Circadian rhythms in circulating T lymphocyte subtypes and plasma testosterone, total and free cortisol in five healthy men. Clin Exp Immunol 71:329–335

Lohrmann H-P, Schreml W (1982a) Cytotoxic drugs and the granulopoietic system. Springer, Berlin Heidelberg New York (Recent results in cancer research, vol 81)

Lohrmann H-P, Schreml W (1982b) Granulopoietic toxicity of cytotoxic agents: pathogenesis, pathophysiology, methods of modulation, and clinical aspects. In: Lohrmann HP, Schreml W (eds) Cytotoxic drugs and the granulopoietic system. Springer, Berlin Heidelberg New York, pp 155–182

Lund-Johansen F, Bjerknes R, Laerum OD (1990) Flow cytometric assay for the measurement of human bone marrow phenotype, function and cell cycle. Cytometry 11:610–616

Maes M, Stevens W, Scharpe S, Bosmans E, Demeyer F, Dhondt P, Peeters D, Thompson P, Cosyns P, Declerck L, Bridts C, Neels H, Wauters A, Cooreman W (1994) Seasonal variation in peripheral blood leukocyte subsets and in serum in-

terleukin-6, and soluble interleukin-2 and -6 receptor concentrations in normal volunteers. Experientia 50:821–829
Malek J, Suk K, Brestak M (1962) Daily rhythms of leukocytes, blood pressure, pulse rate and temperature during pregnancy. Ann NY Acad Sci 98:1018–1091
Marler JR, Price TR, Clark GL, Muller JE, Robertson T, Mohr JP, Hier DB, Wolf PA, Caplan LR, Foulkes MR (1989) Morning increase in onset of ischemic stroke. Stroke 20:473–476
Marshall J (1977) Diurnal variation in the occurrence of strokes. Stroke 8:230–231
Martini E, Muller JY, Doinel C, Gastal C, Roquin H, Douay L, Salmon C (1988a) Disappearance of CD4 lymphocyte circadian cycles in HIV infected patients: early even during asymptomatic infection. AIDS 2:133–134
Martini E, Muller JY, Gastal C, Doinel C, Meyohas MC, Roquin H, Frottier J, Salmon C (1988b) Early anomalies of CD4 and CD20 lymphocyte cycles in human immunodeficiency virus. Presse Med 17:2167–2168
Mauer AM (1965) Diurnal variation of proliferative activity in the human bone marrow. Blood 26:1–7
Mehta J, Agarwal MB (1980) Cyclic oscillations in leukocyte count in chronic myeloid leukemia. Acta Haematol 63:68–70
Mehta J, Malloy M, Lawson D, Lopez L (1989) Circadian variation in platelet alpha2-adrenoceptor affinity in normal subjects. Am J Cardiol 63:1002–1005
Metcalf D (1986) Annotation. Haematopoietic growth factors now cloned. Br J Haematol 62:409–412
Metcalf D, Stevens S (1972) Influence of age and antigenic stimulation on granulocyte and macrophage progenitor cells in the mouse spleen. Cell Tissue Kinet 5:433–446
Miescher PA, Pola W (1986) Haematological effects of non-narcotic analgesics. Drugs 32:90–108
Migliaccio AR, Migliaccio G, Dale DC, Hammond WP (1990) Hematopoietic progenitors in cyclic neutropenia: effect of granulocyte colony-stimulating factor in vitro. Blood 75:1951–1959
Miyawaki T, Taga K, Nagaoki T, Seki H, Suzuki Y, Taniguchi N (1984) Circadian changes of T lymphocyte subsets in human peripheral blood. Clin Exp Immunol 55:618–622
Moldofsky H, Lue FA, Davidson JR, Gorczynski R (1989) Effects of sleep deprivation on human immune functions. FASEB J 3:1972–1977
Moldofsky H, Lue FA, Eisen J, Keyston E, Gorczynsky RM (1986) The relationship of interleukin-1 and immune functions to sleep in humans. Psychosom Med 48:309–318
Morley AA (1966) A neutrophil cycle in healthy individuals. Lancet ii:1220
Morley AA, Baikie AG, Galton DAG (1967) Cyclic leukocytosis as evidence for retention of normal homeostatic control in chronic granulocytic leukemia. Lancet ii: 1320–1323
Morra L, Ponassi A, Caristo G, Bruzzi P, Zunino R, Parodi GB, Sacchetti C (1984) Comparison between diurnal changes and changes induced by hydrocortisone and epinephrine in circulating myeloid progenitor cells (CFU-GM) in man. Biomed Pharmacother 38:167–170
Moskalik KG (1976) Diurnal rhythm of mitotic activity, DNA synthesis, and duration of mitoses in mouse bone marrow cells. Bull Exp Biol Med (USSR) 81:594
Muller JE, Stone PH, Turi SG, Rutherford JD, Czeisler CA, Parker C, Poole WK, Passamani E, Roberts R, Robertson T, Sobel BE, Willerson JT, Braunwald ESG (1985) Circadian variation in the frequency of onset of acute myocardial infarction. N Engl J Med 313:1315–1322
Muller JE, Ludmer PL, Willich N, Tofler GH, Aylmer G, Klangos I, Stone PE (1987) Circadian variation in the frequency of sudden cardiac death. Circulation 75:131–138
Nathan DG (1988) Hematologic diseases. Textbook of medicine. Saunders, Philadelphia
Nelson W, Tong Y, Lee JK, Halberg F (1979) Methods for cosinor rhythmometry. Chronobiologia 6:305–323

Ogawa M (1993) Differentiation and proliferation of hematopoietic stem cells. Blood 81:2844–2853

Paukovits WR, Elgjo K, Laerum OD (1990a) Pentapeptide growth inhibitors. In: Sporn MB, Roberts AB (eds) Peptide growth factors and their receptors II. Springer, Heidelberg Berlin New York, pp 267–295

Paukovits WR, Guigon M, Binder KA, Hergl A, Laerum OD, Schulte-Hermann R (1990b) Prevention of hematotoxic side effects of cytostatic drugs in mice by a synthetic hemoregulatory peptide. Cancer Res 50:328–332

Perpoint B, Le Bousse-Kerdiles C, Clay D, Smadja-Joffe F, Depres-Brummer P, Laporte-Simitsidis S, Jasmin C, Lévi F (1995) In vitro chronopharmacology of recombinant mouse IL-3, mouse GM-CSF and human G-CSF on murine myeloid progenitor cells. Exp Haematol 23:362–368

Petralito A, Mangiafico RA, Gibiino S, Cuffari MA, Miano MF, Fiore CE (1982) Daily modifications of plasma fibrinogen, platelet aggregation, Howell's time, PPT, TT and antithrombin III in normal subjects and in patients with vascular disease. Chronobiologia 9:195–201

Pisciotta AV (1969) Agranulocytosis induced by certain phenothiazine derivatives. JAMA 308:1862–1868

Pizzarello DJ, Witcofski RL (1970) A possible link between diurnal variations in radiation sensitivity and cell division in bone marrow of male mice. Radiology 97:165–167

Rabkin SW, Mathewson FA, Tate RB (1980) Chronobiology of cardiac sudden death in men. JAMA 244:1357–1358

Ramot B, Brok-Simoni F, Chweidan E, Askenazy YE (1976) Blood leukocyte enzymes. III. Diurnal rhythm of activity in isolated lymphocytes of normal subjects and chronic lymphatic leukemia patients. Br J Haematol 34:79–85

Reinberg A, Gervais P, Halberg F, Gaultier M, Roynette N, Abulker CH, Dupont J (1973) Mortalité des adultes: rythmes circadiens et circannuels dans un hopital parisien et en France. Nouv Presse Med 6:289–292

Reinberg A, Schuller E, Delasnerie N, Clench J, Helary M (1977) Rhythmes circadiens et circannuels des leucocytes, proteines totale, immunoglobulines A, G et M. Etude chez 9 adultes jeunes et sains. Nouv Presse Med 6:3819–3823

Reinberg A, Schuller E, Clench J, Smolensky MH (1980) Circadian and circannual rhythms of leukocytes, proteins and immunoglobulins. Recent advances in the chronobiology of allergy and immunology. Pergamon, New York, pp 251–259

Renbourn ET (1947) Variation, diurnal and over longer periods of time, in blood hemoglobin, hematocrit, plasma protein, erythrocyte sedimentation rate and blood chloride. J Hyg 45:455

Ritchie AWS, Oswald I, Micklem HS, Boyd JE, Elton RA, Jazwinska E, James K (1983) Circadian variation of lymphocyte subpopulations: a study with monoclonal antibodies. Br Med J 286:1773–1775

Rivard GE, Infante-Rivard C, Hoyoux C, Champagne J (1985) Maintenance chemotherapy for childhood acute lymphoblastic leukemia: better in the evening. Lancet ii:1264–1266

Rivard GE, Infante-Rivard C, Dresse M-F, Leclerc J-M, Champagne J (1993) Circadian time-dependent response of childhood lymphoblastic leukemia to chemotherapy: a long-term follow-up study of survival. Chronobiol Int 10:201–204

Rocker L, Feddersen HM, Hoffmeister H, Junge B (1980) Jahreszeitliche Veränderungen diagnostisch wichtiger Blutbestandteile. Klin Wochenschr 58:769–778

Roemeling VR (1991) The therapeutic index of cytotoxic chemotherapy depends upon circadian drug timing. Ann NY Acad Sci 618:292–311

Roitman B, Sothern RB, Halberg F, von Mayersbach H, Scheving LE, Haus E, Bartter FC, Delea C, Simpson H, Tavadia H, Fleming K, Hume P, Wilson C, Halberg E (1975) Circadian acrophases for total blood leukocytes counted on different continents. Chronobiologia 2 [Suppl 1]:58

Ross DD, Pollack A, Akman SA, Bachur NR (1980) Diurnal variation of circulating human myeloid progenitor cells. Exp Haematol 8:954–960

Rud F (1947) The eosinophil count in health and in mental disease. A biometrical study. Thesis, Oslo

Sabin FR, Cunningham RS, Doan CA, Kindwale JA (1927) The normal rhythm of white blood cells. Bull Johns Hopkins Hosp 37:14–67

Scheving LE (1959) Mitotic activity in the human epidermis. Anat Rec 135:7–19

Scheving LE (1981) Circadian rhythms in cell proliferation: their importance when investigating the basic mechanism of normal versus abnormal growth. In: von Mayersbach H, Scheving L E, Pauly JE (eds) 11th international congress of anatomy, part C, biological rhythms in structure and function. Liss, New York, pp 39–79

Scheving LE (1984) Chronobiology of cell proliferation in mammals: In: Rdmunds LE (ed) Cell cycle clocks. Implications for basic research and cancer chemotherapy. Cell cycle clocks. Dekker, New York, pp 455–499

Scheving LE, Burns ER, Pauly JE, Tsai TH (1978) Circadian variation in cell division of the mouse alimentary tract, bone marrow, and corneal epithelium, and its possible implication in cell kinetics and cancer chemotherapy. Anat Res 191:479–486

Scheving LE, Burns ER, Pauly JE, Halberg F (1980) Circadian bioperiodic response of mice bearing advanced L1210 leukemia to combination therapy with adriamycin and cyclophosphamide. Cancer Res 40:1511–1515

Scheving LE, Tsai TS, Feuers RJ, Scheving LA (1989) Cellular mechanisms involved in the action of anticancer drugs. In: Lemmer B (ed) Chronopharmacology: cellular and biochemical interactions. Dekker, New York, pp 317–369

Scheving LE, Tsai TH, Scheving LA, Feuers RJ (1991) The potential of using the natural rhythmicity of cell proliferation in improving cancer chemotherapy in rodents. Temporal control of drug delivery. Ann NY Acad Sci 618:182–227

Scheving LE, Feuers RJ, Tsai TH, Scheving LA (1994) Experimental background for cancer chronotherapy. In: Hrushesky WHM (ed) Circadian cancer therapy. CRC Press. Boca Raton, pp 19–40

Schofield R (1978) The relationship between the spleen colony-forming cell and the haematopoietic stem cell. A hypothesis. Blood Cells 4:7–25

Shadduck RK, Winkelstein A, Nunna NG (1972) Cyclic leukemia cell production in CML. Cancer 29:399–401

Sharkis SJ, LoBue J, Alexander PJ, Rakowitz F, Weitz-Hamburger A, Gordon AS (1971) Circadian variations in mouse hematopoiesis. II. Sex differences in mitotic indices of femoral diaphyseal marrow cells. Proc Soc Exp Biol Med 138:494–496

Sharkis SJ, Palmer JD, Goodenough J, LoBue J, Gordon AS (1974) Daily variations of marrow and splenic erythropoiesis, pinna epidermal cell mitosis and physical activity in C57B 1+6J mice. Cell Tiss Kinet 7:381–387

Sharp GWG (1960) Reversal of diurnal leukocyte variations in man. J Endocrinol 21:107–114

Signore A, Cugini P, Letizia C, Lucia P, Murano G, Pozilli P (1985) Study of the diurnal variation of human lymphocyte subsets. J Clin Lab Immunol 17:25–28

Simmons DJ, Loeffelman K Frier C, McCoy R, Friedman B, Melville S, Kahn AJ (1982) Circadian changes in the osteogenic competence of marrow stromal cells. In: Haus E, Kabat HF (eds) Chronobiology 1982–1983, Proceedings 15th International Conference on Chronobiology. S Karger, Basel, pp 37–42

Sletvold O (1987) Circadian rhythms of peripheral blood leukocytes in aging mice. Mech Ageing Dev 39:251

Sletvold O, Laerum OD (1988) Multipotent stem cell (CFU-S) numbers and circadian variations in aging mice. Eur J Haematol 41:230–236

Sletvold O, Laerum OD, Riise T (1988a) Age-related differences and circadian and seasonal variations in myelopoietic progenitor cell (CFU-GM) numbers in mice. Eur J Haematol 40:42–49

Sletvold O, Laerum OD, Riise T (1988b) Rhythmic variations of different hemopoietic cell lines and maturation stages in aging mice. Mech Aging Dev 42: 91-104

Sletvold O, Smaaland R, Laerum OD (1991) Cytometry and time-dependent variations in peripheral blood and bone marrow cells: a literature review and relevance to the chronotherapy of cancer. Chronobiol Int 8:235–250

Smaaland R, Sothern RB (1994) Cytokinetic basis for circadian pharmacodynamics: circadian cytokinetics of murine and human bone marrow and human tumours. In: Hrushesky WJM (ed) Circadian cancer therapy. CRC Press, Boca Raton, pp 119–163

Smaaland R, Lote KOS, Kamp D, Wiedemann G, Laerum OD (1989) Rhythmen in Knochenmark und Blut: Unterschiede wie Tag und Nacht. Dtsch Med Wochenschr 114:845–849

Smaaland R, Laerum OD, Lote K, Sletvold O, Sothern RB, Bjerknes R (1991a) DNA synthesis in human bone marrow is circadian stage dependent. Blood 77:2603–2611

Smaaland R, Lote K, Sletvold O, Bjerknes R, Aakvaag A, Vollset SE, Laerum OD (1991b) Circadian stage dependent variation of cortisol related to DNA synthesis in human bone marrow. Ann N Y Acad Sci 618:605–609

Smaaland R, Svardal AM, Lote K, Ueland PM, Laerum OD (1991c) Glutathione content in human bone marrow and circadian stage in relation to DNA synthesis. J Natl Cancer Inst 83:1092–1098

Smaaland R, Abrahamsen JF, Svardal AM, Lote K, Ueland PM (1992a) DNA cell cycle distribution and glutathione (GSH) content according to circadian stage in bone marrow of cancer patients. Br J Cancer 66:39–45

Smaaland R, Laerum OD, Sothern RB, Sletvold O, Bjerknes R, Lote K (1992b) Colony-forming units – granulocyte/macrophage and DNA synthesis of human bone marrow are circadian stage-dependent and show covariation. Blood 79:2281–2287

Smaaland R, Lote K, Sothern RB, Laerum OD (1993) DNA synthesis and ploidy in non-Hodgkin's lymphomas demonstrate intra-patient variation depending on circadian stage of cell sampling. Cancer Res 53:3129–3138

Sothern RB, Nelson WL, Halberg F (1977) A circadian rhythm in susceptibility of mice to the anticancer drug, adriamycin. Proceedings of the XIIth international conference of the International Society of Chronobiology, pp 433–438

Sothern RB, Smaaland R, Moore JG (1995) Circannual rhythm in DNA synthesis (S-phase) in healthy human bone marrow and rectal mucosa. FASEB 9:397–403

Spivak JL (1984) Normal hematopoiesis. The principles and practice of medicine. Appleton-Century-Crofts, Norwalk

Stoney PJ, Halberg F, Simpson HW (1975) Circadian variation in colony-forming ability of presumable intact murine bone marrow cells. Chronobiologia 2:319

Suda T, Suda J, Ogawa M, Ihle JN (1985) Permissive role of interleukin 3 (IL-3) in proliferation and differentiation of multipotential hemopoietic progenitors in culture. J Cell Physiol 124:182

Swoyer J, Irvine P, Sackett-Lundeen L, Conlin L, Lakatua D, Haus E (1989) Circadian hematologic time structure in the elderly. Chronobiol Int 6:131–137

Swoyer J, Rhame F, Hrushesky WJM, Sackett-Lundeen L, Sothern R, Gale H, Haus E (1990) Circadian rhythm alterations in HIV infected patients. In: Hayes D, Pauly J, Reiter R (eds) Chronobiology: its role in clinical medicine, general biology, and agriculture. Wiley, New York, pp 437–449

Swoyer JK, Sackett-Lundeen L, Haus E, Lakatua DJ, Taddeini L (1975) Circadian lymphocytic rhythms in clinically healthy subjects and in patients with hematologic malignancies. International congress on rhythmic functions in biological systems, pp 62–63

Tavadia HB, Fleming KA, Hume PD, Simpson HW (1975) Circadian rhythmicity of human plasma cortisol and PHA-induced lymphocyte transformation. Clin Exp Immunol 22:190–193

Tofler GH, Brezinski D, Schafer AI, Czeisler CA, Rutherford JD, Willich SN, Gleason RE, Williams GH, Muller JE (1987) Concurrent morning increase in platelet aggregability and the risk of myocardial infarction and sudden cardiac death. N Engl J Med 316:1514–1518

Touitou Y, Touitou C, Bogdan A, Chasselut J, Beck H, Reinberg A (1979) Circadian rhythms in blood variables in elderly subjects. In: Reinberg A, Halberg F (eds) Chronopharmacology: advances in biosciences. Pergamon, New York, pp 283–290

Touitou Y, Touitou C, Bogdan A, Reinberg A, Auzeby A, Beck H, Guillet P (1986) Differences between young and elderly subjects in seasonal and circadian variations of total plasma proteins and blood volume as reflected by hemoglobin, hematocrit and erythrocyte counts. Clin Chem 32:801–804

Touitou Y, Lévi F, Bogdan A, Bruguerolle B (1990) Abnormal patterns of plasma cortisol in breast cancer patients. Annu Rev Chronopharmacol 7:245–248

Umemura T, Hirata J, Kaneko S, Nishimura JSM, Kozuru M, Ibayashi H (1986) Periodical appearance of erythropoietin-independent erythropoiesis in chronic myelogenous leukemia with cyclic oscillation. Acta Haematol 76:230–234

Vacek A, Rotkovska D (1970) Circadian variations in the effect of X-irradiation on the haematopoietic stem cells of mice. Strahlentherapie 140:302–306

Varmus HE, Lowell CA (1994) Cancer genes and hematopoiesis. Blood 83:5–9

Verma DS, Fisher R, Spitzer G, Zander AR, McCredie KB, Dicke KA (1980) Diurnal changes in circulating myeloid progenitor cells in man. Am J Haematol 9:185–192

Verma DS, Spitzer G, Zander AR, Dicke KA, McCredie KB (1982) Cyclic neutropenia and T lymphocyte suppression of granulopoiesis: abrogation of the neutropenic cycles by lithium carbonate. Leuk Res 6:567–576

Vodopick H, Rupp EM, Edwards CL, Goswitz FA, Beauchamp JJ (1972) Spontaneous cyclic leukocytosis and thrombocytosis in chronic granulocytic leukaemia. N Engl J Med 286:284–290

Volini IF, Greenspan I, Ehrlich L, Gonner JA, Felsenfeld O, Schwartz SO (1950) Hemopoietic changes during administration of chloramphenicol (chloromycetin). JAMA 142:1333–1335

von Schulthess GV, Mazer NA (1982) Cyclic neutropenia (CN): a clue to the control of granulopoiesis. Blood 59:27–37

Wheldon TE, Kirk J, Finlay HM (1974) Cyclic granulopoiesis in chronic granulocytic leukemia: a simulation study. Blood 43:379–385

Wide L, Bengtsson C, Birgegard G (1989) Circadian rhythm of erythropoietin in human serum. Br J Haematol 72:85–90

Williams LH, Udupa KB, Lipschitz DA (1986) Evaluation of the effect of age on hemopoiesis in young and old mice. Exp Haematol 14:827–832

Williams RM, Krause LJ, Dubey DP, Yunis EJ, Halberg F (1979) Circadian bioperiodicity in natural killer cell activity of human blood. Chronobiologia 6:172

Wintrobe MM (1981) Clinical hematology. Lead and Febiger, Philadelphia

Wood PA, Sanchez de la Peña S, Hrushesky WJM (1990) Evidence for circadian dependency of recombinant human erythropoietin (rhEPO) response in the mouse. Annu Rev Chronopharmacol 7:173–176

Young N, Mortimer P (1984) Viruses and bone marrow failure. Blood 63:729–737

Zagon IS, Wu Y, Mclaughlin PJ (1994) Opioid growth factor inhibits DNA synthesis in mouse tongue epithelium in a circadian rhythm-dependent manner. Am J Physiol 267:R645–R652

Zinneman HH, Thompson M, Halberg F, Kaplan M, Haus E (1972) Circadian rhythms in urinary Bence-Jones protein excretion. Clin Res 20:798

Zinneman HH, Halberg F, Haus E, Kaplan M (1974) Circadian rhythms in urinary light chains, serum iron and other variables of multiple myeloma patient. Int J Chronobiol 2:3–16

CHAPTER 19
Chronobiology of the Haemostatic System

F. ANDREOTTI and G. PATTI

A. The Haemostatic System and Its Laboratory Evaluation

When a blood vessel is cut or damaged, the blood in its vicinity tends to solidify. The principal events of this process, known as haemostasis, are: (1) the constriction of the damaged vessel, (2) the deposition of circulating platelets over the injured area and (3) the conversion of fluid blood into a gelatinous mass, or fibrin clot (GANONG 1979). The following introductory sections will outline the function and laboratory evaluation of the elements that are thought to govern haemostasis, that is, endothelial cells, vasomotion, platelets, coagulation and fibrinolysis. A simplified depiction of the main events involved in the haemostatic process is shown in Fig. 1.

I. The Endothelium

In undamaged vessels, blood remains fluid because the intact vessels are internally lined by a monolayer of endothelial cells which continuously produce substances with vasodilator and antiplatelet action (nitric oxide, prostacyclin and tissue-type plasminogen activator) or anticoagulant (heparan sulphate and thrombomodulin) and fibrinolytic properties (plasminogen activators) (AZUMA et al. 1986; JAFFE 1987; LOSCALZO and VAUGHAN 1987; MONCADA et al. 1987; THOMPSON and HARKER 1988). Endothelial cells also produce the vasoconstrictor peptide endothelin-1, and the inhibitor of fibrinolysis, plasminogen activator inhibitor-1 (YANAGISAWA et al. 1988; SPRENGERS and KLUFT 1987). Endothelial function can be defined by measurement of: (1) circulating levels of endothelial cell products, (2) vascular responses before and after removal of the endothelial lining, and (3) endothelial reactions to stimuli.

II. Vasomotion

The intermediate layer, or media, of the three layers that make up the arterial wall is the thickest and is mainly composed of multiple tiers of smooth muscle cells which contract or relax in response to neurotransmitters, hormones, or locally produced substances. A persistent degree of smooth muscle cell contraction is present physiologically and is called vascular tone. Vasomotor function can be investigated by measuring: (1) vascular resistance, calculated as

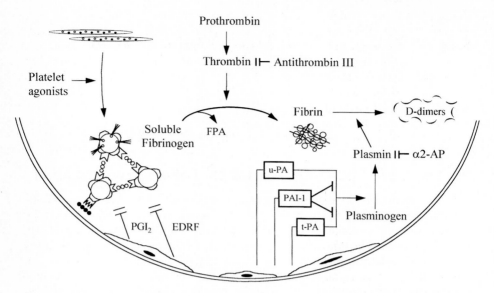

Fig. 1. Simplified scheme of the main events involved in haemostasis. Quiescent platelets (*top left*) are stimulated by agonists or by endothelial cell damage to change into a spherical shape and to release the content of their granules. Activated platelets stick to the subendothelium through the adhesive molecule von Willebrand factor (*four solid circles*) and to one another via fibrinogen bridges (*three open circles*). On the platelet surface, thrombin converts soluble fibrinogen into fibrin strands and releases fibrinopeptide A (*FPA*). The fibrin network consolidates the platelet aggregates. Endothelial cells produce the fibrinolytic enzymes, tissue-type plasminogen activator (*t-PA*) and urinary-type plasminogen activator (*u-PA*), which convert plasminogen into plasmin. Plasmin digests fibrin into soluble degradation products, which typically contain two D domains (D-dimers). The endothelium also produces antiplatelet factors, such as prostacyclin (*PGI$_2$*) and endothelium-derived relaxing factor (*EDRF*). The latter also dilates vascular smooth muscle cells (not shown). Inhibitors of fibrinolysis include the plasminogen activator inhibitor-1 (*PAI-1*), which neutralizes both t-PA and u-PA, and α$_2$-antiplasmin (*α2-AP*). Thrombin is rapidly inhibited by antithrombin III. (Vasoconstriction is not depicted)

the ratio of intravascular blood pressure over blood flow, or (2) the vascular response to constrictor or dilator stimuli.

III. Platelets

Quiescent platelets are circulating, discoid particles, 2–4 mm in diameter, containing secretory granules: the more electron-dense δ-granules contain adenosine diphosphate (ADP) and triphosphate, calcium and serotonin; the less dense α-granules contain growth factors, adhesive and coagulation proteins, the plasminogen activator inhibitor type 1, β-thromboglobulin and platelet factor 4 (WHITE 1974).

Platelets adhere to damaged vascular areas because removal of the endothelium exposes underlying molecules of collagen, fibronectin and von

Willebrand factor, which bind to receptors on the platelet surface (PLOW and GINSBERG 1981; NIEUWENHUIS et al. 1985; MICHELSON et al. 1986). After adhesion to the subendothelium or following the interaction with soluble agonists (ADP, catecholamines, thrombin, serotonin, thromboxane A_2), the platelets are "activated" to release the content of their granules, including ADP, which attracts other platelets (SHATTIL 1981). They are also stimulated to adopt a spherical shape, to extend pseudopods and to stick together through bridges of fibrinogen molecules, forming a compact aggregate, called a haemostatic plug, that seals off the damaged area (BENNET et al. 1982) (Fig. 1).

Platelet agonists bind to receptors on the platelet membrane; these are coupled with guanine nucleotide regulatory proteins which inhibit intraplatelet adenylate cyclase activity and reduce cyclic adenosine monophosphate (AMP) concentrations; this in turn activates phospholipase C, produces inositol triphospate and diacylglycerol, raises intraplatelet calcium concentrations, phosphorylates contractile and other proteins, and causes extrusion of the granule content. Phospholipase A_2 is also activated to form arachidonic acid, which produces thromboxane A_2 and leukotrienes through the action of cyclooxygenase, lipooxygenase and thromboxane synthetase. Platelet α-adrenoceptors, predominantly α_2-receptors, exert part of their agonist effect by inhibiting adenylate cyclase. Inhibition of platelet function is associated, instead, with increased intraplatelet adenylate cyclase activity and cyclic AMP concentrations (NEWMAN et al. 1978; HOFFMANN et al. 1979; BERRIDGE 1984; HANDIN and LOSCALZO 1992).

The assays used to investigate platelet function include the study of: (1) platelet adhesiveness, measured in vitro as the number of retained platelets on a glass bead column (HAUS et al. 1990); (2) platelet serotonin content, which has been inversely related to serotonin secretion (MONTERO et al. 1989); (3) plasma concentrations of β-thromboglobulin, platelet factor 4 or 11-dehydro-thromboxane B_2, considered indices of in vivo platelet activation (KAPLAN and OWEN 1981; BRIDGES et al. 1991); (4) the number of circulating platelet aggregates, assessed as the ratio of platelet count in ethylenediaminetetraacetate (EDTA) plus formalin, over platelet count in EDTA (the method is based on the principle that formalin fixes the aggregates, whereas EDTA on its own breaks them up) (WU and HOAK 1974); (5) platelet aggregation, determined in vitro by measuring the increased intensity of light transmission through a platelet suspension after the addition of agonists such as ADP and noradrenaline (BORN 1962; O'BRIEN 1964); and (6) platelet α_2-adrenoceptor activity, assessed using the radioactive α_2-receptor ligand ^{3}H-yohimbine (MEHTA et al. 1989; WILLICH et al. 1992).

IV. Coagulation and Viscosity

At points of vascular damage, as platelets adhere to the subendothelium, the exposure of subendothelial collagen and of the lipoprotein tissue factor by injured cells also triggers the intrinsic (factor XII initiated) and extrinsic (tissue factor plus factor VII initiated) coagulation pathways, which culminate through a common sequence in the conversion of soluble fibrinogen into insoluble strands of fibrin by action of the final coagulation enzyme thrombin. Thrombin activity is counteracted by inhibitors, such as antithrombin III, which rapidly bind to thrombin, and protein C, which is activated by thrombomodulin and inhibits coagulation cofactors V and VIII (VERSTRAETE and VERMYLEN 1984). The clotting function of blood can be investigated through global tests or by measuring the plasma levels of individual coagulation-related factors.

Global coagulation assays include: (1) the prothrombin time, in which tissue thromboplastin (containing tissue factor) and calcium are added to plasma; the assay screens the extrinsic coagulation pathway and is prolonged by reduced levels of factor VII or of factors of the common pathway (factors X, V, prothrombin or factor II and fibrinogen); (2) the activated partial thromboplastin time (aPTT), which screens the intrinsic coagulation factors (XII, XI, IX, and VIII) and the factors of the common system; the clotting sequence is activated by contact with a finely divided surface and the time to form a clot is measured after adding calcium and phospholipid; (3) the thrombin time, performed by adding a low concentration of exogenous thrombin to plasma; and (4) Howell's time, or recalcification time, performed by adding calcium to platelet-rich plasma (PETRALITO et al. 1982; THOMPSON and HARKER 1988).

Individual coagulation-related factors include single coagulation proteins, as well as certain reaction products, such as the complex between thrombin and antithrombin III, or fibrinopeptide A, released when fibrinogen is converted to fibrin. Thrombin-antithrombin III and fibrinopeptide A reflect recent thrombin generation or activity, as their plasma half-life is approximately 5 min. Coagulation is enhanced by elevated blood and plasma viscosities; these, in turn, are directly related to haematocrit and to the protein concentration of plasma (GANONG 1979). Viscosity is measured as the resistance of blood or plasma to flow.

V. Fibrinolysis

The formation of fibrin is counterbalanced by the fibrinolytic system (ASTRUP 1956), which ultimately leads to the breakdown of the fibrin strands into soluble fragments, or degradation products, through the action of plasmin. In order to be active, plasmin must be converted from its inactive precursor, plasminogen, by specific enzymes, called plasminogen activators (Fig. 1). Plasminogen and plasminogen activators assemble on the fibrin surface and locally generate active plasmin (COLLEN 1985; BACHMANN 1987). The availability of plasminogen activators is rate-limiting and therefore crucial for fibrinolysis.

Excessive fibrinolysis is prevented by the inhibitor of plasmin, α_2-antiplasmin, and by the plasminogen activator inhibitor type 1 (PAI-1), which rapidly and irreversibly neutralizes the two main activators of plasminogen: tissue-type plasminogen activator (t-PA) and urinary-type plasminogen activator (u-PA) (SPRENGERS and KLUFT 1987). Under normal conditions, neither free plasmin nor u-PA are measurable in plasma; for this reason, the capacity of blood to dissolve fibrin has been measured as the activity of t-PA, which in turn is closely regulated by PAI-1 (ANDREOTTI et al. 1988a; KLUFT et al. 1988).

As for coagulation, the assays currently used to detect fibrinolysis in blood can be divided into: (1) global tests and (2) assays that measure individual fibrinolytic factors. The global tests require dilution of the blood sample, as this causes a preferential reduction of the inhibitors of fibrinolysis compared with the active fibrinolytic enzymes (FEARNLEY 1965). They include: (1) the dilute blood-clot lysis-time, in which whole blood is diluted tenfold and then clotted by the addition of thrombin: the shorter the time to lyse the clot, the greater the fibrinolytic activity in the sample (FEARNLEY 1965); (2) the fibrinolytic activity of the euglobulin fraction of plasma, measured using a chromogenic substrate or the clot lysis time (the euglobulin fraction is obtained by plasma dilution and protein precipitation to remove antiplasmins and antiactivators) (DAVIDSON and WALKER 1987); and (3) the fibrin plate assay, in which a standard volume of euglobulin fraction is placed on a fibrin-coated Petri dish: the area of lysis is proportionate to the sample's fibrinolytic activity (ASTRUP 1956).

Virtually all the individual fibrinolytic factors can be measured by sensitive and specific, commercially available kits.

B. Chronobiology of the Haemostatic System

The following sections will focus on the physiological circadian (24-h) or diurnal (daytime) variations of haemostatic function in man. We will mainly refer to observations made in healthy humans, and consider data from patients or experimental animals only when information from normal volunteers is lacking.

I. Circadian Variation of Endothelial Cell Function

1. Endothelium-Related Factors in Plasma

A circadian rhythm in the activity of endothelial cells has been hypothesized to explain the marked fluctuation in the plasma levels of two endothelial cell products, t-PA and PAI-1, during the 24-h cycle. The plasma mass concentration of both these proteins is highest in the morning and lowest in the evening (see Sect. B.V.2 and Fig. 2). This rhythm, if present, however, would involve only certain factors, since other products synthesized and released by endothelial cells, such as von Willebrand factor, show uniform plasma concentrations over the 24 h (JOHANSEN et al. 1990; BRIDGES et al. 1991).

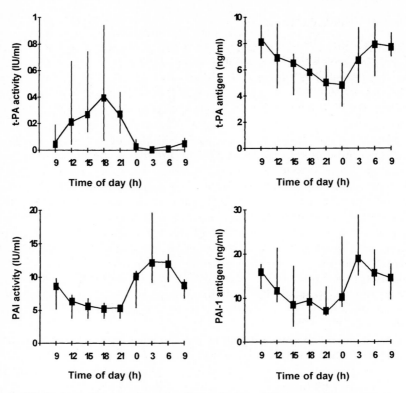

Fig. 2. Fibrinolytic factors of ten healthy volunteers (five men, five women, age 23–80 years, mean 47 years) measured in plasma every 3 h for 24 h. *Boxes* are medians; *bars* are interquartiles. *t-PA*, tissue-type plasminogen activator; *PAI-1*, plasminogen activator inhibitor type 1. (Modified from ANDREOTTI et al. 1988a)

A circadian variation in the plasma concentrations of endothelin-1 has been documented, to our knowledge, only in a group of 12 patients with vasospastic angina. Endothelin levels were measured 1 to 3-hourly between 0800 and 0100 hours and found to be 30%–50% higher at 0800 hours than between 1600 and 2100 hours (ARTIGOU et al. 1993).

A possible circadian rhythm in the endothelial release of prostacyclin has been assessed only indirectly, in ten patients with chronic stable angina, through six consecutive, 4-hourly measurements of plasma prostacyclin stimulating factor (PGI_2 SF) levels. No significant fluctuation in PGI_2 SF was detected over time (BRIDGES et al. 1993). This finding, however, does not entirely exclude the presence of a circadian variation in the secretion of prostacyclin, since a constant stimulatory input may still generate a rhythm of secretion, provided the target organ has a rhythm of responsiveness (ASCHOFF 1979).

2. Effect of Endothelial Removal

To our knowledge, the only information on this topic comes from experiments in rats, where the constrictor effect of noradrenaline on aortic rings was found to vary considerably over the 24-h period (see below). Removal of the endothelial lining enhanced the constrictor response to noradrenaline, but left the circadian rhythm unaffected (GOHAR et al. 1992).

3. Endothelial Reaction to Stimuli

Information on a possible circadian rhythm of the endothelial response to stimuli is scanty. ANGLETON et al. (1989) measured the plasma levels of t-PA activity after 10 min of brachial venous occlusion – considered a stimulus for the endothelial release of t-PA – in 33 healthy men (mean age 31 years) at 0800 and 2000 hours, and found no significant difference in post-occlusion t-PA activity at the two times of day.

Thus, the evidence for a circadian variation in endothelial cell function is still inconclusive.

II. Circadian Variation of Vasomotor Function

1. Coronary Artery Tone

To our knowledge, the only study on the diurnal variation of coronary arterial blood flow has been performed in dogs and deserves some attention. In 21 conscious dogs maintained with constant myocardial metabolic requirements, coronary arterial blood flow was measured by Doppler ultrasonic probe at 0830 and 1630 hours. Flow velocity increased from a mean of 18.8 cm/s in the morning to a mean of 21.2 cm/s in the afternoon ($P<0.001$), while heart rate and blood pressure did not change significantly. The authors conclude that coronary artery vasomotor tone is more pronounced in the morning than in the afternoon, possibly because of consensual changes in sympathetic α-adrenoceptor vasoconstriction (FUJITA and FRANKLIN 1987).

2. Forearm Arterial Tone

In eight normal men (20–38 years old), forearm arterial blood flow was measured both at rest and after submaximal exercise of the forearm muscles at four time points between 0800 and 2330 hours using venous-occlusion plethysmography. Both resting and postexercise blood flow increased by about 45% from 0800 hours in the morning to a maximum in the afternoon. Although arterial blood pressure was not measured in these subjects, the authors argue that the known diurnal increase in blood pressure is of smaller magnitude than their observed changes in flow, and conclude that a fall in peripheral vascular resistance probably occurs from morning to afternoon (KANEKO et al. 1968).

This hypothesis was later confirmed in 12 healthy subjects (mean 44 years old), in whom basal forearm vascular resistance was measured between 0700 and 2100 hours and was found to be 41% greater at 0700 than at 2100 hours

($P < 0.01$). In the same subjects, the infusion of the α-adrenoceptor antagonist phentolamine resulted in an 81% greater vasodilator effect in the morning than in the afternoon and abolished the diurnal changes in forearm vascular tone. In contrast, the infusion of the direct vasodilator nitroprusside did not affect the diurnal variation. Thus, the higher vascular resistance in the morning depends on a greater degree of α-sympathetic vasoconstriction at this time (PANZA et al. 1991). Further confirmation has come from a study in 15 fasting resting men (mean 24 years old), in whom total forearm blood flow was found to increase by 58% from 0900 to 1730 hours ($P < 0.001$), in the absence of a significant increase in arterial blood pressure; flow in the forearm skin microcirculation, however, showed the opposite changes over time, suggesting different regulatory mechanisms for the different vascular beds studied (HOUBEN et al. 1994). A complete 24-h circadian rhythm in peripheral vascular resistance has been recently documented in 12 healthy subjects by continuous monitoring (CUGINI et al. 1993).

3. Response to Constrictor or Dilator Stimuli

A circadian variation in the vascular response to vasoconstrictor agents has been documented, to our knowledge, only in the rat, a nocturnal animal. The constrictor response to noradrenaline of aortic rings from groups of rats kept under a fixed lighting schedule and sacrificed 3-hourly for 48 h was significantly greater at the end, compared with the beginning, of the animals' resting period ($P < 0.001$). The non-selective β-adrenoceptor antagonist propranolol did not alter the response. Since the aorta is sparsely innervated, the authors conclude that this circadian variation is probably determined by changes in the density or affinity of α-adrenoceptors on the aortic vascular smooth muscle cells, rather than by changes in the reuptake of noradrenaline by the nerve endings (GOHAR et al. 1992).

In man, the effect of the direct vasodilator drug nifedipine on epicardial artery diameters and flow was examined at different times of day in 11 subjects with apparently normal coronary arteries at angiography. A 2-mg intracoronary infusion of the drug elicited the same dilator response when given in the morning (six subjects) or in the afternoon (five subjects) (RABY et al. 1993). This finding agrees with the uniform effect throughout the day of nitroprusside, another direct vasodilator, on forearm blood flow (PANZA et al. 1991) and suggests that the action of these drugs is not influenced by the physiological diurnal variation of vasomotor tone.

These studies, therefore, concur in indicating that vasomotor tone in coronary and peripheral muscle arteries and the vasoconstrictor response to noradrenaline are significantly greater in the morning (or at the end of the resting period) than in the afternoon, as a result of enhanced morning α-adrenergic stimuli. The reduced vessel calibre in the morning compared with other times of day may contribute to the occurrence of vascular occlusion at this time, through blood stasis or further vasoconstrictive stimuli.

III. Circadian Variation of Platelet Function

1. Platelet Adhesiveness and Serotonin Content

In ten healthy subjects (mean 31 years old), Haus et al. studied a number of haemostatic variables, including platelet adhesiveness, at six 4-hourly time points distributed over 3 days, starting at 0800 hours and ending at 0400 hours. A statistically significant circadian fluctuation in platelet adhesiveness was detected, with highest values at 0800 hours and lowest values at midnight (HAUS et al. 1990).

The data in the literature concerning platelet serotonin content are somewhat conflicting. In a study conducted in May on ten healthy women (mean 31 years old), no significant difference in the serotonin content of platelet pellets was found by high-performance liquid chromatography in 6-hourly samples over 24 h (MONTERO et al. 1989). This finding was confirmed in a subsequent detailed investigation performed in the winter time on seven healthy men (mean age 27 years), in which blood samples were taken at 0.5–1 hourly intervals for 24 h; however, a significant variation in plasma tryptophan (the amino acid precursor of serotonin) was observed, with maximal levels in the afternoon and minimal ones at night ($P < 0.05$) (EYNARD et al. 1993). In contrast, another study conducted during autumn on 12 healthy men (19–31 years old) recorded a greater than twofold increase in platelet serotonin concentrations, measured 3-hourly for 24 h, from a nadir at 0500 hours to a peak at 1400 hours. In the same subjects, similar changes were observed in the plasma levels of α_1-acid glycoprotein, considered a natural agonist for platelet serotonin uptake (MEYERSON et al. 1989). These apparently conflicting results might be explained, at least in part, by seasonal variations in the circadian rhythmicity of platelet serotonin content (WIRZ-JUSTICE et al. 1977). Changes in platelet serotonin, in turn, might reflect not only variations in δ-granule secretion, but also in the uptake of serotonin or its precursors.

From these data, both the adhesiveness of platelets and, possibly, the secretion of the δ-granule constituent serotonin (based on the one "positive" study) seem to be greatest in the early morning compared to the rest of the day.

2. In Vivo Platelet Activation and Aggregation

The reports in the literature on the circadian variation of plasma indices of platelet activation differ considerably, with some authors finding β-thromboglobulin concentrations highest at 1000 hours (PECHAN et al. 1992), others significantly higher at 1500 hours than at 0800 hours ($P < 0.01$) (JAFRI et al. 1992), and still others comparable in the morning and afternoon (GRIMAUDO et al. 1988). Similarly, the plasma levels of platelet factor 4 have been found by different investigators to be higher at 0800 hours than at 1600 hours ($P < 0.05$) (AKIYAMA et al. 1990), not to vary significantly over the 24 h (JOHANSEN et al. 1990), or to be higher in the afternoon than in the morning ($P < 0.01$) (JAFRI et al. 1992). Finally, the stable metabolite of the short-lived thromboxane A_2, 11-

dehydro-thromboxane B_2, measured in plasma 4-hourly for 24 h in ten healthy men, was found to be higher at 1600 hours than at 0400 hours ($P<0.005$) (BRIDGES et al. 1991). In most of these studies, highest values tended to coincide with periods of physical activity and lowest levels with periods of bed rest; it is possible, therefore, that different schedules of activity are responsible, at least in part, for the conflicting results.

The variation in circulating platelet aggregates between 0600 hours and midnight was investigated in ten healthy subjects (mean 45 years old) by 3-hourly blood samples. A substantial increase (of about 30%) was recorded at 0600 hours, shortly after awakening, compared to nadir levels observed only 3 h later at 0900 hours ($P<0.01$); thereafter, the levels remained more or less unchanged throughout the day. The authors attribute this enhanced early-morning platelet aggregability to increased sympathetic activity and to the physical and mental stress of awaking, and the low levels at 0900 hours to platelet "exhaustion" after a period of hyperactivity (UNDAR et al. 1989).

Thus, the plasma concentrations of platelet-release products do not show a consistent pattern of variation during the day in different studies, possibly because of differences in the experimental protocols. It is not clear to what extent the circadian changes of these factors and of in vivo platelet aggregability, which shows a peak in the early morning, are determined by true endogenous rhythms or by exogenous circumstances, such as awakening and physical activity (see also the following paragraphs).

3. Platelet Aggregation In Vitro

In vitro platelet aggregability has been measured in 15 healthy men (20–35 years old) every 3 h for 24 h. Peak aggregability was recorded at 0900 hours, 1 h after rising, when it increased by approximately 20% in response to ADP and by approximately 100% in response to adrenaline, compared to nadir levels recorded at 0600 hours ($P<0.01$); no other significant change in platelet reactivity was observed during the remaining study period, while plasma adrenaline and noradrenaline increased significantly between 0600 and 0900 hours ($P<0.001$). When the time of awakening was delayed by 4 h, the morning increase in aggregability was not observed. The authors conclude that the peak in platelet reactivity in the morning is related to the concomitant increase in sympathetic activity (TOFLER et al. 1987).

The same group, in two subsequent studies, described a close association between enhanced platelet aggregability, assumption of the upright posture and raised plasma catecholamine concentrations (BREZINSKI et al. 1988). The increase in aggregability produced by standing was found to resolve within 45 min after return to the supine position and to reappear when the upright posture was again assumed. In contrast, exercise on a bicycle did not increase aggregability beyond that determined by orthostatism (WINTHER et al. 1992).

The enhanced morning platelet reactivity to agonists in vitro and the close relation between physical activity and markers of platelet activation have been confirmed by other investigators in diurnal studies performed between 0700 and 2100 hours (MEHTA et al. 1989; JAFRI et al. 1992). The speed of platelet aggregation in response to a fixed concentration of ADP, measured at 1- to 3-hourly intervals between 0800 and 1800 hours in 25 healthy subjects (29–51 years old), was also found to be significantly higher in the morning than in the afternoon ($P < 0.005$). The timing of this variation, however, was not affected by body posture, age, sex or previous night shifts, although peak and trough values tended to be higher in the physically active and in the older and female subjects (JOVICIC and MANDIC 1991).

Two further studies have documented the full circadian behaviour of platelet aggregability by 3- to 4-hourly sampling for 24 h, but are somewhat at variance with the previous data. Petralito et al., in ten healthy subjects performing ordinary activities between 0700 and 2300 hours, observed a peak of aggregability in ADP at 1200 hours and a trough at 2100 hours ($P < 0.01$) (PETRALITO et al. 1982). HAUS et al. (1990), in ten healthy subjects who remained recumbent for at least 30 min before and during each blood sampling, documented a significant sinusoidal fluctuation of ADP-stimulated platelet aggregation, with a peak at 0000 hours and a nadir at 1600 hours.

As for in vivo platelet function, these partly contrasting results may reflect different experimental conditions in the various studies (with special reference to sleep-wake schedule and body posture at the time of sampling) and, consequently, a variable degree of interaction between endogenous and exogenous factors that influence the circadian periodicity of platelet reactivity. Despite these confounding elements, most studies conducted in physically unrestrained subjects have identified a morning peak of in vitro aggregability after awakening, compared with the rest of the day.

4. Platelet α_2-Adrenoceptor Activity

The changes in platelet aggregability in vivo and in vitro over the 24-h period may be related to changes in the circulating levels of platelet agonists (such as catecholamines), as discussed above, or in properties intrinsic to the platelet (such as receptor affinity for agonists, α_2-adrenoceptor number or postreceptor functions). A reduction in the affinity of human platelet α_2-adrenoceptors for agonists has been observed during the 2 h following physiological increases in endogenous catecholamine levels (HOLLISTER et al. 1983). This phenomenon might explain the platelet "exhaustion" hypothesized by UNDAR et al. (1989) after periods of platelet hyperactivity.

The number of α_2-adrenoceptors on the platelet membrane has been measured in seven healthy subjects (aged 30–56 years) and found not to change significantly from 0900 to 2100 hours. In contrast, the dissociation constant for the binding of the α_2-adrenoceptor ligand yohimbine was significantly lower at 0900 than at 1300 or 1700 hours ($P < 0.03$), indicating an increased affinity of platelet α_2-adrenoceptors in the morning. The authors of the study conclude that a circadian rhythm in α_2-adrenoceptor affinity may explain the circadian variation in platelet aggregation in response to adrenergic stimuli (MEHTA et al. 1989). In another investigation conducted in eight fasting healthy men (20–35 years old) between 0600 and 1200 hours, however, no significant changes in either platelet α_2-adrenoceptor number or yohimbine binding affinity over time were observed, despite a significant increase in plasma catecholamine levels and in platelet response to adrenaline after arising. The authors conclude that the higher morning platelet aggregability may be caused by changes in post-receptor mechanisms or in factors extrinsic to the platelet (WILLICH et al. 1992). The apparent divergence in the results by Mehta and Willich may depend on the different duration of the studies or on other conditions which may have led to variable degrees of sympathetic activity. For instance, in the study conducted by Willich, the degree of sympathetic activation may have been inadequate to alter platelet α_2-adrenoceptor activity, as suggested by the authors themselves (WILLICH et al. 1992).

Thus, most studies on platelet function, assessed by a variety of methods – ranging from platelet adhesiveness, to serotonin content, to in vivo or in vitro aggregability, to α_2-adrenoceptor activity – have documented a significant increase in platelet reactivity in the morning between 0500 and 0900 hours, compared with the rest of the day. The precise mechanism governing this process is not clear, although endogenous levels of catecholamines seem to play a part. Given the prominent role of platelets in the composition of arterial thrombi, their higher reactivity in the morning is likely to favour arterial thrombosis at this time.

IV. Circadian Variation of Coagulation

1. Global Measures of Coagulation

a) Prothrombin Time or Quick Test

A statistically significant circadian fluctuation in prothrombin time (PT) has been recorded in ten healthy subjects (five men, mean age 31 ± 11 years) by six 4-hourly samples. The shortest times, indicating highest activity of the extrinsic coagulation system, were at 0800 hours (mean 10 s), and longest times were at 1600 hours (mean 11 s) (HAUS et al. 1990). A significant circadian rhythm in PT has been reported also by MELCHART et al. (1992). AKIYAMA et al. (1990), in 16 healthy Japanese (mean age 37 years, nine men), however, did not find any significant diurnal variation in the activity of either factor VII (specific for PT) or factors II, V and X (common pathway) measured between 0800 and 0000 hours. In contrast, MARCKMANN et al. (1993) in six healthy men (aged 21–30 years) found significantly higher factor VII clotting activity at 1730 and 2130 hours, compared with 0845 hours ($P < 0.01$). The discrepancy between these two studies might be related, at least in part, to differences in the assays used to measure factor VII.

b) Activated Partial Thromboplastin Time

A circadian variation in activated partial thromboplastin time (aPTT) has been documented by 3-hourly sampling for 24 h in ten healthy subjects (mean age 58 years) performing synchronized activities for at least a week. Shortest times, indicating highest intrinsic coagulation activity, were at 0300 hours (mean 25 s) and longest times at 1200 hours (mean 30 s, $P < 0.05$) (PETRALITO et al. 1982). A significant circadian rhythm in aPTT (with minimal levels at 0800 and 1600 hours and maximal values at 2000 hours) and in factor VIII activity (with a peak at 0800 hours and a trough at 2000 hours), but not in factor V, has been reported also by HAUS et al. (1990). AKIYAMA et al. (1990) also documented a diurnal variation in the activity of factor VIII and IX, with acrophases at 0900 hours.

c) Thrombin Time

A significant circadian variation in thrombin time has been reported by both Petralito and Haus; a considerable phase difference, however, is apparent between the results of the two studies, with shortest values occurring at 0900 and 1600 hours, and longest values at 0300 and 0800 hours, respectively (PETRALITO et al. 1982; HAUS et al. 1990). Both investigators also recorded a significant circadian variation in plasma fibrinogen concentrations, with highest levels at 1200 and 0800 hours, respectively, and lowest levels at 0000 hours. AKIYAMA et al. (1990), on the other hand, observed no significant diurnal variation in plasma fibrinogen concentrations.

d) Howell's Time

In the study by PETRALITO et al. (1982), Howell's time was significantly shorter at 0900 hours (89 ± 12 s) than at 2100 hours (124 ± 12 s, $P < 0.01$).

2. Specific Coagulation-Related Factors

Variations in the plasma levels of extrinsic (factor VII), intrinsic (factors VIII and IX) and common (factors II, V, X and fibrinogen) coagulation factors have been described in the previous paragraphs. The effect of a low- or high-fat diet on the circadian plasma levels of coagulation factor VII has been studied by MARCKMANN et al. (1993) in six healthy men served different isoenergetic diets for periods of 2 days. On the 2nd day of each diet, eight blood samples were drawn between 0845 and 0300 hours. The high-fat diets resulted in significantly increased postprandial factor VII clotting activity (using bovine thromboplastin) with peak levels of 131% of the reference value, versus 95% on low-fat diets ($P < 0.01$). The authors suggest that high-fat meals may enhance blood thrombogenicity, through the postprandial lipolysis of triglyceride-rich lipoproteins which may activate factor VII.

No significant circadian or diurnal variation has been observed in the plasma concentration of antithrombin III (PETRALITO et al. 1982; AKIYAMA et al. 1990; HAUS et al. 1990; JOVICIC and MANDIC 1991), thrombin-antithrombin III complexes (DEGUCHI et al 1991; JAFRI et al. 1992), fibrinopeptide A, or von Willebrand factor (AKIYAMA et al. 1990; JOHANSEN et al. 1990; BRIDGES et al. 1991).

Thus, the studies that have investigated coagulation through global functional assays have identified an enhanced clotting activity in the morning; the latter may be explained, at least in part, by concomitant changes in the activity of individual coagulation factors (VIII and IX). Factor VII activity appears to be influenced by exogenous variables, such as the fat content of meals.

3. Blood Viscosity

A significant circadian fluctuation in blood viscosity, running parallel to the daily changes in haematocrit and plasma sialic acid concentrations, was first documented in four healthy subjects (aged 26–59 years, three men) by blood samples collected every 4 h for 72 h. Viscosity peaked at 0800 hours, when the subjects were awake but still in bed and fasting, and reached lowest values at 0400 hours ($P < 0.001$). The authors concluded that this fluctuation was probably the result of an endogenous periodicity interacting with haemodynamic variations (water transfer in and out of the peripheral vascular compartment) caused by changes in posture and muscular activity (SEAMAN et al. 1965). These findings were later confirmed in 11 healthy subjects (aged 18–26 years, eight men) by samples taken 1 to 4 hourly for 24 h. The viscosity of whole blood and plasma again showed circadian fluctuations similar to those of haematocrit and plasma protein concentrations, with a maximum around 1000 hours and a minimum at 0300 hours. The authors conclude that these variations reflect changes in haemoconcentration, with a spontaneous haemodilution occurring during the night (EHRLY and JUNG 1973). Thus, the concomitant enhancement of blood and plasma viscosity and of blood coagulability in the morning are likely to favour blood stasis and the formation of blood clots at this time.

V. Circadian Variation of Fibrinolytic Activity

1. Global Assays of Fibrinolysis

Between the late 1950s and the mid-1970s a number of studies, using global measurements of fibrinolytic activity, demonstrated that the spontaneous capacity of blood to lyse fibrin clots underwent a marked sinusoidal circadian variation, with a fixed phase-relation to external clock-time: fibrinolytic activity increased severalfold from morning to afternoon, reaching a peak at 1800 hours, and then dropped to its lowest between 0300 and 0600 hours (FEARNLEY et al. 1957; CEPELAK and BARCAL 1959; CEPELAK et al. 1966). The variation was thought to depend on changes of a plasminogen activator (FEARNLEY et al. 1957). Although the amplitude of the fluctuation differed from one subject to another, the timing of peak and trough fibrinolytic activity was fairly consistent among individuals; moreover, the range of variation and the 24-h mean level showed good reproducibility within the same subject (ROSING et al. 1970).

Physical activity, body posture or sleep/wake schedule were not responsible for the rhythm, although physical exercise enhanced fibrinolytic activity and elicited a greater response in the evening than in the morning (FEARNLEY et al. 1957; BILLIMORIA et al. 1959; ROSING et al. 1970; JOVICIC and MANDIC 1991). Ageing was found to be associated with a blunted physiological increase of fibrinolytic activity during the day (BUCKELL and ELLIOTT 1959; ROSING et al. 1973). Gender, meals, type of diet, geographical or ethnic origin, and physical training also did not influence the basic features of this rhythm, whereas genetic factors did appear to contribute to its characteristics (CEPELAK and BARCAL 1959; MANN 1967; ROSING et al. 1970; KORSAN-BENGTSEN et al. 1973; SIMPSON et al. 1983; MARCKMANN et al. 1993).

2. Individual Fibrinolytic Factors

More recent studies have established that, of the known components of the fibrinolytic system, only t-PA and its fast-acting inhibitor PAI-1 show a marked circadian variation in plasma (ANDREOTTI et al. 1988b; GRIMAUDO et al. 1988; KLUFT et al. 1988; AKIYAMA et al. 1990; MARCKMANN et al. 1993). The levels of plasminogen, single-chain u-PA, α_2-antiplasmin and a reversible t-PA inhibitor, instead, vary little if at all during the 24 h (NEERSTRAND et al. 1987; ANDREOTTI et al. 1988a; KLUFT et al. 1988; AKIYAMA et al. 1990; HAUS et al. 1990).

The use of quenching antibodies to t-PA and zymographic analysis have demonstrated that the circadian rhythm of fibrinolysis in blood is due exclusively to changes in the activity of t-PA (GRIMAUDO et al. 1988; KLUFT et al. 1983). The 24-h fluctuation of plasma t-PA activity, however, is phase-shifted by several hours compared to the circadian variation of t-PA antigen, but shows a close phase-inversion with respect to the 24-h rhythm of PAI-1 activity and antigen levels (Fig. 2): peak t-PA activity occurs around 1800 hours, when both PAI activity and antigen are at their lowest, whereas lowest t-PA activity is found between 0300 and 0900 hours, when PAI-1 is at its highest (ANDREOTTI et al. 1988a, b; GRIMAUDO et al. 1988; KLUFT et al. 1988; AKIYAMA et al. 1990, MARCKMANN et al. 1993). Thus, plasma t-PA activity, as currently measured in vitro, is closely and inversely related to the levels of PAI-1 throughout the 24-h cycle.

Studies of the daytime variation of fibrinolytic activity in blood are unanimous in finding it lower in the morning than in the afternoon (KLUFT et al. 1988; ANGLETON et al. 1989; SAKATA et al. 1992; MASUDA et al. 1994). In contrast, among studies covering the entire 24-h period, the time of lowest fibrinolysis and of peak PAI-1 levels has varied by as much as 6 h, ranging from 0300 to 0900 hours. Such variability among investigators cannot be entirely explained by differences in the frequency of blood sampling, in assay techniques or in the subjects' age.

Recently, a strong induction of plasma PAI activity has been reported after alcohol consumption, compared to drinking water, with a peak after 5 h (HENDRIKS et al. 1994). This induction might explain, at least in part, the earlier and more marked morning rise of plasma PAI-1 in subjects on unrestricted meals (ANDREOTTI et al. 1988b) – which included wine or beer in the evening – compared to those on alcohol-free diets (MARCKMANN et al. 1993).

In healthy subjects, the circadian fluctuation of t-PA activity is not associated with parallel changes in effective fibrinolysis, assessed as the plasma concentration of fibrin degradation products (AKIYAMA et al. 1990; HAUS et al. 1990; JOHANSEN et al. 1991; JOVICIC and MANDIC 1991; JAFRI et al. 1992). This suggests that in health there is a constant low level of fibrin formation and fibrin degradation (ASTRUP 1956) over the 24-h period, and that t-PA activity – even at its lowest – is sufficient to degrade all the fibrin formed, yielding a constant level of degradation products.

3. Possible Determinants of the Circadian Variation of PAI-1

Since the circadian rhythm of fibrinolysis in blood, as recorded in vitro, is largely determined by changes in PAI-1, we will consider the possible mechanisms regulating the circadian variation of this inhibitor. A number of cell types can synthesize and release PAI-1, including endothelial cells, hepatocytes, vascular smooth muscle cells and monocytes (SPRENGERS and KLUFT 1987); the inhibitor is also contained in platelet α-granules (BOOTH et al. 1988). The fluctuation of PAI-1 in plasma could thus reflect a circadian rhythm in the synthesis and release by endothelial cells of this enzyme. As already stated, however, such a behaviour would be protein specific, as the plasma concentration of von Willebrand factor, another endothelial product, remains stable over the 24 h (AKIYAMA et al. 1990; JOHANSEN et al. 1990; BRIDGES et al. 1991). A circadian variation in the release of PAI-1 from platelet α-granules is also unlikely, as the plasma levels of other α-granule constituents, such as β-thromboglobulin and platelet factor 4, do not vary significantly or show inconsistent changes over the 24-h cycle (GRIMAUDO et al. 1988; AKIYAMA et al. 1990; JOHANSEN et al. 1990; JAFRI et al. 1992; PECHAN et al. 1992). Studies of twins suggest that the 24-h rhythm of PAI-1 may be, to some extent, under genetic control, as monozygotic twins show greater intra-pair concordance in their circadian levels of euglobulin clot lysis time compared with dizygotic twins (CEPELAK and BARCAL 1959). Neither acute β- and α-adrenoceptor antagonism nor chronic β-adrenoceptor antagonism have been found to have any effect on the circadian fluctuation of plasma fibrinolytic activity or of t-PA and PAI-1 levels; these findings tend to exclude an adrenergic regulation of the circadian rhythm of fibrinolysis in blood (HARENBERG et al. 1980; ANDREOTTI et al. 1991).

The circadian rhythm of PAI-1 may be coordinated to that of t-PA antigen, as suggested by the relation between t-PA antigen and PAI-1 levels in many clinical studies (HAMSTEN et al. 1985; PARAMO et al. 1985; HUBER et al.

1988; OLOFSSON et al. 1989; BRIDGES et al. 1993), by the induction of PAI-1 in response to t-PA in human cell cultures (FUJII et al. 1990), by the structural homology in the genetic code of both t-PA and PAI-1 (BOSMA et al. 1988), and by the consensual changes in the plasma mass concentration of the two factors during the 24-h cycle (Fig. 2). Against this hypothesis, however, is the opposite response of t-PA and PAI-1 to interleukin 1 in cultured endothelial cells (SCHLEEF et al. 1988).

Flow-related changes in the clearance of PAI-1 through the liver do not appear to determine the 24-h variation of PAI-1, as hepatic blood flow is maximal around 0800 hours, when PAI-1 levels are high, and minimal around 1600 hours, when PAI-1 levels are low (LEMMER and NOLD 1991). A discordance between the diurnal changes in fibrinolysis and those of plasma corticosteroids, catecholamines and insulin has tended to exclude a relation between these hormones and the circadian rhythm of PAI-1 (MENON et al. 1967; CHANDLER et al. 1990). Activated protein C, which can inactivate and degrade PAI-1 in vitro (DE FOUW et al. 1987), also has no effect on the diurnal variation of PAI-1 when administered in vivo (OKAJIMA et al. 1990). Thus, the mechanism that drives the circadian rhythm of PAI-1 is still elusive; its identification, however, remains an interesting task, as its antagonism might be of therapeutic use in the prevention and treatment of thrombosis-related disorders.

C. Conclusions

The data reviewed in this chapter indicate that the function of most components of the haemostatic system varies physiologically during the 24-h period. These functional circadian variations result in a greater tendency toward vasoconstriction, blood stasis, platelet reactivity, fibrin formation and fibrin accumulation in the morning compared with other times of day.

In the presence of an underlying prothrombotic condition, the overall effect of these changes could theoretically favour the occurrence of vascular thrombotic occlusions in the morning. A number of reports have demonstrated that vaso-occlusive syndromes are indeed more frequent in the morning than during the rest of the 24-h period (WHO 1976; MARSHALL 1977; MULLER et al. 1987; BOGATY and WATERS 1988; COLANTONIO et al. 1989). On the other hand, pathological conditions that increase the prothrombotic tendency beyond the range delimited by the normal circadian fluctuation of the haemostatic system could theoretically lead to thrombosis at any moment of the 24 h. This possibility would explain the substantial number of vaso-occlusive events that occur throughout the 24-h cycle.

The chronobiology of the haemostatic system seems to have important consequences for clinical pharmacology. A greater efficacy of exogenous t-PA and u-PA as thrombolytic drugs for acute myocardial infarction has been reported between 1200 and 0000 hours than between 0000 and 1200 hours (BECKER et al. 1988; FUJITA et al. 1993; KURNIK 1995). This suggests that the physiological morning peak in plasma PAI-1 levels may hinder exogenously induced clot lysis; the greater hepatic blood flow in the morning (LEMMER and NOLD 1991) might also contribute to the reduced thrombolytic efficacy at this time, through an enhanced plasma clearance of exogenous t-PA or u-PA compared with the rest of the day. Similarly, a greater anticoagulant effect of a constant intravenous infusion of heparin has been observed at night (when endogenous coagulation, as assessed by global coagulation tests, exhibits its lowest activity) compared with the morning hours (DECOUSUS et al. 1985).

In conclusion, there is a natural phasic tendency to develop a vaso-occlusive state in the morning, compared with the rest of the day. This phenomenon has clinical implications, as vaso-occlusive events are more frequent, and thrombolytic and anticoagulant drugs less effective, in the morning. Although some of the circadian fluctuations within the haemostatic system may be attributable to changes in sympathetic activity and in haemodynamic conditions, others are driven by endogenous rhythms for which the ultimate mechanism of control is still unknown. Understanding the workings of this "master biological clock" remains a challenge for future research.

References

Akiyama Y, Kazama M, Tahara C, Shimazu C, Otake J, Kamei K, Nakatake T, Sakurai N, Yasumuro Y, Suzuki S, Maeba E, Nishida T (1990) Reference values of haemostasis related factors of healthy Japanese adults. I: circadian fluctuation. Thromb Res 60:281–289

Andreotti F, Davies GJ, Hackett D, Khan MI, de Bart ACW, Dooijewaard G, Maseri A, Kluft C (1988a) Circadian variation of fibrinolytic factors in normal human plasma. Fibrinolysis 2:90–92

Andreotti F, Davies GJ, Hackett DR, Khan MI, de Bart ACW, Aber V, Maseri A, Kluft C (1988b) Major circadian fluctuations in fibrinolytic factors and possible relevance to time of onset of myocardial infarction, sudden cardiac death and stroke. Am J Cardiol 62:635–637

Andreotti F, Kluft C, Davies GJ, Huisman LGM, de Bart ACW, Maseri A (1991) Effect of propranolol (long-acting) on the circadian fluctuation of tissue-type plasminogen activator and plasminogen activator inhibitor-1. Am J Cardiol 68:1295–1299

Angleton P, Chandler WL, Schmer G (1989) Diurnal variation of tissue-type plasminogen activator and its rapid inhibitor (PAI-1). Circulation 79:101–106

Artigou JY, Salloum J, Carayon A, Lechat P, Maistre G, Isnard R, Komajda M, Legrand JC, Grosgogeat Y (1993) Variations in plasma endothelin concentrations during coronary spasm. Eur Heart J 14:780–784

Aschoff J (1979) Circadian rhythms: general features and endocrinological aspects. In: Martini L, Krieger DT (eds) Comprehensive endocrinology series. Endocrine rhythms. Raven, New York, pp 1–61

Astrup T (1956) Fibrinolysis in the organism. Blood 11:781–806

Azuma H, Ishikawa M, Sezizaki S (1986) Endothelium-dependent inhibition of platelet aggregation. Br J Pharmacol 88:411–415

Bachmann F (1987) Fibrinolysis. In: Verstraete M, Vermylen J, Leijnen L, Arnout J (eds) Thrombosis and haemostasis. Leuven University Press, Leuven, pp 227–265

Becker RC, Corrao JM, Baker SP, Gore JM, Alpert JS (1988) Circadian variation in thrombolytic response to recombinant tissue-type plasminogen activator in acute myocardial infarction. J Appl Cardiol 3:213–221

Bennet JS, Vilaire G, Cines DB (1982) Identification of the fibrinogen receptor on human platelets by photoaffinity labeling. J Biol Chem 257:8049–8054

Berridge MJ (1984) Inositol triphosphate and diacylglycerol as second messengers. Biochem J 220:345–360

Billimoria JD, Drysdale J, James DCO, Maclagan NF (1959) Determination of fibrinolytic activity of whole blood with special reference to the effects of exercise and fat feeding. Lancet ii:471–475

Bogaty P, Waters DD (1988) Circadian patterns in coronary disease: the mournfulness of morning. Can J Cardiol 4:5–11

Booth N, Simpson AJ, Croll A, Bennet B, MacGregor IR (1988) Plasminogen activator inhibitor (PAI-1) in plasma and platelets. Br J Haematol 70:327–333

Born GVR (1962) Aggregation of blood platelets by adenosine diphosphate and its reversal. Nature 194:927–929

Bosma PJ, van den Berg EA, Kooistra T, Siemieniak DR, Slightom JL (1988) Human plasminogen activator inhibitor-1 gene. Promotor and structural gene nucleotide sequences. J Biol Chem 263:9129–9141

Brezinski DA, Tofler GH, Muller JE, Pohjola-Sintonen S, Willich SN, Schafer AI, Czeisler CA, Williams GH (1988) Morning increase in platelet aggregability. Association with assumption of the upright posture. Circulation 78:35–40

Bridges AB, McLaren M, Saniabadi A, Fisher TC, Belch JJ (1991) Circadian variation of endothelial cell function, red blood cell deformability and dehydro-thromboxane B_2 in healthy volunteers. Blood Coag Fibrinolysis 2(3):447–452

Bridges AB, McLaren M, Scott NA, Pringle TH, McNeill GP, Belch JJF (1993) Circadian variation of tissue plasminogen activator and its inhibitor, von Willebrand factor antigen, and prostacyclin stimulating factor in men with ischaemic heart disease. Br Heart J 69:121–124

Buckell M, Elliott FA (1959) Diurnal fluctuation of plasma fibrinolytic activity in normal males. Lancet i:660–662

Cepelak V, Barcal R (1959) Fibrinolysis in twins and its variation in circadian rhythm. Plzen Lek Sborn 22:129–131

Cepelak V, Barcal R, Lang N, Cepelakova H (1966) Zum Tag- und Nachtrhythmus der Fibrinolyse. Z Inn Med 21:202–204

Chandler WL, Mornin D, Whitten RO, Angleton P, Farin FM, Fritsche TR, Veith RC, Stratton JR (1990) Insulin, cortisol and catecholamines do not regulate circadian variations in fibrinolytic activity. Thromb Res 58:1–12

Colantonio D, Casale R, Abruzzo BP, Lorenzetti G, Pasqualetti P (1989) Circadian distribution in fatal pulmonary thromboembolism. Am J Cardiol 64:403–404

Collen D (1985) The main components of the fibrinolytic system: biochemical and physiological properties. Eur Heart J 6:193–195

Cugini P, Di Palma L, Di Simone S, Lucia P, Battista P, Coppola A, Leone G (1993) Circadian rhythm of cardiac output, peripheral vascular resistance, and related variables by a beat-to-beat monitoring. Chronobiol Int 10 (1):73–78

Davidson JF, Walker ID (1987) Assessment of the fibrinolytic system. In: Bloom AL, Thomas DP (eds) Haemostasis and thrombosis. Churchill Livingstone, Edinburgh, pp 953–966

Decousus H, Croze M, Levi FA, Jaubert JG, Perpoint BM, DeBonadona JF, Reinberg A, Queneau PM (1985) Circadian changes in anticoagulant effect of heparin infused at a constant rate. Br Med J 290:341–344

de Fouw NJ, van Hinsbergh VWM, de Jing YF, Haverkate F, Bertina RM (1987) The interaction of activated protein C and thrombin with the plasminogen activator inhibitor released from human endothelial cells. Thromb Haemost 57:176–182

Deguchi K, Noguchi M, Yuwasaki E, Endou T, Deguchi A, Wada H, Murashima S, Nishikawa M, Shirakawa S, Tanaka K, Kusagawa M (1991) Dynamic fluctuations in blood of thrombin/antithrombin III complex (TAT). Am J Hematol 38:86–89

Ehrly AM, Jung G (1973) Circadian rhythm of human blood viscosity. Biorheology 10:577–583

Eynard N, Flachaire E, Lestra C, Broyer M, Zaidan R, Claustrat B, Quincy C (1993) Platelet serotonin content and free and total plasma tryptophan in healthy volunteers during 24 hours. Clin Chem 39:2337–2340

Fearnley GR (1965) Fibrinolysis. Arnold, London, pp 28–44

Fearnley GR, Balmforth G, Fearnley E (1957) Evidence of a diurnal fibrinolytic rhythm; with a simple method of measuring natural fibrinolysis. Clin Sci 16:645–650

Fujii S, Lucore CL, Hopkins WE, Billadello JJ, Sobel BE (1990) Induction of synthesis of plasminogen activator inhibitor type-1 by tissue-type plasminogen activator in human hepatic and endothelial cells. Thromb Haemost 64:412–419

Fujita M, Franklin D (1987) Diurnal changes in coronary blood flow in conscious dogs. Circulation 76:488–491

Fujita M, Araie E, Yamanishi K, Miwa K, Kida M, Nakajima H (1993) Circadian variation in the success rate of intracoronary thrombolysis for acute myocardial infarction. Am J Cardiol 71:1369–1371

Ganong WF (1979) Review of medical physiology, 9th edn. Lange Medical Publications, Los Altos

Gohar M, Daleau P, Atkinson J, Gargouil Y-M (1992) Ultradian variations in sensitivity of rat aorta rings to noradrenaline. Eur J Pharmacol 229:69–73

Grimaudo V, Hauert J, Bachmann F, Kruithof EKO (1988) Diurnal variation of the fibrinolytic system. Thromb Haemost 59:495–499

Hamsten A, Wiman B, de Faire U, Blomback M (1985) Increased plasma levels of a rapid inhibitor of tissue plasminogen activator in young survivors of myocardial infarction. N Engl J Med 313:1557–1563

Handin RI, Loscalzo J (1992) Hemostasis, thrombosis, fibrinolysis, and cardiovascular disease. In: Braunwald E (ed) Heart disease. A textbook of cardiovascular medicine. Saunders, Philadelphia, pp 1767–1788

Harenberg J, Weber E, Spohr U, Morl H (1980) Is the diurnal increase of fibrinolytic activity influenced by alpha- or beta-adrenergic blockade? Blut 41:455–458

Haus E, Cusulos M, Sackett-Lundeen L, Swoyer J (1990) Circadian variations in blood coagulation parameters, alpha-antitrypsin antigen and platelet aggregation and retention in clinically healthy subjects. Chronobiol Int 7:203–216

Hendriks HF, Veenstra J, Velthuis-te Wierik EJM, Schaafsma G, Kluft C (1994) Effect of moderate dose of alcohol with evening meal on fibrinolytic factors. Br Med J 308:1003–1006

Hoffmann BB, De Lean A, Wood CL, Schocken DD, Lefkowitz RJ (1979) Alpha-adrenergic receptor subtypes: quantitative assessment by ligand binding. Life Sci 24:1739–1746

Hollister AS, Fitzgerald GA, Nadeau JHJ, Robertson D (1983) Acute reduction in human platelet alpha-2 adrenoreceptor affinity for agonist by endogenous and exogenous catecholamines. J Clin Invest 72:1498–1505

Houben AJH, Slaaf DW, Huvers FC, De Leeuw PW, Nieuwenhuijzen Kruseman AC, Schaper NC (1994) Diurnal variations in total forearm and skin microcirculatory blood flow in man. Scand J Clin Lab Invest 54:161–168

Huber K, Rosc D, Resch I, Glogar DH, Kaindl F, Binder R (1988) Circadian fluctuations of plasminogen activator inhibitor and tissue plasminogen activator levels in plasma of patients with unstable coronary artery disease and acute myocardial infarction. Thromb Haemost 60:372–376

Jaffe EA (1987) Cell biology of endothelial cells. Hum Pathol 18:234–239

Jafri SM, VanRollins M, Ozawa T, Mammen EF, Goldberg AD, Goldstein S (1992) Circadian variation in platelet function in healthy volunteers. Am J Cardiol 69:951–954

Johansen LG, Gram J, Kluft C, Jespersen J (1990) Circadian variations of extrinsic fibrinolytic components in blood from Eskimos. Fibrinolysis 4:35–39

Johansen LG, Gram J, Kluft C, Jespersen J (1991) Chronobiology of coronary risk markers in Greenland Eskimos – a comparative study with Caucasians residing in the same Arctic area. Chronobiol Int 8:400–405

Jovicic A, Mandic S (1991) Circadian variations of platelet aggregability and fibrinolytic activity in healthy subjects. Thromb Res 62:65–74

Kaneko M, Zechman FW, Smith RE (1968) Circadian variation in human peripheral blood flow levels and exercise responses. J Appl Physiol 25:109–114

Kaplan KL, Owen J (1981) Plasma levels of beta-thromboglobulin and platelet factor 4 as indices of platelet activation in vivo. Blood 57:199–202

Kluft C, Jie AFH, Allen AR (1983) Behaviour and quantitation of extrinsic (tissue-type) plasminogen activator in human blood. Thromb Haemost 50:518–523

Kluft C, Jie AFH, Rijken DC, Verheijen JH (1988) Daytime fluctuations in blood of tissue-type plasminogen activator (t-PA) and its fast-acting inhibitor (PAI-1). Thromb Haemost 59:329–332

Korsan-Bengtsen K, Wilhelmsen L, Tibblin G (1973) Blood coagulation and fibrinolysis in relation to degree of physical activity during work and leisure time. Acta Med Scand 193:73–77

Kurnik PB (1995) Circadian variation in the efficacy of t-PA. Circulation 91:1341–1346

Lemmer B, Nold G (1991) Circadian changes in estimated hepatic blood flow in healthy subjects. Br J Clin Pharmacol 32:627–629

Loscalzo J, Vaughan DE (1987) Tissue plasminogen activator promotes platelet disaggregation in plasma. J Clin Invest 79:1749–1755

Mann RD (1967) Effect of age, sex, and diurnal variation on the human fibrinolytic system. J Clin Path 20:223–226

Marckmann P, Sandstrom B, Jespersen J (1993) Dietary effects on circadian fluctuation in human blood coagulation factor VII and fibrinolysis. Atherosclerosis 101:225–234

Marshall J (1977) Diurnal variation in occurrence of strokes. Stroke 88:230–231

Masuda T, Ogawa H, Miyao Y, Yu Q, Misumi I, Sakamoto T, Okubo H, Okumura K, Yasue H (1994) Circadian variation in fibrinolytic activity in patients with variant angina. Br Heart J 71:156–161

Mehta JL, Malloy M, Lawson D, Lopez L (1989) Circadian variation in platelet alpha-2 adrenoreceptor affinity in normal subjects. Am J Cardiol 63:1002–1005

Melchart D, Martin P, Hallek M, Holzmann M, Jurcic X, Wagner H (1992) Circadian variation of the phagocytic activity of polymorphonuclear leukocytes and of various other parameters in 13 healthy male adults. Chronobiol Int 9:35–45

Menon SI, Smith PA, White RWB, Dewar HA (1967) Diurnal variations of fibrinolytic activity and plasma-11-hydroxycorticosteroid levels. Lancet ii:531–533

Meyerson LR, Strano R, Ocheret D (1989) Diurnal concordance of human platelet serotonin content and plasma alpha-1-acid glycoprotein concentration. Pharmacol Biochem Behav 32:1043–1047

Michelson AD, Loscalzo J, Melnick B, Coller BS, Handin RI (1986) Partial characterization of a binding site for von Willebrand factor on glycocalicin. Blood 67:19–26

Moncada S, Palmer RMJ, Higgs EA (1987) Prostacyclin and endothelium-derived relaxing factor: biological interactions and significance. In: Verstraete M, Vermylen J, Leijnen L, Arnout J (eds) Thrombosis and haemostasis. Leuven University Press, Leuven, pp 597–618

Montero D, Ofori-Adjei D, Wagner A (1989) Circadian variation of platelet ^{3}H-imipramine binding, platelet serotonin content, and plasma cortisol in healthy volunteers. Biol Psychiatry 26:794–804

Muller JE, Ludmer PL, Willich SN, Tofler GH, Aylmer G, Klangos I, Stone PH (1987) Circadian variation in the frequency of sudden cardiac death. Circulation 75:131–138

Neerstrand H, Ostergaard P, Bergqvist D, Matzsch T, Hedner MU (1987) tPA inhibitor, tPA:Ag, plasminogen and alpha-2-antiplasmin after low molecular weight heparin or standard heparin. Fibrinolysis 1:39–43

Newman KD, Williams LT, Bishopric NH, Lefkowitz RJ (1978) Identification of alpha-adrenergic receptors in human platelets by (^{3}H)dihydroergocryptine binding. J Clin Invest 61:395–402

Nieuwenhuis HK, Akkerman JWM, Houdjik WPM, Sixma JJ (1985) Human blood platelets showing no response to collagen fail to express glycoprotein Ia. Nature 318:470–472

O'Brien JR (1964) Variability in the human platelets by adrenaline. Nature 202:1188–1190

Okajima K, Koga S, Kaji M, Inoue M, Nakagaki T, Funatsu A, Okabe H, Takatsuki K, Aoki N (1990) Effect of protein C and activated protein C on coagulation and fibrinolysis in normal human subjects. Thromb Haemost 63:48–53

Olofsson BO, Dahlen G, Nilsson TK (1989) Evidence of increased levels of plasminogen activator inhibitor and tissue plasminogen activator in plasma of patients with angiographically verified coronary artery disease. Eur Heart J 10:77–82

Panza JA, Epstein SE, Quyyumi AA (1991) Circadian variation in vascular tone and its relation to α-sympathetic vasoconstrictor activity. N Engl J Med 325:986–990

Paramo JA, Colucci M, Collen D (1985) Plasminogen activator inhibitor in the blood of patients with coronary artery disease. Br Med J 291:573–574

Pechan J, Mikulecky M, Okrucka A (1992) Circadian rhythm of plasma beta-thromboglobulin in healthy human subjects. Blood Coag Fibrinolysis 3:105–107

Petralito A, Mangiafico RA, Gibiino S, Cuffari MA, Miano MF, Fiore CE (1982) Daily modifications of plasma fibrinogen, platelet aggregation, Howell's time, PTT, TT, and antithrombin III in normal subjects and in patients with vascular disease. Chronobiologia 9:195–201

Plow EF, Ginsberg MH (1981) Specific and saturable binding of plasma fibronectin to thrombin-stimulated human platelets. J Biol Chem 256:9477–9482

Raby KE, Vita JA, Rocco MB, Yeung AC, Ganz P, Fantasia G, Barry J, Selwyn AP (1993) Changing vasomotor responses of coronary arteries to nifedipine. Am Heart J 126:333–338

Rosing DR, Brakman P, Redwood DR, Goldstein RE, Beiser GD, Astrup T, Epstein SE (1970) Blood fibrinolytic activity in man: diurnal variation and the response to varying intensities of exercise. Circ Res 27:171–184

Rosing DR, Redwood DR, Brakman P, Astrup T, Epstein SE (1973) Impairment of the diurnal fibrinolytic response in man: effects of aging, type IV hyperlipoproteinemia and coronary artery disease. Circ Res 32:752–757

Sakata K, Hoshino T, Yoshida H, Ono N, Ohtani S, Yokoyama S, Mori N, Kaburagi T, Kurata C, Urano T, Takada Y, Takada A (1992) Circadian fluctuations of tissue plasminogen activator antigen and plasminogen activator inhibitor-1 antigen in vasospastic angina. Am Heart J 124:854–860

Schleef RR, Bevilacqua MP, Sawdey M, Gimbrone MA, Loskutoff DJ (1988) Cytokine activation of vascular endothelium. Effects on tissue-type plasminogen activator and type 1 plasminogen activator inhibitor. J Biol Chem 263:5797–5803

Seaman GVF, Engel R, Swank RL (1965) Circadian periodicity in some physicochemical parameters of circulating blood. Nature 207:833–835

Shattil SJ (1981) Platelets and their membranes in haemostasis: physiology and pathophysiology. Ann Intern Med 94:108–118

Simpson HCR, Mann JI, Meade TW, Chakrabarti R, Stirling Y, Woolf L (1983) Hypertriglyceridaemia and hypercoagulability. Lancet i:786–790

Sprengers ED, Kluft C (1987) Plasminogen activator inhibitors. Blood 69:381–387

Thompson AR, Harker LA (1988) Manual of hemostasis and thrombosis. Davis, Philadelphia

Tofler GH, Brezinski D, Schafer AL, Czeisler CA, Rutherford JD, Willich SN, Gleason RE, Williams GH, Muller JE (1987) Concurrent morning increase in platelet aggregability and the risk of myocardial infarction and sudden cardiac death. N Engl J Med 316:1514–1518

Undar L, Turkay C, Korkmaz L (1989) Circadian variation in circulating platelet aggregates. Ann Med 21:429–433

Verstraete M, Vermylen J (1984) Thrombosis. Pergamon, Oxford, p 31

White JG (1974) Electron microscopic studies of platelet secretion. In: Spaet TH (ed) Progress in haemostasis and thrombosis, vol 2. Grune and Stratton, New York, pp 49–98

WHO International Study Group (1976) Myocardial infarction community registers: public health in Europe. Regional Office for Europe (WHO), Copenhagen 1976; vol 5 (annex II), pp 188–191

Willich SN, Tofler GH, Brezinski DA, Schafer AI, Muller JE, Michel T, Colucci WS (1992) Platelet alpha-2 adrenoreceptor characteristics during the morning increase in platelet aggregability. Eur Heart J 13:550–555

Winther K, Hillegass W, Tofler GH, Jimenez A, Brezinski DA, Schafer AI, Loscalzo J, Williams GH, Muller JE (1992) Effects on platelet aggregation and fibrinolytic activity during upright posture and exercise in healthy men. Am J Cardiol 70:1051–1055

Wirz-Justice A, Lichtsteiner M, Feer H (1977) Diurnal and seasonal variations in human platelet serotonin in man. J Neural Transm 41:7–15

Wu KK, Hoak JC (1974) A new method for the quantitative detection of platelet aggregates in patients with arterial insufficiency. Lancet ii:924–926

Yanagisawa M, Kurihara H, Kimura S, Tomobe Y, Kobayashi M, Mitsui Y, Yazaki Y, Goto K, Masaki T (1988) A novel potent vasoconstrictor peptide produced by vascular endothelial cells. Nature 332:411–415

CHAPTER 20
New Trends in Chronotoxicology

J. CAMBAR and M. PONS

A. Introduction

The rhythmicity of living organisms has been empirically deduced for many centuries, but has been described precisely only during the last decades.

The reality of rhythmicity at different levels of an organism, so-called chronostructures, and temporal changes in functions of whole systems, organs, tissues, or even cells permit us to understand and explain the reality of chronopharmacology and chronotoxicology. Indeed, the demonstration of temporal changes in structure and function allow us to understand why the same exogenous agent, whatever its origin, will not cause the same effect if administered at different times of day or during different months of the year.

An exogenous agent acting upon a rhythmically changing system can cause changes of three major kinds: chronopathology caused by pathological agents, e.g., microorganisms; chronopharmacology in response to drugs or pharmacological agents; and chronotoxicology elicited by toxic agents. Chronotherapeutics, based upon these periodic changes contribute to an optimization of drug use by timed treatment aimed at increasing favourable pharmacological effects and by a decrease in undesired side effects.

It is often difficult, even impossible, to distinguish clearly among the domains of toxicity, tolerance, and efficacy. It is well-known that the difference between pharmacology and toxicology is merely dose-dependent. A chronotoxicological approach can be performed in experimental animals such as rodents, and clinical pharmacological or therapeutical studies can be carried out in volunteers or in patients regarding chronotolerance (and not chronotoxicology). Nevertheless, antitumoral agents can induce toxicological effects either as favorable or undesired effects.

Many reviews have already reported circadian and circannual variations in toxic and pharmacological effects of physical and chemical agents that include many drugs therapeutically used (REINBERG and HALBERG 1971; MOORE-EDE 1973; SCHEVING et al. 1974a, VON MAYERSBACH 1976; CAMBAR and CAL 1985).

Recent studies reporting the latest information on the chronobiological approach to pharmacotoxicology include: CAMBAR et al. 1992; HEINZE et al. 1993; SCHEVING et al. 1993; LEMMER 1994.

The aim of such reviews is, of course, to report the essential classical and more recent studies in this field, but also to encourage the development of a chronobiological approach in toxicology.

This review presents the latest data published on chronotoxicology, in animals as well as in humans. This chapter deals mainly with the chronotoxicological aspect of the use of antibiotics and immunosuppressors. Chronotoxicological aspects related to cancer chronotherapy will not be dealt with in detail since they will be discussed in depth in the review of Levi in this book. Moreover, some mechanistic explanations are put forward for such temporal changes in toxicology, but the role of pharmacokinetics in these mechanisms will be dealt with only briefly in view of the chapter of P.M. BÉLANGER, B. BRUGUEROLLE and G. LABRECQUE in this volume.

B. General Chronotoxicity

I. Chemical Toxic Agents

In toxicology, different types of experiments can be performed regarding the dose of the toxic agent and the number of administrations. Acute (or subacute) toxicity merely considers the damage induced by a single or small number of administrations. Chronic toxicity studies consider the damage induced by doses that are often smaller, but given over days and even weeks or months (namely, for cancer or teratological studies). The dose used can induce either a lethal effect or a nonlethal effect. We present here examples of different types of investigations using treatment with acute lethal, acute sublethal, and chronic sublethal doses and providing different types of information.

The pioneer studies before 1950 reported surprising, contradictory, but exciting data showing that the mortality rate was time-dependent.

The first experiment on circadian variation in murine tolerance to a drug was reported by CARLSON and SERIN (1950), who showed that nikethamide-induced mortality (assessed by the classical median lethal dose LD_{50} test) ranged from 33% at 02.00 h to 67% at 14.00 h. This data, obtained without any particular consideration of chronobiology, showed that temporal considerations were an important variable in experimental medicine.

As reported in an exhaustive review about these pioneer studies (SCHEVING et al. 1974a), HALBERG, HAUS and SCHEVING were the first to systematically and intensively demonstrate the importance of circadian changes in resistance and susceptibility of animals to a variety of different agents.

The general aim of all of these studies was to research, in animals synchronized by a lighting regimen, the influence of administration time on the mortality induced by noxious agents. So, in experiments on mice in which a fixed dose of ouabain (a specific Na/K ATPase inhibitor) was given at 4-h intervals during a 24-h span the number of deaths, recorded 10 min after injection, was circadian-dependent (HALBERG and STEPHENS 1959). A greater percentage of animals (75%) died at the beginning of the light phase and a smaller percentage at the beginning of the dark phase (15%).

These data were confirmed by NELSON et al. (1971) who also reported a circadian variation in susceptibility to the lethal effects of ouabain given subcutaneously and intraperitoneally to mice. In contrast, a very recent paper does not demonstrate significant time-dependent lethality in 576 male mice given ouabain intravenously (FORREST et al. 1993).

Numerous additional studies confirmed that drug toxicity was circadian-phase dependent. Similar conclusions were reached in studies with nicotine in which the same dose lead to 80% deaths at 19.00 h and only 10% at 14.00 h (SCHEVING and VEDRAL 1966) and in studies with 125 mg/kg propranolol inducing death in 50% of animals at 11.00 h and more than 80% at other times of day (LEMMER et al. 1980).

A large number of similar observations have been reported for many types of toxic chemicals such as ethanol (HAUS and HALBERG 1959), local anesthetics such as lidocaine (LUTSCH and MORRIS 1967), anticancer drugs such as cytosine arabinoside (CARDOSO et al. 1970), or Adriamycin (KUHL et al., 1973).

The first pronounced circadian rhythm in lethal toxicity caused by heavy metals was detected by HRUSHESKY et al. (1982) for cis-diamminedichloroplatinum (cisplatin). In six groups of Fisher 344 rats with a single dose (11 mg/kg) of cisplatin at different circadian stages, it was found that cisplatin was tolerated far better late in the animals' active phase.

Our own studies on the chronotoxicity of heavy metals have confirmed such circadian changes in murine tolerance to xenobiotics. The lethal toxicity of mercuric chloride (CAMBAR and CAL 1982) and cadmium sulfate (CAMBAR et al. 1983) was circadian-stage dependent with statistical significance in the case of both metals; tolerance occurred in mice in the middle of the dark period (Fig. 1). The survival time following intoxication supports those for survival rate. Indeed, mice injected with mercuric chloride during the light span died more quickly (day 4) than those injected during the dark span (day 9 to 10). These findings are consistent with those of TSAI et al. (1982) concerning mercuric chloride in BALB/c female mice and those of TAMASHIRO et al. (1986) concerning methylmercury chloride in mice with a reduced mortality around the middle and late activity span.

Moreover, it is interesting to consider the similarity in the mortality rate chronograms of the three heavy metals mercury, cadmium, and cisplatinum (Fig. 1).

Similar chronotoxicological studies carried out in our laboratory with different aminoglycoside antibiotics confirm such temporal tolerance changes. The chronograms of the three aminoglycosides (dibekacin, netilmicin and gentamicin) present quite similar pictures with a minimum during the first part of the night (Fig. 2). Moreover, for the same antibiotic agent, gentamicin, the chronograms of four different doses (250–325 mg/kg) given by a single intramuscular injection in mice are fairly similar (Fig. 3).

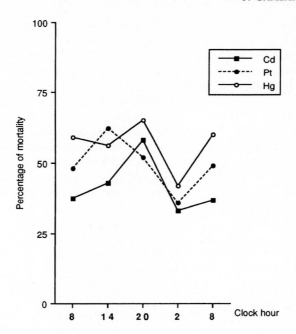

Fig. 1. Time-dependent effect of three heavy metal injections on mortality in mice (*Hg*, mercury; *Cd*, cadmium; *Pt*, platine)

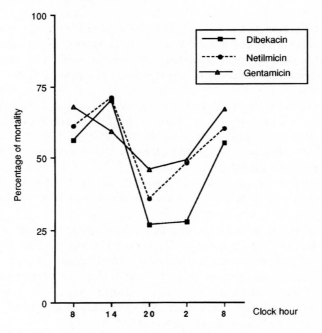

Fig. 2. Circadian changes in the effect of three aminoside antibiotic injections on mortality in mice. (From PARIAT 1986a)

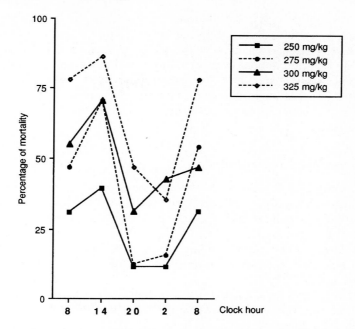

Fig. 3. Time dependence of mortality after a single intramuscular injection of four doses of gentamicin (mg/kg) in mice. (From PARIAT et al. 1988)

These data demonstrate clearly that drugs involving the same targets present a similar general (systemic) chronotoxicity and moreover, as we will explain later, a similar organ (specific) chronotoxicity.

We will develop further these observations in considering the concept of chronesthesy.

Examples presented above expressed circadian changes in mortality rate as a percentage of the animal deaths noted for a given dose of a toxic agent.

Evaluation of temporal changes in the toxicity of drugs assessed by LD_{50} (the dose killing 50% of experimental animals) requires at least four to six different doses. This explains the comparative rarity of such experiments.

Some papers have nevertheless shown circadian changes of LD_{50} in rodents (particularly in mice) for diazepam (Ross et al. 1981), gentamicin (NAKANO and OGAWA 1982), mercuric chloride (CAMBAR and CAL 1982), and procainamide (BRUGUEROLLE 1984).

Recently, circadian changes in LD_{50} have been compared for three different local anaesthetic agents (bupivacaine, mepivacaine, and etidocaine) in mice (PRAT and BRUGUEROLLE 1988; PRAT 1990). In the case of bupivacaine, LD_{50} showed a maximum at 01.00 h (62 ± 2,9 mg/kg) and a minimum at 22.00 h (52.4 ± 0.5 mg/kg). For mepivacaine, LD_{50} was greatest at 10.00 h (130 ± 5 mg/kg) and lowest (maximal toxicity) at 19.00 h (102 ± 3 mg/kg). For etidocaine, LD_{50} occurred during the middle of the night (47.5 ± 2.8 mg/kg) and the maximum at 10.00 h (55 ± 2.3 mg/kg).

It is noteworthy that, for all three anesthetic agents, the lowest LD_{50} (maximal toxicity) was observed during the activity span of the mice.

Likewise, circadian changes in the LD_{50} of three aminoglycoside antibiotics were reported in mice (PARIAT 1986), showing a maximum mortality in the middle of the rest span and a maximum tolerance at 20.00 h. At 14.00 h, LD_{50} was 252 mg/kg for gentamicin, 370 mg/kg for dibekacin, and 123 mg/kg for netilmicin. At 20.00 h, LD_{50} was significantly increased to 340, 405, and 143 mg/kg, respectively.

II. Nonchemical Toxic Agents

Chronotoxicology has also been demonstrated for some nonchemical toxic agents such as bacterial, viral, fungal, and animal toxins. All of these toxins can cause mortality or cellular damage varying greatly as a function of the time of administration.

After *Brucella* endotoxin administration, mice injected at night lived for 75 ± 15 h while those injected during the daytime survived only 37 ± 7 h (HALBERG et al. 1955b).

Similar studies with *Escherichia coli* endotoxin (HALBERG et al. 1980) showed that mortality was 80% when mice were injected at 16.00 h and only 15% at 00.00 h.

Likewise, when the immune system of the mice was stimulated with *Bacillus Calmette-Guérin* (BCG), it was found that 75% of the animals challenged during the middle of the light period survived, whereas only 50% of those challenged during the later part of the dark phase survived (TSAI et al. 1974).

Anisomycine, an inhibitor of protein synthesis isolated from *Streptomyces griseolus*, induces a circadian-dependent phase change in hamster locomotor activity; when given at midnight, the phase change is delayed, and at midday it is advanced (TAKAHASHI and TUREK 1987).

Likewise, tetrodotoxin (TTX), a toxin derived from some fish, for example *Spheroides rubripes*, induces dramatic changes in vasopressin (AVP) release from isolated suprachiasmatic nuclei in the rat; TTX perfused during 12 h of subjective day greatly decreases, even supresses, rhythmic AVP release, whereas perfusion during 12 h of dark did not change this release (EARNEST et al. 1991).

Another toxin from venom of the scorpion, *Heterometrus fulvipes*, induced circadian-dependent toxicity in a tropical mouse, *Mus booduga*, assessed by tissue acetylcholine and acetylcholinesterase levels (SIVA NARAYANA et al. 1984).

Although only indirectly related to toxicology, two physical agents may be circadian-phase dependent, e.g., the induction of seizures by white noise (about 100 dB for 60 s) in a susceptible strain of mice (HALBERG et al. 1955a) and X-ray irradiation showing circadian rhythms in mortality in *Drosophila* and in mice. In the latter, exposure to 555 r at night induced 100% death while during the day it was not lethal.

C. Chronotoxicological Examples in Nonrodents

In most of the studies described above, rodents were used. The choice of such animals is justified by an easy synchronization under well-defined standardized stable conditions and also by a relatively moderate cost, allowing the use of a greater number of animals for toxicological experiments. Nevertheless, chronotoxicological studies have also been carried out in other species, including fish and insects.

For instance, McLeay and Munro (1979) reported circadian variations in zinc tolerance in rainbow trout, and a study in the same species described the influence of season on the lethal toxicity of cyanide (McGeachy and Leduc 1988).

Insects have recently been proposed as simple animal models for chronopharmacological research (Hayes and Morgan 1988). Indeed, such a chronotoxicological approach can optimize the use of pesticides on insects in order to avoid unnecessarily large doses.

The chronotoxicity of malathion and dursban, both pesticides, has been demonstrated in house flies (Frudden and Wellso 1968) and in *Aedes aegypti* larvae (Roberts et al. 1974). All these investigations showed that the same concentration of a pesticide may have a different lethal potency on insects when the time of testing is considered. It has been reported that the same dose of pesticide can kill 90% of insect larvae if they are exposed during the second part of the night and only 20% at the end of the day. Such chronotoxicological considerations may reduce the amounts of pesticides required, and may have important ecological consequences (Thompson et al. 1988).

In conclusion, all these data make it necessary to review critically the significance of toxicity assessed by mortality rate or LD_{50} values when the time of day of their determination is not known. The time has to be taken into account for mortality test evaluations that may have to be performed at two, three, or four different times, if the toxin or the drug is highly patent. An improvement of the use of LD_{50} testing could result from documenting the optimal (highest-tolerance) and worst (highest-toxicity) times of administration. Moreover, such observations point to the relative value of certain toxicological investigations in which differences in LD_{50} values are often used to compare laboratories. It now seems fundamental in toxicology to define the exact hour of the day and even the month of the year and the synchronizing schedule, e.g., the lighting regimen in rodents, to be able to compare toxic doses and to discuss possible conflicting published results.

D. Organ Specific Circadian Chronotoxicity

High lethal doses of potentially toxic material often lead to rapid and often atypical deaths. Moreover, these drastic procedures do not allow any useful predictions applicable to therapeutics. The death of the experimental animal is in most cases the result of damage to different physiological functions and prohibits any hypothesis about the precise organ target responsible for such structural and functional damage.

Therefore, a number of recent studies have used lower nonlethal doses to document more accurately the subacute nonlethal chronotoxicology of various agents in different tissues. As we cannot be exhaustive, only some examples of the chronosusceptibility and chronotolerance of some organs to xenobiotics will be presented.

As already pointed out by Lévi et al. (1988), anticancer drugs cause numerous side effects that can injure nearly all organs. Chapter 11 will focus on the most recent data in chronotolerance of anticancer agents.

We have chosen numerous examples of chronotoxicological studies with drugs or toxic agents inducing specific organ damages such as of the ear, liver, nervous system, or digestive tract. In particular, we shall consider the kidneys because of the relatively large number of nephrotoxic therapeutic agents.

We will end this chapter by referring to some recent clinical studies on the chronotolerance of several widely used drugs in order to point out some practical clinical applications.

I. Liver Chronotoxicity (Chronohepatotoxicity)

Numerous agents have been reported as hepatotoxicants in literature for the past 50 years. Carbon tetrachloride (CCl_4), the most largely used toxicant in experimental hepatitis, causes centrolobular necrosis and fat accumulation in all experimental animals. After a single dose, this necrosis begins to develop with lesions after 12 h and full-blown necrosis after 24 h. Cell necrosis can be observed within 5 to 6 h with dramatic damage to mitochondria and Golgi apparatus.

Other early signs of cell injury include dissociation of ribosomes from the endoplasmic reticulum and disruption of smooth endoplasmic reticulum membranes. Indeed, inhibition of protein synthesis is thought to occur on single unit ribosomes. A blockage in secretion of hepatic triglycerides into plasma is the major mechanism underlying the steatosis induced by CCl_4. The toxicity of CCl_4 involves the cleavage of C-Cl bonds by cytochrome P_{450} to yield trichloromethyl and chlorine free radicals. These radicals attack fatty acids in the membranes of the endoplasmic reticulum, leading to secondary free radicals within the fatty acids, a process termed lipid peroxidation and causing dramatic damage to membranes and enzymes. Temporal differences in hepatic CCl_4 uptake and metabolism may contribute to the circadian susceptibility of the animal liver to toxic injury.

Some reports have been published on temporal variation in the effects of hepatotoxins such as 1,1-dichloroethylene (Jaeger et al. 1973), trichloroethylene (Motohashi et al. 1990a), acetaminophen (Schnell et al. 1983), chloroform (Lavigne et al. 1986; Desgagne et al. 1988), or carbon tetrachloride (Harris and Anders 1980; Bruckner et al. 1984).

LAVIGNE et al. (1986) reported that rats were most susceptible to chloroform-induced hepatotoxicity when treated 2 h after commencement of their active span (21.00 h Fig. 4). The lowest susceptibility was noted at 09.00 h. A more recent study investigated the mechanism of chloroform chronohepatotoxicity in rats (DESGAGNE et al. 1988). The administration of chloroform significantly depressed the hepatic concentration of glutathione both at 21.00 h and 09.00 h, but more at 21.00 h than at 09.00 h. The authors concluded that a lower concentration of glutathione in rat liver at night would result in a greater susceptibility to cellular necrosis induced by metabolites generated from toxic chemicals.

BRUCKNER et al. (1984) reported similar chronohepatotoxicological investigations with carbon tetrachloride (CCl_4) in rats (Fig. 4). The serum activity of four enzymes, glutamic-pyruvic-transaminase (GPT), sorbitol dehydrogenase (SDH), isocitrate dehydrogenase (ICDH), and ornithine-carbamyl-transferase (OCT), assessed at eight different times, was dramatically increased when CCl_4 was given at 18.00 h or 20.00 h, but remained unchanged at 10.00 h. The increases were dramatic, for instance by as much as 400 % for GPT.

It is very interesting to note the similarity in liver chronosusceptibility for the two hepatotoxins, carbon tetrachloride, and chloroform, suggesting the importance of taking into account organ chronesthesy (Fig. 4).

A similar study reported the neurotoxicity of trichloroethylene (MOTOHASHI and MIYAZAKI 1990b), already described as a hepatotoxicant. The chronotoxicity of trichloroethylene, intraperitoneally injected at four different times, was assessed in terms of decreased muscle tone on the inclined plane test in the rat. The percentage rates of rats that slipped and fell were 50% at 03.00 h, 33% at 09.00 h, 7% at 15.00 h and 67% at 21.00 h respectively. Thus, the time of maximal neurotoxic effect of trichloroethylene was in the early active phase.

It is interesting to note that the plasma concentrations of trichloroethylene and two metabolites are maximal at 9.00 h and minimal at 21.00 h; thus, acute neurotoxicity is accentuated during the phase when metabolic hepatic elimination is lowest.

II. Ear Chronotoxicity

A subacute chronotoxicological study of auditory function has been conducted with kanamycin in rats (FISCH et al. 1984). Female Sprague-Dawley rats under light-synchronized conditions received (for six weeks) a daily subcutaneous injection of kanamycin sulfate (225 mg/kg/d) at four different times (08.00 h, 14.00 h, 20.00 h and 2.00 h). After two weeks, the 08.00-h group showed an average hearing loss of 11,5 dB at 32 kHz. The other groups had only minimal changes. After 6 weeks, the 08.00- h and 14.00-h groups (diurnal rest span groups) had a similar dramatic hearing loss (34 dB), while the two other groups had only slight losses. Such considerations can be of value in optimizing the use of aminoglycosides in clinical medicine by timing of their administration.

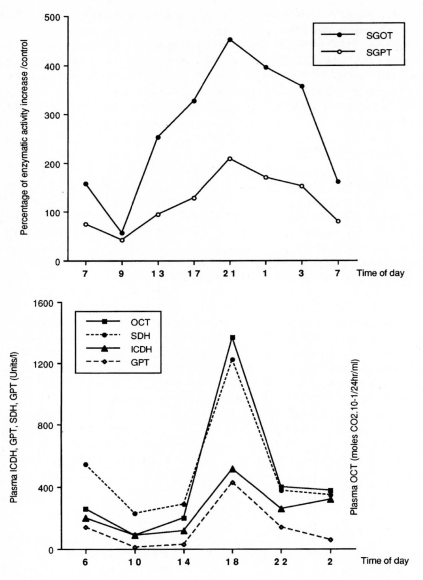

Fig. 4. Circadian changes in chloroform (*above*) and carbon tetrachloride (*below*) induced hepatotoxicity in rats, assessed by different plasma enzymes, ornithine carbamyltransferase (*OCT*), sorbitol dehydrogenase (*SDH*), isocitrate dehydrogenase (*ICDH*), glutamic-pyruvic-transaminase (*GPT*), serum glutamic pyruvic transferase (*SGPT*), and serum glutamic ornithine transferase (*SGOT*). (From BRUCKNER et al. 1984; LAVIGNE et al. 1986)

More recently, similar studies were carried out using gentamicin (100 mg/kg/d) (SOULBAN and YONOVITZ 1986).

It was shown that auditory damage was more dramatic during the day (rest) span that during the night (activity) span. This better nighttime tolerance can be correlated with the peak of body temperature, proposed as a rhythm marker in aminoglycoside-induced intoxication.

Data on renal chronotoxicity of antibiotics will be presented in Section D.IV.

III. Gastric Chronotoxicity

Exciting recent clinical investigations have reported timerelated damage caused by aspirin (ASA) and other nonsteroid anti-inflammatory drugs (NSAID) to the human gastric mucosa. Some studies indicate that nighttime administration of NSAIDs is better tolerated than morning administration (MOORE and GOO 1987; Fig. 5). However, this finding could not be confirmed in a recent study in which the effects of a high (1000 mg) and low dose (75 mg) of ASA on the gastric mucosa were studied at two circadian times (08.00 h vs 20.00 h) in healthy subjects (NOLD et al. 1995). Endoscopic evaluations on mucosal lesions

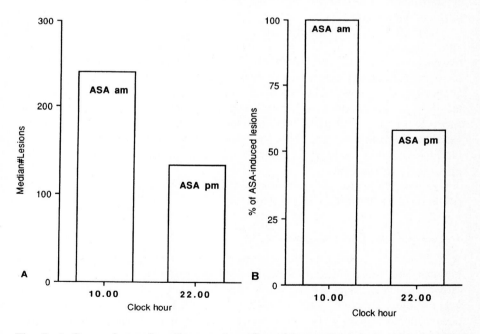

Fig. 5. A Comparison of median number of morning and evening aspirin-induced gastric mucosal lesions and **B** as a percentage of the total number of lesions produced by morning administration of ASA (=100%). (From MOORE and GOO 1987)

from blinded videotapes did not reveal any significant circadian time-dependent variations though a dosage dependency was found.

In a large study of 517 osteoarthritic patients (Lévi et al. 1984), orally administered substained-release indomethacin produced a significantly greater number of undesirable effects when delivered at 08.00 h (30% of patients) than at 20.00 h (9% of patients). Of the 44 therapeutic withdrawals encountered during the study, 29 (66%) were associated with morning ingestion. In an endoscopic study on ten healthy male volunteers, 1300 mg of ASA, administered at 08.00 h or 20.00 h, produced 37% fewer hemorrhagic lesions on the gastric mucosa when administered in the evening.

IV. Renal Chronotoxicity

Renal damage induced by drugs or toxic agents is very frequent, since the kidneys are the major route of elimination of many xenobiotics and receive nearly 25% of the cardiac output. Most drugs are filtered through the glomerular barrier and enter the tubular lumen in the tubular fluid. Many water-soluble substances such as heavy metals, aminoglycoside antibiotics, or cyclosporine can dramatically injure the nephron and alter membrane structures, thus exchanging processes.

In the kidneys, a xenobiotic may reach high intracellular concentrations and produce cell injury by one of three different metabolic pathways. First a toxic agent may enter renal cells and interfere directly with cellular functions (i.e., mitochondrial respiration, membranous transport and permeability properties, lysosomal metabolism, calcium flux, etc.). Second, it may be metabolized in renal cells to a highly reactive intermediate that may bind covalently to protein or initiate lipid peroxidation, resulting in cellular damage. Finally, an extrarenal metabolite that is stable enough to traverse the systemic bloodstream may be toxic to the kidneys either directly or following further intrarenal metabolism.

Renal chronotoxicity of xenobiotics has recently been reviewed in detail (Dorian et al. 1988; Cal et al. 1989a).

1. Heavy Metal Chrononephrotoxicity

A wide variety of toxic manifestations are produced by heavy metals to the kidneys, liver, gastrointestinal tract, heart, testes, pancreas, bones, vessels, etc. However, the kidneys are the most important target organ for heavy metals because these are concentrated mainly in the renal cortex. Their excessive renal accumulation causes well-defined morphological and ultrastructural pathological changes in the proximal tubules.

The immediate response of rat kidneys to the administration of a single dose of metals is massive focal necrosis of the middle third of the proximal tubules, the pars recta. Fragmentation of the microvilli of the brush borders occurs within 3 h of administration of mercury. After 6 h, epithelial cells of the pars racta swell and bulge into the lumen, showing varying degrees of vacuolation of the cytoplasm. After 12 h, the most striking ultrastructural changes are mitochondrial swelling and the increasing number of lysosomal structures. Renal damage extends and leads very rapidly to numerous dead cells, with fragments of dead cells being extruded into the tubules. The epithelium of the pars recta is necrotic, and the cells are in various stages of disruption and dissolution.

For a decade, sublethal chronotoxicological investigations with heavy metals have been carried out with mercury (CAL 1983), cadmium (CAMBAR et al. 1983), and especially platinum derivates, both in rodents (LÉVI 1982a, 1989) and in humans (CAUSSANEL 1989).

LÉVI and collaborators investigated for the first time heavy metal chrononephrotoxicity in cisplatin-treated rats by monitoring serum urea concentration, urinary N-acetyl-glucosaminidase (NAG) activity, and renal histology (LÉVI et al. 1982 b,c). When cisplatin was given at the time of greatest toxicity, a 3.8 fold increase in the 24-h mesor of urinary NAG was observed, in direct relation to the subsequent rise in serum urea. However, when cisplatin was given at a favorable circadian stage (in the middle of the night), the rats showed only a minor NAG rise (1.5 fold) and little renal damage with a minor rise in plasma urea.

We have performed similar studies in our laboratory with sublethal doses of mercuric chloride, the most nephrotoxic of the heavy metals (CAL et al. 1985). Circadian heavy metal nephrotoxicity, induced by a single sublethal injection in rats at four different points of time, could be assessed by the increase in urinary activity of three tubular nephrotoxicity marker enzymes. Single cosinor analysis showed the acrophase in toxicity-induced increase in enzyme excretion at 12.00 h with a mesor of 1485% (about 15-fold the control value). However, this increase reached about 1800% when mercuric chloride was given in the middle of the light span and only 1000% when it was given in the middle of the dark span, i.e., during the time of activity of rats.

The study of mercury distribution revealed that, when rats were treated at the end of the light span, mercury concentration was highest in the kidneys and lowest in urine (CAL 1983; CAL et al. 1985). Similar results were obtained in cisplatin-treated rats by LÉVI (1982a,b), who found that cisplatin was excreted in larger amounts in animals injected during the activity phase than in those injected during the rest phase.

These results in murine studies open the way to chronotherapy with anticancer agents. It is noteworthy that LÉVI (see Chap. 11 on chronotolerance and efficacity of anticancer drugs) can correlate his experimental data with his clinical investigations by considering the inversion of activity rhythm between rodents and humans. The preclinical observations permit the design of clinical chronotherapeutical protocols.

LÉVI et al. (1988) reported investigations in patients with several widely used anticancer agents, such as Adriamycin, 5-fluorouracil, vindesine, or cisplatin. These drugs exhibit circadian stage-dependent pharmacokinetics or pharmacodynamics in cancer patients. In this section on heavy metals we consider only cisplatin chronotolerance and the time-dependent administration schedule of the classical association cisplatin/Adriamycin in patients affected by testicular or ovarian cancers.

For example, murine experiments have shown that Adriamycin was better tolerated in the second half of the host rest span and that cisplatin was least toxic when given in the second half of the host activity span. Patients were therefore randomized to receive either Adriamycin in the morning and cisplatin in the evening (schedule A) or Adriamycin in the evening and cisplatin in the morning (schedule B).

In two independent studies, the necessity of dose reduction and delays in treatment as well as bleeding and infection were more often observed in patients on schedule B than in those on schedule A (HRUSHESKY 1985; MORMONT et al. 1989).

Very recently, an ambulatory treatment with a combination of oxaliplatin, 5-fluorouracil, and folinic acid in patients with colorectal cancer metastases was demonstrated to be more effective and less toxic if drug delivery was chronomodulated rather than constant over time (LÉVI et al. 1992–1994).

It is interesting to point out the similarity in circadian timing of the lethal chronotoxicity of heavy metals with that of circadian sublethal chrononephrotoxicity. We will explain the similar circadian periodicity in heavy metal-induced toxicity in Sect. E on "Chronesthesy as a Mechanistic Approach to Chronotoxicology".

2. Antibiotic Chrononephrotoxicity

The aminoglycosidic group of antibiotics is widely used to treat serious infections by gram-negative bacteria. Nephrotoxicity is a common well-recognized side effect in experimental and clinical investigations. Aminoglycosides damage both glomeruli and tubules reversibly or irreversibly.

Aminoglycosidic nephrotoxicity is generally characterized by depression of the glomerular filtration rate, decreased capacity of urine concentration, increased urinary protein and enzyme excretion, and alterations in tubular organic acid transport. Moreover, ultrastructural damage to the endothelium of the glomerular capillaries and to the tubular epithelium appears to be a characteristic feature of aminoglycosidic nephrotoxicity.

WACHSMUTH (1982) showed, for the first time, a circadian rhythm in the susceptibility of rat kidneys to an antibiotic (cephaloridine) as evaluated by histochemical staining of injured tubules and by the assay of enzymuria and proteinuria. The weakest response was observed when the animals were injected at the onset of the light span and the greatest after the injection at 19.00 h. Similar results were obtained in our group and will be explained in Sect. E "Chronesthesy as a Mechanistic Approach to Chronotoxicology".

We have also reported the circadian susceptibility rhythms of four aminoglycosidic antibiotics that are major nephrotoxic agents (CAL et al. 1985, 1989a; DORIAN et al. 1988).

For example, in November, the same amikacin dose (1,2 g/kg) in rat increased the urinary tubular enzyme activity nearly fivefold when injected at 20.00 h, and only by a factor of 1.2 at 14.00 h (DORIAN et al. 1986).

Likewise, in June, the same 200 mg/kg gentamicin dose increased the urinary NAG activity in rats by 55% at 02.00 h and 180% at 14.00 h and increased urinary AAP activity by 140% at 02.00 h and 22% at 14.00 h. To understand such temporal changes in antibiotic nephrotoxicity, we have investigated a correlation between nephrotoxicity (urinary enzyme excretion and cortex histology) and aminoglycosidic biodisposability (renal and urinary concentration).

For example, at 02.00 h, when enzymuria increase was low, renal gentamicin accumulation was low (345 mg/g kidney weight), and urinary gentamicin concentration (3900 µg/ml) and total excretion (43300 µg/24 h) were very high.

In contrast, at 14.00 h, when the increase in enzymuria was high, renal accumulation was high (442 µg/g kidney weight) and urinary concentration (2517 µg/ml) and excretion (30926 µg/24 h) very low (PARIAT 1986; PARIAT et al. 1988).

A recent, similar study confirms our previous data reporting temporal changes in nephrotoxicity, pharmacokinetics, and subcellular distribution of tobramycin in rats (LIN et al. 1994; Table 1). It compared different parameters following a single injection of 60 mg/kg tobramycin given to two groups of 9–11 week-old female Sprague-Dawley rats at either 14.00 h or 02.00 h.

Nephrotoxicity, assessed by the incorporation of tritiated thymidine into cortical DNA and by NAG urinary excretion, was significantly higher in animals treated at 14.00 h than in those treated at 02.00 h. Moreover, the drug levels in the renal cortex were always higher in animals injected at 14.00 h than in those injected at 02.00 h. These data indicate that the reduction in the clearance of tobramycin during the rest period is in part responsible for higher nephrotoxicity.

Table 1. Cortical H_3-thymidine incorporation, plasma and cortex tobramycin concentrations, and NAG urinary excretion circadian changes after tobramycin injection in rats. (From LIN et al. 1994)

Measured parameter	Group 14.00 h	Group 02.00 h
H_3-thymidine incorporation rate in renal cortex (desintegrations/mn µg DNA)	77,79 ± 41,87 ($p < 0,001$)	25,07 ± 6,65
Tobramycin plasma level (µg/ml)	0,071 ± 0,036 ($p < 0,005$)	0,028 ± 0,024
Tobramycin renal cortex level (µg/g)	382,2 ± 33,7 ($p < 0,2$)	341,8 ± 74,8
NAG urinary excretion (percentage of control)	348 ± 134 ($p < 0,001$)	98 ± 21

NAG, N-acetyl-glucosaminidase.

Table 2. Time dependence of amikacin injection on pharmacokinetics in mice. (From DORIAN, unpublished data)

Administration time	Day		Night	
	08.00 h	14.00 h	20.00 h	02.00 h
Half-life elimination (h)	2.08	2.22	1.38	1.35
Half-life distribution (h)	0.664	0.691	0.56	0.567
Area under the curve (AUC mg.h/l)	68.28	66.85	62.26	62.3
Total plasmatic clearance (l/h/kg)	0.732	0.748	0.803	0.802
Mean residence time (h)	0.887	0.911	0.688	0.676

Parallel to the chronotoxicological studies with amikacin, we have reported chronopharmacokinetic data (DORIAN, unpublished data; Table 2). At 08.00 h and 14.00 h, all pharmacokinetic parameters are higher than at 20.00 h and 02.00 h. Indeed, large and statistically significant circadian variations in elimination half-time (2.1–2.2 h during the day vs 1.3–1.4 at night), distribution volume (0,66–0,69 ml during the day vs 0,56–0,57 ml at night), and mean residence time (0.8–0.9 h during the day vs 0.6–0.7 at night) can be seen. These chronopharmacokinetic data are well correlated with circadian changes in renal haemodynamics and in aminosidic toxicity and nephrotoxicity.

In man, in whom a great number of drugs has been studied (for reviews see BRUGUEROLLE 1987, 1989; BÉLANGER 1987; BÉLANGER and LABRECQUE 1989; WATERHOUSE and MINORS 1989), the chronopharmacokinetics of numerous antibiotics have been reported.

Erythromycin (240 mg/6h) injected orally over three days showed a C_{max} at 11.30 h and the maximal area under the curve (AUC) at 12.00 h (DI SANTO et al. 1975). Likewise, ampicillin (50 mg in a single dose per os at six different times) also presented a C_{max} at 11.00 h (SHARMA et al. 1979). A widely used antifungal agent, griseofulvin, showed circadian-dependent metabolite excretion patterns with a peak at 7.30 h twice as high as that at 12.00 h (KABASAKALIAN et al. 1970).

Cefodizime given i.v. at four different times in eight men and eight women showed dosing-time and sex-related differences in its pharmacokinetics (JONKMAN et al. 1988).

For many years, the administration schedule of aminoglycosides has been discussed in terms of better tolerance by continuous intravenous infusion or by conventional intermittent infusion methods (FELD et al. 1977; BODEY et al. 1975). Some years later, the differences in efficacy and tolerance of aminosides based on once-daily administration or on continuous or discontinuous infusions were considered (POWELL et al. 1983; KOVARIK et al. 1989).

More recently, 50 neutropenic patients receiving continuous infusions of amikacin were reported to have serum levels significantly higher in the early morning than in the evening (ELTING et al. 1990). These authors confirmed circadian changes in renal clearance that are explained in part by temporal changes in glomerular filtration rate.

Temporal changes were also reported in the pharmacokinetics of gentamicin in rat as well as in man (NAKANO et al. 1990; YOSHIYAMA et al. 1992).

In conclusion, the wide use of aminoglycosides in infected patients and the very significant side effects at different organ levels, namely, ears and kidneys make it necessary to optimize their efficacy by chronomodulation. The clinical and experimental data reported in literature support this new temporal approach.

3. Cyclosporine Chronotoxicity

Cyclosporine (CsA) is a potent and clinically useful immunosuppressive agent that has significantly enhanced long-term graft survival after organ transplantation. It has also proved to be beneficial in the treatment of autoimmune diseases. Although CsA has revolutionized transplantation, numerous side effects unfortunately emerged. The most serious of them, nephrotoxicity, has most often been reported as a complication of CsA use (in 30%–50% of the grafted patients), commonly correlated with the CsA blood level.

Indeed, renal damages are numerous, mainly at vascular level but also at tubular level. Renal hemodynamics are first impaired with vasoconstriction of the renal microvessels and dramatic decrease in GFR and RPF are generally reported by the increase in renal vascular resistance. The primary (earlier) vascular target may be the glomerular afferent arteriole, but all of the renal smooth-muscle vasculature is damaged at the endothelial, smooth-muscle, and mesangial levels.

Tubular injuries are secondary to hemodynamic damage, especially of proximal convoluted cells with a permeability dysfunction of water and electrolytes in brush border membranes and with vacuolization and dilatation of endoplasmic reticulum, giant mitochondria and microcalcification in the cytoplasm of necrotic cells.

We will review here the main recent chronotoxicological studies on cyclosporine involving acute toxicology, sublethal toxicity, and pharmacokinetics in experimental animals as well as the recent clinical chronotherapeutic considerations designed to optimize clinical use in immunosuppression.

In Lewis rats treated with CsA (20 mg/kg i.p. daily) at one of six equally spaced times during a 24-h cycle, body temperature decrease and body weight loss were shown to be circadianly dosing-stage dependent, being greater when CsA was given during the dark (activity) span (CAVALLINI et al. 1983; Fig. 6). Survival time for the rats treated during the dark period was significantly shorter than that for animals treated during the light period (27 d vs 43 d).

Likewise, MAGNUS et al. (1985) compared CsA toxicity at six equispaced circadian stages in the rat in order to find treatment schedules that might induce the least damage in animals. CsA was injected daily (20 mg/kg i.p.) at

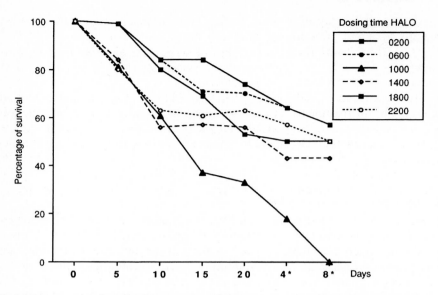

Fig. 6. Percentage of survival over the 21-day period of CsA dosing (40 mg/kg) for six different times of administration. (From MALMARY et al. 1991)

one of six points of time 4 h apart. CsA toxicity in rats, reported by pathologic hypothermia, body weight loss, and survival time, was found to be circadian-stage dependent, being greater for CsA administered during the dark span than during the light span.

A toxic dose of CsA (60 mg/kg) was given to 60 male inbred Lewis rats at one of six equally spaced points of time (BOWERS et al. 1986). Animals were killed 2, 6, 12, and 24 h after CsA administration. Whole blood samples were taken at these times. The toxic effects of CsA were significantly more severe when it was administered during the dark than during the light; weight loss was 20% vs 11% and survival time 26 d vs 42 d. Moreover, the CsA plasma levels were greatly increased (2.3 times higher) in animals treated during the dark period.

Such data demonstrated the correlation between CsA toxicity and plasma levels of the drug.

PATI et al. (1988) attempted a chronopharmacological optimization of oral CsA administration in mice in order to search for a compromise between lowest renal toxicity and highest immunosuppressive effect.

Renal toxicity of CsA can be minimized while achieving a satisfactory immunosuppressive effect when it is administered at the very end of the nocturnal activity span of mice.

CsA nephrotoxicity, assessed by increase in plasma creatinine level, was shown to depend upon the temporal stage of administration of the drug in male Wistar rats treated daily with oral CsA for 21 d (40 mg/kg) at six equally spaced times.

On the third day of treatment, plasma creatinine concentration was enhanced by about 50% in rats treated around the LD transition at 05.00 h (147 mM vs 92 mM for controls) and at 09.00 h (148 mM vs 95 mM for controls).

In contrast, this increase was smaller when CsA was given during the light span at 13.00 h and 17.00 h (104–120 mM vs 93–97 mM for controls). After 16 and 21 days of treatment, serum creatinine levels were dramatically enhanced in animals receiving CsA at the very beginning of the dark activity cycle (21.00 h; with values of 185 mM vs 98 mM for controls) and during the early resting phase (09.00 h; 152 mM vs 95 mM for controls). Rats treated with CsA at the beginning of the dark span (21.00 h) lost body weight compared to control groups, proving a time-dependent effect of CsA on body weight change in rodents (MALMARY et al. 1991).

LUKE et al. (1988) reported markedly greater serum CsA levels in hyperlipidemic obese Zucker rats during the dark cycle.

The glomerular filtration rate, assessed by inulin clearance, was significantly lower when CsA was administered at the onset of the active period of the animals.

Glucose intolerance, described among the adverse metabolic effects of chronic CsA treatment, was investigated in Wistar rats receiving CsA (40 mg/kg) daily by gavage during 21 d, at one of six different circadian stages 4 h apart (MALMARY et al. 1988).

The data from this study demonstrated that the susceptibility of the rodents to the metabolic effects of CsA was circadian-stage dependent, with rapid decline and even disappearance of glucose and insulin circadian rhythms during the first day of treatment. Higher insulin inhibition was evident when rats received CsA during the second half of the dark period and higher hyperglycemia when it was received during the very beginning of the light span, showing a time-dependence in CsA-induced pancreatic toxicity.

Chronopharmacokinetics of intravenously injected CsA at one of four circadian stages were studied in Wistar rats. Differences in area under the curve (AUC) were highly time dependent: 35,275 ± 1185 vs 29,087 ± 752 mg/l h at 15.00 h and 3.00 h respectively. Such data indicate a circadian influence of intravenously administered CsA, with a greater drug exposure during the resting cycle (BATALLA et al. 1994).

Moreover, chronopharmacokinetics of CsA following oral administration was investigated in Wistar rats receiving the same dose of drug at different temporal stages in an attempt to explain the circadian variation in CsA tolerance in rodents.

From these experiments, it appears that plasma maximal concentration (C_{max}) and the maximal area under the curve (AUC) were obtained when CsA was given at the beginning of the resting time (09.00 h) of the animals. Plasma CsA concentrations varied dramatically according to administration time.

High levels were detected when CsA administration occurred during the dark active period of the rodents. The lower plasma concentrations were observed when animals received CsA in the middle of the resting period. (MALMARY et al. 1992; Fig. 7).

Chronic treatment of male Wistar rats with CsA (40 mg/kg for 21 d) was shown to induce great changes in renal temporal structure depending upon the time of administration. Circadian changes of plasma creatinine and urea concentration were demonstrated to be largely disturbed and even abolished, whatever the temporal dosing-stage. Plasma creatinine levels, reflecting glomerular damage, exhibited a greater increase when rodents received CsA in the middle of the resting period (14.00 h) or in the middle of the dark period (2.00 h), correlated to higher plasma and renal CsA concentrations than in treatment at the very beginning of these two spans (8.00 h and 20.00 h). CsA given in the middle of the dark active period (2.00 h) seemed to induce tubular damage earlier with regard to N-acetyl-β-D-glucosaminidase (NAG) and γ-glutamyl-transferase (GGT) levels (BATALLA et al. 1994).

Similar chronobiological data, especially chronopharmacokinetic studies, have been reported in healthy or grafted patients. The CsA blood levels of transplant patients were also shown to vary with administration time, evening treatment inducing higher plasma levels (BOWERS et al. 1988; ALLISON and TRENTON 1987).

VENKATARAMANAN et al. (1986) reported diurnal variations in CsA kinetic parameters in liver transplant patients intravenously infused (140–150 mg) in the morning and at night. The clearance of CsA was found to be higher following nighttime drug administration. (20.00–04.00 h: 479 ml/min and 22.00–

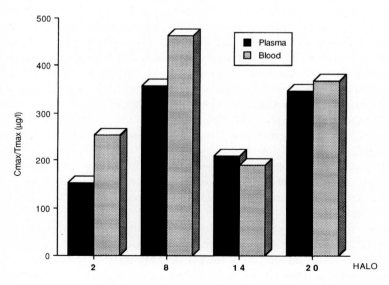

Fig. 7. Time-dependence of C_{max}/T_{max} ratio after daily CsA administration in rat. (From MALMARY et al. 1992)

06.00 h: 536 ml/min) than the clearance obtained during the day (9.00–17.00 h: 369 ml/min and 10.00–18.00 h: 337 ml/min).

CIPOLLE et al. (1988) compared CsA pharmacokinetics in pancreas allografted patients receiving CsA (300 mg) during the resting (21.00 h) and activity (9.00 h) phases. The aim of this investigation was to characterize time-dependent disposition and time variation in drug exposure, suggesting that two-dose CsA regimens improved efficacy and reduced toxicity of the drug. The CsA blood levels were significantly increased in patients receiving the afternoon dose.

Likewise, seven pancreas-transplant patients received CsA orally in two equally divided doses every 12 h at 08.00 h and 20.00 h to evaluate the circadian influences on exposure to CsA. Differences between day-time and nighttime concentrations of CsA were explored. The AUC during the rest cycle in the evening and at night was greater than during the activity phase. The mean residence time, as an index of the average time CsA remained in the body, was greater during the resting phase (5.6 ± 0.4 h) than during the activity phase (4.7 ± 0.6 h). Such data indicates an increased evening and night exposure to CsA (CANAFAX et al. 1988).

Similar findings were obtained by RAMON et al. (1989), showing that CsA plasma concentration was significantly higher after a morning administration (9.00 h; 250 mg/ml vs 800 mg/ml 1 h after drug administration) in kidney allograft recipients.

In spite of these data, there is at present no consensus on the relationship between blood levels and toxic effects of CsA. In clinical practice, such conflicting findings require numerous new studies to open the way to the optimization of CsA therapy; the introduction of chronotherapeutics for CsA can reduce the numerous side effects such as nephrotoxicity and improve efficacy of the drug.

GRUBER et al. (1988) conducted experiments on intrarenal CsA delivery in transplant patients. The pharmacokinetic advantage of local immunosuppression of CsA administered via intra-arterial infusion, using implantable programmable pumps, was examined in renal allograft recipients. The data obtained indicated that a programmable pump could be designed to deliver CsA at variable infusion rates over the 24-h cycle directly into the renal artery to reduce toxic blood levels, while simultaneously a fair immunosuppressive intrarenal blood level of 100 ng/ml is reached.

Nephrectomized dogs bearing an allografted kidney received the same dose of CsA via an external programmable implanted pump at a continuous rate or with sinusoidal schedules. The optimal time established for dogs was quite similar to that found for rats. This approach to chronotherapy demonstrated a significant circadian rhythm in response to sinusoidal CsA treatment (CAVALLINI et al. 1986).

In any attempt to optimize CsA administration, circannual and circaseptan rhythms should also be taken into consideration. An optimal CsA treatment time is pointed out in an experimental paper about segmental pancreatic transplantation in rodents (LIU et al. 1986).

E. Chronesthesy as a Mechanistic Approach to Chronotoxicology

The numerous examples of temporal changes in acute or subacute toxicity of drugs raise the question of the mechanisms through which a drug can be less (or more) toxic at one time than the other.

Originally, studies only reported the crude observations of temporal changes, but more and more frequently recent chronotoxicological reports try to explain temporal changes or at least correlate them (no coincidence!!) with other parameters such as pharmacokinetics or biotransformation processes, for example.

Organs such as the liver or the kidneys undergo a circadian stage-dependent synthesis of their enzymes and even of their nuclear components and present subsequent changes in membrane structure and function. Knowledge of renal or hepatic chronophysiology makes it possible to explain the chronopharmacological variations in hepatic or renal elimination and the metabolism of drugs (BÉLANGER 1987; FEUERS and SCHEVING 1987; CAL et al. 1985 CAL et al. 1989a; CAMBAR et al. 1992).

For example, it is interesting to consider that chronograms of glycogen synthetase and glycogen phosphorylase are a "mirror image" of those of hexobarbital oxidase and barbiturate-induced sleep duration.

Likewise, it is noteworthy that circadian changes in renal hemodynamics and excretion are well correlated with cortical toxic accumulation and elimination.

The general concept of chronesthesy was introduced by REINBERG and HALBERG (1971) and largely revisited by REINBERG (1991) to summarize all these considerations. Chronesthesy can be defined as rhythmic changes in susceptibility or tolerance of a biosystem. This biosystem can be a whole living organism, but also an organ, a tissue, or even a single cell. The chronesthesy of a biosystem to a drug or a toxic agent is all the more precise and well-defined when its structure is well-known and space-limited.

The knowledge of the biorhythmicity in renal function that we have reviewed recently (CAMBAR et al. 1992) permits us to understand the circadian change in the renal susceptibility to different nephrotoxicants (Fig. 8).

Renal decrease of hemodynamics depresses the renal plasma flow and the glomerular filtration rate. The blood and ultrafiltration volumes decrease, thus increasing the intraluminal tubular concentration of the toxic agent. It does greater damage to the brush border membrane and can enter the cell in large amounts to accumulate at the more sensible organelles, for example mitochondria, lysosomes, endoplasmic reticulum, or nucleus.

The intracellular concentration of the toxic agents is greatly increased, causing more damage to the tubular proximal cells. It has also been shown that brush border membrane enzyme activity is higher at the end of the light span. Such circadian changes in renal function can explain biochemical data showing

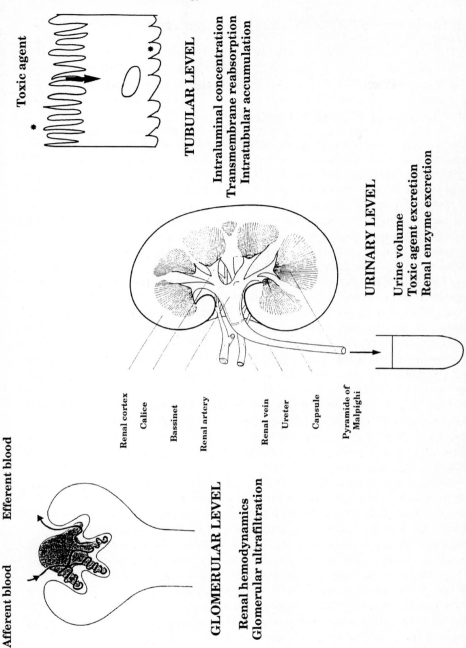

Fig. 8. Renal functions during the light span (rest span of the rat) explaining the daily nephrotoxicity maximum

a daily maximum of renal tubular enzyme excretion and a minimum of urinary volume and toxic excretion.

In the future, it will be interesting to further our knowledge of the chronostructures of the different targets of toxic compounds in normal and pathologic organisms, such as liver, kidneys, and digestive tract.

F. Circannual Chronotoxicology

Some chronotoxicological studies have shown that circadian changes in susceptibility to toxic agents must be examined during different months (or seasons) of the year (CAL et al. 1989b). Indeed, in many cases, it is very surprising to note that the circadian peak time of tolerance found during one season may be totally different from that during another.

Indeed, since the early reports on seasonal variations in the acute toxicity of X rays (FOCHEM et al. 1967) and arabinosylcytosine (SCHEVING et al. 1974b) in rodents and of cyclophosphamide in Chinese hamster and the mouse (PERICIN 1981), only a few recent papers have been concerned with this topic.

The circadian changes in acute toxicity of phenobarbital have been examined in mice during four different months of the year (BRUGUEROLLE et al. 1988). It was found that the same dose (190 mg/kg) given at 16.00 h killed no animal in July but killed 90% in January. Similarly, 270 mg/kg of phenobarbital given at 04.00 h killed all the animals (100%) in October and only 40% in July. The LD_{50} for drug administration at 16.00 h was 204 ± 6;5 mg/kg in January and 254 ± 5;2 mg/kg in July whereas for a 04.00-h administration it was 185 ± 5;5 mg/kg in January and 270 ± 7;7 mg/kg in July. A circadian peak in toxicity was found at 10.00 h in July and at 22.00 h in April and October. Such observations clearly show that in toxicological experiments not only the time of day of drug administration, but also the month of the year must be considered. We can also appreciate the relativity of the LD_{50} value here without any temporal reference!

Similar seasonal changes in murine tolerance of two anthracyclines, doxorubicin and donomycin, have been reported (LÉVI et al. 1988). Tolerance was optimal in autumn and poorest in spring and summer. LÉVI (1982a) had already reported that CDDP tolerance in rats was twice as high in winter as in summer. Documented at the same hour (09.00 h), tolerance was found to be maximal in winter (40%) and low (11%) during the other seasons.

A more recent study in rodents compared the circannual chronotolerance of two platinum derivates, CDDP and 1-OHP (BOUGHATTAS 1989). The circadian mesor of total body weight in intoxicated animals presented a seasonal change with a circannual peak (minimal body weight loss) at the beginning of October for CDPP and at the beginning of February for 1-OHP.

The survival time for 14 mg/kg CDDP was maximal in autumn and minimal in winter, and for 17 mg/kg 1-OHP it was maximal in winter (35 d) and minimal in summer (20 d). Such data may have important consequences in clinical use.

For several years, we have performed investigations on the circannual chronotoxicology of mercuric chloride. Mice received single i.p. injections of mercuric chloride at four different times of day during six months of the year. It has been shown (CAL 1983; CAL et al. 1985) that the same dose of mercury (5 mg/kg) induced a different mortality rate when given in September, in June or in January. The mean mortality rate was maximal in November (82%) and minimal in February (69%).

Another similar experiment, performed with amikacin, showed the non-reproducibility of circadian chronotolerance during two different seasons of the year (DORIAN et al. 1986; Fig. 9). The mortality rate in winter was 48% at 14.00 h and 60% at 02.00 h; in spring, it was 37% at 14.00 h and 23% at 02.00 h. Such observations show that the same toxic doses can reveal a circadian tolerance rhythm with apparently completely reversed acrophases during two different seasons. The acrophase of tolerance on one occasion can become the trough on another. These data show clearly the need to take case before asserting a value for a mortality rate or an optimal time to give a drug.

Similarly marked phase differences were obtained with subacute toxic doses of amikacin (CAMBAR et al. 1992). In autumn, urinary enzyme activity is increased 4.8 times when amikacin is given at 20.00 h and only 1.3 times at 14.00 h; in contrast, during spring this increase is dramatic at 14.00 h (3.5 to 6.5 times) and weak at 20.00 h.

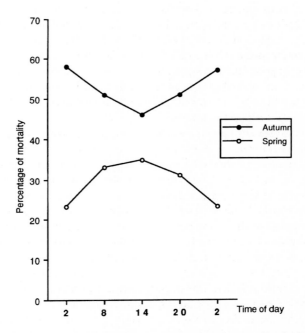

Fig. 9. Circadian change in mortality percentage induced by amikacin injection in rats during two different seasons. (From DORIAN et al. 1986)

Likewise, a comparable influence of the seasons on the severity of kidney damage was reported with gentamicin (PARIAT et al. 1990). In winter (January – February), urinary excretion of all enzymes studied increased about 1.5-to 2fold above the control values with a maximum at the end of the animal rest period at 20.00 h. In spring (March – April), opposite results were obtained with a maximal excretion increase of brush border enzymes after administration of the drug at 08.00 h. In summer (June – July), the maximum increase was at 14.00 h with a trough at 20.00 h. Finally, in autumn (October – November), the chronograms were similar to those obtained during the winter with the maximum again noted at 20.00 h.

In conclusion, all of these studies show clearly that the severity of organ damage produced by the administration of different toxic agents e.g., heavy metals, antibiotics, or anticancer agents, is dependent not only on the circadian time, but also on the season.

These circannual rhythms in tolerance of different xenobiotics could result from various circannual rhythms in organ functions and in hormonal content (testosterone and thyroxin, for example) affecting organ defenses or interfering with them, especially in rodents where they are widely described. However, these hypotheses brought forward remain to be explored in the future.

G. General Conclusions

Toxicologists using experimental animals have often been surprised by the marked discrepancies in results reported in literature. This review has demonstrated the need to consider the exact circadian time and the season of the year the experiment was carried out. Chronotoxicology can thus provide a new complementary approach to toxicology.

Numerous convincing examples have shown the dramatic temporal (circadian and circannual) changes in susceptibility and tolerance of living organisms to toxic substances or drugs. Although the cost of such experiments can be high, the use of drugs presenting severe side effects requires a chronotoxicological approach.

Some studies of underlying mechanisms correlating mortality, organ toxicity and disposition of xenobiotics that were reviewed here open the way for innovatory, exciting experiments.

The circannual changes in the circadian variations in susceptibility or tolerance to drugs warrants caution in any attempt to recommend a definitive optimal time of day or year for treatment with potentially toxic agents. Only a combination of experimental and clinical investigations will allow us to achieve this goal.

Such observations show the dramatic role of time in the toxicity of xenobiotics and indicate clearly that we need to consider the exact time of the investigations of lethal or sublethal toxicological studies and subsequently of clinical ones.

References

Allison TB, Trenton DJ (1987) AM vs PM cyclosporin blood levels. Clin Pharmacol Ther 41:237
Batalla A, Malmary MF, Cambar J, Labat C, Oustrin J (1994) Dosing-time dependent nephrotoxicity of cyclosporin A during 21 days administration to Wistar rats. Chronobiol Int 11, 3:195
Bélanger PM (1987) Chronobiological variation in the hepatic elimination of drugs and toxic chemical agents. Ann Rev Chronopharmacol 4:1–46
Bélanger PM, Labrecque G (1989) Temporal aspects of drug metabolism. In: Lemmer B (ed) Chronopharmacology: cellular and biochemical interactions. Marcel Dekker, New York, pp 15–34
Bodey GP, Chang HY, Rodriguez V, Stewart D (1975) Feasibility of administering aminoglycoside antibiotics by continuous intravenous infusion. Antimicrob Agents Chemother 8:328–333
Boughattas NA (1989) Rythmes de la pharmacocinétique et de la pharmacodynamie de trois agents anticancéreux (cisplatine, oxaliplatine et carboplatine) chez la souris: approche de leurs régulations. Doctoral thesis, Université de Paris VII
Bowers LD, Rabatin JT, Wick M, Canafax D, Hrushesky W, Benson E (1986) Circadian pharmacodynamics of cyclosporin in rats and man. Ann Rev Chronopharmacol 3:219–222
Bowers LD, Canafax DM, Singh J, Seifedlin R, Simmons RL, Najarian JS (1988) Studies of cyclosporin blood levels: analysis, clinical utility, pharmacokinetics, metabolites and chronopharmacology. In: Kahan BD (ed) Cyclosporine – pharmacological aspects. Grune and Stratton, Orlando, pp 137–143
Bruckner JV, Luthra R, Lakatua D, Sackett-Lunden L (1984) Influence of time of exposure to carbon tetrachloride on toxic liver injury. Ann Rev Chronopharmacol 1:373–376
Bruguerolle B (1984) Circadian chronotoxicity of procainamide. IRCS. Med Sci 12:579
Bruguerolle B (1987) Données récentes en chronopharmacocinétique. Pathol Biol 35:925–934
Bruguerolle B, Prat M, Douylliez C, Dorfman P (1988) Are there circadian and circannual variations in acute toxicity of phenobarbital in mice? Fundam Clin Pharmacol 2:301–304
Bruguerolle B (1989) Temporal aspects of drug absorption and drug distribution. In: Lemmer B (ed) Chronopharmacology: cellular and biochemical interactions. M Dekker, New York, pp 3–14
Cal JC (1983) Approche chronobiologique de l'intoxication aigüe par le chlorure mercurique chez les rongeurs. DEA de Nutrition, Université de Bordeaux I
Cal JC, Dorian C, Cambar J (1985) Circadian and circannual changes in nephrotoxic effects of heavy metals and antibiotics. Ann Rev Chronopharmacol 2:143–176
Cal JC, Dorian C, Catroux P, Cambar J (1989a) Nephrotoxicity of heavy metals and antibiotics. In: Lemmer B (ed) Chronopharmacology: cellular and biochemical interaction. Marcel Dekker, New York, pp 655–681
Cal JC, Dorian C, Catroux P, Pariat C, Cambar J (1989b) Les facteurs saisonniers comme source d'erreur en toxicologie expérimentale. Sci Tech Anim Lab 14:121–127
Cambar J, Cal JC (1982) Etude des variations circadiennes de la dose léthale 50 du chlorure mercurique chez la souris. CR Acad Sci 294:149–152
Cambar J, Cal JC, Desmouliere A, Guillemain J (1983) Etude des variations circadiennes de la mortalité de la souris vis-à-vis du sulfate de cadmium. CR Acad Sci 296:949–952
Cambar J, Cal JC (1985) Rythmes biologiques et toxicité des agents physiques, chimiques et médicamenteux. Pharm Biol 158:259–269

Cambar J, L'Azou B, Cal JC (1992) Chronotoxicology. In: Touitou Y, Haus E (eds) Biologic rhythms in clinical and laboratory medicine. Springer, Berlin, Heidelberg New York pp 138–150

Canafax DM, Cipolle RJ, Min DI, Hruseshesky WJM, Rabatin JT, Graves NM, Sutherland DER, Bowers LD (1988) Increased evening exposure to cyclosporin and metabolites. Ann Rev Chronopharmacol 5:5–8

Cardoso SS, Scheving LE, Halberg F (1970) Mortality of mice as influenced by the hour of the day of the drug (araC) administration. Pharmacologist 12:302

Carlsson A, Serin F (1950) Time of day as a factor influencing the toxicity of nikethamide. Acta Pharmacol Toxicol 6:181–186

Caussanel JP (1989) Chronothérapie des cancers par les complexes du platine: étude phase I de l'oxaliplatine. Thèse de Doctorat en Médecine, Université Paris Descartes

Cavallini M, Magnus G, Halberg F, Tao L (1983) Benefit from circadian timing of cyclosporin revealed by delay of rejection of murine heart allograft. Transplant Proc 15:2960–2965

Cavallini M, Halberg F, Cornelissen G, Enrichens F, Margarit C (1986) Organ transplantation and broader chronotherapy with implantable pump and computer programs for marker rhythm assessment. J Contr Release 3:3–13

Cipolle RJ, Canafax DM, Bowers LD, Rabatin JT, Sutherland DER, Hrushesky WJM (1988) Two dose chronopharmacokinetic optimization of cyclosporin in pancreas transplant patients. Ann Rev Chronopharmacol 5:13–16

Desgagne M, Boutet M, Belanger PM (1988) The mechanism of the chronohepatotoxicity of chloroform in rat: correlation between binding to hepatic subcellular fractions and histologic changes. Ann Rev Chronopharmacol 5:235–238

Di Santo A, Chodos D, Halberg F (1975) Chronobio-availability of three erythromycin test preparations assessed by each of four indices: time of peak, peak, nadir and area. Chronobiologia 2, 1:17

Dorian C, Bordenave C, Cambar J (1986) Circadian and seasonal variations in amikacin-induced acute renal failure evaluated by gamma glutamyl transferase excretion changes. Ann Rev Chronopharmacol 3:111–114

Dorian C, Catroux P, Cambar J (1988) Chronobiological approach to aminoglycosides. Arch Toxicol 12:151–157

Earnest DJ, Digiogio SM, Sladek CD (1991) Effects of tetrodotoxine on the circadian pacemaker mechanism in suprachiasmatic explants in vitro. Brain Res Bull 26: 677–682

Elting L, Bodey GP, Rosenbaum B, Fainstein V (1990) Circadian variation in serum amikacin levels. J Clin Pharmacol 30:798–801

Feld R, Valdivieso M, Bodey GP, Rodriguez V (1977) A comparative trial of sisomicine therapy by intermittent versus continuous infusion. Am J Med Sci 135:51

Feuers RJ, Scheving LE (1987) Chronobiology of hepatic enzymes. Ann Rev Chronopharmacol 4: 209–256

Fisch J, Yonovitz A, Smolensky M (1984) Effects of circadian rhythm on kanamycin-induced hearing loss. Ann Rev Chronopharmacol 1:385–388

Fochem K, Michalica W, Picha E (1967) Über die pharmakologische Beeinflussung der tagesrhythmischen Unterschiede in der Strahlenwirkung und zur Frage der jahreszeitlich bedingten Unterschiede der Strahlensensibilität bei Ratten und Mäusen. Strahlentherapie 133:256–261

Forrest AB, Hawley JC, Malone MH (1993) Testing for circadian differences in lethality for intravenous ouabaïn in male mice. J Ethnopharmacol 39:161–166

Frudden L, Wellso SG (1968) J Econ Entomol 61:1692

Gruber SA, Cipolle RJ, Canafax DM, Rabatin JT, Erdmann GR, Hynes PE, Ritz JA, Gould FH, Hrushesky WJM (1988) Circadian-shaped intrarenal cyclosporin delivery. Ann Rev Chronopharmacol 5:29A

Halberg F, Bitiner JJ, Gully RJ, Albrecht PG, Brackney EL (1955a) Proc Soc Exp Biol Med 88:169

Halberg F, Spink WW, Albrecht P, Gully RJ (1955b) J Clin Endocrinol Metabolism 15:887

Halberg F, Stephens AN (1959) Susceptibility to ouabaïn and physiologic circadian periodicity. Proc Minn Acad Sci 27:139–143

Halberg F, Johnson EA, Brown BW, Bittner JJ (1980) Susceptibility rhythms of *Escherichia coli* endotoxin and bioassay. Proc Soc Exp Biol Med 103:142–144

Harris RN, Anders MW (1980) Toxicol Appl Pharmacol 56:191–198

Hayes DK, Morgan NO (1988) Insects as animal models for chronopharmacological research. Ann Rev Chronopharmacol 5:243–246

Heinze W, Kruger S, Kamp HW (1993) Circadian and circannual rhythmicity – effects on toxicity of substances. Monatsh Veterinarmed 48:37

Hrushesky W, Levi F, Halberg F, Kennedy BJ (1982) Circadian stage dependence of 6-diamminedichloroplatinum lethal toxicity in rats. Cancer Res 42:945–949

Hrushesky W (1985) Circadian timing of cancer chemotherapy. Science 228:73–75

Jaeger RJ, Conolly RB, Murphy SD (1973) Res Commun Chem Pathol Pharmacol 6:465–471

Jonkman JHG, Reinberg A, Oosterhuis B (1988) Dosing time and sex-related differences in the pharmacokinetics of cefodizime. Chronobiologia 15:89–102

Kabasakalian P, Katz M, Rozenkrantz B, Towley E (1970) Parameters affecting absorption of griseofulvin in a human subject using urinary metabolic excretion data. J Pharm Sci 59:595–600

Kovarik JM, Hoepelman IM, Vermoef J (1989) Once-daily aminoglycoside administration: new strategies for an old drug. Eur J Clin Microbiol Infect Dos 8:761–769

Kuhl JFW, Grace TB, Halberg F, Rosene G, Scheving LE, Haus E (1973) Effect: tolerance of Adriamycine by Bagg Albino mice and Fisher rats depends on circadian timing of injection. Int J Chronobiol 1:335–343

Lakatua DJ, Thompson M, Haus E, Sackett-Lundeen L (1986) Chronopharmacokinetics of serum gentamicin in Sprague-Dawley male rats. Ann Rev Chronopharmacol 3:365–368

Lavigne JG, Belanger PM, Dore F, Labrecque G (1986) Temporal variations in chloroform induced hepatotoxicity in rats. Toxicology 26:267–273

Lemmer B, Simrock R, Hellenbrecht D, Smolensky MH (1980) Chronopharmacological studies with propranolol in rodents: implications for the management of CODP patients with cardiovascular disease. In: Smolensky MH and Reinberg A (eds), Recent advances in chronobiology of allergy and immunology, Pergamon Press, New York, pp 195–208

Lemmer B (1994) Chronopharmacology – time, a key in drug treatment. Ann Biol Clin 52:1

Lévi F (1982a) Chronopharmacologie de trois agents doués d'activité anticancéreuse chez le rat et chez la souris. Chronoefficacité et chronotolérance. Thèse d'Etat de Sciences Naturelles, Université de Paris VI

Lévi F, Hrushesky WJ, Halberg F, Langevin E, Haus E, Kennedy BJ (1982b) Lethal nephrotoxicity and hematologic toxicity of cis-diamminedichlorplatinum ameliorated by optimal circadian timing and hydration. Eur J Cancer Clin Oncol 18: 471–477

Lévi F, Hrushesky WJ, Blomquist CH, Lakatua DJ, Haus E, Halberg F (1982c) Reduction of cisplatin nephrotoxicity by optimal drug timing. Cancer Res 42:950–955

Lévi F, Boughattas NA, Blazsek I (1988) Comparative murine chronotoxicity of anticancer agents and related mechanisms. Ann Rev Chronopharmacol 4:283–331

Lévi F, Misset JL, Brienza S, Adam R, Metzger G, Itzakhi M, Caussanel J, Kunstlinger F, Lecouturier S, Descorps-Declere A, Jasmin C, Bismuth H, Reinberg A (1992) A Chronopharmacologic phase II clinical trial with 5-fluorouracil, folinic acid, and oxaliplatin using an ambulatory multichannel programmable pump. Cancer 69:893–900

Lévi F, Gastiaburu J, Bismuth A, Jasmin C, Misset JL (1994) Chronomodulated versus fixed infusion rate delivery of ambulatory chemotherapy with oxaliplatin, fluorouracil, and folinic acid (leucoverin) in patients with colorectal cancer metastases: a randomized multi-institutionnal trial. J Natl Cancer Inst 86:1608–1617

Lin L, Grenier L, Bergeron Y, Simard M, Labrecque G, Beauchamp D (1994) Temporal changes of pharmacokinetics, nephrotoxicity and subcellular distribution of tobramycin in rats. Antimicrob Agents Chemother 38:54–60

Liu T, Cavallini M, Halberg F, Cornelissen G, Field J, Sutherland DER (1986) More on the need for circadian, circaseptan and circannual optimization of cyclosporine therapy. Experientia 42:20–22

Lucht F, Tigaud S, Esposito G, Cougnard J, Fargier MP, Peyramond D, Bertrand JL (1990) Chronokinetic study of netilmicin in man. Eur J Clin Pharmacol 39:199–201

Luke DR, Vadiei K, Brunner LJ (1988) Influence of circadian changes in triglyceride concentrations on the pharmacokinetics and experimental toxicity of cyclosporin. Ann Rev Chronopharmacol 5:31–34

Lutsch EF, Morris RW (1967) Circadian periodicity in susceptibility to lidocaïne hydrochloride. Science 156:100–102

McGeachy SM, Leduc G (1988) The influence of season and exercise on the lethal toxicity of cyanide to rainbow trout. Arch Environ Contam Toxicol:17, 313

McLeay DJ, Munro JR (1979) Photoperiodic acclimation and circadian variations in tolerance of juvenile raibow trout to zinc. Bull Environ Contam Toxicol 23:552–557

Magnus G, Cavallini O, Halberg F, Cornelissen G, Sutherland DER, Najarian JA, Hrushesky WJM (1985) Circadian toxicology of cyclosporin. Toxicol Appl Pharmacol 77:181–185

Malmary MF, Kabbaj K, Oustrin J (1988) Circadian dosing-stage dependence in metabolic effects of ciclosporine in the rat. Ann Rev Chronopharmacol 5:35–38

Malmary MF, Kabbaj K, Labat C, Oustrin J (1991) CsA A dosing-time dependent effects on plasma creatinine and body weight in male wistar rats treated for 3 weeks. Chronobiol Int 8:25–34

Malmary MF, Kabbaj K, Labat C, Batalla A, Houti I, Moussamih S, Oustrin J (1992) Chronopharmacokinettics of CsA A in the wistar rat following oral administration. Eur J Drug Metab Pharmacokinet 17:135–144

Mayersbach H von (1976) Time. A key in experimental and practical medicine. Arch Toxicol 36:185–216

Moore JG, Goo RH (1987) Day and night aspirin induced gastric mucosal damage and protection by ranitidine in man. Chronobiol Int 4:111–116

Moore-Ede MC (1973) Circadian rhythms of drug effectiveness and toxicity. Clin Pharmacol Ther 14:925–935

Mormont MC, Boughattas N, Levi F (1989) Mechanisms of circadian rhythms in the toxicity and efficacy of anticancer drugs: relevance for the development of new analogs. In: Lemmer B (ed) Chronopharmacology: cellular and biochemical interactions. M Dekker, New York, pp 395–437

Motahashi Y, Kawakami T, Miyazaki Y, Takano T, Ekataksin W (1990a) Circadian variation in trichloroethylene toxicity under a 12:12 hr light-darkness in rats. Toxicol Appl Pharmacol 104 1:139–148

Motahashi Y, Miyazaki Y (1990b) Temporal variations in acute neurotoxicity of trichloroethylene in the rat. Ann Rev Chronopharmacol 7:177–180

Nakano S, Ogawa N (1982) Chronotoxicity of gentamycin in mice. IRCS Med Sci 10:592–593

Nakano S, Song J, Ogawa N (1990) Chronopharmacokinetics of gentamicin: comparison between man and mice. Ann Rev Chronopharmacol 7:277–280

Nelson W, Kupferberg H, Halberg F (1971) Dose-dependent evaluation of a circadian rhythmic change in susceptibility of mice to ouabaïn. Toxicol Appl Pharmacol 18:335–339

Pariat C (1986a) Etude expérimentale de la chrono-susceptibilité rénale de trois aminoglycosides. Thèse de Doctorat d'Etat es Sciences Pharmaceutiques, Université de Poitiers

Pariat C, Cambar J, Piriou A, Courtois P (1986b) Circadian variation in the nephrotoxicity induced by high doses of gentamicin and dibekacin in rats. Ann Rev Chronopharmacol 3:107–110

Pariat C, Courtois P, Cambar J, Piriou A, Bouquet S (1988) Circadian variations in the renal toxicity of gentamicin in rats. Toxicol Lett 40:175–182

Pariat C, Ingrand P, Cambar J, de Lemos E, Piriou A, Courtois P (1990) Seasonal effects on the daily variations of gentamicin induced nephrotoxicity. Arch Toxicol 64:205–209

Pati A, Florentin I, Lemaigre G, Mechkouri M, Levi F (1988) Chronopharmacologic optimization of oral cyclosporin A in mice: a search for a compromise between least renal toxicity and highest immunosuppressive effects. Ann Rev Chronopharmacol 5:43A

Pericin C (1981) Effect of seasonal changes on the acute toxicity of cyclophosphamide in the chinese hamster and the mouse under laboratory conditions. Experientia 37: 401–402

Powell SH, Thompson WL, Klinger JD (1983) Once-daily vs continuous aminoglycoside dosing: efficacy and toxicity in animal and clinical studies in gentamicin, netilmicin and tobramycin. J Infect Dis 147:918–932

Prat M (1990) Variations circadiennes de la toxicité aigüe et de la pharacocinétique plasmatique, cardiaque et cérébrale de trois anesthésiques locaux (bupivacaïne, étidocaïne et mépivacaïne) chez la souris; approche de leur mécanisme. Thèse d'Université d'Aix-Marseille II

Prat M, Bruguerolle B (1988) Chronotoxicity and chronokinetics of two local anaesthetic agents, bupivacaine and mepivacaine, in mice. Ann Rev Chronopharmacol 5:263

Ramon M, Morel D, Penouil F, Grellet J, Potaux L, Saux MC, Brachet-Liermain A (1989) Variations nyctémérales de la cyclosporine administrée par voie orale à des transplantés rénaux. Thérapie 44:371–374

Reinberg A, Halberg F (1971) Circadian Chronopharmacology. Ann Rev Pharmacol 2:455–492

Reinberg A, (1991) Chronopharmacologie. In: Reinberg A, Labrecque G, Smolensky MH (eds) Chronobiologie et chronothérapeutique. M/S Flammarion, Paris, pp 39–55

Roberts DR, Smolensky MH, Bartholomew P, Hsi P, Scanlon JE (1974) In: Scheving L, Halberg F, Panly J (eds) Chronobiology. Igaku Shoin, Tokyo, p 612

Ross FHN, Sermons AL, Owasoyo JO, Walker CA (1981) Circadian variation of diazepam acute toxicity in mice. Experientia 37:72–73

Scheving L, Von Mayersbach H, Pauly JE (1974a) An overview of Chronopharmacology. J Eur Toxicol 7:203–227

Scheving LE, Cardoso SS, Pauly JE, Halberg F, Haus E (1974b) Variations in susceptibility of mice to the carcinostatic agent arabinosylcytosine. In: Scheving L, Halberg F, Pauly J (eds) Chronobiology. Igaku Shoin, Tokyo, pp 213–217

Scheving LE, Scheving LA, Fewers RJ, Tsai TH, Cope FO (1993) Chronobiology as it relates to toxicology, pharmacology and chemotherapy. Regul Toxicol Pharmacol 17 2:209–218

Schnell CR, Bozigian HP, Davies MH, Merrick B, Johnson K (1983) Circadian rhythm in acetaminophen toxicity: role of nonprotein sulfhydryls. Toxicol Appl Pharmacol 71:353–358

Sharma SD, Deshpande VA, Samvel MR, Vakil BJ (1979) Chronobioavailability of ampicillin. Chronobiologia 6:156

Siva-Narayana RB, Maniraj B, Sasira BK (1984) Impact of scorpion heterometrus fulvipes venom on cholinesterase rhythmicity in the tropical mouse mus booduga. Indian J Physiol Pharmacol 28: 47–52

Soulban G, Yonovitz A (1986) Gentamicin-induced chronotoxicity: use of body temperature as a circadian marker rhythm. Ann Rev Chronopharmacol 3:293–296

Takahashi JS, Turek FW (1987) Anisomycin, an inhibitor of protein synthesis, perturbs the phase of a mammalian circadian pacemaker. Brain Res 405:199–203

Tamashiro H, Arakaki M, Akagi H, Murao K, Kirayama K, Smolensky MH, Hsi B (1986) Chronotoxicity of methyl mercury chloride in BALB/c mice. Ann Rev Chronopharmacol 3:95–98

Tsai TH, Burns ER, Scheving LE (1974) Anat Rec 178:478

Tsai TH, Scheving LE, Pauly JE (1982) Circadian variation in host susceptibility to mercuric chloride and paraquat in Balb-Cann female mice. In: Takahashi R, Halberg F, Walker CA (eds) Advances biosciences. Pergamon, Oxford, pp 249–255

Thompson HH, Wlaker CH, Hardy AR (1988) Avian esterases as indicators of exposure to insecticides. The factor of diurnal urination. Bull Environ Contam Toxicol 41:4

Venkataramanan R, Yang S, Burchart GJ, Ptachcinski RJ, Van Thiel DH, Starzl TH (1986) Diurnal variation of cyclosporine kinetics. Ther Drug Monitor 8:380–381

Wachsmuth ED (1982) Quantification of acute cephaloridine nephrotoxicity in rats: correlation of serum and 24 hour urine analyses with proximal tubular injuries. Toxicol Appl Pharmacol 63:429–445

Waterhouse JM, Minors DS (1989) Temporal aspects of renal drug elimination. In: Lemmer B (ed) Chronopharmacology: cellular and biochemical interactions. M Dekker, New York, p 35–50

Yoshiyama Y, Kobayashi T, Tomonaga F, Nakano S (1992) Chronotoxical study of gentamicin induced nephrotoxicity in rats. J Antibiot (Tokyo) 45:806–808

CHAPTER 21
Chronopharmacology of H_1-Receptor Antagonists: Experimental and Clinical Aspects (Allergic Diseases)

A.E. REINBERG

A. Histamine and H_1-Receptor Antagonists: The Conventional Pharmacological Approach

Histamine is one of the mediators involved in allergic reactions, during which histamine may be released from mast cells of "peripheral" tissues such as skin, upper airways and lung, in response to antigen-antibody reaction (e.g. in specifically sensitized patients). Histamine may also be liberated as a result of mechanical, thermal or chemical injuries of cells. Histamine may be bound to two types of receptors, H_1 and H_2, according to the different pharmacological effects observed. In addition, both type and number of receptors differ from one organ (or tissue) to another in a given species. For example, histamine increases blood pressure in guinea pigs and rabbits since in these species H_1-receptors (leading to vasoconstriction) predominate on large arterial vessels; in humans, cats and dogs, histamine decreases blood pressure (in these species, H_1- and H_2-receptors predominate on small arterial vessels).

Such a conventional approach takes into account, for a given animal model, differences in the type of receptor according to their number at a given site (anatomy in space) but not as yet their number at a given time (anatomy in time). The latter is considered only when a chronopharmacological approach is used, which is the aim of this chapter.

H_1-receptor antagonists (H_1-RAs) antagonize most, but not all, the pharmacological effects of histamine. As such, they reduce in varying degrees the intensity of symptoms in a number of allergic diseases, such as allergic rhinitis, eczema, urticaria, asthma and anaphylactic reactions (CHARPIN 1992). In humans, H_1-receptors are mainly distributed in skin, airways and lung (CHARPIN 1992).

The H_1-RA binds to the histamine H_1-receptors without initiating a response. This is a competitive antagonism since the observed effect(s) depends on the respective amount of histamine and H_1-RA present at a given site, at a given time. H_1-RAs have a preventive rather than a curative clinical effect. In addition, H_1-RAs do not antagonize equally all types of histamine-induced effects and have other properties not related to the major one. In relation to the neuromediator roles of histamine in the brain, H_1-RAs may have sedative effects, and may influence cognitive processes and performance, because they do not reduce the amount of histamine released nor do they prevent further release of histamine. Some H_1-RAs, mainly the older ones, also have m-cholinereceptor blocking activity. Some of these are considered non-desired effects.

This is mainly true for the first generations of H_1-RAs; modern ones such as terfenadine and cetirizine do not (or only to a reduced degree) cross the blood-brain barrier and therefore have sedative effects only in the range as observed with placebo (CAMPOLI-RICHARDS et al. 1990).

The intradermal (i.d.) injection of histamine to human subjects induces local reactions, such as itching, erythema and wheal, resulting from neural stimulation, and changes in both vascular diameter and permeability with transfer of fluids and related oedema. The administration of an H_1-RA results in more or less pronounced inhibition of these local reactions to histamine. Therefore, skin testing in both humans and guinea pigs represents a widely used method for the evaluation of the H_1-RA properties of a molecule.

However, the conventional pharmacological approach to H_1-RA properties does not take into consideration changes related to dosing time(s) for skin testing, quantification of parameters (e.g. in pharmacokinetics) and evaluation of effects (both desired and non-desired ones).

B. Skin Reactivity to Histamine and Allergens

In 1952, CORMIA studied the induction of itching from intradermally (i.d.) injected histamine and showed that the doses required to reach the threshold were about 100 times less at midnight than at 1400 hours. In 1965, REINBERG et al. (REINBERG 1965; REINBERG et al. 1965) proposed a method of standardizing both the manner of skin testing and accurate quantitative measurement of the area of wheal and erythema resulting from i.d. histamine injection. Statistically significant circadian rhythms with large amplitude were thus validated (Fig. 1).

Six healthy adults synchronized with diurnal activity from 0800 to 0000 hours and a nocturnal rest were involved in the initial study. Each subject's profile was obtained by sampling at 4-h intervals during a continuous 24-h span. Skin reactivity was evaluated after i.d. injection of 10 mg histamine dichlorhydrate in 0.1 ml saline exclusively on the flexor surface of both forearms. The injection site was changed each time. The mode of injection was standardized. The objective reaction was measured 15–20 min after the injection of both saline (control) and histamine.

On average, the 24-h mean was 20 cm^2 (limits 7–34 cm^2) for the erythema and 2.5 cm^2 (limits 1.2–5.5 cm^2) for the wheal surface. Both erythema and wheal reactions to histamine varied according to a circadian rhythm; they were synchronized in phase with a peak time at 2300 hours for all the subjects and a trough at either 0700 or 1100 hours (Fig. 1). These findings on the cutaneous chronesthesy to histamine were expanded to include a histamine liberator (48/80) as well as allergens in sensitized patients (REINBERG et al. 1965, 1969). Our findings have been confirmed by others (e.g. SMOLENSKY et al. 1974 and LEE et al. 1977). In addition, SMOLENSKY et al. (1974) documented in spontaneously

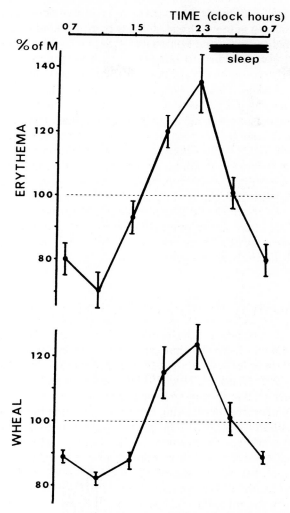

Fig. 1. Circadian rhythms of skin reactions (areas of erythema and wheal) to intradermal injection of histamine 10 μg. Data for each function are expressed as the mean percentile deviation for a group of six healthy subjects (±SEM). (From REINBERG et al. 1965, 1969)

menstruating women an approximately 30-day rhythm of the 24-h mean of both skin reactions documented at 7-day intervals throughout the menstrual cycle. The least cutaneous reactivity occurred around midcycle (presumably at ovulation); the most reactivity occurred on the 1st day of menses. This 30-day rhythm was suppressed in women taking oral contraceptives. SCHEVING et al. (1973) demonstrated the persistence of a circadian rhythm in the cutaneous response of guinea pigs maintained under constant conditions (e.g. continuous illumination). Since 1978, we have performed experiments (REINBERG et al. 1966, 1978, 1984) on skin reactivity, but injecting 2 µg histamine rather than 10 µg/0.1 ml (REINBERG 1965; REINBERG et al. 1965, 1969, 1984). This is in agreement with HÜTHER et al. (1977) that 2 µg histamine is optimal for cutaneous testing.

C. Lung Reactivity to Histamine and Allergens

Timings of the (nocturnal) peak and the (morning) trough of human skin reactivity to i.d. histamine fit well with that of the bronchial reactivity to this substance when inhaled in aerosol form (TAMMELING et al. 1977). Indeed, studies in which the bronchial challenge was documented around the clock with inhaled aerosols of histamine (TAMMELING et al. 1977), allergen extracts (GERVAIS et al. 1977) and acetylcholine (REINBERG et al. 1971) have shown a peak time of the bronchial reactivity occurring at night, with a huge (70%) peak-to-trough difference.

D. Dosing-Time-Dependent Changes in the Acute H_1-Receptor Antagonist Effects of Four Agents

The administration of H_1-RA results in more or less pronounced inhibition of the local response to histamine. At each of several test times for each subject, the cutaneous response after treatment with H_1-RA can be compared with that obtained as a control without an antihistamine (or with a placebo). The change can be expressed at each test time for each individual as a percentage deviation from the corresponding clock-hour and test-time control response.

Four H_1-RAs were tested. In all studies the drug was taken at 0700 hours for one of the profiles and at 1900 hours for the other. The dosing times (Rx) were selected after a preliminary study with Rx times of 0700, 1100, 1900 and 2300 hours; 1100 and 2300 hours correspond, respectively, to trough and peak times in the skin reactivity to i.d.-injected histamine, while 0700 and 1900 hours correspond to the times when drugs are usually taken by patients. Differences in Rx-time-related effects of H_1-RAs were small or nil between 0700 and 1100 hours on the one hand and 1900 and 2300 hours on the other. On an experimental basis it was decided to focus the study on dosing times of 0700 and 1900 hours.

I. Cyproheptadine

In healthy subjects standardized with diurnal activity (from 0700 to 2300 hours) and nocturnal rest, the areas of erythema and wheal evaluated at 15–20 min after i.d.-injected histamine were studied at 4-h intervals during three 24-h spans. A control was obtained for each subject to establish the

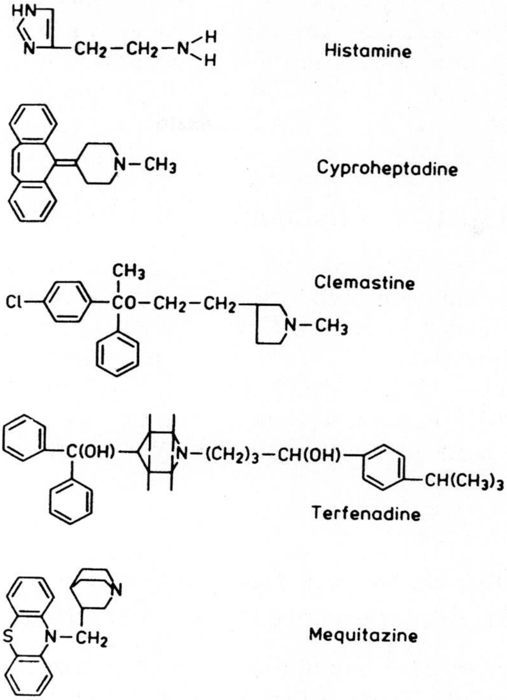

Fig. 2. Molecular structure of histamine and the four H_1-receptor antagonists considered in this chapter

reference circadian variation. Second and third groups of profiles were obtained after a single (4-mg) oral dose of the cyproheptadine H_1-RA (Periactine, Merck, Sharp and Dohme; molecular structure shown in Fig. 2). Each testing span of 24 h was separated from the others by a 7-day span for washout.

The total duration in hours (e.g. return to control values) of H_1-RA effects of cyprophetadine administered at 0700 hours was 15.1 ± 0.9 h ($\pm$SEM) with reference to erythema and 17.5 ± 0.9 h with reference to wheal. Its inhibitor effect lasted only 5.9 ± 1.03 h (erythema) and 8.6 ± 1.32 h (wheal) when the drug was taken at 1900 hours. In other words, the duration of the cyproheptadine activity depends on the timing of its administration.

In addition, the span of time (in hours) to reach the maximum inhibition (equivalent to T_{max}) and the maximum inhibition (as a percentage of appropriate clock-hour control; equivalent to E_{max}) are also dosing time-dependent (Table 1). Rx at 1900 hours was associated with the shorter T_{max} and the larger E_{max} compared to values obtained with Rx at 0700 hours.

Table 1. Dosing-time-dependent changes (0700 vs 1900 hours) of four antihistamines quantified by the inhibition of skin tests with i.d.-injected histamine

Agent[a] (dose) [No. of subjects]	Skin reaction	Dosing time: clock hours	Total duration[b] (h) of inhibitory effects: return to control values	T_{max} time (h)[b] to reach maximum inhibition	E_{max}: maximum inhibition[b] as a percentage of control
Cyroheptadine (4 mg) [6]	E	0700	15.1 ± 0.9***	10.6 ± 0.3***	48.5 ± 3.5*
		1900	5.9 ± 1.0	4.0 ± 0.1	59.6 ± 1.3
	W	0700	17.5 ± 0.9***	12.7 ± 0.6***	47.1 ± 3.2**
		1900	8.6 ± 1.3	4.3 ± 0.3	59.4 ± 1.2
Clemastine (3 mg) [10]	E	0700	25.0 ± 3.6**	18.8 ± 3.8**	54.9 ± 3.7**
		1900	15.0 ± 3.2	4.0 ± 0.8	71.2 ± 3.8
	W	0700	21.0 ± 4.2*	13.5 ± 3.8**	31.3 ± 4.3***
		1900	12.5 ± 3.7	5.5 ± 3.0	53.7 ± 4.3
Terfenadine (60 mg) [10]	E	0700	26.8 ± 2.4***	11.3 ± 3.1**	68.0 ± 5.8**
		1900	19.3 ± 1.2	4.3 ± 1.0	80.3 ± 3.5
	W	0700	23.0 ± 2.5***	9.7 ± 3.3**	38.5 ± 5.2***
		1900	14.6 ± 2.4	4.1 ± 0.9	66.2 ± 3.3
Mequitazine (5 mg) [6]	E	0700	21.6 ± 1.2**	12.0 ± 1.7***	47.3 ± 2.3*
		1900	15.5 ± 1.5	4.9 ± 1.3	56.7 ± 2.2
	W	0700	21.3 ± 1.3***	12.9 ± 1.2***	37.1 ± 2.2*
		1900	13.3 ± 0.8	4.6 ± 0.9	44.9 ± 1.8

E, erythema; W, wheal.
[a] Single oral administration, at least 1 week apart. Healthy subjects were synchronized with diurnal activity from ~0700 to ~2300 hours and nocturnal rest.
[b] Mean $\overline{X}$ ± SEM and t-tested differences, with *$P < 0.05$; **$P < 0.04$ to $P < 0.01$; ***P 0.005 to $P < 0.0001$.

II. Clemastine and Terfenadine

A double-blind placebo-controlled randomized single-dose crossover chronopharmacological study of these two H_1-RAs was undertaken in ten healthy young volunteers (REINBERG et al. 1978). Terfenadine (Teldane, Merrell) 20 and 60 mg, and clemastine (Tavegyl, Sandoz) 1 and 3 mg were tested (molecular structures shown in Fig. 2). Drug or placebo was administered at 0700 or 1900 hours at 1-week intervals with subjects synchronized with diurnal activity (0700 to 2300 hours) and nocturnal rest. Testing procedures were similar to those described for cyproheptadine. Validated dosing time-dependent changes of H_1-RA effects are summarized in Fig. 3 and Table 1.

The time from drug Rx to maximal effect and the duration of effect were longer with both drugs if they were administered at 0700 hours rather than at 1900 hours; also, the degree of maximum inhibition was greater if the drugs were administered at 1900 hours.

Dose-related inhibition of the skin reaction to histamine was obtained with both drugs: 60 mg terfenadine had an approximately equivalent inhibitory

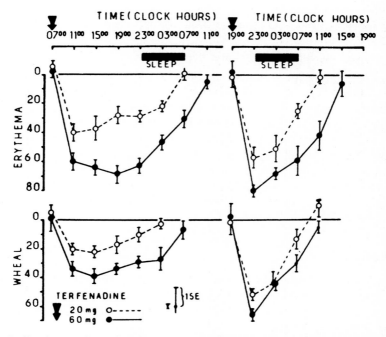

Fig. 3. Daily changes in terfenadine-induced inhibition of local skin reactions to histamine: i.d. 2 µg/0.1 ml. A double-blind and randomized procedure was used to test the effects of placebo and terfenadine (20 and 60 mg, respectively); dosing times: 0700 vs. 1900 hours at weekly intervals. The observed change of each variable/agent/dose/time point/subject was expressed as a percentage of the corresponding value on the control curve (placebo). Time point, means ± SEM. $n = 10$. (From REINBERG et al. 1978)

activity to 3 mg clemastine. However, only clemastine (3 mg) had a significant central depressant effect, as shown by the results of self-rated sleepiness and by the random-number addition test.

III. Mequitazine

The aims of the study by REINBERG et al. (1984a) were to document time-related changes resulting from acute administration of mequitazine (MQZ, Primalan, Pharmuka-Rhône-Poulenc; molecular structure shown in Fig. 2), oral dosing with regard to H_1-RA effects, tolerance and pharmacokinetics. Data gathering and analyses were similar to those already reported. The three sessions of the study consisted of a control without medication, and dosings of 5 mg MQZ at 0700 and 1900 hours.

1. Skin Tests

The time from drug administration to obtain the maximum effect (e.g. T_{max} for wheal) as well as the entire duration of effects (Table 1) was larger when MQZ was given at 0700 hours (21.3 ± 1.3 h) than at 1900 hours (13.3 ± 0.8 h, $P < 0.002$). With regard to E_{max}, the H_1-RA was stronger when MQZ was administered at 1900 hours than at 0700 hours (Table 1).

2. Psychopharmacological Tests and Performance

Statistically significant circadian rhythms were detected during the control span in self-rated sleepiness as well as in the speed to perform the three other tests: random number addition, eye-hand skill and reaction time. Time-related changes with regard to MQZ administration (Rx) were expressed for each time point as the Rx/control measurement ratio and thereafter averaged (mean ± SEM). No statistically significant changes between control, Rx at 0700 hours and Rx at 1900 hours were observed for either sleepiness or the eye-hand skill test. For both the random number addition and reaction time tests, a performance better than control (higher speed) was observed with Rx at 0700 hours but not with Rx at 1900 hours.

IV. Remarks

It is well known that food can greatly modify drug absorption from the gastrointestinal tract. However, in the case of the H_1-RAs being considered here, this does not seem to play a role in their circadian phase-dependent effect since similar phenomena were observed with either spontaneous diet or the equivalent meals standardized for both nutrients and caloric amount.

H_1-RAs most effectively block the histamine-induced smooth muscle constriction (mainly bronchi, gastrointestinal tract and uterus) as well as the histamine-induced increase in permeability of veinlets. Four different molecules (Fig. 2) exhibit similar dosing-time-related changes in their antihistamine properties (Table 1). This is the first of a set of arguments leading one to think that a circadian rhythm in H_1-receptors exists in humans. H_1-receptors presumably reach their maximum number (?) of binding sites in the evening hours, for example, between 1900 and 2300 hours in humans with diurnal activity and nocturnal rest. The question mark associated with maximum number is related to the fact that circadian rhythms experimentally documented in a set of other receptors in rodents (WIRZ-JUSTICE 1987; LEMMER and LANG 1984) and humans (LÉVI et al. 1987; PANGERL et al. 1986) concerned changes in number rather than in affinity. Specific studies on H_1-receptors have not yet been performed in relation to circadian time. In addition, nothing is known as to whether the post H_1-receptor effects of histamine are tied to circadian variations and could thus explain the diurnal variation in the H_1-RA effects involving Ca^{2+} exchanges, Ca-calmodulin and myosin-kinase.

E. Chronopharmacokinetics of Mequitazine

The chronopharmacological study of MQZ is complemented by the study of its pharmacokinetics. The chronokinetics of MQZ has been documented by collecting total urine voidings at fixed clock hours (at 2- to 4-h intervals) during a 24-h span (REINBERG et al. 1984b). Urinary determination of total and conjugated MQZ required the combination of both gas/liquid chromatography and mass spectrometry for high precision, specificity and reproducibility. With the MQZ dosing time at 0700 hours, the urinary drug excretion described a biphasic curve, the first (and highest) peak occurring after 5.5 ± 0.5 hours ($\pm$ SEM) (e.g. between 0400 and 0700 hours) (Fig. 4). Such a curve pattern fits with a two-compartmental model. When MQZ dosing time was 1900 hours, a monophasic curve was found for urinary drug excretion, the peak occurring after 10.7 ± 1.2 hours (e.g. between 0040 and 0700 hours). This other curve pattern fits with a one-compartmental model. This also means that a peak in MQZ urinary excretion is always present during the second half of the rest span, whatever the dosing time.

There were no statistically significant differences in the dosing-time-related 24-h urinary excretion (as expressed by the area under the curve) of total and conjugated MQZ or in C_{max}, the peak concentration. Finally, a biphasic curve pattern was found (FOURTILLAN et al. 1982) in plasma MQZ with Rx at 0700 hours, fitting well with changes in urinary excretion for this dosing time.

The study of the chronokinetics of MQZ has provided one example that dosing-time-related changes may affect curve pattern (REINBERG and SMOLENSKY 1983). At the present time there is no explanation for the time-dependent nature of the pharmacokinetic model for MQZ. Circadian changes in the hepatic metabolism and/or the enterohepatic circulation could well play an important role, but this has not yet been demonstrated.

For both tested dosing times with MQZ, no correlation has been validated between parameters of skin reaction and pharmakokinetic changes. In addi-

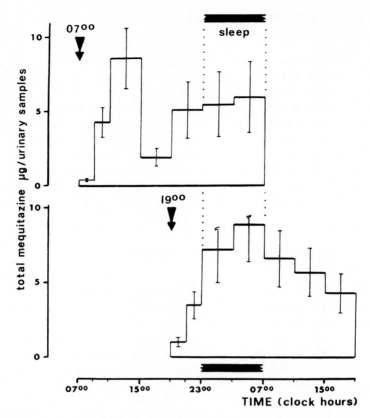

Fig. 4. Circadian time dependency in pharmacokinetics of MQZ. A 5-mg dose of MQZ was given orally to six healthy subjects at 0700 and 1900 hours 1 week apart. They were synchronized with diurnal activity from about 0700 to 2300 hours alternating with nocturnal rest. Determinations of MQZ were made by mass fragmentography techniques in total urine voidings collected at first every 2 h (twice) and thereafter at 4-h intervals during 24 h. The results are graphed as the total MQZ excreted per urinary collection interval, expressed relative to the entire 24-h excretion, following each time-specified treatment, at 0700 or 1900 hours. The kinetics were characterized by two peaks (a two-compartment model) when the antihistamine was ingested at 0700 hours and by one peak (a one-compartment model) only when given at 1900 hours. Means ± SEM. (From REINBERG et al. 1984a)

tion, changes in curve patterns of antihistamine effects have always been monophasic for both dosing times. This provides a second argument in favour of changes in the biosystem chronesthesy (e.g. the skin susceptibility to histamine) being more important than changes in the pharmacokinetics of MQZ.

F. Chronesthesy of Mequitazine with Reference to Chronic Administration

A double-blind placebo-controlled crossover and randomized study involving seven healthy adults was designed to quantify the effects of 7.5 mg MQZ given orally at 0700 hours with placebo at 1900 hours or at 1900 hours and placebo at 0700 hours during two separate 7-day spans. A 3rd week was devoted to placebo only. The order of these three 7-day spans was randomized. Subjects were synchronized with diurnal activity from 0700 hours to midnight and nocturnal rest. Every 2 h during daily activity subjects rated their sleepiness and assessed their oral temperature, peak expiratory flow (PEF; bronchial patency) and the speed at which a random number addition test was completed. At the end (Saturday/Sunday) of each of the 3 weeks, circadian changes in cutaneous reactivity to histamine were assessed at 4-h intervals. In addition, the speed of both reaction time and eye-hand skill were determined every 2 h.

I. Skin Tests

Even during chronic administration of MQZ, responses to skin tests exhibited dosing time-dependent changes, despite the fact that this agent has a half-life of approximately 36 h (FOURTILLAN et al. 1982). When Rx time was 0700 hours, the H_1-AH effects increased progressively, reaching a maximum around 1800 hours (e.g. $T_{max} = 11$ h) and a minimum (return to control) around 0600 hours (Fig. 5). When MQZ was ingested at 1900 hours, the maximum effect occurred more rapidly, between 2100 hours and midnight (T_{max} after 2–4 h); the effect was minimum around 1800 hours. Therefore, both acute and chronic administration of MQZ exhibited similar dosing-time-dependent effects with regard to T_{max} (Table 1; Fig. 5). This means that with Rx at 1900 hours, MQZ is still effective during the subsequent morning and afternoon hours. However, the 24-h adjusted means of the H_1-RA effect were roughly similar for chronic MQZ at 0700 and 1900 hours..

II. Bronchial Tests and Temperature

PEF provides an estimate of large bronchi diameter (bronchial patency). Mean 24-h PEF exhibited a small (5%) but statistically significant increase in comparison to control levels, with dosings at both 0700 and 1900 hours (REINBERG et al. 1984b). The mean 24-h oral temperature decreased slightly with administration of MQZ, particularly with Rx at 1900 hours.

III. Psychopharmacological and Performance Tests

Self-assessed duration and quality of sleep as well as circadian rhythms of self-rated sleepiness were not altered by administration of MQZ. Such findings confirm that MQZ does not induce drowsiness (REINBERG et al. 1984a;

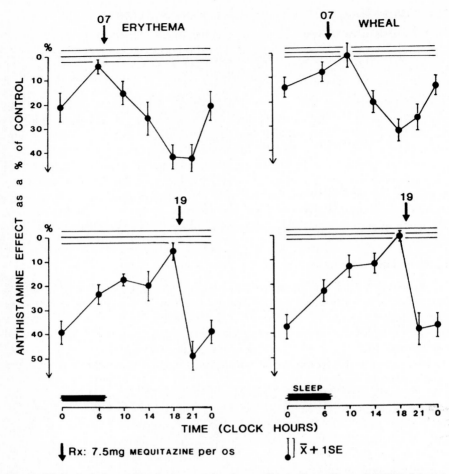

Fig. 5. Biological-time-related changes in the antihistamine effect of MQZ (chronic administration). MQZ-induced inhibition (with regard to control) of the local skin reactions to histamine (i.d. 2 µg/0.1 ml). Changes for each time point/individual/type of Rx are expressed as a percentage of control value (C = 100%) and thereafter averaged (means ± SEM). $n = 6$. (From REINBERG et al. 1984b)

GERVAIS et al. 1975). Both the random-number addition test and reaction time exhibited a small decrease (5% with $P < 0.03$ and $P < 0.01$, respectively) and thus better performance with Rx at 0700 hours. No change was observed with Rx at 1900 hours. The eye-hand skill rhythm was not influenced by administration of MQZ, regardless of whether given at 0700 or 1900 hours.

IV. Remarks

A third argument giving major importance to chronesthesy rather than to chronokinetics of H_1-RA can be added to those already presented. According to FOURTILLAN et al. (1982), total MQZ has a half-life of 36 h; after chronic administration it reaches a plateau with a constant plasma level. Therefore, dosing-time-related changes in H_1-RA effects observed after 7 days are likely to be related to changes in skin chronesthesy (e.g. possibly involving H_1-receptors) since plasma levels were presumably constant over 24 h, though this has not been documented.

G. Chronoepidemiology of a Population of 765 Allergic Patients

Because of the spontaneous evolution of symptoms, the use of patients with allergic rhinitis as their own controls during several weeks of study is not recommended. Therefore, similar groups of patients with different treatment schedules and doses had to be considered as part of a multicentre study (GERVAIS 1984; REINBERG et al. 1985, 1986). This "chronoepidemiological" aspect is critical from a methodological point of view and therefore deserves to be reported here.

I. Circadian and Circannual Rhythms in Allergic Symptoms

During a 24-h span (without drug administration) 765 patients suffering from allergic rhinitis volunteered to self-assess a set of symptoms: sneezing, obstructed nose, runny nose, itchy nose and (if any) cough, dyspnea, etc. (GERVAIS 1984; REINBERG et al. 1985, 1986). With the help of their physician, items such as age, gender, occupation, duration of disease and time of year were taken into account. Symptoms were self-rated four times daily (precise clock hours corresponding to awakening, lunch, dinner, retiring) using sets of visual analogue scales. Statistically significant circadian rhythms were detected. For those patients with diurnal activity and nocturnal rest, major symptoms showed a circadian acrophase (peak time) in the early morning hours (Fig. 6) without age- or gender-related differences. Smoking habits, geographic localization and new versus old cases had no major influence on curve pattern.

However, interindividual differences were observed, leading us to state that symptoms of allergic rhinitis predominate in the morning hours in about 60% of patients (REINBERG et al. 1986, 1987), as reported also by NICHOLSON and BOOGIE (1973).

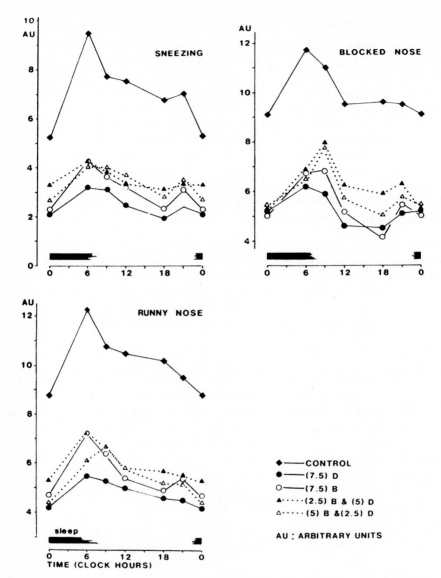

Fig. 6. Chronoeffectiveness of MQZ (mg) given to allergic rhinitis patients at breakfast *B* or dinner *D*. Circadian rhythms of symptoms observed during the control span persisted during the treatment despite the dramatic fall in their intensity. Means ± SEM. $n = 765$. (From REINBERG et al. 1984b)

H. Chronotherapy of Allergic Rhinitis with Mequitazine

The aim of the study (REINBERG et al. 1986, 1988) was to learn when and how to decide how much MQZ should be given. A total of 1052 adult patients suffering from allergic rhinitis, as verified by conventional clinical and biological criteria, volunteered for the study. Four times every 24 h – upon awakening, before lunch and dinner, and when retiring – each patient self-assessed, by means of a visual analogue scale, sneezing, obstructed nose, runny nose, itchy nose and (if any) cough, dyspnea, etc., during a 7-day period.

The first (control) day was without medication. During the following days MQZ was ingested according to one of several administration schedules with respect to the timing – morning (M) or evening (E) – and dose: 5 mg (M) and 2.5 mg (E); 2.5 mg (M) and 5 mg (E); 7.5 (M); 7.5 mg (E); 10 mg (M); 10 mg (E); and 5 mg (M) and 5 mg (E). Patients were randomized with regard to one of the seven treatment schedules. Each patient was requested to take no medication other than MQZ, and most complied. This nationwide (France) study involved 17 centres, in each of which 10 physicians participated. Both before and after entering the study, each patient received at least two medical examinations. Data were gathered from June 1983 to June 1984.

I. Chronotolerance

The estimation of MQZ tolerance by physicians and patients showed good agreement. The occurrence of MQZ side effects was related to the timing of treatment. For example, with 10 mg MQZ each 24 h, the best overall tolerance resulted from evening (91.2% rather than morning (86.8%) administration ($\chi^2 = 6.92$, with $P < 0.01$). A difference in tolerance was found for only one symptom: dry mouth ($\chi^2 = 16.8$, $P < 0001$).

II. Chronoeffectiveness

MQZ proved to be a very powerful H_1-RA agent. The improvement in the 24-h mean of each self-rated symptom (expressed as a percentage of control) was evaluated with regard to both the dose and the scheduling of MQZ. Circadian rhythms of symptoms persisted during MQZ treatment irrespective of dose and timing (Fig. 6). This observation fits with the fact that the inhibitory effect of any H_1-RA is pronounced but not complete (Table 1).

III. Remarks

An evening administration of MQZ more efficiently controlled common symptoms of allergic rhinitis – sneezing, stuffy, runny and itchy nose – than did the morning one. Differences in Rx-time-related efficiency of the drug were especially noted in the subgroup of patients who had predominantly morning symptoms (Fig. 7; REINBERG 1989); this subgroup accounted for 60% of the patient population.

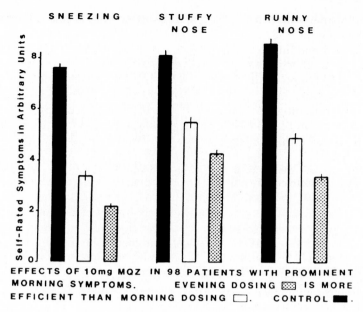

Fig. 7. Twenty-four-hour means of self-rated symptoms with regard to dose and dosing time of MQZ on the subgroups of patients (about 60% of the population) with predominant symptoms in the early morning. Treatments were associated with a dramatic decline in symptoms of sneezing and itchy, runny or stuffy nose. Rather large differences were observed (with $P<0.005$) between dosing-time-qualified effectiveness in the subroup of patients whose peak time of symptoms occurred in the morning hours. Symptoms were also better controlled by evening than by morning administration in the population as a whole (1053 patients); differences were small but statistically significant ($P<0.01$). Moreover, the higher the dose (10 vs. 7.5 mg), the better the control. (From REINBERG et al. 1986)

I. General Comments

Histamine is one of the mediators involved in allergic reactions. Its plasma concentration exhibits a circadian rhythm with a nocturnal peak time in humans (BARNES et al. 1980). Local histamine release (e.g. resulting from intradermal injections of a histamine liberator (REINBERG et al. 1965, 1969) also exhibits a circadian rhythm with a peak time around midnight in humans.

The indirect evidence summarized in this chapter thus leads to the conclusion that H_1-receptors or the H_1-receptor-mediated process in signal transduction are likely to have a circadian rhythmicity. Therefore, circadian rhythms in both histamine release and H_1-receptors have to be taken into account for a better understanding of time-related changes in allergic reactions involving histamine as well as circadian changes in the effectiveness and tolerance of H_1-RA. As an example, circadian changes in the H_1-RA effect of a drug such as MQZ persist (REINBERG et al. 1984b) despite the fact that a steady state in plasma concentration is reached during its chronic administration (FOURTILLAN et al. 1982).

It appears that, in allergic rhinitis, a chronic (once-a-day) dosing at dinner time was more effective than the dosing at breakfast time in those patients having mainly symptoms in the morning, who represent 60%–70% of all cases (Fig. 7; REINBERG et al. 1986; NICHOLSON and BOOGI 1973). In addition, using a chronopharmacological approach, it was possible to show that the cholinergic activity of MQZ – clinically associated with a dryness of the mouth – can be minimized with the evening dosing of the drug (REINBERG et al. 1986).

Acknowledgements. This review has been written with the help of Laurette Véza and Thérèse Trenel-Pontremoli donations for Chronobiologic Research at the Fondation A. de Rothschild, Paris.

References

Barnes P, FitzGerald G, Brown M, Dollery C (1980) Nocturnal asthma and changes in circulating epinephrine, histamine and cortisol. N Engl J Med 303:263–267

Campoli-Richards DM, Buckley M M-I, Fitton A (1990) Cetirizine – a review of its pharmacological properties and clinical potential in allergic rhinitis, pollen-induced asthma, and chronic urticaria. Drugs 40:762–781

Charpin J (1992) Traité d'allergologie, 3rd edn. Flammarion, Paris

Cormia FE (1952) Experimental histamine pruritus. J Invest Dermatol 19:21–27

Fourtillan JB, Girault J, Bouquet S, Ung Hong Ly, Lefebvre MA (1982) Pharmacocinétique de la méquitazine. In: Charpin J (ed) International symposium on histamine. Pharmuka-Spret-Mauchant, Paris

Gervais P (1984) Multicentric chronotherapeutic study of mequitazine in patients with allergic rhinitis. Annu Rev Chronopharmacol 1:65–68

Gervais P, Gervais A, De Beule R, Van der Bijl R (1975) Essai comparé d'un nouvel antihistaminique: la méquitazine et d'un placebo. Acta Allergol 30:286–297

Gervais P, Reinberg A, Gervais C, Smolensky M, DeFrance O (1977) Twenty-four-hour rhythm in the bronchial hyperreactivity to house dust in asthmatics. J Allergy Clin Immunol 59:207–215

Hüther KJ, Renftle G, Barraud N, Burke JT, Koch-Weser J (1977) Inhibitory activity of terfenadine on histamine induced skin wheals in man. Eur J Clin Pharmacol 12:195–199

Lee RE, Smolensky MH, Leach C, McGovern JP (1977) Circadian rhythms in cutaneous sensitivity to histamine and selected antigens including phase relationship to urinary cortisol excretion. Ann Allergy 38:231–236

Lemmer B, Lang PH (1984) Circadian-phase-dependency in ^{3}H-dihydro-alprenolol binding to rat heart ventricular membranes. Chronobiol Int 1:217–223

Lévi F, Benavides J, Touitou Y, Quarteronet D, Canton T, Uzan A, Auzeby A, Guezemy C, Sulon J, Le Fur G, Reinberg A (1987) Circadian rhythm in the membrane of circulating human blood cells: microviscosity and number of benzodiazepine binding sites. Chronobiol Int 4:435–443

Nicholson PA, Boogie W (1973) Diurnal variation of symptoms of hay fever: implication of pharmaceutical development. Curr Med Res Opin 1:395–400

Pangerl A, Remien J, Hean E (1986) β-Adrenoreceptor density on peripheral mononuclear leukocytes depends on the time of day, season and sex. Annu Rev Chronopharmacol 3:331–334

Reinberg A (1965) Hours of changing responsiveness in relation to allergy and the circadian adrenal cycle. In: Aschoff J (ed) Circadian clocks. North-Holland, Amsterdam, pp 214–218

Reinberg A (1989) Chronopharmacology of H₁-antihistamine. In: Lemmer B (ed) Chronopharmacology. Dekker, New York, pp 115–136

Reinberg A, Sidi E (1966) Circadian changes in the inhibitory effects of an antihistaminic drug in man. J Invest Dermatol 46:415–419

Reinberg A, Smolensky MH (1983) Biological rhythms and medicine. Springer, Berlin Heidelberg New York

Reinberg A, Sidi E, Ghata J (1965) Circadian rhythms of human skin reactivity to histamine or allergens and the adrenal cycle. J Allergy Clin Immunol 36:273–283

Reinberg A, Zagulla-Mally Z, Ghata J, Halberg F (1969) Circadian reactivity rhythms of human skin to house dust, penicillin and histamine. J Allergy Clin Immunol 44:293–306

Reinberg A, Gervais P, Morin M, Abulker C (1971) Rythme circadien humain de seuil de la réponse bronchique à l'acétylcholine. C R Acad Sci (Paris) 272:1879–1881

Reinberg A, Lévi F, Guillet P, Burke JT, Nicolaï A (1978) Chronopharmacologic study of antihistamines in man with special reference to terfenadine. Eur J Clin Pharmacol 14:245–252

Reinberg A, Lévi F, Fourtillan JP, Pfeiffer C, Bicakova-Rocher A, Nicolaï A (1984a) Antihistamine and other effects of 5 mg mequitazine vary between morning and evening acute administration. Annu Rev Chronopharmacol 1:57–60

Reinberg A, Lévi F, Bicakova-Rocher A, Blum JP, Ouechni MM, Nicolaï A (1984b) Biologic time-related changes in antihistamine and other effects of chronic administration of mequitazine in healthy adults. Annu Rev Chronopharmacol 1:61–64

Reinberg A, Gervais P, Smolensky M, Lévi F, Ugolini C, Del Cerro L (1985) Chronoepidemiolgic study of allergic rhinitis. Chronobiologia 12:87

Reinberg A, Gervais P, Ugolini C, Del Cerro L, Bicakova-Rocher A (1986) A multicentric chronotherapeutic study of mequitazine in allergic rhinitis. Annu Rev Chronopharmacol 3:441–444

Reinberg A, Gervais P, Ugolini C, Smolensky MH, Del Cerro L, Bicakova-Rocher A (1988) Circadian and seasonal changes of symptoms in allergic rhinitis. J Allergy Clin Immunol 81:52–62

Scheving LE, Sohal GS, Enna D, Pauly JE (1973) The persistence of circadian rhythm in histamine response in guinea pigs maintained under constant illumination. Anat Rec 175:1–7

Smolensky MH, Reinberg A, Lee R, McGovern JP (1974) Secondary rhythms related to hormonal changes in the menstrual cycle: special reference to allergology. In: Ferin M, Vande Wiele RL, Halberg F (eds) Biorhythms and human reproduction. Wiley, Chichester, pp 287–306

Tammeling CJ, De Vries K, Kruyt EW (1977) The circadian pattern of the bronchial reactivity to histamine in healthy subjects and in patients with obstructive lung disease. In: McGovern JP, Smolensky MH, Reinberg A (eds) Chronobiology in allergy and immunology. Thomas, Springfield, pp 139–149

Wirz-Justice A (1987) Circadian rhythms in mammal neurotransmitter receptors. Prog Neurobiol 29:219–259

CHAPTER 22
Local Anaesthetics

B. BRUGUEROLLE

A. Introduction

Local anaesthetic drugs prevent the generation and conduction of the nerve impulse; many drugs are used at present, such as amide-type agents (bupivacaine, articaine, etidocaine, lidocaine, mepivacaine) or esters (procaine, tetracaine, etc.). These drugs differ by their onset and duration of action as well as their toxic effects (cardiovascular depression and central nervous system stimulation). Thus the indication depends on the state of the patient as well as the desired length of the local anaesthetic effect. Chronopharmacological data have now been reported for both animals and humans and clearly demonstrate that efficacy, kinetics and toxicity of local anaesthetic drugs depend on the time of their administration (REINBERG 1986, 1992; BRUGUEROLLE 1989). Local anaesthetics, which are widely used in experimental and clinical practice, are modified by such variations and, thus, possible changes in their toxicity and efficacy are important factors to be taken into account. The present chapter aims to review time dependency of local anaesthetics in both animals and humans.

B. Experimental Chronopharmacology of Local Anaesthetics

I. Time Dependency in Toxicity

The first study describing temporal changes in the toxicity of lidocaine is an experimental work by LUTSCH and MORRIS (1967). These authors investigated the circadian-phase-dependent response of lidocaine on the central nervous system: a single dose of 65 mg lidocaine/kg was given intraperitoneally to different groups of mice every 3 h during a 24-h period. The lidocaine-induced convulsant activity was circadian time dependent with a maximal response (83%) occurring at 2100 hours (beginning of the dark/activity period of the animals). BRUGUEROLLE and PRAT (1988a, c, 1990a) demonstrated the circadian-phase-dependent toxicity of the local anaesthetics bupivacaine, etidocaine, lidocaine and mepivacaine in rodents. As shown in Table 1 and Fig. 1, the LD_{50} values of bupivacaine, etidocaine and mepivacaine were determined at six different hours of the 24-h scale and indicated a circadian time dependency: the highest mortality (i.e. the lowest LD_{50}) occurred during the

Table 1. Chronotoxicological data of some local anaesthetics in animals

Drug	Species (strain)	Variables investigated	Hours of administration	Major findings	Refs.
Bupivacaine	Mice (NMRI)	Mortality, LD_{50}	Every 4 h during 24 h	Maximum mortality at 2200 hours	BRUGUEROLLE and PRAT 1987
Lidocaine alone Lidocaine + flumazenil	Mice (NMRI)	Convulsant actions, mortality	Every 4 h during 24 h	Minimum time to convulse at 2300 hours, partial inverse agonist activity of flumazenil only significantly detected during daytime	BRUGUEROLLE and EMPERAIRE 1993
Etidocaine	Mice (NMRI)	Mortality, LD_{50}	1000, 1600, 1900, 2200, 0100, 0400 hours	Maximum mortality at 0400 hours	BRUGUEROLLE and PRAT 1990
Lidocaine	Mice (albino)	Convulsant actions (65 mg/kg)	Every 3 h during 24 h	Maximum % of convulsions at 2100 hours	LUTCH and MORRIS 1967
Mepivacaine	Mice (NMRI)	Mortality, LD_{50}	1000, 1600, 1900, 2200, 0100, 0400 hours	Maximum mortality at 1900 hours	BRUGUEROLLE and PRAT 1988

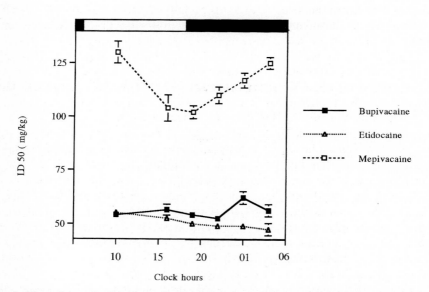

Fig. 1. Circadian variations in bupivacaine, etidocaine and mepivacaine LD_{50} (means ± SEM, mg/kg) in mice ($n = 10$ for each point). (Redrawn from BRUGUEROLLE and PRAT 1987, 1988, 1990)

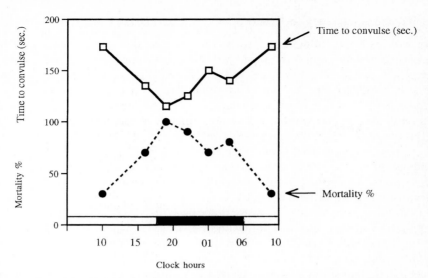

Fig. 2. Circadian variations in percentage mortality (means ± SEM%) and time to convulse (s) observed after a single i.p. 125-mg/kg dose of mepivacaine in mice ($n = 10$ for each time point). (Redrawn from BRUGUEROLLE and PRAT 1988)

dark/activity period at 2200, 0400 and 1900 hours, respectively. Also, with some of these agents, the circadian time dependency of the convulsant activity (index of the central nervous system toxicity) was documented as evaluated by the drug-induced seizures; Fig. 2 shows that the relative mortality and the time elapsed until onset of seizure observed for a single 125-mg/kg i.p. dose of mepivacaine in mice varied along the 24-h scale. Interestingly, the time course for mortality was inversely related to that of seizures (Fig. 2).

Recently a circadian time dependency of the lidocaine-induced toxicity and the influence by pretreatment with the benzodiazepine receptor antagonist flumazenil was reported; again, latency for lidocaine-induced convulsions was shortest at 2300 hours, corresponding to the time of maximal mortality. Also, flumazenil influenced lidocaine-induced toxicity in a circadian-time-dependent way since the partial inverse agonist activity of flumazenil was only significantly detected during daytime but not during night time (BRUGUEROLLE and EMPERAIRE 1993). All these results agree well with those of LUTSCH and MORRIS (1967), indicating that the circadian susceptibility of mice to local anaesthetics is highest during the dark phase and least during the resting phase. These data may indicate that in humans increased susceptibility to the toxic effects of local anaesthetics may occur during man's activity phase, i.e. during the day. If this hypothesis is right (see below), it gives further support to the notion that animal data in rodents can only be compared with human data by adequately taking into account the phase difference in activity period of these two species.

II. Time Dependency in Kinetics

Chronokinetics of lidocaine were investigated in the rat after a single intramuscular dose administered at four different time points within a 24-h period (BRUGUEROLLE and JADOT 1983). The results showed circadian variations in lidocaine kinetics: when lidocaine was given at 1600 hours its elimination half-life was shorter than when given at 1000, 2200 or 0400 hours and the peak drug concentration (C_{max}) was highest. Temporal variations in distribution, metabolism and elimination processes as well as circadian variations of membrane permeability may contribute to both chronergy and chronokinetics of the drug. These findings may explain, at least partly, that the highest susceptibility of mice to lidocaine was observed in the dark span (LUTSCH and MORRIS 1967; BRUGUEROLLE and EMPERAIRE 1993).

Similar chronokinetics to those described for lidocaine were reported for bupivacaine in mice (BRUGUEROLLE and PRAT 1987). After a single i.p. dose of 20 mg/kg at 1000, 1600, 2200 or 0400 hours, pharmacokinetic parameters were found to depend greatly on the hour of administration. At 2200 hours toxicity was highest, coinciding with peak plasma concentration of bupivacaine. Similar data have been reported for plasma chronokinetics of etidocaine and mepivacaine in mice (Table 2). In addition a time dependency in tissue levels in heart and brain of bupivacaine, etidocaine and mepivacaine was documented in mice (BRUGUEROLLE and PRAT 1990a). This point is of particular importance, in the light of the toxicity of these compounds elicited in the central nervous as well as in the cardiovascular system. A significant circadian variation in the penetration of these three local anaesthetics to heart and brain tissues was demonstrated (PRAT and BRUGUEROLLE 1990a, b), with peak values

Table 2. Chronopharmacokinetic changes in some local anaesthetics in animals

Drug	Species	Hours of administration dosing, route	Major findings	Refs.
Bupivacaine	Mice	1000, 1600, 2200, 0400 hours, 20 mg/kg i.p.	Highest C_{max}, C_{max}/t_{max} ratio at 2200 hours longest $T_{1/2}\,\beta$ at 2200 hours	BRUGUEROLLE and PRAT 1987
Etidocaine	Mice	1000, 1600, 2200, 0400 hours	Highest C_{max} at 0400 hours	BRUGUEROLLE and PRAT 1990
Lidocaine	Rats	1000, 1600, 2200, 0400 hours, 50 mg/kg i.m.	Highest plasma levels at 1600 hours, shortest half-life at 1600 hours	BRUGUEROLLE et al. 1991
Lidocaine	Rats	0700, 1600 hours, transcutaneous	Highest plasma levels at 0700 hours shortest half-life at 1600 hours	BRUGUEROLLE et al. 1982
Mepivacaine	Mice	1000, 1600, 2200, 0400 hours	Highest C_{max}/t_{max} ratio and V_d at 2200 hours, longest $T_{1/2}\,\beta$ at 2200 hours	BRUGUEROLLE and PRAT 1988

at 1000 hours in cardiac tissues for the three drugs and at 1000, 1600 and 2200 hours in brain tissues for bupivacaine, etidocaine and mepivacaine, respectively.

Beside circadian variations, it is known that the kinetics of some drugs may also change according to the oestrous or menstrual cycle. Thus, in female rats, the possible changes according to the stage of the oestrous cycle in bupivacaine kinetics were investigated (BRUGUEROLLE 1992). After a single i.p. 20-mg/kg dose during pro-oestrus, oestrus or dioestrus, bupivacaine kinetics did not significantly vary, with the exception of C_{max} being significantly higher during dioestrus.

Finally, the transcutaneous passage of lidocaine has been investigated in rats after cutaneous application of this drug in the morning or in the evening (BRUGUEROLLE et al. 1991), resulting in significantly higher plasma levels after morning dosing. This experiment first documented a time dependency in the transcutaneous passage of drugs.

C. Clinical Pharmacology of Local Anaesthetics

I. Basis for Clinical Chronopharmacology

Circadian variations of nerve conduction velocity have been demonstrated in healthy volunteers resulting in highest conduction velocity at night (FERRARIO et al. 1980). It is now well known that symptoms and onset of certain diseases are not randomly distributed over the 24-h scale but predominate at certain times of day: this implies that the timing of treatment with local anaesthetics has to vary according to the symptoms observed (BOURDALLE-BADIE et al. 1990). Local anaesthetics in man are often used for acute pain treatment. Many physicians intuitively know by observation of their patients that pain is not constant but varies through the day, the month or the year; it has been demonstrated that pain (dental, osteoarthritis, etc.) varies during the 24-h period (PROCCACCI et al. 1973). For instance, PÖLLMAN and HARRIS (1978) have documented circadian variations in dental pain in 543 patients suffering from caries profunda: maximal pain was observed between 0300 and 0600 hours. Other studies have confirmed these data, indicating a minimal pain threshold in the human tooth around 0600 hours (PÖLLMAN 1981). In order to explain these findings it is interesting to note that in rats circadian variations of brain enkephalin levels were described, with a peak occurring at the end of the dark period (WESCHE and FREDERICKSON 1981).

II. Time Dependency in Efficiency

REINBERG and REINBERG (1977) demonstrated in both human teeth and skin a significant circadian rhythm of the duration of local anaesthesia induced by lidocaine and betoxycaine: the longest duration of anaesthesia was found at 1500 hours, with a peak to trough difference of more than 100% of the 24-h

Table 3. Clinical chronopharmacology of local anaesthetics

Drug	Measured parameter	Main observed modification	Refs.
Betoxycaine	Local anaesthetic duration (dental and cutaneous)	Maximum local anaesthetic observed at 1500 hours	REINBERG and REINBERG 1977
Articaine	Sensitivity threshold (area under curve effect)	Local anaesthetic effect maximum at 1400 hours	LEMMER and WIEMERS 1989
Lidocaine	Local anaesthetic duration (dental and cutaneous)	Maximum local anaesthetic observed at 1500 hours	REINBERG and REINBERG 1977
Mepivacaine	Local anaesthetic duration (dental)	Maximum local anaesthetic effect observed at 1500 hours	PÖLLMAN 1981

mean (Table 3). Under conditions of daily dental practice, PÖLLMAN (1981) also found evidence that the duration of the effect of mepivacaine depended on the time of day: mepivacaine-induced duration of local anaesthesia was longest at 1500 hours and both the onset of pain and the disappearance of numbness followed a similar circadian rhythm.

Finally, LEMMER and WIEMERS (1989) studied the stimulus threshold with and without application of the local anaesthetic articaine to the frontal tooth of patients in a dentist's surgery at four different times of day. In accordance with the data mentioned above, articaine plus the vasoconstrictor epinephrine displayed a diurnal rhythm, with the longest duration after dosing at 1400 hours as calculated by peak effect (E_{max}) as well as by the area under the time-effect curve (AUC). Moreover, they also gave evidence for a circadian phase dependency in the dose response curve of the local anaesthetic (LEMMER and WIEMERS 1989).

III. Time Dependency of Pharmacokinetics in Man

A study of lidocaine was performed in order to investigate possible chronokinetics of lidocaine in man during its use in daily dental practice (Table 4). Four groups of six men were injected at 0930, 1230, 1530 and 1830 hours, respectively, with a single dose of lidocaine 0.65 mg/kg during dental surgical interventions (local injection at Spix's spine). The data revealed a significant variation of the area under plasma concentration curves according to the hour of injection, with AUC being greatest at 1530 hours (BRUGUEROLLE and ISNARDON 1985).

The chronokinetics of bupivacaine were also investigated for postoperative pain relief in human subjects receiving a constant-rate peridural infusion of this drug (0.25 mg/kg per hour during 36 h) (BRUGUEROLLE et al. 1988). The data clearly showed that bupivacaine plasma levels were not constant with constant

Table 4. Chronopharmacokinetic changes in some local anaesthetics in humans. Chronopharmacokinetic changes in some general or local anaesthetics in man

Drug	Species	Hours of administration	Major findings	Refs.
Bupivacaine	Man ($n=13$)	Peridural constant rate infusion during 36 h (0.25 mg/kg/h)	Circadian variations in bupivacaine plasma levels in spite of a constant rate infusion. Maximum Cl at 0630 hours	BRUGUEROLLE et al. 1988
Lidocaine	Man ($n=24$)	0930, 1230, 1530, 1830 hours, 0.65 mg/kg acute injection	Greatest area under plasma concentration curve at 1530 hours	BRUGUEROLLE and ISNARDON 1985
Bupivacaine + prilocaine (EMLA)	Children ($n=29$)	0730, 1630 hours, cutaneous application	Highest C_{max} after evening administration	BRUGUEROLLE et al. 1991

i.p., intraperitoneal route; C_{max}, maximum plasma concentration; T_{max}, time to reach C_{max}; Cl, plasma clearance

rate infusion and in addition never reached toxic plasma levels. In spite of the continuous (36 h) and constant infusion rate, the plasma clearance of bupivacaine, however, varied along the 24-h scale with a maximum at 0600 hours (Fig. 3).

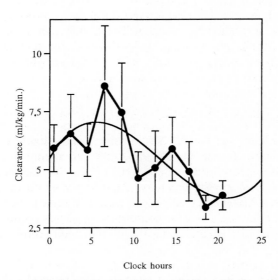

Fig. 3. Circadian variations in bupivacaine plasma clearance (means ± SEM, ml/kg per minute) after a peridural constant-rate infusion in man ($n=13$). (Redrawn from BRUGUEROLLE et al. 1988)

Surprisingly, the transcutaneous passage of drugs has not yet been investigated from a chronokinetic point of view. Recently the influence of the circadian time of cutaneous application on the kinetics of lidocaine was studied in children (BRUGUEROLLE et al. 1991). The data indicate that the lidocaine plasma levels were significantly higher in the evening, having an inverse correlation with pain scores (BRUGUEROLLE et al. 1991). The plasma levels of local anaesthetics can indicate the degree of elimination and thus may be inversely correlated to the amount of the drug applied to the skin.

D. Possible Mechanisms Involved

Local anaesthetic drugs include the amides (such as lidocaine and bupivacaine) and the esters (such as tetracaine). These drugs are weak bases which are combined with a strong acid to provide a soluble salt; they consist of a substituted amino group (hydrophilic) connected by an intermediate chain to an aromatic residue (lipophilic). At tissue pH, local anaesthetic drugs exist in both ionized and un-ionized forms, the relative percentage of each being dependent on the pK_a of the anaesthetic and the pH within the tissue. Only the un-ionized lipid-soluble form of the drug is able to cross membrane barriers before the site of action is reached.

Most local anaesthetic drugs are relatively non-specific in that they exert stabilizing effects on the plasma membranes of all kinds of excitable cells and even non-excitable cells (e.g. red blood cells). Local anaesthetics act on the plasma membranes of excitable cells to inhibit the increase in Na^+ permeability; they are considered to produce this effect in two main ways. Partly, they may exert a non-specific effect associated with their lipid-soluble form (base). Additionally, some local anaesthetic drugs may act more specifically and block Na^+ channels by modifying gating. The non-specific action resulting from accumulation of the drug in the phospholipid bilayer is important for some local anaesthetics (e.g. benzocaine); however, most of them exert their action predominantly through attachment to a receptor site in the Na^+ channel. Recently, RAGSDALE et al. (1994) described the location of the local anaesthetic receptor site in the pore of the Na^+ channel and identified molecular determinants of the state-dependent binding of local anaesthetics.

Thus local anaesthetic drugs act by lipophilicity and blockade of Na^+ channels: as proposed by REINBERG and REINBERG (1977) circadian changes in the effects of local anaesthetics may result from temporal changes in membrane properties. Circadian changes in membrane permeability and in access to channels may explain, in part at least, temporal changes in local anaesthetic efficiency and kinetics. This hypothesis has been addressed in studies using red blood cells.

The red blood cell penetration of drugs may be considered as an "in vitro" model for the study of the passage of drugs through biological membranes. Thus, the red blood cell penetration of lidocaine was investigated in rats; significant circadian changes in concentration of lidocaine in the erythrocytes were found without correlation with circadian variations of total plasma levels (BRUGUEROLLE et al. 1983; BRUGUEROLLE and JADOT 1983). These data clearly indicate a circadian time dependency in the penetration of lidocaine into red blood cells after systemic drug application. Moreover, at different time points, lidocaine levels in red blood cells are higher than the corresponding free plasma levels: thus, the circadian variations of lidocaine accumulation in red blood cells may depend on a passive transport (free drug penetration) and/or active transport. However, a possible active transport has not been further quantified. These findings totally agree with our previous chronokinetic data on lidocaine (BRUGUEROLLE et al. 1983) as well as with data from LUTSCH and MORRIS (1967).

In order to evaluate temporal changes of penetration of local anaesthetics through biological membranes, temporal variations of the passage of bupivacaine, etidocaine and mepivacaine into red blood cells were also investigated in mice (BRUGUEROLLE and PRAT 1988b, 1990b). A circadian variation of the passage of the three anaesthetic agents was demonstrated, with a maximum occurring at 0400 hours for bupivacaine and at 1000 hours for etidocaine and mepivacaine. The respective lipophilicity of each of these compounds may explain differences in the amplitude of the circadian pattern: for instance, the highest amplitude in the circadian rhythm in local anaesthetic penetration into red blood cells was observed with the most lipophilic compound, i.e. bupivacaine. These data may also explain, in part at least, chronokinetic changes of the local anaesthetics.

Chronokinetics of local anaesthetics may also be involved in the time dependency of their efficiency (BRUGUEROLLE and LEMMER 1993; LEMMER and BRUGUEROLLE 1994). As previously mentioned, chronokinetics may be explained by circadian variations in the different steps of the fate of such drugs in the organism, i.e. distribution, protein binding, metabolism and elimination.

Temporal variations in plasma protein binding and drug distribution of local anaesthetics have been documented for lidocaine in rats and bupivacaine and etidocaine and mepivacaine in mice (BRUGUEROLLE and PRAT 1992); a significant circadian variation in protein and tissue binding in the three local anaesthetics was demonstrated. Nevertheless, we did not demonstrate a temporal relationship between the respective free plasma levels and the tissue levels; thus the temporal variations in free plasma, brain and heart levels do not explain the temporal changes in local-anaesthetic-induced mortality as demonstrated by BRUGUEROLLE and PRAT (1987, 1990b). Drugs may also be transported by red blood cells. Time dependency of drug binding to erythrocytes has been reported for local anaesthetics such as lidocaine, bupivacaine, etidocaine and mepivacaine as previously mentioned (BRUGUEROLLE and JADOT 1983; BRUGUEROLLE and PRAT 1988b, 1990b). The time dependency of the passage of drugs into red blood cells provides a strong argument for the existence of temporal variations in the passage of drugs through biological membranes for which red blood cells are often used as a model. Finally, daily variations in drug distribution may also depend on circadian variation in organ blood flow as concluded from studies in rat in which imipramine and its metabolite desmethylimipramine were determined in plasma and in brain after different routes and circadian times of application of imipramine (LEMMER and HOLLE 1991).

Hepatic drug metabolism is generally assumed to depend on liver enzyme activity and/or hepatic blood flow. For drugs with a high extraction ratio such as local anaesthetics, the metabolism depends mainly on hepatic blood flow; circadian variations in hepatic blood flow may thus explain temporal variations in the clearance of local anaesthetic drugs. However, to our knowledge, this hypothesis has not yet been addressed experimentally.

The predictibility of the amplitude and the pattern of temporal variations in drug kinetics has been approached by several authors (BELANGER et al. 1984; LEMMER et al. 1985, 1990). Concerning local anaesthetics, BRUGUEROLLE and PRAT (1988b, c) have underlined the chronokinetic differences of local anaesthetics related to their physicochemical properties, as mentioned above.

E. Conclusions

The pharmacokinetics of local anaesthetics can be significantly influenced by time of day of administration. This observation is rather new and has clinical implications. Moreover, chronokinetics can, but must not always, be responsible for daily variation in drug effects and/or side effects. Thus, temporal variations of local anaesthetics in experimental or clinical practice need to be taken into account because these drugs are very often used. Temporal periodicity analyses of therapeutic/toxic ratios may eventually contribute to a better understanding of the mortality and margin of safety of local anaesthetics.

References

Belanger P, Labrecque G, Dore F (1984) Rate limiting steps in the temporal variations of the pharmacokinetics of some selected drugs. In: Haus E, Kabat H (eds) Chronobiology 1982–1983. Karger, Basel, p 359

Bourdalle-Badie C, Bruguerolle B, Labrecque G, Robert S, Erny P (1990) Biological rhythms in pain and anesthesia. Annu Rev Chronopharmacol 6:155–181

Bruguerolle B (1989) Time dependence of general and local anaesthetic drugs. In: Lemmer B (ed) Chronopharmacology, cellular and biochemical interactions. Dekker, New York, pp 581–596

Bruguerolle B (1992) Bupivacaine kinetic changes during the oestrous cycle in rats. J Pharm Pharmacol 44:440–441

Bruguerolle B, Emperaire N (1993) Flumazenil and lidocaine-induced toxicity: is the inverse agonist type activity circadian-time dependent? J Pharm Pharmacol 45:678–679

Bruguerolle B, Isnardon R (1985) Daily variations in plasma levels of lidocaine during local anaesthesia in dental practice. Ther Drug Monit 7:369–370

Bruguerolle B, Jadot G (1983) Influence of the hour of administration of lidocaine on its intraerythrocytic passage in the rat. Chronobiologia 10:295–297

Bruguerolle B, Lemmer B (1993) Recent advances in chronopharmacokinetics – methodological problems. Life Sci 52:1809–1824

Bruguerolle B, Prat M (1987) Temporal changes in bupivacaine kinetics. J Pharm Pharmacol 39:148–149

Bruguerolle B, Prat M (1988a) Circadian phase dependent pharmacokinetics and acute toxicity of mepivacaine. J Pharm Pharmacol 40:592–594

Bruguerolle B, Prat M (1988b) Temporal variations of membrane permeability to local anaesthetic agents, bupivacaine and mepivacaine, documented by their erythrocytic passage in mice. Annu Rev Chronopharmacol 5:227–230

Bruguerolle B, Prat M (1988c) Chronotoxicity and chronokinetics of two local anaesthetic agents bupivacaine and mepivacaine in mice. Annu Rev Chronopharmacol 5:227–230

Bruguerolle B, Prat M (1990a) Circadian phase dependent acute toxicity and pharmacokinetics of etidocaine in serum and brain of mice. J Pharm Pharmacol 42:201–202

Bruguerolle B, Prat M (1990b) Temporal variations in the erythrocyte permeability to bupivacaine, etidocaine and mepivacaine in mice. Life Sci 45:2587–2590

Bruguerolle B, Prat M (1992) Circadian phase-dependent protein and tissular binding of three local anesthetics (bupivacaine, etidocaine and mepivacaine) in mice: a possible mechanism of their chronotoxicokinetics? Chronobiol Int 9:448–452

Bruguerolle B, Valli M, Bouyard L, Jadot G, Bouyard P (1983) Effect of the hour of administration on the pharmacokinetics of lidocaine in the rat. Eur J Drug Metab Pharmacokinet 8:233–238

Bruguerolle B, Dupont M, Lebre P, Legre G (1988) Bupivacaine chronokinetics in man after a peridural constant rate infusion. Annu Rev Chronopharmacol 5:223–226

Bruguerolle B, Giaufre E, Prat M (1991) Temporal variations in transcutaneous passage of drugs: the example of lidocaine in children and in rats. Chronobiol Int 8:277–282

Ferrario VF, Tredici G, Crespi V (1980) Circadian rhythm in human nerve conduction velocity. Chronobiologia 7:205–209

Lemmer B (1991) Implications of chronopharmacokinetics for drug delivery: antiasthmatics, H_2-blockers and cardiovascular active drugs. Adv Drug Deliv Rev 6:83–100

Lemmer B, Bruguerolle B (1994) Chronopharmacokinetics. Are they clinically relevant? Clin Pharmacokinet 26:419–427

Lemmer B, Holle L (1991) Chronopharmacokinetics of imipramine and desipramine in rat forebrain and plasma after single and chronic treatment with imipramine. Chronobiol Int 8:176–180

Lemmer B, Wiemers R (1989) Circadian changes in stimulus threshold and in the effect of a local anaesthetic drug in human teeth: studies with an electronic pulptester.- Chronobiol Int 6:157–162

Lemmer B, Winkler H, Ohm T, Fink M (1985) Chronopharmacokinetics of beta-receptor blocking drugs of different lipophilicity (propranolol, metoprolol, sotalol, atenolol) in plasma and tissues after single and multiple dosing in the rat. Naunyn Schmiedebergs Arch Pharmacol 330:42–49

Lemmer B, Nold G, Behne S, Becker HJ, Liefhold J, Kaiser R (1990) Chronopharmacology of oral nifedipine in healthy subjects and in hypertensive patients. Eur J Pharmacol 183:521–526

Lutsch EF, Morris RW (1967) Circadian periodicity in susceptibility to lidocaine hydrochloride. Sciences 156:100–102

Pöllman L (1981) Circadian changes in the duration of local anaesthesia. J Interdiscipl Cycle Res 12:187–191

Pöllman L, Harris H (1978) Rhythmic changes in pain sensitivity in teeth. Int J Chronobiol 5:459–464

Prat M, Bruguerolle B (1988) Temporal changes in the acute toxicity of mepivacaine in mice. Med Sci Res 16:715–716

Prat M, Bruguerolle B (1990a) Circadian phase dependency of cardiac tissue levels of three amide type local anaesthetics. Annu Rev Chronopharmacol 7:257–260

Prat M, Bruguerolle B (1990b) Temporal variations of brain tissue levels of the three local anaesthetics in the mouse. Annu Rev Chronopharmacol 7:261–264

Procacci P, Moretti R, Zoppi M, Cappelletti C, Voegelin MR (1973) Rythmes circadiens et circatrigintidiens du seuil de la douleur cutanée chez l'homme. Bull Gr Et Rythmes Biol 5:65–75

Ragsdale DS, McPhee JC, Scheuer T, Catterall WA (1994) Molecular determinants of state-dependent block of Na^+ channels by local anesthetics. Science 265:1724–1728

Reinberg A (1986) Circadian rhythms in effects of hypnotics and sleep inducers. Int J Clin Pharmacol Res 6:33–44

Reinberg A (1992) Concepts in chronopharmacology. Annu Rev Pharmacol Toxicol 32:51–66

Reinberg A, Reinberg MA (1977) Circadian changes of the duration of local anaesthetic agents. Naunyn Schmiedebergs Arch Pharmacol 297:149–152

Wesche L, Frederickson RC (1981) The role of the pituitary in the diurnal variation in tolerance to painful stimuli and brain enkephalin levels. Life Sci 29:2199–2205

CHAPTER 23
Biological Rhythms in Pain and Analgesia

G. LABRECQUE, M. KARZAZI and M.-C. VANIER

A. Introduction

Pain is one of the most common causes for which patients seek advice and help from health professionals. This is a very complex phenomenon always characterized as an unpleasant sensation which often disturbs the normal patterns of a patient's activity, sleep and thoughts. Pain is a subjective phenomenon and factors such as anxiety, fatigue, suggestion or emotion as well as prior experiences can influence its perception. Thus, the patient is the only person who can describe the intensity of his or her pain and clinicians must rely on this subjective information to prescribe analgesic drugs. This is also why it is very difficult to have a good experimental model of pain.

Over the last 20 years, a body of knowledge has accumulated on the nature and mechanisms of pain and on the type of pain. Many algorithms have been developed to define new approaches to pain management with the hope that these new strategies will produce an adequate relief of pain (FOLEY 1989). Research has produced new opioid analgesics more potent than morphine. More recently, new drug formulations (such as slow-release preparations) or new drug delivery systems [such as the patient-controlled analgesia (PCA) pump] have been developed in response to the perceived inadequacies of pain treatment. For example, a patient can use a PCA pump to self-administer small intravenous (i.v.) or subcutaneous (s.c.) doses of opioid analgesics and could hopefully achieve a balance between pain, analgesia and sedation. Unfortunately, pain is still difficult and frustrating to treat, even with these new devices. Large interindividual variations are found from one patient to the next both in pain intensity and in the doses of opioids self-administered by patients. Studies carried out in the last decade have also indicated time-dependent variations in the intensity and in the neurochemistry of pain as well as in the effects of analgesic drugs. The objectives of this paper are to present the main findings in the chronobiology of pain and in the temporal variations of drug-induced analgesia. A secondary objective is to review briefly the physiology of pain and the methods used to evaluate pain level in patients.

B. Physiological Mechanisms of Pain

I. Pain Conduction Pathways

Pain is conveyed by peripheral nerves that communicate with higher centers via the spinal cord. The first studies looked at the discharges induced by mechanical and thermal stimulations of cat's lingual nerve and they reported the presence of pain receptors that are also called the nociceptors. These pain receptors are free nerve endings responsible for the initiation of an impulse by converting mechanical, thermal or chemical stimulation into electrical activity. It is believed that the tissue damage causes the release of a variety of chemicals that sensitize the nociceptors and are primarily responsible for the painful stimulus. Histamine, serotonin, bradykinin, substance P, prostaglandins, and leukotrienes are the main agents involved in the perception and transmission of pain.

The receptors have individual connections with a parent fiber. In humans, the nociceptive stimuli are transmitted from the periphery to the spinal cord predominantly via the myelinated $A\delta$ or the unmyelinated C-primary afferents. Roughly 25% of $A\delta$ and 50% of C fibers are nociceptive. These two groups of nociceptive fibers have been associated with different qualities of pain: the $A\delta$ fibers carry the sharp, stabbing and well-localized pain whereas the dull, burning or diffuse pain is transmitted through the C fibers. Both $A\delta$ and C fibers transmit the pain influx relatively slowly and they end in the dorsal horn of the spinal cord, which is divided into several laminae connected by multiple interneurons. C fibers terminate in laminae II (the substantia gelatinosa) while $A\delta$ fibers end in laminae I and V. The primary afferents synapse with the interneurons of the laminae to activate the ascending nociceptive systems (spinothalamic, spinocervical and spinoreticular tracts). Through relays at the brainstem level, pain signals are sent to the ventral posteromedial and the posterolateral nuclei of the thalamus, which is the first level of conscious pain. From there, they relay to the postcentral gyri of the cerebral cortex.

Transmission and interpretation of pain are not simply a unidirectional process, but are subjected to peripheral and central modifications. The sensory functions of the spinal cord are controlled by various regions of the brain via descending tracts that limit pain transmission by postsynaptic inhibition of the primary afferents (BAUSBAUM and FIELDS 1978; WILLIS 1982). Thus, the firing of dorsal horn neurons in response to noxious skin heating can be inhibited by stimulating the periaqueductal gray matter (PAG), the reticular formation and the thalamus. Furthermore, the electrical stimulation of the ventrolateral portion of the central gray substance of the mesencephalon and the PAG produces profound analgesia without general behavioral depression. These areas are rich in opiate receptors and they exhibit powerful opiate binding (FISHMAN and CARR 1992).

II. Neurotransmitters and Pain Perception

Pain and analgesia are conveyed by well-described primary and secondary intracellular events. Endogenous substances such as H^+ ions, serotonin, histamine, bradykinin and prostaglandins have excitatory effects on nociceptors. Therefore, they have been called the endogenous algesic substances. Experimental evidence suggests that the excitatory effect of these algesic substances is due in part to a direct action on the nociceptive nerve membrane.

In addition to the other algesic substances released from non-neuronal tissues, substance P has been suggested to be a factor in inflammation and pain. This is an undecapeptide originally detected in extracts of gut and brain that has many pharmacological actions including vasodilatation, stimulation of smooth muscles, diuresis and natriuresis. Substance P has also been localized in the central nervous system and in areas such as the dorsal root ganglia and the dorsal horn of the spinal cord. Thus, substance P is a leading candidate to be a neurotransmitter involved in the passage of the nociceptive stimuli from the periphery to the spinal cord and to higher structures (JESSEL 1983; PERNOW 1983). This neurotransmitter is responsible for the pain transmission associated with injury or disease, and drugs depleting substance P are associated with analgesia.

Although morphine has been used as an analgesic for many years, the search for endogenous substances that could bind to the opioid receptor sites and could produce morphine-like effects began about 20 years ago. KOSTERLITZ and HUGHES (1975) were the first to isolate two naturally occurring peptides from the pig brain that were called leucine-enkephalin and methionine-enkephalin. Further research identified three major groups of opioid peptides which are known as the endorphins, the enkephalins and the dynorphins. Each active peptide is obtained from a larger specific precursor molecule which does not possess any intrinsic analgesic activity. The PAG and the nucleus raphe magnus within the rostral ventromedial medulla are the areas of the brain with the largest concentrations of opioid peptides. Many studies suggest that enkephalin and endorphin inhibit certain neuronal mechanisms of pain transmission. For instance, animal studies showed that analgesia can be produced by the electrical stimulation of brain regions such as the PAG, which are known to have large concentrations of peptides (MAYER and LIEBESKIND 1974). These effects are abolished by administration of naloxone, a specific opiate receptor antagonist. Other studies demonstrated also that enkephalins reduced the release of substance P from nerve endings of primary afferent fibers of nociceptors (MUDGE et al. 1979). This observation supports the hypothesis that the opioid peptides are involved in the inhibition of the nociceptive transmission. The fact that small-diameter fibers have opiate receptors also strengthens this hypothesis. In addition, injections of morphine in the PAG relieve pain by activating descending pathways and thus inhibit the primary afferent transmission in the dorsal horn (YAKSH 1978). There is also evidence that the activation of descending pathways involves the nucleus raphe magnus and the related serotonergic nuclei which mediate this inhibition (RIVOT et al.

1984). However, the mechanism by which serotonin inhibits transmission in the dorsal horn is still unknown.

This brief literature review indicates clearly that pain perception is a rather complex phenomenon. Thus, it is not surprising to note that pain intensity is difficult to quantify and many instruments have been designed to evaluate this parameter. The main methods used to measure pain in patients will now be summarized briefly.

III. Measurements of Pain

In the last decade, considerable progress has been made in the methods used to quantify pain. The first studies were carried out a century ago and chemical, electrical, thermal or mechanical stimuli were used to induce pain (WOLFF 1983). As pointed out by WOLFF (1983), the early investigations focused on the evaluation of pain threshold (i.e., the stimulus intensity at which pain is first felt) and pain tolerance (i.e., the level of intensity at which the subject withdraws from the noxious stimulus) because these parameters are easy to use and to understand. However, pain was then considered as a unidimensional sensory experience and the experimental models can be used for acute pain only. Another major problem of the early pain models was the large variability in the responses to painful stimuli (BLITZ and DINNERSTEIN 1968).

It is now believed that pain is a complex multidimensional phenomenon and the sensory, emotional and behavioral components must be considered in the methods used to assess pain. Therefore, pain evaluations can be achieved mainly by quantifying patients' behaviors or by using tools that can quantify the subjective reports of pain by patients. The measurement of the behavioral components is designed to quantify objective behaviors known to accompany the painful experience. The variables most commonly used are: daily activity diaries, sleep patterns, food intake, demands for or the intake of medications, and amount of time spent standing, sitting or reclining. Thus, the observation of patients' behavioral changes provides an objective and practical method for inferring that they are experiencing pain (CHAPMAN et al. 1985). This technique is especially reliable and valid in infants and adults who have difficulties communicating meaningfully (MCGRATH 1990; READING 1989). Although the behavioral procedures are important tools, especially to measure the efficacy of analgesic treatments, subjective techniques such as rating scales and pain questionnaires are still the most widely used tools in clinical settings. These techniques will now be described briefly.

1. The Pain Questionnaires

The McGill Pain Questionnaire (MPQ) is one of the first tests developed to evaluate the multidimensional nature of clinical pain (MELZACK 1975). This test is actually the questionnaire most widely used and it consists of 20 sets of words describing the sensory, affective and evaluative dimensions of the pain experience (MELZACK and TORGERSON 1971; MELZACK 1983). Patients are asked to select the words that are most accurately related to their pain. The MPQ also contains an intensity scale called the Present Pain Intensity scale (PPI) and other items to determine the pain location and to compare the present pain to previous pain experiences. This instrument was shown to be able to discriminate among different pain syndromes and to provide reliable data on both the quantitative and qualitative aspects of pain (GRAHAM et al. 1980; DESCHAMPS et al. 1988). One limitation of the MPQ is that patients may have some difficulty understanding its complex vocabulary. It may also be too long to use for some patients, because the test takes 5–10 min to administer.

2. The Rating Scales

Categorical scales which could be either verbal or nonverbal, numerical scales and visual analogue scales could be used to evaluate pain intensity. These methods are commonly used to assess pain in clinical settings, although they consider pain as a unidimensional experience that varies only in intensity. However, these scales are easy to use, they are well understood by most patients and they have been used effectively in clinical settings to provide reliable information about pain and analgesia.

The categorical scales consist of a list of adjectives describing different levels of pain. A number is associated with each adjective and the patient's pain score is the number associated with the word considered to describe best his or her pain. Categorical scales can be made by numerous points or categories (HUSKISSON 1983) and the basic scale consists of a 4-point scale where pain is graded as slight, moderate, severe or agonizing. MELZACK and TORGERSON (1971) included a verbal scale in their questionnaire with five categories: mild, discomforting, distressing, horrible, excruciating. An 8-point facial expression picture scale has been developed as a tool which can be easily used in children and in people with language difficulties or mental disabilities (MCGRATH 1987). These scales are simple, short and easy to administer, but their use is limited by the lack of sensitivity to detect relatively small changes in the pain intensity.

The numerical scales constitute numbers that patients must choose to quantify pain intensity. Several models of numerical scales are available and 6-point, 11-point or 101-point numerical scales could be used in clinical studies with scales in a written or a verbal form (JENSEN et al. 1986). For instance, the patient is asked to rate pain by stating verbally the number representing pain intensity, whereas he or she must circle the number corresponding to the pain level when the written test is used. This number may vary between 0 and 5, 0 and 10 or 0 and 100.

The Visual Analogue scales (VAS) are other satisfactory subjective methods to measure pain (JENSEN et al. 1986). Usually, a 10-cm straight line is used and words such as *no pain* and *pain as bad as it could be* are placed at both ends of the line. The patient is required to draw a mark on the line at a point which corresponds to the level of pain felt. The distance (in centimeters) from the *no pain* end of the VAS to the patient's mark is used as an indicator of the severity of pain. Descriptive terms and/or numbers could also be placed at intervals along the line of the VAS. These scales have been called the graphic scales and they increased the sensitivity and the accessibility of the scales, especially for patients with mental disabilities (Scott and Huskisson 1976). However, the presence of descriptive terms along the VAS may influence the performance of the test (Scott and Huskisson 1976) by concentrating values around descriptive terms or numbers.

In summary, assessments of human pain have evolved from studies on pain threshold and tolerance determinations in pain-free volunteers to the use of a wide variety of psychophysical, behavioral and physiological methods to measure the multidimensional aspects of clinical pain. With the notable exception of the McGill Pain Questionnaire, clinical pain is usually determined by using either VAS or numerical or categorical scales. These tools are easy to use and to understand by patients and they provided valuable data for studies on pain and on pain relief. Unfortunately, there is no ideal way to quantify pain and the selection of a method must be based on the types of information needed by the investigators and the characteristics of the subject or the group of patients to be evaluated. It must also be pointed out that the measurement technique (Morawetz et al. 1984) and the attitude of investigators toward pain may influence the evaluation of pain in patients. These factors could explain why it is difficult to obtain reproducible data from one study to the next.

C. Chronobiology of Pain

I. Biological Rhythms in Pain Control Mechanisms

While there are no data indicating the existence of biological rhythms in the pathways of pain transmission, brain concentrations of neurotransmitters involved in pain modulation have been shown to fluctuate over a 24-h period. Table 1 summarizes the data presently available.

In mice kept on an L:D 12:12 regimen (light on at 0600 hours), Wesche and Frederickson (1978, 1981) reported that the opioid-like peptide levels in the whole brain were twice as high by the end (2173.3 ± 119.7 pmol/g brain at 1530 hours) than at the beginning of the resting period (1099.6 ± 120.8 pmol/g brain at 0730 hours). Hypophysectomy did not alter the time-dependent changes in the temporal variation in the brain level of this peptide. Time-dependent changes have also been detected in the concentration of β-endorphin in the pituitary gland, the pons, the medulla and the cerebellum of the rat brain. Peak levels of this peptide always occurred during the activity period of the animals, i.e., between 2000 hours and 2400 hours (Kerdelhue et al. 1983). Indirect evidence on the circadian variations of brain enkephalin levels was obtained by Puglisi-Allegra et al. (1982) and Oliverio et al. (1982), who studied the stress-induced analgesia in mice kept on an L:D 12:12 regimen (light on at 0800 hours). These investigators used the tail-flick and the hot-

Table 1. Biological rhythms in the tissue levels of pain neurotransmitters

Species	Location	Parameters	Time of Peak	Time of Trough	References
Mice	Hypothalamus	Substance P	2nd h of activity	2nd h of rest	Kelderhue et al. 1981
Mice	Whole brain	Opioid-like peptide	9th h of rest	1 h of rest	Wesche and Frederickson, 1978, 1981
Rats	Brainstem Cerebellum	Met-enkephalin	5th h of activity	11th h of rest	Kerdelhue et al. 1983
Horses	Plasma levels	β-Endorphin	4th h of activity	Last hour of rest	Hamra et al. 1993
Monkeys	CSF	Opioid-like peptide	3rd h of activity	3rd h of rest	Naber et al. 1981a,b
Humans	CSF				
Healthy volunteers	Plasma levels	β-Endorphin	Morning	Evening	Petraglia et al. 1983
Adult volunteers	Plasma levels	β-Endorphin	Morning	Mid-afternoon	Hindmarsh et al. 1989
33.4-week old infants	Plasma levels	β-Endorphin	Morning	mid-afternoon	Hindmarsh et al. 1989
31.7-week old infants	Plasma levels	β-Endorphin	Morning	mid-afternoon	Sankaran et al. 1989
Pregnant women	Plasma levels	β-Endorphin	Morning	Midnight	Räisänen 1988
Young volunteers	Plasma levels	β-Endorphin	Early morning	Midnight	Rolandi et al. 1992

plate tests and found that pain sensitivity was lower between 0800 hours and 1400 hours (i.e., in the resting period of mice) and significantly higher between 2200 hours and 0400 hours (i.e., during the activity period). The time of peak and low pain sensitivity was slightly different when the animals were kept on constant light as the lowest and highest pain sensitivities were found between 2200 hours and 0200 hours and between 0800 hours and 2200 hours, respectively. These authors conclude that the rhythm of stress-induced analgesia is due to the release of an opioid peptide because naloxone (5 mg/kg) significantly reduced the tail-flick test score.

The data presented above were obtained in rodents, which are nocturnal animals. It is interesting to correlate these data with those obtained by HAMRA et al. (1993) in diurnally active horses kept on the normal seasonal photoperiod (L:D 14:10). The plasma concentrations of immunoreactive β-endorphins were determined in eight mature thoroughbred horses during the months of June and July. The β-endorphin concentrations in the brain varied significantly with time and they were highest at 0900 hours. These data again indicate that plasma endorphin levels are higher during the activity period of the animals, although the clock time for the activity period is different in nocturnal and diurnally active animals.

NABER et al. (1981a) used a radio-immunological assay and quantified the opioid activity in humans and in monkeys. They sampled human plasma and monkey cerebrospinal fluid (CSF) at 2-h intervals over a 24-h period and found an episodic secretion of an opioid substance, with a morning peak (1000 hours) and an evening dip (2200 hours). There was a 40% difference between peak and trough values of human plasma. These data are in rather good agreement with those of PETRAGLIA et al. (1983), who reported simultaneous circadian variation of β-endorphin concentrations in six healthy volunteers, with peak and trough levels occurring at 0800 hours and 2000 hours, respectively.

More recently, HINDMARSH et al. (1989) were able to demonstrate a similar time-dependent variation in β-endorphin levels in the plasma of neonates: the endogenous opioid levels were much higher in the morning than in the afternoon. In this study, the β-endorphin levels found in adult volunteers (mean age 32.5 ± 1.7 years) were compared to the opioid levels found in minimally stressed 33.4-week-old infants who were able to breath spontaneously and had a diagnosis of mild hyaline membrane disease or wet lung disease and in 31.7-week-old infants suffering from severe hyaline membrane disease with or without apnea but under mechanical ventilation. In the three groups included in this study, the plasma concentrations of β-endorphins were always significantly higher in the morning than in the afternoon. In another study, SANKARAN et al. (1989) determined the plasma levels of β-endorphins at three different hours of the day in 17 hospitalized infants with a mean gestational age of 31.7 weeks: again peak and trough plasma peptide levels were found at 0900 hours and 1500 hours, respectively. This difference in endorphin levels is surprising in 32- or 33-week-old infants and one wonders why it was observed. The data found in these studies suggest that the rhythm of endorphin levels is related to the physiology of infants. It could also be argued that the larger morning endorphin levels are due to the enhanced environmental activity or noise usually observed in a neonatalogy ward at this time of day. No evidence was found in the literature to support these hypotheses and further research is needed in this area.

The plasma levels and diurnal variation of β-endorphin were also studied during pregnancy and early puerperium. RÄISÄNEN (1988) used a specific radioimmunoassay for β-endorphin and determined its concentration in the plasma of 62 healthy pregnant women and of 11 healthy nonpregnant women. Compared to those of nonpregnant controls, the mean concentration of β-endorphin was significantly decreased in the first and second trimester, but a twofold rise was detected in the last semester of the pregnancy. The diurnal variation of β-endorphin was studied in five women near term and again in the early puerperium. A circadian pattern was found at 38–41 weeks of pregnancy (before delivery), with highest and lowest values found at 0800 hours and 2400 hours, respectively. This time-dependent variation was not found on the 4th day after delivery. Finally, the chronobiology of β-endorphin was also studied in 6 young (28–37 years) and 6 elderly (78–84 years) healthy male volunteers and in 12 patients suffering from either Alzheimer-type dementia ($n=6$) or multi-infarct dementia ($n=6$), with blood sampling every 4 h from 0800 hours to 2000 hours and every 2 h from 2400 hours to 0600 hours. A circadian rhythm of β-endorphin was found only in the young volunteers: peak and trough values were found at 0600 hours and 2400 hours, respectively (ROLANDI et al. 1992).

It is also interesting to note that NABER et al. (1981b) investigated the diurnal variation of opiate receptor binding. The amount of ^{3}H-naloxone was measured every 4 h across a 24-h period in the forebrain of rats that had been housed under an L:D 12:12 regimen (light on at 0700 hours) for 3 weeks. A significant circadian variation was found, with a peak at 2200 hours (i.e., 3rd h of activity) and a trough at 0200 hours, with an amplitude of 42%. The data also indicated that the difference in binding throughout the day was due to changes in the number of binding sites rather than in the affinity of the drug for the receptor site. These investigators also reported that the rhythm persisted in constant darkness: the peak and nadir values were found at 2200 hours and 0600 hours, respectively, with an amplitude of 78%. To our knowledge, there is no chronopharmacological study on the binding of specific agonists to the μ, κ, δ opioid receptors and to the related σ receptor.

Table 1 also shows that the levels of substance P in the substantia nigra and in the central gray matter of the rat brain were significantly higher between 2000 hours and 0400 hours than at any other hours of the day. In the preoptic area, substance P also increased significantly by 66% between 0800 hours and 1200 hours and the brain concentration remained at this level up to 0400 hours (KERDELHUE et al. 1981).

Very few studies have been carried out for longer than 24 h in order to determine the periodicity in the neurochemistry of pain. VON KNORRING et al.

(1982) identified a circannual variation in the concentrations of endorphins in the CSF of 90 patients with chronic pain syndrome of psychogenic and organic etiology: the highest concentrations were found in January and February and the lowest CSF levels occurred in July and August. To our knowledge, no other data are available on the circannual variation in the levels of opioid peptides.

In summary, the data obtained in laboratory animals and in humans indicate that circadian variations are found in the plasma and brain concentrations of β-endorphin and enkephalins: highest values were always obtained late during the resting period or during the activity period. It must be noted that the acrophases and bathyphases of the different endogenous peptides are best characterized by referring to the activity or sleep periods (i.e., 5th h of activity, 2nd h of rest) rather than to the clock-time itself. Indeed, the clock-time is a useful time-reference value to compare human data obtained in diurnally active individuals, but it will be useless when comparing the biological rhythms of shift workers to those of diurnally active humans or when comparing data obtained in nocturnal animals (such as the rat) and in diurnally active humans.

II. Biological Rhythms of Pain

1. Pain Studies in Experimental Models

The scientific and standardized study of temporal variations in the sensory threshold to painful stimuli began in the early part of the twentieth century. In healthy volunteers, GRABFIELD and MARTIN (1912) and MARTIN and GRABFIELD (1914) showed that the highest irritability to an electrical stimulation occurred at 1030 hours, while lowest irritability was found between 2330 hours and 0100 hours and between 0400 hours and 0500 hours. More recently, PROCACCI et al. (1973, 1974) studied 34 healthy volunteers (age range 25–36 years) and they reported that pain induced by radiant heat was maximal at 0630 hours and minimal 12 h later. This circadian rhythm of pain could not be found in the women included in the study ($n=15$), but a 25-day rhythm was detected in both men and women, with peak pain occurring on day 8 after the beginning of the study. This infradian rhythm (period > 28 h) persisted in menopausal women, but not in women taking oral contraceptives. The morning peak and the evening trough of pain were also detected in the study of DAVIS et al. (1978), who used a psychological test to evaluate pain induced by electric shocks.

The circadian variation of pain sensitivity in pericranial musculature was studied in 12 healthy women and 12 men. Pain was induced at six times over a 24 h period by using a wide, inflatable pressure cuff to reduce blood circulation in the pericranial vessels for 20 s. Pain intensity was measured with a scale consisting of six categories (very weak to very strong), which were further divided into 10 units (GÖBEL and CORDES 1990). The data showed a circadian variation in the sensitivity of very intense headache: it was least around 1400 hours and it increased gradually during the day to reach a peak during the night hours (0200 hours) and during the early morning hours. The data also indicated that there was no significant effect of "time of day" on pain threshold, but a difference was noted according to sex as women were about twice as sensitive as men.

The nociceptive flexion reflex (NFR) can also be used as an objective model of pain and the data obtained with this test correlate well with those reported with the classical measures used in quantitative psychology (WILLER 1977, 1985). BOURDALLÉ-BADIE et al. (1990b) used the NFR in five healthy men synchronized on the usual pattern of diurnal activity and nocturnal sleep (2300 hours to 0800 hours) to study the circadian variations of pain. The electric nociceptive stimulation was made at the level of the sural nerve and the muscular response was registered at the femoral biceps. A verbal scale from 1 to 10 was also used to evaluate the intensity of pain at different hours of the day. The threshold of the NFR was largest at 1700 hours and smallest at 0100 hours; pain intensity was greatest at 0500 hours and lowest between 0900 hours and 1300 hours.

In 18 healthy volunteers, HUMMEL et al. (1992, 1994) studied the chronobiology of pain produced by carbon dioxide and they reported that the intensity of the painful stimuli were significantly higher during the night. In another series of placebo-controlled experiments, KOBAL et al. (1992) produced pain by carbon dioxide stimuli applied to the left nostril of 18 healthy volunteers, who rated the intensity of the painful stimuli with a visual analog scale. Two peaks of pain were found during the 24 h period and they occurred at 1000 hours and 0200 hours. A different pattern was found in dental pain as JORES and FREES (1937), BICHLMAYER (1937) and PÖLLMANN (1984) showed that peak pain threshold (i.e., minimum pain) occurred between 1500 hours and 1800 hours, while it was lowest (highest pain) at midnight.

It must also be pointed out that some investigators could not find any clinically significant circadian variation in experimental pain. Table 2 presents a list of human studies where no circadian variation of pain sensitivity was reported. For instance, STRIAN et al. (1989) examined the diurnal variations in pain and thermal sensitivity thresholds in 11 healthy volunteers tested for 2 days with 7 measurements per day. Thermal sensitivity thresholds to warm and cold stimuli to the right hand and foot were also investigated. Diurnal variations in the pain threshold measures were found for some subjects, but these variations did not have a consistent pattern in all subjects. In most studies, the investigators reported a nonsignificant 5%–8% variation in pain threshold or pain reaction, but these studies were usually done on a small number of volunteers.

It is difficult to explain the contradictory results on the chronobiology of experimentally induced pain. The divergent findings may be due mainly to methodological differences because the type of pain stimuli varied from one study to the next and the nociceptive stimuli could be transmitted by two types

Table 2. Experimental studies reporting lack of circadian variations in pain sensitivity in healthy volunteers. [Adapted and modified from references summarized by Göbel and Cordes (1990)]

Sites of Stimuli	Types of Stimuli	No of Subjects	Stimuli Parameters	References
Hand, foot	Cold	11 men	Detection threshold Pain threshold	Strain et al. 1989
Hand	Current	3 men	Detection threshold	Macht et al. 1916
Forearm	Current	16 men and 16 women	Pain threshold Pain tolerance threshold	Stacher et al. 1982
Forearm	Current	12 Women	Detection threshold Pain threshold Intervention threshold Pain tolerance threshold	Morawetz et al. 1984
Forehead	Heat	3 men	Pain threshold	Hardy et al. 1940
Forehead	Heat	150 men	Pain threshold	Schumacher et al. 1940
Forehead	Heat	15 men	Pain threshold Pain tolerance threshold	Chapman and Jones 1944

of fibers (Aδ and C fibers). It is also interesting to note that pain was measured in different parts of the body and many parameters were used to quantify pain threshold or the intensity of experimental pain. Furthermore, the investigators paid no attention to the differences between sharp and dull pain or between epicritic and protopathic pain. The differences found in these studies cast no doubt on the reliability of human data, but they strongly suggest that further studies must be done on the chronobiology of experimental pain in human volunteers. In these studies, the type of pain, the synchronization of volunteers to sleep-activity pattern as well as the method of inducing pain and the intensity of the painful stimuli should be standardized and described in detail, to be certain that other investigators will be able to use the same methods to study the temporal variations of experimental pain. In most studies, the sleep-activity pattern of the volunteers was not indicated by the authors and there was no information on subjects' morningness or eveningness. This information is important because the sleep-activity pattern is one of the main synchronizers of human biological rhythms and it would be interesting to know whether a correlation could be drawn between the type of activity pattern of the volunteers and the chronobiology of pain in these subjects. Although we would assume that the volunteers were diurnally active and were not drinking coffee or smoking cigarettes during the studies, the chronobiological timing of the subjects cannot be compared exactly. This difference in the time of experimentation could be responsible for some discrepancies in the temporal variations of experimental pain.

It is also well known that the affective reaction to a nociceptive stimulus is a very important component of pain, no matter whether the painful experience is induced by experimental procedures, by disease or by surgery. Thus, it is possible to suggest that the circadian variations in experimental pain are influenced by psychological factors. This assumption is supported by a third study in which the stimulation threshold in the front tooth was delivered by self-measurements in 28 subjects for 4–5 days. Significant daily rhythms were detected in 21 out of the 28 subjects. Lower acrophases were scattered along the 24-h time scale, indicating subjective perception of the stimulus (LEMMER and WIEMER 1989). This argument is important and this is a further reason why the chronobiology of pain must also be studied in animals.

FREDERICKSON et al. (1977) and WESCHE and FREDERICKSON (1981) used the hot-plate test and reported that mice kept on an L:D 12:12 regimen (light on at 0600 hours) exhibited a time-dependent change in their ability to tolerate noxious stimuli. The jump response latency of control mice was highest between 1530 hours and 0430 hours, and lowest between 0430 hours and 1130 hours: thus, the lowest irritability occurred at the end of the resting period and during the first 8 h of the activity period, while highest irritability was found at the end of the activity period and during the first 5 h of the resting period. Hypophysectomy did not alter the time-dependent changes either in the temporal variation in the brain levels of met-enkephalin or in the jump test latencies (WESCHE and FREDERICKSON 1981). Many investigators found similar data in experiments carried out in rodents (BODNAR et al. 1978; BUCKETT 1981; CROCKETT et al. 1977; KAVALIERS and HIRST 1983; KAVALIERS et al. 1983) and in golden hamsters (PICKARD 1987). In a recent study, MARTÍNEZ-GÓMEZ et al. (1994) assessed the changes in pain in relation to time of day and to estrus. Wistar female rats were kept on an L:D 12:12 regimen and the tail-flick method was used to measure pain threshold, which was studied at three times of day: 10 min after the light was on and 2 and 8 h after the light was off. Under these experimental conditions, maximal threshold was found early in the activity period of the rats (i.e., 2 h after the light was off). Estrus was determined by taking vaginal smears and significantly lower thresholds were found during estrus and metestrus, whereas maximal thresholds occurred during proestrus and diestrus. It is also interesting to note that ovariectomy had no effect on the 24-h variations of pain but it abolished completely the cyclicity caused by the estrus cycle. These data suggest that the 24-h variation of pain threshold is not related to the sexual hormones whose plasma levels are known to vary according to time of day. Finally, HAMRA et al. (1993) measured pain in eight thoroughbred horses exposed to the normal summer photoperiod (L:D 14:10). Pain sensitivity was measured using the skin-twitch reflex latency and the hoof withdrawal at 3-h intervals between 0600 hours and 1800 hours and at 2400 hours. Both responses varied significantly over time and were 37% longer at 0900 hours, with a smaller secondary peak at 1500 hours. The lowest value was obtained at 1800 hours. The animal data suggest that the time-dependent variations in the threshold and in the intensity of pain are not only due to psychological reasons. Further research is needed to determine the physiological mechanisms of these variations.

2. Pain Studies in Patients

As it is the case with many other diseases, circadian variations have been described in pain. For instance, the morning peak of chest pain induced by myocardial infarction is well known to clinicians, as the rhythmic appearance of infarction was first reported by MASTER (1960) before being rediscovered recently by MULLER et al. (1985). The exacerbation of pain over a 24-h period is observed everyday in hospital settings, but clinicians are not really aware of the chronobiology of different types of pain and of the meaning of this information in relation to treating their patients. Table 3 summarizes the main data on biological rhythms of clinical pain.

A circadian pattern of pain was found in 543 patients with toothache caused by dental caries (PÖLLMANN 1984). The data indicated that the majority of patients experienced the onset of toothache in the morning hours, with a peak at 0800 hours. Thus, an interesting correlation can be found between the circadian variation of pain induced by electrical or cold stimuli of the teeth (PÖLLMANN 1984) and the temporal pattern of toothache (PÖLLMANN 1984): the threshold of electrical stimulation was lowest between midnight and 0600 hours, while the acrophase of patients' toothache increased from midnight to reach a peak at 0800 hours.

Table 3. Biological rhythms of pain in patients

Causes of pain	Patient Numbers	Time of Peak	Trough	References
Backache	60	Morning,		POWNALL and PICKVANCE 1985
	19	2000 hours	0800 hours	PEDNAULT and PARENT 1993
Biliary colic	50	2300–0300 hours	0900–1300 hours	RIGAS et al. 1990
Cancer	130	1800 hours	0400–1000 hours	SITTL et al. 1990
Intractable pain				GLYNN et al. 1975
	41	2200 hours	0800 hours	FOLKARD 1976
Migraine	15	1000 hours	0000 hours	SOLOMON 1992
	117	0800–1200 hours	0000 hours	WATERS and O'CONNORS 1971
	114	0400–0800 hours	1200 hours	OSTFELD 1963
Myocardial	1229	0900 hours	0500 and 1900 hours	MASTER 1960
infarction	703	0900 hours and 2000 hours[a]	2200 hours	MULLER et al. 1985
Osteoarthritis	20	2200 hours	0200–0600	BELLAMY et al. 1990
Rheumatoid arthritis		0600–0800 hours	hours	KOWANKO et al. 1981, 1982
Toothache	543	0800 hours	1500 hours	PÖLLMANN 1984

[a] In the study by MULLER et al. (1985), the main peak of myocardial infarction was found at 0900 hours, but a secondary peak was detected in the evening (2000 hours).

In 41 patients suffering from intractable pain, GLYNN et al. (1975) and FOLKARD (1976) determined the pattern of pain in patients with visual analog scales every 2 h during the waking span for 7 consecutive days. The intensity of intractable pain increased throughout the day and the highest pain intensity was found at the last evaluation of the day, i.e., around 2200 hours.

RIGAS et al. (1990) studied the circadian rhythm of biliary colic in 50 patients who underwent emergency or elective cholecystectomy for symptomatic biliary tract stones. All patients were interviewed during their first postoperative week and the following aspects were noted: time of onset and pain features such as location, duration, number and time of prior pain episodes, relationship to meals and body position. Patients who could not recall the exact time of pain with confidence were asked to describe their painful episode as occurring during the "daytime" (0600–2000 hours), "night time" (2000–0600 hours) or "anytime". The pain pattern was compared to a control group consisting of 27 patients with renal colic. The data indicated that 76% of the 50 patients with biliary pain were able to report a specific clock time or a 2-h time span during which the painful episodes occurred. Biliary pain displayed a significant circadian rhythm, with a peak between 2300 hours and 0300 hours. In contrast, no periodicity was found in patients with renal pain or in those with nonbiliary abdominal pain secondary to miscellaneous causes.

A circadian pattern was also found recently in the time of onset of migraine (SOLOMON 1992). A total of 214 migraine attacks were reported by 15 patients during the 20-week study period. There was a marked increase in attacks between 0800 hours and 1200 hours, with a peak at 1000 hours and a trough by midnight. These data are in agreement with the early work of WATERS and O'CONNORS (1971) and with the data of OSTFELD (1963), who reported that migraine attacks are commonest at the end of the night (0400 hours) or early morning (0800–1200 hours) and rarest during the night.

The chronobiology of pain reported by arthritic patients is of interest because the occurrence of pain varies according to the cause of arthritis (LABRECQUE et al. 1994). In patients with rheumatoid arthritis (RA); the acrophases of pain were found mostly at the beginning of the day (KOWANKO et al. 1981, 1982), whereas it occurred at the end of the day in osteoarthritic (OA) patients (BELLAMY et al. 1990; LÉVI et al. 1985; JOB-DESLANDRE et al. 1983). Figure 1 presents these data. However, it must be pointed out that important interindividual differences can be found in the time of highest pain intensity. Table 4 indicates that while most OA patients reported pain at the end of the day, others reported that peak pain occurred early in the morning and no rhythm of pain could be detected in a few patients (BELLAMY et al. 1990; LÉVI et al. 1985). These interindividual differences should be taken into account when prescribing pain and anti-arthritic medications to patients.

The chronopathology of backache has been studied by two groups of investigators. The first study was carried out by POWNALL and PICKVANCE (1985) in 60 patients with persistent but recent (less than 1 year) or chronic (3 or more years) low back pain. Patients whose back pain was due to malig-

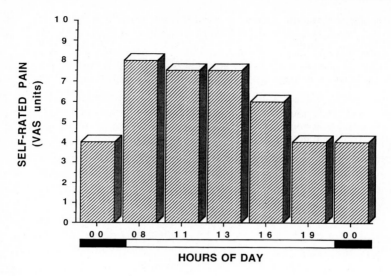

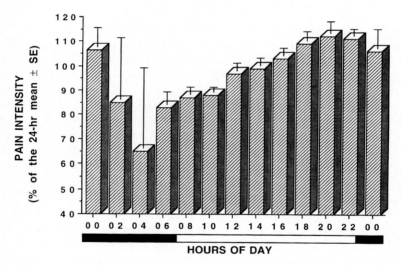

Fig. 1. Circadian variation of pain in patients with rheumatoid arthritis and osteoarthritis. *Upper graph* shows the mean self-rated pain determined at different hours of the day by a patient with RA. *Lower graph* illustrates the pain perceived by 20 OA patients who were self-assessing pain ten times daily for 1 week. Data are mean values ± SE of the data obtained at different hours of the day. (Redrawn from Kowanko et al. 1982 and Bellamy et al. 1990)

Table 4. Interindividual differences in the circadian rhythm in patients with osteoarthritic pain. The data obtained in the two studies indicate that most patients had a single peak of pain/24-h period but few did not have any circadian profile of arthritic pain

Hours of the day	Lévi et al. (1985)	Bellamy et al. (1990)
0800–1400 hours	3	0
1400–2000 hours	19	8
2000–0000 hours	ND	5
0000–0800 hours	8	3
2 peaks/24 h	23	0
No circadian profile	4	4
Number of patients	57	20

ND, not determined in the study

nancy, infection and abnormal metabolism were excluded from the study. After at least 1 week free from nonsteroidal anti-inflammatory use, patients evaluated the circadian variation of their backache before being allocated to a treatment group with ibuprofen. Thirty-three patients indicated that their pain was worst in the morning and 18 in the afternoon compared with only 6 experiencing pain in the evening or night period. Completely different data were obtained by PEDNAULT and PARENT (1993) in a study of 19 outpatients with chronic low back pain caused by discal degeneration, fusion pseudarthrosis or discoidectomy. The patients were synchronized on a diurnal activity pattern (sleep: 0000–0700 hours) and they used a visual analog scale every 4–5 h during the activity span for 7 consecutive days to self-evaluate pain intensity. Although some interindividual differences were noted, data analysis clearly indicated a circadian variation of chronic low back pain, with highest and lowest pain occurring at 2000 hours and 0800 hours, respectively. It is not possible to explain this discrepancy between these two studies, but it may be suggested that there are differences in the causes of pain studied by the two groups of investigators which could account for the differences reported in the pain pattern. These data indicate again why authors must give complete information on the potential causes of pain in the patients included in their studies.

Although pain is reported regularly after surgery and by patients with cancer, the chronobiology of pain itself was rarely studied directly in these patients. To our knowledge, the only study identifying the temporal variation of pain in cancer patients was done by SITTL et al. (1990), who used a visual analog scale (VAS) at eight times during the 24-h period. Their data showed that peak pain occurred at 1800 hours. Most other studies were carried out in patients receiving opioid analgesics and they were mainly concerned with the relief produced by these drugs. Therefore, these data will be reviewed in the section of the review discussing the chronotherapeutics of pain.

The data obtained in patients clearly indicate that investigators must characterize the chronopathology of a disease before beginning to determine drug effectiveness. Further research is needed because the clinical chronobiology of pain is mostly unknown, except for arthritic diseases. The optimal time for drug administration may be quite different depending on when the peak symptoms occur. Few investigators and clinicians consider the chronobiology of pain when preparing pain treatment strategies. This might explain in part why there are reports indicating that pain was still present in many patients undergoing thoracic or abdominal surgery, even though morphine or meperidine was administered continuously after surgery (MARKS and SACHAR 1973; UTING and SMITH 1979).

D. Chronopharmacology of Analgesics

I. Animal Studies

The chronopharmacological studies on pain were carried out mainly in laboratory animals receiving either local anesthetics or morphine-like drugs. As the time-dependent variations in the pharmacological and toxicological effects of local anesthetics are discussed in Chap. 22 of this volume, the chronopharmacology of opioids in rodents will now be described.

MORRIS and LUTSCH (1967) were the first to study the circadian variation in morphine-induced analgesia. In mice synchronized on 12:12 L:D (light on at 0600 hours), morphine (8 mg/kg) was injected intraperitoneally at eight different times of the day. The data indicated that peak analgesia occurred during the 3rd h of darkness (i.e., at 2100 hours) whereas minimal analgesia was found at the 9th h of the resting period (i.e., at 1500 hours). In agreement with these initial data, FREDERICKSON et al. (1977) reported a diurnal rhythm in the responsiveness of mice to nociceptive stimuli, in the analgesic effect of morphine and in the hyperalgesic activity of naloxone. Groups of ten mice synchronized on a 12:12 L:D (light on at 0600 hours) regimen were tested at 1-h intervals over a 24-h period for hot-plate response latencies 15 min after injection of saline or naloxone and 30 min after morphine administration. The data indicated that the latencies in the control group were highest (i.e., low pain) between 1530 hours and 0430 hours and lowest (i.e., high pain) between 0430 hours and 1130 hours. Morphine (2 mg/kg, s.c.) increased the 24-h mean of the jump latency, and its effectiveness was largest between 1430 hours and 2130 hours and lowest at the beginning of the rest period (between 0600 hours and 1100 hours). Finally, naloxone (3 mg/kg, s.c.) dampened the jumping latency rhythm, with the maximal effect occurring at the end of the rest period and most of the activity span (i.e., between 1530 hours and 0430 hours), whereas the effect was smallest and most variable during the resting period of the mice (i.e., between 0630 hours and 1530 hours). KAVALIERS and OSSENKOPP (1988) carried out experiments to determine day-night rhythms of opioid and non-opioid stress-induced analgesia at 2 hours of the day. They reported that the latencies of response (expressed in seconds) to warm stimuli were significantly longer in the mid-dark than in the mid-light period. These data might be most readily explained by the circadian rhythm of endogenous opioid activity, which was shown to be highest when rodents were active.

In classical pharmacology, investigators have determined the action of morphine and related drugs on the adrenergic, serotonergic and cholinergic systems to study the mechanisms of the time-dependent variations in the action of these drugs on the central nervous system. Unfortunately, the chronopharmacology of opioids on these neurotransmitter systems has not been studied extensively. One of these studies was carried out in rats synchronized on an L:D 12:12 regimen (light on at 0600 hours) and the data showed that a small intraperitoneal dose of morphine (1 mg/kg) significantly raised total brain acetylcholine content at the beginning of the activity period (1900 hours), whereas a large dose (31.1 mg/kg) was needed to achieve the same effect at the beginning of the resting period (0800 hours) (LABRECQUE 1973; LABRECQUE and DOMINO 1975). Similar data were obtained when levorphanol was administered at the same 2 hours of the day (LABRECQUE 1973). The circadian rhythm of brain acetylcholine levels (HANIN et al. 1970; LABRECQUE 1973; LABRECQUE and DOMINO 1975; LABRECQUE et al. 1982) may explain these data because smaller doses of morphine were needed to raise brain levels of the neurotransmitter when acetylcholine levels were lowest and vice versa. Further research is needed in this area.

Finally, the circannual variations in the effects of opioids in laboratory animal have not been studied very much. In the first study done in mice, BUCKETT (1981) reported a circannual variation in analgesia, with the peak occurring in winter. BECKMAN et al. (1982) examined the characteristics of morphine physical dependence in ground squirrels. The naloxone-precipitated abstinence syndrome was qualitatively unchanged throughout the year, but the abstinence syndrome showed clear quantitative seasonal differences. These differences were evident in terms of both the frequency of occurrence of particular signs and the percentage of the morphine-dependent population exhibiting them. This rodent hibernator exhibited a strong and characteristic abstinence syndrome during its nonhibernation state.

II. Human Studies

Although clinicians are well aware that it is difficult to establish a dosage regimen for satisfactory relief of pain in their patients, it is surprising to note that very few studies have been carried out on the chronotherapeutics of pain. As reviewed by BOURDALLÉ-BADIE et al. (1990a) and BRUGUEROLLE (Chap. 22, this volume), the early studies showed that the duration of action of local anesthetics such as lidocaine (REINBERG and REINBERG 1977) or carticaine (LEMMER and WIEMERS 1989) was twice as long at 1500 hours as at 0800 hours. Unfortunately, the time-dependent changes in the effects of other analgesics have not been studied extensively.

1. Patients with Arthritic Pain

Morning pain is a very characteristic symptom of rheumatoid arthritis (RA) and the nonsteroidal anti-inflammatory drugs (NSAIDs) are known to produce important side effects on the gastro-intestinal (GI) tract. Thus, it is not surprising that studies were done on the time-dependent variations in the effectiveness and toxicity of NSAIDs. The first and also the largest multicenter chronotherapeutic study on NSAIDs was carried out by French rheumatologists in 517 patients suffering from osteoarthritis (OA) affecting the hip ($n=240$), the knee ($n=240$) or both joints ($n=37$) (BOUCHANCOURT and LE LOUARN 1982; SIMON et al. 1982; LÉVI et al. 1982, 1984, 1985). The objective of these studies was to answer the simple question often asked by patients: *When should I take my once-daily medication?* Each patient was his own control and took a 75-mg sustained-release indomethacin capsule at 0800 hours for 1 week, at 1200 hours for another week and at 2000 hours for the last week. Study 1 was an open crossover design, studies 2 and 4 were an open and randomized crossover design and study 3 was a randomized double-blind crossover design. Each patient was asked to report side effects and to perform self-ratings of pain intensity (visual analog-scale) every 2 h from 0700 hours to 2300 hours for 1–2 days during the washout period on days 4–6 of each test week. In these four studies, the effectiveness of indomethacin was found by both physicians and patients to be very good or excellent. When the data of all patients were grouped per hour of indomethacin ingestion, no significant differences could be found in the effectiveness of indomethacin as a function of the hour of indomethacin administration. Indeed, the morning ingestion time was considered best by 28% of the patients, whereas 22% and 35% of the patients preferred the noon and evening administrations, respectively. However, marked interindividual differences were observed in the time of peak symptoms of arthritis (see Table 4) and when patients were asked to indicate when they preferred to take their NSAID tablets. The data were used to determine the optimal time of drug administration for these drugs: the evening administration was most effective in subjects with a predominantly nocturnal pain, while indomethacin ingestion in the morning or at noon was most effective in subjects with peak pain occurring in the afternoon or evening. The analgesic effect of the drug was increased by about 60% over previous values when the NSAID was taken at the time preferred by the subjects (LÉVI et al. 1985).

In 17 patients suffering from RA, KOWANKO et al. (1981) and SWANNEL (1983) showed that a 100-mg dose of flurbiprofen administered twice a day was more effective than a 50-mg dose ingested four times daily. However, subjective measurements of pain and stiffness indicated that part of the twice daily flurbiprofen dose must be administered at night to control morning stiffness and pain. REJHOLEC et al. (1984) confirmed these data in a larger double-blind multicenter study.

REINBERG et al. (1991) also completed a smaller study on the chronoeffectiveness of tenoxicam in patients with spondylarthritis ($n=11$), OA

($n=7$) and RA ($n=8$). This newer NSAID has interesting pharmacokinetic properties as its half-life is about 3 days and it is highly bound (99.3%) to plasma proteins (HEINTZ 1989). In the double-blind, crossover randomized study, patients were given oral tenoxicam 20 mg for 2 weeks at 0800 hours, 1200 hours or 2000 hours and self-monitored pain and stiffness throughout the day. The results from this preliminary study suggest that optimal effectiveness occurred with the noon administration of tenoxicam. Further research is needed on NSAIDs with similar pharmacokinetic profiles.

Patients taking NSAIDs report GI or neurosensorial side effects that are so troublesome that they sometimes cease to take their medication. Thus, it was not surprising to find that the frequency of side effects was the most striking difference observed in the effects of indomethacin administered at different times of the day. The chronotolerance observed in the double-blind crossover trial with 66 patients and lasting 3 weeks will be discussed (LÉVI et al. 1985). The data indicate that 33% of the patients reported undesirable effects when indomethacin was ingested at 0800 hours in comparison with 7% of the patients reporting these undesirable effects when the drug was taken at 2000 hours. During the 3 weeks of the study, the side effects were consistently higher after the 0800 hours administration of the drug than at any other times of day. About 75% of the undesirable effects were CNS related (vertigo, headache, anxiety) while the others were of GI origin (nausea, gastric pain, diarrhea). There was no difference in the frequency of side effects in males and females. Similar data were obtained in another double-blind trial, with 118 outpatients with OA of the hip or the knee receiving 200 mg slow-release tablets of ketoprofen: GI tolerance was twice better for the evening group of patients taking ketoprofen (BOISSIER et al. 1990). These data are in agreement with the findings of MOORE and GOO (1987), who observed that the number of gastric mucosal lesions produced by oral administration of acetylsalicylic acid (ASA) at 1000 hours was twice as large as those obtained after the 2200 hours dosing. Thus, the data suggest that NSAIDs may be better tolerated when administered in the evening than in the morning to OA patients. However, great interindividual differences were found. In a recent crossover study, 16 healthy male volunteers ingested single oral doses of 75 mg and 1000 mg ASA at either 0800 hours or 2000 hours. Two hours after dosing, the gastric lesions were objectively scored by direct and video endoscopic examinations. No circadian phase dependency could be confirmed for either dose of ASA (NOLD et al. 1995). It is of interest to note that no patient complained about GI pain although lesions were present after the 1000-mg ASA dose. These data may indicate that the intensity of lesions is not be correlated with pain threshold.

Although these studies focused on the chronoeffectiveness and the chronotolerance of NSAIDs in arthritic patients, the data are pertinent to a discussion on pain and analgesia. Indeed, they showed that administration of drug as a function of when pain is greatest leads to the optimization and to the individualization of drug treatment. The data also showed that careful selection of time of administration may also significantly reduce the undesirable effects of drugs. Clinicians should now consider that the time of drug administration is important, to ensure that their patients obtain maximal pain relief and minimal side effects from the analgesic drugs prescribed.

2. Patients with Postoperative and Cancer Pain

Very few investigators studied the temporal variations in the effects of morphine and other opioid analgesics in patients with postoperative pain. In the first chronopharmacological study with patient-controlled analgesia (PCA), GRAVES et al. (1983) found a significant circadian variation in i.v. morphine dosing requirements in patients undergoing different elective abdominal surgery or in obese patients with gastric bypass surgery. These investigators reported that maximal demands for morphine occurred at 0900 hours, whereas the lowest requests for the drug were observed at 0300 hours. In fact, the morning doses were 15% higher than those administered during the night. DUNN et al. (1993) also studied the morphine requirements in 24 diurnally active patients undergoing abdominal surgery. In this PCA study, two peaks of morphine demands were found, with acrophases occurring between 0800 hours and 1200 hours and between 1600 hours and 2000 hours, respectively. Completely different data were obtained in a study done in 55 patients undergoing thoracic or abdominal surgery (LABRECQUE et al. 1988). The objective of this 3-day double-blind study was to compare the analgesia produced by a constant infusion of morphine or by s.c. injections of the drug every 4 h. Pain was assessed with a visual analog scale 1 h after each s.c. injection during 3 days. A circadian variation was detected on day 1 of morphine administration, no matter how the opiate was given. The acrophase of pain was detected between 1830 hours and 1915 hours. It is not yet possible to explain the discrepancies between these studies.

In cancer patients, the first PCA study was done by AUVIL-NOVAK et al. (1988) in 19 postsurgical gynecological cancer patients receiving either morphine sulfate or hydromorphone. Pump records were collected to ascertain the number of PCA attempts at self-medication/4-h intervals, the total dosage and the number of demands undelivered due to the lockout of the pump. Demands were significantly greater between 0400 hours and 0800 hours for both morphine and hydromorphone and they were significantly less between 1200 hours and 1600 hours for morphine and between 0000 hours and 0400 hours for hydromorphone. The upper graph in Fig. 2 shows that the morning doses were about 60% greater than those administered at other hours of the day. There was a temporal variability in drug demands which exceeded that allowed by the pump: peak demands in this situation occurred between 0000 hours and 0400 hours for morphine and between 0400 hours and 0800 hours for hydromorphone. The lowest demands were found between 1200 hours and 1600 hours for morphine or between 1600 hours and 2000 hours for hydromorphone. The other PCA study by AUVIL-NOVAK et al. (1990) was done in 45 postsurgical gynecological patients receiving morphine. These data showed

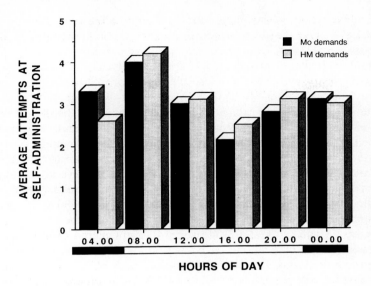

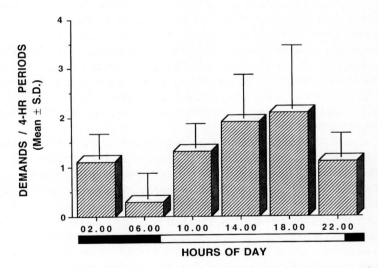

Fig. 2. Doses of morphine administered to cancer patients at different hours of the day. *Upper graph* shows the mean demands for morphine sulfate (*black bars*) and hydromorphone (*gray bars*) self-administered by 19 gynecological cancer patients for the first day after surgery. *Lower graph* illustrates the mean PCA hydromorphone demands ± SE requested by eight cancer patients over a 24-h period. (Redrawn from AUVIL-NOVAK et al. 1988 and VANIER et al. 1992)

that the highest doses of morphine were administered between 0800 hours and 1200 hours, whereas the lowest doses were administered between midnight and 0400 hours, respectively.

WILDER-SMITH et al. (1992) used a verbal rating scale in 20 diurnally active cancer patients and recorded the effect of 16 mg morphine administered at 4-h intervals. The pain intensity scores were highest between 0800 hours and 2000 hours and they were significantly smaller during the night. These investigators also studied the diurnal pattern of pain in 42 cancer patients receiving either slow-release morphine or buprenorphine (WILDER-SMITH and WILDER-SMITH 1992). These data indicate that there is a distinct circadian rhythm of pain intensity during treatment with long-acting opioids: maximal effectiveness was found in the evening (2200 hours) or at night (0200 hours) while the smallest change of pain intensity occurred early in the day (1000 hours).

Finally, VANIER et al. (1992) measured pain intensity in eight cancer patients receiving hydromorphone by basal rate injection associated with PCA demands or continuous infusion of the opioid with a basal rate infusion of saline. Pain intensity was measured with a VAS at 4-h intervals: pain was twice as high at 2200 hours (1.33 ± 0.8 arbitrary pain units) as at 1400 hours (0.64 ± 0.7 arbitrary pain units). The lower graph in Fig. 2 indicates that when the PCA bolus demands were summed with the rescue doses of analgesics, mean requests for the medication were highest between 1800 hours and 2200 hours and lowest between 0200 hours and 0600 hours. Similar data were obtained by BRUERA et al. (1992), who reviewed the circadian distribution in the extra doses of opioids received by 61 patients admitted to a Palliative Care Unit. Extra doses of opioids were determined at 4-h intervals over a 24-h period. A total of 1322 extra doses were administered during 610 patient days. The largest number of extra doses was administered between 1000 hours and 2200 hours and this number decreased by 1.76-fold during the night time or in the early morning (0200–1000 hours). CITRON et al. (1992) also reported that the demands for analgesic were 48% higher during the day than at night.

The large differences found in the intensity of pain and in the amount of opioids administered throughout the day cannot be explained at present. Pain is a complex phenomenon and investigators must always take into account that many factors can influence the intensity of pain and its evaluation by patients. More research is needed before a consensus can be reached on the time of drug administration that should be recommended to obtain optimal pain relief.

E. Conclusions

In this review, the main studies dealing with the time-dependent variations in the neurochemistry of pain and in the chronopharmacology of opioid analgesics have been reviewed. There are circadian variations in the levels of plasma or brain enkephalin or endorphin: peak levels were found at the beginning of the activity span. The intensity of pain also varied throughout the day but the hours of highest and lowest pain intensity are not the same for every disease. These variations were very large during the 24-h period and it is possible that some incertitudes, inadequacies of pain treatment and even drug overdosing or underdosing which often occur in clinical practice partly explain these differences in the chronobiology of diseases.

Further research is needed before the findings on the chronobiology and chronopharmacology of pain can be used daily in clinical practice. Investigators should be very careful in their interpretation of data on the relief of pain produced by the administration of analgesics. For instance, an increase in the doses of opioids administered (or self-administered) to patients at a particular hour of the day may not necessarily mean that the intensity of pain was greater during this period of time. This hypothesis should be accepted only when a clear correlation can be demonstrated between the increases in the doses of analgesics administered and in pain intensity. A recent study by HUMMEL et al. (1994) strongly supported this comment because these investigators showed that dihydrocodeine and tramadol exerted a stronger analgesic effect when administered in the evening. Thus, investigators must be encouraged to quantify pain intensity, pain relief and the amount of analgesic administered throughout the day in the same patient.

In PCA studies, the total number of self-administration attempts by each patient over a 24-h period should be determined and compared to the real number of opioid doses injected. This information might be important because pain is a subjective phenomenon which can be easily altered by many factors including previous experiences or anxiety that may vary throughout the 24-h period. A 24-h rhythm in the level of anxiety could influence both pain intensity and pain relief by analgesics. The importance of psychological factors on the evaluation of pain intensity and pain relief is stressed by the work of PÖLLMANN (1987), who showed the circadian variation of the analgesic potency of placebo administration. Indeed, this investigator reported that a sugar-coated placebo pill can increase pain threshold in healthy teeth by 25%–30% when the pill is ingested during the day (i.e., between 0900 hours and 2100 hours), but it did not produce an effect at night. It can also be suggested that the acrophase of the pain rhythm may vary with the types of surgery or diseases, and the chronopharmacological data obtained in patients suffering from one disease may not be extrapolated to patients with another disease. Therefore, the psychological factors influencing pain intensity must be considered and clearly identified by the investigators who want to characterize the chronobiology of pain in experimental or clinical conditions.

Finally, the studies with local anesthetics indicate that the duration of action of these drugs varies according to time of day (REINBERG and REINBERG 1977; LEMMER and WIEMERS 1989). To our knowledge, this information is not yet available for morphine and related drugs, but such a time-dependent variation would influence the number of drug demands by patients throughout the 24-h period.

More research is thus needed in this area before it will be possible to recommend a chronotherapeutic strategy for the administration of opioid analgesics to patients with pain. The data obtained so far suggest that the practitioners will have to take these findings into account in the immediate future and an important change of attitude will be needed because clinicians may have to modulate the doses of the drugs throughout the day to obtain maximal effectiveness of analgesic drugs.

References

Auvil-Novak SE, Novak RD, Smolensky MH, Kavanagh JJ, Kwan JW, Wharthon JT (1988) Twenty-four hour variation in self-administration of morphine sulfate and hydromorphone by post-surgical gynecologic cancer patient. Annu Rev Chronopharmacol 5:343–346

Auvil-Novak SE, Novak RD, Smolensky MH, Morris MM, Kwan JW (1990) Temporal variation in the self-administration of morphine sulfate via patient-controlled analgesia in post-operative gynecologic cancer patients. Annu Rev Chronopharmacol 7:253–256

Bausbaum AI, Fields HF (1978) Endogenous pain control mechanisms: review and hypothesis. Ann Neurol 4:451–462

Beckman AL, Llados-Eckman C, Stanton TL, Adler MW (1982) Seasonal variation of morphine physical dependence. Life Sci 30:147–153

Bellamy N, Sothern RB, Campbell J (1990) Rhythmic variations in pain perception in osteoarthritis of the knee. J Rheumatol 17:364–372

Bichlmayer A (1937) Lokale Anästhesie und Allgemeinbetäubung in der Zahn-, Mund und Kieferheilkunde. Lehmanns, p 10

Blitz B, Dinnerstein AJ (1968) Effects of different types of instructions on pain parameters. J Abnorm Psychol 73:276–280

Bodnar RJ, Kelly DD, Spiaggia A, Glusman M (1978) Biophasic alterations of nociceptive thresholds induced by food deprivation. Physiol Psychol 6:391–395

Boissier C, Decousus H, Perpoint B, Ollagnier M, Mismetti P, Hocquart J, Queneau P (1990) Timing optimizes sustained-release ketoprofen treatment of osteoarthritis. Annu Rev Chronopharmacol 7:289–292

Bouchancourt P, Le Louarn C (1982) Etude chronothérapeutique de l'indomethacinmétacine à effet prolongé dans l'arthrose des membres inférieurs. Tribune Med [Suppl]:32–35

Bourdallé-Badie C, Bruguerolle B, Labrecque G, Robert S, Erny P (1990a) Biological rhythms in pain and anesthesia. Annu Rev Chronopharmacol 6:155–182

Bourdallé-Badie C, Andre M, Pourquier P, Robert S, Cambar J, Erny P (1990b) Circadian rhythm of pain in man: study by measure of nociceptive flexion reflex. Annu Rev Chronopharmacol 7:249–252

Bruera E, Macmilland K, Kuehn N, Miller MJ (1992) Circadian distribution of extra doses of narcotic analgesics in patients with cancer pain: a preliminary report. Pain 49:311–314

Buckett WR (1981) Circadian and seasonal rhythm in stimulation produced analgesia. Experientia 37:878–879

Chapman CR, Casey KL, Dubner R, Foley KM, Gracely RH, Reading AE (1985) Pain measurement: an overview. Pain 22:1–31

Chapman WP, Jones CM (1944) Variations in cutaneous and visceral pain sensitivity in normal subjects. J Clin Invest 23:81–91

Citron ML, Kaira JM, Seltzer VL, Hoffman M, Walczak ML (1992) Patient-controlled analgesia for cancer pain: a long-term study of inpatient and outpatient use. Cancer Invest 10:335–341

Crockett RS, Bornschein RL, Smith RP (1977) Diurnal variation in response to thermal stimulation: mouse hot-plate test. Physiol Behav 18:193–196

Davis GC, Buchsbaum MS, Bunnew WE (1978) Naloxone decreases diurnal variation in pain sensitivity and somatosensory evoked potentials. Life Sci 23:1449–1460

Deschamps M, Band PR, Coldman AJ (1988) Assessment of adult cancer pain: shortcomings of current methods. Pain 32:133–139

Dunn M, Blouin D, Labrecque G (1993) Time-dependent variations in morphine-requirements in patients with post-operative pain. Proceedings of the 21st conference of the International Society for Chronobiology, Quebec (Canada), abstract no III-10

Fishman SM, Carr DB (1992) Basic mechanisms of pain. Hospital Practice, 15 Oct, pp 63–76

Foley KM (1989) Controversies in cancer pain. Cancer 63:2257–2265

Folkard S (1976) Diurnal variation and individual differences in the perception of intractable pain. J Psychosom Res 20:289–301

Frederickson RCA, Burgis V, Edwards JD (1977) Hyperanalgesia induced by naloxone follows diurnal rhythm in responsivity to painful stimuli. Science 198:756–758

Glynn CJ, Lloyd JW, Folkard S (1975) The diurnal variation in perception of pain. Proc R Soc Med 69:369–372

Göbel H, Cordes P (1990) Circadian variation of pain sensitivity in pericranial musculature. Headache 30:418–422

Grabfield GP, Martin EG (1912) Variations in the sensory threshold for faradic stimulation in normal human subjects I. The diurnal rhythm. Am J Physiol 31:300–308

Graham C, Bond SS, Gercovich MM, Cook MR (1980) Use of the McGill Pain Questionnaire in the assessment of cancer pain:replicability and consistency. Pain 8:377–387

Graves DA, Batenhorst RL, Bennett JG, Wettsetein WO, Griffen BD, Wright BD, Foster TS (1983) Morphine requirements using patient-controlled analgesia: influence of diurnal variation and morbid obesity. Clin Pharm 2:49–53

Hamra JG, Kamerling SG, Wolfsheimer KJ, Bagwell CA (1993) Diurnal variation in plasma ir-beta-endorphin levels and experimental pain thresholds in the horse. Life Sci 53:121–129

Hanin E, Massarelli R, Costa E (1970) Acetylcholine concentrations in rat brain: diurnal oscillation. Science 170:341–342

Hardy DJ, Wolff HG, Goodell H (1940) Studies on pain. A new method for measuring pain: observations on spatial summation of pain. J Clin Invest 19:649–657

Heintz RC (1989) Pharmacocinétique du ténoxicam. In: Gaucher A, Pourel J, Netter P, Kessler M (eds) Actualités en physiopathologie et pharmacologie articulaire. Masson, Paris

Hindmarsh KW, Tan L, Sankaran K, Laxdal VA (1989) Diurnal rhythms of cortisol, ACTH and β-endorphin levels in neonates and adults. West J Med 151:153–156

Hummel T, Hepper M, Kaiser R, Bös R, Liefjold J, Kobal G (1992) Investigation of circadian effects on the plasma levels of dihydrocodeine and tramadol after peroral administration. Proceedings of the 5th International Conference on Biological Rhythms and Medications, Amelia Island (Fl), abstract no X111-4

Hummel T, Kraetsch HG, Lötsch J, Hepper M, Liefhold J, Kobal G (1994) Analgesic effects of dihydrocodeine and tramadol when administered either in the morning or evening. Chronobiol Int 12:62–72

Huskisson EC (1983) Visual analogue scales. In: Melzack R (ed) Pain measurement and assessment. Raven, New York, pp 33–37

Jensen MP, Karoly P, Braver S (1986) The measurement of clinical pain intensity: a comparison of six methods. Pain 27:117–126

Jessel TM (1983) Substance P in the nervous system. In: Iversen LL, Iversen SD, Snyder SH (eds) Neuropeptides. Plenum, New York (Handbook of psychopharmacology, vol 16)

Job-Deslandre C, Reinberg A, Delbarre F (1983) Chronoeffectiveness of indomethacin in four patients suffering from an evolutive osteoarthritis of the hip or knee. Chronobiologia 10:245–254

Jores A, Frees J (1937) Die Tagenschwankungen der Schmerzempfindung. Deutsche Med Wochenschr 63:963–963

Kavaliers M, Hirst M (1983) Daily rhythms of analgesia in mice: effects of age and photoperiod. Brain Res 279:387–393

Kavaliers M, Ossenkopp KP (1988) Day-night rhythms of opioid and non-opioid stress-induced analgesia: differential inhibitory effects of exposure to magnetic fields. Pain 32:223–229

Kavaliers M, Hirst M, Teskey GC (1983) Ageing, opioid analgesia and the pineal gland. Life Sci 32:2279–2287

Kerdelhue B, Palkovits M, Karteszi M, Reinberg A (1981) Circadian variation in substance P, luliberin (LH-RH) and thyroliberin (TRH) contents in hypothalamic and extra hypothalamic brain nuclei of adult male rats. Brain Res 206:405–413

Kerdelhue B, Karteszi M, Pasqualini C, Reinberg A, Mezzy E, Palkovits M (1983) Circadian variations in beta-endorphin concentrations in pituitary and in some brain nuclei in the adult male rat. Brain Res 261:243–248

Kobal G, Hummel T, Kraetsch HG, Lötsch J (1992) Circadian analgesic effects of dihydrocodeine and tramadol. Proceedings of the 5th international conference on biological rhythms and medications, Amelia Island (Fl), abstract no X111-5

Kosterlitz HW, Hughes J (1975) Some thought on the significance of enkephalin, the endogenous ligand. Life Sci 17:91–96

Kowanko IC, Pownall R, Knapp MS, Swannel AJ, Mahoney PGC. (1981) Circadian variations in the signs and symptoms of rheumatoid arthritis and in the therapeutic effectiveness of flurbiprofen at different times of the day. Br J Clin Pharmacol 11:477–484

Kowanko ICR, Knapp MS, Pownall R, Swannel AJ (1982) Domiciliary self-measurement in rheumatoid arthritis and the demonstration of circadian rhythmicity. Ann Rheum Dis 41:453–455

Labrecque G (1973) Acetylcholine antirelease effect of selected narcotic agonists and antagonists. PhD thesis, University of Michigan

Labrecque G, Domino EF (1975) Chronopharmacologie I. Variation de l'effet de la morphine sur l'acétylcholine cérébral en fonction du rythme nycthéméral. Vie Med Can Franc 4:130–135

Labrecque G, Domino EF, Halberg F (1982) Circadian rhythm of total brain acetylcholine content: effect of light and blindness. In: Takahashi R, Halberg F, Walker CA (eds) Toward chronopharmacology. Pergamon, New York, p 19 (Advances in the biosciences, vol 41)

Labrecque G, Lepage-Savary D, Poulin E (1988) Time-dependent variations in morphine-induced analgesia. Annu Rev Chronopharmacol 5:135–138

Labrecque G, Bureau JP, Reinberg A (1994) Biological rhythms in the inflammatory response and in the effects of non-steroidal anti-inflammatory agents. Pharmacol Ther 66:285–300

Lemmer B, Wiemers R (1989) Circadian changes in stimulus threshold in the effect of local anesthetic drug in human teeth: studies with an electronic pulptester. Chronobiol Int 6:157–162

Levi F, Le Louarn C, Peltier A (1982) Etude chronopharmacologique en double-aveugle de 75 mg d'indometacine à effet prolongé dans l'arthrose. Tribune Med [Suppl]:48–53

Lévi F, Le Louarn C, Reinberg A (1984) Chronotherapy of osteoarthritic patients: optimization of indomethacin sustained release (ISR). Annu Rev Chronopharmacol 1:345–348

Lévi F, Le Louarn C, Reinberg A (1985) Timing optimized sustained indomethacin treatment of osteoarthritis. Clin Pharmacol Ther 37:77–84

Macht DJ, Herman NB, Levy CS (1916) A quantitative study of the analgesia produced by opium alkaloids, individually and in combination with each other in normal man. J Pharmacol Exp Ther 8:1–37

Marks RM, Sachar EJ (1973) Undertreatment of medical inpatients with narcotic analgesics. Ann Intern Med 78:173–181

Martin EG, Bigelow GH, Grabfield GB (1914) Variations in the sensory threshold for faradic stimulation in normal human subjects II. The nocturnal rhythm. Am J Physiol 33:415–422

Martínez-Gómez M, Cruz Y, Salas M, Hudson R, Pachero P (1994) Assessing pain threshold in the rat:Changes with estrus and time of day. Physiol Behav 55:651–657

Master AM (1960) The role of effort and occupation (including physicians) in coronary occlusion. JAMA 174:942–948

Mayer DJ, Liebeskind JC (1974) Pain reduction by focal electrical stimulation of the brain: an anatomical and behavioral analysis. Brain Res 68:73–93

McGrath PA (1987) An assessment of children's pain: a review of behavioral, physiological and direct scaling techniques. Pain 31:147–176

McGrath PA (1990) Pain in children: nature, assessment, and treatment. Guilford, New York

Melzack R (1975) The McGill pain questionnaire: major properties and scoring methods. Pain 1:277–299

Melzack R (1983) The McGill pain questionnaire. In: Melzack R (ed) Pain measurement and assessment. Raven, New York, pp 111–117

Melzack R, Torgerson WS (1971) On the language of pain. Anesthesiology 34:50–59

Moore JG, Goo R (1987) Day and night aspirin-induced gastric mucosal damage and protection by ranitidine in man. Chronobiol Int 4:11–116

Morawetz RF, Parth P, Pöppel E (1984) Influence of the pain measurement technique on the diurnal variation of pain perception. In: Bromm B (ed) Pain measurement in man. Elsevier, Amsterdam, pp 409–415

Morris RW, Lutsch EF (1967) Susceptibility of morphine-induced analgesia in mice. Nature 216:494–495

Mudge AW, Leeman SE, Fischbach GD (1979) Enkephalin inhibits release of substance P from sensory neurons in culture and decreases action potential duration. Proc Natl Acad Sci USA 76:526–530

Muller JE, Stone PH, Turin ZG, Rutherford JD, Czeisler CA, Parker C, Poole WK, Passamani E, Robert R, Robertson T, Sobel BE, Willerson JT, Braunwald E (1985) The Milis Study Group: circadian variation in the frequency of onset of acute myocardial infarction. N Engl J Med 313:1315–1322

Naber D, Cohen RM, Pickar D, Kalin NH, Davis G, Pert CB, Bunney WE Jr (1981a) Episodic secretion of opioid activity in human plasma and monkey CSF. Life Sci 28:931–935

Naber D, Wirz-Justice A, Kafka MS (1981b) Circadian rhythm in rat brain opiate receptor. Neurosci Lett 21:45–50

Nold G, Drossard K, Lehman K, Lemmer B (1995) Gastric mucosal lesions after morning versus evening application of 75 mg or 1000 mg acetylsalicylic acid (ASA). Naunyn Schmiedebergs Arch Pharmacol 351:R17

Oliverio A, Castellano C, Puglisi-Allegra S (1982) Opiate analgesia: evidence for a circadian rhythm in mice. Brain Res 249:265–270

Ostfeld AM (1963) The natural history and epidemiology of migraine and muscle contraction headache. Neurology 13:11–15

Pednault L, Parent M (1993) Circadian rhythm of chronic low back pain caused by discal degeneration, fusion pseudoarthrosis or discoidectomy. Proceedings of the 21st conference of the International Society of Chronobiology, Quebec (Canada), abstract no III-9

Pernow B (1983) Substance P. Pharmacol Rev 35:85–141

Petraglia F, Facchinetti F, Parrini D, Micieli G, De Luca S, Genazzi AR (1983) Simultaneous circadian variation of plasma ACTH, β-lipoprotein, β-endorphin and cortisol. Horm Res 17:147–152

Pickard GE (1987) Circadian rhythm of nociception in the golden hanster. Brain Res 425:395–400

Pöllmann L (1984) Duality of pain demonstrated by the circadian variation in tooth sensitivity. In: Haus E, Kabat H (eds) Chronobiology 1982–1983. Karger, Basel, p 225

Pöllmann L (1987) Circadian variation of potency of placebo as analgesic. Funct Neurol 22:99–103

Pownall R, Pickvance NJ (1985) Does treatment matter? A double blind crossover study of ibuprofen 2400 mg per day in different dosage schedules in treatment of chronic low back pain. Br J Clin Pract 39:267–275

Proccaci P, Moretti R, Zoppi M, Cappelletti C, Voegelin MR (1973) Rythmes circadiens et circatrigindiens du seuil de la douleur cutanée chez l'homme. Bull Groupe Et Rythmes Biol 5:65–75

Proccaci P, Dellacorte M, Zoppi M, Maresca M (1974) Rhythmic changes of the cutaneous pain threshold in man. A general review. Chronobiologia 1:77–87

Puglisi-Allegra S, Castellano C, Oliverio A (1982) Circadian variations in stress-induced analgesia. Brain Res 252:373–376

Räisänen I (1988) Plasma levels and diurnal variation of β-endorphin, β-lipotropin and corticotropin during pregnancy and early puerperium. Eur J Obstet Gynecol Reprod Biol 27:13–20

Reading AE (1989) Testing pain mechanisms in persons in pain. In:Wall PD, Melzack R (eds) Textbook of pain, 2nd edn. Churchill Livingstone, Edinburgh, p 269

Reinberg A, Reinberg MA (1977) Circadian changes of the duration of local anesthetic agents. Naunyn Schmiedebergs Arch Pharmacol 297:149–152

Reinberg A, Manfredi R, Khan MF, Chaouat D, Chaouat Y, Delcambre B, Legoff P, Maugars Y, Valat JP (1991) Chronothérapie de ténoxicam. Therapie 46:101–108

Rejholec V, Vitulova V, Vachtenheim J (1984) Preliminary observations from a double-blind crossover study to evaluate the efficacy of flurbiprofen given at different times of day in the treatment of rheumatoid arthritis. Annu Rev Chronopharmacol 1:357–360

Rigas B, Torosis J, McDougall CJ, Vener KJ, Spiro HM (1990) The circadian rhythm of biliary colic. J Clin Gastroenterol 12:409–414

Rivot JP, Weil-Fugazza J, Godefroy F, Bineau-Thurotte M, Ory-Lavollée L, Besson JM (1984) Involvement of serotonin in both morphine and stimulation-produced analgesia: electochemical and biochemical approaches. In: Kruger L, Liebeskind JV (eds) Advances in pain research and therapy, vol 6. Raven, New York, p 135

Rolandi E, Gandolfo C, Franceschini R, Cataldi A, Garibaldi A, Barreca T (1992) Twenty-four-hour beta-endorphin secretory pattern in Alzeimer's disease. Neuropsychobiology 25:188–192

Sankaran K, Hindmarsh KW, Tan L (1989) Diurnal rhythm of β-endorphin in neonates. Dev Pharmacol Ther 12:1–6

Schumacher GA, Goodell H, Hardy JF, Wolff HG (1940) Uniformity of pain threshold in man. Science 92:110–112

Scott J, Huskisson EC (1976) Graphic representation of pain. Pain 2:175–184

Simon L, Hérisson P, Le Louarn C, Lévi F (1982) Etude hospitaliére de chronothe-rapeutique avec 75 mg d'indométacinemétacine à effet prolongé en pathologie rhumatismale dégénérative. Tribune Med [Suppl]:43–47

Sittl R, Kamp HD, Knoll R (1990) Zirkadiane Rhythmik des Schmerzempfindens bei Tumorpatienten. Nervenheilkunde 9:22–24

Solomon GD (1992) Circadian rhythms and migraine. Cleve Clin J Med 59:326–329

Stacher G, Bauer P, Schneider C, Winklehner S, Schmierer G (1982) Effects of combination of oral naproxen sodium and codeine in experimentally induced pain. Eur J Clin Pharmacol 21:485–490

Strian F, Lautenbacher S, Galfe G, Hölzl R (1989) Diurnal variations in pain perception and thermal sensitivity. Pain 36:125–131
Swannel AJ (1983) Biological rhythms and their effect in the assessment of disease activity in rheumatoid arthritis. Br J Clin Pract 38 [Suppl 33]:16–19
Uting JE, Smith JM (1979) Postoperative analgesia. Anaesthesia 34:320–332
Vanier MC, Labrecque G, Lepage-Savary D (1992) Temporal changes in the hydromorphone analgesia in cancer patients. Proceedings of the 5th international conference on biological rhythms and medications, Amelia Island (Fl), abstract no X111-8
Von Knorring L, Almay BGL, Johansson F, Terenius L, Wahlström A (1982) Circannual variation in concentrations of endorphins in cerebrospinal fluid. Pain 12:265–272
Waters WE, O'Connors PJ (1971) Epidemiology of headache and migraine in women. J Neurol Neurosurg Psychiatry 34:148–153
Wesche L, Frederickson RCA (1978) Diurnal differences in opioid peptide levels correlated with nociceptive sensitivity. Life Sci 24:1861–1868
Wesche L, Frederickson RCA (1981) The role of the pituitary in the diurnal variation in tolerance to painful stimuli and brain enkephalin levels. Life Sci 29:2199–2205
Wilder-Smith CH, Wilder-Smith OH (1992) Diurnal patterns of pain in cancer patients during treatment with long-acting opioid analgesics. Proceedings of the 5th international conference on biological rhythms and medications, Amelia Island (Fl), abstract no X111-7
Wilder-Smith CH, Schimke J, Bettiga A (1992) Circadian pain responses with tramadol (T), a short-acting opioid and alpha-adrenergic agonist, and morphine (M) in cancer pain. Proceedings of the 5th international conference on biological rhythms and medications. Amelia Island (Fl), abstract no X111-6
Willer JC (1977) Comparative study of perceived pain and nociceptive flexion reflex in man. Pain 3:69–80
Willer JC (1985) Studies on pain. Effects of morphine on a spinal nociceptive flexion reflex and relation pain sensation in man. Brain Res 331:105–114
Willis WD (1982) Control of nociceptive transmission in the spinal cord. In: Autrum H, Ottoson D, Perl ER, Schmidt RF (eds) Progress in sensory physiology, vol 3. Springer, Berlin Heidelberg New York, pp 1–159
Wolff BB (1983) Laboratory methods of pain measurement. In: Melzack R (ed) Pain measurement and assessment. Raven, New York, pp 7–13
Yaksh TL (1978) Analgesic action of intratechal opiates in cat and primate. Brain Res 153:205–210

Subject Index

ACE inhibitors 265 (table)
acetaminophen (paracetamol) 180, 193, 564
N-acetyl aspartylglutamate (NAAG) 11
acetylcholine 80 (table), 357
acetylsalicylic acid 180 (table), 210, 368–369
N-acetyl serotonin 377
N-acetyl transferase 192 (table), 377, 381, 439
α_1- acid glycoprotein 541
acrophase 513
actinomycin D 508 (table)
activated partial thromboplastin time 536
N-acyl-4-aminomethyl-5-methoxy-4-methyl-1,2,3,4-tetrahydrocarbazoles 403
adenosine diphosphate 517
adenosine monophosphate 535
adenylate cyclase 41, 135–137
   activation by manganese ions 140
   β-adrenergic stimulation 144
   circadian variation in brain 148–149 (table)
   isoprenaline effect 144
   pathway in adrenal glands 151–152
   pathway in epidermis 151
   pathway in liver 151
   pathway in signal transduction in cardiovascular system 138–144
     circadian variations 138, 140
     seasonal variations 140
   pathway in signal transduction in central nervous system 146–147
   pathway in transmembranous signal transduction 135–137
   types I–IV 135–136
adrenal glands 3
   cAMP circadian variation 152
   cortisone concentration 151
   steridogenesis 152
α-adrenoceptor antagonist 145
β-adrenoceptor(s) 263
   antagonists 263–267, 272–278
   blockade 263
   pharmacological characterization 263–267
α-adrenoceptor-G-protein phospholipase C system 135
β-adrenoceptor-G-protein adenyly cyclase complex 135
adrenocorticotrophic hormone (ACTH) 105 (table), 106, 107, 151
   cortisol response 118 (table)
Adriamycin 498, 569–570
affective disorders 135–146, 387
   blunted circadian amplitude 464–465
age-related changes in period in day/night synchronized humans 120 (table)
age-related changes in phase shifting in circadian system 120 (table)
age-related internal desynchronization 120 (table)
aging 95–122
   body temperature 101
   chronobiology 100–116
     approach 96
   clinical chemical variables 113–115
     plasma proteins 114–115
     volemia 114–115
   haematology 115–116
   sleep pattern changes 100–101
albumin 185
albuterol 213, 214, 218
aldosterone 105 (table), 106, 118 (table), 119 (table)
   age-related changes in circannual rhythms 121 (table)
   MESOR decrease in old age 107
alkylating agents 508 (table)
O^6-alkylguanine -DNA-alkyltransferase 305
allergens 590–592
allergic rhinitis 601, 602 (fig.)
   chemotherapy 602 (fig.), 603, 604 (fig.), 605
allyl alcohol 193
alprenolol 195

Alzheimer's disease 106
  DHEA-S 108
amikacin 197, 571, 572–573, 581
γ-aminobutyric acid (GABA) 79, 86 (table), 428
aminoglycosides v 560 (table), 570–571, 572
aminophylline 180 (table)
aminopyrine 192
aminopyrine N-demythylase 189, 190 (table)
amitryptylline 180 (table)
amlodipine 264 (table), 267, 269, 272, 276 (table)
amphetamine 198 (table)
amphetamine sulfate 198 (table)
ampicillin 572
anaemia 492
analgesics 636–643
  animal studies 636–637
  human studies 637–652
    patients with arthritic pain 638–640
    patients with cancer pain 640–642
    patients with postoperative pain 640–642
  patient-controlled 640
Andante (X chromosomal mutant) 66
angina pectoris 261, 272–273
  stable 273, 280
  unstable 281
  variant 273, 279
angiotensin I 270
angiotensin II 87, 270, 338
aniline hydroxylase 189, 190 (table)
6-anilinoquinone-5, 8-quinone 43
anisomycin 445, 562
Anolis caroliensis 375
antacids 361
anthracyclines 302 (table), 508 (table)
antibiotic chrononephrotoxicity 570–573
anticancer agents 299–324, 491–494, 508 (table), 519
  bonemarrow toxicity 519
  clinicopharmocology 307–313
    chronopharmacokinetics 309–313
    rhythms in target tissues 307–309
  clinical validation of chronotherapy 314–324
    chronotherapy of metastatic colorectal cancer 322–324
    phase I trials of chronomudulated chemotherapy 315–321
    phase II trials 321–322
  experimental chronopharmacology 299–307
    circadian rhythms in antitumor efficacy 305–307

    rhythms in susceptibility of target tissue 304–305
    toxicity rhythms 299–303
    summer photoperiod 300
    winter photoperiod 300
anticholinergics 360
antidepressants 471
  light sensitivity blunting 465
S-antigen 7
antihypertensives 271
antimetabolites 302 (table)
antipyrine 191, 192
antithrombin III 536, 545
  complexes 546
antiulcer drugs 355
antrum (distant stomach) 351–352
aplastic anaemia 491
Arabidopsis 73, 74
arabinosylcytosine (Ara-c) 305, 508 (table), 580
arachidonic acid 190 (table)
ardicaine 607, 612 (table)
arterial embolism 261
arterial vessels 259
  nocturnal haemodynamic changes 259
Aschoff's rule 427
aspirin 277 (table), 281, 567–568
asthma 205–241
  acid reflux effect 211
  β_2-adrenergic agonist medications 213–219
  airway caliber 210
  anticholinergic agents 219–212
  chronobiology 208–11
  chronopharmacology/chronotherapy of medications 212–219
  chronotherapy of glucocorticoids 227–239
  chronotherapy of sustained-release tablet β_2-agonist medications 214–217
  circadian rhythm in airways hyperreactivity 210
  circadian rhythm in airways inflammation 208–210
  circadian rhythm in autonomic nervous system function 210
  circadian rhythm in bronchodilator effect of conventional β_2-agonist therapy 213–214
  circadian rhythm in endocrine system function 210
  circadian rhythm in pulmonary function 208
  continuous positive airways pressure 211

day-night differences in airways
secretions, clearance cooling,
and antigen secretion 210
day-night pattern of acute asthma
205–207
deaths 207
early reactions 207–208
effect duration of conventional β_2-
agonist aerosols 213
gastroesophageal reflux 211
late reactions 207–208
mast cell stabilizers 239–240
nocturnal body position 210
pharmacotherapy 211–212
recurrent reactions 207–208
sleep apnea effect 210
sustained effect β_2-agonist aerosol
medications 217–219
atenolol 182, 263, 272, 274 (table), 280, 479
atrial fibrillation 61
atrial natriuretic peptide 145
atropine 220
autonomic nervous system, circadian
rhythm 20, 32

Bacillus Calmette-Guérin 562
backache 633–634
bambuterol 217
barbiturates 101, 105 (table)
basal acid secretion 358–359
basic circadian oscillator 56
analytical approaches 56–57
B cell 121
beclomethasone dipropionate 233
behaviour medication research
tactics 422–425
altering opportunities for
activity 422–423
assessment problems 424–425
confinement 423–424
measuring activity 422
restraint 423–424
testing drugs in vitro 424
benazepril 265 (table), 270
benzene 491
benzodiazepines 101, 379, 418–419
periodicity after single treatment
420
beomycin 511
betoxycaine 612 (table)
bevantol 274 (table)
bile acids 183
biliary colic 632 (table), 633
binding proteins 106
biological age of an individual 96
biological clock 387

biological condition of an individual 96–97
biological response modifiers 508 (table)
biphenyl 4-hydroxylase 190 (table)
birds' eyes 5
bisoprolol 263, 273
blind subject, melatonin
production 390–391
blast cell count 520
β-blockers 275 (table)
blood, peripheral 514
blood pressure 104–106, 121, 251–252, 252–258
age-related change in biologic
rhythm 118 (table)
age-related change in circannual
rhythm 121 (table)
aging patients with essential
hypertension 104–106
circadian fluctuations 252–253
circadian regulation, vascular cAMP
system 144
elderly subjects, children
compared 106
nocturnal decline 52
in:
asthma 254
brainstem infarct 253
cardiac transplantation 254
catecholamine-producing
tumor 254
chronic renal failure 254
congestive heart failure 254
Cushing's syndrome 254
cyclosporine treatment 254
diabetes mellitus 253
exogenous glucocorticoid
administration 254
fatal familial insomnia 253
hyperaldosteronism 254
hypertension with left ventricle
hypertrophy 254
hyperthyroidism 254
neurogenic hypertension 253
pregnancy toxemia 254
pseudohypoparathyroidism 255
recombinant human
erythropoietin therapy 255
severe hypertension 254
Shy-Drager syndrome 253
pathophysiology of regulation 252–258
physiology of regulation 252–258
pre-waking rise 252–253
sympathetic system effect 253
blood viscosity 536, 546
body clock 1
body temperature 101

body temperature (*Contd.*)
    changes in old people 101
    decline 2–3
bone marrow 487–491
    biological rhythms 494
    drug-induced hypoplasia 491–494
    marker rhythms for proliferative
        activity-cytoxicity 513
brachial artery 259
bradyarrhythmias 261
bradykinin 621
brain
    circadian pacemaker for locomotor
        activity rhythm 65–66
    glial cells 66
    neurons 66
    pacemaker function 66
brain infarction, hemodynamic 256
breast cancer 307–308, 512
    temperature in:
        breast surface 308
        contralateral healthy breast 309
        tumor 309
8-bromo-cGMP 43
bronchodilators 213
Brucella endotoxin 562
Bryophyllum phosophenol pyruvate
        carboxylase 74
budesomide 233, 234, 236
Bulla gouldiana 49–50
bupivacains 561–562, 562–563, 667
    animal data 608 (table), 610 (table)
    circadian variations 608 (table)
    + prilocaine 613 (table)
bupranolol 263
buprenorphine 642
busulfan 309, 311 (table), 312 (table)

cadmium 559, 569
    flux 554, 445
cadmium sulfate 559
caffeine 192
calcium 14, 16, 43
    intracellular 145, 450
calcium channel blockers 264 (table),
        267–270, 276–277 (table), 278–
        280
    pharmacological characterization
        267–270
calmodulin 43
calretinin 86 (fig.), 87
cAMP *see* cyclic AMP
cancer
    cell division cycle 29
    chemotherapeutic treatment regimens,
        circadian influence 369–370

chemotherapy *see* anticancer agents
    host tolerance 29
    pain 632 (table), 640–642
captopril 270
    + hydrochlorothiazide 265 (table),
        270
carbachol, intraventricular
        injection 425–426
carbamazine 186 (table), 191, 192
carbon tetrachloride 564, 565, 566
        (fig.)
carboplatin (CBCDA) 300, 301–303,
        309, 310 (table), 508 (table)
cardiovascular active drugs 268
        (table)
cardiovascular disease, drug
        treatments 269
cardiovascular system
    blood pressure *see* blood pressure
    chronobiological mechanisms 252–
        261
    rhythms in signal transduction 138–
        146
        adenylyl cyclase pathway 138–145
        guanylyl cyclase pathway 145
        phospholipase cyclase
            pathway 145–146
cartetol 263
carticaine 637
catalase 78
catecholamines 106, 112–113, 256
    excretion, circannual rhythm 117–121
    tumors producing 256
Cdc2-cyclin B complex 49
Cdc25 49
cefodizine 572
cell division cycle 29–50
    circadian control model 47
    events gating 30
    G_1 phase 31
    G_2/M boundary 31
    'gating' phenomenon 32
    'hatching' 32
    maturation (M-Phase) - promoting
        factor 31
    oscillator-cell cycle coupling, role of
        cAMP 33–47
    regulation 31–32
        downstream pathway: cAMP-
            dependent kinase 43–46
        upstream analysis: oscillation genesis in
            cAMP 41–43
CDC28 31
cell proliferation in mammalian tissues 29
central nervous system rhythms in signal
        transduction 146–151

Subject Index

adenylyl cyclase pathway 146–147
  guanylyl cyclase pathway 147
  phospholipase C pathway 151
cephaloridine 570
cerebellum, cAMP concentration 146, 147
cerebral blood flow 258
cerebral cortex
  cAMP concentration 147
  circadian patterns 147
cerebral infarction 517
cetirizine 590
chemical neuroanatomy of mammalian circadian system 79–83
chemical toxic agents 558–562
chloramphenicol 491
chlordiazepoxide 425
chloroform 193, 194 (figs.), 566 (fig.)
chlorophyll 74
chlorpromazine 379
chronoepidemiology of population of 765 allergic patients 601
  circadian patterns 147
  circadian rhythms in allergic symptoms 601
  circannual rhythms in allergic symptoms 601
chronesthesy 578–580
chronobiotic drugs 403
chronon hypothesis 71
chronostructures 557
chronotherapeutic agents 21, 29
chronotoxicology, new trends 557–582
  examples in nonrodents 563
  organ specific circadian chronotoxicity 563–577
    ear 565–567
    kidney *see* renal chronotoxicity
    liver 564–565, 566 (fig.)
    stomach 567–568
cibenzoline 268 (table)
cimetidine 361
circadian clock 31
  genetics 58–71
  identification criteria 4–5
  intracellular 'hands' 74
  localization in vertebrates 4–9
  molecular biology 58–71
  output signals 73–74
circadian desynchronization 117
circadian genetics 58
circadian oscillators 65
  pineal gland 6–8
  regulation of cell division cycles 29–50
  retina 5–6
circadian rhythms 1–3, 4, 58, 314

absent 117
amplitude decrease 117
internal desynchronization 121
measurement 458–459
circadian system
  age-related changers in phase shifting 120 (table)
  entrainment, melatonin and 18–21
  non-photic entrainment 17–18
  'photon counter' 13
circannual chronotoxicology 580–582
circannual rhythms 117–121
  age-related changes 121 (table)
cisplatin (CDDP) 186 (table), 197, 300, 301–303, 302 (table), 303 (fig.), 508 (table), 559, 569–570
  circadian changes 310 (table)
  renal damage 303
  short intravenous infusion 309
cisplatinum 559
citalopram 467
clemastine 593 (fig.), 594 (fig.), 595–596
clinical chemical variables, aging effect 113–115
  plasma proteins 114–115
  volemia 114–115
clock responsive element 74
clonidine 471
clorgyline 425, 471
coagulation 535–536
  circadian variation 544–546
  drugs affecting 281–282
  global measures 544–546
    activated partial thromboplastin time 545
    Howell's time 545
    prothrombin time (Quick test) 544
    thrombin time 545
  specific coagulation-related factors 545–546
colonic metastatic adenoma 369–370
colony-stimulating factors 490
colorectal cancer metastases 322–324, 570
common carotid artery 259
congenital luxation of hip 98
congenital stenosis of aorta 98
congestive heart failure 271
conidia 60
conidiation 74
converting enzyme inhibitors 270
  pharmacological characteristics 270
core body temperature 2 (fig.), 3, 387
coronary arteries 259
  arteriosclerotic 259
  tone 539

coronary heart disease 261
  chemopharmacology 172–282
  chronopharmacology 272–282
    β-adrenoceptor antagonists 272–282
    calcium channel blockers 278–280
    drugs affecting coagulation 281–282
    organic nitrates 280–281
corticosteriod(s) 3, 98, 227–239
  actions 227
  aerosol forms 227
  aerosol therapy 233–234
  before bedtime administration 233
  circadian rhythm 151, 152
    in pharmacodynamics 228–231
    in pharmacokinetics 228
  tablets 27
  chronotheraphy 231–233
corticosteroid-binding globulin 187
corticosterone 119 (table)
corticotropic axis chronobiology 337–338
corticotropin-releasing hormone 334
cortisol 2 (fig.), 105 (table), 107, 210, 387–388
  age-related change in circannual rhythm 121 (table)
  circadian variations 337–338
  granulocyte mobilization 494
  in old age 107
  peaking 230–231
  rhythms in affective disorder 464
  secretion 337–338
cortisone 98 (table)
  granulocyte mobilization 494
C-peptide 105 (table), 119 (table)
cromolyn sodium (Intal) 239–240
Cushing's syndrome 135, 254, 255
cycles 1
cyclic AMP (cyclic adenosine 3′, 5′-monophosphate 30, 31, 43
  circadian variation 151–152
  oscillation genesis 41–43
  perturbation of cell division rhythm 34–40
  rhythm generation mechanisms 153
  role in oscillator-cell cycling coupling 36–47
cyclic AMP-dependent kinases 43–46
cyclic AMP-response element 14–15
  modulator gene 379
cyclic GMP (cyclic nguanosine 3′, 5′-monophosphate) 145, 442
  concentration variation in brain (rat) 150 (table)
cyclic haemopoiesis 517–519
cyclic neutropenia 517
cyclin 31, 32, 48–50
cyclin B 48

homolog 49
cyclin-cdc2/cdc28 (MPF) system 31
cycloheximide 444
cyclophosphamide 99 (table), 508 (table), 580
cyclosporine 573–577
cyproheptadine 592–593, 594 (table)
  molecular structure 593 (fig.)
cytochrome P-450 190 (table), 509
  isoenzymes 190-191
cytochrome P-450 monooxygenase 189–192
cytosine arabinoside 29, 511, 559

daunorubicin 508 (table)
decarboxylase inhibitors 379
dehydroepiandrosterone (DHEA) 107–108, 11, 118 (table)
dehydroepiandrosterone sulfate 107–108, 117, 118 (table) 119 (table)
dehydropyrimidine dehydrogenase 304
5,7-dehydroxytryptamine b 17
dehydro-thromboxane B_2 542
deoxythymidine kinase 304
depression 457–479
  activity feedback 466–467
  animal models of depression/circadian rhythms 468–478
    ablation-based depression models 475
      olfactory bulbectomy 475
      thyroidectomy 475
    animals models of SAD? 476
    behaviroal feedback effects 476–477
    behaviorally characterized inbred strains 473–474
      inbred mice differing in affective behavior 474
      selection for drug sensitivity 474
      selection for hypertension 473–474
    psychopharmacology of circadian rhythms 468–472
      antidepressants 471
      free-running period 468–470
      pharmacogenic-developmental model 472
      phase-shifting 468–470
    stress-induced animal models 472
  circadian rhythms 459–467
    phase-advanced 459
    psychological models see above
  cortisol rhythms 464
  free-running rhythms 466
  light sensitivity increase 465
  neurotransmitter dysregulation hypothesis 465
  phase-advance hypothesis 460

Subject Index

phase-delayed locomotor activity 467
psychomotor disturbance 466–467
rhythm instability 465
seasonal 461–464
sleep disturbance 460–461
  antidepressives effect 460
social zeitgeber 465–466
switch into mania/hypomania 460
N-desmethyldiazepam 186
desmethylimipramine 379
desynchronization 1
developmental chronobiology 97–100
dexamethasone 99 (table), 107
  absorption 180 (table)
DHEA-S 105 (table)
diabetes mellitus 333
  chronopharmacological approach 339–344
    pulsatility of plasma insulin concentrations 341–342
    retinal degeneration 333
    type II 341–342
1,2-diacylglycerol 138, 139 (fig.)
diazepam 105 (table), 180 (table), 186 (table), 186–187, 561
dibekacin 559, 560 (fig.)
1,1-dichloroethylene 193, 564
diclofenac 180 (table)
digoxin 180 (table), 268
dihydrofolate reductase 304
1,4-dihydropyridine 267, 278
dihyropyridine derivatives 267
dihydropyrimidine dehydrogenase 509, 517
diltiazem 276 (table), 27
6-dimethylaminopurine 49
dipuridamole 180 (table)
disopyramide 186 (table)
DNA
  synthesis 31–32, 307
  synthesis inhibitors 508 (table)
dopamine 80 (table), 105 (table), 112, 118 (table)
dopamine D_1 receptors 20
doxazosin 271, 272
doxorubicin 302 (table), 309, 310 (table), 312 (table), 313, 315, 508 (table)
doxorubicin-cisplatin combination 306
doxorubicin-melphalan combination 305
DRB (reversible transcription inhibitor) 449
Drosophila 21
  genetic/molecular genetics of clock 59–60

per gene 72
single gene mutation 58
X chromosome 59
Drosophila melanogaster 65–66
drugs
  absorption see drugs absorption
  anti-ulcer, proton-pump inhibiting 355
  bioavailability, circadian rhythms infleuncing 355–357
  circadian rhythm parameter in elderly subjects 105 (table)
  circadian variations 178
  cystostatic 29
  cytotoxic 491–494
  disposition following oral administration 177 (fig.)
  distribution see drugs distribution
  effects on circadian rhythms 415–431
drugs absorption 178–184, 355
  drugs absorbed faster in the morning 180 (table)
  factors modifying temporal variation 181–184
    blood flow 183–184
    motility 182–183
    pharmaceutical formulation 182
    secretion 182–183
    solubility 181–182
  inorganic compounds 179
  lipophilic drugs 181–182
  macromolecules 179
  modifying factors 179 (table)
  organic compounds 179–180
  parameters 178
drugs biotransformation 188–196
  conjugations 192–193, 184–195 (figs.)
    to:
      acetate 192–193
      glutathione 193
      glucuronic acid 192–193
      sulfate 192–193
  factors related to biotransformation 195–196
    first pass effect 196
    hepatic blood flow 195
    oxidation: cytochrome P-450 monooxygenase 188–192
drugs distribution 185–188
  erythrocyte bonding 187–188
  plasma protein bonding 185–187
dualism ('in yang') 43
duodenal ulcer disease 358, 359, 362 (fig.)
  healing 367
durstan 563

Dutimelan 232–233, 239
dynorphins 621

E_1 118 (table)
E_2 118 (table)
ear chronotoxicity 565–567
elderly people
　changes in time reaction to environmental synchronizers 95
　failure to adapt to synchronizer changes 95
　vital functions decline 95
embryonic cell cycle 'engine' 30
enalapril 145, 180 (table), 265 (table), 268 (table), 270, 272
endogenous algesic substances 621
endogenous circadian oscillation 71
endogenous rhythms 95
endorphin(s) 621
β-endorphin 626, 627
endothelin-1 533, 537–538
endothelium 490, 533
enkephalin (opioid growth factor) 83, 507
enkephalins 621
entrainment of circadian system 3–4
entrainment pathways 80–85
environmental cues 3
eosinophil leukocytes 115, 516
4'-epidoxorubicin 508 (table)
epinephrine 105 (table), 112, 118 (table), 210, 213–214, 517
epipodophyllotoxins 508 (table)
erythrocytes, drug binding to 187–188
erythromycin 572
erythropoietin 490
Erythrochia coli endotoxin 562
esophageal atresia 98
estradiol 105 (table), 416–417
estrone 111
ethanol 99 (table), 559
　absorption 180 (table)
etidocaine 561–562, 607
　animal data 608 (table)
　circadian variations 608 (table)
　in mice 610 (table)
Euglena gracilis 31, 32–33
　cell division rhythmicity 32–33
　circadian oscillator role 33
　phase delays 33
　ZC mutant 32, 34, 43, 44–45
Euphyllin 223
Euphylong 223, 225, 226 (fig.)
exercise tests 273
extracellular growth factors 490
extracellular growth factors 490

famotidine 361, 363
fatal familial insomnia 255
feeding behavior chronology 338–339
femoral neck fracture in elderly subject 107
fetal biological rhythms 97, 98 (table)
　activity 98 (table)
　bladder volume 98 (table)
　'breathing movements' 98 (table)
　estriol 98 (table)
　heart rate 98 (table)
fetal rhythms of susceptibility to teratogenic agents 97–100
　cortisone 99 (table)
　cyclophosphamide 99 (table)
　dexamethasone 99 (table)
　ethanol 99 (table)
　5-flurouracil 99 (table)
fibronolysis 536–537, 546–549
　circadian variation of PAI-1 548–549
　global assays 546–547
　individual factors 547–548
fibrinpeptide A 536, 546
fibrobastic cells 490
flavoprotein reductase 189
floxuridine 300–301
fluniosolide 233, 234
fluodeoxyuridine 508 (table)
fluopyrimidine(s) 304, 517
5-fluorpyrimidine 369–370
5-fluorouracil 99 (table), 300–301, 302 (table), 309, 313, 319–321, 508, 569–57
　circadian changes 310 (table), 311 (table)
　5-FO-FA-L-OHP 320A21
f-fluorpyrimidine 369–370
flurbiprofen 638
fluticasone 233, 234
fluvoxamine 380 (table)
folionic acid 321 (table), 370
food 696
forearm arterial tone (resistance) 145, 539–540
formeterol 217
forskolin 38, 41, 140
free-running rhythms 466
functional bowel syndromes 351
forosemide 181

galanin 80 (table)
gases 79
gastric acid secretion rhythm 357–367
　pharmacological implications 361–366
　therapeutic implications 367

Subject Index

gastric chronotoxicity 567
gastric mucosal flow 184
gastric ulcer 367
gastrin 357
gastrin-releasing peptide 85
gastro-esophageal reflux 361
gastro-intestinal blood flow 183–184, 258
genes
  cab-1 74
  cab-2 74
  c-fos 14, 15, 16
    promoter 14–15
  cla-1 61
  frq 59, 60, 61, 62–64
  immediate-early 14–17
    phase-shifting 6
  jun-B 14
  nuclear (pea) 73
  per 59, 60, 66, 69, 71
    molecular biology 66–68
  prd-1 61
  prd-2 61
  prd-4 61
  single gene mutations 58
  timeless 68
genetic-environmental interaction 95
geniculothalamic tract 18, 83–84
gentamicin 180, 197, 559, 560 (fig.), 561 (fig.), 561, 571, 573
  renal damage 562
glial fibrillary protein 10
glucocorticoids 302 (table)
  inhaled 234–239
glucose 333
  chronobiology of blood glucose levels 334–335
  chronobiology of glucose tolerance 335
glucose phosphate isomerase 102 (table)
glucose tolerance test 337
glucuronosyltransferase 192 (table)
glutamate 11, 14, 79, 80 (table)
  action on:
    N-methyl-D-aspartate receptors 81
    non-N-methyl aspartate receptors 81
  retinohypothalamic tract neurotransmitters 81
glutamergic antagonists 14
glutamic acid decarboxylase 84
glutathione 193, 195 (fig.), 304–305, 509–510, 565
  reduced 307
glyceroltrinitrate 277 (table), 280
glycine 80 (table)
glycogen phosphorylase 578

glycogen synthetase 578
δ-1-glycoprotein 185
glycoprotein inhibitory factors 488–489
golden hamsters, single gene mutations 58
  tau mutant 69
gonadal steroids 41–418
Goodwin oscillator 71, 73
Gonyaulax poplyedra 49
G-proteins 140, 153, 401
  α-subunit 136
granulocyte 487, 488
granulocytopenia 493
griseofulvin 572
growth hormone 110–111, 119 (table)
guanylyl cyclases 135
  A 138
  B 138
  C 138
  pathway 137–138
    cGMP 152
    epidermis 152
    signal transduction in CNS 147
    signal transduction in CVS 145
    transmembranous signal transduction 137–138
  retinal 138
  soluble 137

H^+ ions 621
haematocrit circadian rhythm 514
haematology in aged subjects 115
haemoglobin circadian rhythm 514
haemopoiesis 487–491
haemopoietic cells, multifrequency time structure 519
haemopoietic progenitor cells 488, 518
haemopoietic stem cells 487, 488
haemopoietic system 491
  chronobiology of peripheral blood in health/disease 513–517
    eosinophil leucocytes 516
    lymphocytes 515–516
    natural killer cells 516–517
    neutrophil leucocytes 515
    platelets 517
    total white blood cells 514–515
  chronopharmacological aspects 508–511
  circadian aspects 494–507
    circadian/circannual proliferative activity in healthy human bone marrow 499–505
    circadian variation in bone marrow DNA synthesis in cancer patients 504–507

haemopoietic system (*Contd.*)
  proliferative activity in murine bone marrow   494–498
  stage-dependent cytotosicty of murine bone marrow   498–499
  cyclic hemopoiesis   517–519
  disease-dependent chronotherapautic effects/consequences   511–512
  implications, scientific/clinical   519–522
  marker rhythms for proliferative acitivity/cytotoxicity of bone marrow   513
  regulation of hemopoietic circadian rhythms   507–508
haemostatic system
  chronobiology   537–549
  circadian variation of coagulation *see* coagulation
  circadian variation of endothelial cell function   537–539
    endothelial reaction to stimuli   539
    endothelial removal effect   539
    endothelium-related factors in plasma   537–538
  circadian variation of fibrinolytic activity *see* fibrinolysis
  circadian variation of platelet function *see under* platelet
  circadian variation of vasomotor function   539–540
    coronary artery tone   539
    forearm artery tone   539–540
    response to constrictor/dilator tone   540
HALO (hours after light onset)   299
head and neck tumor   309
  maximal tumor surface temperature   309
  mitotic index   309
heart
  anginal attacks   261
  atrial fibrillation   261
  atrial rate   260
  bradyarrhythmias   261
  ectopic beats   261
  electrical properties   260–261
  left-sided Kent bundle   260–261
  left ventricular function   261
  left ventricular hypertrophy   261
  output   258
    noradrenaline effect   259
  Q-T interval changes   260
  rate changes   104–106, 251
  refractoriness   260
  sarcolemma   259

sinus node formation   260
sudden death   261
supraventricular extra systoles   261
ventricles (rat)   140
ventricular arrhythmias   260, 261
ventricular premature beats   261
ventricular rate   260
heavy metals   559, 560 (fig.)
Helicobacter pylori   367
heparin   282
Heterometrus fulvipes toxin   562
hexoseminidase   102 (table)
hippocampus   163
histamine   80 (table), 210, 357, 358 (fig.), 589, 604
  H_1 receptor   589–590
    antagonists   361, 362, 589, 592–597
  H_2 receptor   589
  excitatory effect on nociceptors   621
  intradermal injection   590
  lung reactivity   592
  molecular structure   593 (fig.)
  skin reactivity   590–592
H^+K^+-ATPase inhibitors   366, 367
house duct aerosols   210
Howell's time (recalcification time)   536, 545
human seasonality   387
hyaline membrane disease   626
hydrochloric acid   357
hydrochlorothiazide   181, 181 (table)
hydromorphone   640, 642
hydroxizin   105 (table)
18-hydroxy-11-deoxycorticosterone   106–107
8-hydroxy-(di-n-propylamino)-tetralin   165–166, 166, 425
5-hydroxyindoleacetic acid   160–161, 162, 163
hydroxyindole-*O*-methyltransferase   377
17-hydroxyprogesterone   111
5-hydroxytryptamine (serotonin)   17–18, 157–169, 377, 439, 445, 621
  autoreceptor activity   165–168
  coronary blood flow effect   259
  function of circadian rhythm in turnover   168–169
  δ-granulocyte   541
  hippocampal release   163–164
  metabolism   162–164
  release   162–164
  serotonergic neurone firing rate   164, 165
  sites controlling circadian rhythms in serotonergic neurone function   159 (table)

tissue levels 167
uptake inhibitors 379
5-hydroxytryptamine$_{1B}$ autoreceptors 166, 167, 168
5-hydroxytryptamine$_{1D}$ 167
5-hydroxytrytophan 164, 377, 439
5-hydroxytrytophan decarboxylase 161, 164
hydroxyurea 99 (table)
hypertension
  BP changes with aging 104–106
  chronopharmacology 262–271
  'false negatives' BP 106
  secondary 135
hypnotics 101
hypoglycemia 261
hypothalamic-pituitary-adrenal axis 3, 106–107
hypothalamic-pituitary-neurohumeral axes 260
hypothalamic-pituitary-thyroid axis, seasonal variations 113
hypothalamus
  cAMP content 146
  cGMP variations 147
  dorsomedial 12
  neurotransmitters in 334
  paraventricular 12
    nuclei 338
  preoptic 12
  subparaventricular 12
  suprachiasmatic nuclei 8–9, 80–81

indapamide 271
indomethacin 160 (table), 181, 181 (table), 182, 188
  plasma protein binding 186
indoramin 271
infancy rhythms 100, 102–103 (table)
infraradian rhythms 95–96, 117
inositol 138
inositol-1, 4-biphosphate 138
inositol phosphates 135
inositol-1-phosphate 138
inositol-triphosphate 138, 145
inositol-1,4,5-triphosphate 43, 138, 139 (table)
insulin 119 (table)
  annual changes 342–343
  chronobiology of plasma 335–337
  circannual rhythm of basal secretion 344
  pulsatility of plasma concentrations 341–342
insulin-related growth factor-1 (somatomedin C) 111

Intal (cromolyn sodium) 239–240
interferon 488
interferon-α 498–499, 508 (table)
interferon-β 498
interferon-γ 498, 499, 508 (table)
intergeniculate leaflet 83–84, 168–169
  connections 84 (fig.)
  GHT projection 84
  SP+ plexus 84
internal representation of time 4
internal temporal order 1–3
iodopsin cone pigment 5
ipratropium bromide 220
iron 118 (table)
  absorption 179
  serum concentration in elderly 116
irregular working schedule 1
ISDN 268 (table), 277 (table), 281
IS-5-MN 268 (table), 277 (table), 281
3-isobutyl-1-methylxanthine 41–43
isoprenaline 140, 144
isoproterenol 214
isorbide dinitrate 280
isosorbide-5-mononitrate 355
isradipine 256, 264 (table), 267, 267

jet lag 1, 395–397, 398 (fig.)

kanamycin 565
kanamycin sulfate 565
ketanserin 18
ketoprofen 180 (table), 182
kinases 73

lansoprazole 366
lateral geniculate complex 83–84
  intergeniculate leaflet 80
LD$_{50}$ 561–562
lead absorption (rat) 179
left ventricular ejection fraction 273
leucine-enkephalin 621
leukaemia
  acute lymphoblastic 314, 511, 519
  chronic granulocytic 517
  chronic myelocytic 518
leukotrienes 535
lidocaine 559, 607, 612 (table), 612, 637
  animal data 608 (table)
  circadian zeitgeber 381
  rats 610 (table)
  toxicity 607 609
  trancutaneous passage 611
lidocaine + flumenazil, animal data 608 (table)
light-dark cycles 4
  drug action in 415

light effects 80
light therapy 512
linoleic acid 190 (table)
lipophilicity 263
lithium 138, 415, 427, 471
liver 356–357
  blood flow 357
  cAMP concentration 151
  chronotoxicity 564–565, 566 (fig.)
  cytochrome P-450 isoenzymes 190–191
  microsomal enzyme activity 356–357
  microsomal metabolism 189
local anesthetics 186–188, 561–562, 607–616
  clinical pharmacology 611–614
  mechanisms 614–616
  time dependency in efficiency 611–612
  time dependency in kinetics 610–611
  time dependency in toxicity 607–609
  time dependency of pharmacokinetics 612–614
locomotor activity effect 417
lorazepam 180 (table)
losartan 145, 270
luciferin-binding protein 74
lutenizing hormone 109, 111, 119 (table)
  age-related change in circannual rhythm 121 (table)
LY83583 (6-anilinoquinone-5,8-quinone) 43
lymph nodes 514
lymphocytes 115, 119 (table)
lymphoma 809

macrophages 490
malathion 563
mammals 69–70
  circadian system 415, 416 (fig.)
  single-gene mutants 69–70
manganese ions 140, 141
mania 466
mast cells 239–40
  stabilizers 239–240
maturation (M phase)-promoting factor 31
mean corpuscular hemoglobin 116
Medtronic 315
melanoma 309
melatonin 2 (fig.), 3, 5, 7–8, 18–21
  abnormal timing 390
  adrenal gland production 108
  age-related changes in biologic rhythm 118 (table)
  agonists 401–403
  agricultural commercial preparations 385

antagonists 401–403
childhood amplitude 109
chronobiotic effects 425
chronotherapeutic agent 19
circadian amplitude decrease 117
circadian marker rhythm 387–391
circadian profiles (rodents) 389–390
circadian rhythms effect 391–395
core body temperature effect 395
decline with aging 109
dim light melatonin onset (DLMO) 388
entraining agent 7, 18–21
human reproduction role 390
in:
  affective disorder 390
  blind subjects 390–391
  depression 390
  mania 390
  old age 390
human seasonality relationship 387
infusions 420–422
injections 420–422
light control of secretion 380–383
mechanism of action in control of circadian rhythms 398–400
mechanism of action in control of seasonal rhythms 385–387
peak 389, 390
  decline in old age 390
photoperiodic history 383
plasma 388
posture effect 388
production 376–379
receptors 401
saliva 388
seasonal cycles control 385–386
sleep disturbances 392–393
synthesis
  control mechanisms 378 (fig.)
  from tryptophan 377
therapeutic uses 395–398
zeitgeber 393–394
melphalan 508 (table)
membrane depolarization 442
mepivacaine 561–562, 607, 612 (table)
  animal data 608 (table), 610 (table)
  circadian variations 608 (table)
mequitazine 180 (table), 593 (fig.), 594 (table), 596
  allergic rhinitis chronotherapy 602 (fig.), 603, 604 (fig.), 605
  cronesthy 599–601
    bronchial tests 599
    perforance testing 599–600

Subject Index

psychcopharmacological tests 599–600
skin tests 599
temperature 599
6-mercapturine 302 (table), 309, 310 (table), 311 (table), 511
mercuric chloride 559, 561, 581
mercury 559, 569
metabolic mutants 62
metergoline 166
methionine-enkephalin 621
methotrexate 192, 301, 302 (table), 309, 310 (table), 311 (table), 508 (table), 510–511
  toxicity 510
  treatment of acute lymphoblastic leukaemia 314
5-methoxy-3 (1,2,3,6-tetrahydro-4-pyridine)-1H indole (RU24969) 166
N-methyl-D-aspartate receptor 14, 81
6-methyl- prednisolone 228, 229 (fig.), 230, 302 (table)
metoprolol 263, 272, 274 (table)
  + nifedipine 274 (table)
mice
  mutagenizead, semidominant mutation 70
  single gene mutation 58
microsomal monooxygenase 189
  phospholipid fraction 189
microsomal protein 190 (table)
midazalom 180
midbrain raphe nuclei 11
migraine 632 (table), 633
migrating myoelectric complex 183, 352
mitomycin C 508
mitotic index 307, 309
mitoxantrone 300, 302 (table), 321
MK801 428
molecular clock mechanisms 70–71
molluscs 3
monaminergic neurotransmitter systems 457
monocytes 115, 490
morphine 195, 621, 636, 637
  slow-release 642
mortality, human 121
myocardial β-adrenoceptors 141–143
myocardial infarction 261, 272, 517, 632 (table), 633
myocardial ischemia 278

nadolol 274 (table)
naloxone 636
naloxone-precipitated abstinence syndrome 637
napthalenic agonists 403
natural killer cells 516–517
nedocromil sodium (Tilade) 239–240
negative feedback oscillator model 71
netilimicin 559
neuron 79
neuropeptide(s) 79
neuropeptide Y 80, 88 (figs.), 334, 425
  neurons producing 83
Neurospora 21
  genes 74
  genetic molecular genetics of clock 59–64
  metabolic pathways defective 62
  single gene mutations 58
Neurospora crassa 58–59, 72
  circadian rhythm 60
  clock mutants 60–62
  conidia 60
  light perception 72
  long-lived proteins 72
  temperature changes in perception 72
neurotensin 88 (fig.), 334
neurotransmitters 621–622
  small molecule 79, 80 (table)
neutropenia 517, 518
neutrophils 115, 119, 515
newborn, biological rhythms 100, 102–103 (table)
nifedipine 180 (table), 182, 195, 264 (table), 267, 268 (table), 540
  for:
    coronary heart disease 276 (table)
    myocardial ischemia 278–279
nikethamide 558
nitrate(s) 277 (table)
nitrate reductase 73
nitrendipine 180 (table), 264 (table), 265 (table), 267
nitric oxide 17, 79, 81, 135, 145, 280
p-nitroanisole demythalase 189, 190 (table)
$N\omega$-nitro-1-arginine methyl ester 145
nitrosoguanidine 60
nizatidine 361
nociceptive flexion reflex 629
nonchemical toxic agents 562
non-Hodgin's lymphoma 520, 521 (fig.)
non-N-methyl-D-aspartate
  antagonists 14
  receptors 81
non-steroidal anti-inflammatory drugs 379, 491, 567–568, 638, 639
  side-effects 639

noradrenaline 169, 259, 379
  circadian rhythms in turnover 189–170
  hippocampal release 70–71
  in paraventricular nucleus (rat) 170
  release control 171
norephedrine 198 (table)
norephedrine hydrochloride 198 (table)
norepinephrine 80 (table), 105 (table), 106, 112–113
  age-related changes in biologic rhythm 118 (table)
  children 112
  circannual rhythm absence 113
  circannual rhythm age-related changes 121 (table)
  elderly subjects 112, 112–113
  urinary excretion 112
nortriptyline 180
nutritional status, seasonal variations 339 (table)

omeprazole 366
oncology, clinical validation of chronotherapy 314–324
  chronotherapy of metastatic colorectal cancer 322–324
    phase I trials 315–321
    phase II trials 321–322
opiate receptor binding 627
opioid growth factor (enkephalin) 507
orciprenaline 214
organic nitrates 277 (table)
  pharmacological characterization 280–281
orotate phosphoribosyltransferase 304
oscillations 55–56
  amplitude 55
  frequency 55
  genetic control 56
  period length 55
  phase 55–56
  temperature compensation 56
  zeitgebers 56
oscillators, low-temperature 458
osteoarthritis 632 (table), 633, 634 (fig.), 635 (fig.), 638–639
ouabain 558–559
ovarian cancer 309, 314, 512
oxaliplatin 300, 302 (table), 498, 508 (table)
oxaprotiline 379
oxitropium bromide 220
oxprenolol 263, 266 (table), 268 (table)
oxytocin 334, 338

$p34^{cdc2}$cyclin (mitosis cyclin) 34
$p34^{cdc2}$p60 (G_1cyclin) 34
$p34^{cdc2}$protein 31, 32, 34, 48
pain 619
  biological rhythms in pain control mechanisms 624–628
  biological rhythms in patients 632–636
  carbon dioxide production 629
  chronobiology 624–636
  conduction pathways 620
  hot-plate test 631
  measurements 624–626
    pain questionnaires 622–623
    rating scales 623–624
    visual analogue scales 623–624
  neurochemical periodicity 627–628
  perception 621–622
palmitic acid 190 (table)
pantoprazole 366
paracetamol (acetaminophen) 180, 193, 564
paracrine factors 12
Passer domesticus 375
Peptichemio 508 (table)
periaqueductal gray matter 620, 621
peripheral resistance at night 258
peripheral vascular resistance 540
pesticides 563
pharmacodynamics 177
pharmacokinetics 177
phase response curve 4
phase shifting 1
phenobarbital 191, 580
phenothiazine 105 (table)
  chronic users 491
phenylbutazone 181, 181 (table), 191
phenyltoin 186 (table), 191
phocomelia 98
phosphatidylinositol-4,5-biphosphate 138, 139 (table)
phosphodiesterases 41, 137
phosphoenol pyruvate carboxylase 74
phospholipase A_2 535
phospholipase C 135, 139 (fig.), 535
  β-isoenzyme 138
phospholipase C pathway 152
  inositol triphosphate 152
  phospholipids 152
in:
  signal transduction in CNS 151
  signal transduction in CVS 145–146
  transmembranous signal transduction 138
photic effects mimicking/modulation 425–428

behavioural inhibition of photic shifts 427–428
period and Aschoff's rule 427
phase shifts 427
retinohypothalamaic tract transmitter 425–427
sensitivity to light 427
photoentertainment pathway 80–83
 intrinsic retinal pathways 80
 phoreceptors 80
 retinal ganglion cells 80
photoreceptor cells 66
photosystem II 73
phototransduction pathway 7
physical therapy 95
pigment-dispersing hormone 66
pindolol 263, 273, 274 (table)
pineal gland 3, 6–8, 109–110, 375–403
 α-adrenoreceptor stimulation 379–381
 λ-adrenoreceptor stimulation 379–381
 autonomous operation in individual cells 10
 grafting into thymus 109–110
 innervation 377
 melatonin secretion 3, 6, 7–8, 109–110, 259
  aging effect 109–10
  suppression 11
 neuromediators 379
 neuroreceptors 379
 photosensitivity 6
  removal effects 7, 375–376
 secretion inhibition 379
pinealocytes 109
pirarubicin 300, 303–304
pituitary-gonadal axis 111
pituitary-thyroid axis, age-related changes 113
plasma, endothelium-related factors in 537–538
plasma proteins 114–115, 119 (table), 185–187
 patient with advanced cancer 511–512
plasminogen 547
plasminogen activator inhibitor-1 533, 536–537, 548–549
platelets 115, 481, 488, 517, 534–535
 adhesiveness 541
 α-adrenoceptors 535
 α$_2$-adrenoceptors 543–544
 aggregability 281
  inhibition 281
 agonists 535
 'exhaustion' 543
 in vivo activation 541–542

in vivo aggregation 542
 serotonin content 541
platinum 300, 301, 314
 analogs 508 (table)
 derivatives 569, 580
pléthora relative, la 258
portal pressure 260
postvagotomy patients 360
pranofen 180 (table)
prazosin 271
prednisolone 180 (table), 186 (table), 187, 228
prednisone 228–229, 323 (fig.), 233
premenstrual depression 464
Prinzmetal's variant angina 281
procainamide 268 (table), 561
procaine 607
17-OH progesterone 105 (table), 118 (table), 121 (table)
prolactin 105 (table), 111, 118 (table)
 age-related change in circannual rhythm 121 (table)
propranol 180 (table), 182, 186 (table), 263, 266 (table), 268 (table), 271, 479, 540
 anginal attacks aggravation 273
 coronary disease patients 274 (table), 275 (table)
 sudden cardiac death reduction 272
propranol LA 274–275
prostacyclin 538
prostglandins 621
protein 40kDa 446
protein kinase A 72, 136
prolactin 111
prothrombin time (Quick test) 536, 544
proton pump inhibitors 358
pulmonary arterial pressure 259
Proventil Repetabs 215–216
'pulsiogium' 251

Quick test (prothrombin time) 536, 544
quinapril 265 (table)

ramipril 270
ranitidine 361, 362–363, 364 (figs.), 365 (figs), 366 (fig.)
raphe nuclei 11, 17
recalcification time (Howell's time) 536, 545
receptor kinases 7
red blood cells 116, 487, 488
 circadian rhythm 514
 half-life 492
refractoriness 384
renal blood flow 258

renal chronotoxicity 568–577
  antibiotic 570–573
  cyclosporine 573–577
  heavy metal 568–570
renin 118 (table)
renin-angiotensin-aldosterone
      systems 260
reserpine 105 (table), 471
retina 5–6, 435–461
  autonomous operation in individual
      cells 10
  neurons 447
  ganglion cell projections 80
  rhythmicity in *Aplysia*
        californica 440–446
    light-induced phase shifts 442–444
    pacemaker entrainment 442–446
    pacemaker localization 441
    rhythm expression 446
    rhythm generation 441–442
    serotonin-induced phase shifts 444–
      446
  rhythmicity in *Bulla gouldiana* 446–
      451
    pacemaker entrainment 449–450
    pacemaker localization 447
    rhythm expression 450–451
    rhythm generation 447–449
  rhymicity in *Xenopus* 435–440
    pacemaker entrainment 436–439
    pacemaker localization 436
    rhythm expression 439–440
    rhythm generation 436
retinohypothalamic tract 11, 80–83,
      425–427
rheumatoid arthritis 632 (table), 633,
      634 (fig.), 638–640
rhythms
  free-running 96
  infraradian 95
ritanserin 18
mRNA 70
rozatidine 361
RU24969 166

salbutamol 214
salicylic acid 180 (table)
saline injections 428–430
  central 429–430
  peripheral 428–429
salmeterol 217–218
salmeterol xinafoate 218
seasonal affective disorder 390, 461–464
seasonal rhythms 464
seasonal variations of nutritional
      status 339 (table)

second messenger mechanisms 135–153
serotonergic agonists 11
serotonergic neurones firing rate 164, 165
serotonin *see* hydroxytryptamine
serotonin N-acetyl transferase 5
serotonin-induced phase shifts (*Aplysia*)
      444
sexual hormone variation 631
sheep seasonal breeding cycle 384
short life span, animals with 96
Shy-Drager syndrome 253
signal transduction between clocks 29–
      50
single gene mutation 58
skin epidermal carcinoma 309
sleep 460–461
  pattern changes in old people 100–101
  two-process model 461
sleep disturbances 460–461
  antidepressives effect 460
sleep-wakefulness cycle 1–3
social cues 17
social zeitgebers 465–464
sodium nitroprusside 145
somatodendritic autoreceptor 166
somatomedin C (insulin-related growth
      factor-1) 111
somatostatin 80 (table), 87
Sordaria finicola 63
sparrows, pineal gland removal effect 7
spleen 514
S. pompe cycle 32
'splitting' 69
spondylarthritis 638
stearic acid 190 (table)
stem cells 488–489, 490
  pluripotent 492
steoidogenesis activator polypeptide 152
stomach
  hydrogen ion concentration 360
  oxyntic glands 357
  parietal cells 358 (fig.)
  rhythm in mucosal defence 367–369
Streptomyces griseolus 562
stress erosive gastritis 361
styrene 193
subcutaneous blood 258
substance P 81, 85, 627
sudden cardiac death 261, 517
suicide 387
6-sulfatoxymelatonin 109, 388–389
sulfonamides 197
sulfotransferase phenol 192 (table)
sulindac 192
suprachiasmatic nuclei 377, 415, 416–
      417

Subject Index

α-sympathetic vasoconstrictor
  activity   259
synchronisation   1, 95

T_3 113, 121 (table)
  total   113
T_4 105 (table), 113, 121 (table)
  total   113
tau   69–70, 71, 381
T cell   121, 490
temporal isolation   3
tenoxicam   638
terbutaline   214–215
terfenadine   590, 593 (fig.), 594 (table), 595–596
testosterone   105 (table), 111
testosterone 7α-hydroxylase   190 (table)
testosterone 16α-hydroxylase   190 (table)
testosterone 6β-hydroxylase   190 (table)
tetracaine   607
4'-O-tetrahydropyranyl   488, 508 (table)
tetrodotoxin   10, 562
thalamus
  paraventricular nucleus   85
  suprachiasmatic nuclei   8–21, 85–87
    afferent connections   11–12
    astrocytes   10
    autonomous operation in individual cells   10
    circadian function   87
    circadian signal generation   10
    destruction effects   9
    efferents   87
    entrainment   13–14
    gene expression in   14–17
    glutamatergic signals   14–17
    inputs   85
    internal structure   9–10
    lesions   9
    neuroanatomy   9–13
    neuropeptide Y activation   18
    nonphotic inputs   85
    organisation   9–10, 85–87
    outputs   12–13
    pacemaker function   85–87
    peptidergic transmitters   10
    photoresponsive areas   20
    retinal fibres within synapse   11
    subdivisions   85–87
  ventral lateral geniculate nucleus, intergeniculate leaflet   11–12
Theo Dur   216–217, 226 (fig.)
theophylline   180 (table), 182, 188, 213, 214 (fig.), 221–227, 227, 355
  chronoeffectiveness   223
  chronokinetics of SR preparations   221–222

chronotherapy of tablets   223–227
  nocturnal serum concentration   225
third ventricle, transplants in   87
threosulfan   508 (table)
thrombin   536
thrombin time   536
thrombocytopenia   492–493, 517, 518
thromboxane A_2 535
thymus   514
thyroid hormone circannual rhythm   117–121
Tilade (nedocromil sodium)   239–240
time, internal representation   4
'timeless' mutant   66
thyroid-stimulating hormone   113, 119 (table)
tissue blood flow   258
tissue-type plasminogen activation   537, 547, 548
  reversible inhibitor   547
tobramycin   197, 571
tolbutamide   191
tomato plants   73
toothache   632
transcortin   187
transduction   7
transferring growth factor-β   488
transmembranous signaling   135
transmembranous signal transduction   135–138
  adenylyl cyclase pathway   135–137
  guanylyl cyclase pathway   137–138
  phospholipase C pathway   138
triamcinolone   233, 234, 239, 240
triamcinolone acetate aerosol   237–239
triazolam   180 (table)
trichloroethylene   564, 565
tritium release experiments   164
tryptophan   160–161, 541
  availability   160–161
  extracellular concentration   163
  uptake   161
tryptophan hydroxylase   160
tryptophan 5-hydroxylase   377, 439
tumors
  arrhythmic   520
  circadian periodicity   520
  ultraradian periodicity   520
tumor necrosis factor-α   488
tyrosine hydroxylase   170

ultradian rhythms   71, 117
Uniphyl   223
Uniphyllin   223, 224 (fig.)
uridine phosphorylase   304
urinary pH fluctuation   197

urinary-type plasminogen activator 537, 547
 single-chain 547
urokinase 282

valproic acid 10 (table), 182, 186 (table), 187
vasoactive intestinal polypeptide 80 (table), 85, 86 (fig.), 88 (fig.), 260, 379
vasomotion 533–534
vasopressin 86 (fig.), 87, 88 (fig.), 334, 338
vasospasm 279
venous-filling index 260
venous hemodynamics 260
verapamil 182, 195, 268 (table), 269, 279
 sustained-release 267
vertebrate clock 1–21
VGF 87
vinblastine 508 (table)
vinca alkaloids 508 (table)
vincristine 508 (table), 511

vindesine 312 (table), 569–570
virus interaction with drugs 491
viscosity 536, 546
vitamin B_{12} 179, 425
Volmax 216
VP-16 508 (table)

white blood cells 514–515, 518
white noise 562
Willebrand's factor 537, 546
winter depression 58, 387

X chromosome 58
xenobiotics 559
 renal chronotoxicity 568
xipamide 271
X rays 562, 580

'yin yang' (dualism) hypothesis 43
yohimbine 543

ZC mutant 31
zeitgeber 71–73, 381, 458